Modern Biotechnology and its Applications

NIPA® GENX ELECTRONIC RESOURCES & SOLUTIONS P. LTD.
New Delhi-110 034

Modern Biotechnology and its Applications

Editor

Kambaska Kumar Behera (Ph.D, Post.Doc.)
Assistant Professor
Department of Bio-Science and Biotechnolgy
Banasthali University, Banasthali, Rajasthan-304022, India

NIPA® GENX ELECTRONIC RESOURCES & SOLUTIONS P. LTD.
New Delhi-110 034

NIPA® GENX ELECTRONIC
RESOURCES & SOLUTIONS P. LTD.

101,103, Vikas Surya Plaza, CU Block
L.S.C. Market, Pitam Pura, New Delhi-110 034
Ph : +91 11 27341616, 27341717, 27341718
E-mail: newindiapublishingagency@gmail.com
web: www.nipabooks.com

For customer assistance, please contact
Phone: + 91-11-27 34 17 17
Fax: + 91-11- 27 34 16 16
E-Mail: feedbacks@nipabooks.com

ISBN: 978-81-19235-16-2

Composed and Designed by NIPA®.

Dedicated

To

My Mother & My Father

Smt. Dutimani Behera

and

Shri Rama Chandra Behera

FOREWORD

It is a matter of great pleasure and privilege to record the following on the book Modern Biotechnology authored by Dr. Kambaska Kumar Behera, a young faculty member of Banasthali University, Rajasthan.

'Biotechnology', as a field of knowledge, is both a 'science' and a 'technology'--- depending on its particular knowledge area being reckoned and referred to by the readers. It is fairly correct also to say that knowledge generation and applications made or explored under this particular subject area are overwhelmingly huge by volume and in their diversity, may be surpassing any of those of other scientific-cum-technological fields as contemporary and as recent as the Biotechnology is. To prepare and present for the young generation specialized readers, a title Modern Biotechnology and its Applications, therefore seems to be so bold and mega-sized and also exploratory. Encouragingly, Dr. Behera as author has been successful in touching upon almost all the perceivable title points all of which were to be there to give that 'completeness' of the offering he made -- and as could be sought after.

The book is having as many as sixteen chapters and quite noticeably --- there remain a majority which is abiding to those, patterned in the existing good text books and in reference books for academic beginners and commoners. There also remain those chapters with such precise and cutting edge title-areas as are under 5th, 9th, 11th, 15th, and 16th. Together these make the preparation quite variegated and interesting.

The author seems to have tried in as many instances as possible, to present in each of the chapters, the science based information alongside technological supplements, which together may help to intellectually organize any early researcher and the design seems to benefit the readers considerably.

To this narrator, it also occurs that to a steady reader -- willing to cover the full content at one go, the sequencing and succession of knowledge organization needs a constructivist mind for a knowledge continuum and a breezy flow of learning in more harmonious fashion.

At the close it compels this commenter repeat our commonly used Indian words of good-wish… *'ayam arraambha , shuvayya bhabatu'*-it is the beginning, let it be blessed.

Professor. (Dr.) Sunil Kumar Brahmachari
Ex- Vice-Chancellor
Uttar Banga Krishi Viswavidyalaya, West Bengal

ACKNOWLEDGEMENTS

It is my privilege to express my deep sense of gratitude to Prof. Sunil Kumar Brahmachari, Ex-Vice-Chancellor, Uttar Banga Krishi Viswavidyalaya, West Bengal for his kind consent to write the foreword for this volume.

I acknowledge the assistance, encouragement, excitement and efforts of the contributors in organizing and publishing this book. The excellent contents and uniqueness of each individual chapter will make this book as a text book on the use of biotechnology and its advancement for a moderately long term, which usually does not occur with books associated with this genre. I thank all the contributors for their understanding and patience. There are several people who deserve special thanks but few of them are" Prof.(Dr.) Aditya Shastri, Vice Chancellor, Banasthali University and Prof.(Dr.) Santilata Sahoo (HOD),P.G.Department of Botanyl, Utkal University,Vanivihar,Odisha, who have blessed and inspired a lot to complete this task.

Words are not enough to thank to my wellwisher Saudamini Muddali, BajiRout , Sikha Niketan, Sector-3, Rourkela-769002, Odisha and all my friends and colleagues who also inspired a lot to complete this work.

Last but not the least; I would like to assert my intricate appreciation of gratitude, best regards to my loving parents, family members for their incessant support and cooperation.

Editor

PREFACE

Modern biotechnology excels the molecular techniques that use whole or parts of living organisms to produce or improve commercial products and processes. It is a rapidly evolving branch of natural sciences which started with the creation of the first recombinant gene thirty years ago. These techniques are used in many different ways, changing the way we live by improving the foods we eat, the beverages we drink, the clothes we wear, and the medicines we take. They also have enhanced other aspects of our lives through the development of new detection methods for early diagnosis of many diseases such as arteriosclerosis, cancer, diabetes, Parkinson's and, Alzheimer's. The application of biotechnology methods in the food and agricultural industry is one of the many aspects of biotechnology that has great impact on society. By the year 2050, it is expected that more than 10 billion people will be living on this planet, and it is also believed that there may not be enough resources to feed the world population (UNFPA, 1995). Hunger and malnutrition already claims 24,000 lives a day in the developing countries of Asia, Africa, and Latin America (James, 2003). Malnutrition however, is not exclusive to developing nations. Many people in industrialized countries, although mostly well fed, still suffer from lack of proper nourishment that offers the greatest potential to stop hunger today and help to avoid mass starvation in the future. Through biotechnology, scientists can enhance a crop's resistance to diseases and environmental stresses, allowing crops to be grown in relatively unproductive and unsuitable land. Recent developments in biotechnology will allow the production of more nutritious, safer, tastier, and healthier food. Advances in genetic engineering are revolutionizing the way we produce and consume food and it is quite possible that in the next decade a large percentage of the food we eat will be bioengineered. Modern Biotechnology encompass a continuously evolving methods or materials, from molecular techniques for generating energy to non-toxic cleaning products. It is that innovation which reduces waste by changing patterns of production and consumption through DNA engineering. The book Modern Biotechnology and its Applications in carries 32 chapters and covers most of the tools and technology developed by our distinguished scientists mainly focusing, how to save the planet Earth through production and productivity by applying the knowledge of modern technology. The objective of the book is to highlight the exploration and development of biotechnological tools and techniques for the social welfare.

Editor

CONTENTS

Part - II

CHAPTER – 1

Plant Tissue Culture Prospects and Limitation

Kambaska Kumar Behera
Department of Bioscience and Biotechnology, Bansathali University, Banasthali-304022 Rajasthan, India
Email : kambaska@yahoo.co.in

Summary

Plant tissue culture is the technique of growing plant cells, tissues or organs in isolated from the mother plant on artificial media. It has opened up unprecedented opportunities in many areas of basic and applied biological research. This includes techniques and methods used to research into many botanical disciplines and has several practical objectives of plant tissue culture and micropropagation for sustainable development. Indeed, the landmark publication by Gottlieb Haberlandt who is arguably referred to as the "Father of Tissue Culture", is often cited as the origin and emergence of plant tissue culture and its subsequent applications. Success of biotechnological approaches is dependent on regeneration of intact plants following genetic modification, generally by micropropagation. Potential uses of plant tissue culture in biotechnology to further our understanding of plant physiology, how plants function, resolution and expression of use full characters form in vitro to in vivo condition. Agro bacterium-mediated genetic transformation along with other modes of transformation such as microprojectile bombardment has been accomplished mainly through plant tissue culture. Incorporation of disease, stress resistance and other horticulturally important traits are a logical outcome to be anticipated in the future. Use of molecular technologies for identification of genotypes, clones and their ancestors will enable researchers and producers to verify identity and parentage of propagules which are produced by conventional or modern propagation methods. There is no doubt about applications of plant biotechnology which will emerge as the twenty first Century and continues to unfold. The present chapter gives a brief idea about the importance of this technology, types, nutritive requirements, application, advantages and limitations of plant tissue culture.

Introduction

The commercial production of *in vitro* plants is growing worldwide. Its monetary value has significantly increased over the last two decades and there is a great potential for continued further growth in both domestic and international markets (Jain, 2002). Major agronomically, medicinally and horticultural important plants such as *Dioscorea, Curcuma, Zingiber, Saccharum, Stevia, Musa, Begonia, Ficus, Anthurium, Chrysanthemum, Rosa, Saintpaulia*, and *Spathiphyllum* are being produced in the developed countries (Anonymous, 2003; Behera et al.2009). About 212.5 million plants including 157 million ornamental plants amounting to 78% of the total production were reported (Pierik, 1991a,b). The Netherlands dominates export of ornamental plants including pot plants like Begonia, Ficus, Cyclamen, Philodendron, Saintpaulia, *Spathiphyllum* and *Rhododendron* (Anonymous, 2003). About 156 ornamental genera are propagated through tissue culture in different commercial laboratories worldwide. The shares of major producers are:- The Netherlands (33%), Japan (24%), Italy (11%), USA (12%), Thailand (10%) and others (14%). The major exporting countries are:- The Netherlands (59%), Colombia (10%), Italy (16%), Israel (4%), Spain (2%), Kenya (1%) and others (18%). The four leading exporters (The Netherlands, Colombia, Italy and Israel) constitute about 80% of the world market. The share of the developing countries of Africa, Asia and Latin America are less than 20% (Rajagopalan, 2000; Schiva, 2000). Planting material of ornamental plants is in great demand for commercial production as well as for domestic gardens and landscaping. The better quality planting material is a basic need of growers for boosting productivity both in agriculturally and medicinally important plants. Chebet et al. (2003) reported the use of biotechnological approaches to improve horticultural crop production. The objective of plant propagation via tissue culture, termed micropropagation, is to propagate plants true-to-type, that is, as clones. Plants obtained from tissue culture are called microplants and can be derived from tissue cultures in three ways:

1. From pre-existing shoot buds or primordial buds (meristems) which are encouraged to grow and proliferate.
2. Following shoot morphogenesis when new shoots are induced to form unorganized tissues or directly upon explanted tissues of the mother plant.
3. Through the formation of somatic embryos which resemble the seed embryos of intact plants, and which can grow into seedlings in the same way. This process is called somatic embryogenesis.

History in Plant Tissue Culture

Haberlandt (1902) first envisioned the concept of plant tissue culture when he cultivated cells of various species on a medium containing glucose and peptone. Although he did not observe cell division, he described an approach whereby cellular physiology could be followed without the complexity of the whole level intervening. Haberlandt's pioneering research provided the groundwork for the first instances of callus induction and sustainable cell growth nearly forty years after his original experiments. Three independent and nearly simultaneous reports described callus initiation from carrots (Gautheret et al. 1939) and tobacco (White 1939) *in vitro* cultures.

Callus induction from other dicotyledonous species soon followed (Gautheret, 1983) and callus from the first monocot was reported by LaRue (1949) in Zea mays endosperm cultures. One of the reasons for their success in initiating and maintaining these cultures was the use of the plant growth regulators. The discovery of indole acetic acid (IAA) by Went (1926) provided a means to initiate callus from nontumorous tissues. Further more, the positive influence of coconut milk on cell proliferation was attributed to new class of growth regulator, kinetin and other cytokinins (Miller et al.1955). The combined use of IAA and kinetin by Skoog and Miller (1957) illustrated the synergism association with these compounds regarding the development of synchronized shoot-root plant regeneration. It should be pointed out the phrase 'plant tissue culture' is a really a misnomer in that the only tissue of plant origin that can be cultured is wound callus. In other instance of *in vitro* culture where differentiation has occurred into shoots or roots, mixed tissues are involved in the final organs produced. However, Murasigue and Skoog (1962) develop the MS media for plant tissue culture *in vitro* which is now the most widely used media in plant tissue culture.

In a historical perspective, Cocking (1983) described problems and frustrations associated for the initial attempt to isolate and culture protoplasts. However, there have been several advances in this technology since the early 1970s. This includes the first reports of regeneration from protoplasts of *Nicotiana tobaccum*. The work on *Solanum tuberosum* was also a milestone since it represented the first case of protoplast-derived plant regeneration from a major agronomic crop. Advances are occurring in several monocotyledonous species such as *Dactylis glomerata, Oryza sativa, Saccharum* species, *Triticum aestivum and Zea mays*, where plant regeneration is now possible. A survey of historical milestone in plant tissue culture is shown in Table 1.

Table 1 : Historical milestones in plant cell and tissue culture techniques.

Year	Authors	Results	Species
1892	Klercker	First attempts to isolate protoplasts mechanically.	-
1902	Haberlandt	First cultivation experiments with isolated plant cells; cell growth, but no cell division obtained.	*Tradescantia*
1904	Hanning	Establishment of embryo culture for the first time	*Cochleria raphanus*
1909	Kuster	First observation of fusing cells	-
1922	Kotte, Robins	*In vitro* cultivation of root tips, no permanent cultures obtained	*Zea, Pisum*
1924	Dieterich	Embryo rescue-	*Linum*
1925	Laibach	"artificial premature birth"	
1934	Gautheret Nobecourt	First permanent callus culture using B-vitamins and auxins	*Daucus, Nicotiana glauca* × *N. longsdorffii*
1942	Gautheret	Observation of secondary metabolites in plant callus culture	-
1946	Ball	Micropropagation: first development of stem tips and sub adjjacent regions	*Tropaeolum*
1952	Morel and Martin	Stem tips and adjacent regins: plants free of viruses.	*Dahlia*
1954	Morel *et al.*	First suspension cultures of single cells or cell aggregates. Nurse culture	*Tagetes, Nicotiana, Daucus, Picea, Phaseolus*
1955	Mothes and Kala	First reports of secondary metabolite production in liquid media	-
1956	Routien and Nickell	US patent No.2747334 for the production of substances from plant tissue culture.	*Phaseolus*
1958	Wickson and Thimann Reinert, Steward *et al.*	Establishment of axillary's branching Somatic embryogenesis in tissue cultures	*Daucus*
1959	Tukecke and Nickell	First report of large-scale (1341) culture of plant cells: carboy system	*Ginkgo,Lolium, Rosa,llex*
1960	Bergmann	Cell clones obtained from single cultured cells plated in an agar medium	*Nicotiana, Phaseolus*
1960	Jones *et al.*	Hanging drop culture in conditioned medium	*Nicotiana*
1960	Cocking	Method for obtaining large number of protoplasts from plant tissue	*Lycopersicon*
1965	Morel	Clonal multiplication of horticultural plant (orchid) through tissue droplet	*Cymbidium*
1965	Vasil and Hildebrandt	Regeneration of a plant from one single cell cultivated in a hanging droplet	*Nicotiana*

Contd.

1966	Kohlenbach	First cell division and culture of differentiated mesophyll cells.	*Macleaya*
1967	Kaul and stabe	Reports of the yields of certain of differentiated mesophyll cells.	*Ammi*
1967	Bourgin and Nisoh Guha and Maheshwari	*In vitro* production of haploid plant from immature pollen with cultured anthers.	*Nicotiana Datura*
1970	Carlson	Isolation of auxotrophic mutants from cultured cells	*Nicotiana*
1977	Noguchi *et al.*	Cultivation of tobacco cells in 200001reactors	*Nicotiana*
1978	Zenk	Manifold increase in product yields by selection over parent plant documented for a variety of plant metabolites	-
1979	Brodelius *et al.*	Alginate beads used to immobilize plant cells for biotransformation and secondary metabolite production	-
1981	Shuler	Use of hollow reactor for secondary metabolite production	-
1983	Mitsui petrochemi.	First industrial production of secondary plant products by suspension cultures	*Lithospermum*
1983	Barton Brill	Insertion of foreign genes attached to a plasmid	-
1983	Chilton	Production of transformed tobacco plants following single cell transformation	Tobacco
1985	Kohntoco	Somatic hybrids in tobacco mediated by electro fusion	Tobacco
1986	Sundberg glemelius	Somatic hybrids in Brassicaceae	Plant species of Brassicaceae.

Use of Plant Tissue Culture in Agricultural Biotechnology

Green biotechnology is biotechnology applied to agricultural processes. An example is the designing of transgenic plants to grow under specific environmental conditions or in the presence (or absence) of certain agricultural chemicals. One hope is that green biotechnology might produce more environment friendly solutions than traditional industrial agriculture. An example of this is the engineering of a plant to express a pesticide, thereby eliminating the need for external application of pesticides. Another example of this would be Bt corn, Bt brinjal. Green revolution in India is a successful example of green biotechnology, resulting in increased yield and productivity and self sufficiency in food production. Introduction of various biofertilsers and biopesticides are other examples of green biotechnology.

Biotechnology in its broadest sense means the application of all natural sciences and engineering in the direct or indirect use of living organism or parts of organisms in their natural or modified form, in an innovative manner in the production of goods and services and/or to improve existing industrial processes. The market application of the modern biotechnology techniques is typically in the general areas of human health care, agriculture and food production, industrial bio-processing and other public good and environment settings. Thus, biotechnology refers to a set of technologies that involve understanding, mapping, manipulation or change of the genetic characteristics of a living organism. Following is the application of modern biotechnology techniques in various fields ranging from agricultural application to industrial processes.

1. Plant Biotechnology- Significance in Food Production

Green biotechnology which is more commonly known as Plant Biotechnology is a rapidly expanding field within Modern biotechnology. It basically involves the introduction of foreign genes into economically important plant species, resulting in crop improvement and the production of novel products in plants. Use of environment friendly and cost effective alternatives to industrial chemicals such as bio-fuels, bio-fertilizers and bio-pesticides are not only resulting in enhanced crop output, improvement in health and safety standards, these new products are also leading to less environment pollution and use of green technology. The ever increasing demand of agricultural produce has given new impetus to research in the field and has resulted in great benefits for farmers and users alike.

The significance of the application of modern biotechnology in the area of food production and its resultant impact in terms of human health and development can not be undermined. As the world is faced with ever increasing population and more and more food shortage and regional imbalances, new technologies and techniques are being developed to enhance production and increase the shelf life of perishable items. It is in this direction that new research initiatives in the field of green biotechnology are being made to enhance productivity and nutrition value of food items. Foods produced through modern biotechnology can be categorized as follows:

1. Foods consisting of or containing living/viable organisms, e.g. maize.
2. Foods derived from or containing ingredients derived from Genetically Modified Organisms (GMOs), e.g. flour, food protein products, or oil from GM soybeans, wheat etc.
3. Foods containing single ingredients or additives produced by GM microorganisms (GMMs), e.g. colours, vitamins and essential amino acids.
4. Foods containing ingredients processed by enzymes produced through GMMs, e.g. high-fructose corn syrup produced from starch, using the enzyme glucose isomerase (product of a GMM).

The first genetically modified food item - GM food (delayed-ripening tomato) was introduced on the US market in the mid-1990s. Since then, GM strains of maize, soybean, rape and cotton have been adopted by a number of countries and marketed internationally. In addition, GM varieties of papaya, potato, rice, squash and sugar beet have been trialled or released. It is estimated that GM crops cover almost 4% of total global arable land.

The development of GM organisms has revolutionized the scenario of world food production. It has also offered the potential for increased agricultural productivity or improved nutritional value that can contribute directly to enhancing human health and development. From a health perspective, there may also be indirect benefits, such as reduced agricultural chemical use and enhanced farm income, and improved crop sustainability and food security, particularly in developing countries. While the introduction of GM crops has undoubtedly changed the agricultural scenario and led to significant impact on human development, it has also raised various social, cultural and ethical issues and reluctance on the part of various countries and governments to accept the GM food even in times of grave needs such as famine and drought. While some countries have established pre-market regulatory standards for risk assessment of each and every food item before being launched in the market for application, there may be a case of a consistent and uniform international regulatory structure so as to ensure that such food items conform to a set of standards which are fair and equitable. This will also help quell a number of misgivings and doubts in the minds of countries yet to benefit from these food items. For example, recent non-approval of Bt brinjal in India while pesticide residue load in the cultivated egg plant creates more health problems. Today plant biotechnology encompasses the following main areas of research and application.

Status of Plant Biotechnology in India

India is the largest group of tissue culture scientists in the world .Mostly the research is directed to wards the development of improved for agriculture, horticulture and forestry using tissue culture method. Another use of tissue culture methods in some labotaries involves studies in the areas of developmental biology, biochemistry, physiology, genetics and molecular biology for basics research .The Government of India set up the National Biotechnology Board in1982 by realizing and recognizing the importance and potential of biotechnology for sustainable development (Sharma, 2001). The Department of Biotechnology supported the establishment of seven Centres for Plant Molecular Biology throughout the country. Today, there are about 50 public research units in India using tools of modern biotechnology for agriculture, especially techniques for cell and tissue culture. The Indian government allocates an estimated US$15 million annually on plant biotechnology research, while the private sector contributes about US$10 million (Huang et al.2002).

The Indian plant biotechnology research agenda is dominated by tissue culture and micro propagation, exploitation of heterosis vigor, development of new hybrids and planting material with desirable traits, and the genetic enhancement of important crops (Sharma, 2001). India's success in tissue culture and micro propagation techniques led to the development of Micro propagation. Technology Parks, which serve as a platform for the effective transfer of technology to entrepreneurs, including training and the demonstration of technology for mass multiplication of horticulture crops and tress. Considerable research progress has been made with cardamom and vanilla, both important crops. Cardamom yields have increased 40% using tissue culture plants.

A number of multi-institutional projects have also been launched, including the development of transgenic for resistance to geminiviruses in cotton, mungbean, and tomato, resistance to rice tungro disease, resistance to bollworms in cotton, development of a nutritionally enhanced potato with a balanced amino acid composition, and development of molecular methods for heterosis breeding. The first GM crop to be released for commercial cultivation is Bt cotton, developed by the Maharashtra Seed Company (Mahyco) in partnership with Monsanto. The approval, which was given in March 2002, came after several years of field trials following the biosafety procedures laid down by the government. Three cotton hybrids were granted permission for field sowing in six states for three years. For the first season, farmer demand for Bt cotton seed was very high; it is estimated that 44,500 hectares of certified Bt cotton were planted by nearly 55,000 farmers. Other transgenic crops that are awaiting approval for commercial cultivation include transgenic herbicide-tolerant mustard hybrids and nutritionally enhanced potato varieties.

Table 2 : List of Plant Species which have been perfected for Large Scale Propagation

Plant category	Plants
Fruits	Banana, Grapes, Pineapple, Strawberry, Sapota
Cash crops	Sugarcane, Potato
Spices	Turmeric, Ginger, Vanilla, Large Cardamom, Small Cardamom
Medicinal plants	Aloe vera, Geranium, Stevia, Patchouli, Neem
Ornamentals	Gerbera, Carnation, Anthurium, Lily, Syngonium, Cymbidium
Trees	Teak, White teak, Bamboo, Eucalyptus, populus

Plant Production Biotechnology

There is scarcely any aspect of plant production that will not undergo profound changes as a result of the application of biotechnology. Commercial applications of plant genetic engineering have not yet occurred in accelerated speed. At the present time, more traditional aspects of biotechnology such as tissue culture have had an important impact, especially in the acceleration of the breeding process for new varieties and in the multiplication of disease-free seed material.

Realizing the potential of plant tissue culture technology in revolutionizing the commercial agriculture sector by enabling mass propagation of elite, high yielding and disease free plants throughout the year, the Department of Biotechnology (DBT) has identified it as a priority area and initiated a number of programmes aimed at development and commercialization of the technology in an integrated manner. A number of research and development projects in various research institutes and universities have been supported for perfecting protocols of important plant species namely forest trees, horticulture crops and plantation crops both Govt. and private partnership (Table-2 and 4).

Govil and Gupta (1994) reported 21 Indian commercial units in production in 1994 with targeted production capacity of 80 m plants and 14 units in the pipeline with annual production capacity of 60 m plants. Four units had annual targeted production capacity of over 10 m, eight with capacity of 1-5 m plants while nine units with targeted production of 5 m plants each were in the pipeline. Most of the units covered in this article, details of production capacity of 30 Indian commercial units are available. An analysis shows that there are 4 major units with production capacity varying from 15 to 25 m, 3 major units with capacity of 10 m, 12 units with production capacity of 5 to 6.25m, 4 units with production ranging from 1.5 to 3 m, 7 units including 2 pilot plants with production capacity of 1 m, totalling to 190 m plants per annum. A perusal of plants micropropagated by Indian commercial units (Table 3) indicates that plants belonging to the horticulture segment of the ornamental industry are the major items being produced. This is also the international trend as shown earlier (Govil and Gupta, 1994). It is clear that of 14 units engaged in micropropagation of fruit crops, 12 are producing banana, 5 are aseptically multiplying strawberry, while 3 each are engaged in papaya and pineapple production. Only 6 units are producing 7 forest tree species 72 leaving enough scope for expansion. Twenty four units are producing a large number of ornamentals. A few vegetable crops and some plantation crops like cardamom, vanilla, sugarcane and ginger are also being multiplied (Table 3). Indian units need to strike an intelligent balance between product diversity as well as quantity and should try to introduce novelty and explore commercial potential of endemic ornamentals as new entrants. In fact, production by micropropagation units is market-driven and based on advance orders. The production of large number of ornamental and horticultural plants reflects their demand by consumers as well as their willingness to pay the higher cost. Globally, the floriculture market alone is supposed to be of Rs. 750 billion level. While considering cost benefit ratio for identification of products, social and environmental aspects whose monetary value cannot be estimated, should also be given due consideration. With the liberalised industrial policy, incentives provided by the Government and demonstration efforts made by the commercial units, tissue culture products of our units are fast capturing domestic market as well as increasingly becoming popular in international markets. Looking at the brighter prospects of the industry, Government of India had fixed export target to Rs. 1300 million for cutflowers, seeds, live plants and tissue culture products in 1995-96 (Kumar, 1994).

Table 3 : List of Current Commercial Micropropagated Tissue Culture Units in India.

Sl.no.	Name of The Unit	Fruit Crops
1	Harrisons Malayalam Ltd.	Banana, *Citrus, Grapes, Papaya, Pomegranate, Strawberry
2	Godrej Plant Biotech Ltd.	Banana, Pineapple, Strawberry
3	IAHS	Banana, Cherryapple, Peach, Pineapple, Strawberry
4	Nath Seeds	Banana
	Bio Tissue Labs	Banana
	Microplantae,	Banana
	Kumar Gene Tech,	Banana
	SIV Industries Ltd.	Banana
	Maharashtra State Seed Corporation,	Banana
	National Agricultural &Scientific Research Foundation	Banana
5	A.V. Thomas & Co	Banana, Pineapple
6	SPIC	Banana, Datepalm, Mango, Papaya (Female)
7	Enzo-chem	Papaya
8	Agrigene International	Apple
9	Microplantae	Strawberry
	Indrayani Biotech	Strawberry
	Saubhagya Kalpataru Pvt. Ltd.	Banana, Strawberry
Forest Trees		
1	Harrisons Malayalam Ltd.,.,.	Eucalyptus
2	Tata Tea Ltd	Eucalyptus
3	SIV Industries Ltd	Eucalyptus
4	Thapar Corp. R&D Centre,	Eucalyptus
5	IAHS	Teak, Kadamb
6	SPIC	*Calistemon*, *Eucalyptus*, Jackfruit, *Simaruba*
7	Godrej Plant Biotech	*Salavadora*
8	Saubhagya Kalpataru Pvt. Ltd.	Bamboo
Ornamentals		
	Name of the Unit	**Foliage Plants**
1	Harrisons Malayalam Ltd	*Aglaonema*, *Calathea*, *Cordyline*, *Diffenbachia*,*Ficus*, *Philodendron*, *Spathiphyllum*, *Syngonium*

Contd.

	Name of the Unit	Flowering Plants
1		*Alstroemeria*, *Anthurium*, *Callalily*, Carnation,*Cestrum*, *Chrysanthemum*, *Gerbera*, *Impatiens*,*Limonium*, Rose (Miniature)
2	IAHS	*Alpinia*, *Anthurium*, *Calethea*, *Callalily*,*Chrysanthemum*, *Cordyline*, *Croton*, *Diffenbachia*,*Dracaena*, *Eryngium*, *Ficus*, *Freesia*, *Gardenia*,*Geranium*, *Gerbera*, *Gladiolus*, *Hydrangea*, *Iris*,*Impatiens*, *Lilies*, *Maranta*, Orchids, *Philodendron*, Rose, *Spathiphyllum*, *Syngonium*
3	SPIC,	*Anthurium*, SPIC *Anthurium*, *Calathea*, *Cordyline*, *Gerbera*, *Maranta*, Orchids, Rose, *Syngonium*
4	Godrej Plant Biotek	*Anthurium*, *Calathea*, *Ficus*, Lily, OrnamentalBamboo, *Spathiphyllum*, *Syngonium*
5	Agrigene International	*Ficus*, *Spathiphyllum*
	A.V. Thomas & Co.	*Chrysanthemum*, *Gerbera*, Lily, Orchids, Rose
6	Cadilla Laboratories Ltd	*Calathea*, Lily, *Spathiphyllum*, *Syngonium*
7	Kalpataru Irrigation Systems Ltd.	Lily
8	Rallis India	Carnation, Orchids
	COSTFORD promoted Unit	***Dendrobium* (Orchid), *Phaelenopsis***
1	Phytoclone	*Dendrobium*
2	Pocha seed,	Ornamentals
3	Khoday Biotek	Ornamentals
4	Padamjee Paper Mills, ,	Foliage plants
5	Florida Plant Cultures	Foliage plants
	Green Laboratories	Foliage plants
6	National Agricultural & Scientific Research Foundation	*Anthurium*, *Spathiphyllum*
7	Raj Agrobase Pvt. Ltd.	Orchids
8	SIV Industries Ltd.	Orchids
9	Beena Nursery	Orchids
10	Indrayani Biotech	Rose
11	Kumar Gene Tech.	Flowering plants
12	Microplantae	Flowering plants
13	Khushboo Kenibreed Plants	Flowering plants
14	Kenibreed Plants	*Cymbidium*, *Oncidium*, *Phalaenopsis*, *Vanda*
15	Saubhagya Kalpataru Pvt. Ltd	Rose
16	SIV Industries Ltd.	*Chrysanthemum*, *Philodendron*

Contd.

	Name of the Unit	Vegetable Crops
1	Harrisons Malayalam Ltd.	Potato, *Asparagus*
2	SPIC	Potato
3	Proagro	*Brassica napus, Brassica juncea*, *Lycopersicon esculentum* (tomato)
4	Frontier Biotech	Microtubers of Potato (*Solanum tuberosum*)
5	Hindustan Lever Ltd.	Microtubers of Potato (*Solanum tuberosum*
	AgroIndia Biotechnologies	Microtubers of Potato (*Solanum tuberosum*
6	Kalyani Agrotech	Microtubers of Potato (*Solanum tuberosum*
7	IAHS	*Asparagus*, *Cauliflower, Cabbage*, Tomato
8	Agrigene International	*Asparagus officinalis* (exotic vegetable)
	Name of the Unit	**Plantation Crops**
1	Harrisons Malayalam Ltd.	Cardamom, Sugarcane, Vanilla
2	IAHS	Cardamom, Coffee, Ginger, *Miscanthus*, Sugarcane
3	Tata Tea Ltd.	Coffee, Tea
4	SPIC	Cardamom, Vanilla
	A.V. Thomas & Co	Cardamom, Ginger, Turmeric, Vanilla, Black Pepper, Tea
5	Bush Boake Allen	Vanilla
6	Dhampur Sugar Mills,., ,	Sugarcane
7	Frontier Biotech	Sugarcane
8	Saubhagya Kalpataru Pvt. Lt	Sugarcane
9	Western India Tissue Tech Ltd.,	Sugarcane
10	Agri India Biotechnologies	Sugarcane
11	EID Parry Ltd.	Sugarcane
12	Rallis India Ltd	Sugarcane
13	Vasant Dada Sugar Institute	Sugarcane
14	Kalpataru Irrigation Systems Ltd.	Coffee, Tamarind
15	National Agricultural & Scientific Research Foundation	Cardamom

Recommendations

The widespread application of *in vitro* propagation depends on its cost-competitiveness and is profitable only when there is an associated advantage over conventional propagation methods. We should set up modest, commercially viable units at appropriate sites taking into account cost of land, construction, availability of infrastructure such as water, power, sanitation, labour and distance from airport. We should encourage horizontal technology transfer from viable operating indigenous units rather than purchasing technology from abroad. Import of equipment needs to be minimised and costly consumables like agar-agar should be substituted by 'ragi' or 'maida' and sucrose by Indian table sugar. In-house R&D units and interaction with nearby R&D institutes need to be strengthened to optimise production and solve contamination, hardening and related problems. The units should carefully select plant material taking into consideration availability of stocks, reproducible protocol with rapid multiplication rates, demand in domestic and international markets as well as build good market reputation by delivering quality products on time. The micropropagation industry is particularly well suited for a country like India as it is environment friendly and labour-intensive. It can really solve our unemployment problem to a reasonable extent, only if planned judiciously.

Blessed with diverse agro climatic zones and availability of both skilled manpower as well as labour at cost effective rates, combined with liberalised Government policies and incentives provided, Indian tissue culture industry, though a little late starter, is all set to have its presence felt on the international scenario and make a dent in the national economy by earning considerable foreign exchange on a recurring basis.

Table 4 : Distribution of Tissue Culture units vis-a-vis the demand in India

Region	Demand (In thousand nos.)	No of Units
North	10108	5
East	8162	3
West	11129	16
South	39182	21

The table indicates a need for setting up additional units in the northern and eastern regions to cater to the increasing demand for tissue cultured medicinal and horticulture plants. The demand projections are highly conservative. Considering the vast potential that TCPs offer for supplementing and replacing the conventionally propagated plants for improving agricultural productivity, the potential demand is expected to be significantly higher than the projected figures. This potential can be tapped by creating large-scale awareness about the benefits of TCPs and reduction of the cost of production.

Initiating Plant Tissue Cultures

Explants : Tissue cultures are started from pieces of whole plants. The small organs or pieces of tissue that are used are called explants. The part of the plant (the stock plant or mother plant) from which explants are obtained, depends on:

1. The kind of culture to be initiated;

2. The purpose of the proposed culture;
3. The plant species to be used.

Explants can therefore be of many different kinds. The correct choice of explants material can have an important effect on the success of tissue culture. Plants growing in the external environment are invariably contaminated with micro-organisms and pests. These contaminants are mainly confined to the outer surfaces of the plant, although, some microbes and viruses may be systemic within the tissues (Cassells, 1984). Because they are started from small explants and must be grown on nutritive media which is also favourable for the growth of microorganisms. The plant tissue cultures must usually be established and maintained in aseptic conditions. Most kinds of microbial organisms in, particular bacteria and fungi, compete adversely with plant material growing *in vitro*. Therefore, as far as possible, explants must be free from microbial contaminants when they are first placed on a nutrient medium. This usually involves growing stock plants in ways that will minimize infection. So the plant material should be treated with disinfecting chemicals to kill superficial microbes, and sterilizing the tools used for dissection, the vessels and media in which cultures are grown (Cassells and Doyle, 2005). Some kinds of plants can, however, be micropropagated in non-sterile environments.

Isolation and Incubation

The work of isolating and transferring cultured plant material is usually performed in special rooms or inside hoods from which microorganisms can be excluded. Hoods used for isolation can be placed in a draught-free part of a general laboratory, but are much better situated in a special inoculation or transfer room reserved for the purpose. Cultures, once initiated, are placed in incubators or growth rooms where light, temperature and humidity can be properly controlled. The rate of growth of a culture will depend on these factors.

The Cultural Environment

Plant cultures are commenced by placing one or more explants into a pre-sterilized container of sterile nutrient medium. Some explants may fail to grow, or may die, due to microbial contamination. To ensure the survival of an adequate number, it therefore is to initiate several cultures at the same time and each being started from an identical organ or piece of tissue. Explants taken from stock plants at different times of the year may not give reproducible results in tissue culture. This may be due to variation in the level of external contaminants or because of seasonal changes in endogenous (internal) growth regulator levels in the stock plant.

Most *in vitro* studies focus on manipulating of the medium. By varying the nutrient salts, gelling agents and especially plant growth regulators it has been possible to grow callus, stimulate axillary shoot proliferation, rooting, induce shoot organogenesis and somatic embryogenesis. However, *in vitro* growth and morphogenesis are also governed by properties of the explants. It is well known that different genotypes will often not respond in the same way when cultured on the same medium. This has required countless empirical studies on medium optimization for different species or even cultivars of the same species. Within a plant during its growth and development, different parts of the genetic code are expressed differentially in different cells. When different genes are "turned on" or "turned off" in a cell or even a few cells, it can have profound influences on plant growth, development and phase change (Hackett and Murray, 1997). This differential expression is referred to as epigenetic variation. Control of epigenetic variation is the focus of plant propagation (Hartmann et al. 2002) and tissue culture.

Types of Culture Media used in Plant Tissue culture

It is obvious that process operating condition like optimum media composition and physiochemical parameters play a vital role in the improvement of *in vitro* regeneration and multiplications of the explants that have a great influence on the overall process economics. MS is a universal medium for plant tissue culture but now days there are other media which has fulfilled nutritional requirements for growth and multiplication. The rate at which cultures grow and produce shoots during micropropagation can be influenced by the physical nature of the medium. For many purposes it is convenient to make media semi-solid by incorporating a gel such as agar.

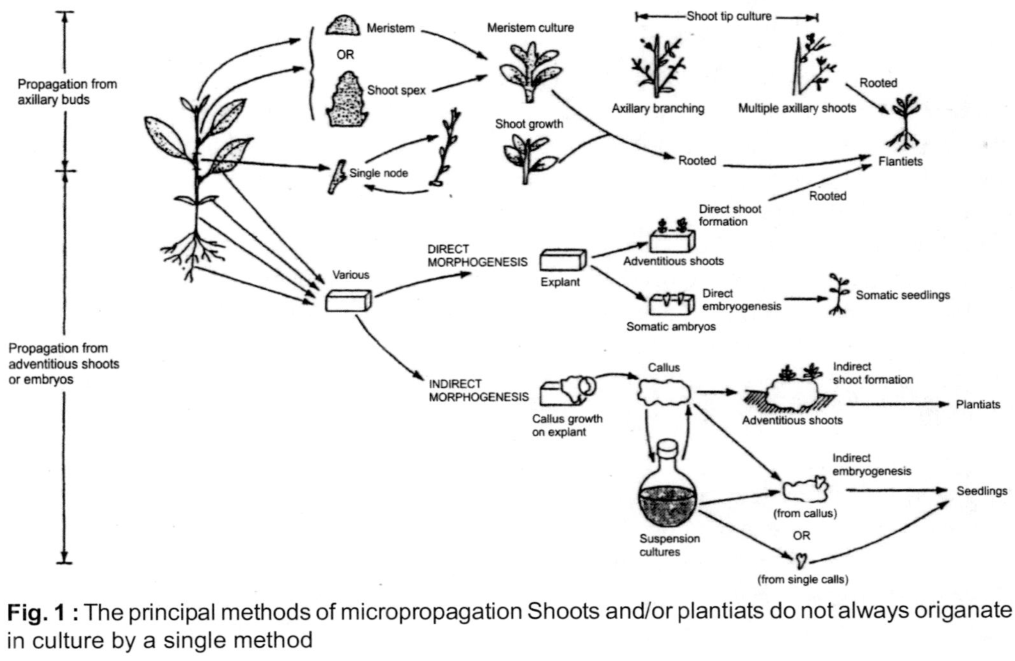

Fig. 1 : The principal methods of micropropagation Shoots and/or plantiats do not always origanate in culture by a single method

Media : Plant material will only grow *in vitro* when provided with specialized media. A medium usually consists of a solution of salts supplying the major and minor elements necessary for the growth of whole plants, together with:-

1. Various vitamins (optional)
2. Various amino acids (optional)
3. An energy source (usually sucrose)
4. And plant growth regulator

Growth and development of plant cultures usually also depends on the addition of plant growth regulators to the medium. Plant growth regulators are compounds, which are used at very low concentration and capable of modifying growth or plant morphogenesis. Many workers define a medium as a completed mixture of nutrients and growth regulators. This is a rather inflexible method, as growth regulators frequently need to be altered according to the variety of plant, or at different stages of culture, whilst the basic medium can stay unchanged. It is therefore recommended that nutritional and regulatory components should be listed separately. Plant material can be cultured either in a liquid medium or on a medium that has been partially solidified with a gelling agent (e.g., Agar). The method employed will depend on the type of culture and its objective.

The most commonly used medium is the formulation of Murashige and Skoog (1962). This medium was developed for optimal growth of tobacco callus and the development involved a large number of dose-response curves for the various essential minerals. Plant culture media are usually pH sensitive and started at pH 5.4-5.8.

Solidified media : It is the media where gelling agents were added, so that they have become semi-solid and are widely used for explant establishment. They are also employed for much routine culture of callus or plant organs (including micropropagation), and for the long-term maintenance of cultures. Agar is the most common solidifying agent, but a gellan gum is also widely used. Cultures grown on solid media are kept static. They require only simple containers of glass or plastic, which occupy little space. Only the lower surface of the explant, organ or tissue is in contact with the medium. This means that as growth proceeds there may be gradients in nutrients, growth factors and the waste products of metabolism, between the medium and the tissues. Gaseous diffusion into and out of the cells at the base of the organ or tissue may also be restricted by the surrounding medium.

Liquid media : Liquid media are essential for suspension cultures, and are preferred for critical experiments on the nutrition, growth and cell differentiation in callus tissues. They are also used in some micropropagation work. Very small organs (e.g. anthers) are often floated on the top of liquid medium and plant cells or protoplasts can be cultured in very shallow layers of static liquid, providing there is sufficient gaseous diffusion. Larger organs such as shoots (e.g. proliferating shoots of shoot cultures) can also often be grown satisfactorily in a shallow layer of nonagitated liquid where part of the organ protrudes above the surface. However, some method of support is necessary for small organs or small pieces of tissue, which would otherwise sink below the surface of a static liquid medium, or they will die for lack of aeration. Systems of support which have been found to be effective and which can be used instead of agar solidified media. Many tissues and organs, small and large, also grow well unsupported in a liquid medium, providing it is aerated by shaking or moving. Some kind of agitation is essential for suspension cultures to prevent cells and cell aggregates settling to the bottom of the flask. Other purposes served by agitation include: the provision of increased aeration, the reduction of plant polarity, the uniform distribution of nutrients and the dilution of toxic explant exudates (Lim-Ho, 1982). There are several alternative techniques. Plant cell suspensions can be cultured very satisfactorily when totally immersed in a liquid culture medium, providing it is shaken (by a rotary or reciprocating shaking machine) or stirred (e.g. by a magnetic stirrer) to ensure adequate aeration. This method may also be used for culturing organs of some plants (e.g. proliferating shoot cultures), but the fragmentation, which occurs, can be disadvantageous. Periodic immersion may be achieved by growing cultured material in tubes or flasks of liquid medium which are rotated slowly. Steward and Shantz (1956) devised so-called 'nipple flasks' for this purpose which had several side-arms. They were fixed to a wooden wheel, which was rotated so that tissue in the arms of each flask was alternately bathed in medium and drained or exposed to the air (Fig. 1.1). This technique ensured that callus tissue for which they were used was well aerated. The medium usually became turbid as cells dissociated from the callus and started a cell suspension. Flasks of this sort are seldom used to-day because of their cost. A similar alternating exposure can be achieved by placing calluses in vessels, which are rotated slowly. An alternative to the costly rotating systems to achieve periodic immersion of the cultures, is the increasingly popular temporary immersion system in which static vessels are periodically or temporarily flooded with culture medium (Teisson and Alvard, 1995). Medium is pumped from a reservoir container into the culture vessel for experimentally determined time intervals repeated over a 24 hour cycle. This system prevents anoxia and has the advantage that the medium can easily be changed in the reservoir.

Liquid medium in flasks or column bio-reactors (fermentors) can be circulated and at the same time aerated, by the introduction of sterile air. Shearing forces within air-lift reactors are much less than in mechanically-stirred vessels so that plant cell suspensions suffer less damage. Bio-reactors are used in the pharmaceutical industry to produce high value plant secondary products and to carry out substrate conversions. Low cost bio-reactors developed for micropropagation have been described in detail in Hvoslef-Eide and Preil (2005) Rather than immersing callus or organ cultures, liquid medium may be slowly dripped onto the growing tissues or applied as a mist and afterwards the liquid drained or pumped away for recirculation (Weathers and Giles, 1987). A particular advantage of this technique is the ability to grow cultures in a constant and non-depleted medium; nutrients can be varied frequently and rapidly and their availability controlled by altering either concentration or flow rate. Toxic metabolites, which in a closed container might accumulate and inhibit growth, can be removed continuously. As complicated apparatus is needed, the method has not been widely used. The relative merits of solid and liquid media are:-

Advantage of Solid Media

1. Small explants are easily seen and recovered
2. Explants retain the same orientation throughout culture
3. Plant material is held above the medium so that no special means of aerating the culture is used
4. Shoots and roots grow in a more orderly fashion because the medium is stationary.

Disadvantages of Semi-solid Media.

1. Some agars contain inhibitory substances which may prevent morphogenesis in certain cultures (Powell and Uhrig, 1987), rates of growth can be slow, toxic exudates from explants do not diffuse away quickly.
2. Poor oxygen diffusion to roots developing in the medium: Low oxygen tensions will affect both root growth and function and is probably one of the reasons that plantlets from culture often show lesions in root functioning after transplanting, with potentially very serious consequences for plant water relations and the quality of plants produced (Verslues et al.1998).
3. Gel from solid media adhering to roots may cause problems for propagators when they are transferring plantlets to soil.
4. Use of semi-solid media increases the time taken to clean containers and glassware for re-use. Vessels may need to be autoclaved to re-melt the agar scrubbed and hand rinsed before being placed in a dishwasher.

Advantage of Liquid Media

1. Their use often results in faster rates of growth than are possible on semi-solid media. This is because a greater surface area of the explant is in contact with a liquid, and, providing the medium is agitated, the diffusion gradients for nutrients and for gases between it and the explant are reduced. These two factors combined, enable a more efficient uptake of nutrients and growth regulators.

2. In addition, toxic metabolites, which may accumulate in the vicinity of the tissue are effectively dispersed.

3. The partial pressure of oxygen in the root tissues will be increased. Because plant material must be supplied with oxygen, the uses to which a static liquid medium can be put are very limited. Aeration may not be a problem in cultures contained in very small volumes of liquid (e.g. hanging drops), or in those plant parts submersed in shallow layers of liquid medium. A supply of oxygen for submerged cells, tissues or organs in large volumes of liquid media is normally obtained by shaking, rotating the containers, stirring the medium, or by introducing a flow of sterile air (or a controlled mixture of gases) into the culture vessel.

4. There may also be direct effects of low oxygen tensions on membrane functioning and on water and ion uptake capacity (Jackson et al. 2003). Small organs or small pieces of tissue can be made to float more effectively on liquid media if its density is increased by the addition of Ficoll.

Disadvantage of Liquid media

1. Maize cells in suspension culture produce a mucilage, which interferes with cell growth: the formation of a very large amount of mucilage similarly hampered the culture of *Phragmites australis* embryogenic callus (Straub et al. 1988).

2. The motion within liquid cultures may damage delicate tissue. The growth of 'meristemoidlike' green globular bodies (produced from *Asplenium nidus* rhizome segments) was unsuccessful on rotated or shaken liquid medium. Higuchi and Amaki (1989) suggested that this might have been due to the meristems on the surface of the globular bodies suffering from collision damage.

3. One serious disadvantage of using liquid media for shoot growth and multiplication is that shoots, which are perpetually submerged in liquid cultures, may have a water-soaked appearance and will then be useless for micropropagation.

Problems of Establishment of Culture

Phenolic Oxidation

Some plants, particularly tropical species, contain high concentrations of phenolic substances that are oxidized when cells are wounded or senescent. Isolated tissue then becomes brown or black and fails to grow. The prevention of blackening, which can be a serious problem in tissue culture and can be avoided by adding some supplements like Vitamin C, and other antioxidants.

Minimum Inoculation Density

Certain essential substances can pass out of plant cells by diffusion. Substances known to be released into the medium by this means include alkaloids, amino acids, enzymes, growth substances and vitamins (Street, 1969). The loss is of no consequence when there is a large cluster of cells growing in close proximity or where the ratio of plant material to medium is high. However, when cells are inoculated onto an ordinary growth medium at a low population density, the concentration of essential substances in the cells and in the medium can become inadequate for the survival of the culture. For successful culture initiation, there is thus a minimum size of explants or quantity of separated cells or protoplasts per unit culture volume. Inoculation density also affects the initial rate of growth *in vitro*. Large explants generally survive more frequently and grow more rapidly at the outset than very small ones. In practice, minimum inoculation density varies according to the genotype of plant being cultured. For commencing suspension cultures it is commonly about 1-1.5 x 104 cells/ml. The minimum cell density phenomenon is sometimes called a 'feeder effect' because deficiencies can often be made up by the presence of other cells growing nearby. Suspension cultures can be started from a low density of inoculums by 'conditioning' a freshly prepared medium - i.e. allowing products to diffuse into it from a medium in which another culture is growing actively, or adding a quantity of filter-sterilized medium which has previously supported another culture. It is possible to overcome the deficiencies of plant cells at low starting densities by adding small amounts of known chemicals to a medium. For example, Kao and Michayluk (1975) have shown that *Vicia hajastana* cells or protoplasts can be cultured from very small initial inoculum or even from individual cells. A standard culture medium was supplemented with growth regulators, several organic acids, additional sugars (apart from sucrose and glucose), and in particular, casein hydrolysate (casamino acids) and coconut milk. There is often a maximum as well as a minimum plating or inoculation density for plant cells or protoplasts. In a few cases the effective range has been found to be quite narrow.

Patterns of Growth and Differentiation

A typical un-organized plant callus, initiated from a new explant or a piece of a previously-established culture, has three stages of development, namely:

1. The induction of cell division.
2. A period of active cell division during which differentiated cells lose any specialized features they may have acquired and become dedifferentiated.
3. A period when cell division slows down or ceases and when, within the callus, there is increasing cellular differentiation.

These phases are similarly reproduced by cell suspensions grown in a finite volume of medium (a batch culture), where according to a variety of different parameters that can be used to measure growth (e.g. cell number, cell dry weight, total DNA content) and S-shaped growth curve is generally obtained.

The growth phases are:

1. A lag phase
2. A period of exponential and then linear growth (Log phase)
3. A stationary phase when growth comes to a halt (Stationary phase)
4. A period when the rate of growth declines (Death phase)

Some differentiation of cells may occur in cell cultures during the period of slowed and stationary growth, but generally it is less marked and less complete than that which occurs in callus cultures. Cultures cannot be maintained in stationary phase for long periods. Cells begin to die and, as their contents enter the nutrient medium, death of the whole culture accelerates. Somewhat similar patterns of growth also occur in cultures of organized structures. These also cease growth and become moribund as the components of the medium become exhausted.

Subculturing

Once a particular kind of organized or unorganized growth has been started *in vitro*, it will usually continue if callus cultures, suspension cultures, or cultures of indeterminate organs are divided to provide new explants for culture initiation on fresh medium. Subculturing often becomes imperative when the density of cells, tissue or organs becomes excessive; to increase the volume of a culture; or to increase the number of organs (e.g. shoots or somatic embryos) for micropropagation. The period from the initiation of a culture or a subculture to the time of its transfer is sometimes called a passage. The first passage is that in which the original explant or inoculum is introduced. Suspensions regularly subcultured at the end of the period of exponential growth can often be propagated over many passages. However, many cultures reach a peak of cell aggregation at this time and aggregation often becomes progressively more pronounced in subsequent passages (Street, 1977). Subculture is therefore more conveniently carried out during the stationary phase when cell aggregation is least pronounced. Rapid rates of plant propagation depend on the ability to subculture shoots from proliferating shoot or node cultures, from cultures giving direct shoot regeneration, or callus or suspensions capable of reliable shoot or embryo regeneration. A further reason for transfer, or subculture, is that the growth of plant material in a closed vessel eventually leads to the accumulation of toxic metabolites and the exhaustion of the medium, or to its drying out. Thus, even to maintain the culture, all or part of it must be transferred onto fresh medium. Callus subcultures are usually initiated by moving a fragment of the initial callus (an inoculum) to fresh medium in another vessel. Shoot cultures are subcultured by segmenting individual shoots or shoot clusters. The interval between subcultures depends on the rate at which a culture has grown: at 25°C, sub-culturing is typically required every 4-6 weeks. In the early stages of callus growth, it may be convenient to transfer the whole piece of tissue to fresh medium, but a more established culture will need to be divided and only small selected portions used as inocula. Regrowth depends on the transfer of healthy tissues. Decontamination procedures are theoretically no longer necessary during sub culturing, although sterile transfer procedures must still be used. However, when using shoot or node cultures for micropropagation, some laboratories do re-sterilize plant material at this stage as a precaution against the spread of contaminants. Cultures which are obviously infected with micro-organisms should not be used for sub culturing and should be autoclaved before disposal.

Subculturing Hazards

There are several hazards in sub culturing. This may cause the characteristics of the culture to be altered and may mean that some of the plants regenerated from the culture will not be the same as the parent plant. Cultures may also inexplicably decline in vigour after a number of passages, so that further subculture becomes impossible. Several kinds of callus may arise from the initial explant, each with different morphogenic potential. Strains of callus tissue capable of giving rise to somatic embryos and others without this capability can, for instance, arise simultaneously from the culture of grass and cereal seed embryos. Careful selection of the correct strain is therefore necessary if cultures capable of producing somatic embryos are ultimately required. Timing of the transfer may also be important, because if left alone for some while, non-embryogenic callus may grow from the original explant at the expense of the competent tissue, which will then be obscured or lost. Although subculturing can often be continued over many months without adverse effects becoming apparent, cultures of most unorganised cells and of some organised structures can accumulate cells that are genetically changed.

Acclimatization and Field Establishment of the *in vitro* Plant

Production of plantlets with profuse rooting *in vitro* is important for successful establishment of regenerated plants in soil (Kumar and Nair, 1979; Ohyama, 1970).Rooted micropropagules were removed from culture tubes and the roots were washed under running tap water to remove agar. The plantlets were transferred to 2.5 cm sterile poly pots (each small plastic cup containing one plantlet) containing 200-250 gm vermiculite inside a growth chamber at 28^0C and 70-80% RH. After 3 weeks they were transplanted to earthen pots containing a mixture of soil + sand + manure in a 1: 1: 1 ratio and kept under a shade house for a period of 3 weeks for acclimatization. The potted plants were generally irrigated with Hoagland's solution (50 ml per pot) every 3 days for a period of 3 weeks. Survival rate of the plantlets were recorded after 3 weeks (Behera et al. 2009). Plants started producing fresh shoots and roots after one week of transplanting. Later they were transferred to the field condition.

Micropropagation

It is the practice of rapidly multiplying stock plant material to produce a large number of progeny plants by using modern plant tissue culture methods. It has been used to multiply novel plants, such as those have been genetically modified or bred through conventional plant breeding methods. It is also used to provide a sufficient number of plantlets for planting from a stock plant which does not produce seeds, or does not respond well to vegetative propagation.

The process begins with the selection of plant material to be propagated. The clean stock material for propagation is important in the production of the healthiest plants. Once the plant material is chosen for culture, the collection of explants begins and is dependent on the type of tissue to be used, and can include stem tips, anthers, petals, pollen and others.

Complete propagules of *Curcuma longa* L. in plastic pot after hardening

Hardening of rooted plant lets of *Saccharum officinarum* L. in plastic trays

The explants material is then sterilized on their surfaces, usually in multiple courses of bleach and alcohol washes and finally rinsed in sterilized water. This small portion of plant tissue is placed on a growth medium, typically containing sucrose as an energy source and with one or more plant growth regulators (plant hormones). Usually the medium is thickened with agar to create a gel which supports the explants during growth. Some plants are easy to grow on simple mediums but other plants need more complicated mediums before they will successively grow. The medium is sterilized during preparation to prevent fungal and bacterial contaminations that can outgrow and smother the explants. Autoclaves and filter sterilization are used to remove potential contaminates. The plant tissue should now begin to grow and differentiate into new tissues. For example, media containing cytokinins are used to create branched shoots from plant buds.

Multiplication : Multiplication is the taking of tissue samples produced during the first stage and increasing there number. Following the successful introduction and growth of plant tissue, the establishment stage is followed by multiplication.

In vitro plantlets of *Dioscorea alata* L.ready for hardening

In vitro plantlets of *Curcuma longa* L. ready for hardening

Through repeated cycles of this process, a single explant sample may be increased from one to hundreds or thousands of plants. Depending on the type of tissue grown, multiplication can involve different methods and mediums. If the plant material grown is callus tissue, it can be placed in a blender and cut into smaller pieces and re-cultured on the same type of culture medium to grow more callus tissue. If the tissue is grown as small plants called plantlets, hormones are often added that cause the plantlets to produce many small offshoots that can be separated and sub-cultured.

Pre-transplant

This stage involves treating the plantlets/shoots produced to encourage root growth and "hardening." It is performed *in vitro*, or in a sterile "test tube" environment. Root growth does not always occur in the earlier stages in plant cell culture and it is of course a requirement for successful plant growth after the micropropagation procedure. It is often performed *in vitro* by transferring the plantlets to a growth medium containing auxins which stimulates root initiation. The pre-transplant stage is not always performed. Some plants are micropropagated and grown in culture.

"Hardening" refers to the preparation of the plants for a natural growth environment. Until this stage, the plantlets have been grown in "ideal" conditions, designed to encourage rapid growth. Due to lack of necessity, the plants are likely to be highly susceptible to disease and often do not have fully functional dermal coverings and will be inefficient in their use of water and energy. *In vitro* conditions are high in humidity and plants grown under this condition and do not form a working cuticle and stomata that keep the plant from drying out, when taken out of culture the plantlets need time to adjust to more natural environmental conditions. Hardening typically involves slowly weaning the plantlets from a high-humidity, low light, warm environment. This is done by moving the plants to a location high in humidity, such as a green house with regular mist watering.

Transfer from culture : In the final stage of plant micropropagation, the plantlets are removed from the plant media and transferred to soil or (more commonly) potting compost for continued growth by conventional methods. This stage is often combined with the "pre-transplant" stage.

Advantages

Micropropagation has a number of advantages over traditional plant propagation techniques:

1. The main advantage of Micropropagation is the production of many plants that are clones of each other.
2. Micropropagation can be used to produces disease-free plants.
3. Micropropagation produce rooted plantlets ready for growth, and saves time for the grower, instead of seeds or cuttings.
4. It has an extraordinarily high fecundity, producing thousands of propagules in the same time it would take a conventional technique to produce tens or hundreds
5. It is the only viable method of regenerating genetically modified cells or cells after protoplast fusion.
6. It is a good way of multiplying plants which produce seeds in uneconomical amounts (if at all) or whose seed can't be stored (vgr. recalcitrant seeds).
7. Micropropagation often produces more robust plants, leading to accelerated growth compared to similar plants produced by conventional methods.

Disadvantages

Micropropagation may appear to be the perfect means of multiplying plants, but it has associated problems**:**

1. It is *very* expensive, and can have a labor cost of more than 70%.
2. An infected plant sample can produce infected progeny. This is uncommon if the stock plants are carefully screened and vetted to prevent culturing plants infected with virus or fungus.
3. Not all plants can be successfully tissue cultured, often because the proper medium for growth is not know or the plants produce secondary metabolic chemicals that stunt or kill the explants.
4. Sometimes plants or cultivars do not come true to type after being tissue cultured, this is often dependent on the type of explants material utilized during the initiation phase or the result of the age of the cell or propagule line.
5. Some plants are very difficult to disinfect of fungal organisms.

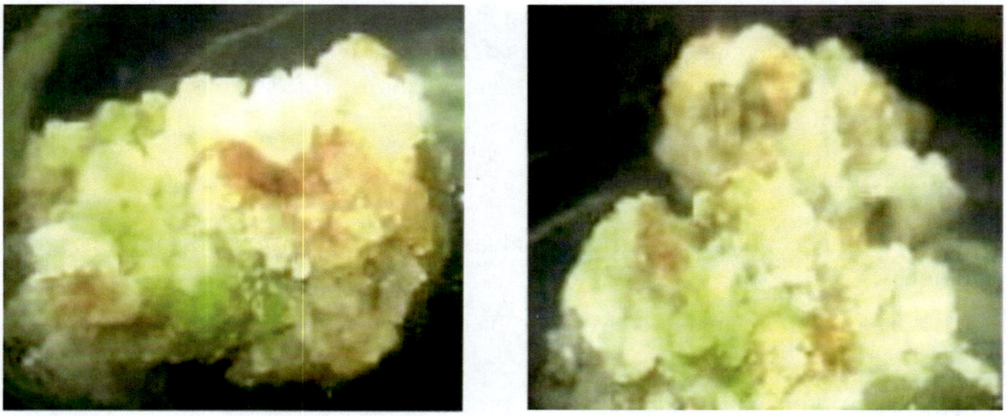

Callus regeneration of *Hypericum mysorense*

The greatest limitation is the cost of production, and for most plants seeds, which are normally disease free and produced in good numbers readily produce plants in good numbers at a lower cost. For this reason, many plant breeders do not utilize micropropagation because of the prohibitive cost and other breeders use it to produce stock plants that are then used for seed multiplication. Mechanization of the process would eliminate most of the labour cost associated, but this has proven difficult so far despite active attempts to develop this technology.

Callus-Mediated Organogenesis

The induction of callus growth and subsequent differentiation and organogenesis is accomplished by the differential application of growth regulators and the control of conditions in the culture medium. With the stimulus of endogenous growth substances or by addition of exogenous growth regulators to the nutrient medium, cell division, cell growth and tissue differentiation are induced. There are many reports on the regeneration of various medicinal plants via callus culture. Satheesh et al. (1988) have reported the regeneration of shoots from callus of *Plumbago rosea* using appropriate concentrations of auxins and cytokinins. Mantell and Hugo(1989) have also reported a high frequency of shoot, root, and microtuber production from *Dioscorea alata* depending on the culture medium used, the type of explant from which the calli originated, and the photoperiod . Plant regeneration has been achieved from leaf callus of *Cephaelis ipecacuanha* on Murashige and Skoog medium supplemented with 4.5 mg/L kinetin and 0.1 mg/L α-Naphthaleneacetic acid (NAA) (Jheng et al.2006) . Ghosh and Sen (1991) established plant regeneration via callus cultures from different explants of *Asparagus cooperi* (Karkonen, 2001). The relative importance of genotype of the explants and their interactions for *in-vitro* plant regeneration via organogenesis in *Solanum melongena* has been investigated (Sharma et al. 1995). Basu and Chand (1996) achieved shoot bud differentiation from root-derived callus of *Hyoscyamus muticus* in

MS medium supplemented with 0.05 mg/L NAA and 0.5 mg/L BA. Saxena et al.(1997) reported plant regeneration via organogenesis from callus cultures derived from mature leaves, stems, petioles and roots of young seedlings of *Psoralea corylifolia* . *In-vitro* organogenesis of *Zingiber officinale* via callus culture has been described by Rout and Das (1997). Further Shasany et al.(1998) reported the influence of different growth regulators on high frequency plant regeneration from internodal explants of *Mentha arvensis* . Successful plant regeneration was reported from stem and leaf-derived callus of *Centella asiatica* on MS medium supplemented with 4.0 mg/L BA, 2.0 mg/L kinetin, 0.25 mg/l NAA, and 2.0 mg/L adenine sulfate. Rapid regeneration of *Plumbago zeylanica* in callus culture was achieved on MS medium supplemented with 2.0 mg/L BA and 0.5 mg/l IAA. The efficient systems for *in-vitro*regeneration of *Solanum laciniatum* and *Echinacea pallida* have been established from leaf explants on medium supplemented with BA and NAA, Pandey et al.(2002) have also reported the *in-vitro* regeneration of *Lepidium sativum* from various explants on medium supplemented with BA and NAA.

Suspension culture : When the pieces of calli are transferred to liquid medium and the medium is continuously agitated to obtain isolated single cells, the culture is called suspension culture. Basically there are two types of suspension culture viz.Batch culture and continuous culture.

***In vitro* Rice Seedling**

Batch culture: When cultures are continuously propagated by routinely taking a small aliquot of the suspension and transferring it to a fresh medium and maintained by sub culturing, the suspension culture is called batch culture.

Continuous culture: When cultures are grown in culture vessels under steady state for prolonged periods by adding fresh mediums and draining out the used medium, the suspension culture is called continuous culture.

Anther culture: It is one of the very popular methods for production of haploid plants through culturing anthers or microspores on artificial culture medium. This leads to the growth of microspore in to saprophytes. After the initial reports of successful production of haploids from anther culture of Datura (Guha and Maheswari,1964). Haploids have been reported in more than 150 species belonging to 23 families of angiospermic plant taxa of economically important species (Maheswari et al.1958). Haploid plants may be obtained from pollen grains by placing anthers or isolated pollen grains on a suitable culture medium. The anthers may be taken from plants grown in the field or in pots, but ideally these plant should be grown under controlled temperature, light and relative humidity ; the optimum condition may differ from species to species. Often the capacity for haploid production declines with the age of donor plants.

Floral buds of the appropriate developmental stage are collected, surface sterilized and their anthers are excised and placed horizontally on culture medium .Floral buds with small anthers may themselves be cultured and in some cases the entire inflorescence has been cultured .Care should be taken to avoid injury to anthers since it may induce callus formation from anther walls. Alternatively, pollen grains may be separated from anthers and cultured on a suitable medium.

Anther Culture in Rice

Rice is consumed by nearly half of the world's population and it is the staple food for the largest number of people on the earth thereby providing one quarter of global per capita energy. It is grown on 125 million hectares world-wide and rice production was 600 million tons in 2000. The 2.3 folds increase in rice production over the last 40 years has been largely contributed by intensive research leading to better varieties able to grow in shorter period and showing higher yield potential as well as resistance/ tolerance to various pests, disease and abiotic stresses. In India, rice is grown in 44 million hectares which is 28% of total arable land. About 74 kg of milled rice per capita was consumed in 1999 in India providing about 30% of total calories/capita/day. The demand of rice in India is expected to increase from 84 million tons in 2000 to 114 million tons in 2030, an increase of about 35%. This increase in production is to be achieved without major increase in arable land and with rapidly degrading ecosystem. Such a challenge requires deployment of all possible means of higher production, which to large extent involves development of better varieties and efficient agricultural management.

To produce high yielding varieties of rice suited to specific locations; it is imperative to understand intrinsic yield-limiting factors at molecular level to help deployment of natural and engineered genetic variability most effectively. Advent of molecular biotechnology in seventies opened a new possibility for genetic improvement of rice. Genes associated with specific traits could be mapped and introgressed leading to molecular breeding and genetic engineering which allowed movement of genes across the incompatibility barriers (Tyagi et al. 1999). Simultaneously, availability of efficient gene tagging systems led to application of the concepts of functional genomics in rice. In 1998, an International Rice Genome Sequencing Program (IRGSP) was initiated to sequence the entire genome of rice, estimated to be about 430 Mb, drawing upon the information already available on rice genetics and generated by the Rice Genome Program (RGP) in Japan. The components of IRGSP have developed an integrated physical and genetic map as well as a comprehensive transcript map of rice. The high quality (>lOX) sequence of rice has already been declared in December 2004. India is actively participating in this public sector international venture. Monsanto and Syngenta from private sector have also produced a draft sequence of the variety being investigated by IRGSP and made available their data to IRGSP. An *indica* variety genome has been sequenced in China. Notwithstanding the enormous value of these reports as a significant step in sequencing complete rice genome, the lower sequence reads in any particular area and loose relationship of the genes to each other and to the genetic map, as reported, makes the application of information cumbersome and required consistent efforts by IRGSP to produce highly accurate sequence to serve as golden standard (International Rice Genome Sequencing Project 2005). It is estimated that the rice genome may' contain 37,544 genes. Functional analysis of these genes and their useful alleles would be a challenge for rice scientists in near future and require global attention and initiatives by various countries. India, being a major rice producer and consumer, obviously need to take a lead in the functional genomics of rice.

Anther of Rice

Callus of Rice

Success in genome sequencing and transgenics is paving the way for preparing a road map of functional genomics which is expected to correlate action of a gene to a trait in cellular and organismal context. Realizing the immense potential, various nations have provided huge in-put to researchers to initiate investigation on functional genomics of plants. Prominent among these are Plant Genomics Programs supported by the National

Science Foundation USA, European Functional Genomics Program, Australian Centre for Plant Functional Genomics, Rice Simulator Program of Japan and Chinese Program on Functional Genomics of rice. The International Rice Research Institute (IRRI) has helped form an International Rice Functional Genomics Working Group. Most of these programs concern with development and application of tools and resources for genome-wide understanding of gene functions. This includes analysis of expression profiles, generation of mutants and tagged lines or target-specific silencing of genes, functional validation in transgenics, allele mining, comparative genomics and virtual plant concept.

The initiation point of these investigations is either selected traits or a group of genes and in certain cases the whole transcriptome. Most of these approaches ultimately aim at providing input for marker-assisted breeding or genetic engineering of new crop varieties. In appreciation of the potential and need of India, a meeting of Indian Rice Scientists and Scientists from IRRI was organized in New Delhi in May 2002. Deliberations of two days led to crystallization of the idea of forming a National Consortium for Functional Genomics of Rice (NCFGR). It was also decided to develop programs on the functional genomics of indica rice in selected areas. In parallel with 'Impact Research and Delivery' approach and the 'Strategic Research' approach are expected to unravel the function of most crucial group of genes from rice, i.e. signal transduction and transcription components. The analysis of the rice genome reveals that such genes might be represented by as much as 8000 genes. Their activity is responsible for perceiving environmental (abiotic and biotic) and intrinsic (developmental and hormonal) signals and prepare the plant to respond to adverse environment to survive and follow a set pattern of life-cycle. The program, therefore, starts with analysis of expression and data mining to select useful genes which would be target for functional validation. The "Strategic Research" area project involves five institutions and thirteen scientists having prior experience in rice genetics, transgenics and genomics. At least four scientists from Delhi University South Campus (UDSC) and Indian Institute of Science (IISc) would work together to procure/design/develop a microarray system to analyze signal transduction (e.g. kinases, receptors) and transcription factor genes. This would require in silico analysis of the whole genome and transcriptome-related information to be undertaken by UDSC and IISc. The selection of target genes will be based on their expression pattern during critical developmental windows of the rice plant (including seed development, meristem determination, floral initiation and reproductive organ development). For this, existing germplasm as well as germplasm developed by other network programs will be utilized and useful alleles will be identified. Selected crucial genes will be functionally validated in transgenics by over-expression and functional knockout approaches, to be-taken at UDSC, IISc, Madurai Kamraj University (MKU), Osmania University (OU) and the Directorate of Rice Research (DRR). Simultaneously, vectors for artificial target-specific gene silencing will also be developed and utilized.

Specific technology development for functional genomics, viz, (a) targeted gene integration by homologous recombination (MKU), (b) efficient transformation of Indian genotypes (OU), and (c) germplasm resource, diversification and trials (DRR) would also be undertaken. It is expected that microarray analysis would narrow down to about 50 genes for functional analysis. For each gene, one over-expression and one knockout (antisense or RNAi) construct is to be made. In addition, there would be at least 50 constructs for stage-specific promoters. Thus, each group at UDSC and groups at DRR, IISc, MKU and OU would work with about 20 constructs to generate at least 10-20 independent transgenic lines for each and analyze them in network mode at To and TI generations. It is expected that 'Strategic Research' program would identify several useful genes for improvement of rice and related crops as well as validate their function for crop improvement.

Stigma of Rice

Lema and Palea of Rice

The IRRI platform would help development of international collaboration for functional validation and go on to resource and information sharing. Agreement of Scientists from Japan to be advisors and contributors of genomics resources has also been obtained. It may also be mentioned that functional genomics is a frontier area undergoing rapid development and would require regular moderation and flexibility to achieve the goals most rapidly. The 'Strategic Research' program would also share microarray platform and bioinformatics logistics with other areas of NCFGR. Looking at the potential of groups involved and a minimum need to achieve the objectives, a budget of Rs 670 millions has been approved. The present proposal has been developed by in-puts from participating labs which can take shape of a virtual institution in the area of 'Strategic Research' for gene discovery in rice. It is expected that in-put from

IRRI, Philippines, and NIAS, Japan, would greatly supplement our initiative, while efforts would be made to take advantage of other international programmes as well.

Somaclonal Variation

It is the term used to describe the variation seen in plants that have been produced by plant tissue culture. Chromosomal rearrangements are an important source of this variation. Somaclonal variation is not restricted but is particularly common in plants regenerated from callus. The variations can be genotypic or phenotypic, which in the later case can be either genetic or epigenetic in origin. Typical genetic alterations are: changes in chromosome numbers (polyploidy and aneuploidy), chromosome structure (translocations, deletions, insertions and duplications) and DNA sequence (base mutations). Typical epigenetic related events are: gene amplification and gene methylation. If no visual, morphogenic changes are apparent, other plant character screening procedures must be applied. There are both benefits and disadvantages to somaclonal variation. The phenomenon of high variability in individuals from plant cell cultures or adventitious shoots has been named somaclonal variation .

Advantage : The major advantage of somaclonal variation is in plant improvement which leads to the creation of additional genetic variability. The major Characteristics of somaclonal mutants can be enriched during *in vitro* culture include resistance to disease pathotoxins, herbicides and tolerance to environmental or chemical stress, as well as for increased production of secondary metabolites.

Disadvantages : A serious disadvantage of somaclonal variation occurs in operations which require clonal uniformity, as in the horticulture and forestry industries where tissue culture is employed for rapid propagation of elite genotypes. It is well known that increasing numbers of subculture increases the likelihood of somaclonal variation, so the number of subcultures in micropropagation protocols should be kept to a minimum. Regular reinitiating of clones from new explants might reduce variability over time. Another way of reducing somaclonal variation is to avoid 2,4-D in the culture medium, as this hormone is known to introduce variation.

Plant embryogenesis: Plant embryogenesis refers to the process of development of plant embryos, being either a sexual or asexual reproductive process that forms new plants. It may occur naturally in the plant as a result of sexual fertilization, and these embryos are called **zygotic embryos.** The zygotic embryo develops into seeds, which can germinate and give rise to seedlings. Plant cells can also be induced to form embryos in plant tissue culture; these embryos are called **somatic embryos**. An embryo is made up of actively growing cells and the term is normally used to describe the early formation of tissue in the first stages of growth. It can refer to different stages of the sporophyte and gametophyte plant. In both gymnosperms and angiosperms, the young plant contained in the seed, begins as a developing egg-cell formed after fertilization (sometimes without fertilization in a process called apogamy) and becomes a plant embryo. This embryonic condition also occurs in the buds that form on stems. The buds have tissue that has differentiated but not grown into complete structures. They can be in a resting state over winter or when conditions are dry, and then commence growth when conditions become suitable. Before they start growing into stem, leaves, or flowers, the buds are said to be in an embryonic state.

Zygotic Embryogenesis :- The zygotic embryo is formed following double fertilisation of the ovule, forming the plant embryo and the endosperm which together go on to develop into a seed. Seeds may also develop without fertilization through pathways referred to as apomixis.

Somatic Embryogenesis :- Somatic embryogenesis is a process by which asexual or somatic cells are induced to form embryos in culture. Somatic embryogenesis is a multi-step regeneration process starting with formation of pro-embryogenic masses, followed by somatic embryo formation, maturation and regeneration. The important processes involved in the somatic embryogenesis lead to the formation of somatic embryos which resemble the zygotic embryos of intact seeds and can grow into seedlings on suitable medium. Plant regeneration via somatic embryogenesis from single cells,that can be induced to produce an embryo and then a complete plant, has been demonstrated in many medicinal plant species. Arumugam and Bhojwani (1990) noted the development of somatic embryos from zygotic embryos of *Podophyllum hexandrum* on MS medium containing 2 μM BA and 0.5 μM IAA (Basu et al.1996). Ghosh and Sen (1991) reported regeneration and somatic embryogenesis in *Asparagus cooperi* on MS medium having 1.0 mg/L NAA and 1.0 mg/L kinetin. Embryogenic calluses and germination of somatic embryos in nine varieties of *Medicago sativa* has been achieved by using a medium containing 2,4-dichlorophenoxyacetic acid (2,4-D) and TDZ. Zhou et al.(1994) have achieved the induction of somatic embryogenesis in cells from *Cayratia japonica*.Somatic embryogenesis and subsequent plant regeneration from callus derived from immature cotyledons of *Acacia catechu* has also been achieved on medium supplemented with 13.9 μM kinetin and 2.7μM NAA. Gastaldo and Caviglia (1996) induced somatic embryogenesis from bark callus of *Aesculus hippocastanum* on MS medium supplemented with 2.0 mg/L kinetin, 2.0 mg/L 2,4-D and 2.0 mg/l NAA.

High frequency somatic embryogenesis and plant regeneration from suspension cultures of *Acanthopanax koreanum* have been reported by Choi et al.(1997) on a medium containing 4.5 μM 2,4-D. Das et al. (1999) reported high frequency somatic embryogenesis in *Typhonium trilobatum* on medium containing 1.0 mg/L kinetin and 0.25 mg/L NAA. The suspension culture of *Catharanthus roseus* from stem and leaf explants on medium containing NAA and kinetin has been established by Zhao et al. (2002). Chand and Sahrawat. (2002) have reported the somatic embryogenesis of *Psoralea corylifolia* L. from root explants on medium supplemented with NAA and BA. Efficient development and germination of somatic embryos are prerequisites for commercial plantlet production. Lowering of growth regulator concentrations in culture media has improved embryo development and germination of many medicinal plants (Kumar, 1992).Germination of the somatic embryos is achievable on MS medium without the growth regulator. However, Arumugam and Bhojwani (1990) noted that inclusion of BA (2μM) and gibberellic acid (GA3, 2.8 μM) in the medium stimulated embryo development of *Podophyllum hexandrum*, although 75% of the embryos germinated on basal MS medium devoid of growth regulator. Similar results were reported on the germination of embryos of *Psoralea corylifolia* (Kunitake et al. 1997). Wakhlu et al. (1990) have reported that the somatic embryos of *Bunium persicum* matured and germinated on the basal MS medium supplemented with 1.0 mg/L kinetin.

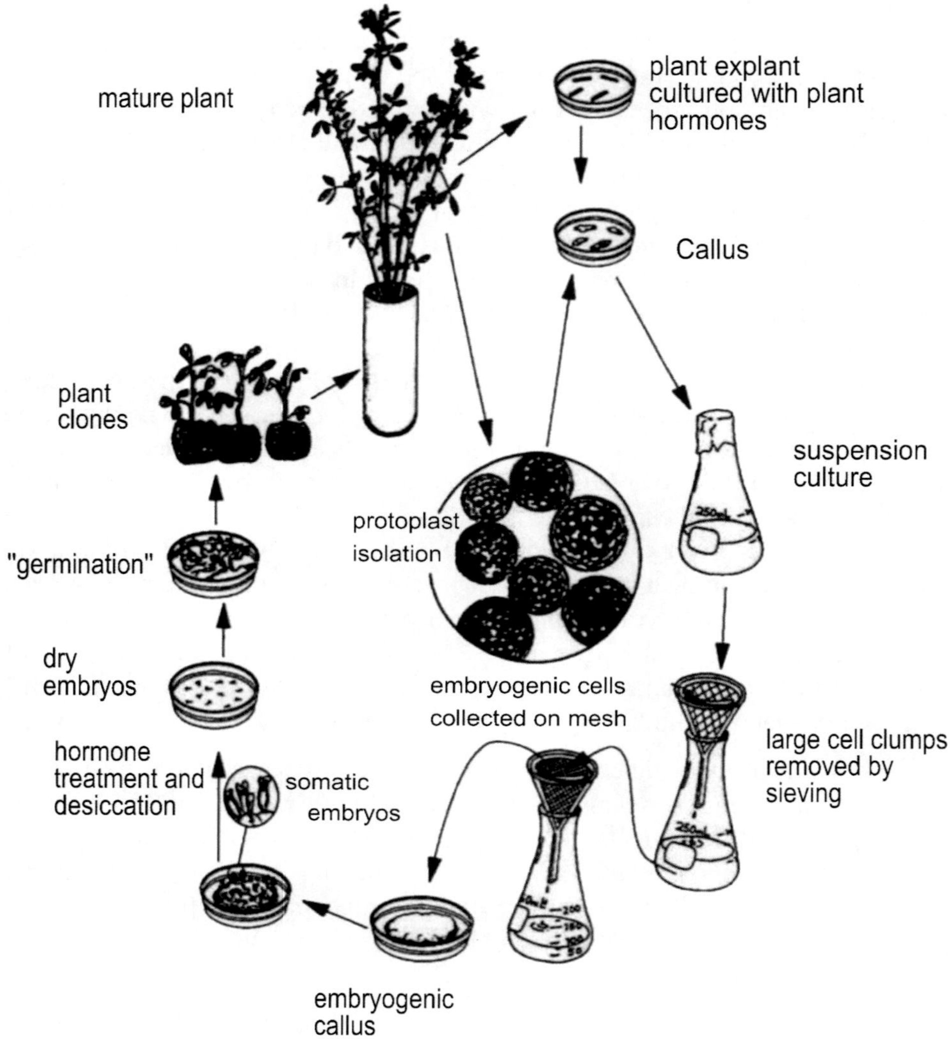

Fig.2 : Schekmatic Representation of Callus Mediated Organogenesis

Somatic embryos are formed from plant cells that are not normally involved in the development of embryos, i.e. ordinary plant tissue. No endosperm or seed coat is formed around a somatic embryo. Applications of this process include: clonal propagation of genetically uniform plant material; elimination of viruses; provision of source tissue for genetic transformation; generation of whole plants from single cells called protoplasts; development of synthetic seed technology. Cells derived from competent source tissue are cultured to form an undifferentiated mass of cells called a callus. Plant growth regulators in the tissue culture medium can be manipulated to induce callus formation and subsequently changed to induce embryos to form from the callus. The ratio of different plant growth regulators required inducing callus or embryo formation varies with the type of plant.

The Stages of Embryogenesis

Zygotic and somatic embryos share a number of characteristic developmental stages; however, the very early steps in their development are not well correlated. Following fertilization, the zygote undergoes an asymmetrical cell division that gives rise to a small apical cell that becomes the embryo and a large basal cell (called the suspensor) that functions to provide nutrients from the endosperm to the growing embryo. Asymmetrical cell division also seems to be important in the development of somatic embryos, and while failure to form the suspensor cell is lethal to zygotic embryos, it is not lethal for somatic embryos. From the eight cell stage (octant) in the zygotic embryo, 'embryo patterning' is apparent; however, somatic embryos at this stage may be quite variable, therefore zygotic and somatic embryos become most comparable from the globular stage. In the globular stage, the embryo develops radial patterning through a series of cell divisions, with the outer layer of cells differentiating into the 'protoderm.' The globular embryo can be thought of as two layers of inner cells with distinct developmental fates; the apical layer will go on to produce cotyledons and shoot meristem, while the lower layer produces the hypocotyl and root meristem. Bilateral symmetry is apparent from the heart stage; pro-vascular cells will also differentiate at this stage. In the subsequent torpedo and cotyledonary stages of embryogenesis, the embryo completes its growth by elongating and enlarging .

Synthetic Seeds : Synthetic seeds are defined as artificially encapsulated somatic embryos, shoot buds, cell aggregates, or any other tissue that can be used for sowing as a seed and that possess the ability to convert into a plant under *in vitro* or *ex vitro* conditions, and that retain this potential also after storage. The synthetic seed technology has been developed to use somatic embryos and/or other micropropagules as seed analogues successfully in the field or greenhouse, and their mechanical planting at a commercial level. The technology provides methods for preparation of seed analogues called synthetic seeds or artificial seeds from the micropropagules like somatic embryos, axillary shoot buds, apical shoot tips, embryogenic calli as well as protocorm or protocorm like bodies. For the last decades, intensive research efforts have been made on synthetic seed production in a number of plant species.

Embryoids of *Abrus precatorius*

Isolated somatic embryos of *Euphorbia pulcherrima*

Germination of somatic embryos of *Euphorbia pulcherrima*

Despite these researches, practical implementation of the technology is yet to be fully realized due to limitations encountered with the production, development, maturation and subsequent conversion of the micropropagules into plantlets under *ambient* conditions. In the conventional plant tissue culture for clonal propagation, storage and transportation of propagules for transplantation is a major problem. To over come this problem in recent years the concept of synthetic or artificial seeds has become popular, in which somatic embryos are encapsulated in a suitable matrix (e.g., sodium alginate), along with substances like mycorrhizae, insecticides, fungicides and herbicides. In India this technique of synthetic seeds is being standardized and practiced for sandalwood and mulberry at BARC under the leader ship of Dr. P.S. Rao.

Advantages of Synthetic Seeds

1. Longer duration of storage without loss of viability.
2. Useful as units of delivery and easy to handle.
3. They can be directly shown in the soil like natural seeds and do not need hardening in green house.

Disadvantages of Synthetic Seeds

1. High cost of their production
2. Only feasible in commercial basis

Protoplast Culture, Somatic Hybridization and Somatic Cybridization

One of the most significant developments in the field of plant tissue culture, witnessed during the last few decades, is the isolation, culture and fusion of protoplasts.

A more recent achievement is the manipulation and regeneration of these cultured or fused protoplasts into whole plants. Since in plant cells, the plasma membrane is bound by a rigid cellulose wall (unlike animal cells), it has been relatively difficult to handle plant cells. Only in 1960, it was demonstrated by E.C. Cocking at the University of, Nottingham (U.K.), that naked cells called protoplasts can be obtained through enzymatic degradation of cell walls. This led to significant developments in the field of somatic cell genetics in higher plants. Cultured protoplasts can be used not only for somatic cell fusions, but also for taking up foreign DNA, cell organelles, bacteria and virus particles. In view of this, the isolation and culture of protoplasts has become a very important area of research, within the realm of plant biotechnology.

Protoplast Culture and Regeneration of Plants

The culture methods for isolated protoplasts are similar to those for single cells earlier. The culture method selected also depends upon the objective of the experiment. For instance protoplasts can be cultured following 'Bergmann's cell plating technique. Both semi-solid medium and liquid medium can be used, although the liquid medium is preferred. Protoplasts can be suspended in a liquid medium in Erlenmeyer flasks without shaking and can be cultured in small quantities in 'hanging drops' or in 'microchambers . Multidrop array (MDA) techniques may also be used. Within 2-4 days, protoplasts start developing cell walls, which can be detected by staining with 0.1% calcofluor white (CPW) fluorescent stain.While the presence of a proper wall is essential for a regular division, it is not always a prerequisite for nuclear division; not all protoplasts with cell walls formed, embark upon division.

Exploitation of genetic variability is an essential strategy in plant breeding programme for the genetic improvement of crop plants. Improvements of crop plants by conventional breeding are time consuming, expensive and labour intensive. Unconventional techniques can shorten the process. One of these techniques is somatic hybridization. Somatic hybridization through protoplast fusion has been envisaged for the production of Para sexual hybrids to overcome sexual incompatibility.The ultimate objective to transfer desirable traits from distantly related taxa. This para-sexual hybridization is a means of increasing genetic variability of several crops gene pool, not only by overcoming sexual incompatibility but it can produce nuclear cytoplasmic combinations that are difficult or impossible to obtain by conventional sexual crossing. Klericker (1892) first isolated plant protoplasts of *Stratiotes aloides* by plasmolysing and subsequently slicing the tissue. Protoplasts obtained by this procedure were used mainly in physiological studies. However, limited yields, excessive damage, and restriction of the technique to specific tissues, were the major drawbacks. Large scale isolation of plant protoplasts, using cell wall degrading enzymes (Cocking, 1960), gave impetus to research in protoplasts. Initially, a two step procedure was adopted, first producing single cells and then protoplasts. However, with further experimentation, a mixed enzyme procedure was developed, involving maceration and protoplast isolation in a single step from intact tissues.

The major activity of the cell wall degrading enzymes is in the digestion of pectins, cellulase and hemicellulases (xylans) that largely constitute the plant cell wall. Pectinases degrade the galacturonic acid residues of pectin's that confer the cell to cell adhesion, and apparently macerate the tissue to single cells; cellulases digest the cellulose component, conferring the spherical shape to the protoplasts, while hemicellulases assist in the breakdown of xylans. Incubation in a mixture of the three types of enzymes in suitable ratios facilitates release of a large number of protoplasts. The impurities present in the commercially supplied crude enzymes reduce the protoplast viability and enzyme purification by gel filtration can be beneficial for sustained division of protoplasts. The pH of the enzyme mixture and temperature of incubation are some of the factors that influence the release of protoplasts. In certain cases, vacuum assists initial penetration of the enzyme and subsequent digestion.

Following protoplast release, filtering through an autoclavable mesh removes undigested tissue, while repeated washing by centrifugation, removes enzymes and finer debris. Separation of protoplasts from debris is also achieved using sucrose floatation. A two-phase system using dextran and PEG or sodium metrizoate or a discontinuous density gradient system incorporating percoll are some of the other methods used to purify plant protoplasts from debris.

Protoplast Isolation

Mesophyll tissues of leaves are a convenient source for a large number of uniform cells for protoplast isolation. The following points require close attention prior to selecting leaves for protoplast isolation.Growth conditions in terms of nutrition, humidity, light and temperature are critical in determining the yield of stable protoplasts.Over-nourished or undernourished plants are not suitable . Age of the plants from which the leaves are excised for protoplast isolation determines their stability. Generally plants between 6-7 weeks are suitable in protoplast isolation. Position of the leaf on the source plant also influences the yield and quality of the protoplasts. A well expanded 3rd and 4th leaf is suitable. *In vitro* raised axenic cultures are favored over the field/glass-house grown plants for obtaining leaves for protoplast isolation.

Cell wall formation is evident by the change of shape of the protoplasts within 24-48 hrs of plating. Divisions occur by 3-7 days of culture initiation. For sustained growth of protoplasts the osmoticum of the culture medium has to be lowered by sequential dilutions periodically. For protoplasts that are embedded in agar or agarose and alginate beads, small blocks containing the dividing protoplasts are transferred to the surface of the same medium but with lower osmoticum. The same can further be shifted again to a medium with a lower osmotic pressure. Once the protoplasts have grown up to a callus stage, the procedure for regeneration of plants there from is essentially the same as that employed in regular tissue culture techniques.

Mechanism of Protoplast Fusion

Major constituents of the cell membrane are phospholipids and proteins. Lipid molecules consist of a polar head group that are hydrophilic, to which is attached one or two long hydrocarbon chains of tails which are hydrophobic. The phospholipids are arranged in a bilayer into which peripheral or integral structural proteins are embedded in a mosaic-like fashion. Exogenous chemical stimuli cause disturbances in the intra-membranous proteins and glyco-proteins, increasing thereby the membrane fluidity and resulting in fusion of membranes at the point of contact. Such fluidity is related to the proportion of saturated phospholipids in the membrane and is influenced by temperature. The intra-membranous phosphate groups are postulated to induce a negative surface charge on the protoplasts. This tends to repel the protoplasts from one another (Carlson.1973).

The underlying mechanism of fusion by chemical methods and also electric field induced fusion of plant protoplasts, remains similar to that suggested by Poste and Allison, 1971 for animal cells. According to these workers, membrane proximity of the fusion partners is an essential pre-requisite followed by an induction phase, whereby the charges induced in electrostatic potential of the membrane result in fusion. It was the early part of the century, that Kuster (1910) demonstrated the ability of onion epidermal protoplasts to fuse upon de-plasmolysis. Also Michel (1937) showed that protoplasts could be fused by using potassium nitrate as the plasmolyticum. However, fusions were rare and the culture of fusion products was difficult. In the light of enzymatic procedures for protoplast isolation, interest developed in protoplast fusion with an aim to genetic manipulation. Power et al. (1971) reported sodium nitrate induced fusion of root protoplasts from various species. This method met with success in the production of the first somatic hybrid. However, fusion frequency following this method was low and the sodium nitrate treatment also affected the division capability of the protoplasts. This stimulated interest in other fusogens, and as a consequence rapid advances in the technique of protoplast fusion have been made in the last decade . Fusion methods that have gained maximum attention are the use of high pH combined with high Ca++ in solution, polyethyleneglycol (PEG) and more recently the electric field induced electrofusion. Several other methods have also been reported to induce protoplast fusion. These include manipulation with a perfusion micropipette, sea water, lysozyme and inactivated viruses.

Action of high pH and Ca^{++}

A calcium solution buffered at high pH induces aggregation of the protoplasts and their fusion. The addition of Ca^{++} causes the potential of the surface negative charge on protoplasts to be reduced, facilitating protoplast adhesion. The high alkalinity (pH 9.5 - 10.4) induces the formation of intramembranous lysophospholipids such as lysolecithin and lysophosphatidyl -ethanolamine that increase membrane fluidity that results in fusion. This method has been successfully used in the production of numerous somatic hybrid plants. Protoplasts derived from cell suspension cultures generally are not as tolerant to the use of high pH Ca++ method as the mesophyll derived protoplasts.

Action of Polyethyleneglycol (PEG)

PEG has been most widely used to fuse plant protoplasts. Both, the concentration and molecular weights of PEG are important in relation to fusion. PEG, whose general formula is HOCH2-(CH2-O-CH2) -CH2OH is a water soluble compound whose other linkages make the molecule slightly negative in charge. Thus, addition of Ca++ (CaCl2 or Ca (NO3)2 links the compound with membrane surfaces. The high molecular weight of the polymer acts as a bridge connecting the protoplasts together. A strong affinity of PEG for water causes local membrane dehydration and increased fluidity. This in combination with the reduction of an exclusion volume between adjacent protoplasts causes diminishing mutual membrane electrostatic repulsion. The redistribution of glycoprotein and glycocalyx macromolecules, causes fusion. PEG of molecular weight 400 to 6,000 was found to be active in fusion, whereas PEG 200 and 20,000 was almost inactive. Thus factors, other than purely chemical structure determine fusogenicity. Compounds structurally related to PEG, namely, polyvinyl alcohol, polyvinyl pyrrolidone and polyglycerol are known to induce fusion.(Cocking et al .1981) Gelatin, and dextran sulphate also induce fusion to varying degrees. PEG has been used in combination with other treatments to enhance fusion frequency. An increase in the frequency of heterokaryon formation was observed when protoplasts were pre-incubated in lysozyme. Also, combination of PEG with high pH Ca^{++} solution, or the addition of DMSO or concanavalin-A gave rise to higher fusion frequencies in comparison to treatments with PEG alone. It is generally observed that in fusion experiments involving PEG as the fusogen, the elution of PEG subsequent to its application is a critical step in determining the frequency of stable fusion products. A rapid change of osmotic pressure subsequent to fusion in the process of transferring to a culture medium results in excessive damage, particularly when mesophyll protoplasts are involved in the fusion process. (Cocking 1972).

Prospective of plant tissue culture: Plant tissue culture is used, widely in plant science and it has a number of commercial applications which are :-

1. Tissue culture systems provide the necessary interface between molecular biology and bioengineering and these will ultimately permit a fuller understanding of plant development, growth, productivity etc.
2. Plant tissue culture offer several options for clean "management" of biomaterials and enables a "Modular approach" to both basic and applied research.
3. Micro-propagation is used to produce large numbers of identical individuals with in a short span of time and space
4. It is widely used in forestry, horticulture including many agronomy and medicinally important plants.
5. It can also be used in to conserve rare plant species by means of seed and gene. Bank through modern techniques.
6. A plant breeder may use tissue culture to screen cells, rather than plants for advantageous characters, e.g., herbicide resistance/tolerance etc.
7. Large-scale growth of plant cells in liquid culture in bioreactors as a source of secondary products, like biopharmaceuticals including molecular farming.
8. To crossing distantly related species by protoplast fusion and regeneration of the novel hybrid in human welfare.
9. For production of dihaploid plants from haploid cultures to achieve homozygous lines more rapidly in breeding programmes, usually by treatment with colchicines which causes doubling of the chromosome number.
10. As a tissue for transformation, followed by either short-term testing of genetic constructs or regeneration of transgenic plants.
11. The techniques of tissue culture provides year round supply of planting material for off season cultivation.

Limitation of Plant Tissue Culture

1. The cost of production per plant by tissue culture is more in comparison to conventional methods.
2. Plant tissue culture experiments must be handled by highly trained people because it requires special care and diligent observation.
3. Chances of reduction of genetic diversity are high in tissue culture plants.
4. Instance of lethal mutation may be induced during the *in vitro* culture.
5. Labor-intensive as "tissue culture" is generally and currently practiced. But effort beginning to be automated.
6. All species are not yet equally amenable to the full range of manipulations of tissue culture technologies.
7. The technology is not feasible for laymen.

Conclusion

Plant tissue culture is a practice used to propagate clones of the plants. There are various reasons this may be done, among them to create exact copies of plants that produce particularly good flowers, fruits, or otherwise desirable traits for some purpose. It produces large number of plantlets which multiples in the absence of seeds or necessary pollinators. Plant tissue culture can also be used to regenerate whole plants from plant cells that have been genetically modified. It relies on the fact that all plant cells have the ability to generate a whole plant (totipotency). Single cells (protoplasts), pieces of leaves and roots can be used to generate a new plant on culture media by providing the requisite nutrients and plant growth regulators. Most in vitro studies focus on manipulating the medium which can lead to profound effects on explants growth and development. However, not all explants respond well in vitro and the response of success can be increased by considering the plants that are the sources of explants. Simply the time of year or portion of the stock plant from which explants are collected can have great influences on *in vitro* performance. Additionally, it may be necessary to rejuvenate or otherwise treat the stock plant or donor tissues prior to harvesting explants to place in culture.

Many commercial laboratories and national institutes worldwide use *in vitro* culture system for rapid plant multiplication, germplasm conservation, elimination of pathogens, genetic manipulations, and for secondary metabolite production. Annually, millions of *in vitro* plants are routinely produced in world wide. The great potential of micropropagation for large-scale plant multiplication can be tapped by cutting down the cost of production per plant by applying low-cost tissue culture, which is to adopt practices and proper use of equipment and resources to reduce the unit cost of micropropagule and plant production without compromising the quality. Bioreactor technology may cut down the cost of plant production provided proper precautions are taken to prevent contamination. Somatic embryogenesis facilitates cryopreservation, synseed development, mutations, and genetic transformation. Plant transformation methods and enhanced gene silencing technology can effectively be used to evaluate and authenticate newly discovered endogenous genes to characterize their function in plants as well as to genetically manipulate trait quality and productivity. Recent progress in genetic manipulation of plant cells has opened new possibilities for improvement of quality traits in important plant species.

References

Anonymous.(2003).Omzettabel Kamerplanten.Vakbl. Bloemist, 21: 136–137.

Aoshima, Y. (2005).Efficient embryogenesis in the callus of tea (*Camellia sinensis*) enhanced by the osmotic stress or antibiotics treatment. Plant. Biotechnol, 22: 277-280.

Arumugam, N and Bhojwani, S.S.(1990).Somatic embryogenesis in tissue cultures of *Podophyllum hexandrum*. Can. J. Botany, 68: 487-491.

Arya, S., Kalis. R.K.and Arya, I.D.(2000).Induction of somatic embryogenesis in *Pinus roxburghii* Sarg. Plant. Cell. Rep, 19: 775-780.

Baez, E.M. and Hernandez, Z.(1997).Protein changes associated with plant regeneration in embryogenic calli of sugarcane (*Saccharum* sp.). Plant Cell Tissue Organ Cult., 51:153-158.

Basu, P. and Chand, S.(1996). Regeneration of plantlets from root-derived callus of *Egyptian Henbane*. Cell Chromosome Res, 19: 31-34.

Behera, K.K., Sahoo, S. and Prusti, A.B. (2009). Efficient In Vitro Micropropagation of Greater Yam (*Dioscorea alata* L. cv. Hinjilicatu) Through Nodal vine Explant. Indian J. Plant physiology,14(3): 250-256.

Behera, K.K., Sahoo, S., Prusti, A.B. (2009). Regeneration of Plantlet of Water Yam (*Dioscorea oppositifolia* L.) through *In Vitro* Culture from Nodal Segments. Not. Bot. Hort. Agrobot. Cluj, 37: 94-102.

Blochl, A., March, G.G., Sourdioux, M., Peterbauer, T. and Richter, A. (2005).Induction of raffinose oligosaccharide biosynthesis by abscisic acid in somatic embryos of Alfalfa (*Medicago sativa* L.). Plant Sci, 168: 1075-1082.

Capuana, M. and Debergh, P.C.(1997). Improvement of the maturation and germination of horse chestnut somatic embryos. Plant Cell Tissue Organ Cult, 48: 23-29.

Carlson, P.S.(1973). The use of protoplast for genetic research. Proc.Natl.Acad.Sci, 70:598-602.

Cassells, A.C. and Doyle, B.M.(2005). Pathogen and biological contamination management: the road ahead. In: Loyola- Vargas, V.M., Vázquez-Flota, F., (Eds.). *Plant Cell Culture Protocols,* Humana Press, New York, pp. 35-50.

Cassells, A.C. and Plunkett, A.(1984). Production and growth analysis of plants from leaf cuttings, and from tissue cultures of disks from mature leaves and young axenic leaves of African violet (*Saintpaulia ionantha* Wendl.). Sci. Hortic, 23:361–369.

Castillo, P., Marquez, J., Rubluo, A., Hernandez, G. and Lara ,M. (2000). Plant regeneration from callus and suspension cultures of *Valeriana edulis* ssp. *Procera* via simultaneous organogenesis and somatic embryogenesis. Plant Sci, 151: 115-119.

Chand, S. and Sahrawat, A.K.(2002). Somatic embryogenesis and plant regeneration from root segments of *Psoralea corylifolia* L. an endangered medicinally important plant. *In vitro* Cell Dev. Biol, 28: 33-38.

Chebet, D.K., Okeno, J.A. and Mathenge, P.(2003). Biotechnological approaches to improve horticultural crop production. Acta. Hortic,625: 473–477.

Choi, Y.E., Kim, J.W. and Soh, W.Y. (1997).Somatic embryogenesis and plant regeneration from suspension cultures of *Acanthopanax koreanum*. Plant Cell Rep., 17: 84-89.

Choi, Y.E., Ko, S.K., Lee, K.S. and Yoon, E.S.(2002a).Production of plantlets of *Eleutherocossus sessiliflorus* via somatic embryogenesis and successful transfer to soil. Plant Cell Tissue Organ Cult., 69: 201-204.

Choi, Y.E., Lee, K.S., Kim, E.Y., Kim, Y.S., Han, J.Y., Kim, H.S., Jeong, J.H. and Ko, K.S. (2002b). Mass production of Siberian ginseng plantlets through large-scale tank culture of somatic embryos. Plant Cell Rep., 21: 24-28.

Cocking, E. C.(1983).The enzymatic isolation of plant protoplasts: historical perspective. Int. Rev. Cytol. 16:1–3.

Cocking, E.C.(1972).Plantcell protoplast isolationand development. Anna. Rev. Plantphysiology, 23: 29-50.

Cocking, E.C., Davey, M.R., Pental, A. and Power, J.B. (1981). Aspect of plant genetic manipulation. Nature, 293: 265-270.

Das, A.B., Das, P.,(1992). Rapid multiplication of *Musa acuminata* 'Dwarf Cavendish' through meristem culture and evalua-tion of chromosomal stability. *Orissa Journal of Horticulture 20*: 22-27.

Das, P., Palai, S.K., Patra, A., Samantaray, S. and Rout GR.(1999). In vitro somatic embryogenesis in *Typhonium trilobatum* Schott. Plant Growth Regulation 27:193–197.

Gastaldo, P. and Caviglia, A.M.(1996). Somatic embryogenesis and esculin formation in calli and embryoids from bark explants of Aesculus hippocastanum L. Plant Sci. Shannon, 119: 157-62.

Gautheret, R.J. (1939).Sur la possibilite′ de realiser a culture indefinite des tissues de tubercules de carotte. C R Hebd Seances Acad Sc., 208:118–120.

Gautheret, R.J. (1983). Plant tissue culture: A history. Botanical Magazine, Tokyo 96: 393–410.

Ghosh, B.E. and Sen, S.(1991). Plant regeneration through somatic embryogenesis from spear callus culture of *Asparagus cooperi* Baker. Plant Cell Rep., 9: 667-670.

Govil S, Gupta SC (1994) Plantmicropropagation industry in India retrospects and prospects. Jour. Sci. Ind. Res., 53: 768–777.

Guha,S. and Maheswari, S.C.(1964). Invitro production of embryos from anthers of *Datura*. Nature , 204: 497.

Haberlandt, G. (1902). Kulturversuche mit isolierten Pflanzenzellen. Sitzungsber K Preuss Akad Wiss Wien. Math Naturwiss 111:69–92.

Hackett, W.P. and Murray, J.R.(1997).Approaches to understanding maturation or phase change. in: Geneve, R.L., Preece, J.E., Merkle, S.A., (eds.) Biotechnology of Ornamental Plants. CAB International, Wallingord, UK, pp. 73-86.

Hartmann, H. T., Kester, D.E., Davies, F.T. Jr. and Geneve, R.L.(2002). Hartmann and Kester's Plant Propagation Principles and Practices. Prentice-Hall, Upper Saddle River, N.J. pp 880.

Huang, J., Pray, C. and Rozelle, S.(2002). Enhancing the crops to feed the poor. Nature, 418: 678-684.

Hvoslef-Eide, A.K. and Preil, W.(2005). Liquid culture systems for *in vitro* plant propagation. Springer, Dordrecht.

Jackson, M.B., Saker, L.R., Crisp, C.M., Else, M.A. and Janowiak, F.(2003). Ionic and pH signalling from roots to shoots of flooded tomato plants in relation to stomatal closure. Plant Soil, 253: 103-113.

Jain, S.M.(2002). Feeding the world with induced mutations and biotechnology. Proc. Int. Nuclear Conference 2002 – Global trends and Perspectives. Seminar 1: agriculture and bioscience. Bangi, Malaysia: MINT; p. 1–14.

Jain, S.M.(2002). Tissue culture and induced mutations useful tools for floriculture industry. *In Vitro* Cellular & Developmental Biology-Plant, 38: 643.

Jheng, F.Y., Do, Y.Y., Liauh, Y.W., Chung, J.P.and Huang, P.L.(2006). Enhancement of growth and regeneration efficiency from embryogenic callus cultures of *Oncidium* 'Gower Ramsey' by adjusting carbohydrate sources. Plant Sci, 170: 1133-1140.

Kao, K.N. and Michayluk, M.R.(1975).Nutritional requirements for growth of *Vicia hajastana* cells and protoplasts at a very low population density in liquid media. Planta, 126: 105-110.

Karkonen, A.(2001).Plant tissue cultures as models for tree physiology: somatic embryogenesis of *Tilia cordata* and lignin biosynthesis in *Picea abies* suspension cultures as case studies. Ph.D Thesis, University of Helsinki, Finland.

Klericker, J.A.(1892). Eline Methode zur Isoliering lebender protoplasten. Oefuers. Vetenskaps Adad. Stockholm, 9: 463-471.

Kumar K B (1994) Commercialization of micropropagation. In: Mukhopadhyay S (ed) Commercialization of Micropropagated Plants in India (pp 25-29). *Naya Prakash, Calcutta*

Kumar, A.(1992).Somatic embryogenesis and high frequency plantlet regeneration in callus cultures of *Thevetia peruviana*. Plant Cell Tissue Organ Cult, 31: 47-50.

Kumar, C.R.M. and Nair, G.M.(1979). Note on the effect of planting material and levels of farmyard manure on the yield of *D. alata* J. Root. Crops, 5: 67-68.

Kunitake, H. and Mii, M.(1997).Somatic embryogenesis and its application for breeding and micropropagation in Asparagus (*Asparagus officinalis* L.). Plant. Biotechnol, 15: 51-63.

LA RUE, C.D. (1949). Cultures of the endosperm of maize. Am. J. Bot. 36, 798.

Lim-Ho, C. L.(1982) .Tissue culture of local orchid hybrids at the Singapore Botanic Gardens. in: Rao A. N. (ed.), pp. 295-300.

Mahaswari, N. and Lal,M.(1958). *In vitro* culture of Iberis amara. Nature,181: 631-632.

Mantell, S.H. and Hugo, S.A.(1989). Effects of photoperiod, mineral strength, inorganic ammonium, sucrose and cytokinin on root, shoot and microtuber development in shoot cultures of *Dioscorea alata* L and *D. bulbifera* L yams. Plant Cell Tiss Org Cult,16: 23–37.

Miller, C.O., Skoog, F., Von Saltza, M.H. and Strong, F.M. (1955). Kinetin, a cell division factor from deoxyribonucleic acid. J Am Chem Soc 77:1392-1396.

Murashige, T. and Skoog, F. (1962). A revised medium for rapid growth and bioassays with tobacco tissue cultures. Physiol Plant 15:473–497

Ohyama, K.(1970). Tissue culture in mulberry tree. Jpn. Agric. Res. Quart, 5: 30-34.

Pandey, D., Malik, S., Bora, M. and Srivastava, P.S.(2002).Rapid protocol for in-vitromicropropagation of *Lepidium sativum* L. and enhancement in the yield of lepidine. In-vitroCell Dev. Biol- Plant, 38: 451-455.

Pierik, R.L.M.(1991a). Micropropagation of ornamental plants. Acta. Hortic, 289: 45–53.

Pierik, R.L.M.(1991b). Commercial micropropagation in Western Europe and Israel. in: Debergh, P.C., Zimmerman, R.H., (Eds). Micropropagation. Kluwer Acad. Publ, Netherlands, pp. 155–65.

Powell, W. and Uhrig, H.(1987). Anther culture of *Solanum* genotypes. Plant Cell Tissue Organ Cult, 11: 13-24.

Power, J.B. and Cocking, E.C.(1971). Fusion of plant protoplast. Sci.Progr,59: 181-198.

Rajagopalan, C.(2000). Export potential of Indian floriculture and need of policy environment. Floric Today, 9: 29–33.

Rout, G.R. and Das, P.(1997). *In-vitro* organogenesis in ginger (*Zingiber officinale* Rosc.). J. Herbs. Spices and Med Plants, 4: 41-51.

Satheesh, K. K. and Bhavanandan, K.V.(1988). Micropropagation of Plumbago rosea Linn. Plant Cell Tissue Organ Cult, 15: 275- 278.

Saxena, C., Palai, S.K., Samantaray, S., Rout, G.R. and Das, P.(1997). Plant regeneration from callus cultures of *Psoralea corylifolia* L. Plant Growth Reg, 22: 13-17.

Schiva, T.(2000). Strategies for development of commercial floriculture in Asia and the Pacific. Report of the APO seminar, 2nd – 6th May, New Delhi, India, pp. 27–38.

Sharma, M.(2001). Persley, G.J., MacIntyre, L.R., (Eds.), Agricultural biotechnology: Country case studies. CABI Publishing, Wallingford, UK, pp. 51.

Sharma, M., Charak, K.S. and Ramanaiah, T.V.(2003). Agricultural biotechnology research in India: Status and policies. Current Science, 84: 297-302.

Sharma, P. and Rajam, M.V.(1995). Genotype, explant and position effects on organogenesis and somatic embryogenesis in eggplant (*Solanum melongena* L.). J. Exp. Botany. 46: 135- 141.

Shasany, A.K., Khanuja, S.P.S., Dhawan, S., Yadav, U., Sharma, S. and Kumar, S.(1998). High regenerative nature of *Mentha arvensis* internodes. J. Biosci, 23: 641-646.

Skoog, F. and Miller, C.O. (1957). Chemical regulation of growth and organ formation in plant tissues cultured *in vitro*. Symp Soc Exp Biol 11:118–131.

Steward, F.C. and Shantz, E.M.(1956). The chemical induction of growth in plant tissue cultures. in: Wain and Wightman, (Eds.) The Chemistry and Mode of Action of Plant Growth Substances. Butterworth Scientific Publications Ltd, London, pp. 165-186.

Street, H.E.(1969).Growth in organized and unorganized systems - knowledge gained by culture of organs and tissue explants. in: Steward, F. C., (Ed.). Plant Physiology - a Treatise. 5B, Academic Press, New York, pp. 3-224.

Street, H.E.(1977). Cell (suspension) cultures - techniques. in: Street, H.E., (Ed.) 1977a (q.v.), pp. 61-102.

Teisson, C. and Alvard, D.(1995).A new concept of plant *in vitro* cultivation in liquid medium: temporary immersion. in: Terzi, B., Cella, R., Falavigna, A., (Eds.). Current issues in plant molecular and cell biology. Academic Publishers, Dordrecht, pp. 105-109.

Tyagi, A. K., Mohanty, A., Bajaj, S., Choudhary, A. and Maheshwari, S. C.(1999). Transgenic rice: A valuable monocot system for crop improvement and gene research. Crit. Rev. Biotechnol, 19: 41-79.

Verslues, P.E., Ober, E.S.and Sharp, R.E.(1998). Root grwoth and oxygen relations at low water potentials. Impact of oxygen availability in polyethylene glycol solutions. Plant Physiol, 116: 1403-1412.

Wakhlu, A.K., Nagari, S. and Barna, K.S.(1990).Somatic embryogenesis and plant regeneration from callus cultures of *Bunium persicum* Boiss. Plant Cell Rep, 9: 137-138.

Weathers, P. and Giles, K.L.(1987). Regeneration of many varieties of plants using a novel method of *in vitro* cell culture. *In Vitro* Cell Dev. Biol, 23: 69A (Abst.).

Went. (1926). Growth accelerating substance in coleoptile of *Avena sativa*. Proc. K. Akad Wet Amsterdam.30:10-19.

White, P.R. (1939). Potentially unlimited growth of excised plant callusin an artificial nutrient. Am J Bot 26:59–64.

Zhou, J., Ma, H., Guo, F. and Luo, X.(1994). Effects of thidiazuron on somatic embryogenesis of *Cayratia japonica*. Plant Cell Tissue Organ Cult, 36: 73-79.

❑❑❑

CHAPTER – 2

in vitro Clonal Propagation of Forests Trees : By Bud Culture Technique

Afaque Quraishi
Department of Biotechnology, Pt. Ravishankar Shukla University, Raipur (C.G.) India
Email : dr_afaque@rediffmail.com

Summary

Tropical forests are declining sharply and continuously growing demand of wood leads to removal of trees from their natural habitats. Clonal propagation of superior trees is must for further exploration and conservation of these genetic resource. Tissue culture technique offers rapid and large scale propagation of desired traits. However, micropropagation of adult trees is difficult; many problems are associated like- maturity, contamination, phenolic leaching, season, variants, long life cycle, etc. Shoot bud culture is the easiest and successful tissue culture technique for the toughest material to clone-mature trees. The chapter deals with the bud culture method of heartwoods and associated problems at different stages – explant establishment, shoot multiplication, rooting and acclimatization. Also the possible solutions of the problems are discussed on the basis of previous reports.

1. Introduction

World's green cover by tropical forests is declining at a rapid pace for the demand of timber, pulp and other products; however, the vast depletion of tropical forests is for fuel wood. Reforestation requires tree improvement and production of genetically superior trees for producing fast growing trees with better wood quality, disease resistance and stress tolerance (Durzan 1988). Hence it is necessary to protect present existing resources.

For the purpose these genetic gains can be maintained by clonal propagation (Durzan 1988). Cloning of mature trees is must as it is not possible to determine the desired qualities in seedlings. While conventional propagation through natural regeneration and vegetative cuttings may be slow and cumbersome. Moreover, cuttings and rooted stock involves lengthy delays due to quarantine procedures to control the spread of disease.

In contrast, the *in vitro* culture offers a sustainable and viable tool for rapid propagation and storage of germplasm. Clonal propagation by tissue culture is an alternative to cuttings and has the potential to provide high multiplication rates of uniform genotypes, resulting in short-term gains (Durzan 1988). Also these aseptic *in vitro* cultures would reduce the chances of introducing foreign pathogens (Jones 1986) and allow for the exchange of desirable clones *in vitro* to various countries for selection work (Jones et al. 1991). However, use of tissue culture technique to clone the tropical trees is less frequent.

India is one of the world's top 12 mega-diversity countries (Singh and Chowdhery 2002). Increasing demand for wood leading to continuous depletion of forest land is continuously losing its natural flora at an alarming pace – 1.5 m. ha every year (Shivarajan and Balachandran 1999). Thus propagation and conservation of the fast depleting genetic diversity of forest trees is one of the major concerns faced today in India; and immediate efforts are required in this regard. *In vitro* clonal propagation has the capacity to produce large numbers of plants within limited space and time irrespective of the season, under controlled conditions of temperature, light intensity, and photo period. It also offers an alternative cloning procedure for forest trees that do not respond well to conventional methods of propagation, either sexual or vegetative (Bonga 1987).

2. Micropropagation by Bud Culture

Micropropagation of mature trees is difficult due to various factors, like maturity, contamination, presence of phenolic compounds, long life cycle, great genetic variations, etc. Bud culture is the easiest and successful method to propagate toughest materials- 'mature trees'; because the small shoot is already present in the form of small bud. Thus, for complete plant regeneration only elongation of this bud and then rooting is required; the technique has been applied to a wide range of mature Indian tree species (Table 1). We will discuss the associated problems and solutions as well, step by step for different micropropagation stages.

2.1. Explant Selection

It is well established that *in vitro* propagation of plants is also influenced by source of initial tissue which is related to their endogenous hormonal status (George 1993). Also maturational gradients exist in trees with maturation increasing towards the top of the tree (Bonga 1982). It is reported that coppice shoot explants rather than explants from crown branches of mature neem tree are better source for micropropagation

(Quraishi et al. 2004). For *Aegle marmelos* tree, 11 - 15th nodal stem segments of twigs proved to the best explants to initiate cultures (Pati et al. 2008). Pe´rez-Tornero et al. (2010) took shoots for explants from rapidly-extending new branches of lemon tree during spring. Nodal explants of *Ochreinauclea missionis* tree were more responsive compared to shoot tips (Dalal and Rai 2001). For collecting nodal explants, some of the branches of sandalwood tree were pruned periodically to obtain new sprouts (Sanjay et al. 2006). Position of the twig on the mother plant also influenced the response of the nodal segments; nodal explants of vertically oriented twigs of *Crataeva adansonii* exhibited better response than the explants of angular branches (Sharma et al. 2003).

Healthy current-year shoots were collected in May from scions of superior teak trees (Shirin et al. 2005). These were grafted and on the root stock for 5 years. Nodal segments of these grafted shoots were used as explants.

2.2. Explant Establishment

Generally nodal segments with single axillary bud are used as explants to initiate shoot culture. Establishment of cultures *in vitro* using mature tissues is problematic. Mature trees are often recalcitrant to vegetative reproduction (Jones and van Staden 1997). As it is easier to propagate juvenile tissues (Thorpe et al. 1991), rejuvenation and the establishment of juvenile characteristics prior to *in vitro* culture have proved to be important (Jones and van Staden 1997). When rejuvenating mature tissues, the most juvenile tissue and the smallest explant possible should be used so as to disrupt the controlling forces of the mother plant. These controlling forces can be reduced by pretreatment of the plant with hormone applications and the use of repeated culture (Bonga 1987).

2.2.a. Surface sterilization

In tree tissue culture, contamination of cultures is the major problem. Since the mature trees can't be grown in green houses, as in case of crops and flowering plants. Different strategies have been adopted for microbial free explant to initiate culture; in general mercuric chloride, ethanol, different fungicides and antibiotics have been used for the purpose. For example coppice shoots of *Acacia mearnsii* (Beck et al. 1998) were pretreated for 7 days with fungicides, prior to taking explants. Coppice material was surface sterilized in 70% ethanol and soaked in fungicide, combined with boric acid. Surface disinfestations of the explants was done using $HgCl_2$, then the explants were thoroughly washed in sterile distilled water.

The nodal explants of *Aegle marmelos* (Pati et al. 2008) were washed thoroughly with Tween-20 and were rinsed in running tap water for 30 minutes. The explants were treated with fungicide and antibiotic. The explants were treated with $HgCl_2$ and washed with sterile distilled water and dried on pre sterilized filter paper sheets.

Surface sterilization was done by treating the explants of *Holarrhena antidysenterica* (Kumar et al. 2005) with fungicide- Bavistin, followed by 70% ethanol and $HgCl_2$, antibiotic solution under laminar hood.

2.2.b. Season

Generally for tropical trees growing season during spring is found suitable to establish the newly growing contamination free buds and plants are at correct physiological state. As is *Holarrhena antidysenterica* trees, nodal explants exhibited maximal morphogenic response from May to July (Kumar et al. 2005); similarly April for Neem (Arora et al. 2010), March for *Melia azedarach* (Husain and Anis 2009) and May for Teak (Tiwari et al. 2002).

2.2.c. Phenolic leaching

Most of the tree explants released excessive phenolics into the culture medium from their cut ends and may die; another big hurdle for tree culture as observed in Teak (Tiwari et al. 2002), Neem (Arora et al. 2010), *Pterocarpus santalinus* (Prakash et al. 2006), *Terminalia arjuna* (Pandey et al. 2006), *Cleistanthus collinus* (Quraishi and Mishra 1998), etc. This problem was overcome by frequent transfer of explants to the fresh medium (Tiwari et al. 2002; Prakash et al. 2006); suspending explants in absorbent-Polyvinylpyrollidone (PVP) and/or antioxidants solution and also incorporation of these chemicals to the explant establishment medium (Purohit and Tak 1992; Quraishi et al. 2004). For *Ochreinauclea missionis,* activated charcoal is incorporated in the medium to check the phenolic exudates, as effective absorbent (Dalal and Rai 2001).

2.2.d. Rejuvenation

Sometimes in order to search most juvenile tissue within the mature tree for its clonal propagation, explants from the coppice shoots exhibited better regeneration capacity *in vitro.* Probably because of an increased availability of hormones and other chemical factors produced by the roots (Bonga 1987). Improved *in vitro* propagation from these basal sprout explants has been reported in Neem (Quraishi et al. 2004), *Acacia mearnsii* (Beck et al. 1998), *Artocarpus heterophyllus* (Amin and Jaiswal 1993) and *Eucalyptus tereticornis* (Das and Mitra 1990).

2.2.e. Medium

As observed in herbs, medium developed by Murashige and Skoog (MS) (1962) is the most successful medium for micropropagation of trees also. It is considered as properly balanced medium for both inorganic and organic nutrients to fulfill the growth requirements. Amongst the five different basal media assayed (B5 (Gamborg et al. 1968), Nitsch (Nitsch 1969), Woody Plant Medium (WPM) (Lloyd & McCown 1980), MS, and Knop's (Knop, 1865) media. MS was the most effective for *in vitro* shoot multiplication of nodal explants from *Holarrhena antidysenterica* trees (Kumar et al. 2005). Husain and Anis (2009) also reported MS medium was found best for the cultures of *Melia azadarch* tree; followed by ½ MS, WPM and B5 media in decreasing order. Some other media have also been used like- Driver and Kuniyuki medium (DKW) (1984) for lemon (Pe´rez-Tornero et al. 2010).

2.2.f. Plant growth regulators

Plant hormones- cytokinin, auxin and gibberellic acid (GA), play key role to give directions to the culture growth. A higher cytokinin to auxin ratio in the medium promotes shoot formation and a higher auxin to cytokinin ratio induced root differentiation (Skoog and Miller 1957). The requirement of hormones depends on endogenous levels of cultured tissue. Synthetic substances which mimic plant hormones- plant growth regulators; are generally more effective than natural plant hormones like 6-benzyladenine (BA) for shoot proliferation and indole-3-butyric acid (IBA) to induce rooting. Probably because of these plant growth regulator are more stable in plant systems than the natural ones.

Generally explant establishment medium of trees contain a cytokinin; as the cytokinin influenced cell division, change in apical dominance, shoot differentiation, etc. The mostly used cytokinin is BA ranging from 0.5 to 2 mg/l levels. However, other cytokinins also effective for particular species, as Ahmad and Anis (2007) reported best bud break response in presence of Thidiazuron (TDZ), in *Vitex negundo*. Sometimes combination of cytokinin and much lower concentration of an auxin was found effective for shoot bud establishment: BA & 1-naphthelene acetic acid (NAA) in *Terminalia arjuna* (Pandey et al. 2006), Teak (Shirin et al. 2005), *Holarrhena antidysenterica* (Mallikarjuna and Rajendrudu 2007), *Santalum album* (Sanjaya et al. 2006); BA and indole-3-acetic acid (IAA) in *Aegle marmelos* (Pati et al. 2008), *Murraya koenigii* (Rout 2005). However, in lemon, the best results regarding shoot productivity were obtained with the combination of BA and GA (Pe´rez-Tornero et al. 2010).

2.3. Shoot Multiplication

Shoot proliferation is the second stage of micropropagation; and is very important for its commercial feasibility. As higher shoot production rate cut down the production cost via tissue culture technique. This can be achieved by excision of elongated shoot from the original explant aseptically and divided into nodal segments followed by subsequent transfer in multiplication medium. Incubation in multiplication medium for 4 to 6 weeks, shoots has been regenerated. Again these shoots have been divided in nodal segments and transferred to fresh multiplication medium for further shoot proliferation, called subculture cycles. Generally with increasing subculture cycles enhanced the shoot multiplication rate may be due to the *in vitro* rejuvenation. Shoot multiplication medium may be similar to the explant establishment medium or little manipulation is required for the PGR requirements. Further, if leaching of phenolics is absent at this stage then the antioxidants and absorbents can be withdrawn from the multiplication medium. In some reports, incorporation of adjuvant in multiplication medium have been found beneficial for shoot production, Adenine sulfate, casein hydrolysate (CH), polyvinyl pyrollidone (PVP), etc. The effectiveness of CH could be attributed to different amino acids present which affect nitrogen assimilation, alleviation of phosphate deficiency, replacement of toxic ammonium ions and chelation of divalent metals; hence the overall response (George 1993). Adenine in the form of adenine sulfate can stimulate cell growth and greatly enhance shoot formation (Murashige 1974).

Vitrification is another constraint encountered during multiplication. Vitrified shoots have glassy hydrated appearance, become thick and brittle with abnormal leaves. These shoots are with poor growth, become necrotic and subsequently die. Generally liquid medium causes vitrification than semisolid medium. Similarly continuous culture in high cytokinin may promote vitrification. Vitrification can be reduced by using semisolid medium or by increase the concentration of gelling agent; lowering the cytokinin level; or by improving gaseous exchange of culture vessels.

2.4. Rooting and Acclimatization

For rooting, i*n vitro* regenerated shoots in multiplication medium are excised aseptically and individual shoots transferred to rooting medium in order to produce complete plants. For *in vitro* rooting, usually the salt concentration is reduced to half than multiplication medium. And this rooting medium is supplemented with a suitable auxin- indole-3-butyric acid (IBA), NAA or IAA. Maximum rooting percentage recorded in IBA containing medium in lemon (Pe′rez-Tornero et al. 2010), *Terminalia arjuna* (Pandey et al. 2006), *Pterocarpus santalinus* (Prakash et al. 2006); in NAA supplemented medium in Teak (Shirin et al. 2005); in IAA enriched medium in *Holarrhena antidysenterica* (Kumar et al. 2005); or with the combination of these auxins.

For primary hardening, usually plantlets with well developed roots removed from the culture vessels and washed in running tap water to remove medium. These plantlets treated with fungicide and/or antibiotics and then transplanted in small pots or seed trays filled with vermiculite or coconut husk or sand; kept in greenhouse with high relative humidity and 30°C ±2 temperature for 4-6 weeks. These acclimatized plants then transferred to poly-bags filled with soil mixture under partial shade for secondary hardening. Well developed plants are then ready for the plantation in field.

Ex vitro rooting of regenerated shoot is another option, which is advantageous because of simultaneous rooting and hardening of plants. It also reduced the production cost considerably. Further, these plants exhibited a high percentage of survival (Moncousin 1991). For this, *in vitro* grown shoots removed from culture and exposed to quick dip of high auxin concentration and then transferred to pots or seed trays filled with potting mixture and kept in green house condition – high humidity and ambient temp. *Ex vitro* rooting and acclimatization has been reported in *Melia azedarach* (Husain and Anis 2009), Teak (Tiwari et al. 2002), *Ochreinauclea missionis* (Dalal and Rai 2001), etc.

3. Somaclonal Variations

The occurrence of genetic variation in micropropagated plants is matter of great concern and its success depends only on the clonal uniformity (Bonga and Durzan 1987). Generally, culture conditions (Thorpe 1990), genotype (Mohmand and Nabors 1990), and ploidy level and *in vitro* culture duration (Lee and Phillips 1987) are known to be associated with the occurrence of somaclonal variation.

Therefore it is important to determine the particular micropropagation protocol for a particular plant for the production of genetically uniform plants without somaclonal variations; especially for trees with long life span (Rani and Raina 2000). A multidisciplinary approach at morphological, chromosomal, biochemical and molecular levels would be best for the complete characterization of tissue culture derived plants. If protocol induced variants then it should be modified in the light of above mentioned causes. So make sure that there is no incidence of variants in tissue cultured plants or at least minimized at acceptable levels.

Table 1 : List of some mature Indian trees propagated via the bud culture technique

Tree species	References
Aegle marmelos	Pati et al. (2008)
Artocarpus heterophyllus	Amin and Jaiswal (1993)
Azadirachta indica	Arora et al. (2010), Quraishi et al. (2004)
Bauhinia vahlii (climber)	Dhar and Upreti (1999)
Bauhinia variegata	Mathur and Mukunthakumar (1992)
Citrus limon	Pe´rez-Tornero et al. (2010)
Cleistanthus collinus	Quraishi and Mishra (1998)
Commiphora wightii	Barve and Mehta (1993)
Crataeva adansonii	Sharma et al. (2003)
Dalbergia sissoo	Gulati and Jaiwal (1996)
Dalbergia latifolia	Raghava Swamy et al. (1992)
Ochreinauclea missionis	Dalal and Rai (2001)
Feronia limonia	Purohit and Tak (1992)
Holarrhena antidysenterica	Mallikarjuna and Rajendrudu (2007), Kumar et al. (2005)
Melia azedarach	Husain and Anis (2009)
Murraya koenigii	Rout (2005)
Pongamia pinnata	Sujatha and Hazra (2007)
Psidium guajava	Amin and Jaiswal (1987)
Pterocarpus santalinus	Prakash et al. (2006)
Santalum album	Sanjaya et al. (2006)
Tectona grandis	Shirin et al. (2005), Tiwari et al. (2002),
Terminalia arjuna	Gupta et al. (1980), Devi et al. (1994)
Vitex negundo	Pandey et al. (2006), Ahmad and Anis (2007)

4. Conclusion

Demand for timber and pulp is continuously growing all over the world and natural forests are under great pressure. Micropropagation of trees is good option to exploit genetic gain of desired traits; as tissue culture technique has been applied to a wide range of tree species. It is very important method which can play an important role in forest tree improvement programs via improving the quality of the trees that are produced. However *in vitro* tree propagation is lagging far behind. Lot of challenges season, contamination, maturity, phenolic leaching, genetic variations, long life cycle, big resources and lands for field trials, etc. have to overcome. Nevertheless conservation and utilization of existing valuable resources is must and to make sure that this valuable wealth should available for future as well. Thus the micropropagation can be used for mass propagation and conservation of forest trees and would also helpful in different forestry programs and further future tree improvement using genetic transformation technology.

References

Ahmad N and Anis M (2007) Rapid clonal multiplication of a woody tree, *Vitex negundo* L. through axillary shoots proliferation. Agroforest Syst 71:195–200

Amin MN and Jaiswal VS (1993) *In vitro* response of apical bud explants from mature trees of jackfruit (*Artocarpus heterophyllus*). Plant Cell Tissue Organ Cult 33:59–65

Amin MN and Jaiswal VS (1987) Rapid clonal propagation of guava through *in vitro* shoot proliferation on nodal explants of mature trees. Plant Cell, Tissue and Organ Culture 9: 235-243

Arora K, Sharma M, Srivastava J, Ranade SA and Sharma AK (2010) Rapid *in vitro* cloning of a 40-year-old tree of *Azadirachta indica* A. Juss. (Neem) employing nodal stem segments. Agroforest Syst 78:53–63

Barve DM and Mehta AR (1993) Clonal propagation of mature elite trees of *Commiphora wightii.* Plant Cell, Tissue and Organ Culture 35: 237-244

Beck SL, Dunlop R and van Staden J (1998) Rejuvenation and micropropagation of adult *Acacia mearnsii* using coppice material. Plant Growth Regulation 26: 149–153

Bonga JM (1987) Clonal propagation of mature trees: problems and possible solutions. In: Bonga JM, Durzan DJ (eds) Cell and tissue culture in forestry, vol 3. Martinus Nijhoff, Dordrecht, pp 249–271

Bonga JM (1982) Vegetative propagation in relation to juvenility, maturity and rejuvenation. In: Bonga JM & Durzan DJ (eds) Tissue Culture in Forestry (pp. 387–412). Nijhoff/Junk, The Hague

Bonga JM and Durzan DJ (1987) Cell and tissue culture in forestry. Vol. 1. Dordrecht: Martinus Nijhoff

Dalal NV and Rai RV (2001) *In vitro* propagation of *Ochreinauclea missionis* (Wall. Ex G. Don), an ethnomedicinal endemic and threatened tree. *In vitro* Cell. Dev. Biol.- Plant 37: 820-823

Das T and Mitra GC (1990) Micropropagation of *Eucalyptus tereticornis*. Plant Cell Tissue Organ Cult 22:95–103

Devi YS, Mukherjee BB and Gupta S (1994) Rapid cloning of elite teak (*Tectona grandis* Linn.) by *in vitro* multiple shoot production. Indian J. Exp. Bio. 32: 668–671

Dhar U and Upreti J (1999) *In vitro* regeneration of a mature leguminous liana (*Bauhinia vahlii* Wight & Arnott) Plant Cell Reports 18: 664–669

Driver JA and Kuniyuki AH (1984) *In vitro* propagation of paradox walnut rootstock. HortSci. 19: 507–509

Durzan, DJ (1988) Applications of cell and tissue culture in tree improvement. In: Bock, G., Marsh, J., eds. Applications of plant cell and tissue culture (Ciba Foundation Symposium 137). Chichester: Wiley: 36–59

Gamborg OL, Miller RA and Ojima K (1968) Nutrient requirements of suspension cultures of soybean root cells. Exp. Cell Res. 50: 151–158

George EF (1993) Plant propagation by tissue culture, parts 1 and 2. Basingstoke: Exegetics Ltd.

Gulati A and Jaiwal PK (1996) Micropropagation of *Dalbergia sissoo* from nodal explants of mature trees. Biologia Plantarum 38(2): 169-175

Gupta PK, Nadgir Al, Mascarenhas AF and Jaganathan V (1980) Tissue culture of forest trees – Clonal multiplication of *Tectona grandis* (teak) by tissue culture. Plant Sci. Lett. 17: 259–268

Husain MK and Anis M (2009) Rapid *in vitro* multiplication of *Melia azedarach* L. (a multipurpose woody tree). Acta Physiol Plant 31: 765–772

Jones C (1986) Getting started in micropropagation of Tasmanian blackwood (*Acacia melanoxylon*). Proc. Int. Plant Prop. Soc. 36:477–481; 1986

Jones C, Smith D, Gifford H and Nicholas I (1991) Field trial results of *Acacia melanoxylon* from tissue culture. Proc. Int. Plant Prop. Soc. 41:116–119; 1991

Jones NB and van Staden J (1997) Micropropagation of Eucalypts. In: Bajaj YPS (ed) Biotechnology in Agriculture and Forestry, Vol. 39, High-Tech and Micropropagation, V. Springer-Verlag, pp 286–329

Knop W (1865) Quantiative undersuchung iiber die emahrungsprocess der Pflanzen. Landw. VerStnen. 7: 93–107

Kumar R, Sharma K and Agrawal V (2005) *In vitro* clonal propagation of *Holarrhena antidysenterica* (L.) Wall. Through nodal explants from mature trees. *In vitro* Cell. Dev. Biol.- Plant 41: 137-144

Lee M and Phillips RL (1987) Genomic rearrangements in maize induced by tissue culture. Genome 29: 122-128

Lloyd G and McCown B (1980) Commercially feasible micropropagation of mountain laurel, *Kalmia latifolia*, by use of shoot tip culture. Proc Int Comb Plant Prop Soc 30:421–427

Mallikarjuna K and Rajendrudu G (2007) High frequency *in vitro* propagation of *Holarrhena antidysenterica* from nodal buds of mature tree. Biologia Plantarum 51(3): 525-529

Mathur J and Mukunthakumar S (1992) Micropropagation of *Bauhinia variegata* and *Parkinsonia aculeata* from nodal explants of mature trees. Plant Cell, Tissue and Organ Culture 28: 119-121

Mohmand AS and Nabors MW (1990) Somaclonal variant plants of wheat derived from mature embryo explants of three genotypes. Plant Cell Rep. 8: 558-560

Moncousin CH (1991) Rooting of *in vitro* cuttings. In: YPS Bajaj (ed) Biotechnology in Agriculture and Forestry, Vol. 17. High-Tech and Micropropagation. Springer-Verlag, Berlin, Heidelberg, Germany

Murashige T (1974). Plant propagation through tissue culture, Annu. Rev. Plant Physiol., 25:135–166.

Murashige T and Skoog F (1962) A revised medium for rapid growth and bioassays with tobacco tissue cultures. Physiol. Plant. 15: 473–497

Nitsch JP (1969) Experimental androgenesis in *Nicotiana*. Phytomorphology 19: 389–404

Pandey S, Singh M, Jaiswal U and Jaiswal VS (2006) Shoot initiation and multiplication from a mature tree of *Terminalia arjuna* Roxb. *In Vitro* Cell. Dev. Biol.—Plant 42:389–393

Pati R, Chandra R, Chauhan UK, Mishra M and Srivastava N (2008) *In vitro* clonal propagation of bael (*Aegle marmelos* Corr.) CV. CISHB1 through enhanced axillary branching. Physiol. Mol. Biol. Plants 14(4): 337-346

Pe´rez-Tornero O, Tallo´n CI and Porras I (2010) An efficient protocol for micropropagation of lemon (*Citrus limon*) from mature nodal segments. Plant Cell Tiss Organ Cult 100: 263–271

Prakash E, Khan PSSV, Rao TJVS and Meru ES (2006) Micropropagation of red sanders (*Pterocarpus santalinus* L.) using mature nodal explants. J For Res 11:329–335

Purohit SD and Tak K (1992) *In vitro* propagation of an adult tree *Feronia limonia* L. through axillary branching. Indian J Exp Biol 30:377–379

Quraishi, Afaque, Koche V, Sharma P and Mishra SK (2004) *In vitro* clonal propagation of neem (*Azadirachta indica*). Plant Cell, Tissue and Organ Culture 78: 281–284

Quraishi A and Mishra SK (1998) Micropropagation of nodal explants from adult trees of *Cleistanthus collinus*. Plant Cell Reports 17: 430–433

Raghava Swamy BV, Himabindu K and Lakshmi Sita G (1992) *In vitro* micropropagation of elite rosewood *(Dalbergia latifolia* Roxb.) Plant Cell Reports 11:126-131

Rani V and Raina SN (2000) Genetic fidelity of organized meristem-derived micropropagated plants: A critical reappraisal. *In Vitro* Cell. Dev. Biol.-Plant 36: 319-330

Rout GR (2005) Direct plant regeneration of Curry Leaf Tree (*Murraya koenigii* Koenig.), an aromatic plant. *In Vitro* Cell. Dev. Biol.—Plant 41:133–136

Sanjaya, Muthan B, Rathore TS and Rai VR (2006) Micropropagation of an endangered Indian sandalwood (*Santalum album* L.) J For Res 11: 203–209

Sharma PK, Tyagi P, Sharma KC and Kothari SL (2003) Clonal micropropagation of *Crataeva adansonii* (DC.) Prodr.: A multipurpose tree. In Vitro Cell. Dev. Biol.—Plant 39: 156–160

Shivarajan VV and Balachandran I (1999) Ayurvedic Drugs and Their Plant Sources. Oxford and IBH Publishing Co. Pvt. Ltd., New Delhi, India, pp. 9.

Singh NP and Chowdhery HJ (2002) Biodiversity conservation in India. In: Das AP (ed), Perspective of Plant Biodiversity. Department of Botany, North Bengal University, West Bengal, India, pp 501–527.9–11

Shirin F, Rana PK and Mandal AK (2005) *In vitro* clonal propagation of mature *Tectona grandis* through axillary bud proliferation. J For Res 10:465–469

Skoog F and Miller CO (1957) Chemical regulation of growth and organ formation in plant tissues cultured *in vitro*. Symp. Soc. Exp. Biol. 11: 118-131

Sujatha K and Hazra S (2007) Micropropagation of mature *Pongamia pinnata* Pierre. *In Vitro* Cell.Dev.Biol.—Plant 43: 608–613

Thorpe TA (1990) The current status of plant tissue culture. In: Bhojwani, S. S., ed. Plant tissue culture: applications and limitations. Amsterdam: Elsevier Science: 1-33

Thorpe TA, Harry IS and Kumar PP (1991) Application of micropropagation to forestry. In: Debergh PC and Zimmerman RH (eds) Micropropagation. Dordrecht: Kluwer Academic Publishers, pp 311–335

Tiwari SK, Tiwari KP and Siril EA (2002) An improved micropropagation protocol for teak. Plant Cell, Tissue and Organ Culture 71: 1–6

❑❑❑

CHAPTER – 3

Evaluating Phytoremediation using *in vitro* Plant Cultures

*Indira P. Sarethy[1] and Kailash Paliwal[2]
[1]Department of Biotechnology, Jaypee Institute of Information Technology, Noida - 201307, India.
Email : Indirap.sarethy@jiit.ac.in
[2]Indian Institute of Advance Research, Gandhinagar, Gujarat- 382 007, India.

Summary

Human activities involving rapid industrialization and urbanization without concomitant environmental planning have resulted in a situation where substantial parts of our land and water resources have been polluted with organic and inorganic toxicants. Environmental engineering principles have been utilized to detoxify polluted areas; however, the efforts, adopting predominantly physico-chemical methods, have addressed the issue to a restricted extent, focusing on solutions for specific categories of pollutants and applicable for limited areas, in addition to being expensive. Hence cost-effective sustainable alternatives in the form of bioremediation (involving microorganisms) and phytoremediation (involving plants alone or along with associated rhizospheric microorganisms) are gaining public acceptance. Considerable scientific evidence has proved that many plant species can detoxify the pollutants utilizing various mechanisms. In this background, in vitro plant cultures have served as invaluable tools in identifying appropriate plant species for phytoremediation. Better understanding of the complex interactions underlying phytoremediation has been facilitated using in vitro cultures where the only variable is the pollutant, in contrast to field experiments where there are many confounding variables. While straightforward extrapolation of in vitro data can be erroneous, these systems provide the basic framework for translating laboratory-scale experiments to field-scale technologies. This chapter addresses the role of phytoremediation as a measure for removing various categories of environmental pollutants, the types of phytoremediation mechanisms involved, their benefits and limitations and the role of in vitro plant cultures in understanding how phytoremediation can be used for practical applications.

1. Introduction

Anthropogenic activities have resulted in relentless squalor of land and water due to severe and uncontained exploitation of natural resources and discharge of pollutants. Pollution originates from point sources like discharge of liquid and solid waste from industries or from vehicle exhaust or activities such as mining. The non-point sources pertain to the usage of fertilizers, insecticides, and herbicides as part of agricultural practices (Nriagu and Pacyna, 1988; Schalscha and Ahumada, 1998; McGrath et al., 2001). It is estimated that approximately 52 million hectares, comprising 16% of the total land area, is affected by soil pollution to various extents (Memon and Schroeder, 2009). The most heavily polluted areas are those in the industrialized regions. The problem is acute in developed countries; however, developing countries, including India, which face haphazard industrial growth and urbanization without proper attention being paid to planned development, face the danger of pollution-related damage in a shorter time span.

Current pollution-abatement practices utilize environmental engineering principles based on physico-chemical processes to undertake remediation measures and combat the ever-increasing soil and water pollution levels. However, these methods are difficult to implement on a large scale due to the expense involved as well as production of toxic compounds during the remediation itself. Hence, efforts are being made to develop sustainable and cost-effective technologies utilizing bioremediation techniques. Phytoremediation, a type of bioremediation, involves the use of plants and microorganisms associated with their rhizosphere, to eliminate, contain, or detoxify the pollutants. Many plant species have been found to carry out effective detoxification of different categories of pollutants. The conversions of the toxicant to harmless or less toxic compound occurs due to inherent ability of certain plants to either uptake and transform the toxic compound or accumulate or volatilize it or, in some cases, simply make the toxicant non-bioavailable.

2. Types of Environmental Pollutants

Xenobiotics (chemicals released due to human/anthropogenic activities) include organic and inorganic chemicals (Schwitzguebel, 2004). The various categories of environmental pollutants are summarized in the following sections.

2.1 Organic Pollutants

2.1.1 Petroleum-based products

Petroleum spillages (accidental or during extraction) occur in soil as well in water bodies and require imperative measure for proper treatment. Crude oil spills are a major problem worldwide and contaminate the soil and water to large extents, requiring substantial expenses for clean-up (Bundy et al., 2002). This is because petroleum comprises a mixture of different types of hydrocarbons with diverse properties and thus needs remediating technologies to address each of them. The total petroleum hydrocarbons (TPH) are hazardous because they can also accumulate in food chains (Heitkamp and Cerniglia, 1988; McElroy et al., 1989). TPH are divided into two major categories: Gasoline consists of small-chain alkanes or volatile aromatics (C6-C10) with relatively low boiling point (60-170^{0}C). Isopentane, n-butane, n-pentane, the two-ringed PAH naphthalene and BTEX (benzene, toluene, ethylbenzene and xylene) belong to this group. Gasoline is comprised of 40-70% alkanes and cycloalkanes, about 10% alkenes, and 20-50% aromatics (AEHS, 1998a and 1998b). The benzene group compounds are recognized to be mutagens and carcinogens (Mortelmans et al., 1986]. The diesel group contains long-chain alkanes (C10-C40) and the polyaromatic hydrocarbons (PAHs), with boiling points between 200-300^{0}C. This group is comprised of 60-90% alkanes and cycloalkanes, about 5% alkenes and 10-30% aromatics (AEHS, 1998a; Trapp et al., 2001). Some of the PAHs have been identified as priority pollutants by the US Environmental Protection Agency (EPA). PAHs can also arise due to incomplete coal combustion, oil or organic matter (Kanaly and Harayama, 2000; ATSDR, 2004).

2.1.2 Polychlorinated biphenyls

The polychlorinated biphenyls (PCB) are lipophilic and differ in the number of chlorine atoms bound to biphenyl ring. These are used for filling transformers and as paint additives. However their lipophilic nature results in their accumulation in the food chain as parts of fat deposits. While their production has been stopped in most countries due to concerns about their toxicity, many volatile PCBs are still cases for concern since they contaminate large areas (Mackova et al., 2009).

2.1.3 Nitroesters

The nitrate ester group is made of different compounds such as nitroglycerine (NG) and Ethylene glycol (EGDN) (Ye et al., 2004). NG is used for manufacturing dynamite and rocket propellants. It also has a therapeutic use for heart problems. EGDN is an important component of dynamite along with NG. These compounds persist in the environment and cannot be easily degraded using chemical methods. They are used by construction industries and as plasticizers for solid propellants (Christodoulatos et al.,

1997; Marshall and White, 2001). The degradation products also are toxic. These nitrate esters have exhibited toxicity in many animals (Ellis et al., 1984; Zalta et al., 1985; Imoto et al., 1986; Johansson et al., 1987; Tai and Tsuruta, 1997; Lowry et al., 2001). In man, toxicity symptoms are severe headaches and cardiovascular problems (Fukuchi, 1981; Kanerva et al., 1991; Stayner et al., 1992; Lamm et al., 1993; Santoro et al., 2001). NG also causes phytotoxicity (Hannink et al., 2002). Trinitrotoluene (TNT) is a nitroaromatic compound found around ammunition factories. It also presents considerable environmental problems due to its toxicity and persistence in soil (Won et al., 1976; Spain, 1995).

2.1.4 Dyes

It is estimated that at least 10^6 tons of 10,000 types of dyes or pigments are produced annually on a global scale (Schwitzguebel et al., 2002a). A significant quantity (280,000 tons) is also discharged as waste (Willetts and Ashbolt, 2000). Synthetic anthraquinone is a component of dyes and its sulphonated derivatives have a wide variety of uses. For example, benzene sulphonate (BS), *p*-toluene sulphonate (*p*-TS) and naphthalene sulphonic acids are used in manufacture of pickling agents, insecticides, wetting agents, dispersants and anti-oxidants. Due to the presence of sulphonic group or sometimes a nitro group as side chains, these compounds, by and large, cannot be easily subjected to biodegradation; however, BS and *p*-TS are still moderately biodegradable (Khlebnikov et al., 1997). The organosulphonate group, comprising of the carbon-sulphur bond that is thermodynamically stable, confers stability and prevents its degradation (Kertesz et al., 1994). Hence this group of compounds requires urgent attention for biologically-mediated degradation systems (Riediker et al., 2000).

2.1.5 Trichloroethylene

Trichloroethylene (TCE) is a volatile organic compound and has been shown to degrade to form compounds such as trichloroethanol, trichloroacetic acid, and dichloroacetic acid in hybrid poplars (Newman et al., 1997). It is a common groundwater pollutant. It causes hepatotoxicity and is carcinogenic too.

2.1.6 Phenols & Anilines

Phenols and anilines are alcohol and amine derivates of benzene respectively and are soluble in polar as well as non-polar solvents (Harvey et al., 2002). These are primarily produced as by-products during bleaching of pulp and are used as wood preservative (Keith and Telliard, 1979). Pentachlorophenol (PCP) is used as an herbicide and is found in wood preservative treatment sites (Crosby, 1981). Derivatives of phenol such as chlorophenols are highly recalcitrant and carcinogenic too (Waidyanatha et al., 1996).

Anilines are components of synthetic dyes, herbicides, and anti-oxidants in the rubber industry (Ellenhorn and Barceloux, 1988). Around 10% is discharged into industrial wastewater effluents (Vaidya and Datyre, 1982). They are extremely stable compounds and can resist extremes of light and temperature. Azo derivatives are carcinogenic under anaerobic conditions.

2.1.7 Pharmaceuticals & personal care products (PPCPs)

Emerging micropollutants, comprising the organic as well as pharmaceuticals represent new threats to the environment. The group of compounds making up the pharmaceuticals and personal care products (PPCPs) are of significance in terms of their contribution to environmental pollution. These compounds are primarily manufactured for daily use and are popular ingredients in medicines and cosmetics. They exhibit poor biodegradability. Environmental concerns center around their ever-increasing usage and improper disposal since their utility is primarily domestic. Manufacturing units pose another area of concern though they predominantly conform to environmental legislations regarding waste disposal. Recent studies show that the pharmaceutically active compounds occur in ng/l to µg/l range (Kolpin et al., 2002; Schröder et al., 2007). The other group of PPCPs i.e. housekeeping chemicals and personal care products (such as sun-screen creams) are also discharged into water bodies. Additional ingredients present in them such as solvents or fragrances pose a further hazard. These compounds can also show endocrine-disrupting activity also, altering hormone levels. In man, these compounds can together lead to increased incidences of breast and testicular cancer, decreased sperm counts and more reproductive abnormalities. The major pharmaceuticals discharged into the environment are analgesics (ibuprofen, diclofenac), psycholeptics (diazepam), anti-epileptics (carbamazepine), hypolipidemics (clofibrate), cardiovascular drugs (beta-blockers, ACE inhibitors), hormones, many antibiotics, and veterinary drugs (Barber et al., 2006).

2.2 Inorganic Pollutants: Heavy Metals & Metalloids

The major metals contributing to environmental pollution are copper (Cu), lead (Pb), cadmium (Cd), zinc (Zn), mercury (Hg) and nickel (Ni) – all belonging to the heavy metal category- and arsenic (As) and selenium (Se) which are metalloids. These metals can cause toxicity to plants by affecting photosynthesis and can even lead to death (Garbisu and Alkorta, 2001). They can also be accumulated in plants and form a part of human nutrition contributing to toxicity on excessive intake. It is known that Cd and Zn cause gastrointestinal, heart and brain damages (Schmidt, 2003; Schwartz et al., 2003).

3. Current Technologies for Environmental Remediation and Their Drawbacks

Many strategies are used for remediation of the different categories of pollutants (Saleh et al., 2004). Methods predominantly utilize excavation of the impacted site and transferring the soil to a landfill followed by chemical washing. While the methods address the situation at the given point of time, they are costly and merely remove the pollutant to a different site. *In situ* bioremediation techniques have been tried as alternatives to the 'dig and dump' strategy. Tilling is one method which can bring the pollutant to the surface and expose it for oxidative degradation (Huang et al., 2004a, 2004b and 2005). However, the method is suitable for volatile organic compounds only. Microbial degradation also contributes to removal of organic compounds (Stapleton et al., 1998; Dua et al., 2002). However, microbial biomass needs to reach a critical level before it can be used successfully in the field (Brookes and McGrath, 1984).

4. Why Phytoremediation?

The industrialized countries have thousands of hectares requiring urgent remediation efforts. It is estimated that the United States of America (USA) has around 50,000 sites contaminated with metals (Ensley, 2000) and Germany has 80,000 such areas (Franzius, 1994). Estimates for developing countries (including India) are yet to be quantified comprehensively. Technical systems utilizing physico-chemical methods are suitable for smaller sites but focus on remediation of specific pollutants. Here phytoremediation can be a sustainable alternative and may be up to 1000-fold cheaper than the estimated remediation costs of $7 to $8 billion annually (Bennett et al., 2003). Moreover it offers *in situ* remediation solutions (Salt et al., 1998).

The 'green liver' concept emphasizes that plants have the capacity to metabolize and degrade xenobiotics. Many woody plants, shrubs, and annuals have been reported to have pronounced capabilities for detoxification of pollutants. Their ability to consume large quantities of water and release them by evapo-transpiration prevents movement of toxicants to ground water. The same phenomenon also facilitates transportation of pollutants into the plant system. Associated microorganisms in the root zone also contribute to breakdown of pollutants (Siciliano & Germida, 1999). These microbes can be stimulated by plant exudates to further enhance the degradation process.

4.1 Mechanisms of Phytoremediation

Plants utilize certain broad strategies for protecting themselves from the toxic effects of pollutants (Salt et al., 1998). These mechanisms ultimately affect the concentration of pollutants left and thereby, its toxicity (EPA, 2000). Figure 1 provides a diagrammatic representation of these events.

1. **Phytoextraction :** Some plants are capable of removing metals or organics from soil and concentrate them in harvestable parts such as leaves. If such plants are used for detoxification of a site, it is essential to remove these plants from that site. The advantage of the method is that even if the plants have to be land filled, mass of the harvest is still much less when compared to the original

contaminated soil. This method is ideally suited for metal-contaminated soil (EPA, 2000). It is noteworthy that the first phytoremediation-related patent in USA was related to phytoextraction (McCutcheon and Schnoor, 2003).

2. **Phytovolatilization :** The plants take up the pollutant into their system; subsequently, the pollutant or its volatile degradation product is transpired along with water vapour from the leaves (EPA, 2000). Some quantity of the pollutant can also diffuse from other parts of the plant (such as stem) through which translocation occurs before reaching the leaves. The method is effective with pollutants in soil or water. Organic compounds are effectively degraded in this manner. TCE and some volatile forms of Se, Hg and As have been found to be degraded in this manner (EPA, 2000).

3. **Phytodegradation :** Contaminants are taken up into the plant body and subsequently metabolized. It is essential that the compound is soluble to some extent so that it can be taken up inside the plant cell. This holds true for the first two methods also. Chlorinated solvents and some herbicides can be degraded in this manner. The method is suitable for soil and contaminated groundwater too.

4. **Rhizodegradation :** In this method, the rhizosphere-associated microorganisms play a major part in degradation of the contaminants. According to McCutcheon and Schnoor (2003), the number of microorganisms in rhizosphere can be up to 100 fold-higher when compared to non-rhizosphere soil. Plant exudates such as sugars, amino acids, and enzymes play a major role in stimulating presence and growth of microorganisms. This method is successful for breaking down organic contaminants such as PAHs, chlorinated solvents, pesticides, PCBs and BTEX (EPA, 2000).

5. **Rhizofiltration :** Contaminants are taken up by the plant roots and concentrated there itself; subsequently, these can be removed when the plant is harvested. This method is ideal for removing organic compounds or metal contaminants present in groundwater, surface water, or waste water.

6. **Phytostabilization :** The plant growing in a particular area influences the local soil geochemistry environment and this results in adsorption of contaminants on the roots. There may also be precipitation of metals on the roots. There is reduction in mobility (any soil or wind erosion) of the contaminant due to the stable presence of plants. This method has been found suitable for metals and other inorganic contaminants.

   The detoxification events inside plants occur incorporating one or more of the following series of reactions:

a) **Phase I : Conversion :** Oxidations, reductions or hydrolysis reactions start the detoxification process.

b) **Phase II : Conjugation :** The products formed from phase I are linked to glutathione, sugars or amino acids. Now the solubility and toxicity levels of the compound are altered.

c) **Phase III: Compartmentation :** Subsequent to conjugation, the compounds are removed as waste or segregated into vacuoles or bound to the cell wall material (hemicellulose/lignin). In this form, the compounds can no longer be extracted in their original form since they are now bound (Schwitzguebel et al., 2002b).

In the case of organic pollutants, the extent of phytodegradation and phytovolatilization depend on actual movement of the compound into plant roots and translocation into the plant body (Schroeder et al., 2002). Since organic compounds are hydrophobic, they may preferentially bind to soil particles. Hence the bioavailability of the pollutant determines the success of any method used for phytoremediation of soil (Cunningham et al., 1996). Plant exudates such as organic acids could encourage adsorption of compounds on to plant roots from soil particles [Liste and Alexander, 2000). In the case of compounds such as benzopyrene, a carcinogen, just absorption to roots could serve to limit its environmental toxicity even without further translocation and metabolization.

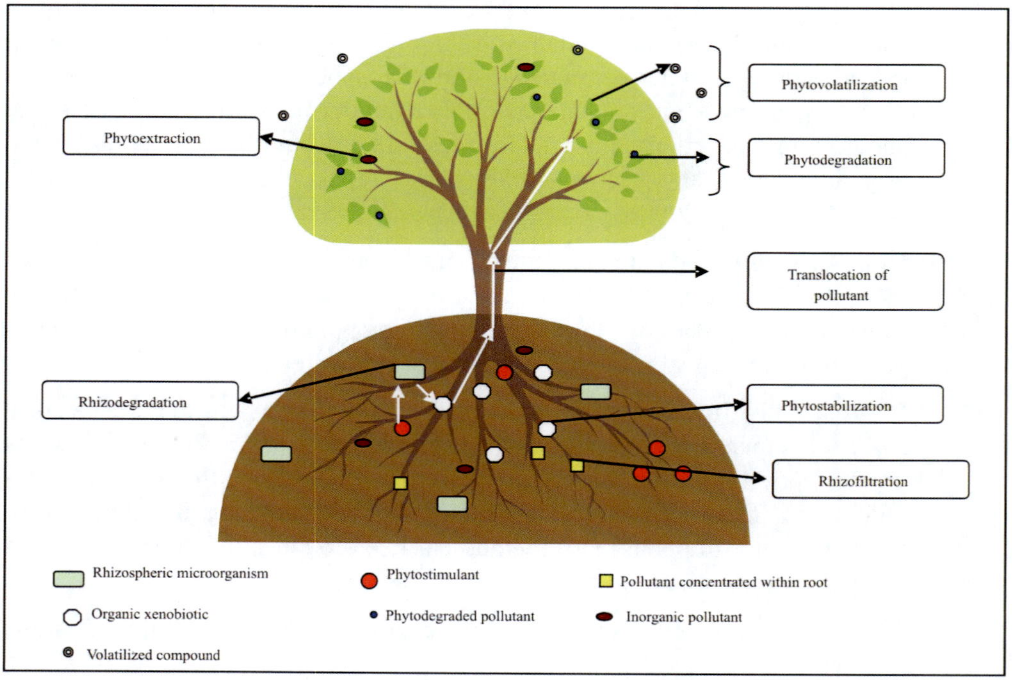

Fig. 1 : Various mechanisms involved in phytoremediation of organic and inorganic pollutants (Adapted from Black 1999).

Generally, a mixture of grasses and legumes has been found suitable for surface soils polluted with hydrocarbons. Studies have shown that degradation of petroleum-based compounds has been greater when species such as *Festuca arundinacea* Vill., *Medicago sativa* L., *Juncus roemerianus* Scheele or *Panicum maximum* Jacq. have been grown on the contaminated soils (Ferrera-Cerrato et al., 2006; Sangabriel et al., 2006).

In the case of PPCPs, Rosal and group (2010) had undertaken a survey of sewage treatment plants (STPs) and found compounds such as acetaminophen, N-formyl-4-amino-antipiryne and galaxolide were the main individual pollutants generally found at levels over 20 ppb. Commonly prescribed medicines auch as those belonging to the beta-blockers group (atenolol, metoprolol and propanolol), lipid regulating drug (fenofibric acid), antibiotics (erythromycin, sulfamethoxazole and trimethoprim), anti-inflammatory drugs (such as diclofenac and mefenamic acid), the antiepileptic medication carbamazepine and the antiacid omeprazole could be removed at a level below 20% in the STP treatment. Hydroxyl radical reactions were shown to be the important contributors towards oxidative transformation of these compounds.

When considering inorganic pollutants, plants utilize two types of mechanisms for better survival; they can either induce pH changes in the rhizosphere so that metal uptake is prevented or they can accumulate the metals in their harvestable parts. The chelating and accumulating capacity of a plant also depends on production of chelatins, more commonly known as phytochelatins (Lips et al., 2000; Mejare and Bulow, 2001). The stress hormone abscisic acid stimulates enzymes acting on reactions converting cysteine to glutathione and phytochelatins. Hence stress conditions such as heavy metal presence also stimulate phytochelatin production (Cobbett, 2000).

Whatever be the method to be adopted depends on the outcome of screening tests and a preliminary knowledge of plant characteristics, their local availability, knowledge of their resistance to specific pollutants, rhizospheric microorganisms etc. (Trapp & Karlson, 2001).

4.2 Benefits

Phytoremediation has many advantages (Glass, 1999; EPA, 2000; Dietz and Schnoor, 2001; Schwitzguebel et al., 2002b; Wilson, 2004), some of which are listed below:

1. Low cost compared to chemical treatment methods and hence suitable for rural areas also
2. Negligible disturbance to environment since it is basically an in situ technique
3. Applicable to many pollutants (organic and/or inorganic)
4. Less generation of secondary wastages that can possibly cause further toxicity

5. No movement of contaminants to other areas, better soil and water quality in treated areas
6. Top soil can be reclaimed for agricultural purposes since less soil erosion
7. Less technical know-how required and hence lowered complexities since it is primarily powered by solar energy
8. Provides for maintenance of habitat and biodiversity
9. Can serve as carbon sink and maintain carbon balance
10. Better public acceptance
11. Selection of plants that have a secondary use such as energy (fuel wood, biodiesel etc.) or for pulp production could provide value-added marketable products.

4.3 Limitations

A few factors still preclude large-scale use of phytoremediation technologies. It is expected that the technology must be preferably continued for longer time, which may not be possible with some annual plant species. Prior knowledge of tolerance levels of plants to specific pollutants, required time frame for clean-up and suitable agronomic practices have to be established. Consideration must also be given to post-remediation factors such as proper disposal of harvestable plant parts. If a specific pollutant is not bioavailable to the plant, treatment may not be successful. Transfer of pollutant to food chain is still possible if animals or insects feed on the candidate plant. Attention must also be given to the toxicity potential of the phyto-transformed product.

5. Plant Cell Cultures as Tools for Studying Phytoremediation

Phytoremediation is not an independent event mediated by the plant system alone; rather it involves the complex interaction between the plants and the microbial community in the rhizosphere. The complexities can be partially or completely unraveled when using controlled systems where the only variable is the pollutant. Hence, *in vitro* plant culture systems have considerably enhanced our knowledge in understanding the effect of the environmental toxicants on the plant systems and the mechanisms by which they are detoxified or otherwise metabolized.

The major advantages (Golan-Goldhirsh et al., 2004) provided by plant tissue cultures are:

1. Once initiated, the cultures can be maintained in the same state (producing plantlets/callus) for long periods for time.
2. The interference from confounding factors is non-existent. Since cultures are maintained under sterile conditions, microbial growth is precluded. Hence the

effect of the pollutant on the axenic plant culture can be evaluated thoroughly. This will also conclusively prove the remediation capacity of the plant and strengthen its usage for the purpose while removing any uncertainties of microbial influence on remediation.

3. The controlled conditions maintain physiological activities at levels which permit effective evaluation.
4. Analysis of metabolites is easier.
5. Cell cultures provide a functional and practical structure for possible metabolic engineering and are more cost-effective.

However, care needs to be exercised when extrapolating the *in vitro* data to the plant that is growing in field conditions and being subjected to extraneous environmental variations. The review by Golan-Goldhirsh and co-workers (2004) provides insights into the requirement for undertaking a holistic approach in phytoremediation studies. However, in recent times when phytoremediation technologies are garnering attention for their advantages in comparison to other in situ methods, the contribution of *in vitro* plant cultures has been significant and has played a key role in translating the laboratory-scale results to field-scale technologies.

5.1 Initiating *in vitro* Cultures : Suitable Explants

Different explants may be used for when starting phytoremediation studies. There can be many objectives (as shown below) when undertaking such investigations. Accordingly, it is essential to keep the overall aim of the particular study in mind before deciding upon the explant. The same is shown in Figure 2 as a flowchart to facilitate easy decision making.

1. If there is a necessity to undertake a preliminary investigation, then a screening study could be started for purpose of understanding which type of explant would best serve the purpose. Various explants (leaves, root, internodes, nodes etc.) could be inoculated on a range of concentration of the pollutant and phytotoxicity assessed. The best survivors could be considered for further studies.
2. There may be a need to micropropagate a hyper-accumulating genotype to generate genetically uniform plants since there may not be enough planting material available in the wild. In this case nodal explants containing axillary buds would be preferred.
3. If the objective is merely to understand the uptake kinetics of a particular pollutant, cell suspension cultures, or differentiated tissues could be used as the starting material.
4. When the intention is to understand gene expression patterns, initial information about the plant part primarily contributing to removal of the pollutant may be

necessary. For example, it may be known that a particular heavy metal primarily accumulates in leaves. In this case, the ideal explant would be the leaf or shoot.

Primary determinants for successful achievement of any objective are availability of suitable explants, their response to culture conditions, and their field response when compared to standard culture conditions. Accordingly various studies have utilized many types of explants. More details are provided in the section on case studies.

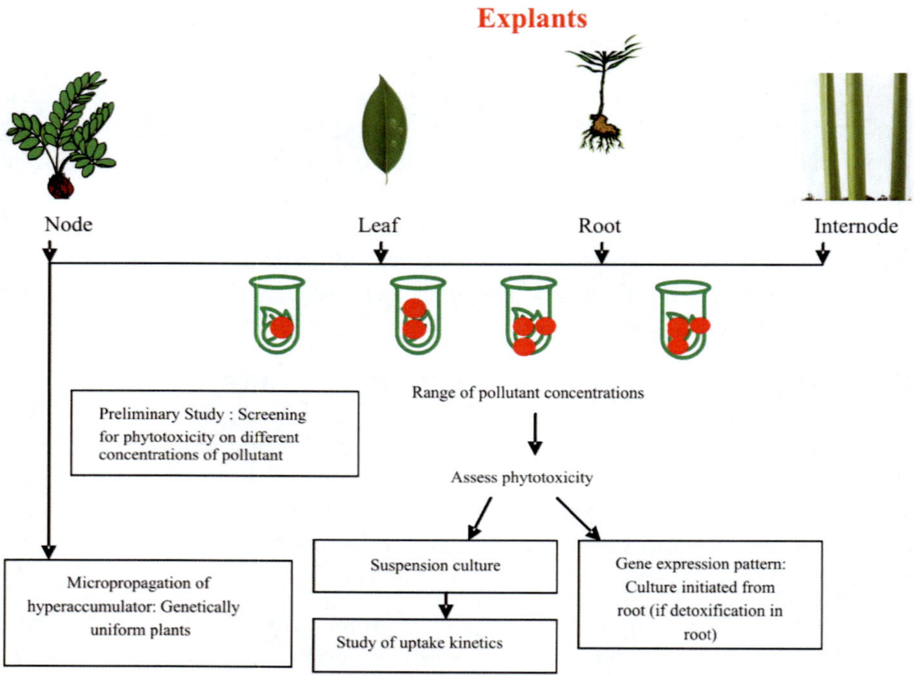

Fig. 2 : Flowchart of events in *in vitro* phytoremediation studies.

5.2 Analyses

After deciding on the most suitable explant, it would be necessary to include assessment parameters with respect to phytoremediation. The essential pre-requisite here is to understand the phytotoxic range of the pollutant being studied. The plant may not exhibit any symptoms of phytotoxicity in the field at a particular concentration of the pollutant but an explant may be unable to survive at the same concentration. Hence the highest concentration of the test compound which facilitates explant survival should be determined. It should be kept in mind that the highest concentration may not be the best one in enhancing our understanding of any phytoremediation event occurring. Hence a judicious approach is required where a sub-lethal dose or a lower concentration may serve to provide a better perspective on the mechanism actually operating in the plant cells.

5.2.1 Evaluation of *in vitro* phytoremediation

The overall study structure pertaining to *in vitro* phytoremediation follows the points being articulated in the following sub-sections. There may be variations (inclusion/exclusion of specific steps) for which the case studies provide further inputs.

When studying the phytoremediation capability, the chosen explant is first cultured on media containing a range of concentrations of the test compound. The results will provide information for choosing the working concentration of the test compound. Cultures may also be screened incorporating several of the test compounds in the same medium since in contaminated soil, there is generally a mixture of several pollutants (Kucerova et al., 2001). Generally, plant growth regulators (PGRs) are incorporated in the medium to facilitate growth of the explants. Many reports suggest influence of these PGRs on the response of the explant to a test compound. In view of this, it is essential to include a control (medium excluding the test compound) for comparison purposes. Statistical analyses support the results further and are critical for interpretation of the results obtained. The common assessment parameters are:

5.2.1.1 Survival percentage of explants

This requires assessment of the number of explants surviving the pollutant stress and calculating the percentage of surviving explants. Comparison with control (pollutant not included in medium) will provide a broad perspective for choosing the working concentration of the pollutant for further assessments.

5.2.1.2. Morphological parameters

These are standard parameters that convey the suitability of the culture conditions and their appropriateness for the purpose. These are accordingly given weightage in the overall assessment structure.

1. Shoot length and number
2. Root length and number
3. Surface area
4. Diameter
5. Leaf number and area including any abnormalities
6. Any other changes deviating from the normal morphological response

In this context, it may be mentioned that changes in roots need to be studied carefully, since these are the first structures which are likely to encounter the pollutant in the field and hence may display toxicity symptoms much earlier, leading to even death of the plant. Assessment of toxicity only on above-ground shoot parts, excluding the roots, may lead to erroneous assessment and may not provide true reflection of actual field situation. Hairy root cultures (explained in section 5.3) facilitate in this type of assessment.

In case of suspension cultures, growth of the culture with the test compound is noted. The growth may be assessed in terms of fresh weight/dry weight. Quantitation of the test compound before and after culturing (inoculation of suspension culture followed by growth in the medium for specified period) will provide the data of any biotransformation or any other type of phytoremediation being carried out by the plant cells.

5.2.1.3. Anatomical features

These include histological studies targeting changes in cell structure, morphology, diameter, stele parameters etc.

5.2.1.4. Assessment of phytoremediation

The initial and final quantities of the test pollutant in the medium are carried out using specific assays or other techniques. For example, High Performance Liquid Chromatography (HPLC) may be used to detect and quantify nitroglycerine after extracting the compound from the plant tissue or medium sample. For arsenic, acid digestion of the sample is required followed by quantitation using an atomic absorption spectrophotometer. Hence, the investigators should familiarize themselves with standard protocols for detection and quantitation and apply the same.

5.2.1.5. Acclimatization and field studies

After the data has been analyzed on suitability of a particular explant to remediate a pollutant, the regenerated plantlet has to be acclimatized and checked for similar remediation capacity under field conditions. Two strategies may be adopted here:

5.2.1.5.1 Hydroponic culture

The *in vitro* regenerated plantlet (with properly formed roots) is now grown in a hydroponic culture (without solid substrate but containing essential nutrients dissolved in water). Standard formulations available such as Hoagland solution are suitable. Addition of plant growth regulators may be avoided at this stage. The test compound can be incorporated either on transplantation or shortly after roots have stabilized in the hydroponic culture. Similar assessment parameters (5.2.1.1-5.2.1.5.) may be utilized to assess phytoremediation capability of the acclimatized plantlet.

5.2.1.5.2. Pot culture

The *in vitro* grown plantlets may be grown in pots containing soil with added test compound, followed by similar assessment studies. Care should be taken to ensure same number of plants in all replicate pots. It is not necessary for the soil to be sterile. Hence the pots can serve as a microcosm and provide matching conditions as in field in terms of rhizospheric microorganisms and other common elements present in soil, thereby providing a more correct picture of the remediation capacity of the *in vitro* regenerated plant.

5.3. Transgenics/Hairy Root Cultures

A well-studied phenomenon is that many types of bacteria can transfer genes to higher plant species (Broothaerts et al., 2005). The chief amongst these organisms are *Agrobacterium tumefaciens* and *A. rhizogenes*. Investigations on these soil bacteria have contributed to the field of genetic engineering and gene transfer of agriculturally significant traits into target plant species. In phytoremediation studies, *A. rhizogenes* has contributed in a noteworthy manner to the overall understanding of mechanisms operating during toxicant uptake by plants. The crux lies in that the gene transfer event leads to the formation of adventitious roots or "hairy roots". While plants can differ in their susceptibility pattern to this bacterium, the system has offered an excellent system to understand the role of roots in toxicant uptake and further metabolism.

The hairy root produced as a result of infection contains the T-DNA segment of the Ri-plasmid (root-inducing plasmid) within its nuclear genome (Chilton et al., 1982). Since the transfer event occurs in trans, it is easier to select for transgenic plants from screened hairy roots (Christey, 2001) utilizing binary vectors and disarmed plasmids (lacking the tumor-forming capability). Hairy root cultures have facilitated large-scale production of therapeutically important secondary metabolites and other phytochemicals (Shanks and Morgan, 1999; Choi et al., 2000) as well as in undertaking phytoremediation studies (Nedelkoska and Doran, 2000).

Production of transgenic plants using *A. tumefaciens* has been successful in many species and has resulted in the plants over-expressing the genes responsible for metabolizing or detoxifying the pollutant. Field studies have also shown their enhanced remediation capability; however, concerns on release of GM plants into the environment and possible gene transfer to non-target organisms has precluded their commercial use in phytoremediation technologies. Many factors can influence the *Agrobacterium*-mediated transformation process such as the genotype, wounding pattern, stimulation of synthesis of phenolic products which can stimulate the vir gene of the *Agrobacterium* strain, promotion of bacterial attachment to host plant cell, T-DNA transfer and stable integration of the transgene in the host plant genome (Gelvin, 2000).

While using *Agrobacterium*-mediated transformation systems for phytoremediation studies, the factors/conditions for facilitating stable transformation require to be optimized. These conditions are:

1. Optimal growth (cell number) of the *A. rhizogenes* strain and thereby, the medium composition in which the strain is grown
2. Explant which requires to be transformed- type, size, physiological state and number of explants per culture vessel
3. Effective inoculum density of bacterial culture for transformation
4. Time period of incubation for the transforming event to occur, temperature, photoperiod and medium composition
5. Antibiotic and its concentration to be added after transformation to kill the Agrobacterium cells
6. Proof of transformation using appropriate and accurate primers
7. Selection of transformants exhibiting production of hairy roots
8. Appropriateness of the hairy roots for phytoremediation- tolerance to toxicant at the concentrations tested and capability to remediate the toxicant.

5.4. Limitations of Using *in vitro* Cultures

The major drawback is that intricate interactions in the field between the plant and its rhizosphere are not well-represented under *in vitro* culture conditions. The synergistic interaction between these systems resulting in accelerated remediation of the pollutant cannot be predicted in these systems and data extrapolations may provide an inaccurate picture of the remediation event (Macek et al., 2000).

6. Case Studies

The experimental studies described below broadly follow the pattern suggested in section 6. However, the variations incorporated during individual experiments provide an insight into the strategies that may be adopted to understand and evaluate different aspects of phytoremediation.

6.1 Organic Pollutants

6.1.1 PAH & PCBs

The capacity of *in vitro* cultured cells from different plant species (hairy root clone of black nightshade and callus of wheat, barley, soybean, tomato, mulberry and birch) to tolerate the presence of PAHs and PCBs was compared in a study conducted by Kucerova and co-workers (2001). Six types of PAHs and a mixture of PCBs were tested for growth inhibition. The plant cells were grown as submerged cultures in 250 ml flasks containing 100 ml of medium. Now 0.5 ml solutions of different PAHs (naphthalene, acenaphthene, phenanthrene, anthracene, fluoranthene and pyrene) in methanol were added to the medium so that there was a final concentration of 3 mg L^{-1} of each of the component. PCBs were added in methanol solutions (0.5 ml each) to give a final concentration of 50 mg L^{-1}. After 14 days of culture, data analysis revealed that there was no specific correlation between resistance of individual plants to the toxicants and the ability to metabolize the toxicants. Dead cells (killed by boiling) were used as controls. Growth of all cells (except nightshade) was affected in the presence of the toxicants. However, the cells were able to metabolize the toxicants to different levels (wheat cells could successfully remove approximately 20% of PCBs and 90% of PAHs from nutrient media in two weeks) thus showing that the plant cells themselves have an inherent capacity to metabolize xenobiotics in the absence of microorganisms. It was also found that differentiated and transformed plant cell cultures degraded PCBs with much better efficacy than callus or non-transformed tissues obtained from the same species (Macková et al., 1997a and 1997b).

6.1.2 Nitroesters

Experiments were conducted as part of a pilot project directed towards construction of a biological waste disposal plant which could decontaminate the waste water generated during the production process of explosives such as nitroglycerine (NG) and Ethylene glycol dinitrate (EGDN) which are components of dynamite. Phytotoxicity of these nitroesters was evaluated on in vitro cultivated regenerants of *Juncus inflexus* and *Phragmites australis* and suspension cultures of flax (*Linum usitatissimum*). For the *in vitro* regenerants, EGDN was added in different concentrations (100, 500 and 1,000 mg L^{-1}). At concentrations higher than 500 mg L^{-1}, EGDN induced wilting and browning and subsequent death of the regenerants. Stem and leaf segments of flax were used to generate callus and further suspension cultures. NG was added as a methanol solution (100 μl per flask). The initial concentration of NG in the medium was kept at 300 mg L^{-1}. The tolerance and viability of these cells was assessed using the reduction of 2,3,5-triphenyltetrazolium chloride (TTC) to water-insoluble red formazan. However, these cells absorbed NG and transformed it to both dinitroglycerine 1,2- DNG and 1,3- DNG isomers. While waste water contains 1-2 g L^{-1}, the regenerants showed phytotoxicity even at 200 mg L^{-1}. The degradation products also could have possibly contributed towards toxicity to the cultures. A possible solution is that, knowing that these plants are able to transform such explosives, waste water could be fed to them for detoxification but in diluted levels where pollutant concentrations are higher than 100 mg L^{-1} (Podlipna et al., 2008).

6.1.3 TNT, NG and PETN

Phytoremediation of the explosives trinitrotoluene (TNT), nitroglycerine (NG) and pentaerytritoltetranitrate (PETN) was studied using *in vitro* cultures of *Rheum palmatum, Saponaria officinalis and Populus simonii*. In the case of rhubarb *(R. palmatum)*, callus was initiated from petiole explant of in vivo grown plants. The callus was inoculated into a suspension culture containing 10 mg L^{-1} naphathaleneacetic acid (NAA) after 4 weeks. Seed explants served as the starting material for initiation of suspension culture of Saponaria officinalis (medium containing 10 mg L^{-1} 2,4-dichlorphenoxyacetic- 2,4-D, 1mgL^{-1} of benzylaminopurine- BAP and 1mgL^{-1} of kinetine). These cultures were maintained for 15 days in the dark. Poplar (*Populus simonii* L.) and aspen (*Populus tremula* × *tremuloides*) were grown hydroponically on liquid MS medium and sub-cultured every 12 weeks on liquid MS medium inside Magenta GA7-3 vessels without membrane rafts. Four concentrations of TNT (25, 50, 75, 100 mg L^{-1}) was incorporated in the medium. The concentration causing 50% inhibition in growth was evaluated. The xenobiotics were degraded to less toxic products by cell cultures as well as hydroponically grown plants thus substantiating that similar metabolization patterns occur in cells as well as in plants (Vanek et al., 2003).

6.1.4 Trichloroethylene (TCE)

Investigations to understand the role of hybrid poplars (*Populus trichocarpa* × *Populus deltoids*) to degrade TCE were carried out. Trees were found to remove more than 99% of the TCE (Newman et al., 1999). Negligible amounts of TCE were transpired. To discount possible role of microorganisms, poplar cell suspensions were subjected to TCE and it was found that same metabolites were produced under culture conditions as seen in the whole plant (Newman et al., 1997; Shang et al., 2001; Shang and Gordon, 2002).

6.1.5 Paraquat (A non-selective herbicide)

This is a commonly used herbicide and is a common pollutant of soils and in agricultural water run-off which also ends up polluting the ground water. In a study, leaf discs were prepared from the wild-type poplar hybrid *Populus canescens* and its two transgenic lines over-expressing γ-glutamylcysteine synthetase in the cytosol/ chloroplasts. These poplar clones were micropropagated using shoot cultures initially. These discs were then subjected to different concentrations of paraquat (4×10^{-9} to 4×10^{-6} M). GST activity showed a decrease with increasing paraquat concentrations. While lower concentrations were non-toxic, at 4×10^{-6} M, there was bleaching of the leaf discs and at 4×10^{-7} M there was inhibitory effect on chloroplast. When sucrose levels were increased in the medium, paraquat induced GST activity. Light also played a role where photosynthetic efficiency was shown to decrease in leaf discs incubated in continuous darkness in medium supplemented with paraquat (Gullner et al. 2005).

In another study by the same group, paraquat-tolerant poplar (*Populus x canescens*) clones were selected using nodal segments cultured at sub-lethal paraquat concentrations (0.4 μmol L^{-1}). The regenerated plantlets were further acclimatized and their tolerance tested in contaminated soil. These clones showed 20-fold higher GST activity when grown on at 0.4 µmol L^{-1} paraquat and 40-fold higher GST activity when grown on at 1 µmol L^{-1} paraquat than wild-type after analysis by quantitative reverse transcriptase real time PCR (qRT-PCR). Additionally lipooxygenase and glutathione reductase activities were found to be enhanced in these clones. Hence subjection to sub-lethal toxicity levels followed by micropropagation was suggested to serve as an effective alternative to usage of controversial GM plants, especially where oxidative stress is involved (Bittsansky et al. 2009).

6.1.6 Dye precursors

To conclusively identify the role of plant systems, exclude any microbial interaction, and better understand the role of each component, cacti were utilized. Since cacti are known to produce protective compounds such as alkaloids using aromatic hydrocarbons as the substrate, it was hypothesized that these plant cells would have a better capability to transform xenobiotics belonging to aromatics group. *In vitro* cultures of these cacti were used to evaluate any biodegrdative/biotransformation events using dye precursors (Bruhn and Bruhn, 1973). Cultures of *Epithelantha micromeris* and *Strombocactus discifomis* were grown on an MS medium containing 2,4-D and different dye precursors: 2-hydroxy-4-sulfo-6-nitro- 1-naphtalenediazonium (Compound I), 2-chloro-5-nitrobenzene- sulfonic acid (Compound 11), 2-hydroxy-4-sulfo-naphtalenediazonium (Compound 111) and 2-amino-4-nitrophenol (Compound IV), each at 1 mM. HPLC analysis showed that both cultures showed variations in removal of the compounds in terms of days. Some of the accumulated degradation products were also toxic to the cultures as deduced from their respective growth curves.

6.1.7 Sulfonated anthraquinones

Since plants like *Rheum, Rumex* and *Rubia* contain anthraquinones, it was considered that their cells would contain the enzymes for degradation of sulphonated anthraquinones. *Rheum palmatum* cells were grown in shake flasks and bioreactors containing up to 700-800 mg/L^{-1} anthraquinones having sulphonate groups in different positions. Various degradation/biotransformation events such as simple uptake from the medium, transformation, desulphonation, conjugation, or partial degradation occurred depending on the position of the sulphonated moiety on the aromatic ring structure. (Schwitzguebel et al., 1995; Schwitzguebel, 1996; Schwitzguebel et al., 1996; Duc et al., 1999). These cells could also accumulate other sulphono-aromatics such as 2-chloro-5-nitro-benzene sulphonate 2-hydroxy-4-sulpho- and 2-hydroxy- 4-sulpho-6-nitro-naphthalene-diazonium produced by the dye industry.

6.2 Inorganic: Heavy Metals and Metalloids

6.2.1 Heavy metals

In order to study the tolerance of *Ailanthus altissima*, a fast-growing and contamination-resistant species, to heavy metals, a micropropagation protocol was developed using nodal segments. Subsequently these shoots were cultured in medium containing different concentrations of heavy metals ($MnSO_4.1H_2O$: 0.2, 0.4, 0.6, 0.8 and 1.6 mM; $ZnSO_4{\cdot}7H_2O$: 0.06, 0.12, 0.24, and 0.48 mM; $CuSO_4{\cdot}5\ H_2O$: 0.02, 0.05, 0.1 and 0.2 mM). These cultures showed tolerance to heavy metals which was comparable to that of cell lines of *Acer pseudoplatanus* (obtained from seedlings growing on metal contaminated soils - Watmough and Dickinson 1995). Zinc tolerance was comparable to *in vitro* cultures of Populus (Kalisova-Spirochova et al. 2003). Based on these studies, poplar is widely favored for phytoremediation of contaminated soil (Rockwood et al. 2004).

6.2.2 Arsenic

Gametophyte of *Pteris vittata* and callus induced from it were subjected to five concentrations of arsenate (0, 0.2, 0.5, 1, and 2 mM final concentration) in half-strength MS solid medium and *in vitro* tolerance was then evaluated (Yang et al., 2007). In another sub-set, callus of *P. vittata* and *Arabidopsis thaliana* were used in a suspension culture experiment also containing different concentrations of arsenate (0.02, 0.05, 0.2, 0.5, 1, and 2 mM). Arsenate (As) concentrations up to 2 mM did not cause cell death of *P. vittata* callus while *A. thaliana* callus was unable to tolerate even 0.2 mM concentrations. Accumulation of As was three times more at 0.2 mM level compared to *A. thaliana* callus. In the suspension cultures, about 60% of the total As was removed when 7.5 g of *P. vittata* callus was grown on medium containing 150 ml of half-strength MS liquid medium containing 450 lg As for 2 days. The *P. vittata* callus, sporophytes, and gametophytes all tolerated and grew well on 1 mM of arsenate and accumulated As when grown for 15 or 30 days on 2 mM arsenate. Hence it was postulated that presence of undifferentiated cells, ease of culture and hyperaccumulating characteristic make *P. vittata* callus a good system to study the mechanism it adopts towards phytoremediation.

In a similar study using *P. vittata*, media supplemented with upto 0.5 mM sodium arsenate (incorporating 0, 0.5, 1.0, 2.0, 3.0 mM in media) accelerated growth of the gametophyte and callus, while higher levels inhibited it (Zheng et al. 2008). However, As accumulation in gametophytes and callus increased up to a level of 763.3 and 315.4 mg kg^{-1}, respectively. These cultures showed enhanced copper accumulation also when cultured on medium containing different concentrations of $CuSO_4$ (0, 0.5, 1.0, 2.0, 3.0 mM). Hence heavy metal tolerance could be studied effectively using micropropagation systems. These results suggested that the high efficiency propagation system could be a useful and rapid means to identify other heavy metal tolerance and accumulation.

6.2.3 Copper

An ecotype of *Brassica nigra* was obtained from a contaminated site and after analysis, was found to have significant quantity of copper in its shoots. Callus from this ecotype was used to regenerate plantlets. These regenerated plantlets were grown in soil culture containing 200 ppm Cu. It was found that the shoots accumulated Cu to greater extent than roots (Memon et al., 2006). The γ-glutamylcysteine (γ-EC) expression in the shoots of Cu-treated plants was around 3.5 times that of control plants (Memon et al., 2008). Hence *in vitro* regeneration could serve the purpose of producing plantlets which have better phytoremediating capacity.

6.2.4 Cadmium/Zinc

A mining ecotype of *Sedum alfredii* was found to be a Cd/Zn hyper-accumulator and since availability of planting was a limitation, a micropropagation protocol was developed. Leaves were used as starting explants for development of callus from which plantlets were developed. The *in vitro* developed plantlets were acclimatized using hydroponic culture and grown on 200 M cadmium and found to hyper-accumulate it. The Cd content in leaves was highest and reached 11,676 mg kg^{-1} DW at 400 μM Cd. The roots were more sensitive to Cd than field-grown plants and they were smaller in size too (Zhao et al., 2009).

6.3 Transgenics

6.3.1 Hairy root cultures

6.3.1.1 PCBs

A hairy root clone of *Solanum nigrum* was used to understand phytoremediation of polychlorinated biphenyls (PCBs) present at an initial concentration of 10 mg in 100 ml medium. The culture inoculum was 4 g in 100 ml of medium. Five different growth regulator combinations were tested along with incorporated PCBs. The PCBs were transformed by the hairy root culture. However, presence of the growth regulators had significant influence on the transformation of the PCBs. Growth of the culture in basal media was very poor. Highest biomass increase was seen when NAA+Kin was added in the media. The cells could transform PCBs even after growth cessation. Similar amounts of different ages of cultures were tested for their ability to transform PCBs. Older inocula (37, 68 days) were more effective indicating that transformation is stimulated in older cultures (Mackova et al., 1997b).

6.3.1.2 Chlorophenols

A carrot hairy root culture was evaluated for its ability to tolerate and metabolize phenols and chlorophenols (de Araujo et al., 2002). Carrot discs were prepared from roots. Standard sized discs were co-cultivated with *A. rhizogenes* and transformation was confirmed by PCR analysis. Two clones with highest and lowest growth indices were selected and evaluated for their tolerance to phenols by inoculating 50 mg each of the clones into media supplemented with PGRs and final concentrations of 50, 500, 1000, 2000 and 5000 mmol/l of phenol (P), 2-chlorophenol (CP), 2,6-dichlorophenol (DCP) and 2,4,6-trichlorophenol (TCP). Based on tolerance levels, 1000 $mmol^{-1}$ of P, and 50 $mmol^{-1}$ of CP, DCP, and TCP were used further and were added during exponential phase of growth (21 days). The comparison was carried out using different controls: untreated root cultures, root cultures inactivated by autoclaving for 1 hour and the culture medium without roots but containing P (1000 $mmol^{-1}$) and TCP(50 $mmol^{-1}$). The clones could remove more than 90% of the phenolic compounds from the culture medium within 120 h of inoculation and were metabolized in the root tissue.

6.3.1.3 Pharmaceuticals

Cell cultures of *Armoracia rusticana and Linum usitatissimum* and and hydroponically grown plants of *Lupinus albus, Hordeum vulgaris, and Phragmites australis* were used to study their potential to detoxify different concentrations of some pharmaceutical compounds such as diclofenac, ibuprofen, and acetaminophen that are commonly found as contaminants in the environment (Kotyzaab et al., 2010). *A. rusticana* hairy roots were able to remove 100% of acetaminophen within eight days of culture. *L. usitatissimum* successfully removed ibuprofen (1 day) and diclofenac (6 days). *Lupinus* was able to remove acetaminophen in 4 days while ibuprofen was removed by *Phragmites*. All plants studied displayed lowered effectiveness in removal of diclofenac. The removal percentage also depended on the initial starting amounts of the test compounds in the medium.

In another study (Huber et al., 2009), hairy root cultures of *A. rusticana* were grown with acetaminophen (1 mM) in a liquid medium. After incubation for 3 hours, roots were transferred into fresh growth medium. Samples were studied for metabolites after 0, 2, 4 and 6 hours. The 6 hour sample showed presence of paracetamol (18%) paracetamol-glucoside (64%), paracetamol glutathione (17%) and cysteine conjugate (1%). A similar detoxification but involving glucuronization rather than glucosidation mechanism is found in mammalian systems. Glutathione conjugate is also present in mammalian cells. Considering these factors, it was proposed that these hairy root cultures of *A. rusticana* could be developed as a model system to further study detoxification mechanisms in plants and their potential to remediate pharmaceutical pollutants.

6.3.2 Transgenic plants

There has been considerable research on improving the capability of plants to remove environmental pollutants by incorporating appropriate genes from microorganisms, animals and other plants and over-expressing them resulting in genetically modified plants with enhanced phytoremediation potential. The *Agrobacterium*-mediated system provides a convenient platform for achieving the same. The excellent and exhaustive review by Doty (2008) provides considerable insight into the development of transgenic tailor-made for phytoremediation. While there is considerable number of reports, only a few representing some of the major pollutants are presented below.

French's group incorporated the pentaerythritol tetranitrate (PETN) reductase from *Enterobacter cloacae* PB2, which can utilize the explosives as a sole nitrogen source (Binks et al., 1996), into tobacco cultures which resulted in increased tolerance to trinitroglycerin and TNT (French et al., 1999). This enzyme is the only known one which can remove nitrate from TNT and convert it into a harmless compound. In a similar strategy, genes containing a flavodoxin reductase (xplB) and a fused flavodoxin cytochrome P450 (xplA) were isolated from an RDX-utilizing bacterium and transformed in *Arabidopsis* (Rylott et al., 2006). These transgenics showed enhanced potential to remove RDX at 40 mg L^{-1} within 5 days, without any phytotoxicity, while a non-transgenic plant could not do so.

The *CYP2E1* gene was over-expressed in hybrid poplar (*Populus tremula x Populus alba*). The metabolism of TCE was found to be enhanced more than 100-fold compared with control plants (Doty et al., 2007). Amonsgt the transformed clones, the one showing highest expression of CYP2E1 transgene was able to remove TCE fastest.

When tobacco plants were transformed with an extracellular fungal laccase (Sonoki et al., 2005), the transgenic plants were able to secrete the laccase into the rhizosphere, and could remove the pollutants bisphenol A (BPA) and pentachlorophenol. The transgenic line showing best efficiency removed 10-fold more BPA than the controls in 1 week.

Dhanker and co-workers (2002) constructed a transgenic *Arabidopsis* which had γ-ECS gene and arsenate reductase C (*ArsC*) gene to ensure that mobility and sequestration of As was controlled. The transgenics showed only 3-fold higher accumulation of As; however, this was a success in converting a non-accumulator species into a hyper-accumulator. *Spartina alterniflora* was transformed with *merA* and *merB* genes and showed enhanced tolerance to mercuric chloride and phenylmercuric acetate (Czako et al. 2006). The best line could tolerate up to 500 μM $HgCl_2$, which was twice the concentration which inhibited the wild-type plants.

7. Conclusions

The developing as well as developed countries are looking at decontamination of thousands of hectares of polluted land. Contaminated water requires further efforts. Methods currently approved and available cannot meet these demands. It was estimated that even partial cleaning of contaminated sites in the European Union would require

US$ 1000 billion. For the US, it was estimated to be nearly one trillion dollars (Saier and Trevors, 2010). The phytoremediation industry has shown encouraging growth in recent years (Glass, 2000) and is comprised of companies focusing solely on phytoremediation technologies and those which offer consultation services for providing phytoremediation solutions. The academia and research institutes are also actively involved by providing the initial experimental data required to develop new technological know-how (Pilson-Smits, 2005).

Focus has shifted to not only identifying and utilizing appropriate plants suitable for phytoremediation in situ but also harnessing their biomass (which is otherwise left unutilized) for bioenergy (Schröder et al., 2008), which would provide an economically feasible solution. Growing plants with remediation capacity but which also have additional value-added products (such as oil, fiber or fragrance) would also serve a similar role and can have potential to bring down the cost of metal phytoremediation technology which is high compared to that of organic pollutants. Recovery of the metal, especially those which are expensive, from the plant parts after phytoremediation can also be an option. Trees with phytoremediation capacity can be used for pulp production. In this context, the review by Vangronsveld and co-workers (2009) suggests that cash crops like tobacco, which are also known to be good accumulators of metals, could be used for fast-track breeding and subsequent plantation on degraded soil and, thereby, facilitate soil amelioration. Tobacco *in vitro* clones were assessed in two types of soils (pH KCl 5.0-5.5 and pH KCl 6-6.8). In the first site, the metal extraction gain was found to increase by many folds for Cd, Zn, Pb and Cu when compared with non-modified clone. Based on the study, it was proposed that the clean-up time using this *in vitro* bred clone would be reduced to about 29 years compared to the originally estimated 58 years.

A good number of phytoremediation-related genes could be present in plants that are expressed in the presence of environmental pollutants. The complete genome sequences of the model dicot *A. thaliana*, monocot *Oryza sativa* and the tree *Populus* are available in public domain. Data mining has revealed the presence of many genes coding for enzymes or transporters suggesting their utility in detoxifying organic or inorganic toxicants in the environment. Multidisciplinary research teams are required to work in a holistic manner utilizing the knowledge culled from basic processes of plant biology, an understanding of rhizospheric biology and chemistry and scaling the information obtained from lab-based studies to actually design field-scale technology adopting environmental engineering principles (Schwitzguebel et al., 2002b; Schwitzguebel, 2004).

More studies are required in the Indian scenario which will help us to identify plants growing in waste/degraded soil that is considered unfit for agricultural purposes and thereby facilitate their utilization for phytoremediation. Candidate plants such as *Solanum nigrum* and *Convolvulus* sp., which generally grow in soil with poor nutrient status, have been documented to be suitable for uptake of pollutants. However, there is urgent requirement to translate these studies into field-level applications.

In vitro plant culture experiments have considerably added to knowledge on phytoremediation mechanisms occurring in plants and have provided crucial data for scaling up the information. However, straight forward extrapolation has not always been successful, underscoring the fact that judicious interpretation of data is necessary. Nevertheless, use of *in vitro* cultures in the initial steps followed by hydroponic and pot cultures has provided a more clear picture and significant information on the overall mechanisms operating in plants which can help develop successful phytoremediation technology.

References

AEHS (Association for Environmental Health and Sciences) TPH Criteria Working Group. (1998b). Volume 2: Composition of Petroleum Mixtures, Wade Weisman (ed.). Available at: http://www.aehs.com/publications/catalog/contents/Volume2.pdf

AEHS (Association for Environmental Health and Sciences) TPH Criteria Working Group. (1998a). Volume 1: Analysis of Petroleum Hydrocarbons in Environmental Media, Wade Weisman (ed.). Available at: http://www.aehs.com/publications/catalog/contents/Volume1.pdf

ATSDR (Agency for Toxic Substances and Disease Registry). (2004). Toxicological Profile for Polycyclic Aromatic Hydrocarbons (PAHs. Available at: http://www.atsdr.cdc.gov/toxprofiles/tp69.html.

Barber, L.B., Murphy, S.F., Verplanck, P.L., Sandstrom, M.W., Taylor, H.E., and Furlong, E.T. (2006). Chemical loading into surface water along a hydrological, biogeochemical, and land use gradient-A holistic watershed approach: Environ. Sci. Technol. 40 (2), 475-486. doi: 10.1021/es051270q.

Bennett, L.E., Burkhead, J.L., Hale, K.L., Terry, N., Pilon, M. and Pilon-Smits, E.A.H. (2003). Analysis of transgenic Indian mustard plants for phytoremediation of metal contaminated, mine tailings. J. Environ. Qual. 32, 432-440.

Binks, P.R., French, C.E., Nicklin, S. and Bruce, N.C. (1996). Degradation of pentaerythritol tetranitrate by *Enterobacter cloacae* PB2. Appl. Environ. Microbiol. 62, 1214-1219.

Bittsánszky, A., Gábor, G., Gullner, G., Kiss, J., Szabó, Z., Kátay, G., Heszky, L. and Kömíves, T. (2009). *In vitro* breeding of grey poplar (*Populus × canescens*) for phytoremediation purposes. J. Chem. Technol. Biotechnol. 84 (6), 890-894.

Black, H. (1999): Phytoremediation: A growing field with some concerns. The Scientist. 13, 5-6.

Brookes, P.C. and McGrath, S.P. (1984). Effects of metal toxicity on the size of the soil microbial biomass. J. Soil Sci. 35, 341-346.

Broothaerts, W., Mitchell, H.J., Weir, B., Kaines, S., Smith, L.M.A., Yang, W., Mayer, J.E., Roa-Rodriguez, C. and Jefferson, R.A. (2005). Gene transfer to plants by diverse species of bacteria. Nature. 433, 629-633.

Bruhn, J.G. and Bruhn, C. (1973). Alkaloids and ethnobotany of Mexican peyote cacti and related species. Econ. Bot. 27, 241-251.

Bundy, J.G., Paton, G.I. and Campbell, C.D. (2002). Microbial communities in different soil types do not converge after diesel contamination. J. Appl. Microbiol. 92, 276-288.

Chilton, M-D., Tepfer, D.A., Petit, A., David, C., Casse-Delbart, F. and Tempe, J. (1982). *Agrobacterium rhizogenes* inserts T-DNA into the genomes of the host plants root cells. Nature, 295, 432-434.

Choi, S.M., Son, S.H., Yun, S.R., Kwon, O.W., Seon, J.H. and Paek, K.Y. (2000). Pilot-scale culture of adventitious roots of ginseng in a bioreactor system. Plant Cell Tiss. Organ Cult. 62 (3), 187-193.

Christey, M.C. (2001). Use of Ri-mediated transformation for production of transgenic plants. In Vitro Cell. Dev. Biol.-Plant, 37 (6), 687-700.

Christodoulatos, C., Bhaumik, S. and Brodman, B.W. (1997). Anaerobic biodegradation of nitroglycerin. Water Res. 31(6), 1462-1470. doi:10.1016/S0043-1354(96)00390-9

Cobbett, C.S., (2000). Phytochelatins and their roles in heavy metal detoxification. Plant Physiol. 123, 825-832.

Crosby, D.G. (1981). Environmental chemistry of pentachlorophenol. Pure & Appl. Chem. 53, 1051-1080.

Cunningham, S.D., Anderson, T.A., Schwab, A.P. and Hsu, F.C. (1996). Phytoremediation of soils contaminated with organic pollutants. In Advances in Agronomy, Sparks, D.L. (Ed), vol. 56, pp 55-114, London: Academic Press.

Czako, M., Feng, X., He, Y., Liang, D. and Marton, L. (2006). Transgenic *Spartina alterniflora* for phytoremediation. Environ. Geochem. Health. 28, 103-110.

de Araujo B.S., Charlwood, B.V., Pletsch, M. (2002). Tolerance and metabolism of phenol and chloroderivatives by hairy root cultures of *Daucus carota* L. Env. Pollut. 117, 329-335.

Dhankher, O.P., Li, Y., Rosen, B.P., Shi, J., Salt, D., Senecoff, J.F., Sashti, N.A. and Meagher, R.B. (2002). Engineering tolerance and hyperaccumulation of arsenic in plants by combining arsenate reductase and gamma-glutamylcysteine synthetase expression. Nature Biotechnol. 20, 1140-1145.

Dietz, A.C. and Schnoor, J.L. (2001). Advances in phytoremediation. Environ. Health Perspect. 109, 163-168.

Doty, S.L. (2008). Enhancing phytoremediation through the use of transgenics and endophytes. New Phytol. 79, 318-333.

Doty, S.L., James, C.A., Moore, A.L., Vajzovic, A., Singleton, G.L., Ma, C., Khan, Z., Xin, G., Kang, J.W., Park, J.Y. et al. (2007). Enhanced phytoremediation of volatile environmental pollutants with transgenic trees. Proc. Nat. Acad. Sci., USA. 104, 16816-16821.

Dua, M., Singh, A., Sethunathan, N. and Johri, A.K. (2002). Biotechnology and Bioremediation: Successes and Limitations. Appl. Microbiol. Biotechnol. 59, 143-152.

Duc, R., Vanek, T., Soudek, P., Schwitzgutbel, J.P. (1999). Accumulation and transformation of sulfonated aromatic compounds by rhubarb cells (*Rheum palmatum*). Int. J. Phytoremediation. 1, 255-271.

Ellenhorn, M.J. and Barceloux, D.G. (1988). Medical toxicology diagnosis and treatment of human poisoning. New York, NY, Elsevier Science Publishing Co., Inc.

Ellis, H.V., Hong, C.B., Lee, C.C., Dacre, J.C. and Glennon, J.P. (1984). Subacute and chronic toxicity studies of trinitroglycerin in dogs, rats, and mice. Fundam. Appl. Toxicol. 4, 248-260. doi:10.1016/0272-0590(84)90126-X

Ensley, B.D. (2000). Rationale for use of phytoremediation. In: Phytoremediation of toxic metals: using plants to clean up the environment, Raskin, I., Ensley, B.D. (eds), pp. 3-11, J. Wiley & Sons, New York, USA.

EPA. (2000). Introduction to Phytoremediation. EPA 600-R-99-107, Office of Research and Development. Available at: http://cluin.org/download/remed/introphyto.pdf

Ferrera-Cerrato, R., Lópes-Ortiz, C., Alarcón, A., Delgadillo-Mártinez, J., Trejo-Aguilar, D., Mendoza-López, R., et al. (2006). Environmental biotechnology and engineering. Proceedings of the Second International Meeting on Environmental Biotechnology End Engineering (2IMEBE). México City. México.

Franzius, V. (1994). Aktuelle entwicklungen zur altlastenproblematik in der Bundesrepublik Deutschland. Umwelt Technologie Aktuell. 6, 443-449.

French, C.E., Rosser, S.J., Davies, G.J., Nicklin, S. and Bruce, N.C. (1999). Biodegradation of explosives by transgenic plants expressing pentaerythritol tetranitrate reductase. Nature Biotechnol. 17, 491-494.

Fukuchi, Y. (1981). Nitroglycol concentrations in blood and urine of workers engaged in dynamite production. Int. Arch. Occup. Environ. Health. 48, 339-346. doi:10.1007/ BF00378682

Garbisu, C. and Alkorta, I. (2001). Phytoextraction: A cost effective plant-based technology for the removal of metals from the environment. Biores. Technol., 77 (3), 229-236. [doi:10.1016/S0960-8524(00)00108-5]

Gelvin, S.B. (2000). *Agrobacterium* and plant genes involved in T-DNA transfer and integration. Ann. Rev. Plant Physiol. Plant Mol. Biol., 51, 223-256.

Glass, D.J. (1999). US and International Markets for Phytoremediation, 1999-2000, D. Glass Associates, Inc, Needham, MA, USA.

Glass, D.J. (2000). Economic potential of phytoremediation. In: Phytoremediation of toxic metals: using plants to clean up the environment (Raskin, I. and Ensley, B.D. Eds.), Wiley, New York, pp 15-31.

Golan-Goldhirsh, A., Barazani, O., Nepovim, A., Soudek, P., Smrcek, S., Dufkova, L., Krenkova, S., Yrjala, K., Schröder, P. and Vanek, T. (2004). Plant response to heavy metals and organic pollutants in cell culture and at whole plant level. J. Soils & Sediments. 4 (2), 133-140.

Gullner, G., Gyulai, G., Bittsánszky, A., Kiss, J., Heszky, I. and Komives, T. (2005). Enhanced inducibility of glutathione S-transferase activity by paraquat in poplar leaf discs in the presence of sucrose. Phyton. 45 (3), 39-44.

Hannink, N.K., Rosser, S.J. and Bruce, N.C. (2002). Phytoremediation of explosives. Crit. Rev. Plant Sci. 21 (5), 511-538. doi: 10.1080/0735-260291044340

Harvey, P.J., Campanella, B.F., Castro, P.M.L., Harms, H., Lichtfouse, E., Schaeffner, A. R., Smrcek, S. and Werck- Reichhart, D. (2002). Phytoremediation of polyaromatic hydrocarbons, anilines and phenols. Environ. Sci. Pollut. Res. 9, 29-47.

Heitkamp, M. A. and Cerniglia, C. E. (1988). Mineralization of polycyclic aromatic hydrocarbons by a bacterium isolated from sediment below an oil field. Appl. Environ. Microbiol. 54, 1612-1614.

Huang, X.D., El-Alawi, Y., Gurska, J., Glick, B.R. and Greenberg, B.M. (2005). A multi-process phytoremediation system for decontamination of persistent total petroleum hydrocarbons from soils. Microchem. J. 81, 139-147.

Huang, X.D., El-Alawi, Y., Penrose, D.M., Glick, B.R. and Greenberg, B.M. (2004a). Responses of three grass species to creosote during phytoremediation. Environ. Poll. 130, 453-463.

Huang, X.D., El-Alawi, Y., Penrose, D.M., Glick, B.R. and Greenberg, B.M. (2004b). A multi-process phytoremediation system for removal of polycyclic aromatic hydrocarbons from contaminated soils. Environ. Poll. 130, 465-476.

Huber, C., Bartha, B., Harpaintner, R. and Schröder, P. (2009). Metabolism of acetaminophen (paracetamol) in plants-two independent pathways result in the formation of a glutathione and a glucose conjugate. Environ. Sci. Pollut. Res. 16, 206-213.

Imoto, S., Kuramoto, M., Iwabuchi, K., Nagai, H. and Shimpo, K. (1986). Percutaneous chronic toxicity study of 10% nitroglycerin (NT-1 ointment) in rabbits. J. Toxicol. Sci. 11 (2), 31-57.

Johansson, P., Ehrenstrom, F. and Ungell, A.L. (1987). Study on adrenergic function after development of tolerance to Ethylene Glycol Dinitrate (EGDN) in rats. Pharmacol. Toxicol. 61 (3), 172-181.

Kalisová-Spirochová, I., Puncochárová, J. and Kafka, Z. (2003). Accumulation of heavy metals by *in vitro* cultures of plants. Water Air Soil Poll. 3, 269-276.

Kanaly, R.A. and Harayama, S. (2000). Biodegradation of high-molecular weight polycyclic aromatic hydrocarbons by bacteria. J. Bacteriol. 182 (8), 2059-2067.

Kanerva, L., Laine, R., Jolanki, R., Tarvainen, K., Estlander, T. and Helander, I. (1991). Occupational allergic contact dermatitis caused by nitroglycerin. Contact Derm. 24, 356-362. doi: 10.1111/j.1600-0536.1991.tb01750.x

Keith, L.H. and Telliard, N.A. (1979). ES&T special report: Priority pollutants I - A perspective view. Environ. Sci. Technol. 13, 416-423.

Kertesz, M.A., Cook, A.M. and Leisinger, T. (1994). Microbial metabolism of sulfur- and phosphorus-containing xenobiotics. FEMS Microbiol. Rev. 15, 195-215.

Khlebnikov, A., Zhoukov, V. and, Peringer, P. (1997). Comparison of p-toluenesulphonic acid degradation by two *Comamonas* strains. Biotechnol. Lett. 19, 389-393.

Kolpin, D.K., Furlong, E.T., Meyer, M.T., Thurman, M,E., Zaugg, S.D., Barber, L.B., and Buxton, H.T. (2002). Pharmaceuticals, hormones, and other organic wastewater contaminants in US streams, 1999-2000: A national reconnaissance. Env. Sci. Tech. 36, 1202-1211.

Kotyzaab, J., Soudeka, P., Kafkab, Z., Vanka, T. (2010). Phytoremediation of Pharmaceuticals-Preliminary Study. Int. J. Phytorem. 12 (3), 306 - 316.

Kucerova, P., in der Wiesche, C., Wolter, M., Macek, T., Zadrazil, F. and Mackova, M. (2001). The ability of different plant species to remove polycyclic aromatic hydrocarbons and polychlorinated biphenyls from incubation media. Biotechnol. Lett. 23, 1355-1359.

Lamm, S.H., Grumski, K.S. and Fethers, R.F. (1993). Short-term exposures to ethylene glycol dinitrate concentrations greater than 0.4 mg/m3 show no evidence of nitrate-induced headaches. Am. J. Epidemiol. 138, 653.

Lips, S.H., Fediuc, E., Omarov, R. and Erdei, L. (2000). A comparative study of sulfurylating enzymes involved in the biosynthesis of phytochelatins. In Abstracts of the Inter-COST Workshop on Bioremediation (Sorrento, Italy), 15-18 November, pp 84-86.

Liste, H.-H. , Alexander, M. (2000). Accumulation of phenanthrene and pyrene in rhizosphere soil. Chemosphere. 40, 11-14.

Lowry, B.P., Bradfield, J.F., Carroll, R.G., Brewer, K. and Meggs, W.J. (2001). A controlled trial of topical nitroglycerin in a New Zealand white rabbit model of brown recluse spider envenomation. Ann. Emerg. Med. 37 (2), 161-165. doi: 10.1067/mem.2001.113031

Macek, T., Macková, M. and Kás, J. (2000). Exploitation of plants for the removal of organics in environmental remediation. Biotechnol. Adv. 18, 23-34.

Mackova, M., Macek, T., Burkhard, J., Ocenaskova, J., Demnerova, K. and Pazlarova, J. (1997a). Biodegradation of polychlorinated biphenyls by plant cells. Int. Biodeterior. Biodegrad. 39, 317-325.

Mackova, M., Macek, T., Kucerova, P., Burkhard, J., Pazlarova, J. and Demnerova, K. (1997b). Degradation of polychlorinated biphenyls by hairy root culture of *Solanum nigrum*. Biotechnol Lett. 9, 787-790.

Mackova, M., Prouzova, P., Stursa, P., Ryslava, E., Uhlik, O., Beranova, K., Rezek, J., Kurzawova, V., Demnerova, K. and Macek, T. (2009). Phyto/rhizoremediation studies using long-term PCB-contaminated soil. Environ. Sci. Pollut. Res. 16, 817-829.

Marshall, S.J. and White, G.F. (2001). Complete denitration of nitroglycerin by bacteria isolated from a washwater soakaway. Appl. Environ. Microbiol. 67 (6), 2622-2626. doi:10.1128/ AEM.67.6.2622-2626.2001

McCutcheon, S.C. and Schnoor, J.L. (2003). Overview of phytotransformation and control of wastes. In: Phytoremediation: Transformation and Control of Contaminants, S. McCutcheon and J. Schnoor (eds.), John Wiley & Sons, Inc., Hoboken, NJ.

McElroy, E., Farrington, J. W. and Teal, J. M. (1989). In: U. Varanasi (ed.) Metabolism of polycyclic aromatic hydrocarbons in the aquatic environment, CRC Press Inc., Boca Raton, Florida.

McGrath, S.P., Zhao, F.J. and Lombi, E. (2001). Plant and rhizosphere process involved in phytoremediation of metal-contaminated soils. Plant Soil. 232 (1/2), 207-214.

Mejare, M. and Bulow, L. (2001). Metal-binding proteins and peptides in bioremediation and phytoremediation of heavy metals. Trends Biotechnol. 19, 67-73.

Memon, A.R. and Schröder, P. (2009). Implications of metal accumulation mechanisms to phytoremediation. Environ. Sci. Pollut. Res. 16, 162-175.

Memon, A.R., Yildizhan, Y. and Demirel, U. (2006). Cu tolerance and accumulation in *Brassica nigr*a and development of *in vitro* regeneration system for phytoremediation. COST action 859: phytotechnologies to promote sustainable land use and improve food safety. WG2 and WG3 workshop -omics approaches and agricultural management: driving forces to improve food quality and safety? Universite Jean Monnet et Ecole Nationale Superieure des Mines, Saint-Etienne, France, pp 37-38.

Memon, A.R., Yildizhan, Y. and Keskin, B.C. (2008) Enhanced Cu tolerance in Brassica nigra (L.) is associated with increased transcription level of γ-glutamylcysteine synthatase (γ-ECS) and phytochelatin synthase (PCS). COST action 859: genes and proteins involved in steps of phytoextraction and degradation of pollutants, workshop WG2: exploiting "-omics" approaches in phytotechnologies. University of Verona, Verona, Italy, p 68.

Mortelmans.K., Haworth.S., Lawlor.X, Speck,W., Tainer.B. and Zeiger.E. (1986). *Salmonella* mutagenicity tests II. Results from the testing of 270 chemicals. Environ. Mutagen. 8 (7), 1-119.

Nedelkoska, T.V. and Doran, P.M. (2000). Hyperaccumulation of cadmium by hairy roots of Thlapsi caerulescens. Biotechnol. Bioeng. 67 (5), 607-615.

Newman, L., Strand, S., Choe, N., Duffy, J., Ekuan, G., Ruszaj, M., Shurtleff, B.B., Wilmoth, J., Heilman, P. and Gordon, M.P. (1997). Uptake and biotransformation of trichloroethylene by hybrid poplars. Env. Sci. Technol. 31, 1062-1067.

Newman, L., Wang, X., Muiznieks, I., Ekuan, G., Ruszaj, M., Cortellucci, R., Domroes, D., Karscig, G., Newman, T., Crampton, R.S. et al. (1999). Remediation of trichloroethylene in an artificial aquifer with trees: a controlled field study. Env. Sci. Technol. 33, 2257-2265.

Nriagu, J.O. and Pacyna, J.M. (1988). Quantitative assessment of worldwide contamination of air water and soils by trace metals. Nature. 333 (6169), 134-139. [doi:10.1038/333134 a0]

Pilson-Smits, E. (2005). Phytoremediation. Annu. Rev. Plant Biol. 56, 15- 39.

Podlipna, R., Fialova, Z. and Vanek, T. (2008). Toxic effect of nitroesters on plant tissue cultures. Plant Cell Tiss. Organ Cult. 94, 305-311. DOI 10.1007/s11240-008-9407-8

Riediker, S., Suter, M.J.F. and Giger, W. (2000). Benzene and naphthalenesulfonates in leachates and plumes of landfills. Water Res. 34, 2069-2079.

Rockwood, D.L., Naidu, C.V., Carter, D.R., Rahmani, M., Spriggs, T.A., Lin, C., Alker, G.R. and Segrest, S.A. (2004). Short rotation woody crops and phytoremediation: Opportunities for agroforestry? Agroforest. Syst. 61-62, 51-63.

Rosal, R., Rodríguez, A., Perdigón-Melón, J.A., Petre, A., García-Calvo E., Gómez, M.J., Agüera, A. and Fernández-Alba, A.R. (2010). Occurrence of emerging pollutants in urban wastewater and their removal through biological treatment followed by ozonation. Water Res. 44 (2), 578-588.

Rylott, E.L., Jackson, R.G., Edwards, J., Womack, G.L., Seth-Smith, H.M., Rathbone, D.A., Strand, S.E. and Bruce, N.C. (2006). An explosive-degrading cytochrome P450 activity and its targeted application for the phytoremediation of RDX. Nature Biotechnol. 24, 216-219.

Saier, M.H. and Trevors, J.T. (2010). Phytoremediation. Water Air Soil Pollut. 205 (Suppl 1), S61-S63.

Saleh, S., Huang, X.-D., Greenberg, B.M. and Glick, B.R. (2004). Phytoremediation of persistent organic contaminants in the environment. In Applied Bioremediation and Phytoremediation, (Series: Soil Biology,Vol. 1), A. Singh and O. Ward (Eds.), pp 115-134. Springer-Verlag, Berlin.

Salt, D.E., Smith, R.D. and Raskin, I. (1998). Phytoremediation. Annu. Rev. Plant Physiol. Plant Mol. Biol. 49, 643-668.

Sangabriel, W., Ferrera-Cerrato, R., Trejo-Aguilar, D., Mendonza-López, M. R., Cruz-Sánches, J. S., López-Ortiz, C., et al. (2006). Internacional de Contaminación Ambiental. 22, 63-73.

Santoro, A., Rovati, L.C. and Setnikar, I. (2001). Assessment of skin safety of a new glyceryl trinitrate transdermal patch. Arzneimittelforschung. 51, 29-37.

Schalscha, E. and Ahumada, I. (1998). Heavy metals in rivers and soils of central chile. Water Sci. Technol. 37 (8), 251-255. [doi:10.1016/S0273-1223(98)00255-8]

Schmidt, U. (2003). Enhancing phytoremediation: The effect of chemical soil manipulation on mobility, plant accumulation, and leaching of heavy metals. J. Environ. Qual. 32, 1939-1954.

Schröder, P., Herzig, R., Bojnov, B., Ruttens, A., Nehnevajova, E., Stamatiadis, S., Memon, A.R., Vassilev, A., Caviezel, M. and Vangronsveld, J. (2008). Bioenergy to save the world-novel plants for bioenergy production. Environ. Sci. Pollut. Res. Int. 15, 196-204.

Schröder, P., Navarro-Aviñó, J., Azaizeh, H., Golan-Goldhirsh, A., DiGregorio, S., Komives, T., Langergraber, G., Lenz, A., Maestri, E., Memon, A.R., Ranalli, A., Sebastiani, L., Smrcek, S., Vanek, T., Vuilleumier, S. and Wissing, F. (2007). Using phytoremediation technologies to upgrade waste water treatment in Europe. Env. Sci. Pollut. Res. 14 (7), 490 - 497.

Schroeder, P., Harvey, P.J. and Schwitzgukbel, J.P. (2002). Prospects for the phytoremediation of organic pollutants in Europe. Environ. Sci. Pollut. Res. 9, 1-3.

Schwartz, C., Echevarria, G. and Morel, J.L. (2003). Phytoextraction of cadmium with *Thlaspi caerulescens*. Plant Soil. 249 (1), 27-35. [doi:10.1023/A:1022584220411]

Schwitzguebel, J.P. (1996). Biodegradation of xenobiotics by plants. In: Kurzfassungen der DECHEMA-Jahrestagungen '96, Band I, DECHEMA, Frankfurt am Main, pp 442-443.

Schwitzguebel, J.P. (2004). Potential of phytoremediation, an emerging green technology: European Trends and Outlook, Proc. Indian Natn Sci Acad. B70 (1) pp 131-152.

Schwitzguebel, J.P., Aubert, S., Grosse, W. and Laturnus, F. (2002a). Sulphonated aromatic pollutants: Limits of microbial degradability and potential of phytoremediation. Environ. Sci. & Pollut. Res. 9 (1), 62 - 72.

Schwitzguebel, J.P., Thirot, J.L., Mathieu, N., Novotny, M., Vanek, T. (1996). Biotransformation of sulfonated aromatic compounds by plant cells. Experientia. 52, A22.

Schwitzguebel, J.P., van der Lelie, D., Baker, A., Glass, D.J. and Vangronsveld, J. (2002b). Phytoremediation: European and American Trends. J. Soils & Sediments. 2 (2), 91-99.

Schwitzguebel, J.P., Vanek, T., Mathieu, N., Thirot, J.L. and Novotny, M. (1995). Biotransformation of sulfonated xenobiotic compounds by plant cells. In: Biosorption and Bioremediation, (Eds. Macek, T., et al.), Prague, pp L 2-10.

Shang, Q.T. and Gordon, M.P. (2002). Transformation of [14C] trichloroethylene by poplar suspension cells. Chemosphere. 47, 957-962.

Shang, Q.T., Doty, S.L., Wilson, A.M., Howald, W.N. and Gordon, M.P. (2001). Trichloroethylene oxidative metabolism in plants: the trichloroethanol pathway. Phytochem. 58, 1055-1065.

Shanks, J.V. and Morgan, J. (1999). Plant 'hairy root' culture. Curr. Opinion Biotechnol., 10 (2), 151-155.

Siciliano, S.D. and Germida, J.J. (1999). Enhanced bioremediation of chlorobenzoates in rhizosphere soil. Soil Biol. Biochem. 31, 299-305.

Sonoki, T., Kajita, S., Ikeda, S., Uesugi, M., Tatsumi, K., Katayama, Y., Iimura, Y. (2005). Transgenic tobacco expressing fungal laccase promotes the detoxification of environmental pollutants. Appl. Microbiol. Biotechnol. 67, 138-142.

Spain, J.C. (1995). Biodegradation of nitroaromatic compounds. Annu. Rev. Microbiol. 49, 523-555.

Stapleton, R.D., Savage, D.C., Sayler, G.S. and Stacey, G. (1998). Biodegradation of aromatic hydrocarbons in an extremely acidic environment. Appl. Environ. Microbiol. 64 (11), 4180-4184.

Stayner, L.T., Dannenberg, A.L., Thun, M., Reeve, G., Bloom, T.F., Boeniger, M., et al. (1992). Cardiovascular mortality among munitions workers exposed to nitroglycerin and dinitrotoluene. Scand. J. Work Environ. Health. 18, 34-43.

Tai, T. and Tsuruta, H. (1997). The effects of nitroglycol on rat isolated cardiac muscle. Ind. Health. 35, 515-518. doi: 10.2486/indhealth.35.515

Trapp, S. A., Koehler, L.C., Larsen, K.C., Zambrano and Karlson, U. (2001). Phytotoxicity of fresh and weathered diesel and gasoline to willow and poplar trees. J. Soils & Sediments. 1 (2), 71-76.

Trapp, S. and Karlson, U.J. (2001). Aspects of phytoremediation of organic pollutants. J. Soils & Sediments. 1 (1), 37 -43.

Vaidya, A.A. and Datyre, K.V. (1982). Environmental pollution during chemical processing of synthetic fibres. Colourate. 14, 3-10.

Vanek, T., Nepovím, A., Podlipná, R., Zeman, S. and Vágner, M. (2003). Phytoremediation of selected explosives. Water Air Soil Poll. Focus 3, 259-267.

Vangronsveld, J., Herzig, R., Weyens, N., Boulet, J., Adriaensen, K., Ruttens, A., Thewys, T., Vassilev, A., Meers, E., Nehnevajova, E., van der Lelie, D. and Mench, M. (2009). Phytoremediation of contaminated soils and groundwater: lessons from the field. Environ. Sci. Pollut. Res. 16, 765-794.

Waidyanatha, S., Lin, P., and Rappaport, S.M. (1996). Characterization of chlorinated adducts of hemoglobin and albumin following administration of pentachlorophenol to rats. Chem. Res. Toxicol. 9, 647-653.

Watmough, S.A., Dickinson, N.M. (1995). Multiple metal resistance and co-resistance in *Acer pseudoplatanus* L. (sycamore) callus cultures. Ann. Bot. 76, 465-472.

Willetts, J. and Ashbolt, N. (2000). Understanding anaerobic decolourisation of textile dye wastewater: mechanism and kinetics. Water Sci. Technol. 42 (1-2), 409-415.

Wilson, M.A. (2004). Ecosystem Services at Superfund Redevelopment Sites. Prepared for U.S. EPA, Office of Solid Waste and Emergency Response, Policy Analysis and Regulatory Management Staff.

Won, W.D., DiSalvo, L.H. and Ng, J. (1976). Toxicity and mutagenicity of 2,4,-6-trinitrotoluene and its microbial metabolites. Appl. Environ. Microbiol. 31, 576-550.

Yang, X., Chen, H., Xu, W., He, Z. and Ma, M. (2007). Hyperaccumulation of arsenic by callus, sporophytes and gametophytes of *Pteris vittata* cultured *in vitro*. Plant Cell Rep. 26, 1889-1897.

Ye, J., Singh, A., Ward, O.P. (2004). Biodegradation of nitroaromatics and other nitrogen-containing xenobiotics. World J. Microbiol. Biotechnol. 20, 117-135. doi:10.1023/B:WIBI.0000021720.03712.12

Zalta, A.H., Karlin, K.M. and Zimmerman, T.J. (1985). The effect of nitroglycerin ointment on the external ocular structures of rabbits. J. Ocul. Pharmacol. 1 (1), 71-77.

Zhao, S-J., Zhang, Z-C., Gao, X., Tohsun, G. and Qiu, B-S. (2009). Plant regeneration of the mining ecotype *Sedum alfredii* and cadmium hyperaccumulation in regenerated plants. Plant Cell Tiss. Organ Cult. 99, 9-16.

Zheng, Y., Xu, W., He, Z. and Ma, M. (2008). Plant regeneration of the arsenic hyperaccumulator *Pteris vittata* L. from spores and identification of its tolerance and accumulation of arsenic and copper. Acta Physiol. Plant. 30, 249-255.

❑❑❑

CHAPTER – 4

Synthetic Seed : A New Concept in Plant Biotechnology

Kambaska Kumar Behera
Department of Bioscience and Biotechnology, Banasthali University, Banasthali Rajasthan-304022, India
Email:kambaska@yahoo.co.in

Summary

Synthetic seed technology is an alternative to traditional micropropagation for the production and delivery of cloned plantlets. Several aspects of the technique are still underdeveloped and hinder its commercial application. This technology may be of greater value in breeding programs and allows the propagation of many elite genotype-derived plants in a short time and would provide an easy and novel propagation system for the elite as well as difficult to root species. Also, encapsulation of propagules that were produced in vitro could reduce the cost of micropropagation of plantlets for commercialization and final delivery to the farmer field. This technology also employed for germplasm storage and exchange purposes.

Introduction

Synthetic seed production is a well documented applied technology that capitalizes on the capacity for rapid plant multiplication via somatic embryogenesis (Castillo et al. 1998). Synthetic seed is also defined as a somatic embryo, or any other vegetative propagule, with an artificial, biodegradable coating entrapped in a nutrient medium, supplying carbon sources, mineral nutrients, vitamins and growth regulators which is engineered for use in the commercial propagation of plants (Gray and Purohit, 1991; Redenbaugh, 1993). The entrapping matrix or coating should not have any adverse effect on the embryo or propagule and should allow germination under both *in vitro* or *ex vitro* conditions. The term therefore, reflects the artificial nature of the seed coat as well as somatic origin of the encapsulated propagule (Malabadi and Nataraja, 2002). At present, Alfalfa and Sugar cane synthetic seed systems are the most advanced and have progressed to direct field planting of synthetic seeds (Fuji et al. 1987; Nieves et al. 2003). Propagation by synthetic seeds that contain somatic embryos has advantages over propagation by organogenesis (Ghosh and Sen, 1994). Although there has been steady research in the area of synthetic seed production, adequate attention has not been directed to storage of encapsulated propagules (Fujii et al. 1992). In recent years, significant progress in biotechnological research has opened up unprecedented opportunities in many areas of basic and applied biological research. The application of biotechnology and conventional recombination can help in solving the propagation and preservation of highly valuable plants through formation of synthetic seeds.

Case Study of Synthetic Seed

Parts of a Typical Synthetic Seed

A typical synthetic seed has the following parts such as: (a) **Plant propagule** (somatic embryo or shoot bud) :- Somatic embryos are bipolar structures with both apical and basal meristematic regions, which are capable of forming shoot and root, respectively. (b) **Matrix** (synthetic gametophyte) :- This is the method and system of manufacturing artificial seed coat. The encapsulating matrix has the ability to incorporate nutrients, biofertlizers, pesticides, nitrogen - fixing bacteria, antibiotics or other essential additives. The direct delivery of encapsulated material will save many subcultures to obtain plants and also eliminate the difficult stage of acclimatization of *in vitro* plants (c) **Seed shell** :- These are the artificial seed coats prepared with complex mixture of alginate-gelatin which was used to develop the coat system for encapsulation. The concept of artificial seed technology has been applied, commercially and successfully in crops like *Azadirachta indica, Alfalfa, Gossypium hirsutum, Santalum album* and in some orchids (*Dendrobium, Spathoglottis plicata, Cymbidium, Phalaenopsis*) etc.

Different Steps for a Successful Production of Synthetic Seeds

The propagules (embryos/ axillary buds/ shoot tips) are carefully isolated from aseptic *in vitro* cultures and blot dried on sterilized filter paper, and is then mixed in sodium alginate prepared in nutrient medium. The propagules are then picked up manually by forceps and dropped into a solution of calcium chloride for about 40 minutes. After the incubation period, the beads (synthetic seeds) are recovered by decanting the calcium chloride solution and washing them in sterile water for 3 to 4 times before culturing on nutrient medium or on different substrates such as filter paper, cotton or soil for their growth and conversion to plants.

Steps for Production of Synthetic Seed

1. Selection of plant materials
2. Establishment of somatic embryogenesis
3. Selection of mature somatic embryo
4. Synchronization of embryos maturation/ desiccation
5. Mass production of somatic embryos
6. Encapsulation
7. Mass production of synthetic seed
8. Synthetic seed development
9. Storage or Plantlet development
10. Acclimatization in Greenhouse and field planting

Encapsulation and Addition of Adjuvants

Synthetic seeds can be prepared with or without a coating. Based on this, four types of synthetic seeds have been proposed:

1. Uncoated desiccated somatic embryos (dry artificial seeds)
2. Coated, desiccated embryos
3. Encapsulated, hydrated embryos
4. Hydrated embryos in a fluid-drilling gel (Redenbaugh et al. 1990)

Several gels like agar, alginate, polyco2133, carboxymethyl cellulose, carrageenan, gelrite, guargum, sodium pectate, tragacanth gum, etc., were tested for synthetic seed production. Alginate hydrogel is frequently selected as a matrix for synthetic seed because of its moderate viscosity and low spinnability of solution, low toxicity for somatic embryos and quick gellation, low cost and bio-compatibility characteristics (Redenbaugh, 1993). Based on the gel-forming procedure, the encapsulation agents generally are categorized into three groups.

1. Gel formation by ionic cross-linking of a charged polymer (alginate, chitosan, carrageenan, etc.).
2. Gel formation by cooling a heated polymer (agar, agarose, carrageenan, etc.).
3. Gel formation by a chemical reaction.

Sometimes the encapsulation matrix may be incorporated by fungicides, pesticides, antibiotics, nutrients etc. for prevention of embryos from desiccation and mechanical injury. The best example is activated charcoal; by addition of this the alginate breaks up that increases respiration of somatic embryos. In addition, charcoal retains nutrients within the hydrogel capsule and slowly releases them to the growing embryo.

Fig 1 : (A) Artificial seeds encapsulated with sodium alginate; (B) Germination of somatic embryo; (C) Ruptured beads showing sprouting of shoots

Nutritive Requirements

The culture medium, should meet the nutritive and metabolic requirements of the embryo as *in vivo*, to have proper growth of the embryo. However, it may be cautioned that this suggestion although is logical and convincing, we still do not have adequate knowledge of the requirements of embryos *in vivo*, at various stages of development in different plant species. The artificial nutritive media should contain certain mineral salts, micro nutrients, vitamins, growth regulators and other organic additives. As a source of energy a carbon source is also essential. Sucrose is the most commonly used source. The vitamins and the growth regulators in the medium depend on the species used for experimentation and the stage of development of the embryo at the time of culturing. Carbohydrate as a source of carbon energy for continued growth of cultured embryo is on the most important ingredient. The concentration of the source should be appropriate. Nitrogen as a source of protein synthesis is another requirement. The embryos receive and utilize nitrogen in an organic form and synthesize the required protein from inorganic nitrates and sucrose in the medium. The biochemical events involved in the utilization is not known. Various amino acids have been experimented in growing embryos. Asperagine and glutamine are widely used with satisfactory results.

Case study-I: [Manjkhola et al. 2005]

Arnebia euchroma (Royal) Jonst (Boraginaceae), a critically endangered medicinal plant o f the Himalaya in which artificial seed production was studied. Manjkhola et al. (2005) reported the protocol for encapsulation and germination of synthetic seeds in *Arnebia euchroma*. After several proliferation and subcultures, early cotyledonary-stage somatic embryos (5–6mm long) were taken and surface sterilized and put in to sodium alginate (1 to 4%) and calcium nitrate (100mM for 25 min) solution for encapsulation. The alginate beads containing somatic embryos were washed three times with sterilized distilled water and wiped dry with filter paper. Encapsulated and non-encapsulated somatic embryos were inoculated in full or half-strength MS medium with 3% sucrose and 0.7% agar, supplemented with IBA (0, 2.5μM) in combination with BA (0, 2.5 μM). Data on the percentage of germination and secondary embryogenesis were recorded after 6 week of inoculation. Low concentrations (1–2%) of sodium alginate beads were too soft to handle, and at higher concentration (4%) the beads were too hard and hindered the emergence of roots and shoots. When encapsulated and non-encapsulated somatic embryos were kept for germination, the maximum germination percentage of encapsulated (60.6%) and of non-encapsulated somatic embryos (73.8%) occurred on hormone-free MS medium. However, the germination percentage was not significantly different from that of half-strength MS hormone-free medium. In MS medium as well as in half-strength MS medium supplemented with 2.5 μM IBA and 2.5 μM BA, the encapsulated embryos showed a significantly lower germination percentage. In this combination, secondary embryogenesis was observed on both MS medium as well as half-strength MS medium. Non-encapsulated somatic embryos showed faster growth than encapsulated somatic embryos. Root emergence in encapsulated somatic embryos occurred after 2 week of inoculation and after 1 week in non-encapsulated somatic embryos. The slower growth of encapsulated embryos can be attributed to internal diffusional limitations, which have been demonstrated for many immobilized biocatalysts (Klein et al. 1984). This is the first report on development of synthetic seed for *Arnebia euchroma*. Application of these protocols would be helpful in promoting conservation of natural populations. Plant regeneration via somatic embryogenesis is advantageous for genetic transformation to obtain higher rates of production of naturally occurring pigments in *A. euchroma*. Additionally, synthetic seeds can be useful for storing valuable germplasm.

Case study-II [*Hyoscyamus muticus* L. By Pandey and Chand, 2005]

Encapsulation of Somatic Embryos: Sodium alginate (CDH, Mumbai) was added in the range of 2.5 to 4% (w/v) in MS medium with or without growth regulators and pH was adjusted to 5.8. For encapsulation, 75 mM calcium chloride was prepared either in MSO (Murashige and Skoog medium without any growth regulator) or MS basal medium containing growth regulators. Both the gel matrix and complexing agent were autoclaved at 15 lb pressure at 121°C for 15 min. Encapsulation was accomplished by mixing the somatic embryos (developed through callogenesis or isolated directly from soacked seeds) in to the sodium alginate solution and dropping these into the calcium chloride solution. Each drop contained one somatic embryo. The drops set as small, white bead when left in $CaCl_2$ solution for 30 min. The beads containing the entrapped somatic embryos were retrieved and washed 2-3 times with sterilized distilled water. Encapsulated somatic embryos were cultured on MS medium with or without growth regulators.Under the *in vitro* conditions, encapsulated somatic embryos were cultured on MS medium supplemented with MS + NAA (2.64 μM) + BAP (2.2 μM), MSO, ½ MSO, and MS + NAA (1.34 μM) + BAP (1.11 μM), for conversion into plantlets. After sprouting, the encapsulated shoot buds were transferred to the same medium. Four-week-old shoots with fully expanded leaves emerging through the ruptured capsule were placed on MSO or MS medium fortified with IBA (4.92 μM) to trigger root induction.

Effect of Alginate Bead Composition and Storage on Plant Recovery: The effect of storage and alginate bead composition on plant recovery, somatic embryos of *Hyoscyamus muticus* were encapsulated in alginate matrix containing 2, 3, 3.75, and 4% sodium alginate. Encap- sulated somatic embryos were stored for 15, 30, 45 and 60 days at 4°C in the dark. Stored encapsulated embryos were cultured on MSO medium for plant regeneration.

The calcium alginate beads differed morphologically in respect to texture, shape and transparency with different combinations of sodium alginate and $CaCl_2$. An encapsulated matrix of 3% sodium alginate with 75 mM $CaCl_2.2H_2O$ was found most suitable for the formation of ideal beads. Sodium alginate concentration of 1-2.5% was not suitable for encapsulation because the resulting beads were of undefined shape and were too soft to handle, whereas at higher concentration the beads were hard enough to cause considerable delay in sprouting. Lower concentration of $CaCl_2$ also caused delay in the preparation of the gel matrix in addition to affecting the bead quality.

In vitro germination encapsulated somatic embryos sprouted within 8- 10 days on MS medium containing growth regulators (NAA, 2.64 µM and BAP, 2.2 µM). Sprouting took 10-12 days on MSO medium. The percentage of germination was 68.8, 53 and 40% on MSO, MS+NAA (2.64 µM) + BAP (2.22 µM), MS+NAA (1.34 µM)+BAP (1.11 µM) media, respectively. Roots developed from the sprouted buds on MSO and MS+IBA (4.92 µM) within 4-7 weeks of cultures. Encapsulated somatic embryos cultured on MS + NAA (2.64 µM) + BAP (2.22 µM) exhibited multiple shoot induction (6-8 shoots per embryo) whereas only single shoot emergence was noted in the case of synthetic seeds cultured on MSO medium.

Following storage at 4°C for 15, 30, 45 and 60 days encapsulated beads were sown on MSO medium. Retrieval of plants after storage period was feasible only when the encapsulating matrix was prepared using MS nutrients. The use of distilled water in making sodium alginate and $CaCl_2$ solution failed to support germination, perhaps, because of poor nutrition. The rates of sprouting of encapsulated somatic embryos declined with longer storage periods and up to 17% of synthetic seeds were capable of germination following 60 days of storage.Many plant systems produce somatic embryos *in vitro* and these proceed through similar developmental stages as those of zygotic embryos.

Applications

Vegetative propagation is very much beneficial because of the capability of long term storage. Synthetic seeds have many diverse applications in agriculture (Redenbaugh et al. 1990; Redenbaugh, 1993). A similar system could be developed in other crops although the specific commercial applications will undoubtedly differ. In plant breeding programs of cross pollinating species, this technology might provide an alternative way to store germplasm instead of a field or greenhouse nursery. Synthetic seeds also offer the opportunity to store genetically heterozygous plants or other plants with unique gene combination that cannot be maintained by conventional seed production due to genetic recombination that occurs at each generation of seed increase.

Many plants are sterile and do not yield seed. Somatic embryogenesis would be preferred in many instances to make cuttings as a means to propagate these plants. Other plants, particularly some tropical species, produce recalcitrant seeds that cannot be dried. Therefore, long term storage of these species in seed banks is not currently possible. Somatic embryos might be an alternative that would enable long-term storage of these species, but the success of the somatic embryo approach will depend on the cause of the seed's recalcitrance. If its inability to tolerate drying is due to a genetic trait that makes the seed unable to respond to the inductive signals that trigger the expression of desiccation tolerance (e.g. they lack ABA receptors), then the synthetic seed approach will not work. However, if the inability is due to a genetic trait that prevents the plant from producing the required signal (e.g. block in ABA synthesis), then the synthetic seed approach might work.

The technology of Synthetic seed may also have application in the short term storage of propagules to enable the synchronized planting of commodities in a greenhouse. One of the limitations of current micropropagation procedures is that the tissue culture facility and the greenhouse production facility must be physically linked; as well their production must be synchronized to meet peak market demands. Synthetic seeds could be a means to uncouple this linkage enabling the micropropagation to be done year round at a site distance from the production facility. The exact application of synthetic seeds will vary from species to species. In self-pollinated crops that currently have good seed production systems, synthetic seeds will not have any practical applications, but in cross pollinating species, especially those where seed production is difficult and expensive, synthetic seeds offer many advantages and opportunities.

Till date a lot of research works have been carried out on different valuable plant species and the process has gained considerable success as presented in the table.

Table 1 : List of plant species in which encapsulation technology is used to produce synthetic seed.

Botanical name of the plant	Common name	Propagule used for encapsulation
Asparagus cooperi	Satabari	Somatic embryos (Gosh and Sen,1994)
Arachis hypogaea	Ground nut	Somatic embryos
Adhatoda vasica	Basanga	Shoot buds , Anand and Bansal (2002)
Brassica campestris	Mustard	Shoot buds
Carica papaya	Amrutabhanda	Somatic embryos (Castllo et al.1998).
Coffea arabica L	Coffe	Shoot buds (Abla, 2003)
Cucumis sativus	Kakudi,Khera	Hypocotyls (Tabassum et al.2010)

Contd.

Cymbidium giganteum	Orchid	Protocorm-like bodies(Corrie and Tandon,1993)
Daucus carota	Carrot	Somatic embryos (Latif et al.2007).
Dalbergia sisso	Shisham	Nodal Segments (Chand and Singh, 2004).
Dioscorea spp.	Yam	Nodal segments(Hasan and Takagi,1995)
Gypsophila paniculata	Baby's Breath	Shoot tips, (Rady and Hanafy,2004)
Hordeum Vulgare	Barley	Microspore derived embryos (Datta and Potrykus ,1989).
Ipomea batata	Sweet potato	Shoot tips(Pennycooke and Towil, 2001)
Mangifera indica	Mango	Somatic embryos(Ara and Jaiswal, 1999).
Medicago sativa	Alfalfa	Somatic embryos(Senaratna et al.1990)
Morus indica	Mulberry	Shoot buds(Machii&Yamanouchi,1993).
Oryza sativa	Rice plant	Somatic embryos(Suprasanna et al.1996)
Phaius tankervillae	Orchid	Protocorm-like bodies (Fukai et al.1997)
Psidium guajava	Guava	Somatic embryos(Rai et al.2008)
Rauvolfia tetraphylla	Patala garuda	Nodal segment (Faisal et al.2006)
Santalum album	Sandalwood	Somatic embryos(Rao & Bapat,1993)
Simmondsia chinensis	Jojoba	Apica and axillary buds(Hassan ,2003)
Solanum melongena	Brinjal	Somatic embryos (Rao and Singh,1991)
Solanum tubersum	Potato	Shoot tips (Fabre and Dereuddre,1990).
Sarumccharumofficinarum	Sugarcane	Somatic embryos(Nieves et al.2001)
Syringa vulgaris	Lilac	Axillary buds(Refauvelet et al.1998)
Vitis vinifera	Grape	Somatic embryos(Nirala et al.2010)
Zingiber officinale	Ginger	Shoot buds (Sundararaj et al.2010)

Advantages of Artificial Seeds

Progress in biotechnological research during the last two decades has opened up unprecedented opportunities in many areas of basic and applied biological research. The application of biotechnology and conventional recombination can help in solving the propagation and preservation of highly valuable plants through formation of synthetic seeds. Merits of artificial or synthetic seed are listed below :-

1. **Easy handling and Inexpensive transport :** As the synthetic seeds are small in size hence it is easier to store, transport and planting.
2. **Storage life :** Synthetic seeds posses long storage life and also the seed viability remains good for longer period of time.
3. **Product uniformity :** As somatic embryos are used for the production of artificial seeds hence most of the seeds are identical in there uniformity.

4. **To avoid extinction of endangered species and seed less plants :** The most important advantages of synthetic seed are such as it helps in conservation of endangered species e. g. in hedgehog cacti (Echinocereus sp.) and seed less varieties e.g. grapes.
5. **Large scale propagation :** After the standardization of protocol it is very much suitable for large scale monoculture.
6. **Germplasm conservation :** A synthetic seed plays an important role in germplasm conservation.
7. **Elite plant genotypes :** Artificial seed technology preserves, protects and permits economical mass propagation of elite plant genotypes such as orchids.
8. **Dependbility on environmental conditions :** The technologies of synthetic seeds production are not a season dependent as these are prepared inside the laboratory.
9. **Permits direct field use :** For tissue culture raised plants rooting, hardening is necessary but in case of synthetic seeds direct field sowing can give good yield.
10. **Facilitates study :** Basic steps involved in the synthetic seed technology involves seed coat formation, function of endosperm in embryo development and seed germination, somaclonal variation provides wide open facility for study.
11. **Supply of beneficial adjuvants :** Beneficial adjuvants like plant nutrients, plant growth regulators, microorganisms, fungicides, mycorrhizae, antibiotics can be made available to the developing plant embryo as per the requirement as these can be added in to the matrix.
12. **Hybrid production :** Synthetic seed production technology can be used for production of hybrids which have unstable genotypes or show seed sterility such as not susceptible towards infection. It can be used in combination with embryo rescue technique. The rescued embryo can be encapsulated with this technique to form synthetic seeds.

Disadvantages

Several intensive researches in the field of synthetic seed technology were applied in propagating and conserving a number of plant species, but practical implementation of the technology is limited due to the following main reasons.

1. Production and storage of synthetic seed is cost effective hence the production technique itself becomes costlier.
2. Production of viable micropopagules useful in synthetic seed production is less.

3. Anomalous and asynchronous development of somatic embryos.
4. All the embryos cannot mature at a time hence that makes them inefficient for germination and conversion in to normal plants.
5. Seed dormancy is considered as a problem but due to lack of dormancy and stress tolerance in somatic embryos the storage of synthetic seeds are limited.
6. The technology of synthetic seed is limited due to poor conversion of even apparently normally matured somatic embryos and other micropopagules into plantlets. (Klein et al.1984; Ara et al. 2000).

Conclusion

Considerable improvement is seen in the case of artificial seed production during the last twenty years, but several problems remain associated with the commercialization of synthetic seed. The most important requirement about the practical application of artificial seed is the production of high quality micropropagules and large-scale production of seeds. Among tree species, regeneration of viable plantlets from somatic embryos is a frequently encountered problem, including germination, maturation, rooting of shoots or acclimatization and shoot apex elongation. Occurrence of high levels of somaclonal variations in tissue culture is another aspect to be considered seriously while recommending the use of artificial seeds for clonal propagation. Direct sowing of synthetic seeds in the soil does not need acclimatization often required for the tissue cultured plants; thus it provides an ideal delivery system and flexibility in handling and transport instead of large parcels of seeds or plants. Commercialization of synthetic seeds still requires some additional processes such as improvement in the tissue culture practices for the generation of adequate propagules, standardization of methods for synchronization followed by automation of the whole process of sorting; harvesting, encapsulation and germination of the coated propagules can enhance the pace in the production of synthetic seeds.

. The production of synthetic seed must either reduce production cost or increase crop value. Improvement in the practices related to sorting of matured embryos, harvesting, encapsulation and germination can give higher yield and improvement of the quality of synthetic seed. However further study and research related to the improvement of technology can help in the global acceptance of synthetic seeds. One of the future usages of synthetic seeds would be in germplasm conservation through cryopreservation. Either hydrated calcium alginate-based or desiccated polyoxy-ethylene glycol-based artificial seeds might be used, but it is likely that some degree of drying before cryopreservation would be beneficial.

References

Aitken-Christie, J., Kozai, T and Smith, M.A.L.(1995). *Automation and Environmental Control. In:* Plant Tissue Culture. Kluwer Academic Publ., Dordrecht, The Netherlands.

Ara, H., Jaiswal, U. and Jaiswal, V.S. (2000). Synthetic seed: Prospect and limitations. Current Science, 78 (12): 1438:1444.

Capuano, G., Piccioni, E. and Standardi, A.(1998). Effect of different treatments on the conversion of M26 apple rootstock synthetic seeds obtained from encapsulated apical and axillary micropropagated buds. J. of Hort Sci and Biotech. 73: 299-305.

Castillo, B., M.A.L. Smith and U.L. Yadava.(1998). Plant regeneration from encapsulated somatic embryos of *Carica papaya* L. Plant Cell Rep., 17: 172-176.

Fuji, J., D. Slade, K. Redenbaugh and K. Walker.(1987). Artificial seeds for plant propagation. Trends Biotech., 5: 335-339.

Fujii, J., Slade, D. and Aguirre-Rascon, J. (1992). Field planting of alfalfa artificial seeds. In vitro Cell Devel. Biol. Plant, 28: 73-80.

Ghosh, B. and S. Sen.(1994). Plant regeneration from alginate encapsulated somatic embryos of Asparagus cooperi baker. Plant Cell Rep., 13: 381-385.

Gray D.J. and Purohit A.(1991). Somatic embryogenesis and development of synthetic seed technology. Crit. Rev. Plant Sci. 10: 33-61.

Griesbach, R.J. (1987). Methods in Plant tissue culture *In:* Food Biotechnology. Marcel Dekker Publ. New York. 271.

Klein, J.,Vorlop, K. D and Wagner, F.(1984). The implication of reaction kinetics and mass transfer on the design of biocatalytic processes with immobilized cells. Enzyme Engng 7:437–449; 1984.

Malabadi, R.B. and K. Nataraja, (2002). *in vitro* storage of synthetic seeds in *Clitoria ternatea* L. Phytomorphology, 52: 231-243.

Manjkhola, S., Dhar, U and Joshi, M. (2005). Organogenesis, Embryogenesis, and Synthetic eeed Production in *Arnebia euchroma in vitro* Cell. Dev. Biol.Plant 41:244–248.

Nieves, N., Y. Zambrano, R. Tapia, M. Cid, D. Pina and R. Castillo.(2003). Field performance of artificial seedderived sugarcane plants. Plant Cell Tiss. Organ Cult., 75: 279-282.

Pandey, A. and S. Chand.(2005). Efficient plant regeneration from encapsulated somatic embryos of Hyoscyamus muticus L. Ind. J. Biotechnol., 4: 546-550.

Pandey,A.and Chand,S.(2005).Efficient plant regeneration from encapsulated somatic emberyos of *Hyoscyamus muticus* L.Indian J.of Biotechnology,4:546-550.

Redenbaugh, K. and Walker, K.(1990). in *Plant Tissue Culture: Applicationsand Limitations* (ed. Bhojwani, S.), Elsevier, Amsterdam: 102:135.

Redenbaugh, K., Fujii, lA.A., and Slade,D.(1993). Hydrated coatings for synthetic seeds. In K. Redenbaugh, (ed.). Synseeds. CRC Press, Boca Raton. pp. 35-46.

Saiprasad, G.V.S. (2001) Artificial seeds and their applications. Resonance.: 39:47.

❑❑❑

CHAPTER – 5

Plant Secondary Metabolites : *in vitro* Production Strategies

Sharad Vats and Kambaska Kumar Behera
Department of Bioscience and Biotechnology, Banasthali University, Banasthali Rajasthan-304 022, India
Email : vats_yahoo.co.in

Summary

Plant cell culture systems produce myriads of valuable compounds which cannot be synthesized by microbial cells or chemical synthesis. Many of the drugs sold today are simple synthetic modifications or copies of the naturally obtained substances. The burgeoning demand for secondary metabolites in recent years as therapeutic agent has resulted in a great interest, in secondary metabolism, and particularly in the possibility to alter the production of bioactive compounds using cell culture technology. Various strategies such as cell line selection, use of precursors and elicitors, mass culture in bioreactors and others have proved effective in enhancing the production of these metabolites so as to meet the market demand but space for improvement still exists. In order to stretch the boundary, recent advances, new directions and opportunities in plant cell based processes are being examined in this section.

Introduction

Plant Secondary Metabolites (bioactive compounds, phyto-pharmaceuticals, phytoceuticals and natural products) are relatively low molecular weight compounds with restricted distribution in taxonomic groups. They are not needed for the growth and development of plants but have more of ecological role. Earlier these metabolites were regarded as waste products with little significance to plant metabolism. In the due course of time scientists realized the importance of Plant Secondary Metabolites (PSM) in plant defense against biotic and abiotic stresses, as attractant of pollinators and as signal compounds. Now it is a well accepted fact that secondary metabolites are integral part of basic metabolism of plants.

Importance of Plant Secondary Metabolites

Secondary metabolites are characterized by an enormous chemical diversity viz., alkaloids, anthocyanins, flavonoids, rotenoids, quinones, lignans, steroids and terpenoids. More than 100 000 secondary metabolites have been discovered, with the estimated total number in plants exceeding 500 000 (Hadacek, 2002). The knowledge that we have gathered is restricted to few plant species and it indicates that there is a large variation in the biochemical pathways involved and in their regulation. This demands an intensive investigation to unearth the intricate network of metabolic pathways which leads to metabolic diversity and to unveil its myriad of applications.

PSM has also made a huge impact on mankind with its diverse applications. These metabolites have been used as food additives, pigments, dyes, insecticides, cosmetics, perfumes and medicine since ages. Some of the early plant derived natural products include drugs such as morphine (painkiller- *Papaver somniferum*), codeine (anticough- *Papaver somniferum*), reserpine (tranquilizer- *Rauvolfia serpentina*), vinblastine and vincristine (anticancer- *Catharanthus roseus*), atropine (spasmolytic- *Atropa belladonna*), pilocarpine (antiglucoma- *Pilocarpus jaborandi*), taxol (antitumour- *Taxus baccata*), digoxin (cardiotonic- *Digitalis*), cradenolide glycosides (cardiotonic- *Asclepias* sps), certain safe plant origin insecticides such as rotenoids and pyrethrins. Many of these compounds being used today and are with no useful synthetic substitutes having same efficacy and pharmacological specificity (Balandrin & Klocke, 1988). Viewing this WHO realized the significance of plant-derived drugs and advocated the use of phytotherapy, complementary and traditional medicines with proper scientific validation (WHO, 2004).

More than 80% of the world's population still depends on traditional medicine for their primary healthcare needs (Muthu et al. 2006). The botanical market in USA is estimated to be US$1.6 billion per annum. Europe, annually, imports about 400,000 tonnes of medicinal plants (values of US$ 1 billion) from Africa and Asia. This is indicative of the renewed interest in herbal medicine with least residual toxicity to human system (Hoareau & Da Silva, 1999).

Production of bioactive compounds by plants is governed by many factors. Plants produce secondary metabolites in less quantity, making their extraction and purification difficult. As a result, secondary metabolites are generally high value-low volume products (Balandrin & Klocke, 1988). Moreover, due to structural complexity their chemical synthesis in sufficient amount is rather difficult. For example some plants are difficult to cultivate, necessitating collection in the wild and thus the risk of the plant getting extinct. Thus, the only commercially viable sources of these compounds are naturally growing plants or their cell and tissue culture.

There are various advantages of producing valuable secondary products in plant cell culture, rather than *in vivo* in the whole crop plant. These include the following:

1. Production of useful compounds under controlled conditions free from environmental constraints
2. Cell cultures are useful in cases where plants have long life cycle
3. Conservation of biodiversity
4. Production can be more reliable, simpler and more predictable using characterized cell lines
5. Isolation of the phytochemicals can be rapid and efficient, as compared to extraction from complex whole plants
6. Interfering compounds that occur in the field-grown plant can be avoided in cell cultures
7. Yield of metabolite of interest can be enhanced using metabolic engineering

Selection of High Yielding Cell Lines

Plants with high content of desired product (elite plant) form the best source of explant to initiate a callus culture with an objective to obtain high yielding cell lines. Cells, thus obtained, in culture represents a heterogeneous population which may differ in size, shape, karyotype and metabolic activities. The heterogeneity in the biochemical activity existing within cells can be explored to select cell lines having higher levels of metabolite of interest.

Cell cloning provides a promising way of cell line selection. The callus obtained from explant is maintained by periodic subculturing. This makes the callus friable enough for raising a suspension culture. Then the callus is transferred in liquid medium continuously agitated by suitable device. Agitation exerts pressure on cell aggregates, breaking them into free cells or small clumps. Further, uniform distribution of cells and cell clumps is maintained in the medium together with proper gaseous exchange between culture medium and air. Single cells are then taken from the suspension culture and grown on conducive medium to obtain population of cells representing clones of the single cell. Several such populations can thereby be screened for enhanced yield of desired metabolite.

Another approach is using cell aggregates for cloning. Callus initiated from explant on solid media is used to establish suspension culture to obtain cell aggregates, which is further cultured again on solid media. The callus thus obtained is divided in to two parts. One is used for quantitative analysis of desired metabolite, which helps out in the selection of high yielding callus and other is subcultured. This procedure might be repeated several times (clonal reselection) to get cell lines producing maximum content of secondary metabolite of interest.

Clonal reselection plays a significant role in cell line selection. Deus-Neumann & Zenk (1984) reported deterioration accumulation of ajmalicine and serpentine with time in clones initiated from cells of *C. roseus*. They advocated clonal reselection to overcome this problem. A strain of *Euphorbia milli* was also recognized to accumulate about 7 times higher amounts of anthocyanins than that of the parent strain after 24 selections (Yamamoto et al. 1982). Statistical and cell-pedigree analysis proved that production of the red pigments was stable. Instability in the production/accumulation of PSM has been observed in other cultures also. Nicotine content of *Nicotiana rustica* callus was reduced by serial sub-culture and reached a meager amount by the end of the ninth culture passage (Tabata and Hiraoka, 1976). Hirotani and Furuya (1977) could not detect certain cardenolides with time in the callus culture of *Digitalis purpurea*. However, instability in the production of PSM is not always the case. High-yielding cell-lines of tobacco produce significant levels of nicotine for one year without reselection (Ogino et al. 1978). The most impressive example of cell line stability was reported by Yamada and Hashimoto (1990), wherein shikonin production by selected cell lines of *Lithospermum erythrorhizon* was quite stable over several months or even years. Moreover, regular screening to maintain production was not essential.

There are several examples wherein the above techniques were used for enhanced production of secondary metabolites – *Daucus carota* (β- carotene), *Beta vulgaris* (betalain), *Lavandula vera* (biotin), *Coptis japonica* (Berberine), *Vitis hybrid* (Anthocyanin), *C. roseus* (indole alkaloids- Serpentine and ajmalacine) and others.

Visual Selection

Selection can be easily achieved if the product of interest is a pigment. Isolation of callus cell lines of *D. carota* with high level of β-carotene was reported by Eichenberger in 1951. Stability in the production of β-carotene was maintained in cell line for more than 10 yrs. Differences in anthocyanin content in *Zea mays* endosperm callus led to isolation of cell lines with different pigment level. Similarly, *Haplopappus gracilis* culture with high level on anthocyanin was isolated. Significant level of red betalain pigment was selected from white callus of *B. vulgaris* by visual selection (Staba, 1980). Mizukami et al. (1978) selected high pigment producing cell lines in *L. erythrorhizon* callus cultures for shikonin. Shikonin derivatives are bright red, and so visual selection of high-yielding calli was feasible.

Chemical Analysis

In this the desired metabolite is measured chemically to identify high producing cell lines. Zenk et al (1977) used RIA (Radioimmunoassay) to select high yielding strains of *C. roseus* producing serpentine and ajmalicine. The method proved to be highly sensitive, specific and rapid. Detection as low as 0.1 ng/0.1ml sample was possible. Deus and Zenk (1982) selected ajmalicine and other major heteroyohimbine alkaloids producing cells of *C. roseus* using fluorescence assay. ELISA (Enzyme Linked Immunosorbent assay) was used to select high yielding capsaicin cell lines of *Capsicum frutescens* (Aitken and Yeoman, 1986). ELISA makes the screening process comparatively quite quick and sensitive. Simultaneous assay of many cell lines can be done in microtitre plate without any prior extraction of medium. Yamamoto and Yamada (1987) used UV-light for the detection of alkaloid fluorescence in callus cultures of *Rauwolfia serpentina* as callus produce colorless alkaloids. Differential fluorescence under 365 nm UV-light was observed as yellow-green and blue-white colonies. These two kinds of fluorescent colonies were separated from the heterogeneous callus cultures and analyzed for reserpine by TLC. The fluorescence of the reserpine spot that appeared on the TLC plates under 365 nm UV-light was very intense in the yellow-green fluorescent colonies as compared to blue-white fluorescent colonies. Thus, former colony was chosen for high producing cell lines. They also confirmed the results using HPLC. Selection can also be done by supplementing the media with some chemicals which react with the desired compound giving a particular colour.

Resistance Selection

In this method, a large population of cells in culture is exposed to a toxic inhibitor or environmental stresses so that only resistant cells are able grow. Amino acid analog have been widely used for resistant cell line selection. These analog bring about false feedback inhibition of biosynthetic enzyme. The selected cell lines are resistant to these inhibitions and thus overproduce primary metabolites. This overproduction could lead to synthesis of PSM as primary metabolites form precursors of secondary metabolites. Many cell lines and plants have been selected that are resistant to toxic Tryptophan analogs such as 5-methyltryptophan (5 MT) and in most cases the resistance is due to the presence of a feedback-insensitive form of Anthranilate Synthase (AS). This leads to increased level of free Tryptophan. Tryptophan is one of the least abundant amino acids (16– 80 mM) present in plants and is the sole donor of indole ring for the synthesis of auxins, glucosinolates, nicotinic acid, phytoalexins and alkaloids (Radwanski and Last, 1995).

The 5-MT resistant *D. carota* cell lines have been reported to contain 30 times the normal level free tryptophan and significant amount of IAA (Staba, 1980). Callus lines showing differential tolerance to 5MT were isolated in two accessions of *C. roseus* ('Nirmal' and 'C-6'), at 0.22, 0.27 and 0.54 mM level of 5MT stress. Selected lines showed higher crude alkaloid content in the presence of the respective 5MT stress. Presence of both catharanthine and vindoline was also indicated in presence of analogue stress, indicating a positive influence of tryptophan accumulation on diverting the metabolic flux towards synthesis of monomeric indole alkaloids (Seth and Mathur, 2005).

p-fluorophenylalanine (PFP) is an analog of amino acid phenylalanine. PFP toxicity might be due to incorporation into proteins to yield dysfunctional products. Resistance to PFP could be achieved by minimized uptake of the analogue, by metabolic detoxification or through enhanced rates of endogenous synthesis of phenylalanine. The latter response would lead to overproduction of normal amino acids thereby reducing the chances of PFP becoming incorporated into proteins. Phenylalanine is deaminated by phenylalanine ammonia-lyase (PAL) to produce cinnamic acid. This acid is subsequently modified to yield phenylpropanoid and flavonoid phenolic compounds. As PAL is also capable of deaminating PFP to yield fluorinated cinnamic acid derivatives (Berlin et al. 1981), this reaction forms a detoxification mechanism for reducing intracellular PFP concentrations. Gathercole and Street (1976) reported elevated levels of both PAL and soluble polyphenols in PFP resistant *Acer psedoplatanus* suspension culture line. Hyper accumulation of aromatic compounds in suspension cultures of *N. tabacum* (Palmer and Widholm, 1975; Berlin et al.1981) *D. carota* (Palmer and Widholm, 1975) has also been observed. Two callus lines *Lavandula vera* MF and *L. vera* PF resistant to m-F-D,L-phenylalanine and p-F-D,L-phenylalanine, respectively, were obtained, which produced 1.95 and 1.71 times more rosmarinic acid in comparison with the parent culture (Georgiev et al.2006). After adaptation of the selected callus lines in liquid media the yields of rosmarinic acid were 2808.4 mg/L in *L. vera* MF cell suspension and 2594.6 mg/L in *L. vera* PF cell suspension.

Higher levels of flavonoids than parent line was produced in UV resistant *Rosa damascena* cultures (Murphy and Hamilton, 1979). The synthesis and accumulation of phenolic metabolites was suggested to be a likely basis for the increased resistance to the selection agent. In other study *Nicotiana* and *Anchusa* cultures responded to two selective agent UV and p-fluorophenylalanine with an enhanced production of aromatic compounds (Quensel and Ellis, 1989).

Large Scale Production of PSM

After the selection of high yielding cell lines the next important measure is to find ways for the mass cultivation and thereby production of desired metabolite industrially so as to meet the burgeoning demand for PSM. However, many of the distinctive characteristics of plant cells in culture like sensitivity to shear, slow growth rates and low oxygen requirements are manifested in complex ways at large scale cultivation. As the scale of operation increases, mixing inside the bioreactor vessel becomes difficult, resulting in non-uniform concentration of the nutrients and limited oxygen transfer to respiring cells. Changes in the rheological nature of the fluid, wall growth, and clumping of cells resulting in sedimentation, lead to suboptimal utilization of the bioreactor. These problems necessitate a more meticulous analysis when scaling-up from flasks to bioreactors.

Bioreactors

Bioreactors act as a biological factory for the production of high value products and provide many advantages:

1. Larger working volume
2. Better control of culture conditions like pH, nutrient media, temperature, aeration etc. for enhanced metabolite production
3. Homogeneous culture due to stirring mechanism
4. Better contact between the plant biomass and the medium
5. Ability to manipulate the plant biomass in relation to the medium volume
6. Reproducible yields of desired product
7. Comparatively simpler and faster harvest of cells and metabolites
8. The opportunity to perform biosynthetic and/or biotransformation experiments related to metabolite production with enzyme availability

Different bioreactors have been used for providing optimum conditions for effective cell growth and secondary metabolite production. These bioreactors are classified as mechanically agitated, pneumatically driven, disposable and non-agitated. The relative advantages and selection criteria for various bioreactors need to be considered according to the specific metabolite and characteristics of the host plant cell. General criteria for choosing a suitable bioreactor and bioreactor design should consider adequate oxygen mass transfer to cells, low shear stress to cells and adequate mass transfer.

One of the most common bioreactor is the **Stirred Tank Bioreactor (STR)**. The first commercial application of large scale cultivation of plant cells was carried out in STR of 200 L and 750 L capacities to produce shikonin by cell cultures of *Lithospermum erythrorhizon* (Payne et al.1987). Cells of *C. roseus* (Ten Hoopen et al.1994), *Dioscorea deltoidea* (Drapeau et al.1986), *D. lanata* (Fulzele et al. 1992), and *Podophyllum hexandrum* (Chattopadhyay et al.2002) have also been cultured in these reactors with suitable modifications for product formation. STR provides high volumetric mass transfer coefficients and a homogeneous environment enabling production to be controlled with ease. The impeller system is the most important element. Mixing promotes better growth by enhancing the transfer of nutrients from liquid and gaseous phases to cells and the dispersion of air bubbles for effective oxygenation. Plant cells have higher tensile strength as compared to microbial cell but their shear sensitivity to hydrodynamic stresses restricts the use of high agitation. Although proper mixing can be achieved by high shear rate and shear time, it reduces the mean aggregate size and has unfavourable effect on cell viability. In order to minimize the shear forces, numerous modifications have been developed by employing a variety of impeller designs and seals (Fig. 1, i). Theoretical engineering analysis of impeller type and geometry for plant cell culture in a $10m^3$ bioreactor indicated that a pitchedblade turbine (axial flowpattern) with the upward-pumping mode offered advantages compared to Rushton turbines (resulting in radial flow pattern) for solid suspension and for reducing shear stress to plant cells when the power input was restricted by cell damage considerations (Doran, 1999). However, in highly viscous liquid culture the oxygen mass transfer ability of upward-pumping pitch-blade impeller is minimized (Kieran, 2001) as compared to operating the same impellerd in the downward-pumping mode (Junker et al.1998). Impeller systems like marine impeller, pitched blade and helical ribbon impeller are considered suitable for plant cell cultures, which exhibit slow-moving axial flow patterns with low tip speed (Amanullah, 2004). Huang et al (2002) advocated the use of dual impeller system consisting of a flat-blade turbine and pitched-blade turbine, which provided a high oxygen transfer rate and was beneficial to the cell growth and L-DOPA production (Fig. 2A, B).

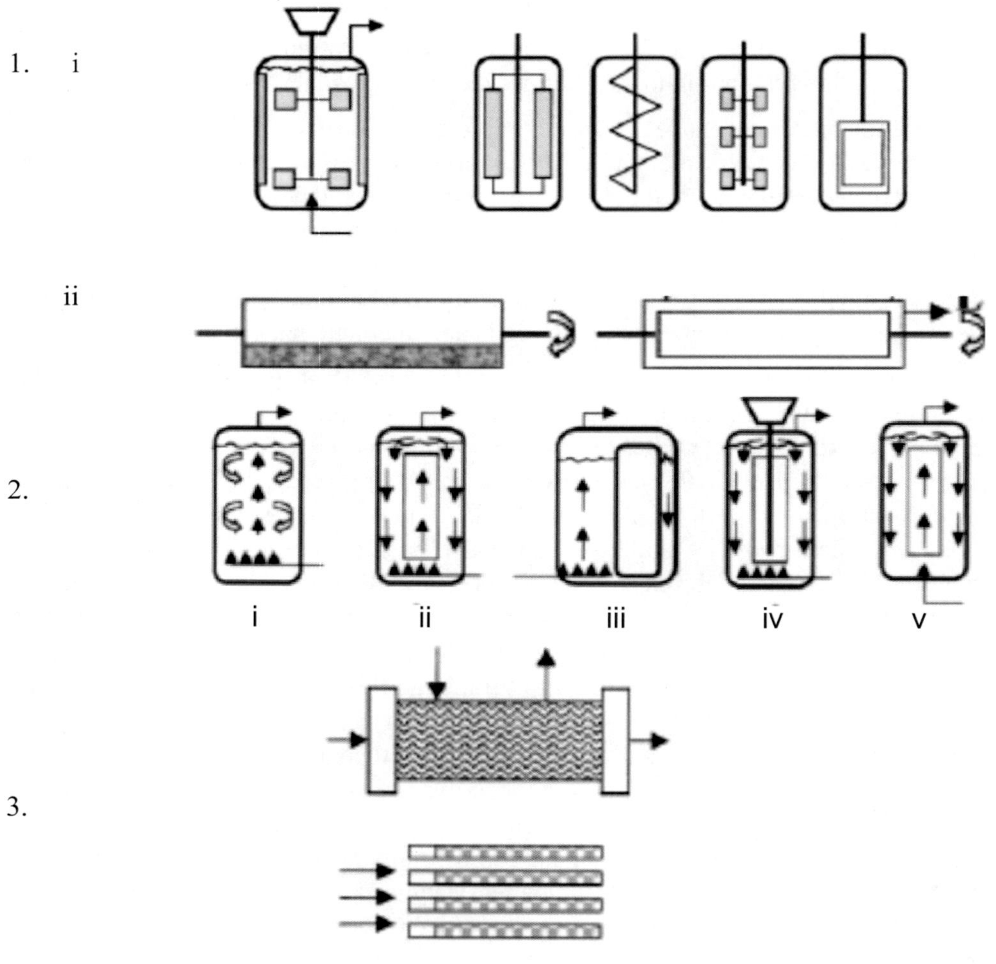

Fig. 1: Types of bioreactors for plant cultures 1) Mechanically agitated bioreactors i. Stirred tank reactor equipped with various propellers (spin, helix, bladed, paddle) ii. Rotary drum tank reactor; 2) Pneumatically driven bioreactors i. bubble column, ii. Internal loop airlift, iii. External loop airlift, iv. Propeller loop, v. Jet loop; 3) Membrane reactor (Sajc et al. 2000)

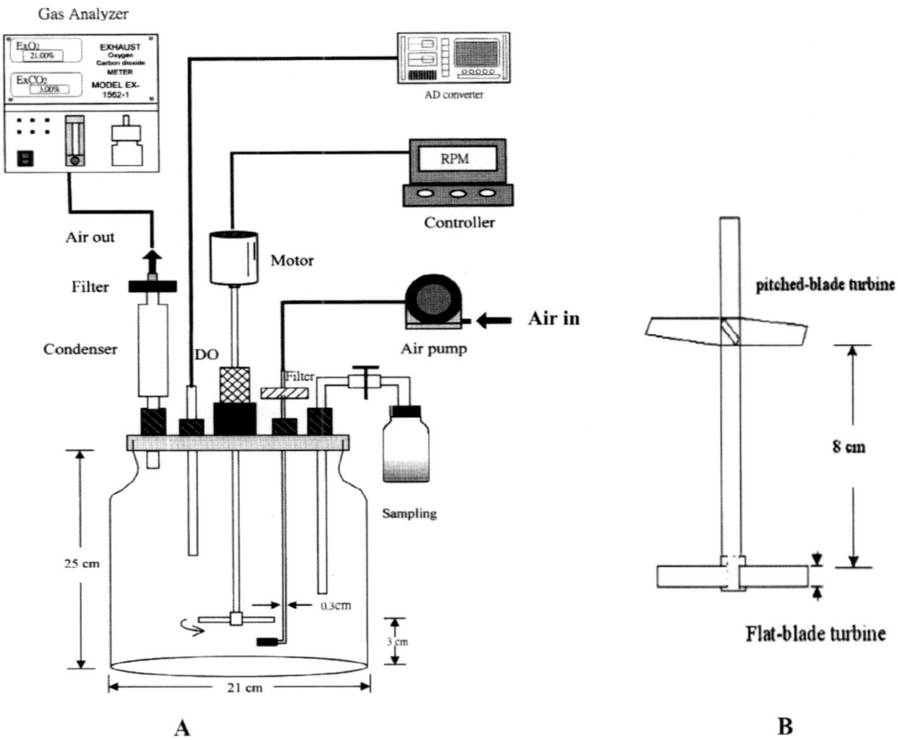

Fig 2A : Schematic diagram of a stirred tank bioreactor; B: Dual impeller system (Huang et al. 2002)

Rotary drum reactors (Fig. 1, ii) have higher surface area to volume ratio than other reactor types. As a result, mass transfer is achieved with comparably less power consumption. Rotary drum reactors used for the cultivation of high-density plant suspensions have shown advantages in terms of suspension homogeneity, low shear environment and reduced wall growth, over either airlift or stirred tank reactors (Sajc et al.2000). The disadvantage is their comparatively high energy consumption in industrial scale operations.

Pneumatic bioreactor (bubble column, air-lift and modified air-lift : It is cylindrical shaped, in which the reaction medium is mixed and aerated by the introduction of compressed air at the bottom through nozzles, perforated plates or a ring sparger (Fig. 1, i-v). An airlift/ modified air-lift bioreactor has internal draught tubes or external loops, which separates the reactor volume into gassed and degassed regions generating a vertically circulating flow (IUPAC, 1997). This configuration prevents bubble coalescence by directing them in one direction, improves oxygen mass transfer by increasing the number of bubbles, distributes shear stress evenly and promotes the cyclical movement of fluid leading to shorter mixing times. These reactors are less suitable for the processes, which involves high cell density or highly viscous liquids as it restricts proper mixing of contents leading to non-homogenous biomass, nutrient, oxygen and pH, and extensive foaming (resulting from extracellular polysaccharides, proteins, fatty acids and high superficial gas velocity). However, low capital costs, uncomplicated mechanical configurations and relatively less operational costs due to low energy requirements are some of the advantages associated with it (Wang et al. 2002). Also, relatively low shear stress environment in pneumatic bioreactors is desirable for shear-sensitive plant cells. The cells of *C. roseus* (Zhao et al. 2001), *D. lanata* (Reinhard et al. 1989), *L. erythrorhizon* (Tanaka, 1987), *Taxus chinensis* (Pan et al, 2000) *Panax ginseng* (Thanh et al. 2006, Fig. 3) have been successfully cultivated in air-lift bioreactors for production of secondary metabolites.

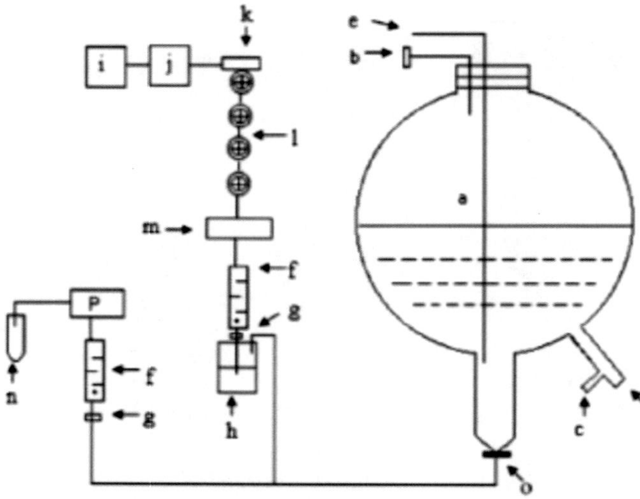

Fig. 3 : Schematic diagram of the balloon-type bubble bioreactor culture system: a – balloon-type bubble reactor, b – air vent, c – inoculum port (5mm diameter), d – sampling port (10mm diameter), e – medium exchange port, f – air flow meter, g – membrane filter, h – water column, i – air compressor, j – air reservoir, k – air cooler, l – filter system, m – air dryer, n – oxygen tank, o – sparger, p – pressure gauge (Thanh et al. 2006).

Disposable bioreactor is a modern alternative to traditional bioreactors. These bioreactors consist of FDA approved sterile biocompatible plastics such as polyethylene, polystyrene and polypropylene, and not glass or stainless steel, as was the case in traditional designs. The plastic chamber that is partially filled with media (10% to 50%), inoculated with cells and discarded after harvest. Usually, the disposable bioreactors eliminates the need for cleaning or sterilization, are low cost, simple to operate, and guarantee high process security. Disposable bioreactors like Wave are characterized by rocking driven wave motion where the mass and heat transfer are regulated by rocking rate, rocking angle and medium filling volume (Fig. 4). There is continuous renewal of the medium in the bag due to rocking effect. Also, bubble-free surface aeration takes place resulting in oxygenation and bulk mixing with less shear stress to cultivated cell.

Few drawbacks that restrict the use of disposable bioreactors are renewal of the disposable cultivation container, which increases the expense on solid waste disposal and consumables. There is limited commercial availability of reliable, disposable sensors for the measurement and control of the main process parameters and peripheral elements, such as valves, sampling systems, and couplers. The most controversial issue has been the detection of extractables and leachables which can interact with the product (Eibl et al. 2010). However, scientists have shown the potential of the wave bioreactor for cultivating mammalian, tobacco, grape and apple suspension cells (Singh, 1999; Eibl and Eibl, 2006; Terrier et al. 2007). Eibl and Eibl achieved high plant cell (*V. vinifera*) biomass productivities of 40 g-FCW/(L-day) with a doubling time of 2 days and observed that there was no significant change in cell morphology when compared to cultivations in stirred-tank bioreactors (Eibl and Eibl, 2006).

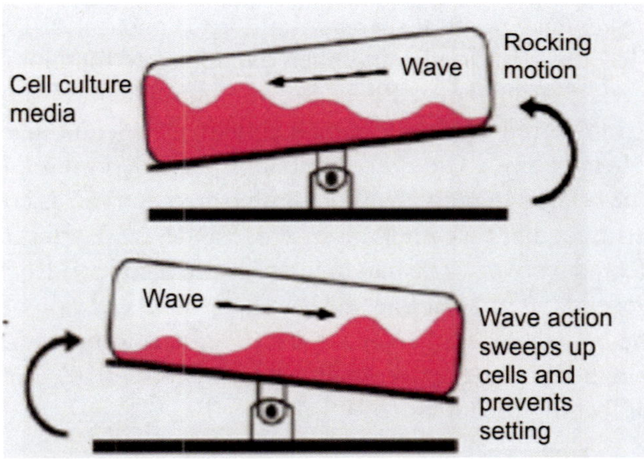

Fig. 4 : Wave bioreactor

Precursor Feeding and Biotransformation

This concept is based on the idea that any compound, which is an intermediate or precursor of a biosynthetic pathway, stands a good chance of enhancing the yield of the final product. An exogenous supply of a biosynthetic precursor to plant culture medium is transformed to a product of economic importance (Biotransformation). Factors such as the concentration and the time of addition of the precursor are to be considered when applying the precursor to the cell culture medium. This approach is more helpful when the precursors are inexpensive. Attempts to enhance the production of PSM by supplying precursor or intermediate compounds have been effective in many cases. Phenylalanine administered to *Coleus blumei* was transformed to rosmarinic acid. The addition of phenylalanine has also been reported to increase the accumulation of hydroxycinnamoyl esters in apple culture, of polyphenols in tobacco cells and of napthoquinones in

Lithospermum cell cultures (Ibrahim, 1988). The addition of phenylalanine has also been reported to increase the accumulation of hydroxycinnamoyl esters in apple culture, of polyphenols in tobacco cells and of napthoquinones in *Lithospermum* cell cultures (Ibrahim, 1988). Increased amount of diosgenin has been reported in the tissue culture of *Dioscorea deltoidea* and *Momordica charantia* using cholesterol (Kaul & Staba, 1968; Kaul et al. 1969; Stohs et al. 1969; Tomita & Uomori, 1973). Production of trigonelline increased on feeding nicotinic acid (Khanna & Jain, 1972; Kamal et al.1996). Production of secologanin and indole alkaloids increased by the addition of loganin, tryptophan and tryptamine in *C. roseus* suspension cultures (Cotin et al.1999; Moreno et al. 1993). Phenylalanine and other potential paclitaxel side-chain precursors (*e.g.* benzoic acid, *N-benzoylglycine* and serine) increased the production of paclitaxel almost six times in the cell culture of *Taxus cuspidata* (Fett-Neto, 1996). Cholesterol, a precursor of alkaloid biosynthesis, was reported to enhance the production of conessine in *Holarrhena antidysenterica* cell suspension culture (Panda et al.1992). Increase in rotenoid content has been reported using phenylalanine and methionine as precursors (Kamal & Mehra, 1995; Kamal et al.1997). Six months old calli of *Cassia occidentalis* and *C. tora* were transferred to MS medium supplemented with three treatment doses of amino acids; phenylalanine (Phe) and methionine (Met) in the ratio of 0.025mM, 0.05mM and 0.1mM. The MS media without amino acids served as control. The callus was later harvested at regular time intervals of 2, 4, 6 and 8 week after fresh subculturing to evaluate the rotenoid recovery. The maximum rotenoid recovery (1.65%) was observed in the treatment dose of 0.05mM of Phe and Met in 6 weeks old calli of *C. occidentalis.* In *C. tora* also, the highest level of rotenoids (1.58%) was found in 0.05mM of amino acids supplemented media in the 6th week callus tissue (Vats, 2010). Table-1 summarizes some examples of feeding precursors of PSM.

Table 1 : Effect of feeding precursor on production of secondary metabolites (Namdeo et al. 2007)

S. No.	Culture	Precursor	Product
1.	*Aesculus californica*	Isoleucine	2-Amino-4-methylhex-4-enoic acid
2.	*Ageratum houstonianum*	^{14}C-labelled precocene I	Precocene II
3.	*Allium cepa*	S-alk(en)yl-cysteine sulphoxides	Glutamyl peptides
4.	*Atropa belladona*	Tropic acid, tropinone, tropanol, tropinone and tropanoltropic acid	Tropane alkaloid
5.	*Azadirachta indica*	Sodium acetate, squaline, IPP, GPP	Azardirachtin
6.	*Camptotheca acuminata*	Tryptophan	Camptothecin
7.	*Capsicum annuum*	L-tyrosine, DL-DOPA	Capsaicin
8.	*Capsicum frutescens*	Phenyl alanine, valine, vanillyl amine, Isocarpic acid, Coumaric acid, Cinnamic acid, Ferulic acid	Capsaicin, vanillin

Contd.

9.	*Catharanthus roseus*	Tryptophan, Secologanin, loganin acid, methyl jasmonate	Ajmalicine , Strictosidine, Terpenoid indole alkaloid (TIA)
10.	*Ceanothus americanus*	Amino acid	Macrocyclic alkaloids
11.	*Celosia plumosa*	Betanidin and betanin	Amaranthin
12.	*Cephaelis ipecacuanha*	L-tyrosine, Phenylalanine, Shikmik acid	Cephaeline
13.	*Cicer arietinum*	4,2, 4-trihydroxychalcone and 7,4-dihydroxyflavanone	Flavanoids
14.	*Cistanche deserticola*	I-Phenylalanine, I-tyrosine, sodium acetate, phenylacetic acid	Phenylethanoid glycosides
15.	*Coleus blumei*	Phenylalanine	Rosmaric acid
16.	*Convallaria majalis*	(R,S)-1-^{14}C, 15N) Methionine	Azetidine-2-carboxylic acid
17.	*Coreopsis tinctoria*	Phenylalanine	1'-Isobutyryloxyeugenol-4-isobutyrate
18.	*Coryphantha macromeris* var *runyonii*	N-methyltyramine	N-methyl-3,4-dimethoxy-β-hydroxyphenethylamine
19.	*Datura tabula*	Tropic acid	Tropic acid
20.	*Datura stramonium*	Cinnamoyl-[2-^{14}C]-tropine-[N-methyl-^{14}C]	Atropine, Scopolamine
21.	*Daucus carota*	Sinapic acid	Anthocyanin
22.	*Delonix regia*	A, γ-Diaminobutryric acid	L-Aazetidine-2-carboxylic acid
23.	*Digitalis lantana*	Cholesterol, sitosterol and stigmasterol	Pregnane derivatives
24.	*Digitalis purpurea*	Glucose	Digitoxose
25.	*Dioscorea deltoidea*	Chlolestrol	Diosogenin

Elicitation

Elicitation is yet another approach to increase the yield of desired metabolite. Plants have been known to accumulate secondary metabolites as a defense mechanism against invading microbes, the phytoalexins (Wolters & Eilert, 1983). It has also been shown that the compounds from pathogens have similar effect in plants as pathogens. These compounds are called as elicitors (Eilert, 1987), and together with the endogenous or constitutive elicitors from the host plant (Dixon et al. 1989), comprise the group of biotic elicitors. Physical and chemical stresses like UV radiation, exposure to cold or heat, fungicides, antibiotics, heavy metals or high salt concentrations are abiotic elicitors (Eilert, 1987). Elicitation can improve the efficiency of product accumulation in cell cultures reducing the time required to obtain a product, considerable amount of metabolite may excrete into the medium along with formation of metabolites within the cell.

Elicitation leads to the induction of various enzymes and additional pathway of metabolism (Wijnsma et al.1985). Various biotic and abiotic elicitors (Table 2 & Table 3) can be used to increase the yield of bioactive compounds (Verpoorte et al.1991; Flores & Curtis, 1992). Catharanthine using Vanadium sulfate (Smith et al. 1987) tryptamine and ajamalicine using fungal elicitors, sodium nitroprusside and arachidonic acid (Kamal et al. 1995; Moreno et al. 1995) from *C. roseus*. Sesquiterpenoids from *Datura stramonium* using Al^{3+}, Cr^{3+}, Co^{2+}, Zn^{2+} (Threfall & Whitehead, 1988). Tropane alkaloids from *A. belladonna* using Cu^{2+} and Cd^{2+} (Lee et al.1998). Capsaicin from *Capsicum frutescens* using Xantha (Johnson et al. 1991). Chitinase from *D. carota* using Salicylic acid (Muller et al. 1994). Activation biosynthesis in *C. roseus* cell suspension culture by using salicylic acid as an elicitor was reported by Kamal et al.(1995). Phytoalexin from *Hyoscyamus albus* using Copper sulphate (Mader, 1999). *G. sylvestr*e cultures were subjected to elicitation using different treatment doses of salicylic acid in suspension culture. Recovery of gymnemagenin was 2.22 times as compared to the control using 0.025 mM of salicylic acid (Vats, 2010).

Table 2 : Effect of biotic elicitor of secondary metabolite (Namdeo, 2007)

Sl. No.	Plant	Product	Biotic elicitor
1.	*Arabidopsis* sp.	Camalexin, Indole glucosinolates	*Erwinia carotovora*
2.	*Bidens pilosa*	Phenylheptaryn	Fungal culture filtrate
3.	*Brugmansia candida*	Hyosujamine	Hemicellulases
4.	*Capsicum annuum*	Capsidiol	Cellulose
5.	*Catharanthus roseus*	Indole alkaloids	Fungal elicitor
6.	*Catharanthus roseus*	N-acetyl-tryptamine	Pythium aphanidermatum
7.	*Catharanthus roseus*	Ajmalicine	Tricoderma viridae
8.	*Catharanthus tictorius*	Polyacetylenes	Fungal polysaccharides
9.	*Cicer arientium*	Medicarpin, Maackiain	Ascochyta rabiei
10.	*Cupressus lusitanica*	Â-thujaplicin	Fungal elicitor

Table 3 : Effect of abiotic elicitor of secondary metabolite (Namdeo, 2007)

Sl. No.	Plant	Abiotic elicitor	Product
1.	*Arabidopsis*	Oxidative stress, amino acid starvation	camalexin
2.	*Atropa belladona*	Cu^{2+}, Cd^{2+}	Tropane alkaloids
3.	*Catharanthus roseus*	Diethyl amino ethyl dichloro phenyl ether, Vanadium sulphate	Indole alkaloids, Catharanthine
4.	*Capsicum annuum*	Arachidonic acid	Capsidiol, Rishitin
5.	*Capsicum frutescens*	Curdlan, Xanthan	Capsaicin

Contd.

6.	*Coleus blumei*	Methyl jasmonate	Rosamarinic acid
7.	*Coleus forskolin*	Methyl jasmonate	Forskolin
8.	*Cupressus lusitanica*	Methyl jasmonate	β-thujaplicin
9.	*Datura stramonium*	Metal ions: Al^{3+}, Cr^{3+}, Co^{2+}, Ni^{+2}, Cu^{2+}, Zn^{2+}	Sesquiterpenoids
10.	*Daucus carota*	Salicylic acid	Chitinase

Hairy Root Culture

The major constraint with cell cultures is that they are genetically unstable and cultured cells tend to produce low yields of secondary metabolites. *Agrobacterium rhizogenes*, the causative agent of hairy root disease in plants, has been used successfully for enhancing secondary metabolite production. The fast growing hairy roots are unique in their genetic and biosynthetic stability and their fast growth offers an additional advantage. These fast growing hairy roots can be used as a continuous source for the production of valuable secondary metabolites. These roots can also synthesize more than a single metabolite and therefore prove economical for commercial production purposes. Transformed roots of several plant species have been studied for the *in vitro* production of secondary metabolites (Table 4). Synthesis PSM in transformed roots is influenced genetically but the effect of other factors such as growth hormone, carbon source, light, temperature etc. can not be undermined (Rhodes et al.1994; Nussbbaumer et al. 1998). Boitel et al (1996) reported enhanced release of alkaloids in *Datura innoxia* hairy root culture using Tween 20. Treating the transformed *Belladonna* roots with 5 mM H_2O_2 induced a transient release of tropane alkaloids (Lee et al.1998). Effect of pH, sonication, temperature, oxygen starvation and osmotic stress on betalaines recovery from *B. vulgaris* was studied by Thimmaraju et al.(2003). Introducing a mixture of two adsorbants (alumina and silica 1:1) in the medium recovered 97.2% of the betalaine effluxes from red beet hairy roots (Thimmaraju et al. 2004). Production of commercially important coumarins, flavonoids and phenolic acid were observed in transformed roots of *Pelargonium sidoides* (Colling et al. 2010). The scaling-up of hairy root cultures using bioreactors is another venture to meet the burgeoning demand for PSM globally.

Table 4: Secondary metabolite production from hairy root culture (Giri and Narasu, 2000)

Sl. No.	Plant	Secondary metabolite
1.	*Aconitum heterophyllum*	Aconites
2.	*Ajuga replans* var. *atropurpurea*	Phytoecdysteroids
3.	*Ambrosia spp.*	Polyacetylenes and thiophenes
4.	*Amsonia elliptica*	Indole alkaloids
5.	*Anisodus luridus*	Tropane alkaloids
6.	*Armoracia laphthifolia*	Peroxidase, Isoperoxidase, Fusicocin
7.	*Artemisia obsinthum*	Essential oils
8.	*Artemisia annua*	Artemisinin
9.	*Astragalus mongholicus*	Cycloartane saponin
10.	*Atropa belladona*	Atropine
11.	*Azadirachta indica*	Azadirachtin
12.	*Beta vulgaris*	Betalin pigments
13.	*Bidens* spp.	Polyacetylenes & Thiophenes
14.	*Brugmansia candida*	Tropane alkaloids
15.	*Calystegia sepium*	Cuscohygrine
16.	*Campanula medium*	Polyacetylenes
17.	*Carthamus* sp.	Thiophenes
18.	*Cassia obtusifolia*	Athraquinone polypeptides pigments
19.	*Catharanthus roseus*	Indole alkaloids, Ajmalicine
20.	*Catharanthus tricophyllus*	Indole alkaloids
21.	*Centranthus ruber*	Valepotriates
22.	*Chaenatis douglasis*	Thiarubrins
23.	*Cinchona ledgeriana*	Quinine
24.	*Coleus forskohlii*	Forskolin
25.	*Coreopsis* spp.	Polyacetylene
26.	*Datura candida*	Scopolamine, Hyoscyamine
27.	*Datura stramonium*	Hyoscyamine, Sesquiterpenes
28.	*Daucus carota*	Flavonoids, Anthocyanin
29.	*Digitalis purpurea*	Cardioactive glycosides
30.	*Duboisia myoporoides*	Scopolamine

Conclusion

This chapter attempts to bring together a number of ideas already published and how these may be used to develop new approaches to the perennial problem of low product yield from plant cell cultures. The need of the hour is to integrate various approaches including focused molecular methodologies, biochemical engineering, downstream processing and others to come out with processes which enable spectrum of phytocompounds to be manufactured to meet the global demand.

References

Aitken, M.M. and Yeoman, M.M. 1986. A rapid screening technique for selection of high yielding capsaicin cell lines of *Capsicum frutescens*. In: Secondary metabolism in plant cell cultures. Morris P (Ed.), Cambridge University Press. pp. 224-229.

Amanullah A, Buckland BC and Nienow AW. 2004. Mixing in the Fermentation and Cell Culture Industries. In: Handbook of Industrial Mixing: Science and Practice. Paul EL, Atiemo-Obeng VA and Kresta SM (Eds.) John Wiley & Sons, Inc., Hoboken, NJ, USA. pp. 1071—1170.

Balandrin MF and Klocke JA. 1988. Medicinal aromatic and industrial materials from plants. In: Biotechnology in Agriculture and Forestry. Bajaj YPS (Ed.). Vol. 40. Springer Verlag. Berlin, Geramny. pp 1-35.

Berlin J, Witte L, Hammer J, Kukoschke KG, Zimmer A and Pape D. 1981. Metabolism of p-fluorophenylalanine in p-fluorophenylalanine sensitive and resistant tobacco cell cultures. Planta 155: 244-250.

Boitel CM, Gontier E, Laberche JC, Ducrocq C and Sangvan-Norreel BS. 1996. Inducer effect of Tween 20 permeabelization treatment used for release of stored alkaloids in *Datura innoxia* Mill. Hairy root cultures. Plant Cell Reports 16: 241–4.

Chattopadhyay S, Srivastava AK, Bhojwani SS and Bisaria VS. 2002. Production of podophyllotoxin by plant cell cultures of *Podophyllum hexandrum* in bioreactor. Journal of Bioscience and Bioengineering 93: 215-220.

Colling J, Groenewald JH, Makunga NP.2010. Genetic alterations for increased coumarin production lead to metabolic changes in the medicinallyimportant *Pelargonium sidoides* DC (Geraniaceae). Metabolic Engineering 12: 561–572.

Contin A, van der Heijden R and Verpoorte R. 1999. Effects of alkaloid precursor feeding and elicitation on the accumulation of secologanin in a *Catharanthus roseus* cell suspension culture. Plant Cell Tissue Organ Culture 56: 111- 119.

Deus B and Zenk MH. 1982. Exploitation of plant cells for the production of natural compounds. Biotechnology and Bioengineering 24: 1965-1974.

Deus-Neumann B and Zenk MH. 1984. Instability of indole alkaloid production in *Catharanthus roseus* cell suspension cultures. Planta Medica 50, 427-431.

Dixon RA, Jennings AC, Davies LA, Gerrish C and Murphy DL. 1989. Elicitor-active components from French bean hypocotyls. Physiological and Molecular Plant Pathology 34: 99-115.

Doran PM. 1999. Design of mixing systems for plant cell suspensions in stirred reactors. Biotechnology Progress 15: 319–335.

Drapeau D, Blanch HW and Wilke CR. 1986. Growth kinetics of *Dioscorea deltoidea* and *Catharanthus roseus* in batch culture. Biotechnology and Bioengineering 28: 1555-1563.

Eibl R and Eibl D. 2006. Design and use of the wave bioreactor for plant cell culture. Plant Tissue Culture Engineering 6: 203–227.

Eibl R, Kaiser S, Lombriser R and Eibl D. 2010. Disposable bioreactors: the current state-of-the-art and recommended applications in biotechnology. Applied Microbiology and Biotechnology 86:41–49

Eilert U. 1987. Elicitation: methodology and aspects of application. In:. Cell Culture and Somatic Cell Genetics of Plants. Constabel F (Ed.). Academic Press, New York, London. pp.153-196.

Fett-Neto AG and DiCosmo F. 1996. Production of paclitaxel and related taxoids in cell cultures of *Taxus cuspidata:* perspectives for industrial applications. In: Plant Cell Culture Secondary Metabolism Toward Industrial Application. DiCosmo and M. Misawa (Eds). CRC Press, New York, USA. pp. 139-166.

Flores HE and Curtis WR. 1992. Approaches to understanding and manipulating the biosynthetic potential of plant roots. In: Biochemical Engineering. Pederson H, Mutharasan RD and Biasio (Eds.). Vol 8. The New York Academy of Science. New York, USA. pp 188-209.

Fulzele D, Kreis W and Reinhard E. 1992. Cardenolide biotransformation by cultured *Digitalis lanata* cells: Semi-continuous cell growth and production of deacetyllanatoside- C in a 40-L stirred tank bioreactor. Planta Medica 58: A601-A602.

Gathercole RWE and Street HE. 1976. isolation stability and biochemistry of a p-flurophenylalanine resistant cell line of *Acer pdeoduplatanaus* L. New Phytology 77:29.

Georgiev M, Pavlov A and Ilieva M. 2006. Selection of high rosmarinic acid producing *Lavandula vera* MM cell lines. Process Biochemistry 41: 2068–2071

Giri A and Narasu ML. 2000. Transgenic hairy roots: recent trends and applications. Biotechnology Advances 18; 1–22.

Hadacek F. 2002. Secondary metabolites as plant traits: current assessment and future perspectives. CRC Critical Review in Plant Science 21(4): 273–322.

Hirotani M and Furuya T. 1977. Restoration of cardenolide-synthesis in redifferentiated shoots from callus cultures of *Digitalis purpurea.* Phytochemistry*,* 16, 610-611.

Hoareau L and Da Silva EJ. 1999. Medicinal plants: a re-emerging health aid. Plant Biotechnology 2(2): 56-70.

Huang SY, Shen YW and Chan HS. 2002. Development of a bioreactor operation strategy for L-DOPA production using *Stizolobium hassjoo* suspension culture. Enzyme and Microbial Technology 30: 79–791.

Ibrahim RK. 1988. Regulation of synthesis of phenolics. In: Cell Structure and Somatic Cell Genetics of Plants, Cell Culture in Phytochemistry. Vol. 4. Constable F and Vasil IK (Eds.). Academic Press Inc. San Diego, USA. 77-95.

IUPAC. 1997. Compendium of Chemical Terminology, 2nd edition.

Johnson TS, Ravishankar GAV and Venkataraman L. 1991. Elicitation of capsaicin production in freely suspended cells and immobilized cell cultures of *Capsicum frutescens*. Food Biotechnology 5: 197-205.

Junker BH, Stanik M, Barna C, Salmon P, Buckland BC. 1998. Influence of impeller type on mass transfer in fermentation vessels. Bioprocess Engineering 19: 403–413.

Kamal R and Mehra P. 1995. *In vitro* influence of precursor amino acids on growth and rotenoid biosynthesis in *Medicago sativa* tissue culture. In: 3rd Dutch German Workshop of Plant Cell and Tissue Culture. Kerkrade, The Netherlands.

Kamal R, Mehra P and Yadav R. 1996. *In vitro* influence of nicotinic acid on production of trigonelline from *Trigonella polycerata*. Indian Journal of Experimental Biology 34: 1147-1149.

Kamal R, Moreno PRH and Verpoorte R. 1995. Stimulation of enzymatic activity and production of indole alkaloids by salicylic acid in suspension cultured cells of *Catharanthus roseus*. In: Symposium on Secondary Products as Physiologically active compounds. Organised by The Society for Medicinal Plant Research and Martin Luther University. Halle-Saale, Wittenberg, Germany.

Kamal R, Yadav R and Mehra P. 1997. Rotenoid biosynthesis potentiality of *Trigonella* species *in vivo* and *in vitro*. Journal of Medicinal and Aromatic Plant Science 19: 988-993.

Kaul B and Staba EJ. 1968. *Dioscorea* Tissue Cultures. I. Biosynthesis and isolation of diosgenin from *Dioscorea deltoidea* callus and suspension cells. Lloydia 31(2): 171-179.

Kaul B, Stohs SJ and Staba EJ. 1969. *Dioscorea* Tissue Cultures. III. Influence of various factors on diosgenin production by *Dioscorea deltoidea* callus and suspension cultures. Lloydia 32(3): 347.

Khanna P and Jain SC. 1972. Effect of nicotinic acid on growth and production of trigonelline by *Trigonella foenum-graecum* L tissue cultures. Indian Journal of Experimental Biology 10(3): 248-249.

Kieran PM. 2001. Bioreactor design for plant cell suspension cultures. In: Multiphase Bioreactor Design. Cabral J, Mota M and. Tramper J. (Eds.). Taylor & Francis Ltd., Routledge, USA. pp. 391–426.

Lee JH, Kim YS, Lee CK, Lee HK and Han SS. 1998. Antiviral activity of some flavonoids on Herpes simplex viruses. Korean Journal of Pharmacognosy 30: 34-39.

Lee KT, Yamakawa T, Kodama T and Shimomura K. 1998. Effects of chemicals on alkaloid production by transformed roots of *Belladonna*. Phytochemistry 49:2343–7.

Mader CJ. 1999. Effects of jasmonic acid silver nitrate and L-AOPP on the distribution of free and conjugated polyamines in roots and shoots of *Solanum tuberosum in vitro*. Journal of Plant Physiology 154: 79-88.

Mizukami H, Konoshima M. and Tabata M. 1978. Phytochemistry 17: 95-97

Moreno PRH, van der Heijden R and Verpoorte R. 1993. Effect of terpenoid precursor feeding and elicitation on formation of indole alkaloids in cell suspension cultures of *Catharanthus roseus*. Plant Cell Report 12: 702- 705.

Moreno PRH, van der Heijden R and Verpoorte R. 1995. Cell and tissue culture of *Catharanthus roseus:* a literature survey II. Plant Cell Tissue Organ Cult. 42: 1-25.

Muller SS, Kurosaki F and Nishi A. 1994. Role of salicylic acid and intrercelluar Ca^{2+} in the induction of chitinase activity in carrot suspension culture. Physiology and Molecular Plant Pathology 45: 101-109.

Murphy TM and Hamilton CM. 1979. A strain of *Rosa damascena* L. cultured cells resistant to ultraviolet light. Plant Physiology 64: 936-941.

Muthu C, Ayyanar M, Raja N and Ignacimuthu S. 2006. Medicinal plants used by traditional healers in Kancheepuram District of Tamil Nadu, India. Journal of Ethnobiology and Ethnomedicine 2: 43.

Namdeo AG, Jadhav TS, Rai PK, Gavali S. and Mahadik KR. 2007. Precursor feeding for enhanced production of Secondary metabolites: A review. Pharmacognosy Reviews 1 (2): 227-231.

Nussbbaumer P, Kapetanidis I, Christen P. 1998. Hairy roots of *Datura candida* X *D. aurea*: Effect of culture medium composition on growth and alkaloid biosynthesis. Plant Cell Reports 17: 405–9.

Ogino, T., Hiraoka, N., and Tabata, M. 1978. Selection of high-producing cell lines of tobacco callus by single-cell cloning. Phytochemistry 17: 1907-1910.

Palmer JE and Widholm J. 1975. Characterization of carrot and tobacco cell cultures resistant to p-fluorophenylalanine. Plant Physiology 56: 233-238.

Pan ZW, Wang HQ and Zhong JJ. 2000. Scale-up study of suspension culture of *Taxus chinensis* cells for production of taxane diterpene. Enzyme Microbial Technology 27: 714-723.

Panda AK, Bisaria VS and Mishra S. 1992. Alkaloid production by plant cell cultures of*Holarrhena antidysenterica*: II. Effect of precursor feeding and cultivation in stirred tank bioreactor. Biotechnology and Bioengineering 39: 1052–1057.

Payne GF, Shuler ML and Brodelius P. 1987. Plant cell culture. In: Large Scale Cell Culture Technology. Lydensen BK (Ed.). Hanser Publishers, New York, USA. pp. 193-229

Quensel AA and Ellis BE. 1989. Comparison of UV irradiation and p-fluorophenylalanine as selective agents for production of aromatic compounds in plant cell cultures. Journal of Biotechnology 10: 27-38.

Radwanski ER and Last RL. 1995. Tryptophan biosynthesis and metabolism: Biochemical and molecular genetics. Plant Cell 7: 921–934.

Reinhard E, Kreis W, Barthlen U and Helmbold U. 1989. Semicontinuous cultivation of *Digitalis lanata* cells: production of â-methyldigoxin in a 300 l airlift bioreactor. Biotechnology and Bioengineering 34: 502-508.

Rhodes MJC, Parr AJ, Ceielutt A and Aird ELH. 1994. Influence of exogenous hormone on the growth and secondary metabolite formation in transformed root cultures. Plant Cell Tissue and Organ Culture 38: 143–51.

Sajc L, Grubisic D and Vunjak-Novakovic G. 2000. Bioreactors for plant engineering: an Outlook for further research. Biochemical Engineering Journal 4: 89-99.

Seth R and Mathur AK. 2005. Selection of 5-methyltryptophanresistant callus lines with improved metabolic flux towards terpenoid indole alkaloid synthesis in *Catharanthus roseus*. Current Science 89 (3): 544-548.

Singh V. 1999. Disposable bioreactor for cell culture using wave-induced agitation. Cytotechnology 30: 149–158.

Smith JI, Smart NJ, Misawa M, Kurz WGW, Tallevi SG and DiCosmo F.1987. Increased accumulation of indole alkaloids by some cell lines of *Catharanthus roseus* in response to addition of vanadyl of vanadyl sulphate. Plant Cell Report 6: 142-145.

Staba EJ. 1980. Secondary metabolism and biotransformation. In: Plant Tissue Culture as a Source of Biochemicals. Staba EJ. (Ed.) CRC Press. Boca Raton, Florida, USA. pp. 59-97.

Stohs SJ, Kaul B and Staba EJ. 1969. The metabolism of cholesterol by *Dioscorea deltoidea* suspension cultures. Phytochemistry 8: 1679.

Tabata M and Hiraoka N. 1976. Variation of alkaloid production in *Nicotiana rustica* callus cultures. Physiologia Plantarum 38: 19-23.

Tanaka H. 1987. Large-scale cultivation of plant cells at high density: A review. Process Biochemistry 22: 106-113.

Ten Hoopen HJG, Van Gulik WM, Schlatman JE, Moreno PRH, Vinke J, Heijnen JJ and Verpoorte R. 1994. Ajmalicine production by cell cultures of *Catharanthus roseus*: from shake flask to bioreactor. Plant Cell Tissue Organ Culture 38: 85-91.

Terrier B, Courtois D, Henault N, Cuvier A, Bastin M, Aknin A, Dubreuil J and Petiard V. 2007. Two new disposable bioreactors for plant cell culture: the wave and undertow bioreactor and the slug bubble bioreactor. Biotechnology and Bioengineering 96: 914–923.

Thanh NT, Murthy HN, Yu KW, Jeong CS, Hahn EJ and Paek KY. 2006. Effect of oxygen supply on cell growth and saponin production in bioreactor cultures of *Panax ginseng*. Journal of Plant Physiology 163: 1337—1341

Thimmaraju R, Bhagyalakshmi N and Ravishankar GA. 2004. *In situ* and *ex situ* adsorption and recovery of betalains from hairy root cultures of *Beta vulgaris*. Biotechnology Progress 20:777-785.

Thimmaraju R, Bhagyalakshmi N, Narayan MS and Ravishankar GA. 2003. Kinetics of pigment release from hairy root cultures of *Beta vulgaris* under the influence of pH, sonication, temperature and oxygen stress. Process Biochemistry 38: 1069 - 1076

Threfall DR and Whitehead IM. 1988. The use of biotic and abiotic elicitors to induce the formation of secondary plant products in cell suspension cultures of Solanaceous plants. Biochemical Society Transaction 16: 71-75.

Tomita Y and Uomori A. 1973. Biosynthesis of isoprenoids. 3. Mechanism of alkylation during biosynthesis of stigmasterol in tissue cultures of higher plants. Journal of Chemical Society (Perkin 1). 22:2656-2659.

Vats S. 2010. Studies on *in vitro* regulation of bioactive phytoceuticals from selected medicinal plants. Ph.D. Thesis, Dept. of Botany. University of Rajasthan. Jaipur, India.

Verpoorte R, Van der Heijden R, Van Gulik WM and Hooken GJH. 1991. Plant biotechnology for production of alkaloids: Present status and prospects. In: The alkaloids. Brossi A (Ed.). Academic Press. San Diego, USA. pp. 1-8.

Wang PM, Huang TK, Cheng HP, Chien YH and Wu WT. 2002. A modified airlift reactor with high capabilities of liquid mixing and mass transfer. Journal of Chemical Engineering of Japan. 35: 354–359.

WHO. 2004. World Health Report. New WHO guidelines to promote proper use of alternative medicines. USA.

Wijnsma R, Van Weerdon G, Harkes IN, Verpoorte R and Svendsen AB. 1985. Anthroquinones as phytoalexins in cell tissue culture of *Cinchona* sps. Plant Cell Report 11: 241-244.

Wolters B and Eilert U. 1983. Elicitoren-Ausloser der Akkumulation von Pflanzenstoffen. Deutsche Apotheker Zeitung 123: 659-667.

Yamada Y and Hashimoto T. 1990. Possibilities for improving yield of secondary metabolites in plant cell cultures. In: Current Plant Science and Biotechnology in Agriculture. Progress in Plant Cellular and Molecular Biology. Nijkamp HJJ, Van der Plas LHW and Van Aartrijk J (Eds.). Kluwer, The Netherlands. pp.547-556.

Yamamoto O and Yamada Y. 1987. Selection of a reserpine-producing cell strain using UV-light and optimization of reserpine production in the selected cell strain. Plant Cell Tissue Organ Culture 8: 125-133.

Yamamoto Y, Mizuguchi and Yamada Y. 1982. Selection of a high and stable pigment - producing strain in cultured *Euphorbia millii* cells. Theoretical and Applied Genetics 61: 113-116.

Zenk MH, El-Shagi H, Arens H, Stockigt J, Weiler EW, Deus B. 1977. In: Plant Tissue Culture and Its Biotechnological Application. Barz W, Reinhard E Zenk MH (Eds.) Springer, Berlin. pp. 27.

Zhao J, Zhu WH and Hu Q. 2001. Enhanced catharanthine production in *Catharanthus roseus* cell cultures by combined elicitor treatment in shake flasks and bioreactors. Enzyme Microbial Technology 28: 673-681.

❑❑❑

CHAPTER – 6

Plant Defense Priming Role of Allelochemicals

R. Aswati Nair[1], Geethu Chellappan[1], Padmesh P[2] and K. Manikantan Menon[2]
[1]School of Biotechnology, NIT-Calicut, Calicut-673 601, Kerala
[2]Biotechnology & Bioinformatics Division, Jawaharlal Nehru Tropical Botanic Garden and Research Institute (JNTBGRI), Palode, Thiruvananthapuram - 695 562, Kerala
Email : aswati@nitc.ac.in

Summary

This review addresses various defense strategies evolved by plants to resist pathogen attack.Of the different defensive strategies, pathogen attack either induces or enhances constitutively the synthesis of a spectrum of diverse chemical compounds in various plant species. These chemical compounds also called secondary metabolites are classified broadly as nitrogen compounds, terpenoids and phenolics and have been documented to possess broad range antifungal, antimicrobial and pesticidal activities. Such defensive compounds could be extracted and developed as eco-friendly biopesticides/microbicides because of their natural occurrence, ephemeral nature and biodegradability. These will be favorable alternative strategies compared to chemical pesticides that are fraught with various harmful side effects.

1. Introduction to Plant Defense

Like animals, plants have to overcome a plethora of biotic stresses caused by a wide variety of pathogens that include bacteria, viruses, fungi, nematodes, insects and many herbivorous animals. Despite being sessile, plants have developed an elaborate and sophisticated innate defense system against these microbial invaders in consequence of the long process of host-pathogen co-evolution (Dodds and Rathjen, 2010). The multiple layers of defense mechanisms evolved by plants ranges from passive-mechanical or preformed chemical barriers, which provide nonspecific/non-host protection (Yun and Loake, 2002), to active responses that provide host- or cultivar-specific resistance (Jackson and Taylor, 1996; Hammond-Kosack and Parker, 2003). Passive-mechanical barriers include the various structural features that are characteristic of different plants like thorns, prickles, spines or trichomes, many of these may sometimes also contain irritants or poisons. In addition to these structures, other mechanical barriers to microbial invasion include the waxy outer layer called cuticle, the secondary protective tissue namely the periderm in addition to wax, resins, latex (Agrawal and Konno, 2009), lignins etc that prevent pathogen ingress. Many plants have non-protoplasmic inclusions primarily in epidermal cells like calcium oxalate crystals (Goldblatt et al. 1984; Rudall, 1994), starch grains, mucilage (slime) and silica bodies (10-30 microns). In the plants, they serve as variety of purposes that includes structural rigidity (Kaufman et al. 1979), lodging resistance (Ma et al. 2006), mechanical resistance that deters predators (Namaganda et al. 2009), tolerance to pathogens (Djamin and Pathak, 1967) and overcoming various abiotic stresses like drought, radiation, nutrient imbalance, temperature fluctuations and osmotic imbalances (Hodson et al. 2005; Ma and Yamaji, 2006).

Table 1 : Summary of the various mechanical and chemical defense barriers in plants

Pre-formed defenses	Mechanisms	Mode of action	Examples	Reference
Physical barriers				
Structural features	Thorns, prickles, spines, trichomes, thick cell wall etc	Deterrents to invading organisms	Thorns on the stem of raspberry plant, serve as a mechanical defense against herbivory.	Fernandes, 1994
Physiological features	Wax, cuticle, latex, resins, silica bodies, crystals and other inclusions	Inhibits/ interferes with pathogen attack	Silicification in grass leaves Larrea tridentate resin serves as an antidessicant, solar radiation filter and also as a herbivore deterrent	McNaughton and Arrants, 1983 Rhoades, 1977
Chemical barriers: Secondary metabolites				
Nitrogen compounds	Alkaloids	Alters the metabolic system of the invading organism, Enzyme inhibition	Aporphine alkaloid liriodenine from Liriodendron tulipifera against Candida albicans	Clark, 1987
	Cyanogenic glycosides		Cyanogenic glycosides of Sorghum bicolor impart resistance to Locusta migratoria	Woodhead and Bernays, 1978

Contd.

	Glucosinolates		Glucosinolate breakdown products negatively affects insect respiratory system	Clauss et al. 2006; Muller et al. 2001
Terpenoids		Inhibits enzymes, Membrane disruption, as pathogenic deterrents, pollinator attractants, plant-plant signaling	Monoterpenes like pyrethroid from Chrysanthemum spp. ; alpha pinene and myrcene from Pinus spp., carvone from Carum carvi with fungicidal activity	Ikeda et al. 1980; Raffa et al. 1985; Hartmans et al. 1995;
			Diterpenes like abietane and seco-abietane from Salvia prionitis	Chen et al. 2002
			Sesquiterpenes confer resistance to Colorado potato beetle in tomato	Carter et al. 1989
Phenolics	Tannins	Membrane destabilization, Antioxidant, inhibition of microbial enzymes, inactivation of microbial adhesins	Catechol (I) and protocatechuic acid (II) accumulation in onion scales imparts resistance to Colletotrichum circinans; lignans in disruption of endocrine system	Link et al. 1929
	Lignans			Harmatha and Dinan, 2003
	Phenolic acid			Wallis et al. 2008; Bonello et al. 2006
	Flavanoids			Treutter, 2006

Pathogen can overcome these physical barriers by way of force penetration using specialized structures, entry through stomata or wounded regions or by production of degradative enzymes (Walton 1994). But since plants produce wide range of secondary metabolites many of which possess antimicrobial activities, the pathogen will be encountered by this second line of defense (Edreva et al. 2008). The various defensive secondary metabolites in plants can be classified as terpenoids, phenolics and nitrogen compounds (Harborne 1999). The chemical barrier has evolved as a consequence of natural selection and plays decisive role in plant resistance against the invading microbes (Ingham 1973; Osbourn 1996). However successful pathogens have evolved to circumvent this chemical arsenal (Morrisey and Osbourn 1999; Papadopoulou et al. 1999). Once the pathogen breaches the preformed mechanical barriers it comes under the robust surveillance of host resistance/ active defense which finally results in resistance or susceptibility of host to the invading pathogen (Jones and Jones 1997; Hammond-Kosack and Jones 1997; Moffett et al. 2002; Morita-Yamamuro 2005).

In resistant plants, pathogen avirulence (avr) gene encoded elicitors (Dangl and Holub 1997; Nimchuk et al. 2001) are perceived by the plant R-gene encoded receptors that "guard" the host proteins (Dangl and Jones 2001; Jones 2001; Rathjen and Moffett 2003). This molecular recognition originally proposed by Flor (1956; 1971) and elaborated in the "**gene-for-gene model**" underlies the molecular basis of defense response and subsequently triggers a hierarchy of defensive molecular events in the plant. After Flor's classic work, it became evident that the outcome of many host-parasite interactions is governed by matching gene pairs [resistance (R) and avirulence (avr) genes, respectively] (Crute and Pink, 1996). R-proteins localized in the plasma membrane or cytoplasm impart resistance to the host plant either by direct detection of elicitors or indirectly through detection of any pathogen mediated alterations to the "guarded" host proteins (Dangl and Jones 2001). Following the R-Avr gene product interaction, a signaling cascade is activated that includes generation of reactive oxygen intermediates (ROIs), transient ion fluxes, intracellular pH changes, cell wall strengthening, release of secondary signaling molecules like nitric oxide (NO), salicylic acid (SA), jasmonic acid (JA) and ethylene (ET) and activation of MAPK (Mitogen activated protein kinase) cascades (Martin 1999; McDowell and Dangl 2000; Heath 2000). All these molecular events eventually result in the transcriptional activation of batteries of plant defense genes, that includes pathogenesis-related (PR) genes. The cumulative effect of all these cellular events arrests microbial proliferation through collapse of challenged plant cells by a localized programmed cell death (PCD), termed the 'hypersensitive response' (HR) (Keen et al. 1993) and shares analogy to animal apoptosis (Lam et al. 2001). HR results in establishment of systemic acquired resistance (SAR) (Ross 1961) that immunizes the entire plant against secondary attack by a broad spectrum of pathogens (Ryals et al. 1996). SAR is characterized by an increase in endogenously synthesized signaling molecules like SA (Malamy et al. 1990; Metraux et al. 1990), concomitant activation of PR genes (Bol et al. 1990 and Ward et al. 1991) and heightened plant resistance (Staskawicz 1995). Other signaling molecules like JA and ET have also been observed to induce expression of defense genes (Wasternack and Parthier 1997; van Wees et al. 2000; Hammond-Kosack and Parker 2003) that are not activated by SA (Penninckx et al. 1996). Apparently it has been proven that JA/ET act antagonistically to SA pathway but global gene expression profiling supports existence of substantial cross-talk between the SA, JA and ET signaling pathways (Glazebrook et al. 2003; Schenk et al. 2000; Thomma et al. 2001). The defense genes activated by these secondary signaling molecules include proteases (Pechan et al. 2000), protease inhibitors or PIs (Azarkan et al. 2004; Kehr 2006; Walz et al. 2004), chitinases (Howard and Glazer 1969; Azarkan et al. 2004; Kim et al. 2003), oxidases like polyphenol oxidase (PPO), peroxidase (POD) (Saby et al. 2003) and lipoxygenases (LOX) (Walz et al. 2004) in addition to phosphatase (Lynn & Clevette-Radford 1987) and lipases (Gandhi & Mukherjee 2000).

2. Chemical Defense in Plants

Among the many different mechanisms employed by plants for combating diseases, the ability to synthesize an arsenal of low-molecular weight volatile and non-volatile chemicals helps them to interact with the ever-changing physical environment (Firn and Jones, 2009). These defensive organic compounds called secondary metabolites have been a major counter defense tactics evolved by plants to withstand pathogen attack (Boller, 1995). Unlike primary metabolites which are found in every dividing cell, secondary metabolite biosynthesis is regulated by environmental factors, genotype of the plant as well as age (Kossel, 1891). There is also difference in distribution with particular metabolites being confined to phylogenetically related taxa while others having much broader distribution (Futuyma and Agrawal, 2009; Dixon, 2001).

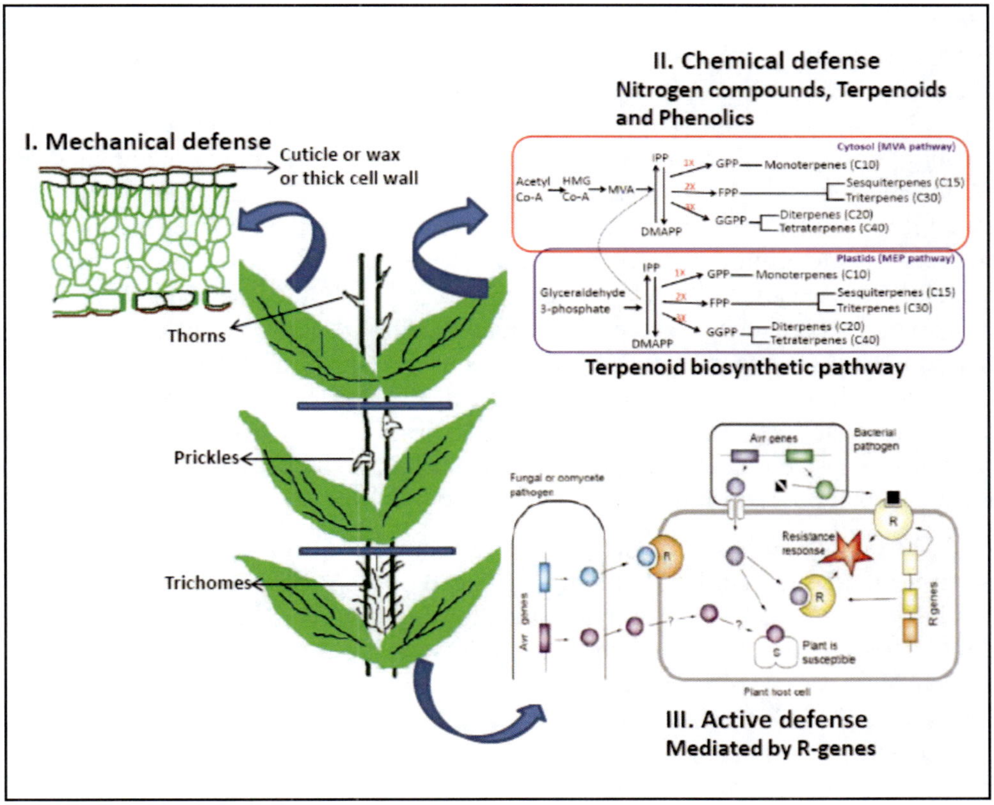

Fig. 1 : An overview of the different defense mechanisms employed by plants; with adaptations from Tyler, 2001.

Even in any particular plant the distribution of secondary metabolites may not be uniform but it will be constitutive with an induction seen in response to pathogen attack

(Zong and Wang, 2006). Constitutive/basal distribution of secondary metabolites, which is central to the optimal defense theory of plant-pathogen interactions (McKey, 1979; Rhoades, 1979; Strauss et al. 2002; Holland, 2009), is of significance considering the crucial role of speed of the host response in determining resistance to the invading pathogen.

The various secondary metabolites synthesized in the plants can be classified into three groups: nitrogen compounds, terpenoids, and phenolics (Harborne, 1999). The nitrogen compounds include alkaloids, cyanogenic glycosides and glucosinolates. Alkaloids widespread among different plant taxa are derived from various amino acids and are basic organic compounds with heterocyclic nitrogen, the position of which varies according to the plant families (Harborne, 1993; Wink, 2004). More than 12,000 different types of alkaloids have been discovered from over 300 plant families (Zwenger and Basu, 2008). The defensive effects of alkaloids are manifested by their ability to affect cell membrane/cytoskeletal structure, as inhibitors of glycosidases and sugar-metabolizing enzymes (Asano et al. 2000) or by inhibition of protein/DNA synthesis (Schmeller et al. 1997). Cyanogenic glycosides composed of an alpha-hydroxynitrile type aglycone and sugar moiety (mostly D-glucose) are stored in inactive forms in plant vacuoles among a wide range of plant taxa (Vetter 2000). Upon enzymic hydrolysis, cyanogenic glycosides release cyanohydric acid (HCN) or prussic acid (Harborne, 1993) which is extremely toxic to a wide spectrum of organism due to its ability to chelate metal ions functioning as co-factors to many key enzymes involved in metabolic processes (McMahon et al. 1995; Francisco and Pinotti, 2000). In the unaffected plant the enzyme and cyanogenic glycosides remain separate and are brought in contact upon pathogen attack (Gruhnert et al. 1994). Sulfur containing glucosinolates or thioglucosides produced widely by plants belonging to the genera Cruciferae have well-documented antimicrobial activity (Fenwick et al. 1983). Hydrolysis of glucosinolates liberates D-glucose, sulphate ion and a series of compounds such as isothiocyanate (ITC), thiocyanate and nitrile that posseses fungitoxic activity (Mithen, 1992; Wink, 2004).

Most numerous and structurally diverse group of organic secondary metabolites are terpenoids (isoprenoids) derived from five carbon isoprene units and having diverse physiological, metabolic and structural functions in addition to their significant role in plant defense (Rees and Harborne, 1985; Sessa et al. 2000; Dussourd, 2003). All terpenoids are derived by isomerization of isopentenyl diphosphate (IPP) to dimethylallyl diphosphate (DMAPP) by isopentenyl diphosphate isomerase. IPP and DMAPP originate from either the plastidial methyl-erythritol 4-phosphate (MEP) pathway or the cytosolic mevalonate (MVA) pathway (Chappell, 1995; Bochar et al. 1999). DMAPP condenses with one, two or three units of IPP catalyzed by prenyltransferases to give geranyl diphosphate (GPP; C10), farnesyl diphosphate (FPP; C15), and geranylgeranyl diphosphate (GGPP; C20) respectively. These three acyclic prenyl diphosphates are converted by a very large group of enzymes called the terpene (terpenoid) synthases (TPSs) to corresponding monoterpenoid (C10), sesquiterpenoid (C15), and diterpenoids

(C20). Subsequent hydroxylation and oxidation of terpenes catalyzed by cytochrome P450 (CYP450) enzymes further diversifies the terpenes (Zulak and Bohlmann, 2010). Gene duplication and neofunctionalization of the enzymes that synthesize and subsequently modify terpenes are primarily responsible for the terpenoid diversity in response to different physiological conditions (Keeling et al. 2008; Trapp and Croteau, 2001).

Phenolics with broad distribution in the plant kingdom are aromatic compounds with hydroxyl groups. Phenolics range from simple plant phenols like hydroxy-benzoic acid to phenylpropanoids (with three carbon side chains) and complex phenyl propanoids like flavonoids that give pigmentation to plants and constitute a complex group called tannins (Harborne, 1993). Flavanoids are further sub-divided as flavanols, anthocyanidins and chalcones and perform a wide range of biological activities that includes their role as attractants or feeding deterrents (Snook, 1994), biotic and abiotic stress protective agents (Tattini et al. 2004; Moore et al. 2005; Schlösser, 1994; Inderjit and Gross, 2000). Antimicrobial activities of flavanoids are mediated by their ability to inhibit microbial cell wall degrading enzymes and chelation of metals necessary for enzyme activity (Skadhauge et al. 1997). Phenolics generally have antioxidant properties and are thus of therapeutic significance (Dai and Mumper, 2010). Systemic induction and accumulation of low molecular weight phenolics and phenolic polymers like lignin and suberin is observed in response to various diseases (Bonello et al. 2006; Wallis et al. 2008; Cvikrova et al. 2006). Inhibition of pathogen colonization by phenolics is mediated via protein precipitation and iron depletion (Scalbert, 1991).

3. Molecular Evolution of Allelochemical Diversity

Secondary metabolite chemistry of plants is dynamic and subject to changes depending on the type of environmental stress (Kliebenstein, 2004; Hammerschmidt 2005; Mary Ann Lila 2006) with upregulation of key genes observed in response to pathogen attack (Baldwin, 1998; Sirvent and Gibson, 2002). Pathogen attacks impose natural selection on plants to evolve defensive strategies (Agrawal, 2007) that in turn depends on the plants life history, genetic attributes and mating pattern (Futuyma and Agrawal 2009; Johnson et al. 2009). Reduced recombination and allelic segregation as a consequence of prolonged asexual reproductive strategies negatively affects defense gene evolution as proposed in the Recombination-Mating hypothesis (Levin, 1975). It is hypothesized that asexual reproduction not only results in erosion of genetic variation but also is susceptible to Muller's Ratchet according to which sexual reproduction purges the deleterious mutations that might affect primary or secondary metabolic processes (Muller, 1964; Paland and Lynch, 2006). As per the Red Queen hypothesis also, maintenance of sexual reproduction is of prime significance in the host-pathogen co-evolutionary arms race (Jaenike, 1977), thereby equipping plants to synthesize an enormous reservoir of natural chemical diversity (Agrawal, 2007). Natural selection pressures will select for alleles that enhance or retain secondary metabolite diversity taking into consideration the fitness cost on the plant (Agrawal and Fishbein, 2008). Phylogenetic analysis has revealed the role of recombination, mutations, gene duplications and reshuffling in maintaining allelic diversity in the secondary metabolite as well defense genetic pool (Wagner, 1998; Gierl and Frey, 2001; Kondrashov et al. 2002). Molecular studies in genes associated with secondary metabolite biosynthesis like polyketide synthases (PKS), genes associated with phenyl propanoid (Milkowski and Strack, 2004; Stehle et al. 2006) and alkaloid (Ober and Hartmann, 1999) biosynthetic pathway, cytochrome P450 (Scott and Wen, 2001) and terpene synthases (TPS) (Bohlmann et al. 1998; Rohdich et al. 2005) have revealed that environmental selection pressures must have favored duplication followed by functional divergence in generating/ maintaining a rich diversity of metabolites that are distinguishable from their biogenetic starter units. Studies have indicated that terpenoid phytochemical complexity is contributed by a multitude of phylogenetically diverse TPS and CYP450 that have been subject to repeated duplication and divergence (Zulak and Bohlmann, 2010). Even single amino acid changes have been observed to alter substrate specificity, product profiles and kinetic efficiency of these enzymes (Jez et al. 2002; Keeling and Bohlmann, 2006). Nucleotide sequence diversity analysis of such functional genes from resistant and susceptible sources have thus provided a framework for understanding the variations in deployment of metabolites and defensive proteins in plant-pathogen responses (Shonle and Bergelson, 2000; Bishop et al. 2000).

Secondary metabolite biosynthetic genes can also evolve by divergent evolution and by domain swapping (O'Brien and Herschlag, 1999; Schmidt et al. 2003; Katoh et al. 2004). Simultaneous analysis of diversity at both neutral and defense-related functional loci has helped in elucidating the evolutionary forces shaping resistance evolution (Burdon and Thrall, 1999; de Meaux et al. 2003). Adaptive evolution at key residues involved in detection of variable pathogen ligands and recombination events generating paralogues have been suggested as primary mechanisms responsible for generation of diversity at these loci (Ehrlich and Raven, 1964; Michelmore and Meyers, 1998; Futuyma and Agrawal, 2009). Besides sequence diversity analysis have also revealed the role of balancing selection in maintaining high nucleotide diversity levels at key resistance loci reflecting "trench warfare" (Stahl et al. 1999) or "recycling polymorphism" (Holub, 2001) between host-pathogen genotypes (Bergelson et al. 2001).

Unfortunately, domestication and breeding procedures have eliminated many of the traits contributing to secondary metabolite diversity in crop plants (Lebot et al. 2005; Olsen and Gross, 2008) with a consequence that has led to heavy reliance on chemical pesticides/insecticides (Holdren and Ehrlich, 1974; Nash et al. 2010). This increased susceptibility in cultivars attributed to "domestication bottleneck" eventually reduces overall genetic diversity in the cultivars as against their resistant wild counterparts (Tang and Knapp, 2003). Hence to gain a comprehensive understanding of the molecular processes governing host-pathogen interactions, diversity should also be sampled in genes involved in secondary metabolite biosynthesis of both cultivars as well in the relatively undisturbed wild relatives of crop plants (Hawkes, 1977). The adverse impact of pesticides on environment and biodiversity compounded by the susceptibility of cultivars of many crop plants has sparked an interest in understanding the mechanisms underlying the evolution and nature of this chemical diversity in developing biologically and ecologically friendly yet sustainable resistance strategies by making use of the plants own innate immune system.

4. Allelochemicals as Biocontrol Agents for Plant Diseases

Growing global awareness about safe food and increasing public consciousness on the adverse impact of pesticides both on human health and environment has witnessed a boom in organic farming and has fuelled the search for greener alternatives in the agricultural sector (Fravel, 2005; Slusarenko et al. 2008). Plant derived allelochemicals or secondary metabolites provide better alternatives than pesticides because of low toxicity, biodegradability and hence reduced risk to environment and human health (Tesar et al. 1988; Jespers and de Waard, 1993; Tripathi and Dubey, 2004) and thus serve as leads for new agrochemicals. Plant derived antimicrobial agents that include essential oils, alkaloids, phenols, flavanoids, tannins and sterols (Burt, 2004; Halama and van Haluwin, 2004) can contribute to disease control because of their enormous diversity. Many allelochemicals have been identified and characterized so far and includes Heliannuols and guaianolides from H. annuus (Macias et al. 1999); 14,15[beta]-Epoxy-prieurianin from bark of Guarea guidona (Lukacova et al. 1982); Valtrate from Valeriana capense (Fuzzati, 1996), insecticidal pyrethrin I and pyrethrin II from Tanacetum cinerariifolium (Staudinger and Ruzicka, 1924); Nicotine from Nicotiana tabacum (Elliott et al. 1974), Azadirachtin from Azadirachta indica (Butterworth and Morgan, 1968; Abbasi et al. 2005; Rani et al. 2006); rotenone from Derris urucu (Geoffrey, 1895), glucosinolates like glucoraphenine ITC and Allyl-isothiocyanate (AITC) from mustard and horseradish (Ishiki et al. 1992; Delaquis and Mazza, 1995).to name a few. Even though performance of these allelochemicals is considered inferior to non-natural chemical pesticides, but the environment friendly and low toxic nature far outweighs the damaging effects of chemical pesticides.

Conclusion

A comprehensive understanding of the chemical complexities seen in different taxa can thus be utilized towards developing measures for crop protection. Encouraging results have been reported regarding the crucial role played by these antimicrobial secondary metabolites in plant-pathogen interactions.Development of these non-chemical/synthetic plant based formulations on a commercial scale is gaining considerable momentum considering the merits that includes negligible risk to human health and environment as against the chemical pesticides. As many of these secondary metabolites have proven biological activities, scale-up production of active principles/

preparations of secondary metabolites, from resistant plants can be optimized and exploited for development of environment friendly biopesticides/microbicides. Alternatively, strategies can be developed through metabolic engineering to mediate the synthesis of these defensive secondary metabolites in susceptible cultivars to combat the invading pathogens.

Acknowledgement

ANR and GC thank NIT-Calicut for the financial support. PP and MM express their sincere thanks to Jawharlal Nehru, Tropical Botanic Garden and Research Institute for the facilities and support received.

References

Abbasi P.A., Riga, E., Conn K.L. and Lazarovits, G. (2005). Effect of neem cake soil amendment on reduction of damping-off severity and population densities of plant-parasitic nematodes and soilborne plant pathogens. Can. J. Plant Pathol., 27: 38-45.

Agrawal, A.A. (2007). Macroevolution of plant defense strategies. Trends in Ecol. And Evol. 22: 103-109

Agrawal, A.A. and Fishbein, M. (2008) Phylogenetic escalation and decline of plant defense strategies. Proc Natl Acad Sci USA 105: 10057–10060.

Agrawal, A.A. and Konno, K. (2009). Latex: A Model for understanding mechanisms, ecology, and evolution of plant defense against herbivory. Annu. Rev. Ecol. Evol. Syst. 40:311–31.

Asano, N. Nash, R.J., Molyneux, R.J. and Fleet, G.W.J. (2000). Sugar-mimic glycosidase inhibitors: natural occurrence, biological activity and prospects for therapeutic application. Tetrahedron: Asymmetry 11:1645–80.

Azarkan, M. Wintjens, R. Looze, Y. and Baeyens-Volant, D. (2004) Detection of three wound-induced proteins in papaya latex. Phytochemistry 65:525–34.

Baldwin, I.T. (1998) Jasmonate-induced responses are costly but benefit plants under attack in native populations. Ecology 95: 8113-8118.

Bergelson, J. Kreitman, M. Stahl, E. and Tian, D. (2001) Evolutionary dynamics of plant R-genes. Science 292: 2281-2285.

Bishop, J.G., Dean A.M. and Mitchell-Olds T (2000) Rapid evolution in plant chitinases: molecular targets of selection in plant–pathogen coevolution. Proc Natl Acad Sci USA 97: 5322–5327.

Bochar, D.A., Friesen, J.A. Stauffacher, C.V., Rodwell, V.W. (1999) Biosynthesis of mevalonic acid from acetyl-CoA. In: Cane DE, ed. Isoprenoids, including carotenoids and steroids, Vol. 2. Comprehensive natural products chemistry. London, UK: Elsevier, 15–44.

Bohlmann, J. Meyer-Gauen, G. and Croteau, R. (1998) Plant terpenoid synthases: Molecular biology and phylogenetic analysis. Proc Natl Acad Sci USA 95:4126–4133.

Bol, J.F., Linthors, H.J.M. and Cornelissen, B.J. (1990). Plant pathogenesis-related proteins induced by virus infection. Annual Review of Phytopathology 28: 113–38.

Boller, T. (1995) Chemoreception of microbial signals by plant cells. Ann. Rev. of Plant Physiol.and Plant Mol. Biol. 46: 189–214.

Bonello, P. Gordon, T.R., Herms, D.A., Wood, D.L. and Erbilgin, N. (2006) Nature and ecological implications of pathogen-induced systemic resistance in conifers: a novel hypothesis. Physiol. and Mol. Plant Pathol. 68: 95-104.

Burdon, J.J. and Thrall, P.H. (1999) Spatial and temporal patterns in coevolving plant and pathogen associations. Am Nat 153: S15–S33.

Burt, S. (2004). Essential oils: their antibacterial properties and potential applications in foods- a review. Int. J. Food Microbiol. 94(3): 223-53.

Butterworth, J.H. and Morgan, E.D. (1968). Isolation of a substance that suppresses feeding in locusts. J. Chem. Soc. Chem. Commun. 23-24.

Carter, C.D., Gianfagna, T.J. and Sacalis, J.N. (1989). Sesquiterpenes in glandular trichomes of a wild tomato species and toxicity to the Colorado potato beetle. J. Agric. Food Chem. 37(5):1425–1428.

Chappell, J. (1995). The biochemistry and molecular biology of isoprenoid metabolism. Plant Physiology 107: 1– 6.

Chen, X., Ding, J., Ye, Y.M, Zhang, J.S. (2002). Bioactive abietane and seco-abietane diterpenoids from Salvia prionitis. J Nat. Prod. 65 (7):1016-20.

Christina, J.P., Paula, J.R. and Mary, G. (2003) Systematics and biology of silica bodies in monocotyledons. The Botanical Rev. 69(4):377-440.

Clark, A.M., Watson, E.S., Ashfaq, M.K. and Hufford, C.D. (1987) In vivo efficacy of antifungal oxaaporphine alkaloids in experimental disseminated candidiasis. Pharm. Res. 4(6): 495-498.

Clauss, M.J., Dietel, S., Schubert, G., Mitchell-Olds, T. (2006). Glucosinolate and trichome defenses in a natural Arabidopsis lyrata population.J Chem Ecol. 32(11):2351-73.

Crute, I. and Pink, D. (1996) The genetics and utilization of pathogen resistance in plants. Plant Cell 8 : 1747-1755.

Cvikrova, M. Mala, J. Hrubcova, M. and Eder, J. (2006). Soluble and cell wall-bound phenolics and lignin in Ascocalyx abietina infected Norway spruces. Plant Sci. 170: 563–570.

Dai, J. and Mumper, R.J. (2010). Plant Phenolics: Extraction, analysis and their antioxidant and anticancer properties. Molecules 15: 7313-7352.

Dangl, J.L. and Holub, E. (1997) La dolce vita: a molecular feast in plant-pathogen interactions. Cell 91: 17-24.

Dangl, J. and Jones, J. (2001) Plant pathogens and integrated defence responses to infection. Nature 411: 826–833.

De Meaux, J., Cattan-Toupance, I., Lavigne, C., Langin, T. and Neema. C. (2003). Polymorphism of a complex resistance gene candidate family in wild populations of common bean (Phaseolus vulgaris) in Argentina: comparison with phenotypic resistance polymorphism. Mol Ecol 12: 263–273.

Delaquis, P.J. and Mazza, G. (1995). Antimicrobial properties of isothiocyanates in food preservation. Food Technol. 49: 73–84.

Dixon, R.A. (2001) Natural products and plant disease resistance. Nature 411: 843-847.

Djamin, A. and Pathak, M.D. (1967) Role of silica in resistance to the Asiatic rice borer, Chilo suppressalis (Walker), in rice varieties. J. Econ. Entomol. 60: 347–351.

Dodds, P.N. and Rathjen, J.P. (2010) Plant immunity: towards an integrated view of plant–pathogen interactions. Nature Reviews Genetics 11: 539-548

Dussourd, D.E. (2003). Chemical stimulants of leaf-trenching by cabbage loopers: natural products, neurotransmitters, insecticides, and drugs. J. Chem. Ecol. 29:2023–47.

Edreva, A. Velikova, V., Tsonev, T., Dagnon, S., Gürel, A., Akta°, L. and Gesheva, E. (2008) Stress-protective role of secondary metabolites: diversity of functions and mechanisms. Gen. Appl. Plant Physiol. 34 (1-2): 67-78.

Ehrlich, P.R. and Raven, R.H. (1964). Butterflies and plants: A study in coevolution. Evolution (Lawrence, Kans) 18: 586- 608.

Elliott, M., Farnham, A.W., Janes, N.F., Needham, P.H. and Pulman, D.A. (1974). Synthetic insecticide with a new order of activity. Nature 248: 710-711.

Fenwick, G.R., Heaney, R.K., Mullin, W.J. (1983). Glucosinolates and their breakdown products in food and food plants. CRC Crit. Rev. Food Sci. Nutr. 18: 123–201.

Fernandes, G.W. (1994). Plant mechanical defenses against insect herbivory." Revista Brasileira de Entomologia 38 (2): 421–433.

Firn, R.D. and Jones CG (2009) A Darwinian view of metabolism: molecular properties determine fitness. Journal of Exp. Botany 60(3); 719–726.

Flor, H. (1956). The complementary genetic systems in flax and flax rust. Adv. Genet. 8: 29-54.

Flor, H. (1971). The current status of gene for gene concept. Annu. Rev. Phytopathol. 9: 275-296.

Francisco, I.A. and Pinotti, M.H.P. (2000) Cyanogenic glycosides in plants. Braz Archives Biol Ttechnol 43(5): 487-492.

Fravel, D.R. (2005). Commercialization and implementation of Biocontrol. Ann. Rev. of Phytopathol. 43: 337-359.

Futuymaa, D.J. and Agrawal, A.A. (2009) Macroevolution and the biological diversity of plants and herbivores. PNAS 106(43): 18054–18061.

Fuzzati, N. Wolfender, J.L., Hostettmann, K., Msonthi, J.D., Mavi, S., Molleyres, L.P. (1996) Isolation of antifungal valepotriates from Valeriana capense and the search for Valepotriates in crude Valerianaceae extracts. Phytochem. Anal. 7(2): 76-85.

Gandhi, N.N. and Mukherje (2000). Specificity of papaya lipase in esterification with respect to the chemical structure of substrates. J. Agric. Food Chem. 48:566–70.

Geoffrey, E. (1895). Contribution a l'etude du Robinia Nicou Aublet, au point de vue botanique, chimique et physiologique. Ann. Inst. Colon. Marseille 2: 1-7.

Gierl, A. and Frey, M. (2001) Evolution of benzoxazinone biosynthesis andindole production in maize. Planta 213:493–498.

Glazebrook, J. Chen, W., Estes, B., Chang, H., Nawrath, C., Metraux, J., Zhu, T. and Katagiri, F. (2003) . Topology of the network integrating salicylate and jasmonate signal transduction derived from global expression phenotyping. Plant J. 34: 217-228.

Goldblatt, P., Henrich, J.E. and Rudall, P. (1984). Occurrence of crystals in Iridaceae and allied families and their phylogenetic significance. Ann. Missouri Bot. Gard 71:1013–1020.

Gruhnert, C., Biehl, B. and Selmar, D. (1994) Compartmentation of cyanogenic glucosides and their degrading enzymes. Planta 195: 36-42.

Halama, P. and van Haluwin, C. (2004). Antifungal activity of lichen extracts and lichenic acids. Biocontrol 49: 95-107.

Hammerschmidt, R. (2005). Phenols and plant–pathogen interactions: the saga continues. Physiol. Mol. Plant Pathol. 66: 77–78.

Hammond-Kosack, K.E. and Jones, J.D.G. (1997). Plant disease resistance genes. Annu Rev Plant Physiol Plant Mol Biol 48: 575–607.

Hammond-Kosack, K. and Parker, J.E. (2003). Deciphering plant-pathogen communication: fresh perspectives for molecular resistance breeding. Current Opinion in Biotechnology 14: 177-193.

Harborne, J.B. (1993). Introduction to Ecological Biochemistry (Fourth Edition). Academic Press, London. 318 pp.

Harborne, J.B. (1999). Classes and Functions of Secondary Products From Plants. Chemicals From Plants. Ed. NJ Walton, De Brown. ICP: London. 1-26.

Harmatha, J. and Dinan, L. (2003). Biological activities of lignans and stilbenoids associated with plant-insect chemical interactions. Phytochem. Rev.2(3): 321-330

Hartmans, K.J., Diepenhorst, P., Bakker, W. and Gorris, L.G.M. (1995). The use of carvone in agriculture, sprout suppression of potatoes and antifungal activity against potato tuber and other plant diseases. Ind. Crops Prod. 4: 3–13.

Hawkes, J.G. (1977). The importance of wild germplasm in plant breeding. Euphytica 26: 615—621.

Heath, M. (2000). Hypersensitive response-related death. Plant Mol. Biol. 44: 321–334.

Hodson, M.J., White, P.J., Mead A and Broadley, M.R. (2005). Phylogenetic variation in the silicon composition of plants. Ann. Bot. (Lond.) 96: 1027–1046.

Holdren, J.P. and Ehrlich, P.R. (1974). Human population and the global environment. Am Sci. 62(3):282-292.

Holub, E.B. (2001). The arms race is ancient history in Arabidopsis, the wildflower. Nat Rev Genet 2: 516–527.

Howard, J.B. and Glazer, A.N. (1969). Papaya lysozyme: terminal sequences and enzymatic properties. J. Biol. Chem. 244:1399–409.

Ikeda, T., Enda, N., Yamane, A., Oda, K. and Toyoda, T. (1980). Attractants for the Japanese pine sawyer, Monochamus alternatus Hope (Coleoptera: Cerambycidae). App.Entomol.and Zool. 15: 358-361.

Inderjit, S. and Gross, E. (2000). Plant phenolics: potential role in aquatic and terrestrial ecosystems. In Polyphenols 2000 (Martens S, Treutter D and Forkmann G eds.), Freising, pp. 206–234.

Ingham, J.L. (1973). Disease resistance in higher plants. The concept of pre-infectional and post-infectional resistance. Phytopathol. Z. 78:314–335.

Ishiki, K., Tokuora, K., Mori, R., Chiba, S. (1992). Preliminary examination of allyl isothiocyanate vapour for food preservation. Biosci. Biotechnol. Biochem. 56: 1476–1477.

Jackson, A. and Taylor, C. (1996). Plant–microbe interactions: Life and death at the interface. Plant Cell 8: 1651–1668.

Jaenike, J. (1977). An hypothesis to account for the maintenance of sex within populations. Evol. Theory 3: 191–194.

Jespers, A.B.K. and de Waard, M.A. (1993) Natural products in plant protection. Neth. J. PI. Path. Supplement 3:109-117.

Jez, J.M., Bowman, M.E. and Noel, J.P. (2002). Expanding the biosynthetic repertoire of plant type III polyketide synthases by altering starter molecule specificity. Proc. Natl. Acad. Sci. 99(8): 5319-5324.

Johnson, M.T., Smith, S.D. and Rausher, M.D. (2009) Plant sex and the evolution of plant defenses against herbivores. Proc. Natl. Sci. USA. 106(43): 18079- 18084.

Jones, D. and Jones, J. (1997). The role of leucine-rich repeat proteins in plant defenses. Adv. Bot. Res. 24: 90-167.

Jones, J.D. (2001). Putting knowledge of plant disease resistance genes to work. Curr Opin Plant Biol. 4: 281-287.

Katoh, S., Hyatt D. and Croteau, R. (2004) Altering product outcome in Abies grandis (“)-limonene synthase and (“)-limonene/(“)-á-pinene synthase by domain swapping and directed mutagenesis. Arch Biochem Biophys 425:65–76.

Kaufman, P.B., Takeoka, Y., Carlson, T.J., Bigelow, W.C., Jones, J.D., Moore, P.H. and Ghosheh, N.S. (1979). Studies on silica deposition in sugarcane (Saccharum spp.) using scanning electron microscopy, energy dispersive X-ray analysis, neutron activation analysis, and light microscopy. Phytomorphology 29: 185–193.

Keeling, C.I., Weisshaar, S., Lin, R.P.C. and Bohlmann, J. (2008). Functional plasticity of paralogous diterpene synthases involved in conifer defence. 105: 1085- 1090

Keeling, I.C. and Bohlmann, J. (2006). Genes, enzymes and chemicals of terpenoid diversity in the constitutive and induced defence of conifers against insects and pathogens. New Phytologist 170: 657–675.

Keen, N., Bent, A. and Staskawicz, B. (1993). Biotechnolgy in Plant Disease Control. (New York).

Kehr, J. (2006). Phloem sap proteins: their identities and potential roles in the interaction between plants and phloem-feeding insects. J. Exp. Bot. 57:767–74.

Kim, J.S., Kim, Y.O., Ryu, H.J., K., wak, Y.S., Lee, J.Y. and Kang H. (2003). Isolation of stress-related genes of rubber particles and latex in fig tree (Ficus carica) and their expression by abiotic stress of plant hormone treatments. Plant Cell Physi. 44:412–14.

Kliebenstein, D.J. (2004). Secondary metabolites and plant/environment interactions: a view through Arabidopsis thaliana tinged glasses. Plant Cell Environ. 27: 675–684.

Kondrashov, F.A., Rogozin, I.B., Wolf, Y.I., Koonin, E.V. (2002). Selection in the evolution of gene duplications. Genome Biol 3:1–9.

Kossel, A. (1891). Archives of Analytical Physiology, Physiol Abteilung, 181–186.

Lam, E., Kato, N. and Lawton, M. (2001). Programmed cell death, mitochondria and the plant hypersensitive response. Nature 411: 848–853.

Lebot, V., Ivancic, A. and Abraham, K. (2005). The geographical distribution of allelic diversity, a practical means of preserving and using minor root crop genetic resources. Expl Agric. 41: 475–489.

Levin, D.A. (1975), Pest pressure and recombination systems in plants. Am. Nat. 109:437–451.

Link, K.P., Dickson, A.D. and Walker, J.C. (1929). Further observations on the occurrence of Protocatechuic acid in pigmented onion scales and its relation to disease resistance in the onion. J. Biol. Chem. 84: 719-725.

Lukacova, V., Polonsky, J., Moretti, C., Pettit, G.R. and Schmidt, J.M. (1982). Isolation and structure of 14,15α-epoxyprieurianin from the South American tree Guarea guidona. J. Nat. Prod. 45(3); 288-294.

Lynn, K.R. and Clevette-Radford, N.A. (1987). Biochemical properties of latices from the Euphorbiaceae. Phytochemistry 26:939–44.

Ma, J.F., Tamai, K., Yamaji, N., Mitani, N., Konishi, S., Katsuhara, M., Ishiguro, M., Murata, Y. and Yano, M. (2006). A silicon transporter in rice. Nature 440: 688–691.

Ma, J.F. and Yamaji, N. (2006). Silicon uptake and accumulation in higher plants. Trends Plant Sci. 11: 392–397.

Macias, F.A., Galindo, J.C.G., Molinillo, J.M.G., Castellano, D., Velasco, R.F. and Chinchilla, D. (1999). Developing new herbicide models from allelochemicals. Pestic. Sci. 55: 662-665.

Malamy, J., Carr, J., Klessig, D. and Raskin, I. (1990). Salicylic acid: A likely endogenous signal in the resistance response of tobacco to viral infection. Science 250: 1002-1004.

Martin, G. (1999). Functional analysis of plant disease resistance genes and their downstream effectors. Curr. Opin. Plant Biol. 2: 273-279.

Mary Ann Lila (2006). The nature-versus-nurture debate on bioactive phytochemicals: the genome versus terroir. J. Sci. Food Agric. 86: 2510–2515.

McDowell, J. and Dangl, J. (2000) Signal transduction in the plant immune response. Trends Biochem Sci. 25: 79-82.

McKey, D. (1979). The distribution of secondary compounds within plants. Herbivores: Their Interaction with Secondary Plant Metabolites (eds G.A. Rosenthal & D.H. Janzen), pp. 56–134. Academic Press, Orlando, FL.

McNaughton, S.J. and Tarrants JL (1983) Grass leaf silicification: natural selection for an inducible defense against herbivores. Proc. Natl. Acad. Sci. 80: 790-791.

Metraux, J.P., Signer, H., Ryals, J., Ward E, Wyss-Benz, M., Gaudin, J., Raschdorf, K., Schmid, E., Blum, W. and Inverardi, B. (1990). Increase in salicylic acid at the onset of systemic acquired resistance in cucumber. Science 250: 1004-1006.

Michelmore, R.W. and Meyers, B.C. (1998). Clusters of Resistance genes in plants evolve by divergent selection and a Birth-and-Death process. Genome Res. 8: 1113-1130.

Milkowski, C., Strack, D. (2004). Serine carboxypeptidase-like a cyltransferases. Phytochemistry 65:517–524.

Mithen, R. (1992). Leaf glucosinolate profile and their relationship to pest and disease resistance in oilseed rape. Euphytica 63(1-2): 71-83.

Moffett, P., Farnham, G., Peart, J. and Baulcombe, D. (2002). Interaction between domains of a plant NBS-LRR protein in disease resistance-related cell death. EMBO J. 21: 4511-4519.

Moore, J.P., Westall, K.L., Ravenscroft, N., Farrant, J.M., Lindsey, G.G. and Brandt, W.F. (2005). The predominant polyphenol in the leaves of the ressurrection plant Myrothamnus flabellifolius, 3,4,5-tri-O-galloylquinic acid, protects membranes against desiccation and free radical induced oxidation. Biochem. J. 385: 301-308.

Morita-Yamamuro, C., Tsutsui, T., Sato, M., Yoshioka, H., Tamaoki, M., Ogawa, D., Matsuura, H., Yoshihara, T., Ikeda, A., Uyeda, I. and Yamaguchi, J. (2005). The Arabidopsis Gene CAD1 Controls Programmed Cell Death in the Plant Immune System and Encodes a Protein Containing a MACPF Domain. Plant Cell Physiol 46: 902-912.

Morrisey, J.P. and Osbourn, A.E. (1999). Fungal resistance to plant antibiotics as a mechanism of pathogenesis. Microbiol Mol Biol Rev 63: 708-724.

Muller, C., Agerbirk, N., Olsen, C.E., Boeve, J.L., Schaffner, U.R.S. and Brakefield, P.M. (2001) . Sequestration of host plant glucosinolates in the defensive hemolymph of the sawfly Athalia rosae. J. of Chem. Ecol. 27(12): 2505-2516.

Muller, H.J. (1964). The relation of recombination to mutational advance. Mut Res. 1:2–9.

Namaganda, M., Krekling, T. and Lye, K.A. (2009). Leaf anatomical characteristics of Ugandan species of Festuca L. (Poaceae). S. Afr. J. Bot 75: 52–59.

Nash, M.A., Hoffmann, A.A., Thomson, L.J. (2010). Identifying signature of chemical applications on indigenous and invasive nontarget arthropod communities in vineyards. Ecol Appl. 20(6): 1693-1703.

Nimchuk, Z., Rohmer, L., Chang, J. and Dangl, J. (2001). Knowing the dancer from the dance: R-gene products and their interactions with other proteins from host and pathogen. Curr. Opin. Plant Biol. 4: 288-294.

O'Brien, P.J. and Herschlag, D. (1999). Catalytic promiscuity and the evolution of new enzymatic activities. Chem Biol 6:R91–R105.

Ober, D., Hartmann, T. (1999). Homospermidine synthase, the firstpathway-specific enzyme of pyrrolizidine alkaloid biosynthesis, evolved from deoxyhypusine synthase. Proc Natl Acad Sci USA 96:14777–14782.

Olsen, K.M. and Gross, B.L. (2008). Detecting multiple origins of domesticated crops. Proc. Natl. Acad. Sci. 105(37) 13701-13702.

Osbourn, A.E. (1996). Preformed antimicrobial compounds and plant defense against fungal attack. Plant Cell 8:1821–1831.

Paland, S. and Lynch, M. (2006). Transitions to asexuality result in excess amino acid substitutions. Science 311:990–992.

Papadopoulou, K., Melton, R.E., Leggett, M., Daniels, M.J. and Osbourn, A.E. (1999). Compromised disease resistance in saponin-deficient plants. Proc Natl Acad Sci USA 22: 12923-12928.

Pechan, T., Ye, L., Chang, Y., Mitra, A. and Lin, L. (2000). A unique 33-kD cysteine proteinase accumulates in response to larval feeding in maize genotype resistant to fall armyworm and other Lepidoptera. Plant Cell 12:1031–40.

Penninckx, I., Eggermont, K., Terras, F., Thomma, B., De Samblanx, G., Buchala, A., Métraux, J.P., Manners J. and Broekaert, W. (1996). Pathogen-induced systemic activation of a plant defensin gene in Arabidopsis follows a salicylic acid–independent pathway. Plant Cell 8: 2309-2323.

Raffa, K.F., Berryman, A.A., Simasko, J., Teal, W. and, Wong. B.L. (1985). Effects of grand fir monoterpenes on the fir engraver, Scolytus ventralis (Coleoptera: Scolytidae). Enviorn. Entomol. 14: 552-556.

Rani, A., Bhat, M.N. and Singh, B.P. (2006). Efficacy of neem formulations against late blight of potato in sub-tropical plains. Crop Research 31: 179-180.

Rathjen, J.P. and Moffet, P. (2003). Early Signal Transduction Events in Specific Plant Disease Resistance. Current Opinion in Plant Biology. 6: 300-306.

Rees, S.B. and Harborne, J.B. (1985). The role of sesquiterpene lactones and phenolics in the chemical defense of the chicory plant. Phytochemistry 24:2225–2231.

Rhoades, D.F. (1977), Integrated antiherbivore, antidesiccant and ultraviolet screening properties of creosotebush resin. Biochem. Syst. Ecol. 5: 281-290.

Rhoades, D.F. (1979). Evolution of plant chemical defense against herbivores. Herbivores: Their Interaction with Secondary Plant Metabolites (eds G.A. Rosenthal & D.H. Janzen), pp. 4–55. Academic Press, Orlando, FL.

Rohdich, F., Bacher, A. and Eisenreich, W. (2005). Isoprenoid biosynthetic pathways as anti-infective drug targets. Biochem Soc Trans 33:785-791.

Ross, A.F. (1961), Systemic acquired resistance induced by localized virus infections in plants. Virology 14: 340–358.

Rudall, P.J. and Caddick, L.R. (1994). Investigation of the presence of phenolic compounds in monocotyledonous cell walls, using UV fluorescence microscopy. Ann. Bot 74:483–491.

Ryals, J.A., Neuenschwander, U.H., Willits, M.G., Molina, A., Steiner, H.Y. and Hunt, M.D. (1996). Systemic acquired resistance. Plant Cell 8: 1809-1819.

Saby, J.K., Bhat, S.G. and Prasada, Rao. U.J.S. (2003). Biochemical characterization of sap (latex) of a few Indian mango varieties. Phytochemistry 62:13–19.

Scalbert, A. (1991). Antimicrobial properties of tannins. Phytochemistry, 30: 3875–3883.

Schenk, P., Kazan, K., Wilson, I., Anderson, J., Richmond, T., Somerville, S. and Manners, J. (2000) . Coordinated plant defense responses in Arabidopsis revealed by microarray analysis. Proc Natl Acad Sci USA 97: 11655–11660.

Schlösser, E. (1994) Preformed phenols as resistance factors. In Natural Phenols in Plant Resistance. Acta Horticulturae 381 (Geibel M, Treutter D and FeuchtW eds.), Freising, pp.615–630.

Schmeller, T., Latz-Bruning and Wink, M. (1997) Biochemical activities of berberine, palmatine and sanguinarine mediating chemical defense against microorganisms and herbivores. Phytochemistry 44:257–66.

Schmidt, S., Sunyaev, S., Bork, P. and Dandekar, T. (2003). Metabolites: A helping hand for pathway evolution? Trends Biochem Sci 28:336–341.

Scott, J.G. and Wen, Z. (2001). Cytochromes P450 of insects: The tip of the iceberg. Pest Manag Sci 57:958–967.

Sessa, R., Benett, M.H., Lewin, M.J., Mansfirls, J.W. and Beale, M.H. (2000). Metabolite profiling of sesquiterpene lactones from Lactuca species. J. Biol. Chem. 275: 26877–84.

Shonle, I. and Bergelson, J. (2000). Evolutionary ecology of the tropane alkaloids of Datura stramonium L. (Solanaceae). Evolution 54: 778–788.

Sirvent, T.M. and Gibson, D.M. (2002). Induction of hypericins and hyperforin in Hypericum perforatum L. in response to biotic and chemical elicitors. Physiological and Molecular Plant Pathology 60: 311-320.

Skadhauge, B., Thomsen, K. and vonWettstein, D. (1997). The role of barley testa layer and its flavonoid content in resistance to Fusarium infections. Hereditas 126: 147–160.

Slusarenko, A.J., Patel, A. and, Portz. D. (2008). Control of plant diseases by natural products: Allicin from garlic as a case study. Eur. J. of Plant Pathol. 121(3): 313-322.

Snook, M.E. (1994). Characterization and quantification of hexadecyl, octadecyl and eicosyl esters of p-coumaric acid in the vine and root latex of sweet potato [Ipomoea batatas (L.) Lam.]. J. Agric Food Chem. 42:2589–95.

Stahl, E.A., Dwyer, G., Mauricio, R., Kreitman, M. and Bergelson, J. (1999). Dynamics of disease resistance polymorphism at the Rpm1 locus of Arabidopsis. Nature 400: 667–671.

Staskawicz, B., Ausubel, F., Baker, B., Ellis, J. and Jones, J. (1995). Molecular genetics of plant disease resistance. Science 268: 661-667.

Staudinger, H. and Ruzicka, L (1924) "Über die wirksamen Bestandteile des dalmatinischen Insektenpulvers". Helv Chim Acta. 7: 177.

Stehle, F., Brandt W, Milkowski C and Strac D (2006) Structure determinants and substrate recognition of serine carboxypeptidase-like acyltransferases from plant secondary metabolism. FEBS Lett 580:6366–6374

Strauss, S.Y., Rudgers, J.A., Lau, J.A. and Irwin, R.E. (2002). Direct and ecological costs of resistance to herbivory. Trends in Ecol. and Evoln. 17: 278–285.

Tang, S. and Knapp, S.J. (2003). Microsatellites uncover extraordinary diversity in native American land races and wild populations of cultivated sunflower. Theor Appl Genet.; 106(6):990-1003.

Tattini, M., Galardi, C., Pinelli, P., Massai, R., Remorini, D. and Agati, G. (2004). Differential accumulation of flavonoids and hydroxycinnamates in leaves of Ligustrum vulgare under excess light and drought stress. New Phytologist 163: 547–561.

Tesar, M.B. and Marble, V.L. (1988). Alfalfa establishment. In Alfalfa and Alfalfa Improvement. Hanson AA, Barnes DK and Hill RR Jr., Am. Soc. Agron. Monogr. Madison, WI. 29: 303-322.

Thomma, B., Pennickx, I., Broekaert, W. and Cammue, B. (2001) The complexity of disease signalling in Arabidopsis. Curr. Opin. Immunol. 13: 63-68.

Trapp, S.C. and Croteau, R.B. (2001). Genomic Organization of Plant Terpene Synthases and Molecular Evolutionary Implications. Genetics, 158: 811- 832.

Treutter, D. (2006). Significance of flavonoids in plant resistance: a review. Environmental Chem. Letters. 4(3): 147-157.

Tripathi, P. and Dubey, N.K. (2004). Exploitation of natural products as an alternative strategy to control postharvest fungal rotting of fruit and vegetables. Postharvest Biology and Technology 32: 235–245.

Tyler, B.M. (2001). Genetics and genomics of the oomycete–host interface. Trends in Genetics 17(11): 611-614.

van Wees, S., de Swart, E., van Pelt, J., van Loon, L. and Pieterse, C. (2000). Enhancement of induced disease resistance by simultaneous activation of salicylate- and jasmonate-dependent defense pathways in Arabidopsis thaliana. Proc Natl Acad Sci USA 97: 8711-8716.

Vetter, J. (2000). Plant cyanogenic glycosides Toxicon. 38(1):11-36.

Wagner, A. (1998). The fate of duplicated genes: Loss or new function? Bioessays 20: 785–788.

Wallis, C., Eyles, A., Chorbadjian, R., McSpadden, Gardener, B., Hansen, R., Cipollini, D., Herms, D.A. and Bonello, P. (2008). Systemic induction of phloem secondary metabolism and its relationship to resistance to a canker pathogen in Austrian pine. New Phytol. 177: 767–778.

Walton, J.D. (1994). Deconstructing the Cell Wall. Plant Physiol. 104: 11 13-1 118

Walz, C., Giavalisco, P., Schad, M., Juenger, M., Klose, J. and Kehr, J. (2004). Proteomics of cucurbit phloem exudates reveals a network of defense proteins. Phytochemistry 65:1795–804.

Ward, E.R., Uknes, S.J., Williams, S.C., Dincher, S.S., Wiederhold, D.L., Alexander, D.C., Ahl-Goy, P., Metraux, J.P. and Ryals, J.A. (1991). Coordinate gene activity in response to agents that induce systemic acquired resistance. Plant Cell 3: 1085-1094.

Wasternack, C. and Parthier, B. (1997). Jasmonate-signaled plant gene expression. Trends Plant Sci. 2: 302-307.

Wink, M. (2004), Phytochemical diversity of secondary metabolites. Encyclopedia of Plant and Crop Science; Taylor and Francis. 915-919.

Woodhead, S. and Bernays, E.A. (1978). The chemical basis of resistance of Sorghum bicolor to attack by Locusta migratoria.Entomologia Experimentalis et Applicata. 24(2):123-144.

Yun, B.Y. and Loake, G.J. (2002). Plant Defence Responses: Current Status and Future Exploitation. J. of Plant Biotechnology 4: 1-6.

Zong, N. and Wang, C. (2006). Larval feeding induced defensive responses in tobacco: comparison of two sibling species of Helicoverpa with different diet breadths. Planta 226: 215-224.

Zulak, K.G. and Bohlmann, J. (2010). Terpenoid Biosynthesis and Specialized Vascular Cells of Conifer Defense. Journal of Integrative Plant Biology 52 (1): 86–97.

Zwenger, S. and Basu, C. (2008). Plant terpenoids: Applications and Potentials. Biotechnology and Molecular Biology Reviews 3: 1-7.

❑❑❑

CHAPTER – 7

Metagenomics Progress, Opportunities and Challenges

Sarvjeet Kaur
National Research Centre on Plant Biotechology, IARI Campus, New Delhi 110012.
Email : dr_sarvjeetkaur@yahoo.com

Summary

Metagenomics is a rapidly developing field of research that enables the study of microbes that have not been cultured or cannot be cultured easily. It has greatly advanced the understanding of ecology and diversity of natural microbial communities and has led to the discovery of many new genes and geneproducts. The various aspects of metagenomics – history, techniques for identification, different approaches of data analysis, comparative and functional genomics, applications in the areas of medicine, agriculture, biotechnology, ecology, bioprospecting, bioremediation, future advances and challenges and issues related to intellectual property, have been discussed.

1. Introduction

The microbes constitute an enormous genetic pool, the potential of much of which is still unexplored. The obstacle with majority of the microbes is the difficulty of cultivation under laboratory conditions. Such microbes are called *unculturables.* This has severely limited the knowledge about physiology and ecology of microbial communities (Schloss and Handelsman 2005). Only a small percentage of the microbes in nature can be cultured. Bacteria may be not amenable to cultivation because of lack of symbionts, nutrients, presence of inhibitory compounds, inappropriate conditions of temperature, pressure, accumulation of toxic waste products from their own metabolism and intrinsic slow growth rate. Furthermore, a clonal culture may not give the correct representation with respect to microbe interaction, functions and population diversity in the environment. It is now possible to detect and exploit the unculturable microbes through modern molecular and computational methods.

Metagenomics is a rapidly developing field of research that enables the study of microbes and microbial communities that have not been cultured or that cannot be cultured easily. The microbes that cannot be grown in the laboratory, are now accessible employing the tools of metagenomics, as it bypasses the requirement for obtaining pure culture. Thus, the development of metagenomics in the recent years has greatly facilitated the genomic investigation of procaryotes in environmental samples, irrespective of the culturability of these organisms (Cowan *et al*. 2005; Commitee on Metagenomics 2007; Foerstner *et al.* 2006; Wooley *et al.* 2010). Metagenomics incorporates the tools of genomics, bioinformatics, and systems biology for the genomic analysis of entire communities of microbes, circumventing the isolation and culture of individual community members.

Metagenomics has been defined as the application of shotgun sequencing to DNA obtained directly from an environmental sample or series of related samples, producing at least 50 Mbp of randomly sampled sequence data (Kunin *et al.* 2008). It employs high-throughput sequencing technology and new algorithms for sequence assembly to facilitate simultaneous genome analysis of many microorganisms and provide information on interaction among members of a microbial community. It is accomplished by direct isolation of DNA from the environmental sample without culture of individual microbial species, followed by cloning into a suitable vector, transformation into a host and screening of the transformants for phylogenetic markers or expression of specific traits (Fig. 1).

1.1 Development of Metagenomics

The earlier investigations on microbes have concentrated on single species grown in pure laboratory cultures, irrespective of their ecological context. Bergey's Manual stated that no organism could be classified without being cultured (Society of Microbiologists, 1923). Therefore, the understanding of microbial communities has lagged behind the understanding of individual members of these communities. Non-traditional methods have since been developed to study the microbes in their complex communities. The true extent of microbial diversity in different environments is now being revealed with the help of such culture-independent molecular methods.

The study of microbial genomes started in the late 1970s, with the sequencing of the genomes of bacteriophages, MS2 (Fiers *et al.* 1976) and -X174 (Sanger *et al.* 1978). The first bacterial genome sequenced was of *Haemophilus influenza (*Fleischmann *et al.* 1995). Subsequently, genomes of 916 bacterial, 1,987 viral, and 67 archaeal species have been sequenced and have opened up the fields of comparative genomics and systems biology (Wooley *et al.* 2010). The ribosomal RNA (rRNA) genes have frequently been used as phylogenetic anchors for assessment of microbial diversity, due to very high sequence conservation among these genes. Direct analysis of 5S and 16S rRNA gene sequences in an environmental sample without culturing revealed the presence of highly diverse unculturable microbes (Pace *et al* 1986). The first large-scale metagenomic studies were on ocean surface water (Venter *et al.* 2004) and an acid mine drainage biofilm (Tyson *et al.* 2004). Since then, metagenomic analyses of diverse environments such as hot springs, deep sea vents, soil, animal rumen and gut, human oral activity and intestine has been carried out to understand the phylogenetic diversity and function of microbial communities in these environments, a task which would have taken much longer and yielded much less information with the traditional methods of analysis (Handelsman 2004; Warnecke and Hugenholtz 2007).

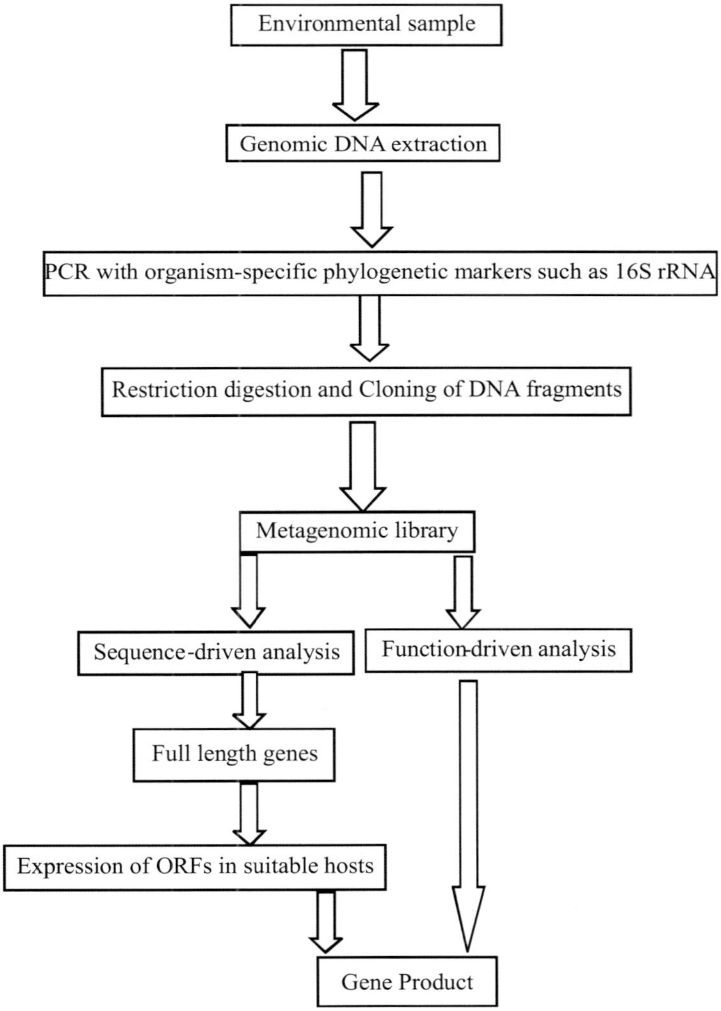

Fig. 1. Schematic representation of the steps involved in construction and screening of a metagenomic library.

1.2 Distinction Between Genomics and Metagenomics

The term *genome* describes the total hereditary material contained in an organism. More specifically, *genome* is used to describe the DNA present in a haploid set of chromosomes in an organism. Genomics has provided genome sequences for fungi, nematodes, fruit flies, *Arabidopsis,* rice, dog, cow, chimpanzee, and several other eukaryotic and prokaryotic organisms. Genome analysis studies on a large and phylogenetically diverse number of microbial species have revealed extremes in genome sizes (<500 kb to almost 10 Mbp) with no clear boundaries between bacterial, archeal, fungal, protest and viral genomes (Fraser-Ligget, 2005).

Metagenomics takes into consideration the entire genetic complement of a microbial community and generates collective genomic information on a habitat. The progress in metagenomics has been enabled by the technological advances and bioinformatics tools for the development of computational methods for the analysis of massive data generated in such studies. For genomics studies, the DNA is extracted from pure cultures of organisms grown in the laboratory. In metagenomics studies, the starting material is a mixture of DNA from a community of diverse types of microorganisms , many of which may be unculturable.. The emphasis is on understanding of community composition and function rather than mere generation of the complete genomes data. Methods based on phylogenetic anchors such as 16S rRNA sequence are employed to reveal information on microbial diversity and abundance and comparative analyses of communities. Cultivation-independent methods such as random shotgun sequencing of DNA extracted directly from environmental samples are used to generate information about the gene content in the community. Strategies such as BAC libraries are used to sub-clone the DNA for sequencing. The function-driven metagenomic analysis involves expression in a host, sequencing and phylogenetic analysis. Thus, a particular community can be mined for target genes of interest by assessing their biochemical functions in suitable hosts (Rondon *et al.* 2000).

The sequence assembly and annotation is tractable in genomic data as it comes from an identifiable clone of a cultured microbe. In contrast, the metagenomic data is partial and noisy as it is derived from heterogeneous microbial communites with thousands of species (Wooley *et al.* 2010). Thus, the difference between genomics and metagenomics is that in metagenomics, with the exception of dominant populations, there is no fixed end point of having produced finished genomes and metagenomes may not progress beyond draft assembly (Kunin *et al.* 2008). Therefore, sequence analysis of metagenomic data requires greater care and expertise.

Many additional methodologies such as metatranscriptomics, metaproteomics and metabolomics, which use high-throughput systems, have been developed to aid in information retrieval (Table 1). With new cost-effective sequencing technologies, metagenomics is increasingly being used by researchers the world over (Commitee on Metagenomics, 2007).

2. Approaches to Metagenomic Analysis

Metagenomics involves direct DNA extraction of DNA from an environmental sample and construction of a DNA library containing microbe genome sequences and its screening by sequence analysis or determination of function of the cloned genes. Microbial communities comprise combinations of bacteria, archaea, microbial eukaryotes and viruses (Kunin *et al.* 2008). To estimate the fraction of species sequenced, rarefaction curves are typically used. A rarefaction curve plots the number of species as a function of the number of individuals sampled. The curve usually begins with a steep slope, which at some point begins to flatten as fewer species are being discovered per sample: the gentler the slope, the less contribution of the sampling to the total number of operational taxonomic units (OTUs). For microbial samples, different OTUs are typically characterized by 16S (prokaryotic) or 18S (eukaryotic) rDNA, and are also referred to as ribotypes (Wooley *et al.*2010). The 16S rRNA sequence can reveal significant information about microbial diversity and abundance.

Table 1 : New methodologies for identification of target genes and their functions

Nomenclature	**Field of study**	**Application**
Proteomics	The total set of proteins in an organism, tissue, or cell type	Useful for identification of novel proteins
Transcriptomics	The total set of RNAs found in an organism, tissue, or cell type	Useful for elucidation of spatially and temporally expressed genes
Metabolomics	The entire complement of metabolites that are generated in an organism	Useful for identification of novel metabolic enzymes and products
Interactomics	The entire set of molecular interactions in an organism	Useful for elucidation of metabolic interactions in cells

2.1 Sample Collection

Selection of appropriate environment and use of correct procedure of sampling is important for obtaining useful and accurate data from a metagenomic study. The composition of a microbial community is influenced by its habitat. Information about the physical, chemical, and ecological aspects of the habitat is required for data interpretation. The sample should be representative of the diverse microbial communities present in a particular environment. The environment selected either should have natural enrichment for target genes, or should be highly genomically diverse or an extreme environment, depending on the objectives of the metagenomic study.

The variability of the habitat and the host over space and time may present constraints in sampling. Some habitats may show seasonal variation such that results from resampling experiments may show differences. Habitats may also differ in the complexity of microbial species. A few species may dominate in some environments while in others no single species may be abundant. The diversity of species is much higher in soil as compared to ocean water. For complex microbial communities such as those from soil samples, full representation of all the community members needs to ensured by replicate sampling.

Differences in experimental protocols can also lead to sampling bias. For example, the use of size filters to select against small viruses as well as large eucaryotic cells in the Sargasso sea metagenomics project, simplified the prokaryotic diversity, but has to be taken into account for comparison of data to other samples (Foertsner *et al.* 2006). Thus, the elucidation of phylogenetic composition of communities is a challenge as some species may be overrepresented while others may be underrepresented in an environmental sample. The protocols for read quality filtering, assembly, and prediction may also vary from one laboratory to another rendering the comparison among data from different laboratories difficult.

2.1.1 Record of metadata

Metadata are descriptions of geographic location, general and specific features of the sample site, such as pH and salinity, temperature, time when the sample was taken, and host condition, method of sampling, size of sample, and sample preparation. Metadata are useful for resampling if required and help provide appropriate context for metagenomic data interpretation. An ocean sample metadata, for example, includes information about sampling date and time, depth, salinity, light intensity, geographical coordinates, pH and soluble gases in the ocean water sample. Correct recording of sample 'metadata' including the biochemical data such as pH, temperature, salinity, geographical data such as height, depth and global positioning coordinates, sample collection data and method of DNA extraction and library construction is important (Kunin *et al.* 2008).

2.2 Nucleic Acids Extraction methods

Extraction and purification of sufficient quantity of high quality DNA from environmental samples presents a challenge in metagenomic studies. Several methods have been developed for efficient extraction of nucleic acids from environmental samples (Krsek and Wellington 1999). The protocols take into consideration the requirement of vigorous extraction with minimal shearing of DNA. Although mechanical bead beating improves recovery of DNA from diverse species, the chemical lysis is considered a better method for the recovery of high molecular weight DNA (Cowan *et al.* 2005).

The constraints in extraction of DNA include too little yield of DNA, presence of compounds that may interfere with DNA extraction procedures, and contaminating DNA from other species which are not intended to be the part of a particular metagenomic study. Furthermore, the presence of contaminants affects the effective concentration of bacterial DNA and the probability of recovering less abundant members of the microbial community. The nucleic acid extraction protocols need to be optimized particularly for samples that have community members that are more difficult to lyse, or are in different physiological or physical states, or that are rare. This is necessary for ensuring that observed abundance of a particular microbial species is not because of the sampling bias.

Therefore, sample enrichment protocols to improve the yield of DNA by facilitating enrichment of bacterial DNA *vis a vis* eukaryotic DNA, are often adopted.

2.2.1 Sample enrichment techniques

The discovery of target genes, which may represent a small fraction of the total nucleic acid content in a metagenomic screening study, can be improved by selective enrichment of target genomes. The uneven abundance of microbial communities further compounds the problem of obtaining entire genome data of community members, due to sampling of the most abundant organisms again and again while missing the rare members. Enrichment methods to remove the redundant DNA facilitate detection of the rare members in a community.

The sample enrichment techniques include filtration, centrifugation, or chromatography for separation of target cells from other contaminants and to increase the proportion of active clones in the metagenomic libraries. These methods often make use of the physical properties for example, the size, of the target cells. In the Sargasso Sea genome sequencing project, size-selective filtration was used to select against small viruses and eukaryotic cells (Venter *et al.* 2004). Phylogenetically novel populations of ultra-small archaea were enriched by using microfilters (Baker *et al.* 2006). Differential centrifugation was used for enrichment of cells of *Buchnera aphidicola*, an obligate symbiont of aphids, from which the first complete genome sequence for an uncultured bacterium was obtained and *Cenarchaeum symbiosum* a archaeal symbiont of a marine sponge, by removing them from their hosts (Schloss and Handelsman 2003). Floorescence-activated cell sorting (FACS) can also be used to sort specific cells on the basis of cell size, DNA content, phytosynthetic pigments or fluorescently labelled probes targeting the cells (Warnecke and Hugenholtz 2007). Physical fractionation of a diverse microbial community by splitting into simpler subsets helps in reduction of complexity within the resulting fractions, thereby aiding in the assembly of individual sequence reads (singletons) into longer stretches (contigs) (Bergeron *et al.* 2007). Apart from facilitating contig identification, it also ensure greater genomic coverage of microbial populations (Warnecke and Hugenholtz 2007).

Organisms and phyla differ in nucleotide compositional aspects such as GC content, dinucleotide and hexanucleotide frequencies, and codon usage (Foerstner *et al.* 2006). The representation of GC rich genomes of some organisms such as Actinomycetes and Acidobacteria in the library can be enhanced by ultracentrifugation to enrich for high GC content DNA (Schloss and Handelsman 2003). The enrichment of metabolically active organisms can be affected by feeding labelled Bromodeoxyuridine to bacteria in soil sample and recovery of labelled DNA by immunocapture. Laboratory enrichment cultures of clones from environmental samples carrying the target trait such as biotin synthesis, or enzyme activity can be obtained by feeding the appropriate substrate to the samples and incubating them prior to DNA isolation (Entcheva *et al.* 2001, Steele *et al.* 2009).

Treatment with nucleases derived from the host (an insect gut) for digestion of host DNA before lysis of bacteria was found to be effective (Guan *et al.* 2006). Physical separation of bacteria from the rest of the soil matrix before lysis has been found to be useful for minimizing contamination with numerous inhibitors of the enzymes used for cloning found in soil (Berry *et al.* 2003). Subtractive hybridization of microbial DNA to immobilized or labeled copies of eukaryotic DNA, and subsequent separation of unbound microbial DNA has also been used for sample enrichment (Holben and Harris 1995).

The sequencing of communities containing eucaryotes, particularly the protists, is expensive because of their enormous genome sizes and low gene coding densities (Kunin *et al.* 2008). In case of eucaryotes, obtaining the molecular data at RNA (metatranscriptomics) or protein level (metaproteomics) bypasses the problem of large amounts of noncoding sequence data. The technologies for recovering RNA from environmental samples are largely similar to those used for DNA isolation, and are modified to optimise the yield of intact RNA by minimising degradation (Cowan *et al.* 2005).

2.3 Metagenomic Data Generation and Analyses

Metagenomic DNA library construction and analysis involves generation of suitably sized DNA fragments, their cloning into an appropriate vector, and screening for target genes of interest. Shotgun clone libraries for genome sequencing are typically prepared using different average sizes of cloned DNA ranging from 3 to 40 kbp (Kunin *et al.* 2008). Cosmid, fosmid and bacterial artificial chromosome (BAC) vectors which allow cloning of larger segments of genomic DNA, have been extensively used for the construction of metagenomic libraries (Entcheva *et al.* 2001; Rondon *et al.* 2000; Cowan *et al.* 2005; Kim *et al.* 2006).

2.3.1 Sequence-based analysis of metagenomic data

The sequence-based analysis makes use of conserved DNA sequences to design hybrization probes or PCR primers for screening of metagenomic libraries for target genes of interest. Sequence-based analysis involves complete sequencing of clones employing phylogenetic anchors which may indicate the probable taxonomic group of source of DNA. Random sequencing of clones and flanking DNA can also provide a phylogenetic linkage with the functional gene (Handelsman 2004). The sequence-based metagenomics approach has been applied to samples from different environments, such as the ocean, coral reefs, thermal vents, hot springs and various types of soils. The reconstruction of the genomes of uncultured organisms in a community in acid mine drainage (Tyson *et al.* 2004) and the Sargasso sea (Venter *et al.* 2004) illustrated the power of large-scale sequencing efforts to enrich the understanding of uncultured communities. Some of the sequence-based data analyses approaches are-

2.3.1.1. 16S rRNA-based surveys

The PCR amplification of 16S rRNA genes using primers corresponding to highly conserved regions of rRNA genes of different microbes serves as a rapid and cost-effective method for phylogenetic typing for the assessment of microbial composition, diversity and abundance in a sample. It is an efficient technique for community sampling and has led to discovery of previously unknown microbial species (Handelsman 2004). Sequence analysis of cloned 16S rRNA PCR products provides a semiquantitative phylogenetic profile of a microbial community. Pyrosequencing of 16S rRNA genes by providing 100 to 200bp 'pyrotags' provides high resolution (Kunin *et al.* 2008).

There are some limitations of PCR screening of microbial communities for the presence of 16S rRNA genes. The sample may have uneven abundance of microbial species. Some rRNA genes may not amplify well with the set of primers used, leading to mismatch between estimates of diversity based on the PCR-based 16S rRNA gene analysis and those based on shotgun sequencing. In some studies though, the estimates were found to be similar (Liles *et al*. 2003; Tyson *et al*. 2004). The multiple and non-identical copies of 16S rRNA genes present in many procaryotes may also pose a challenge in correct estimation of the relative abundance of a species.

Nevertheless, the phylogenetic markers are useful for initial identification and taxonomic affiliation. Additional single-copy phylogenetic markers have been developed to overcome the limitations with rRNA genes (Handelsman 2004). FISH (fluorescence in situ hybridization) oligonucleotide probes and flow cytometry have also been used for analysis of microbial community composition (Wagner *et al.* 2003; Sekar *et al.* 2004).

2.3.1.2 DNA Microarrays

Microarrays are used for assessment of diversity of microbial communities in a large number of samples and for high-throughput robotic screening of target genes. Microarrays based on 16S rRNA and other phylogenetic marker genes have been developed for analysis of microbial communities (Wu *et al.* 2006a). The phylogenetic arrays carry thousands of spots bearing synthetic oligonuceotides which match with the rRNA gene sequences found in databases or are present in samples. DNA microarrays are also useful for assessment of variations in population structure and community function over time and space.

The microarray chips carry information about thousands of unique database sequences. For example, the Affymetrix-based PhyloChip has been used to characterize complex environments such as soil, aquifers, and urban air (Brodie *et al.* 2007; Desantis *et al.* 2007) Low cost microarrays with increased sensitivity and specificity are being developed. However, although microarrays provide useful information with regard to known species, these can not be used to identify the yet-to-be-discovered species.

2.3.1.3 Whole genome shotgun sequencing

In metagenomics, shotgun sequencing is done in the same manner as in genomics. It is sometimes termed as environmental shotgun sequencing (ESS), since the raw genomic material is derived from a community of microbes. Sanger shotgun sequencing has been used for sequencing of prokaryotic genomes (Sanger *et al.* 1977). The DNA of a genomic clone is sheared into random fragments (hence the name "shotgun.") and the library is constructed by cloning the fragments into plasmid vectors and the clones are then sequenced by the dye-termination method. The sequences are then assembled into the whole genome by using computational software. Theoretically, any genome shorter than 5 Mbp can be assembled this way. Regions with large repeats are often not incorporated into the whole genomic picture, thus leaving some gaps (Wooley *et al.* 2010).

The shotgun-sequencing is a good strategy for obtaining information on diversity and function of microbial communities such as bacteriophage and other viruses, eukaryotic species, bacteria and archaea. However, since a large number of genes do not have homologs of known function in the databases, the sequence data need to be supplemented coupled with other analyses for detailed information about the structure and function of genes in microbial communities.

Determination of the complete genome sequence of community members requires adequate depth of sequence coverage for each organism. This may vary depending on level of complexity, abundance and size of genomes of different members of the community. Sequence assembly is easier for simple community as there would be sufficient overlapping fragments of the dominant members as compared to a complex community where the sequence fragments of the more prevalent members will be diluted and the sequences of less prevalent or rare members would only be a very small fraction of the total sequence data. In such cases, the use of sequence data of reference genomes, if available, can help in data inference. Second generation sequencing methods such as pyrosequencing wherein sequencing is performed by polymerase extension of a primed template have been developed (McPherson 2009). Third generation sequencing methods wherein long sequences can be sequenced without amplification are in advanced stage of development (Branton *et al.* 2008).

2.3.1.4 Gene-centric analysis of complex data

Sequencing data from highly complex microbial communities is more difficult to assemble as compared with the data from communities which have dominant populations. However, due to high coding density and the average gene size in most bacterial and archeal genomes, most reads are still able to capture a coding sequence (Kunin *et al.* 2008). Thus, the data analysis in such cases is gene-centric, treating the community as a whole rather than concentrating on contributions from individual species. The gene sequences are mapped to different gene families to get an estimation of their relative abundance and functional differences in metagenomic datasets. Such unassembled reads, which are short DNA sequences containing fragments of functional genes, corresponding to different members in a community are termed Environmental gene tags (EGTs). The (EGTs) provide information about the genes which are essential and represented more frequently in multiple environments. The relative abundance of EGTs provides clues about the metabolic capabilities of members of a particular community. EGTs obtained from Sargasso Sea metagenomic analysis revealed the prevalence of genes that encode rhodopsin-like proteins as compared with ocean environments that receive less sunlight (Venter *et al.* 2004). However, it may not always be the case that high relative abundance signifies metabolic activity of a particular gene family. Gene family of *pufB,* encoding a subunit of a light harvesting protein, were found to be highly expressed but poorly represented in the metagenomic data from ocean surface water (Frias-Lopez *et al.* 2008).

2.3.2 Functional metagenomics

The data of genome content of a microbial community does not provide information on the expression, dynamics and time scale of expression of genes in that community. Statistically significant sequence similarity between genes discovered by metagenomics and the known genes in the databases may suggest functionality of the new sequences. Detailed databases of sequences of target genes and their functions are therefore needed.

However, the sequence comparisons may suggest incorrect inference about function in case of gene sequences having less homology to a known gene. Furthermore, many genes have unknown functions. Discovery of proteins with a defined function may be accomplished more efficiently by functional metagenomics. Functional metagenomics involves expression and sequence and biochemical analysis in a host to identify the clones that express a desired trait. Functional metagenomics has been used to identify novel antibiotics as well as proteins involved in antibiotic biosynthesis, antibiotic resistance, biodegradation of environmental contaminants, vitamin production and signal-transduction pathways (Handelsman 2004). Complementation of an *E. coli* mutant deficient in biotin production led to the isolation of seven new operons for biotin synthesis from enrichment cultures derived from samples of soil or horse excrement (Entcheza *et al.* 2001). Gene expression patterns can provide significant insight into microbial processes active in the environment. For example, the genes involved in microbial phototrophy were among the most highly expressed genes in the microbial community from ocean surface water (Frias-Lopez *et al.* 2008). The success of the functional metagenomics is dependent on the faithful expression of target gene of interest, or the cluster of all the genes needed for that function in the host cell and availability of an efficient assay to detect the function. Several approaches have been developed to tackle these requirements.

2.2.2.1 Heterologous gene expression systems

Functional metagenomics involves expression of genes from unknown microbes in a substitute host, that is, heterologous gene expression. Heterologous gene expression systems, which make use of shuttle vectors that facilitate screening of target genes of interest in different host species, have been developed (Schloss and Handelsmann 2003). The shuttle vectors can be maintained in a number of host species and are useful for expansion of range of gene expression and cloning.

Most genes may not be expressed in the chosen host due to lack of transcription and translation factors required for expression of metagenomic gene sequences in the host. Genes from many microbes may not be expressed in *E. coli*. A survey of 32 procaryotic genomes revealed that only 40 per cent of the genes could be expressed in *E. coli* (Gabor *et al.* 2004). In addition to *E. coli*, several other bacteria such as the actinomycetes, have been exploited as hosts. The *Streptomyces, belonging to actinomycetes,* has GC-rich genome and is useful as host for expression of genes from genes from other high GC content genomes (Wang *et al.* 2000). *Bacillus subtilis* has also been used for heterologous gene expression, particularly for the expression of extracellular proteins (Li *et al.* 2004).Different positive clones were detected in three different hosts- *E. coli*, *Pseudomonas putida* and *S. lividans,* used to screen a metagenomic library, thus indicating the usefulness of employing a wide range of hosts in screening (Martinez *et al.* 2004). The weak ability of the host organism in which the metagenomic library is being maintained to express genes could be a limitation. It is essential to develop techniques for expression of a greater array of genes in *E. coli* as well as in other bacteria. Wide host range plasmid, cosmid and BAC cloning vectors have been developed to screen metagenomic libraries in a range of bacterial hosts (Wexler and Johnston 2010).

2.3.2.2 Functional screens

The genes encoding for the enzymes involved in the biosynthetic pathway of a secondary metabolite are generally clustered contiguously. Yet, to obtain a DNA fragment large enough that contains the entire gene complement required for the synthesis of a particular antibiotic can be challenging due to contamination of DNA particularly from soil samples. Since the frequency of clones that express any given activity is likely to be low, efficient screening methods are required for discovery of new gene products.

Detection of rare, active clones in large libraries has been facilitated by functional screens that are able to detect low levels of activity, for identification of novel antibiotics, antibiotic resistance gene, and many hydrolytic, metabolic and biosynthetic enzymes (Schloss and Handelsmann 2003). The functional screening methods generally involve plating metagenomic clones on indicator agar and detection of activity based on change in colour or appearance of a halo around the colonies and may differ in their sensitivity. Functional gene arrays have also been designed with probes corresponding to target genes (Wu *et al.* 2006a). Further improvements in sensitivity, specificity and speed of these arrays are required for metagenomics.

The phylogenetic origin of the clones may be ascertained by sequencing of the flanking DNA of clones provided the flanking sequences contain a highly conserved gene such as 16S rRNA, RecA or DNA polymerase. Such phylogenetic anchors are useful for drawing inferences on the relationship between phylogenetic affiliation and function.

The methods for accurate representation of community mRNA need to be improved. Metatranscriptomics and proteomics are particularly significant approaches for full utilization of information for discovery of new gene products in a metagenomic dataset. The separation of mRNA from dominanat rRNAs is a technical problem though, with transcriptomics of bacteria and archaea, as the polyA tail which is present in eucarytic mRNAs, is not present in bacterial and archaeal mRNAs (Warnecke and Hugenholtz 2007). Metatranscriptomic analyses is useful for association of expression dynamics with the phylogenetic origin of the expressed patterns. Transcriptomic analyses of microbial communities in ocean surface water revealed transcripts highly similar to previously known genes as well as unique transcripts indicating the presence of new genes (Frias-Lopez *et al.*2008)._It is also possible to develop community-interaction maps for communities with known genomes.

Strain-specific proteomic analysis of expressed proteins in acidophilic bacteria from mine to identify protein variants which differ in only a single acid, was facilitated by mass spectrometry (Lo *et al.* 2007).

Several "hypothetical proteins" have been discovered by functional screening of metagenomic libraries. Development of new heterologous gene expression systems will expand the repertoire of genes that can be expressed and help in gene discovery.

3. Metagenomic Data Analysis

The nature, size and complexitiy of metagenomic datasets make the interpretation of information a formidable challenge. and require state-of-art computational tools for analysis. Appropriate methods of data analysis are necessitated for full exploitation of metagenomics due to the sheer volume and complexity of data generated. Metagenomics requires high-throughput, low-cost sequencing technologies and new sophisticated computational tools for identification of genes and proteins and prediction of their secondary structure from diverse members with different levels of abundance in microbial communities. Computer algorithms and data models which have been developed for analysis of metagenomic sequences are-

3.1 Clustering

Clustering is a primary annotation method in which, a large dataset is divided on the basis of the nucleotide or predicted protein sequence similarity into subsets to identify groups that share a phylogenetic origin. Clustering does not perform well with sequences which are present in low abundance.

3.2 Binning

Th process of association of a sequence with its taxonomic origin is called Binning. It is a clustering method that uses sequence composition features such as oligonucleotide frequencies, GC content and codon usage to divide the sequences into groups that belong to specific genomes or groups of genomes. Binning is used as a preliminary step in metagenomic projects in which the goal is to know not only who populates the sample, but also what the different OTUs are doing. There are different binning strategies such as composition-based binning and phylogenetic binning (McHardy and Rigoutos 2007, Wooley *et al.* 2010).

3.3 Gene Annotation

Genome annotation is a form of clustering, where the predicted genes are assigned to known gene families. The sequence reads are assembled into longer contiguous sequences or contigs based on sequence similarity between reads, for assembly of the whole genome. The fragmentary nature of metagenomic sequences presents challenges for gene annotation. Longer sequence reads are more informative and help in robust annotation. Assembly of whole genomes with shotgun sequencing is generally incomplete such that many sequence reads remain as singletons rather than being joined in contigs (Wooley *et al.* 2010). In addition to annotation by homology, methods for context annotation such as genomic neighbourhood, gene fusion, phylogenetic profiles and coexpression are also used to infer specific functions of metagenomic datasets (Harrington *et al.* 2007). However, since in metagenomics a considerable number of sequences may not be classified or assembled into longer reads, the annotations may only be at the preliminary stage.

3.4 Gene Prediction

Gene prediction is the analysis of genomic DNA sequences to predict their biological functions. Gene sequence databases are very useful for gene annotation and prediction metagenomic datasets. The information is rapidly expanding as the new methods of sequencing and bioinformatics tools are being developed (Kunin *et al.* 2008). Gene prediction involves identification of DNA and protein sequences among the assembled or partially assembled reads by either homology search which identifies similar sequences or by *ab initio* gene prediction approach which examines intrinsic features of sequences to distinguish between coding and noncoding sequences (Kunin *et al.* 2008). For genes which have known homologs, using BLAST (Basic Local Alignment Search Tool) against known databases yields information about the existence of gene family in the metagenomic data. The nucleic acid sequence databases: GenBank, funded by the National Institutes of Health (NIH); EMBL-Bank, funded by the European Molecular Biology Laboratory; and the DNA Databank of Japan and the protein sequence databases: The Protein Information Resource and Swiss-Prot have been extensively used by researchers for both genomic and metagenomic data analysis. Gene prediction on assembled sequences can be as accurate as 85% of the originally predicted genes in the constituting genomes from a high complexity metagenomic dataset and up to 90% for a low complexity dataset (Mavromatis *et al.* 2007).

4. Applications

The metagenomics databases enable gene prediction from complex communities and provide an understanding of microbial populations and ecological processes in natural habitats. **A detailed understanding of microbial communities is useful for sustainable environment.** With the advent of metagenomics, the vast genomic potential of microbial communities has become accessible. Metagenomics has accelerated the discovery of novel genes and their products from the enormous biodiversity available in microbes in their natural habitats. It has provided information about stability and activity of proteins in a wide range of environmental conditions. There is enormous potential for discovery of new genes and their application in biotechnology. It is estimated that there are about 10 trillion genes in 1 g of soil and at least 1 million *different* genes in 1 g of soil (Schloss and Handelsman 2006).

Most of the microbial gene products in use have been obtained from bacteria grown in pure cultures. Access to natural microbial populations for exploration of useful products has been made possible by the culture-independent methods of metagenomics. Metagenomics can aid in developing appropriate culture protocols for the microbes detected through PCR-based phylogenetic analysis by using rRNA sequences as probes. There are several applications of metagenomics in biomedicine, agriculture, and environment. Following are some of the important potential contributions emanating from metagenomic studies-

4.1 Novel Enzymes and Biocatalysts

A wide range of biocatalysts and enzymes such as amylases, cellulases, chitinases, lipases, having novel substrate or product ranges and high stability under extreme conditions and thus having potential applications in food and chemical industry have been discovered in metagenomic studies (Steele *et al.* 2009).

4.2 Biosensors and Diagnostics

Metagenomics is useful for development of novel diagnostics of disease-causing pathogens. Novel sequence-based, culture-independent diagnostic tools that incorporate microbial community profiling, metagenomics and single-cell genomics are envisaged (Pallen *et al.* 2010). Microbiome-based diagnosis will be help in treatment, preventive medicine and development of probiotic therapy based on understanding of the human microbiome, a complex ecosystem composed mainly of uncultured bacteria (Tesse *et al.* 2010). Metagenomics can help discover novel biosensors for development of vaccines and therapeutics against potential pathogens. It can help in identification of microbes that can be used as potential biological weapons and development of biosensors for monitoring and control of such pathogens.

4.3 Novel Pharmaceuticals and Antibiotics

Detection and characterization of many pharmaceuticals and antibiotics from naturally occurring microbes which are otherwise recalcitrant to cultivation in laboratory, has been made possible through metagenomics (Steele *et al.* 2009). Novel genes encoding the enzymes for biosynthesis of biotin were identified in a metagenomic library from forest soil (Entcheva *et al.* 2001). Broad spectrum antibiotics active against both Gram-positive and gram-negative bacteria were identified in a metagenomic study with soil microbial DNA (Gillespie *et al.* 2002). Insight about the diverse variety of natural antibiotic resistance mechanisms present in microbes has also been gained through metagenomics (Riesenfield *et al.* 2004). Many novel antibiotics and their resistance mechanisms, cation transporters and degradative enzymes have been discovered using the function-based metagenomics (Handelsman 2004; Steele *et al.* 2009). Bioactive small molecules, antibacterially active enzymes and small molecule biosynthetic gene clusters have also been discovered using metagenomic approaches (Banik and Brady 2010).

4.4 Bioenergy and Bioremediation

Metagenomics has potential in development of microbial bioenergy resources for sustainable environment by discovery of new cellulases for generation of bioethanol from agricultural wastes such as corn fiber, corn stalks and wheat straw. Habitat-related halotolerant characteristics were observed in cellulases detected in metagenomic screening of samles from extreme environment of a soda lake in Africa (Grant *et al.* 2004). However, apart from extreme environments, the highly genetically diverse environments also contain a range of cellulases. A cellulose enzyme with high haloteolerance and stablility over a wide pH range was characterized by functional screening of a metagenomic library from a soil sample from a non-extreme environment (Voget *et al.* 2006).

Metagenomics can be applied to devise new technologies for environmental monitoring and sustainability such as microbe-enabled oil recovery. Metagenomics will aid in the development of ecosystem models by providing insight into microbial community genomics and biogeochemical interactions. Metagenomics can improve bioremediation by development of microbe-based methods for restoration of damaged ecosystems such as by discovery of novel microbial strains for degradation of xenobiotic compounds and treatments of contaminated sites. Metagenomic screening of microbial communities that drive the acidic outflows produced by oxidation of sulfide minerals (acid mine drainage) have been carried out for exploration of their diverse metabolic pathways and toleration mechanisms (Allen and Banfield 2005; Tyson *et al.* 2004). Enzymes functioning at alkaline pH and used in the detergent industry as additives to improve the stain removal properties have been isolated from Soda lakes host dense populations of alkali-loving microbes (Grant and Heaphy 2010).

4.5 Agriculture and Biotechnology

Metagenomics has considerable potential in agriculture in development of methods for early detection of crop and animal diseases and dangerous microbial food contaminants. Metagenomics is useful for studying phylogenetically distant microbial symbionts living in the same host. By binning the ESS data from bacterial symbionts living in the glassy-winged sharpshooter, an insect that lives solely on tree sap, a nutrient poor diet, it was inferred that one symbiont synthesizes amino acids for the host insect, while another synthesizes cofactors and vitamins (Wu *et al.*2006b). Metagenomic studies have revealed that upto 60% of the psbA genes, a component of Photosystem I, in surface water are of cyanophage origin and the existence of whole photosynthetic cassettes in cyanophages, which may increase host fitness by supplementing cyanobacterial photosystems (Sharon *et al.* 2009).

Metagenomics studies of gut populations in poultry, pigs, and other food animals will help to formulate more effective probiotic mixtures. Metagenomic studies can provide more insight into interactions between microbes and plants. For example, the role of crenarchaeotes and other plant root- colonizing microbes s and their contribution to carbon and nitrogen cycles and the interaction of plant and their microbial symbionts with other antagonistic microbes.

There is enormous potential for discovery of new genes and their application in biotechnology. It is estimated that there are about 10 trillion genes in 1 g of soil and at least 1 million *different* genes in 1 g of soil (Schloss and Handelsman 2006). Metagenomics has enabled discovery of novel biosynthetic and hydrolytic enzymes of microbial communities for biotechnological applications in industrial, food, feed and human health products. Functional characterization of entire proteomes of model microbes will enhance the rate of discovery of novel enzymes.

5. Future Prospects

Metagenomics provides an understanding of distribution, diversity and collective interactions of microbial communities. Full utilization of its potential depends on improvements in techniques of extraction of nucleic acids, library construction and analyses, affordable gene chips, efficient functional screens, improved bioinformatics tools and completion of more reference microbial genome sequences for model microbes. Generation and interpretation of the massive and complex metagenomic data for assembly of complete genomes of community members and information on their metabolic activities is an enormous task. New sequencing technologies for generating DNA sequence at a lower cost and deeper sequence coverage are being developed McPherson (2009). The metagenomics data has potential for bias due to extraction and quantitation methods. Some cloned genes are toxic, and so may be underrepresented in clone libraries used for sequencing. Some viral DNAs are chemically modified and difficult to clone.

These challenges can be addressed by careful analyses using multiple approaches. Cloning-independent approaches such as single cell genome sequencing and cell-sorting methods that readily permit recovery of even unique individual cells have been developed. The single cell genomics approach allows the culture-independent genome analysis of individual community members by isolation of the single cells from the environment, amplification of genome by Multiple Displacement Amplification (MDA) and random genome shotgun sequencing. The complete genome from *Candidatus Sulcia mulleri*, has been sequenced by this approach (Woyke *et al.* 2010). Additional use of metatranscriptomic and metaproteomic approaches such as whole-community RNA amplification, and imaging methods such as FISH using oligonucleotide probes will enhance the usefulness of metagenomic data. Improvements in microarray sensitivity will help in detection of rare members of a microbial community.

The microbial communities from habitats such as tropical, temperate, desert and cold soils; deep sea thermal vents, frozen lakes and acidic hot springs; rhizosphere and phyllosphere; symbiotic associations; and eukaryotic hosts can be exploited for discovery of novel genes for energy production, food security, human health and environment sustainability. Metagenomics will eventually expand beyond viruses, bacteria and archaea, to include population biology and biogeography of microbial eukaryotes such as protists, algae and fungi.

The enormous commercial potential of metagenomics may generate some intellectual property issues. The patents in bioprospecting and metagenomic methods may affect the growth of further metagenomic studies (Commitee on Metagenomics, 2007). Meticulous documentation of research and careful award of patents will be necessitated.

References

Allen, E. E., and Banfield J. F. (2005). Community genomics in microbial ecology and evolution. *Nat. Rev. Microbiol* 3:489-98.

Baker, B. J., Tyson, G.W., Webb, R.I., Flanagan, J., Hugenholtz, P., Allen, E.E. and Banfield, J.F. (2006). Lineages of acidophilic archaea revealed by community genomic analysis. *Science* 314: 1933-1935.

Banik, J.J. and Brady, S.F. (2010). Recent applications of metagenomic approaches toward the discovery of antimicrobials and other bioactive small molecules. *Curr. Opin. Microbiol.* 13: 603-609.

Bergeron, A., Belcaid, M., Steward, G.F. and Poisson, G. (2007). Divide and conquer: enriching environmental sequencing data. *PLoS One* 9: e830.

Berry, A. E., C. Chiocchini, T. Selby, M. Sosio, and E. M. Wellington. (2003). Isolation of high molecular weight DNA from soil for cloning into BAC vectors. *FEMS Microbiol Lett* 223 (1):15-20.

Branton, D., Deamer, D.W., Marziali, A., Beyley, H., Benner, S.A. *et al.* (2008). The potential and challenges of nanopore sequencing. *Nat. Biotech.* 26: 1146-1153.

Brodie, E. L., T. Z. DeSantis, J. P. Parker, I. X. Zubietta, Y. M. Piceno, and G. L. Andersen. (2007). Urban aerosols harbor diverse and dynamic bacterial populations. *Proc Natl Acad Sci USA* 104 :299-304.

Committee on metagenomics. (2007). The New Science of Metagenomics: Revealing the Secrets of Our Microbial Planet. National Academies press. 500 Fifth St. N.W. Washington, D.C. 20001. USA. p158.

Cowan, D., Meyer, Q., Stafford, W., Muyanga, S., Cameron, R. and Wittwer, P. (2005). Metagenomic gene discovery: past, present and future. *Trends Biotech.* 23: 321-329.

Desantis, T. Z., E. L. Brodie, J. P. Moberg, I. X. Zubieta, Y. M. Piceno, and G. L. Andersen. (2007). High-density universal 16S rRNA microarray analysis reveals broader diversity than typical clone library when sampling the environment. *Microb Ecol.* 53: 371-383.

Eisen, J.A. (2007). Environmental Shotgun Sequencing: Its Potential and Challenges for Studying the Hidden World of Microbes. *Plos Biol.* 5:e82.

Entcheva, P., Liebl, W., Johann, A., Hartsch, T. And Streit, W.R. (2001). Direct cloning from enrichment cultures, a reliable strategy for isolation of complete operons and genes from microbial consortia. *Appl. Environ. Microbiol. 67: 89-99.*

Fiers W, Contreras R, Duerinck F, Haegeman G, Iserentant D, *et al.* (1976). Complete nucleotide sequence of bacteriophage ms2 RNA: primary and secondary structure of the replicase gene. *Nature* 260:500–507.

Fleischmann RD, Adams MD, White O, Clayton RA, Kirkness EF, *et al.* (1995). Whole-genome random sequencing and assembly of *Haemophilus influenzae* rd. *Science* 269:496–512.

Foerstner, K. U., vonMering, C. And Bork, P. (2006). Comparative analysis of environmental sequences: potential and challenges. *Phil. Trans. R. Soc. B* 361: 519-523.

Fraser-Ligget, C. M. (2005). Insights on biology and evolution from microbial genome sequencing. *Genome Res.* 15: 1603-1610.

Frias-Lopez, J., Shi, Y., Tyson, G.W., Coleman, M. L., Schuster, S.C., Chisholm, S.W. and Delong, E.F. (2008). Microbial community gene expression in ocean surface waters. *Proc. Natl. Acad. Sci. USA* 105: 3805-3810.

Gabor, E.M., Alkema, W.B.L. and Janssen, D.B. (2004). Quantifying the accessibility of the metagenome by random expression cloning techniques. *Environ. Microbiol.* 6:879-886.

Gillespie, D.E., Brady,S.f., Bettermann, A.D., Cianciotto, nz.P., Liles, M.R., Rodon, M.R., Clardy, J., Goodman, R.M. and Handelsmann, J. (2002). Isolation of antibiotics turbomycin A and B from a metagenomic library of soil microbial DNA. *Appl. Environ. Microbiol. 68: 4301-4306.*

Grant, S., Sorokin, D.Y., Grant, W.D., Jones, B.E. and Heaphy S. (2004). A phylogentic analysis of Wadi el Natrun soda lake cellulose enrichment cultures and identification of cellulose genes from these cultures. *Extremophiles* 8: 421-429.

Grant WD, Heaphy S. (2010). Metagenomics and recovery of enzyme genes from alkaline saline environments. *Environ. Technol.* 31:1135-43.

Guan, C., B. Borlee, J. Ju, L. Williamson, B. Shen, K. Raffa, and J. Handelsman. (2007). Signal mimics derived from a metagenomic analysis of gypsy moth gut microbiota. *Appl. Environ. Microbiol.* 73: 3669–3676

Handelsman, J. (2004). metagenomics: application of genomics to uncultured microorganisms. *Microbiol. Mol. Bio. Rev.* 68: 669-685.

Harrington, E.D., Singh, A.H., Doerks, T., Letunic, I., von Mering, C., Jensen, L. J., Raes, J. and Bork, P. (2007). Quantitative assessment of protein function prediction from metagenomics short sequences. *Proc. Natl. Acad. Sci. USA* 104: 13913-13918.

Holben, W. E., and D. Harris. (1995). DNA-based monitoring of total bacterial community structure in environmental samples. *Mol Ecol* 4:627-31.

Kim, J-N., Choi,G-S., Kim, S-B., Yoon, G-S., Kim, Y-S. and Ryu, R.W. (2006). Screening and characterization of a novel esterase from a metagenomic library. *Protein Expr. Purif.* 45: 315-323.

Krsek, M. and Wellington, E.M.H. (1999). Comparison of different methods for the isolation and purification of total community DNA from soil. *J. Microbiol. Methods* 39:1-16.

Kunin, V., Copeland, A., Lapidus, A., Mavromatis, K. And Hugenoltz, P. (2008). A bioinformatician's aide to metagenomics. *Microbio. Molec. Biol. Rev.* 72: 557-578.

Lo, I., Denef. V.J., VerBerkmoes, N. C., shah, M.B., Goltsamn, D., DiBartolo, G., Tyson, G.W., Allen, E.E., ram, R.J. *et al.* (2007). Strain-resolved community proteomics reveals recombining genomes of acidophilic bacteria. *Nature* 446: 537-541.

Li, W., X. Zhou, and P. Lu. (2004). Bottlenecks in the expression and secretion of heterologous proteins in Bacillus subtilis. *Res. Microbiol.* 155:605-10

Liles, M. R., B. F. Manske, S. B. Bintrim, J. Handelsman, and Goodman R.M. (2003). A census of rRNA genes and linked genomic sequences within a soil metagenomic library. *Appl. Environ. Microbiol.* 69 :2684-91.

Martinez, A., Kolvek, S.J., Yip, C.L.T., Hopke, J., Brown, K.A., MacNeil, I. A. and Osburne, M.S. (2004). Genetically modified bacterial strains and novel bacterial artificial chromosome shuttle vectors for constructing environmental libraries and detecting heterologous natural products in multiple expression hosts. *Appl. Environ. Microbiol.* 70 :2452-2463.

Mavromatis K, Ivanova N, Barry K, Shapiro H, Goltsman E, *et al.* (2007) Use of simulated data sets to evaluate the fidelity of metagenomic processing methods. *Nat Methods*.4:495–500.

McHardy, A.C. and Rigoutos, I. (2007). What's in the mix: phylogenetic classification of metagenome sequence samples. *Curr. Opin. Microbiol.* 10: 499-503.

McPherson, J.D. (2009). Next-generation gap. *Nat Methods* 6:S2–S5

Pace, N.R., Stahl, D.A., Lane, D.J. and olsen, G.J. (1986). The analysis of natural microbial populations by ribosomal RNA sequences. *Adv. Microb. Eco.* 9:1-55.

Pallen M.J., Loman N.J., Penn C.W. (2010). High-throughput sequencing and clinical microbiology: progress, opportunities and challenges *Curr. Opin. Microbiol.* 13:625-31

Riesenfeld, C. S., Schloss, P. D. and Handelsman, J. (2004). Metagenomics: Genomic analysis of microbial communities. *Ann. Rev. Genet.* 38: 525-552

Rondon, M. R., P. R. August, A. D. Bettermann, S. F. Brady, T. H. Grossman, M. R. Liles, K. A. Loiacono, B. A. Lynch, I. A. MacNeil, C. Minor, C. L. Tiong, M. Gilman, M. S. Osburne, J. Clardy, J. Handelsman, and R. M. Goodman (2000). Cloning the soil metagenome: a strategy for accessing the genetic and functional diversity of uncultured microorganisms. *Appl. Environ. Microbiol.* 66:2541-7.

Sanger, F., Nicklen, S. And Coulson, A.R. (1977). DNA sequencing with chain terminating inhibitors. *Proc. Natl. Acad. Sci. USA* 74: 5463-5467.

Sanger F, Coulson AR, Friedmann T, Air GM, Barrell BG, *et al.* (1978). The nucleotide sequence of bacteriophage phix174. *J. Mol. Biol.* 125:225–246.

Schloss, P. D., and J. Handelsman (2003). Biotechnological prospects from metagenomics. *Curr. Opinion Biotech.* 14: 303-310.

Schloss, P. D., and Handelsman, J. (2005). Metagenomics for studying unculturable microorganisms: cutting the Gordian knot. *Genome Biol.* 6: 229-233

Schloss, P. D., and Handelsman, J. (2006). Toward a census of bacteria in soil. *PLoS Comput Biol* 2 :e92.

Sekar, R., Fuchs, B.M., Amman, R. and Pernthaler, J. (2004). Flow sorting of marine bacterioplankton after fluorescence in situ hybridization. *Appl. Environ. Microbiol.* 70: 6210-6219.

Sharon I, Alperovitch A, Rohwer F, Haynes M, Glaser F, *et al.* (2009). Photosystem i gene cassettes are present in marine virus genomes. *Nature* 461:258–262

Society of American Bacteriologists (1923). Bergey's manual of dterminative bacteriology. Williams and Wilkins, Baltimore, Madison.

Steele H. L., Jaeger, K-E., Daniel, R. and Streit, W.R. (2009). Advances in recovery of novel biocatalysts from metagenomes. *J. Mol. Microbiol. Biotechnol.* 16:25-37.

Tasse L, Bercovici J, Pizzut-Serin S, Robe P, Tap J, Klopp C, Cantarel BL, Coutinho PM, Henrissat B, Leclerc M, Doré J, Monsan P, Remaud-Simeon M, Potocki-Veronese G. (2010). Functional metagenomics to mine the human gut microbiome for dietary fiber catabolic enzymes. *Genome Res.* Sep 14

Tyson, G. W., J. Chapman, P. Hugenholtz, E. E. Allen, R. J. Ram, P. M. Richardson, V. V. Solovyev, E. M. Rubin, D. S. Rokhsar, and Banfield, J. F. (2004). Community structure and metabolism through reconstruction of microbial genomes from the environment. *Nature* 428 :37-43.

Venter, J. C., K. Remington, J. F. Heidelberg, A. L. Halpern, D. Rusch, J. A. Eisen, D. Wu, I. Paulsen, K. E. Nelson, W. Nelson, D. E. Fouts, S. Levy, A. H. Knap, M. W. Lomas, K. Nealson, O. White, J. Peterson, J. Hoffman, R. Parsons, H. Baden-Tillson, C. Pfannkoch, Y. H. Rogers, and Smith, H. O. (2004). Environmental genome shotgun sequencing of the Sargasso Sea. *Science* 304:66-74.

Voget, S., Steele, H.L. and Streit, W.R. (2006). Characterization of a metagenome-derived halotolerant cellolase. *J. Biotechnol.* 126: 26-36.

Wagner, M., Horn, M. and Daims, H. (2003). Flourescence in situ hybridisation for the identification and characterisation of prokaryotes. *Curr. Opin. Microbiol.* 6: 302-309.

Wang, G. Y., E. Graziani, B. Waters, W. Pan, X. Li, J. McDermott, G. Meurer, G. Saxena, R. J. Andersen, and Davies, J. (2000). Novel natural products from soil DNA libraries in a streptomycete host. *Org Lett* 2:2401-4.

Warnecke, F. and Hugenholtz P. (2007). Building on basic metagenomics with complementary technologies. *Genome Biol.* 8:231-235

Wexler, M and Johnston A.W. (2010). Wide host range cloning for functional metagenomics *Methods Mol. Biol.* 668: 77-96.

Wooley, J.C., Godzik, A. and Friedberg, I. (2010). A Primer on Metagenomics. *PLoS Comput Biol.* 6: e1000667.

Wu, L., X. Liu, C. W. Schadt and Zhou, J. (2006a). Microarray-based analysis of subnanogram quantities of microbial community DNAs by using whole-community genome amplification. *Appl Environ Microbiol* 72:4931-41.

Wu ,D., Daugherty S.C., Van Aken S.E., Pai G.H., Watkins K.L., Khouri H., Tallon L.J., Zaborsky J.M., Dunbar H.E., Tran P.L., Moran N.A. and Eisen J.A. (2006b). Metabolic complementarity and genomics of the dual bacterial symbiosis of sharpshooters. *PLoS Biol.* 4:e188.

❑❑❑

CHAPTER – 8

Expression of Genetic Information From Transcription to Translation

Nidhi Sharma and Veena Garg
Department of Bioscience and Biotechnology, Banasthali University, Rajasthan - 304 022 India
Email : nidhisharma2006@gmail.com

Summary

Genes are discrete elements that govern the appearance of specific traits. Morgan Sturtevant and their coworkers demonstrated that genes have specific address. They reside at particular locations on particular chromosomes and these addresses remain constant from one individual of the species to the next. ***Transcription*** *is the first step leading to gene expression. It is the process of creating a complementary RNA copy of a sequence of DNA. Both RNA and DNA are nucleic acids, which use base pairs of nucleotides as a complementary language that can be converted back and forth from DNA to RNA by the action of the correct enzymes. During transcription, a DNA sequence is read by RNA polymerase, which produces a complementary, antiparallel RNA strand.* ***translation*** *is the second stage of protein biosynthesis (part of the overall process of gene expression). In translation, messenger RNA (mRNA) produced by transcription is decoded by the ribosome to produce a specific amino acid chain, or polypeptide, that will later fold into an active protein. In Bacteria, translation occurs in the cell's cytoplasm, where the large and small subunits of the ribosome are located, and bind to the mRNA. In Eukaryotes, translation occurs across the membrane of the endoplasmic reticulum in a process called vectorial synthesis. The ribosome facilitates decoding by inducing the binding of tRNAs with complementary anticodon sequences to that of the mRNA. The tRNAs carry specific amino acids that are chained together into a polypeptide as the mRNA passes through and is "read" by the ribosome.*

1. Introduction

1.1. Relationship Between Genes and Proteins

Archibald Garrod, a Scotish physician, in 1908 reported that the symptoms exhibited by persons with certain rare inherited diseases were caused by the absence of specific enzymes. Garrod investigated alkaptonuria, a disease in which urine becomes dark on exposure to air. He found that persons with alkaptonuria lacked an enzyme in their blood that oxidized homogentisic acid, a compound formed during the breakdown of the amino acids phenylalanine and tyrosine. As homogentisic acid accumulates, it appears in the urine and darkens in color when oxidized by the air. Garrod discovered the relationship between a genetic defect, a specific enzyme, and a specific metabolic condition. He called such diseases as **'inborn errors of metabolism'**.

Clarification of the actual function of genes came from research in the 1940s on *Neurospora* by George Beadle and Edward Tatum, who later received a Nobel Prize for their work. Before describing their actual experiments, let's jump ahead and examine some aspects of **biosynthetic pathways,** on the basis of our current understanding. We now know that molecules are synthesized as a series of steps, each one catalyzed by an enzyme. For instance, a biosynthetic pathway might have four steps, where 1 is the starting material and 5 is the final product:

$$1 \xrightarrow{A} 2 \xrightarrow{B} 3 \xrightarrow{C} 4 \xrightarrow{D} 5$$

Each step is catalyzed by an enzyme: A, B, C, or D. In turn, each enzyme is specified by a particular gene. We might say that gene *A* specifies enzyme A, gene *B* specifies enzyme B and so forth. Therefore, if we inactivate the gene responsible for an enzyme, we eliminate one required step and the pathway is interrupted.

In the following diagram, enzyme B is eliminated owing to a mutation in gene *B*:

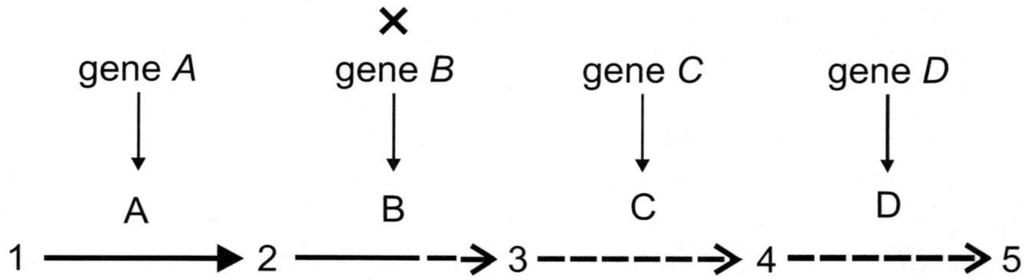

Now the cell cannot carry out the reaction that converts compound 2 into compound 3. It is blocked at compound 2, and cannot go further. But what happens if we add back different intermediate compounds? Suppose that, for example, we feed the cell compound 3 or 4. Can the cell then synthesize the final product, 5? Yes, with either compound 3 or 4 it can, because subsequent steps are not blocked. What if we add more of compound

1? No, adding compound 1 will not allow the synthesis of product 5, because one of the subsequent steps is blocked.

2. Experiments of Beadle and Tatum

Beadle and Tatum analyzed mutants of *Neurospora,* a fungus with a haploid genome. They first irradiated *Neurospora* to produce mutations and then tested cultures from ascospores for interesting mutant phenotypes. They detected numerous **auxotrophs** strains that cannot grow on a minimal medium unless the medium is supplemented with one or more specific nutrients. In each case, the mutation that generated the auxotrophic requirement was inherited as a single-gene mutation: each gave a 1:1 ratio when crossed with a wild type. Figure 1 depicts the procedure that Beadle and Tatum used.

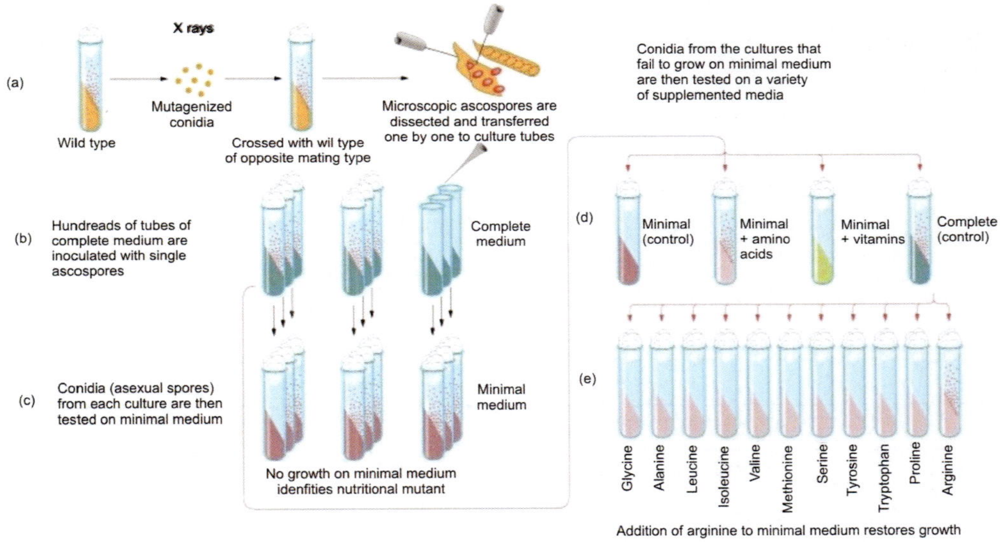

Fig. 1 : Summary of the procedure used by Beadle & Tatum experiments on *Neurospora*

One set of mutant strains required arginine to grow on a minimal medium. These strains provided the focus for much of Beadle and Tatum's further work. First, they found that the mutations mapped into three different locations on separate chromosomes, even though the same supplement (arginine) satisfied the growth requirement for each mutant. Let's call the three loci the *arg-1, arg-2,* and *arg-3* genes. Beadle and Tatum discovered that the auxotrophs for each of the three loci differed in their response to the chemical compounds ornithine and citrulline, which are related to arginine (Fig. 2). The *arg-1* mutants grew when supplied with ornithine, citrulline, or arginine in addition to the minimal medium. The *arg-2* mutants grew on either arginine or citrulline but not on ornithine. The *arg-3* mutants grew only when arginine was supplied. We can see this more easily by looking at Table 1.

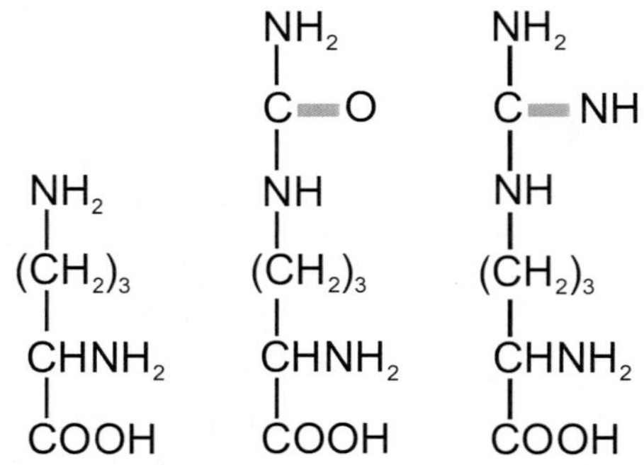

Omithine Citrulline Arginine

Fig. 2 : Chemical structures of arginine and the related compounds citrulline and ornithine. In the work of Beadle and Tatum, different *arg* auxotrophic mutants of *Neurospora* were found to grow when the medium was supplemented with citrulline or with ornithine as an alternative to arginine.

It was already known that cellular enzymes often interconvert related compounds such as these. On the basis of the properties of the *arg* mutants, Beadle and Tatum and their colleagues proposed a biochemical model for such conversions in *Neurospora:*

$$\text{precursor} \xrightarrow{\text{enzyme X}} \text{ornithine} \xrightarrow{\text{enzyme Y}} \text{citrulline} \xrightarrow{\text{enzyme Z}} \text{arginine}$$

Table 1 : Growth of arg Mutants in Response to Supplements

	SUPPLEMENT		
Mutant	**Ornithine**	**Citrulline**	**Arginine**
arg-1	+	+	+
arg-2	"	+	+
arg-3	"	"	+

Note: A plus sign means growth; a minus sign means no growth.

The above relation easily explains the three classes of mutants as shown in Table 1. The *arg-1* mutants have a defective enzyme X, so they are unable to convert the precursor into ornithine as the first step in producing arginine. However, they have normal enzymes Y and Z, and so the *arg-1* mutants are able to produce arginine if supplied with either ornithine or citrulline. The *arg-2* mutants lack enzyme Y, and the *arg-3* mutants lack enzyme Z. Thus, a mutation at a particular gene is assumed to interfere with the production of a single enzyme. The defective enzyme, then, creates a block in some biosynthetic pathway. The block can be circumvented by supplying to the cells any compound that normally comes after the block in the pathway.

We can now diagram a more complete biochemical model:

arg-1$^+$ → enzyme X; *arg-2*$^+$ → enzyme Y; *arg-3*$^+$ → enzyme Z

precursor $\xrightarrow{\text{enzyme X}}$ ornithine $\xrightarrow{\text{enzyme Y}}$ citrulline $\xrightarrow{\text{enzyme Z}}$ arginine

Note that this entire model was inferred from the properties of the mutant classes detected through genetic analysis. Only later were the existence of the biosynthetic pathway and the presence of defective enzymes demonstrated through independent biochemical evidence.

This model, which has become known as the **one-gene–one-enzyme hypothesis,** was the source of the first exciting insight into the functions of genes: genes somehow were responsible for the function of enzymes, and each gene apparently controlled one specific enzyme. Other researchers obtained similar results for other biosynthetic pathways, and the hypothesis soon achieved general acceptance. for, The one gene–one-enzyme hypothesis became one of the great unifying concepts in biology, because it provided a bridge that brought together the concepts and research techniques of genetics and biochemistry.

3. Flow of Information Through the Cell: An Overview

It is now well known that there is an intermediate between a gene and its polypeptide; the intermediate is messenger RNA (mRNA) whose existence was first demonstrated in 1961 by Francois Jacob and Jacques Monod. The formation of RNA from DNA template is called **transcription**. Because its nucleotide sequence is complimentary to that of the gene from which it is transcribed, the mRNA retains the same information as the gene itself. mRNA reaches the cytoplasm where it can serve as a template to direct the incorporation of amino acids in a particular order encoded by the nucleotide sequence of the DNA and mRNA. One DNA molecule can serve as the template in the formation of many mRNA molecules, each of which can be used in the formation of a large number of polypeptide chains. Proteins are synthesized in the cytoplasm by a highly complex process called **translation**. Translation requires the participation of different components including ribosomes. Ribosomes consist of a large and a small subunit. The ribosome is assembled from these subunits at the time it begins to synthesize a particular polypeptide and then is disassembled when synthesis has been completed. Ribosome consists of RNA (rRNA) and protein. rRNAs are also transcribed from one of the DNA strands of the gene. The transfer RNAs (or tRNAs) constitute a third major class of RNA that is required during protein synthesis. tRNAs are required to translate the information in the mRNA nucleotide code into the amino acid 'alphabet' of a polypeptide.

4. Transcription : The Basic Process

Transcription is the enzymatic synthesis of RNA on a DNA template with the help of enzymes called RNA polymerases. This is the first stage in the overall process of gene expression and ultimately leads to synthesis of the protein encoded by a gene. Transcription is catalyzed by an RNA polymerase which requires a double stranded DNA template as well as the precursor ribonucleotides ATP, GTP, CTP and UTP. RNA synthesis always occurs in a fixed direction, from 5' - to the 3' - end of the RNA molecule. Usually only one of the two strands of DNA becomes transcribed into RNA. One strand is known as the **sense** strand. The sequence of RNA is a direct copy of the sequence of deoxynucleotides in the sense strand of DNA (with U in place of T). The other strand is known as the antisense strand. This strand may also be called the template strand since it is used as the templates to which ribonucleotides base-pair for the synthesis of the RNA. The site on the DNA to which an RNA polymerase molecule binds prior to initiating transcription is called the promoter. In addition to providing a binding site for the polymerase, the promoter contains the information that determines which of the two DNA strands is transcribed and the site at which transcription begins (Karp, 2002).

The polymerase moves along the template DNA strand towards its 5' end (i.e. 3' to 5' direction). As the polymerase progresses the DNA is temporarily unwound, and the polymerase assembles a complementary strand of RNA that grows from its 5' terminus in a 3' direction (Fig. 3).

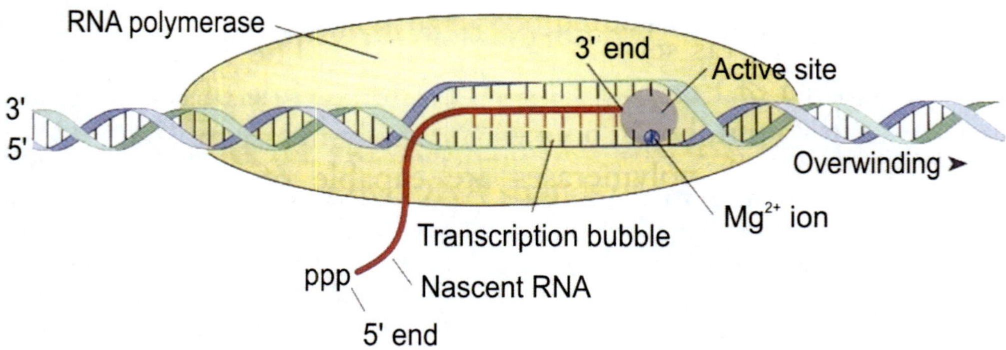

Fig. 3 : A schematic model of the elongation of a newly synthesized RNA molecule during transcription. The polymerase covers approximately 35 base pairs of DNA, the transcription bubble composed of single stranded (melted) DNA contains about 15 base pairs, and the segment present in a DNA-RNA hybrid includes about 9 base pairs. The enzyme generates overwound (positively supercoiled) DNA ahead of itself and underwound (negatively supercoiled) DNA behind itself.

The following reaction is catalyzed by RNA polymerase.

$$RNA + NPPP \rightarrow RNA_{n+1} + PPi$$

In the above reaction ribonucleoside triphosphate precursors (NPPs) are hydrolyzed into nucleoside monophosphates as they are polymerized into a covalent chain. The second reaction is catalyzed by an enzyme pyrophosphatase, in which the pyrophosphate (PPi) produced in the first reaction is hydrolyzed to inorganic phosphate (Pi). The hydrolysis of pyrophosphate releases a large amount of free energy and makes the incorporation of nucleotides essentially irreversible. As the polymerase moves along the DNA template, it incorporates complimentary nucleotides into the growing RNA chain. A nucleotide is incorporated into the RNA strand if it is able to form a proper (Watson-crick) base pair with the nucleotide in the DNA strand being transcribed. This can be seen in the Figure (Fig. 4) where the incoming ATP pairs with the thymine containing nucleotide of the template. Once the polymerase has moved past a particular stretch of DNA, the double helix reforms (Fig. 3). Consequently, the RNA chain does not remain associated with its template as a DNA-RNA hybrid (except for a few nucleotides just behind the site where the polymerase is operating).

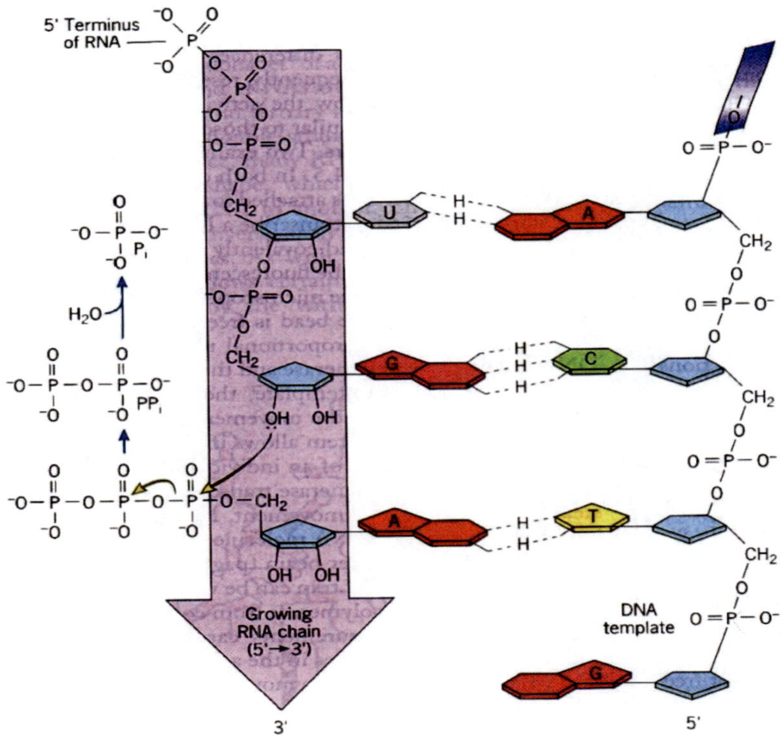

Fig. 4 : Chain elongation occurs as a result of an attack by the 3' OH of the nucleotide at the end of the growing strand on the 5' α-phosphate of the incoming nucleoside triphosphate. The pyrophosphate released is subsequently cleaved, which drives the reaction in the direction of polymerization.

RNA polymerase is *processive*. It remains attached to the DNA over long stretches of the DNA template to produce long RNAs. At the same time, the enzyme must be associated loosely enough so that it can move from nucleotide to nucleotide on the template. For this, the polymerase undergoes the change in conformation, which allows the enzyme to move step-by-step along the DNA template. The energy required to drive the movements of the polymerase is presumably derived from the hydrolysis of the nucleoside triphosphate precursors as they are incorporated into the growing chain.

4.1 Transcription in Prokaryotes

Prokaryotic cells contain a single type of RNA polymerase composed of 5 subunits. These are alpha (α), beta (β), beta prime (β'), omega (ω) and sigma (σ) subunits. In the complete polymerase called the holoenzyme, there are two α subunits and one each of the other four subunits (i.e. $\alpha_2\beta\beta'\ \omega\sigma$). The complete enzyme is required for transcription initiation. However, the σ factor is not required for the transcription elongation and is released from the transcription complex after the transcription initiation. The remaining enzyme, which translocates along the DNA, is known as the core enzyme. The attachment of the σ factor to the core enzyme increases the affinity of the enzyme for promoter sites in the DNA and decreases its affinity for the DNA in general. The σ factor may also promote the separation (or melting) of the DNA strands of the promoter DNA helix, which is required for transcription to begin.

Bacterial promoters are located in the region of a DNA strand just preceding the initiation site of RNA synthesis (Fig 5). The nucleotide at which transcription is initiated is denoted as +1 and the preceding nucleotide as -1. Those portions of the DNA preceding the initiation site (towards the 3' end of the template) are said to be upstream from that site. Those portions of the DNA succeeding it (towards the 5' end of the template) are said to be downstream from that site. Promoters contain conserved sequences which are required for specific binding of RNA polymerase and transcription initiation. The promoter region extends for around 40 bp. Within this sequence, there are short regions of extensive conservation which are critical for promoter function. One sequence is found approximately 10 bases upstream. This -10 sequence is a 6 bp region recognizable in most promoters. This hexamer is generally 10 bp upstream from the start site. The consensus -10 sequence is TATAAT. The second sequence is centered at approximately 35 bases upstream from the initiation site. The -35 sequence is a further 6 bp region recognizable in most promoters. This hexamer typically occurs as the sequence TTGACA (Fig 5). This site in the promoter is the sequence recognized by the ó factor associated with the polymerase. Bacterial cells possess a variety of different ó factors that recognize different versions of the -35 sequence. The primary σ factor, σ^{70} factor, initiates transcription of most genes. The base at the start site is almost always a purine. G is more common than A. Just as the transcription is initiated at specific points in the chromosome, it also terminates when a specific nucleotide sequence is reached. In some cases, a protein called the *rho* factor is required for the termination, whereas in most cases, the polymerase stops transcription and releases the completed RNA chain without requiring additional factors.

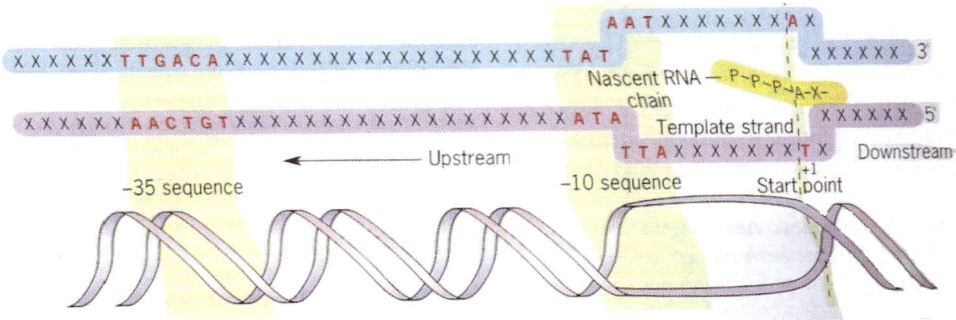

Fig. 5 : The basic elements of a promoter region in the DNA of *E. coli*

Eukaryotic promoters can be divided into core elements and regulatory elements (Sumimoto et al., 1991). Core promoter elements define the site for assembly of the transcription preinitiation complex (PIC) and include a TATA sequence, located upstream of the transcription start site, and an initiator sequence (Inr), encompassing the start site. Promoters can include a TATA box, an Inr sequence, or both of these control elements. A third core element, the downstream promoter element (DPE), was initially described in *Drosophila* and is located about 30 bp downstream of the start site (Burke et al., 1996). The DPE appears to function, in conjunction with the Inr element, as a TFIID binding site at TATA-less promoters.

TATA elements in *S. cerevisiae* are typically located 40 to120 bp upstream of the transcription initiation site. This is in contrast to other eukaryotes, including *Schizosaccharomyces pombe*, where the TATA element is almost always located at a fixed distance of 25 to 30 bp from the start site (Struhl, 1995). The TATA sequence is the binding site for the TATA binding protein (TBP). TBP-TATA association nucleates the assembly of an approximately 4-MDa transcription preinitiation complex, a step that can be rate limiting for transcription initiation in vivo (Klein and Struhl, 1994).

Mutational analysis and random selection for functional TATA elements defined TATAAA as the consensus TATA sequence in yeast (Chen and Struhl, 1988; Singer et al., 1990; Wobbe and Struhl, 1990). Many derivatives of this sequence also confer TATA function, albeit with diminished activity. One derivative, TGTAAA, eliminated TATA function and was used to select for TBP derivatives with altered binding specificity (Strubin and Struhl, 1992). TBPm3, described below, allowed transcription from TGTAAA promoters but not from certain other single-nucleotide derivatives of TATAAA (Strubin and Struhl 1992). This mutant demonstrated the importance of specific interactions between the TATA element and TBP for efficient initiation. Functional analysis of mutated TATA elements revealed that yeast and human TBP have nearly identical TATA sequence requirements, underscoring the evolutionary conservation of the TBPTATA interaction (Wobbe and Struhl 1990; Hampsey 1998).

In order for the antisense strand to become accessible for base pairing, the DNA duplex must be unwound by the polymerase. Negative supercoiling enhances the transcription of many genes, since this facilitates unwinding by the polymerase. However, some promoters are not activated by negative supercoiling, implying that differences in the natural DNA topology may affect transcription. DNA gyrase is responsible for negative supercoiling of the *E. coli* genome and so this may serve as an elegant feedback loop for DNA gyrase protein expression. The initial unwinding of the DNA results in formation of an open complex with the polymerase; and this process is referred to as tight binding.

4.2 RNA Chain Initiation

Unlike DNA synthesis, RNA synthesis can occur without a primer (Fig 6). The chain is started with a GTP or ATP, from which synthesis of rest of the chain is initiated. The polymerase initially incorporates the first two nucleotides and forms a phosphodiester bond between them. The first nine bases are added without the enzyme movement along the DNA. After each of these first 9 nucleotides is added to the chain, there is a significant probability that the chain will be aborted.

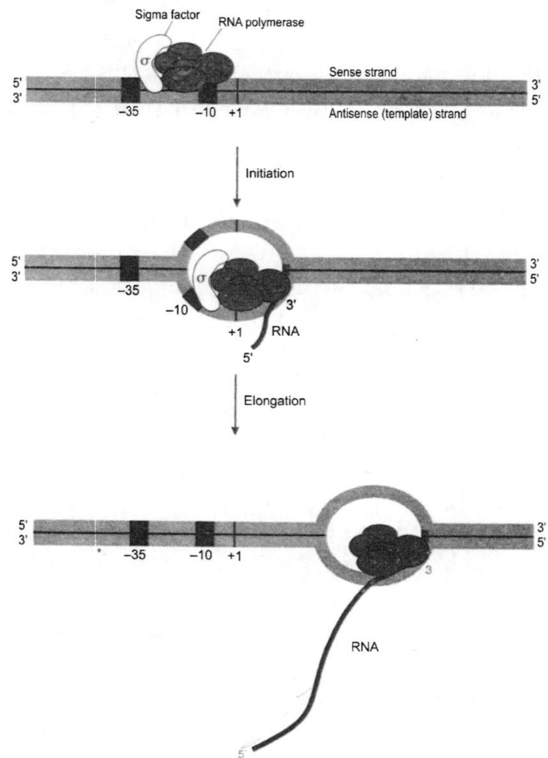

Fig. 6 : Formation of the transcription complex: initiation and elongation.

4.3 RNA Chain Elongation

When initiation succeeds, the enzyme releases the ó factor and forms a ternary complex of polymerase-DNA-nascent (newly synthesized) RNA, causing the polymerase to progress along the DNA (promoter clearance). The region of unwound DNA, which is called the transcription bubble, appears to move along the DNA with the polymerase. The size of this region of unwound DNA remains constant at round 17 bp (Fig 7), and the 5' end of the RNA forms a hybrid helix of around 12 bp with the antisense DNA strand. This corresponds to just less than one turn of the RNA-DNA helix. The *E. coli* polymerase moves at an average rate of 40 nt per sec. Maintenance of the short region of unwound DNA indicates that the polymerase unwinds DNA in front of the transcription bubble and rewinds DNA at its rear.

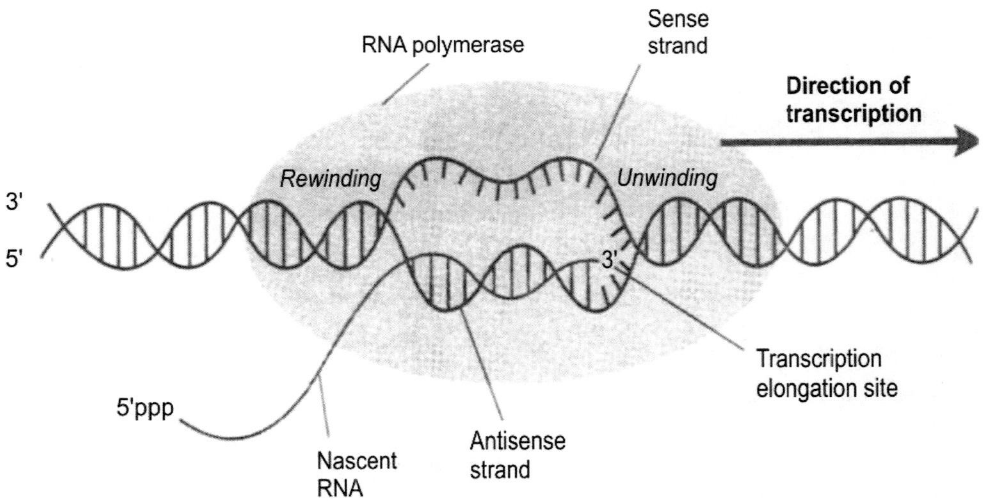

Fig. 7 : Schematic structure of the transcription bubble during elongation.

4.4 RNA Chain Termination

The RNA polymerase remains bound to the DNA and continues transcription until it reaches a **terminator sequence (stop signal)** at the end of the transcription unit (Fig 8). The more common stop signal is an RNA hairpin loop. Self-complimentary sequences at the 3' end of the genes form hairpin structures in the RNA which act as terminators. The stem of the hairpin often has a high content of G-C base pairs giving it a high stability (due to additional stability of G-C base pairs over A-U base pairs), causing the polymerase to pause. The hairpin is often followed by four or more Us which result in weak RNA-antisense DNA strand binding. This favors dissociation of the RNA strand, causing transcription termination.

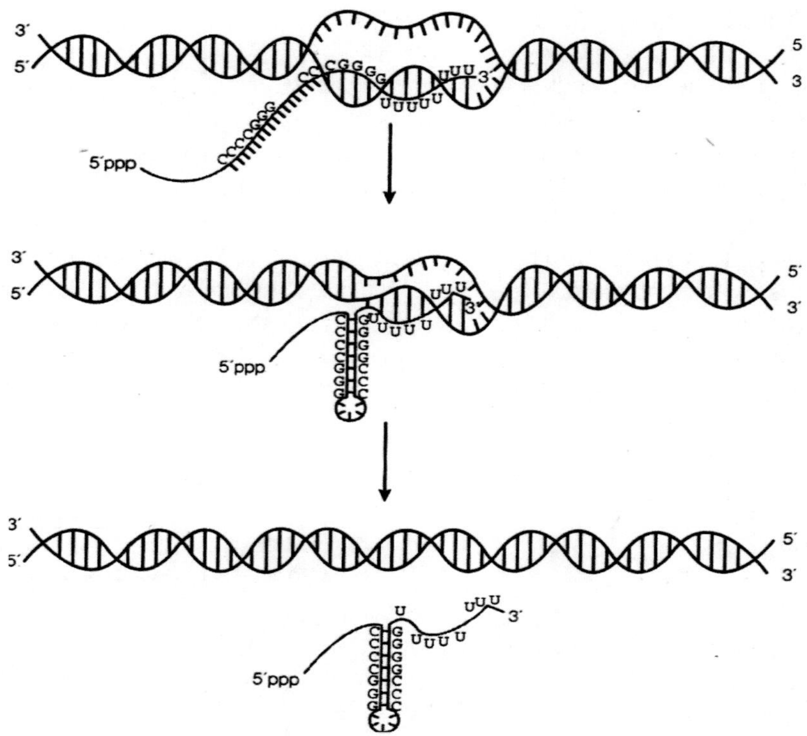

Fig. 8 : Schematic diagram of rho-independent transcription termination.

4.5 Rho-Dependent Termination

While the RNA polymerase can self-terminate at a hairpin structure followed by a stretch of U residues, other known terminator sites may not form strong hairpins. They use an accessory factor, the rho protein (ρ) to mediate transcription termination. Rho binds to specific sites in single stranded RNA. It hydrolyzes ATP and moves along the RNA towards the transcription complex, where it enables the polymerase to terminate transcription at rho-dependent transcriptional terminators. Like rho-independent terminators, these signals are recognized in the newly synthesized RNA rather than in the template DNA. Sometimes the rho-independent terminators are hairpin structures which lack subsequent stretch of U residues which are required for rho-independent termination.

4.2 Transcription in Eukaryotes

Eukaryotic cells have three distinct transcribing enzymes, each responsible for synthesizing a different group of RNAs. **RNA polymerase I** synthesizes the larger (28S, 18S and 5.8S) ribosomal RNAs. It is located in the nucleoli and is insensitive to α-amanitin. **RNA polymerase II** synthesizes mRNAs and most small nuclear RNAs. It is locatd in the nucleoplasm and is very sensitive to α-amanitin; and **RNA polymerase III** synthesizes low molecular weight RNAs and the 5S ribosomal RNA. It is located in the nucleoplasm and is moderately sensitive to α-amanitin.

All three major types of RNAs – mRNAs, rRNAs and tRNAs – are derived from precursor RNA molecules that are considerably longer than the final RNA product. The initial precursor is equivalent in length to the full length of the DNA transcribed and is called the primary transcript, or **pre-RNA**. The corresponding segment of DNA from which a primary transcript is transcribed is called a **transcription unit**. Primary transcripts do not exist in cell as naked RNA but becomes associated with proteins even as they are synthesized. Primary transcripts typically have a fleeting existence, being processed into smaller, functional RNAs by a series of 'cut and paste' reactions. RNA processing requires a variety of small RNAs (90-300 nucleotides long) and their associated proteins. The RNAs are called small nuclear RNAs (snRNAs).

During the past decade, numerous ncRNAs (noncoding RNAs) have been identified as regulators of transcription (Prasanth and Spector, 2007). Short (21–23 nt) double-stranded RNA molecules, such as siRNAs (silencing RNAs) and shRNAs (small hairpin RNAs), are today among the most well characterized ncRNAs (Rana, 2007). siRNAs and shRNAs have become very useful tools for studying gene inactivation, and have revolutionized genetics. Endogenous short (19–25 nt), natural ncRNAs, which are known as miRNAs (microRNAs), appear to play an important role in gene regulation during development (Bushati and Cohen, 2007; Zhao and Srivastava, 2007). Short, as well as long, noncoding transcripts are involved in epigenetic regulation (Yang and Kuroda, 2007; Zaratiegui et al., 2007).

4.2.1 Transcription of ribosomal RNAs

Eukaryotic cells contain millions of ribosomes, each of which contains several molecules of rRNAs together with dozens of ribosomal proteins. The composition of a mammalian ribosome is shown in Fig. 9. Ribosomes are so numerous that more than 80% of the RNA in most cells consists of ribososmal RNA. To furnish the cells with such a large number of transcripts, the DNA sequences encoding rRNA are normally repeated hundreds of times. This DNA, called rDNA, is typically clustered in one or a few regions of the genome. In a non-dividing (interphase) cell, the clusters of rDNA are gathered together as part of one or more irregularly shaped nuclear structures, called nucleoli (singular, nucleolus) that function as ribosome producing organelles.

The bulk of the nucleolus is composed of ribosomal subunits that give the nucleolus a granular appearance (Fig 10). Embedded within the granular mass are one or more rounded cores consisting primarily of fibrillar material. Nucleoli disappear during mitosis and then reappear in the nuclei of the daughter cells around those parts of the genome that contain rRNA genes.

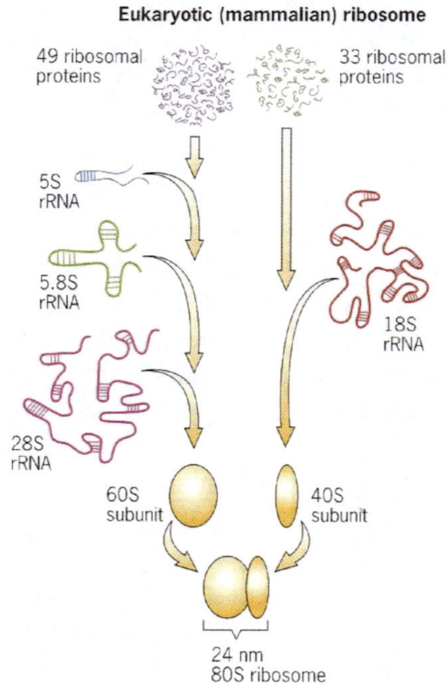

Fig. 9 : The macromolecular composition of a mammalian ribosome. The drawing shows the components present in each of the subunits of a mammalian ribosome.

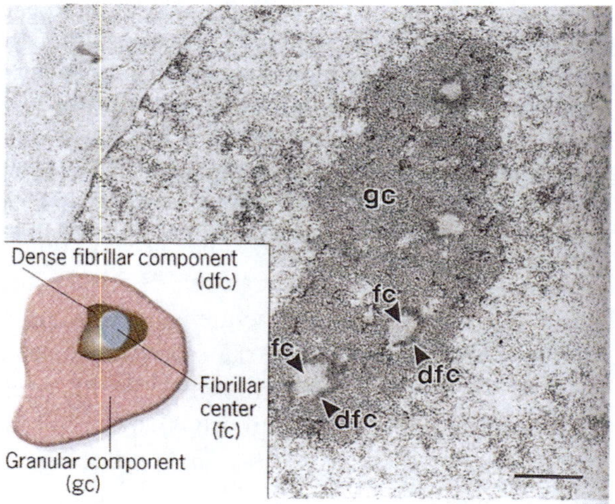

Fig. : 10 : The structure of the nucleolus. The bulk of the nucleolus consists of a granular component (gc). Embedded within the granules are fibrillar centers (fc) that surrounded by a denser fibrillar component (dfc). The fc contains the DNA that codes for ribosomal RNA, and the dfc contains the nascent pre-rRNA transcripts and associated proteins.

Consequently, regions of the chromosome that contain rDNA have been called *nucleolar organizers*. During active rRNA synthesis, the pre-rRNA transcripts are packed along the DNA and stick out perpendicularly from the DNA. Short transcripts can be seen at the start of the gene, which get longer until the end of the transcription unit, which is indicated by the disappearance of the RNA transcripts (Fig. 11). The length of DNA between the shortest and longest RNA fibrils corresponds to a single transcription unit. The promoter lies just upstream from the transcription initiation site. Mammalian pre-rRNA gene promoters have a bipartite transcription control region (Fig 12). The core elements includes the transcription start site and encompasses bases -31 to +6. This sequence is essential for transcription. An additional element of around 50-80 bp named the **upstream control element (UCE)** begins about 100 bp upstream from the start site (-100). The UCE is responsible for an increase in transcription of around 10- to 100-fold compared with that from the core element alone.

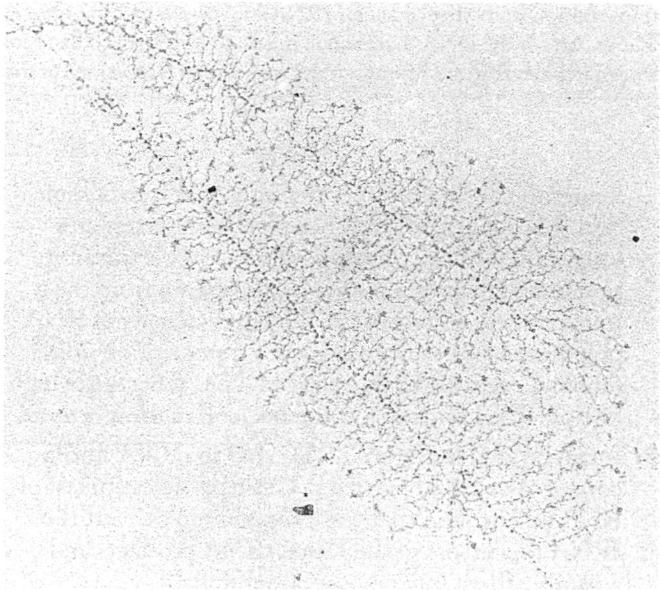

Fig. 11 : View of two nucleolar genes being transcribed. The length of the nascent rRNA primary transcript increases with increasing distance from the point of initiation. Dots at the base of each fibril shows the presence of RNA polymerase molecules.

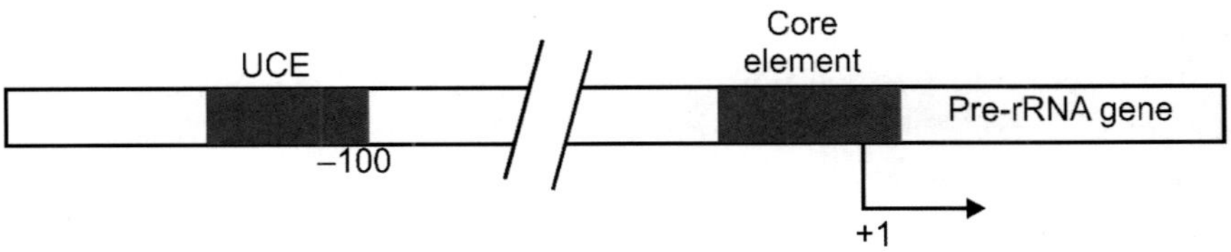

Fig. 12 : Structure of a mammalian pre-rRNA promoter.

A specific DNA binding protein, called **upstream binding factor**, or **UBF**, binds to the UCE. As well as binding to the UCE, UBF binds to a sequence in the upstream part of the core element. One molecule of UBF is thought to bind to each sequence element. The two molecules of UBF may then bind to each other through protein-protein interactions, causing the intervening DNA to form a loop between the two binding sites (Fig 13). A low rate of basal transcription is seen in the absence of UBF, nad this is greatly stimulated in the presence of UBF.

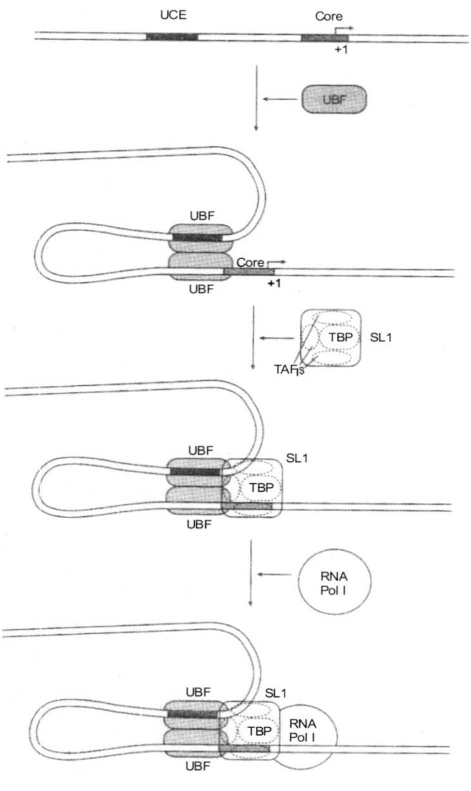

Fig. 13 : rRNA transcription initiation

An additional factor, called selectivity factor (SL1) binds to and stabilizes the UBF-DNA complex and interacts with the free downstream part of the core element. SL1 binding allows RNA Pol I to bind to the complex and initiate transcription, and is essential for rRNA transcription. SL1 has now been shown to contain several subunits, including a protein called TBP (TATA-binding protein). TBP is required for initiation by all three eukaryotic RNA polymerases and seems to be a critical factor in eukaryotic transcription. The other three subunits of SL1 are referred to as TBP-associated factors or TAFs, and those subunits required for RNA Polymerase I transcription are referred to as TAF_1S.

4.2.2 Processing the rRNA precursor

Eukaryotic ribosomes have four distinct rRNAs, three in the large subunit and one in the small subunit. In humans, the large subunit contains a 28S, 5.8S and 5S RNA molecule, and the small subunit contains an 18S RNA molecule. Three of these rRNAs (28S, 18S and 5.8S) are carved by various nucleases from a single primary transcript (called the pre-rRNA). The 5S rRNA is synthesized from a separate RNA precursor outside the nucleolus.

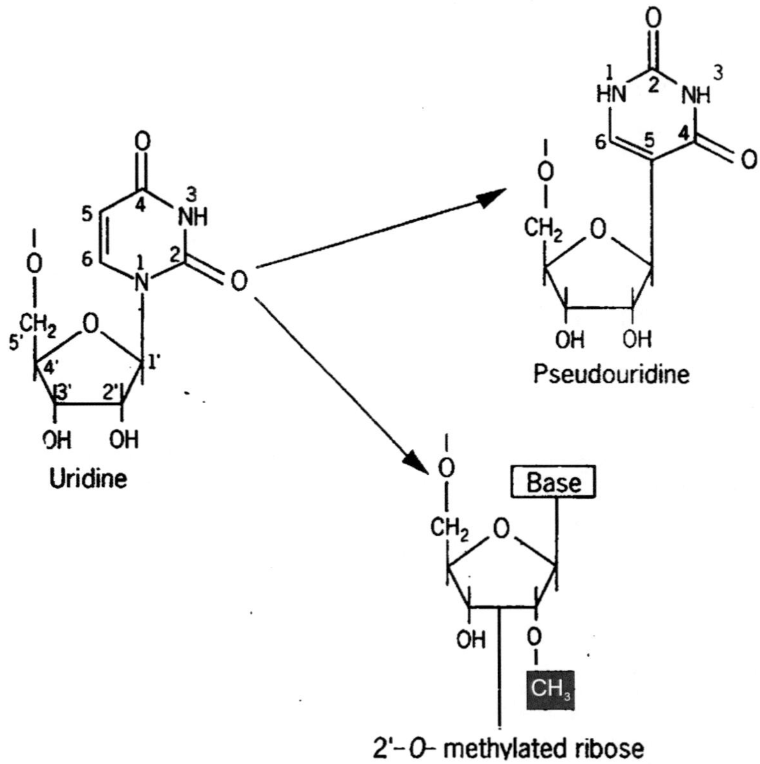

Fig. 14 : Modification of pre-rRNA. The most frequent modifications to nucleotides in a pre-rRNA are the conversion of a uridine to a pseudouridine and the methylation of a ribose at the 2' site of the sugar.

Two of the peculiarities of the pre-RNA, as compared with other RNA transcript are the large number of methylated nucleotides and pseudouridine residues. By the time the human pre-rRNA precursor is first cleaved, over 100 methyl groups have been added to ribose groups in the molecule, and approximately 95 of its uridine residues have been chemically converted to pseudouridine (Fig. 14). All of these modifications occur after the nucleotides are incorporated into the nascent RNA, i.e., *posttranscriptionally*.

The altered nucleotides are located at specific positions and are clustered within portions of the molecules. All of the nucleotides in the pre-rRNA that are altered remain as part of the final products, while the unaltered sections are discarded during processing. The function of the methyl groups and pseudouridines is unclear. It is speculated that these modified nucleotides may protect parts of the pre-rRNA from enzymatic cleavage, promote the folding of the ribosomal RNAs into their final three-dimensional structures, and/or promote interactions of the rRNAs with other molecules.

Because rRNAs are so heavily methylated, their synthesis can be followed by incubating cells with radioactively labeled methionine, a compound that serves in most cells as a methyl group donor. The methyl group is transferred enzymatically from methionine to nucleotides in the pre-rRNA. When [^{14}C]-methionine is provided to cultured mammalian cells for a short period of time, a considerable fraction of the incorporated radioactivity is present in a 45S RNA molecule, which corresponds to a length of about 13,000 nucleotides. The 45S RNA is cleaved into smaller molecules that are subsequently trimmed down to the 28S, 18S and 5.8S rRNA molecules. The combined length of the three mature rRNAs is approximately 7,000 nucleotides, or a little more than half of the primary transcript.

4.2.3 Synthesis and processing of the 5S rRNA.

A 5S rRNA, about 120 nucleotides long, is present as part of the large ribosomal subunit of both prokaryotes and eukaryotes. The 5S rRNA molecules are encoded by a large number of identical genes that are separate from the other rRNA genes and are located outside the nucleolus. The 5S rRNA genes are transcribed by RNA polymerase III. The 5' end of the primary transcript is identical with that of the mature 5S rRNA, but the 3' end usually contains extra nucleotides that are removed during processing. Following synthesis, the 5S rRNA is transported to the nucleolus to join the other components involved in the assembly of ribosomal subunits.

RNA polymerase III is unusual among the three polymerases in that it binds to a promoter site located within the transcribed portion of the gene rather a site upstream from the gene's 5' flank.

4.2.4 Transfer RNAs

Plant and animal cells are estimated to have approximately 50 different species of transfer RNA, each encoded by a DNA sequence that is repeated a number of times within the genome. The degree of repetition varies with the organism. Transfer RNAs are synthesized from genes that are found in small clusters scattered around the genome. A single cluster typically contains multiple copies of different tRNA genes, and conversely, the DNA sequence encoding a given tRNA is typically found in more than one cluster. The DNA within a cluster (tDNA) consists largely of non-transcribed spacer sequences with the tRNA coding sequences situated at irregular intervals in a tandemly repeated arrangement (Fig.15).

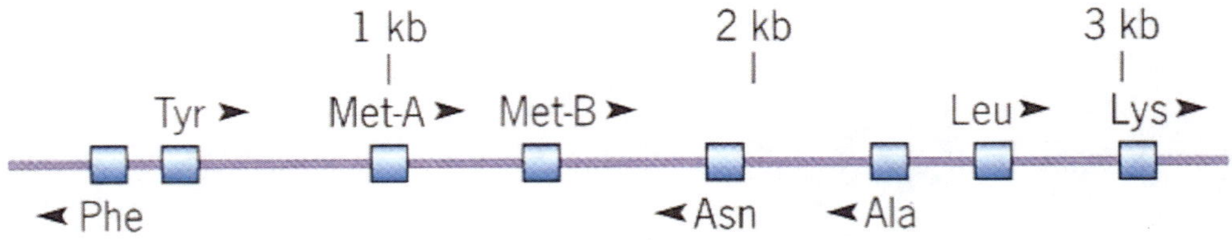

Fig. 15 : The arrangement of genes that code for tRNAs in *Xenopus*.

Like the 5S rRNA, tRNAs are transcribed by RNA Polymerase III (RNA Pol III), and the promoter sequences lies within the coding section of the gene rather than being located at its 5' flank. RNA Pol III has 16 or more subunits. The enzyme is located in the nucleoplsm and apart from synthesizing the precursors of 5S rRNA and tRNAs, it also synthesizes other small nuclear and cytosolic RNAs. The primary/initial transcripts produced from tRNA genes are precursor molecules which are processed into mature RNAs The primary transcript of a tRNA molecule is larger than the final product, and pieces on both the 5' and 3' sides of the precursor tRNA (and an interior piece in some cases) must be trimmed away. One of the enzymes involved in pre tRNA processing is an endonuclease called P, which is present in both bacterial and eukaryotic cells and consists of both RNA and protein subunits. It is the RNA subunit of ribonuclease P that catalyzes the cleavage of the pre-tRNA substrate. All mature tRNAs have the triplet sequence CCA at their 3' end. These three nucleotides are added enzymatically after the tRNA is processed, and they play a key role in protein synthesis. The transcription control regions of tRNA genes lie after the transcription start site within the transcription unit. There are two highly conserved sequences within the DNA encoding the tRNA, called the A box (5' -TGGCNNAGTGG-3') and the B box (5' -GGTTCGANNCC-3'). These sequences also encode important sequences in the tRNA itself, called the D-loop and the TψC loop (Fig 16). This means that highly conserved sequences within the tRNAs are also highly conserved promoter DNA sequences.

Two complex DNA-binding factors have been identified which are required for tRNA transcription initiation by RNA Pol III (Fig. 17). **TFIIIC** binds to both the A and B boxes in the tRNA promoter. **TFIIIB** binds 50bp upstream from the A box. TFIIIB consists of three subunits, one of which is TBP, the general initiation factor required by all three RNA polymerases. The second is called BRF (TFIIB-related factor, since it has homology to TFIIB, the RNA Pol II initiation factor). The third subunit is called B" 2 . The TFIIIB has no sequence specificity and therefore its binding site appears to be determined by the position of TFIIIC binding to the DNA. TFIIIB allows RNA Pol III to bind and initiate transcription. Once TFIIIB has bound, the TFIIIC can be removed without affecting transcription. TFIIIC is therefore an assembly factor for the positioning of the initiation factor TFIIIB.

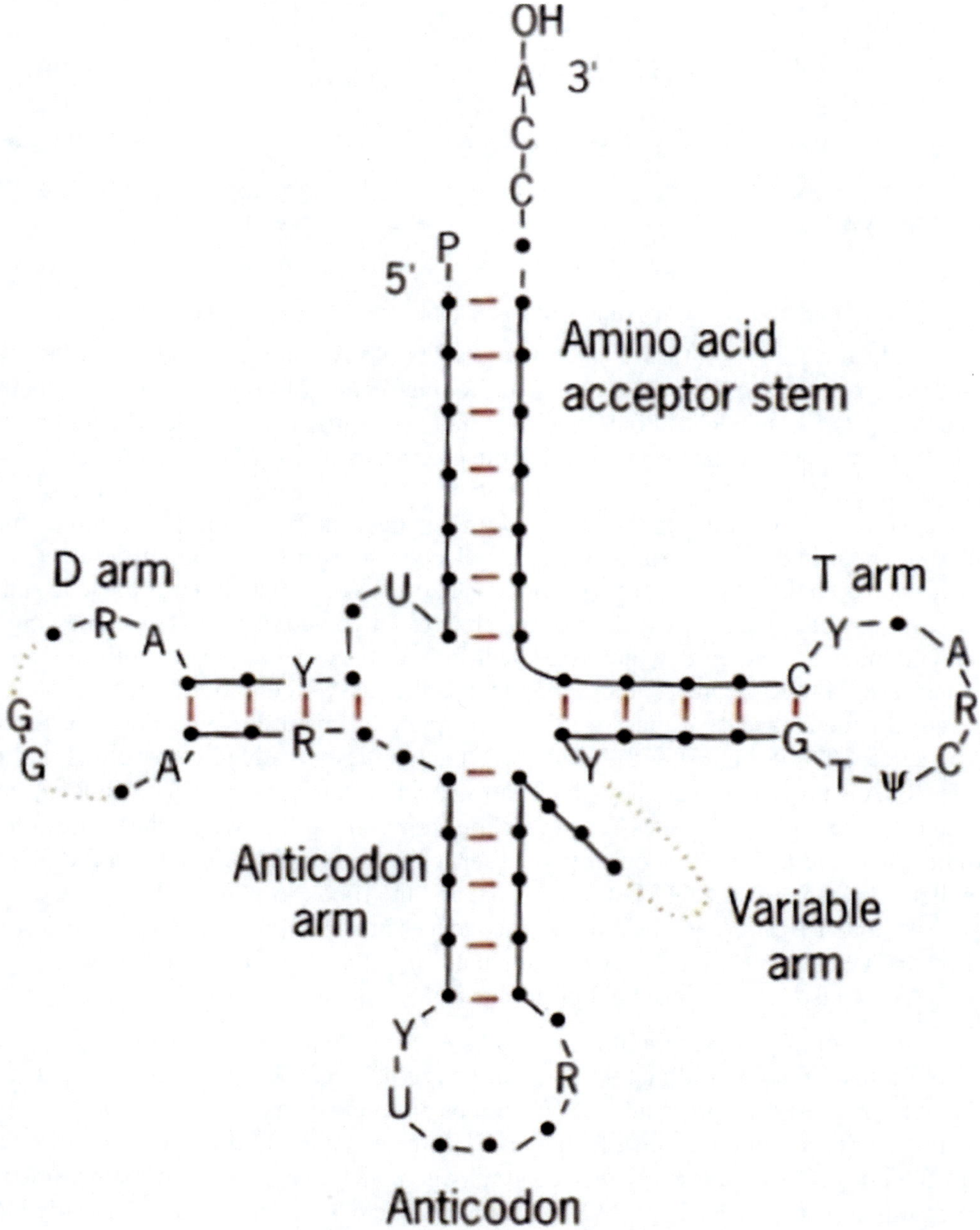

Fig. 16 : Two dimensional structure of tRNAs.

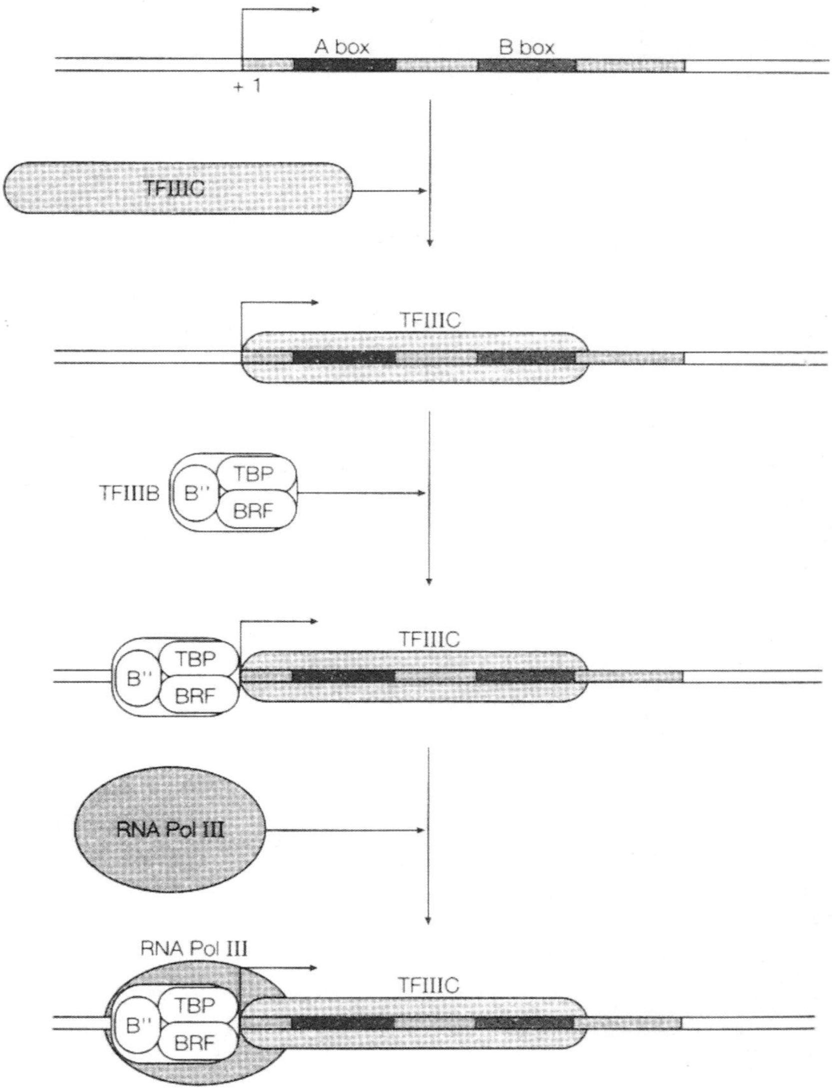

Fig. 17 : Initiation of transcription at a eukaryotic tRNA promoter.

A number of RNA Pol III promoters are regulated by upstream as well as downstream promoter sequences. Other promoters require only upstream sequences, including the TATA box and other sequences found in RNA Pol II promoters. Termination of transcription by RNA Pol III appears only to require polymerase recognition of a simple nucleotide sequence. This consists of a cluster of dA residues whose termination efficiency is affected by surrounding sequence. Thus the sequence 5' -GCAAAAGC-3' is an efficient termination signal in the *Xenopus borealis* somatic 5S rRNA gene.

In prokaryotes, mature tRNAs are generated by processing longer pre-tRNA transcripts, which involves specific exo- and endonucleolytic cleavages by RNases D, E, F and P followed by base modifications which are unique to each particular tRNA type. Following the initial 32 -cleavage by RNase E or F, RNase D can trim the 32 -end to within 2 nt of mature length. RNase P can then cut to give the mature 52 -end. RNase D finally removes two 32 -residues, and base modifications takes place.

Many eukaryotic pre-tRNAs are synthesized with an intron as well as extra 5' - and 3' - nucleotides which are all removed during processing. In contrast to prokaryotic tRNA, the 3' terminal CCA is added by the enzyme tRNA nucleotidyl transferase. Many base modifications also occur. Being composed (in *E. coli*) of a 377 nt RNA and a single 13.7 kDa protein, RNase P is a simple RNP. In both prokaryotes and eukaryotes, its function is to cleave 5' -leader sequences off pre-tRNAs. The RNA component alone can cleave pre-tRNAs *in vitro* and hence is a catalytic RNA, or ribozyme. Several biochemical reactions can be carried out by RNA enzymes, or ribozymes. These catalytic RNAs can cleave themselves or other RNA molecules, or perform ligation or self-splicing reactions. They can work alone, but are often complexed with protein(s) *in vivo*, which enhance their catalytic activity.

4.2.5 Messenger RNAs

When eukaryotic cells are incubated for a short period (1 to 5 minutes) in [^{3}H]uridinne or [^{32}P]phosphate and immediately killed (Fig. 18a), most of the radioactivity is incorporated into a large group of RNA molecules that have the following properties: (1) they have large molecular weights (up to about 80S, or 50,000 nucleotides); (2) as a group, they are represented by RNAs of diverse (heterogeneous) nucleotide sequence; and (3) they are found only in nucleus. Because of these properties, these RNAs are referred to as heterogeneous nuclear RNAs (hnRNAs, Fig. 18a). After a brief pulse when the above cells were placed into an unlabelled medium and chased for an hour so before they are killed, and the RNA extracted (Fig. 18b), the radioactivity in the large nuclear RNAs drops sharply and appears instead in much smaller mRNAs found in the cytoplasm. This suggests that the large, rapidly labeled hnRNAs were precursors to the smaller cytoplasmic mRNAs.

Most of the RNA in the cell is present as 18S and 28S RNA (along with tRNAs and snRNAs). Neither the hnRNA nor the mRNA constitutes a significant fraction of the RNA in the cell. Even though the mRNAs (and their heterogeneous nuclear RNA precursors) constitute only a small percentage of the total RNA of most eukaryotic cells, they constitute a large percentage of the RNA that is being synthesized by that cell at any given moment. These RNAs are degraded after relatively brief periods of time. This is particularly true for the hnRNAs, which have half lives of only few minutes. In contrast, the rRNAs and the tRNAs have half-lives that are measured in days or weeks and thus gradually accumulate to become the predominant species in the cell. The half lives of mRNAs vary depending on the particular species, ranging from about 15 minutes to a period of days.

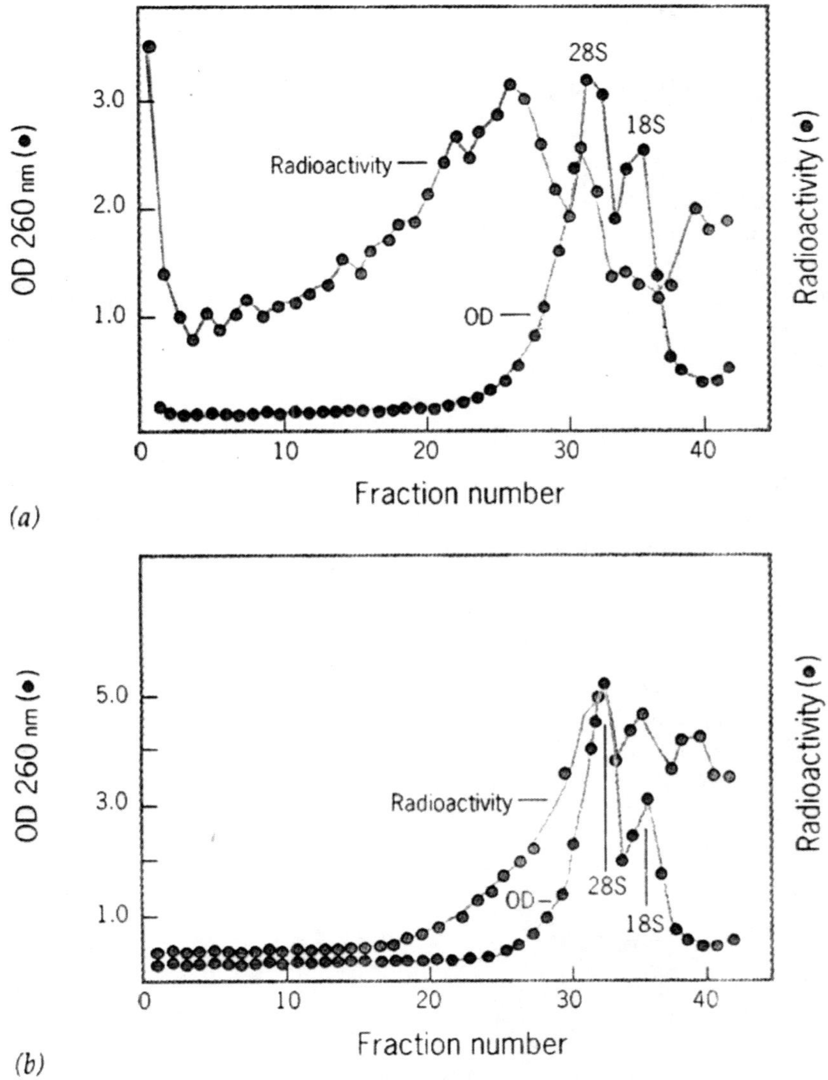

Figure 18 (a and b) : The formation of heterogeneous nuclear RNA (hnRNA) and its conversion into small mRNAs.

4.2.6 Machinery for mRNA transcription

All mRNA precursors are synthesized by RNA polymerase II in cooperation with a number of general transcription factors (GTFs). Promoters that bind RNA polymerase

II lie to the 5' side of each transcription unit. In the vast majority of the genes studied, a critical portion of the promoter (called the *core promoter* element) lies between 24 and 32 bases upstream from the site at which transcription is initiated (Fig. 19a). This region contains a consensus sequence that is either identical or very similar to the oligonucleotide 5' -TATAAA-3' and is known as *TATA box*. The TATA box of the DNA is the site of assembly of a **preinitiation complex** that contains the GTFs and the polymerase. The preinitiation complex must assemble before the transcription of the gene can be initiated.

There are two views regarding the assembly of the preinitiation complex. According to one view, the complex is assembled at the promoter piece by piece. Another view depicts that most of the GTFs are already associated with the polymerase at the time it binds to the promoter. In fact, both the mechanisms may operate simultaneously at different promoters. According to both views, the first step in assembly of the preinitiation complex is the binding of a protein, called, called *TATA-binding protein (TBP)* that specifically recognizes the TATA box of eukaryotic promoters. Thus, as in prokaryotic cells, a purified eukaryotic polymerase is not able to recognize a promoter directly and cannot initiate accurate transcription on its own.

The DNA-binding domain of TBP consists of a 10-stranded â sheet curved into a saddle-shaped structure that sits astride the DNA (Fig. 19b). TBP does not act by itself but is present as a subunit of the protein TFIID (transcription factor for polymerase II, protein D).

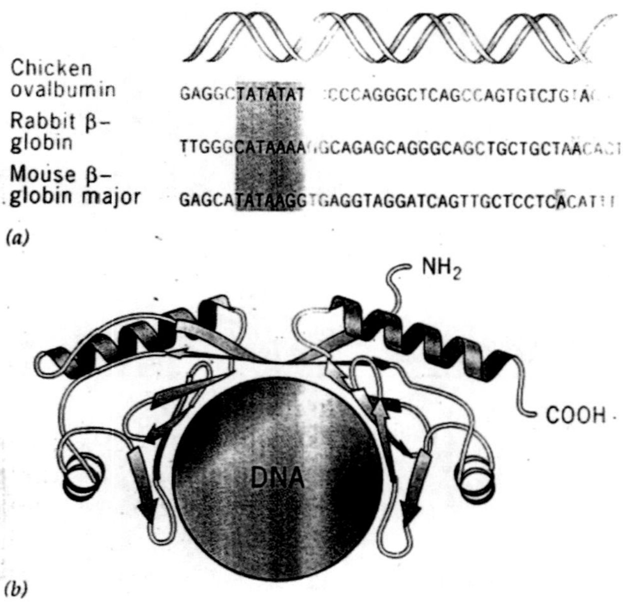

Figure 19 : (a) Nucleotide sequence of the region just upstream from the site where transcription is initiated in three different eukaryotic genes (**b**) Three-dimensional structure of the TATA-binding protein.

Binding of TFIID sets the stage for the assembly of the complete preinitiation complex, which may occur in a stepwise manner as depicted in Figure 20, or in a single step by the binding of a large, preassembled complex (or holoenzyme) that contains the polymerase and the remaining GTFs. Once transcription begins, some of the GTFs (TFIID and possibly TFIIB) may be left behind at the promoter, while others may remain associated with the polymerase as it proceeds down the template (Fig. 21). As long as TFIID remains bound to the promoter, additional RNA polymerase molecules may be able to attach to the promoter site and initiate additional rounds of transcription without delay.

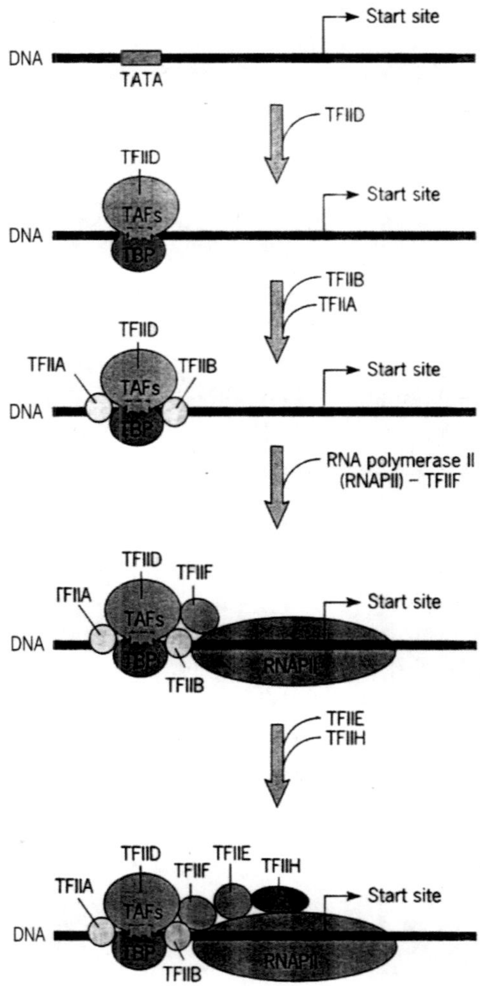

Fig. 20 : Step-wise assembly of the preinitiation complex for RNA Polymerase II at the TATA box of the eukaryotic promoter.

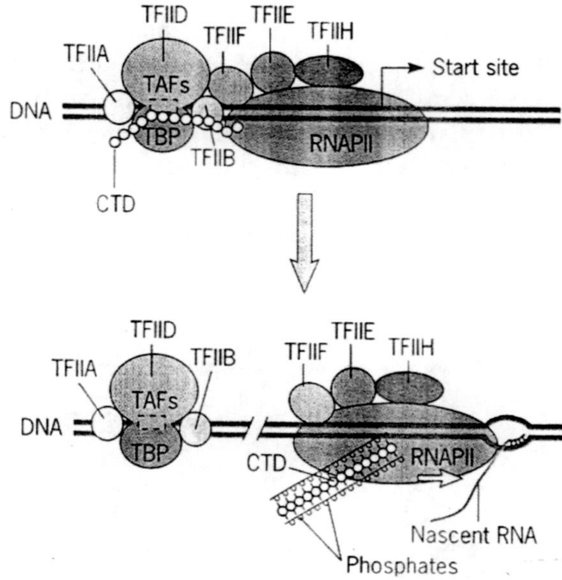

Fig. 21 : Initiation of transcription by RNA polymerase II is preceded by phosphorylation of the C-terminal domain (CTD).

The carboxyl-terminal domain (CTD) of the largest subunit of RNA Polymerase II has a very unusual structure; it consists of a sequence of seven amino acids (-Tyr-Ser-Pro-Thr-Ser-Pro-Ser-) that is repeated over and over. Of the seven amino acids, all except the two proline are prime candidates of phosphorylation by protein kinases. RNA polymerases that assembles into the preinitiation complex is not phosphorylated, whereas the same enzyme engaged in transcription is heavily phosphorylated; all of the phosphate groups are localized in the CTD (Fig. 21). RNA polymerase II is phosphorylated just prior to initiating transcription. Phosphoryaation of the polymerase is catalyzed by a subunit of one of the general transcription factors (TFIIH), which is also protein kinase. Phosphorylation of the polymerase may act as a trigger that uncouples the enzyme from transcription factors that remain bound to the TATA box, allowing the enzyme to move down the DNA template. It is not the transcriptional machinery that moves along the template, but the template that moves through an immobilized transcriptional machine.

4.2.7 Structure of mRNAs

mRNAs contain a continuous sequence of nucleotides encoding a specific polypeptide. They are found in cytoplasm and are attached to ribosomes when they are translated. Most mRNAs contain a significant noncoding segment, that is, a portion that does not direct the assembly of amino acids. For example, approximately 25 percent of each globin mRNA consists of noncoding, nontranslated regions, noncoding portions are found on both the 5' and 3' ends of a messenger RNA and contain sequences that have important regulatory roles. Eukaryotic mRNAs have special modifications at their 5' and 3' termini that are not found on prokaryotic mRNAs nor on tRNAs and rRNAs. The 5' end of eukarytic mRNAs has a methylated guanosine cap, while the 3' end has a string of 50 to 250 adenosine residues that form a poly(A) tail (Fig. 22).

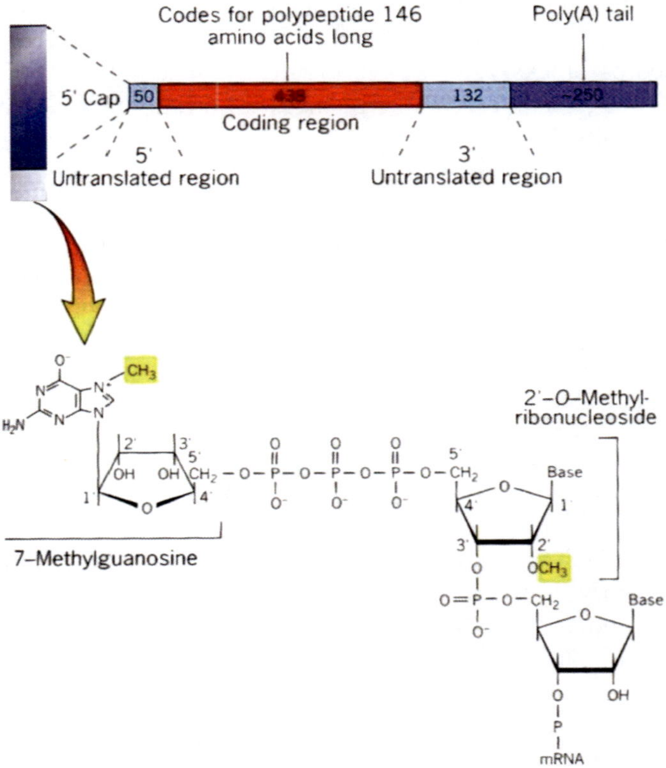

Fig. 22 : Structure of the human α-globin mRNA.

4.2.8 Formation of mRNAs

hnRNAs were proposed to be the precursors for cytoplasmic mRNAs. However, there was huge difference in the size of the two RNA populations, hnRNA being several times larger than the mRNA. Why would cell synthesize huge molecules that were precursors to much smaller versions? Until 1977, it was assumed by molecular biologists that a continuous sequence of nucleotides in an mRNA is complementary to a continuous sequence of nucleoides one strand of the DNA of a gene. Later it was found that the mRNAs were transcribed from segments of DNA that were separated from one another along the template strand. The first observations of major importance were made during the analysis of transcription of the adenovirus genome. Adenovirus is a pathogen capable of infecting a variety of mammalian cells. It was found that a number of different adenovirus mRNAs had the same 150- to 200- nucleotide 5' terminus. One might expect that this leader sequence represents a repeated stretch of nucleotides located near the promoter region of each of the genes for these mRNAs. However, further analysis revealed that the 5' leader sequence is not complementary to a repeated sequence and, moreover, is not even complementary to a continuous stretch of nucleotides in the template DNA. Instead, the leader is transcribed from three distinct and separate segments of DNA (represented by blocks x, y and z in Figure 23). The regions of DNA between these blocks, referred to as intervening sequences (I_1 to I_3 in Figure 23) are missing in the corresponding RNA. Intervening sequences were soon found in other genes, and it became apparent that the presence of genes with intervening sequences-called **split genes**-is more the rule than exception. Those parts of a split gene that contribute to the mature RNA product are called **exons**, whereas the intervening sequences are called **introns**. Split genes are widespread among eukaryotes. The introns of simpler eukaryotes, however, tend to be fewer in number and smaller in size than those of plants and animals. Introns are found in all types of genes, including those that code for tRNAs, rRNAs and mRNAs.

The discovery of genes with introns immediately raised the question as to how such genes were able to produce mRNAs lacking these sequences. One likely possibility was that cells produce a primary transcript that corresponds to the entire transcription unit, and that those portions of the RNA corresponding to the introns in the DNA are somehow removed, which suggests that the segments corresponding to the introns should be present in the primary transcript. It would also provide a reason why hnRNA molecules are so much larger than the mRNA molecules that they ultimate produce. The mRNA formation in eukaryotic cells occurs by the removal of internal sequences of ribonucleotides from a much larger pre-mRNA.

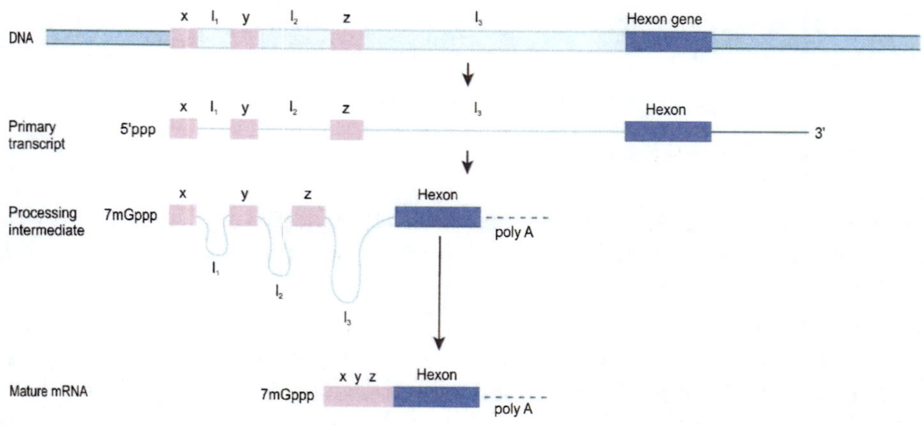

Fig. 23 : Intervening sequences (introns).

4.2.9 The processing of Eukaryotic messenger RNAs

There appears to be little or no processing of mRNA transcripts in prokaryotes. In fact, ribosomes can assemble on, and begin to translate mRNA molecules that have not yet been completely synthesized. Prokaryotic mRNA is degraded rapidly from the 5' - end and the first cistron (protein coding region) can therefore only be translated for a limited amount of time. Some internal cistrons are partially protected by stem-loop structures that form at the 5' - and 3' - ends and provide a temporary barrier to exonucleases and can thus be translated more often before they are eventually degraded.

Because eukaryotic RNA Pol II transcribes such a wide variety of different genes, the collection of products made by this enzyme is referred to as **heterogeneous nuclear RNA (hnRNA)**. Those transcripts that will be processed to give mRNAs are called **pre-mRNAs**. Pre-mRNA molecules are processed to mature mRNA by 5' -capping, 3' -cleavage and polyadenylation, splicing and methylation.

The hnRNA synthesized by RNA Pol II is mainly pre-mRNA that rapidly becomes covered by proteins to form **heterogeneous nuclearribonucleoprotein (hnRNP)**. RNA Pol II transcripts (hnRNA) complex with the three most abundant hnRNP proteins, the A, B and C proteins, to form hnRNP particles. These contain three copies of three tetramers and around 600-700 nucleotides of hnRNA. They assist in RNA processing events.

The first steps in the process of conversion of pre-mRNA to mRNA include the addition of a 5' cap and a 3' poly(A) tail.

5' capping : Very soon after RNA Pol II starts making a transcript, and before the RNA chain is more than 20-30 nt long, the 5' end is chemically modified by the addition of a 7-methylguanosine (m^7G) residue (Fig. 24). This 5' modification is called a **cap** and occurs by addition of a GMP nucleotide to the new RNA transcript in the reverse orientation compared with the normal 3' -5' linkage, giving a 5' -5' triphosphate bridge. The reaction is carried out by an enzyme called mRNA guanyltransferase and there can be subsequent methylations of the sugars on the first and second transcribed nucleotides, particularly in vertebrates. The cap structure forma a barrier to 5' -exonucleases and thus stabilizes the transcript, but the cap is also important in other reactions undergone by pre-mRNA and mRNA, such as splicing, nuclear transport and tramslation.

3' Cleavage and polyadenylation : In most pre-mRNAs, the mature 3' -end of the molecule is generated by cleavage followed by addition of a run, or tail, of A residues which is called the **poly(A) tail**. This feature has allowed the purification of mRNA moleculesfrom the other types of cellular RNAs, permitting the construction of cDNA libraries.

The cleavage and polyadenylation reaction requires that specific sequences be present in the DNA and its pre-mRNA transcript. These consists of 5' -AAUAAA-3' , the **polyadenylation signal**, followed by a 5' -YA-3' , where Y = pyrimidine, in the next

11-20 nt (Fig. 25). Downstream a GU rich sequence is often present. Collectively, these sequence elements make up the requirements of a **polyadenylation site**.

A number of specific protein factors recognize these sequence elements and bind to the pre-mRNA. When the complex has assembled, cleavage takes place and then one of the factors, poly(A) polymerase (PAP), adds up to 250 A residues to the 3' - end of the cleaved pre-mRNA. The poly(A) tail on pre-mRNA is thought to help stabilize the molecule since a poly(A)-binding protein binds to it which should act to resist 3' - exonuclease action. In addition, the poly(A) tail may help in the translation of the mature mRNA in the cytoplasm. Histone pre-mRNAs do not get polyadenylated, but are cleaved at a special sequence to generate their mature 3' - ends.

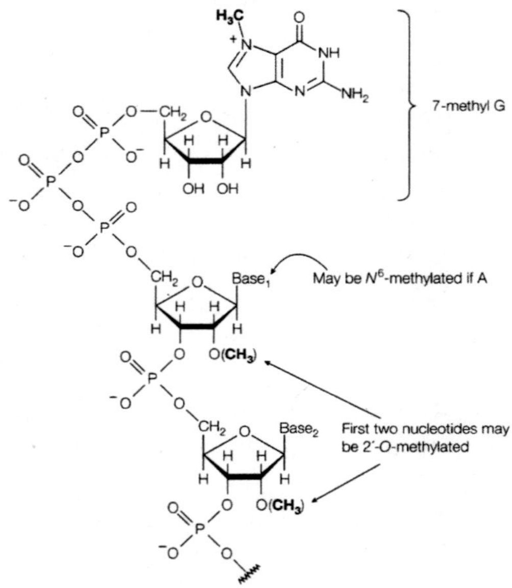

Fig. 24 : The 5' cap structure of eukaryotic mRNA.

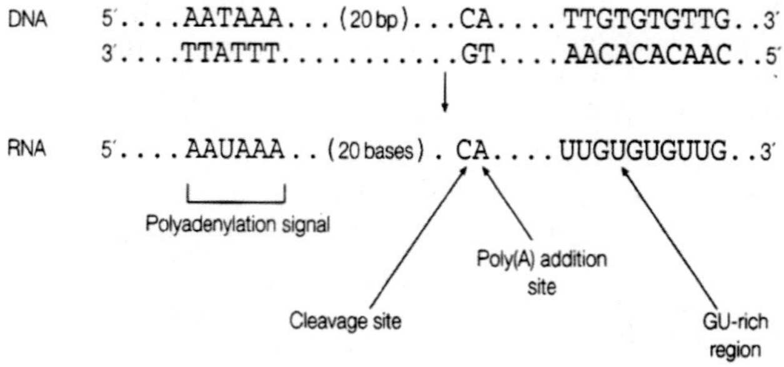

Fig. 25 : Sequence of typical polyadenylation site.

4.2.10 Splicing the primary transcript: The removal of introns

Most protein-coding genes in eukaryotes are interrupted by intervening sequences (introns) which are spliced from the precursor mRNA to generate mature mRNA by cis splicing (removal of introns from colinear transcripts; Green, 1986; Padgett et al., 1986; Sharp, 1987). In kinetoplastida and several nematode species, the exons of a single mRNA can be encoded by different genes located on separate chromosomes.

In addition to the formation of 5' cap and poly(A) tail, parts of primary transcript that corresponds to the intervening DNA sequences (the introns) must be removed by a complex process known as **RNA splicing** (Haung and Ploeg, 1991; Buratti and Baralle 2004).

To splice an RNA, breaks in the strand must be introduced at the 5' and 3' edges (or **splice sites**) of each intron, and the exons situated on either side of the splice sites must be covalently joined (ligated). It is imperative that the splicing process occur with absolute precision, because the presence or absence of a single abnormal nucleotide at any of the splice junctions would cause the resulting mRNA to be mistranslated. In eukaryotes, the splicing machinery recognizes the exon-intron boundaries by the presence in the splice sites (at junctions between exons and introns) of a highly conserved nucleotide sequence of ancient evolutionary origin. The sequence most commonly found at the exon-intron borders within hnRNA molecules is shown in Figure 26. The G/GU at the 5' end of the intron (*the 5' splice site*) and the AG/G at the 3' end of the intron (*the 3' splice site*) are present in the vast majority of the eukaryotic pre-mRNAs. The AG at the 3' -end is preceded by a pyrimidine rich sequence called the **polypyrimidine tract**. About 10-40 residues upstream of the polypyrimidine tract is the sequence called the **branchpoint sequence** which is 5' -CURAY-3' in vertebrates, where R = purine and Y = pyrimidine, but in yeast it is the more specific sequence 5' -UACUAAC-3' (Fig. 27). A single base substitution in the DNA sequence within these invariant sequences blocks the excision of the intron, leading to the accumulation of unspliced precursors. In most cases, unspliced RNAs are recognized and degraded so that they are not shipped to the cytoplasm and translated into nonfunctional polypeptides.

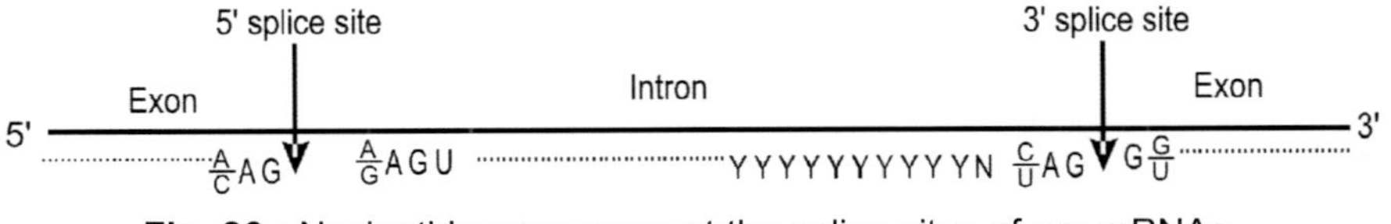

Fig. 26 : Nucleotide sequences at the splice sites of pre mRNAs.

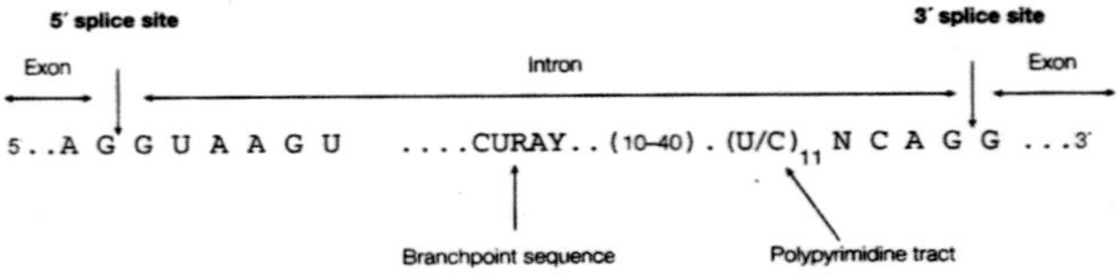

Fig. 27 : Sequences of the splice site consensus.

Splicing has been shown to take place in **two-step process** (Fig 28a). first, the bond in front of the G at the 5' end of the intron at the so called **5' -splice site** is attacked by the 22 -hydroxyl group of the A residue of the branchpoint sequence to create a tailed circular molecule called a **lariat** and free exon I. In the second step, cleavage at the **3' splice site** occurs after the G of the AG. As the two exon sequences are joined together. The intron is released in the lariat form and is eventually degraded.

The splicing process is catalyzed by the U1, U2, U4, U5 and U6 **snRNPs**, as well as other splicing factors. The RNA components of these snRNPs form base pairs with the various conserved sequences at the 5' - and 3' -splice sites and the branchpoint (Fig 28b). Early in splicing, the 5' -end of the U1 snRNP binds to the 5' -splice cite and then U2 binds to the branchpoint. The tri-snRNP complex of U4, U5 and U6 can then bind, and in doing so the intron is looped out and the 5' - and 3' -exons are brought into close proximity. The snRNPs interact with one another forming a complex which folds the pre-mRNA into the correct conformation for splicing. This complex of snRNPs and pre-mRNA which forms to hold the upstream and downstream exons close together while looping out the intron is called a **spliceosome**. After the spliceosome forms, a rearrangement takes place before the two-step splicing reaction can occur with release of the intron as a lariat.

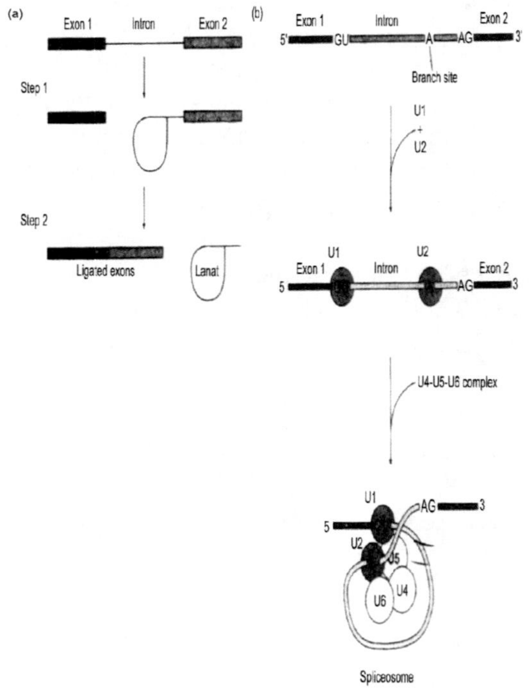

Fig. 28 : Splicing of eukaryotic pre-mRNA. (a) The two step reaction; (b) involvement of snRNPs in splicosome.

4.2.11 Pre-mRNA methylation

The final modification or processing event that many pre-mRNAs undergo is specific methylation of certain bases. In vertebrates, the most common methylation event is on the N6 position A residues, particularly when these A residues occur in the sequence 5' -RRACX-3' , where X is rarely G. Up to 0.1% of pre-mRNA A residues are methylated, and the methylations seem to be largely conserved in the mature mRNA, though their function is unknown.

5. Genetic Code

5.1 Encoding genetic information

Once the structure of DNA had been revealed, it became evident that the sequence of amino acids in a polypeptide was specified by the sequence of nucleotides in the DNA of a gene. It seemed very unlikely that the DNA could serve as a direct, physical template for the assembly of a protein. Instead, it was widely assumed that information stored in the nucleotide sequence was present in some type of **genetic code**. With the discovery of mRNA as an intermediate in the flow of information from DNA to protein, attention turned to the manner in which a sequence written in a ribonucleotide 'alphabet' would code for a sequence in an 'alphabet' consisting of amino acids.

5.2 The Properties of Genetic Code

The genetic code is the correspondence between the sequence of the four bases in the nucleic acids and the sequence of the 20 amino acids in proteins. The code was reported to be triplet, where three nucleotides encode one amino acid, and this agrees with mathematical argument as being the minimum necessary $[(4^2 = 16) < 20 < (4^3 = 64)]$. However, since there are only 20 amino acids to be specified and potentially 64 different triplets, most amino acids are specified by more than one triplet and hence the genetic code is said to be **degenerate**, or to have redundancy. From a fixed start point, each group of three bases in the coding region of the mRNA represents a **codon**, which is recognized by a complimentary triplet, or **anticodon**, on the end of a particular tRNA molecule. The triplets are read in **nonoverlapping** groups and there is no punctuation between the codons to separate or delineate them. They are simply decoded as adjacent triplets once the process of decoding has begun at the correct start point (initiation site). As more gene and protein sequence information has been obtained, it has become clear that the genetic code is very nearly, but not quite universal.

6. Deciphering the Code

By 1961, the general properties of the code were known, but the coding assignment of the specific triplets was not explored out. Marshall Nirenberg and Heinrich Matthaei developed a technique to synthesize their own artificial genetic messages and then determine what kind of a protein they encoded. The first message they tested was a polyribonucleotide consisting exclusively of uridine; the message was called *poly(U)*. When *poly(U)* was added to a test tube containing a bacterial extract with all 20 amino acids plus the materials necessary for protein synthesis (ribosomes and various soluble factors), the system followed the artificial messenger's instructions and manufactured a polypeptide. The assembled polypeptide was analyzed and found to be polyphenylalanine-a polymer of the amino acid phenylalanine. Nirenberg and Matthaei had thus shown that the codon UUU specifies phenylalanine.

Later, the synthetic mRNAs were constructed to test the amino acid specifications for all 64 possible codons. The result was the universal decoder chart for the genetic code shown in Figure 29. The decoder chart lists the nucleotide sequence for each of the 64 possible codons in an mRNA. Instructions for reading the chart are provided in the accompanying legend. The primary exceptions to the universality of the genetic code occur in the codons of mitochondrial mRNAs. For example, in human mitochondria, UGA is read as tryptophan rather than stop. AUA is read as methionine rather than isoleucine, and AGA and AGG are read as stop rather than arginine.even with these discrepancies, the similarities of the mitochondria and nuclear genetic codes far outweigh their differences, and it is believed that one is derived from the other and that all organisms present on Earth today share a common evolutionary origin.

The chart indicates that the amino acids assignments are distinctly non-random. It shows that the codons for a specific amino acid tend to be clustered within a particular portion of the chart. Clustering reflects the similarity in codons that specify the same amino acid. As a result of this similarity in codon sequence, spontaneous mutations causing single base changes in a gene often will not produce a change in the amino acid sequence of the corresponding protein. Codon assignments are such that the similar amino acids tend to be specified by similar codons. For example, the codons of various hydrophobic amino acids are clustered in the first two columns of the chart. Consequently, a mutation that results in the base substitution in one of these codons is likely to substitute one hydrophobic residue for another. In addition, the greatest similarities between amino acid-related codons occur in the first two nucleotides of the triplet, whereas the greatest variability occurs in the third nucleotide. For example, Glycine is encoded by four codons, all of which begin with the nucleotides GG. The explanation of this phenomenon will be reveled in the section of tRNAs.

6.1 tRNA - Primary Structure

tRNAs are the adapter molecules that deliver amino acids to the ribosome and decode the information in mRNA. Their primary structure (i.e. the linear sequence of nucleotides) is 60-95 nucleotides long, but most commonly 76. They have many modified bases sometimes accounting for the 20% of the total bases in any one tRNA molecule.

Four of these, ribothymidine (T), which contains the base thymine not usually found in RNA, pseudouridine (ψ), dihydrouridine (D) and inosine (I), are very common in tRNA, all but the last being present in all tRNA molecules in similar positions in the sequences.

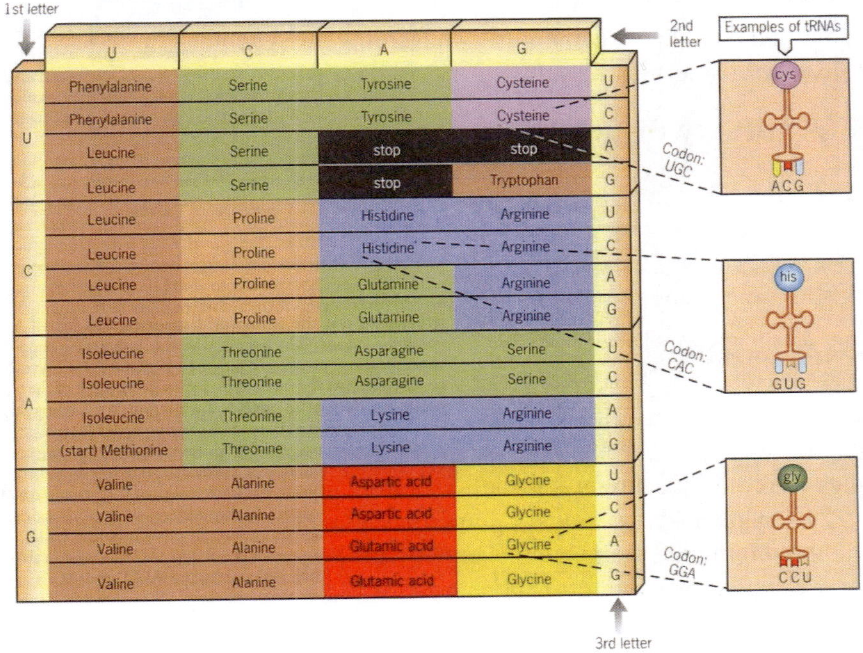

Fig. 29 : The universal decoder chart representing the genetic code listing each of the 64 possible mRNA codons and the corresponding amino acid specified by that codon.

6.2 tRNA - Secondary Structure

All tRNAs have a common secondary structure (i.e. base pairing of different regions to form stems and loops), the **cloverleaf** structure shown in Figure 30a. This structure has a 5' -phosphate formed by RNaseP cleavage, not the usual 5' triphosphate. It has a 7 bp stem formed by base pairing between the 5' - and 3' -ends of the tRNA; however, the invariant residues 74-76 (i.e. the terminal 5' -CCA-3') which are added during processing in eukaryotes are not included in this base pairing region. This stem is called the **amino acid acceptor stem**. Another secondary structural feature is called the D-arm which is composed of a 3 or 4 bp stem and a loop called the **D-loop** (DHU-loop) usually containing the modified base dihydrouracil. The next structural feature consists of a 5 bp stem and a seven residue loop in which there are three adjacent nucleotides called the **anticodon** which are complementary to the codon sequence (a triplet sequence in mRNA) that the tRNA recognizes. The presence of inosine in the anticodon gives a tRNA the ability to base-pair with more than one codon sequence. Next there is a **variable arm** which can have between 3 and 21 residues and may form a stem of up to 7 bp. The other positions of length variation in tRNAs are in the D-loop shown as dashed lines in Figure 30a. The final major feature of the secondary structure is the **T-arm** or TψC arm which is composed of a 5 bp stem encoding in a loop containing the invariant residues GTψC. Majority of the invariant residues in tRNA molecules are in the loops and do not play a major role in forming the secondary structure. Several of them help to form a tertiary structure.

6.3 tRNA Tertiary Structure

There are 9 hydrogen bonds that help to the 3-D structure of tRNA molecules. They mainly involve base pairing between several invariant bases (Fig. 30b). Base pairing between residues in the D-and T-arms fold the tRNA molecules into an **L-shape**, with the anticodon at one end and the amino acid acceptor site at the other. (Fig. 30c).

6.4 tRNA Function

tRNAs are joined to amino acids to become **aminoacyl-tRNAs** (charged tRNAs) in a reaction called aminoacylation. It is these charged tRNAs that are the adaptor molecules in protein synthesis. Special enzymes called aminoacyl-tRNA synthetases (Seryl-tRNA synthetase for serine, Leucyl-tRNA synthetase for leucine and likewise) carry out the joining reaction which is extremely specific (i.e. a specific amino acid is joined to a specific tRNA to form cognate pairs).

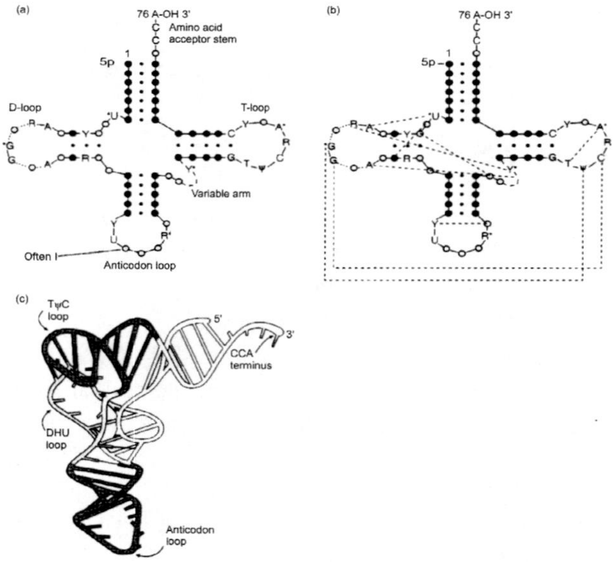

Fig. 30 : tRNA structure. (a) Cloverleaf structure showing the invariant and semi-variant nucleotides, where I = inosine, ψ = pseudouridine, R = purine, Y = pyrimidine and * indicates a modification. (b) Tertiary hydrogen bonds between the nucleotides in tRNA are shown as dashed lines. (c) The L-shaped tertiary structure of yeast $tRNA^{Tyr}$.

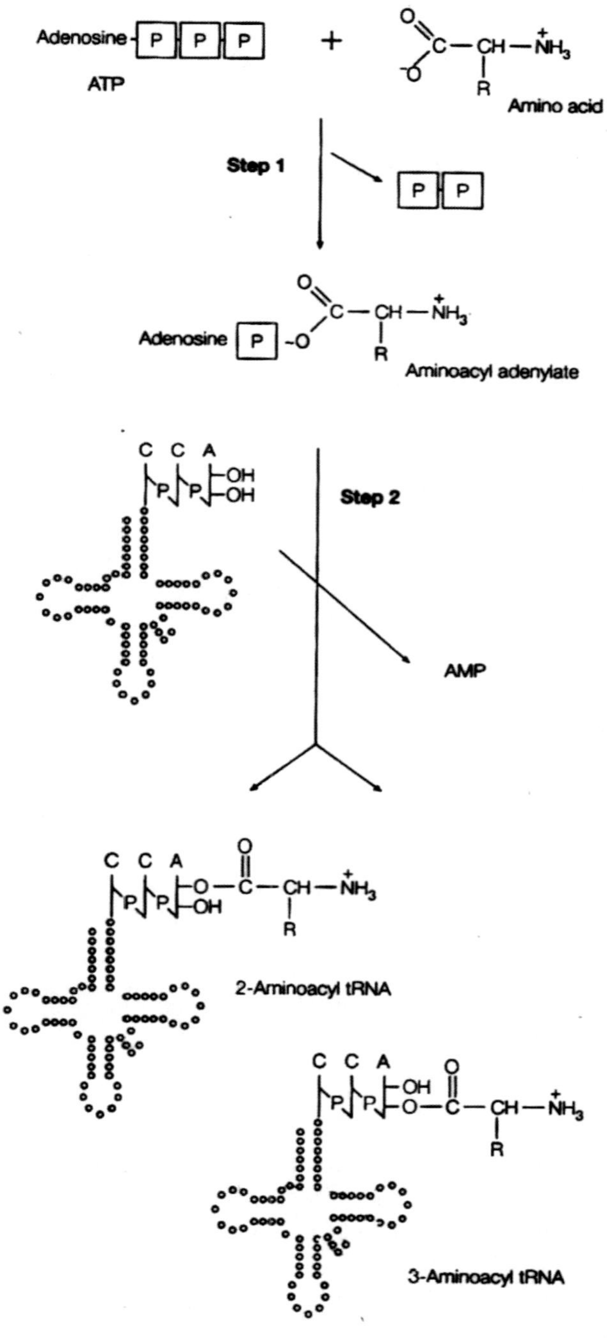

Fig. 31 : Formation of aminoacyl-tRNA.

6.5 Aminoacylation of tRNAs

It is a two-step process (Fig. 31). In the first step, AMP is linked to the carboxyl group of the amino acid giving a high-energy intermediate, called an **aminoacyl adenylate**. The hydrolysis of the pyrophosphate released (to two molecules of inorganic phosphate) drives the reaction forward. In the second step, the aminoacyl adenylate reacts with the appropriate uncharged tRNA to give the aminoacyl-tRNA and AMP. Some synthetases join the amino acid to the 2' - hydroxyl of the ribose and some to the 3' - hydroxyl, but once joined the two species can interconvert. The formation of an aminoacyl-tRNA helps to drive protein synthesis as the aminoacyl-tRNA bond is of higher energy than a peptide bond and thus peptide bond formation is a favorable rection once this energy-consuming step has been performed.

6.6 Proofreading

Some synthetase enzymes that have to distinguish between two chemically similar amino acids can carry out a proofreading step. If they carry out step 1 of the aminoacylation reaction with the wrong amino acid, then the step 2 will not be carried out by these enzymes. Instead they will hydrolyze the amino acid adenylate. This proofreading ability is only necessary when a single recognition step is not sufficiently discriminating. Discrimination between the amino acids Phe and Tyr can be achieved in one step because of the –OH group difference on the benzene ring, so in this case there is no need of proofreading.

7. Translation

Protein synthesis or translation is considered as the most complex synthetic activity in the cell. The assembly of a protein requires all the various tRNAs with their attached amino acids, ribosomes, mRNA, a number of proteins having different functions, cations and GTP.

The synthesis of a polypeptide chain can be distinguished into three activities: initiation, elongation and termination of the chain (Harvey, 2003).

7.1 Initiation

Once it attaches to an mRNA, a ribosome always moves along the mRNA from one codon to the next, i.e., in consecutive blocks of three nucleotides. To ensure that the proper triplets are read, the ribosome attaches to the mRNA at a precise site, termed the initiation codon and specified as AUG, which automatically puts the ribosome in the proper reading frame. The steps involved in the initiation of translation in prokaryotic cells are illustrated in Figure 32.

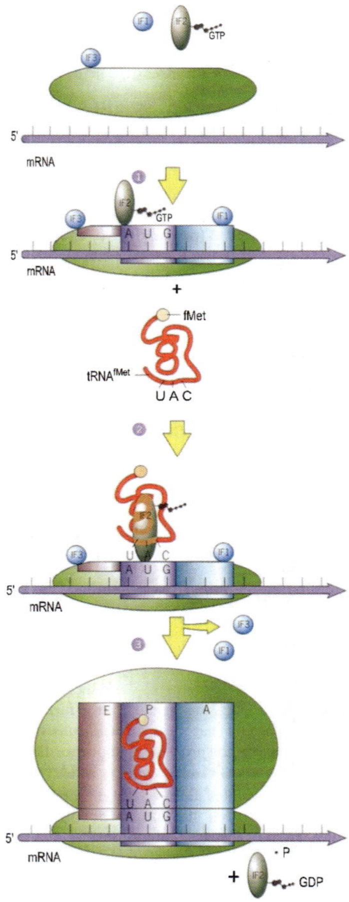

Fig. : 32: The initiation of Protein synthesis in prokaryotes.

Step 1: Bringing the small ribosomal subunit to the initiation codon: An mRNA does not bind to an intact ribosome, but to the small and large subunits in separate stages. The first major step of initiation is the binding of the small ribosomal subunit to the first AUG sequence in the message, which serves as the initiation codon. How does the small subunit select the initial AUG codon as opposed to some internal one? Bacterial mRNAs possess a specific sequence of nucleotides (called the Shine-Dalgarno sequence after its discoverers) that resides 5-10 nucleotides before the initiation codon. The Shine-Dalgarno sequence is complementary to the sequence of nucleotides near the 3' end of 16S rRNA of the small ribosomal subunit. Attachment of the 30S subunit at the AUG initiation codon results from the interaction between these complementary sequences on the mRNA and rRNA. In eukaryotes, the small ribosomal subunit first recognizes the 5' end of the message, which bears the methyl guanosine cap and then it scans along the mRNA until it reaches a recognizable sequence of nucleotides that contains AUG triplet.

Prokaryotic cells require three initiation factors, two of which (IF1 and IF3) are thought to bind to the 30S ribosomal subunit and facilitate its attachment to the mRNA (step 1, Fig. 32). eukaryotic cells require at least 10 initiation factors comprising a total of atleast 25 polypeptide chains. One of these factors, eIF4E, is thought to bind to the cap at the 5' end of the eukaryotic mRNA and mediate interaction between the message and the small ribosomal subunit. An associated factor (eIF4A) moves along the 5' end of the message removing any double stranded regions that would interfere with the movement of the small ribosomal subunit as it scans for the initiation codon (Fig. 32).

Step 2 : Bringing the first aa-tRNA into the ribosome. Methionine is always the first amino acid to be incorporated at the N-terminus of a nascent polypeptide chain. In prokaryotes, the initial methionine bears a formyl group, making it N-formylmethioine. Two types of methionyl-tRNAS are possessed by the cell – one used to initiate protein synthesis (designated $tRNA_i^{Met}$) and a different one (designated $tRNA^{Met}$) to incorporate methionyl residues into the anterior of a polypeptide. Before the association of initiator tRNA with the mRNA-30S subunit complex, it should come in contact with two components GTP and initiation factor 2 (IF2), finally leading to the formation of GTP-IF2-$tRNA_i^{Met}$. Release of IF3 is accompanied with the association of the initiator tRNA complex with the AUG codon of the mRNA.

Step 3 : Assembling the Complete Initiation Complex. Subsequent to the binding of the initiator tRNA to the AUG codon, the large subunits joins the complex. It is followed by the hydrolysis of GTP and release of the soluble factors IF1 and IF2. GTP hydrolysis is responsible for imparting the conformational change in the ribosome that is required for initiating translation.

7.2 The Role of Ribosomes

During translation, a ribosome undergoes a repetitive cycle of mechanical changes that is driven by energy released by GTP hydrolysis. The information stored in mRNA determines the sequence of aminoacyl-tRNAs the ribosome uses in translation. They also play an important role by recognizing various molecules (tRNAs and protein factors) and catalyzing a key reaction. There are three sites in each ribosome with which, it associates with tRNAs. The three sites viz. A (aminoacyl) site, the P (peptidyl) site and the E (exit) site, receive each tRNA in successive steps of elongation cycle.

7.3 Elongation

Step 1: Entry of Aminoacyl-tRNAs. The first step in elongation is the entry of the second amino-acyl-tRNA into the vacant A site, while the charged initiator tRNA is placed within the P site (Fig 33a, step 1). Association of a GTP-protein elongation factor (Tu in prokaryotes and eEF1 in eukaryotes) with the aminoacyl-tRNA is mandatory before it can effectively bind to mRNA in the A site. Tu is required to deliver the aminoacyl-tRNA to the ribosome. Although many aminoacyl-tRNA-Tu-GTP complex can enter the A site, only one anticodon is complimentary to the mRNA codon is situated situated in the A site is able to become trapped there by specifically binding to the mRNA. GTP hydrolysis takes place subsequent to the binding of the proper aminoacyl-tRNA-Tu-GTP complex to the mRNA codon thus leading to the release of Tu-GDP. GTP hydrolysis is responsible for causing the newly arrived aa-tRNA to undergo a pivoting movement that brings the bound amino acid into proper position within the A site of the 50S subunit. Another elongation factor, Ts is required for the regeneration of Tu-GTP from the released Tu-GDP.

Step 2: Peptide Bond formation. In this step, a peptide bond is formed between two amino acids that are juxtaposed in such a position that they can react with one another. (Fig 33, step 2). Peptide bond formation is accomplished as the amino group of the aa-tRNA in the A sites react with the carboxyl group of the aa-tRNA of the P site, displacing the P-site tRNAS (Fig 34). As a consequence of this reaction, the tRNA in the A site has a bound dipeptide, whereas the tRNA in the P site is deacylated.Peptide bond formation is catalyzed by peptidyl transferase, which is a component of the large subunit of ribosome.

Step 3 : Translocation. The formation of the first peptide bond leaves one end of the tRNA molecule of the A site still attached to its complimentary codon on the mRNA and the other end of the molecule attached to a dipeptide. The tRNA of a P site is now devoid of any linked amino acid. In the next step called translocation, a complex of EF-G (translocase) and GTP binds to the ribosome and, in an energy consuming step, the discharged tRNA is ejected from the P site, the peptidyl-tRNA is moved from the A-site to the P site and the mRNA moves by one codon relative to the ribosome. GDP and EF-G are released the latter being reusable. A new codon is now present in the vacant A-site. The elongation factor is eEF2 in eukaryotes. Elongation factors are responsible for converting the chemical energy released by GTP hydrolysis into mechanical energy required for the movement of the ribosome along the mRNA.

Step 4: Releasing the deacylated tRNA: Here the deacylated tRNA leaves the ribosome, thus vacating the E site.

In the entire elongation process, two molecules of GTP are hydrolyzed: one during aminoacyl-tRNA selection and one during translocation.

Once the peptidyl-tRNA has moved to the P site by translocation, the A site is once again open to the entry of another aminoacyl-tRNA, in this case one whose anticodon is complimentary to the third codon (Fig 33, step 4). Once the third charged tRNA is associated with the mRNA in the A site, the dipeptide from the tRNA of the P site is displaced by the aa-tRNA of the A site, forming the second peptide bond and, as a result, a tripeptide attached to the tRNA of the A site. The tRNA in the P site is once again devoid of an amino acid. Peptide bond formation is followed by translocation of the ribosome to the fourth codon and release of the deacylated tRNA, whereupon the cycle is ready to begin again.

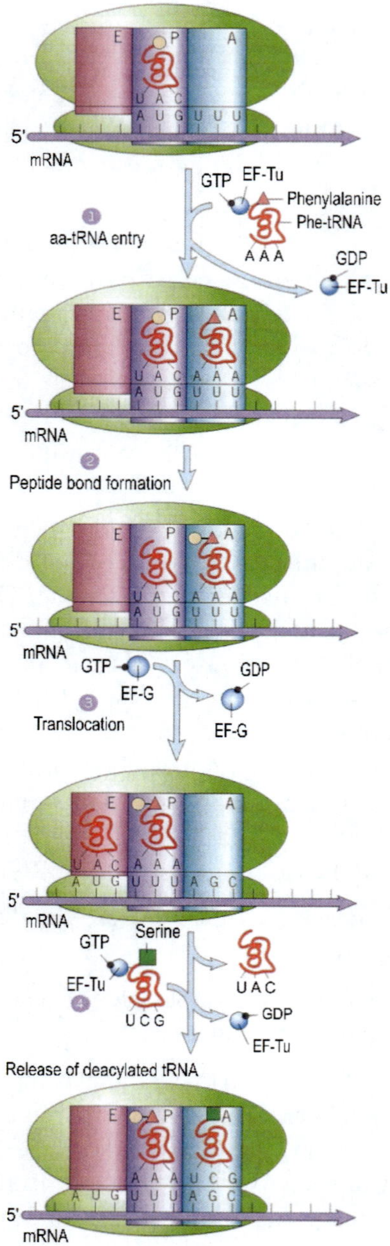

Fig. 33 : Elongation of nascent polypeptide during prokaryotic translation.

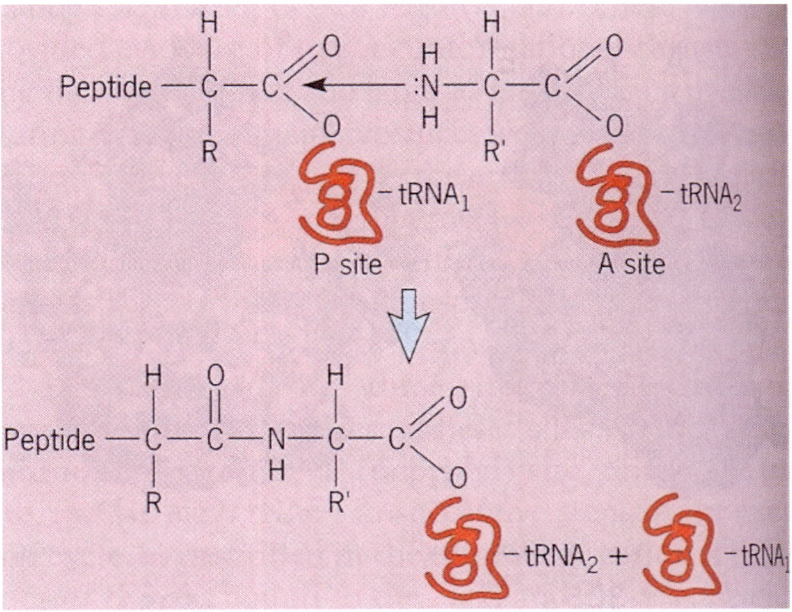

Fig. 34 : Peptide bond formation and the subsequent displacement of the deacylated tRNA.

7.4 Termination

There are no tRNA species that normally recognize stop codons. Instead, protein factors called release factors interact with these codons and cause release of the completed polypeptide chain. RF1 recognizes the codons UAA and UAG, and RF2 recognizes UAA and UGA. RF3 helps either RF1 or RF2 to carry out the reaction. The release factors make peptidyl transferase transfer the polypeptide to water rather than to the usual aminoacyl-tRNA, and thus the new protein is released. To remove the uncharged tRNA from the P-site and release the mRNA, EF-G together with ribosome release factor are needed for the complete dissociation of the subunits. IF3 can now bind the small subunit to prevent inactive 70S ribosome forming.

References

Buratti,E. and Baralle,F.E. (2004). Influence of RNA Secondary Structure on the Pre-Mrna. Splicing Process. Molecular and Cellular Biology. 24, 10505–10514.

Burke,T.W. and Kadonaga J.T. (1996). Drosophila TFIID binds to a conserved downstream basal promoter element that is present in many TATA-box-deficient promoters. Genes Dev. 10, 711–724.

Bushati,N. and Cohen,S.M. (2007). microRNA Functions. Annu. Rev.Cell Dev. Biol. 23, 175–205.

Chen,W. and Struhl,K. (1988). Saturation mutagenesis of a yeast his3 "TATA element": genetic evidence for a specific TATA-binding protein. Proc. Natl. Acad. Sci. USA. 85, 2691–2695.

Green,M.R. (1986). Pre-mRNA splicing. Annu. Rev. Genet. 20, 617-708.

Hampsey,M. (1998). Molecular Genetics of the RNA Polymerase II General Transcriptional Machinery. Microbiology and Molecular Biology Reviews, 63, 465-503.

Harvey,L. Arnold,B. Paul,M. Chris,A.K. Monty,K. Matthew,P.S. Lawrence,S.Z. Darnell,J. (2003). Molecular Cell Biology. 5th edn. W.H. Freeman & Company. New York.

Haung,J. and Ploeg,L.H.T.V. (1991). Maturation of Polycistronic Pre-mRNA in Trypanosoma brucei: Analysis of trans Splicing and Poly(A) Addition at Nascent RNA Transcripts from the hsp7O Locus. Molecular and Cellular Biology. 11, 3180-3190.

Karp,G.. (2002). Cell & Molecular Biology. Concepts & Experiments. 2nd edn., New York, John Wiley.

Klein,C. and Struhl,K. (1994). Increased recruitment of TATA-binding protein to the promoter by transcriptional activation domains in vivo. Science. 266, 280–282.

Padgett,R.A. Grabowski,P.J. Konarska,M.M. Seiler,S. and Sharp,P.A. (1986). Splicing of messenger RNA precursors. Annu. Rev. Biochem. 55, 1119-1150.

Prasanth,K.V. and Spector,D.L. (2007). Eukaryotic regulatory RNAs: an answer to the 'genome complexity' conundrum. Genes Dev. 21, 11–42

Rana,T.M. (2007). Illuminating the silence: understanding the structure and function of small RNAs. Nat. Rev. Mol. Cell. Biol. 8, 23–36.

Sharp,P.A. (1987). Splicing of messenger RNA precursors. Science 235, 766-771.

Singer,V.L., Wobbe,C.R. and Struhl,K. (1990). A wide variety of DNA sequences can functionally replace a yeast TATA element for transcriptional activation. Genes Dev. 4, 636–645.

Strubin,M. and Struhl,K. (1992). Yeast and human TFIID with altered DNA-binding specificity for TATA elements. Cell 68, 721–730.

Struhl,K. (1995). Yeast transcriptional regulatory mechanisms. Annu. Rev. Genet. 29, 651–674.

Sumimoto,H. Ohkuma,Y. Sinn,E. Kato,H. Shimasaki,S. Horikoshi,M. and Roeder,R.G. (1991). Conserved sequence motifs in the small subunit of human general transcription factor TFIIE. Nature 354, 401–404.

Wobbe,C.R. and Struhl,K. (1990). Yeast and human TATA-binding proteins have nearly identical DNA sequence requirements for transcription in vitro. Mol. Cell. Biol. 10, 3859–3867.

Yang,P.K. and Kuroda,M.I. (2007). Noncoding RNAs and intranuclear positioning in monoallelic gene expression. Cell. 128, 777–786.

Zaratiegui,M. Irvine,D.V. and Martienssen,R.A. (2007). Noncoding RNAs and gene silencing. Cell. 128, 763–776.

Zhao,Y. and Srivastava,D. (2007). A developmental view of microRNA function. Trends Biochem. Sci. 32, 189–197.

❑❑❑

CHAPTER – 9

Expressed Sequence Tag and its Potential

Pramod Katara
Department of Bioscience and Biotechnology, Banasthali University, Rajasthan,
Email : pmkatara@gmail.com

Summary

Expressed sequence tags are small pieces of DNA sequence (usually 200 to 500 nucleotides long) that are generated by sequencing either one or both ends of an expressed gene. Currently various EST sequencing projects are underway for numerous organisms, generating millions of short, single-pass nucleotide sequence reads, accumulating in EST databases. Extensive computational strategies and various databases have been developed to organize and analyse both small- and large-scale EST data for gene discovery, functional annotation of putative gene products as well as transcript and single nucleotide polymorphism analysis. Present chapter provides an overview of the significance of ESTs, their properties, databases, analysis software and the applications of ESTs.

1. Introduction

To understand the behaviour of complex biological organization and processes in terms of their molecular constituents (Kirschner, 2005), we must not only identify, catalogue and assign the function of all of its genes and gene products, but also understand regulatory interconnections between DNA, RNA and proteins. Following on from significant advancement in high-throughput technologies (microarrays, automated sequencing and mass spectrometry), transcriptomics, the global study of transcription, together with genomics and proteomics, have undoubtedly contributed to a systems biology approach. These technologies have generated a deluge of data. Fortunately, efficient computational tools (intelligent data networks, query, retrieval, analysis and visualization tools) have now optimized data mining, accelerating the process of discovery.

Expressed sequence tag (EST) and complementary DNA (cDNA) sequences provide direct evidence for all the sampled transcripts and they are currently the most important resources for transcriptome exploration. ESTs are short (200–800 nucleotide bases in length), unedited, randomly selected single-pass sequence reads derived from cDNA libraries. High-throughput ESTs can be generated at a reasonably low cost from either the 5' or 3' end of a cDNA clone to get an insight into transcriptionally active regions in any organism. In 1991, ESTs were used as a primary resource for human gene discovery (Adams et al., 1991). Thereafter, there has been an exponential growth in the generation and accumulation of EST data in public databases for myriad organisms. At present, ESTs enable gene discovery, complement genome annotation, aid gene structure identification, establish the viability of alternative transcripts, guide single nucleotide polymorphism (SNP) characterization and facilitate proteome analysis (Jongeneel, 2000; Dong et al., 2005 ; Rudd 2003).

Whole genome sequencing is currently impractical and expensive for organisms with large genome sizes. Such an approach is unlikely to be applied extensively, irrespective of the significance of such genome data in human and animal health, agriculture, ecology and evolution. In addition, genome expansion, as a result of retrotransposon repeats, makes whole genome sequencing less attractive for plants such as maize (Bennetzen, 2002). In this scenario, EST data sets have been utilized to complement genome sequencing or as an alternative to the genome sequencing of many organisms, earning the label, the 'poor man's genome' (Rudd, 2003). It must be noted that ESTs are subject to sampling bias resulting in under-representation of rare transcripts, often accounting for only 60% of an organism's genes (Bonaldo et al., 1996). However, ESTs in combination with reduced representation sequencing strategies, such as methylation filtration and high C_0t selection, have enabled the successful examination of the gene pool in plants like maize (Barbazuk et al., 2005).

There are several steps in EST analysis and an overwhelming number of tools available for each step. These methods have different strengths and attempt to extract biological information systematically from ESTs, in spite of their error-prone nature. However, there exists confusion in choosing the right tool for each different step of EST analysis and the subsequent downstream annotation at DNA or protein level. The confusion is compounded by the ability of some tools to handle high-throughput EST data, while others cannot.

This review briefly describes how ESTs are generated, to get an idea of the possible sources of error in the sequences obtained; where they are deposited (EST data resources) and the bottlenecks in EST analysis. We also list their proven utility in different application areas, the general methods and protocols being followed by different groups for EST analysis.

2. EST Generation

Messenger RNA (mRNA) sequences in the cell represent copies from expressed genes. As RNA cannot be cloned directly, they are reverse transcribed to double-stranded cDNA using a specialized enzyme, the reverse transcriptase. The resultant cDNA is cloned to make libraries representing a set of transcribed genes of the original cell, tissue or organism. Subsequently, these cDNA clones are sequenced randomly from both the directions in a single-pass run with no validation or full-length sequencing to obtain 5' and 3' ESTs. These ESTs usually range in size from 100–800bp. The resultant set of ESTs is redundant, as the cDNA template used can be of partial or full length. Bonaldo et al., (1996) provide a detailed description of cDNA library construction and normalization applied to remove redundancies.

Dias et al., (2000), have developed a novel cost-effective method for generating high-throughput ESTs called ORESTES (open reading frame expressed sequence tags). This method differs from conventional EST generation by providing sequence data from the central protein coding region, and thus the most informative and desired portion, of transcripts. ORESTES representing highly, moderately and rarely expressed transcripts has been derived from several species with more than a million human sequences and thousands from other species such as cow and honey bee deposited in the Expressed Sequence Tags database, dbEST (da Mota et al., 2004; Camargo et al., 2001; Nunes et al., 2004).

3. Errors Associated with EST Generation

A typical EST sequence (Figure 1A) is only a very short copy of the mRNA itself and is highly error prone, especially at the ends. The overall sequence quality is usually significantly better in the middle (Figure 1B). Vector and repeat sequences either in the end or rarely in the middle are excised during EST pre-processing.

As ESTs are sequenced only once, they are susceptible to errors. Generally, the quality of base reads in individual EST sequences is initially poor (upto 20% or 50–100bp), gradually improves and then diminishes once again towards the end (Aaronson et al., 1996). The overall sequence quality is usually significantly better in the middle ('highly informative length,' Figure 1B). Phred scores provide a measure of sequence quality with higher values corresponding to better sequence quality. Phred examines the peaks around each base call to assign a quality score to each base call that are logarithmically linked to error probabilities. Quality scores range from 4 to 40 and are estimated as $q = -10\times \log_{10}(P)$ where q is the Phred quality score and P is the estimated error probability of that call. A Phred score of 20 represents 1/100 chance of being incorrect or 99% accurate base calling while $q = 30$ denotes 1/1000 chance of being incorrect or 99.9% base calling accuracy. Phred scores can therefore be used to extract either entire sequences or segments of specified quality, based on the biological question being addressed in subsequent analysis.

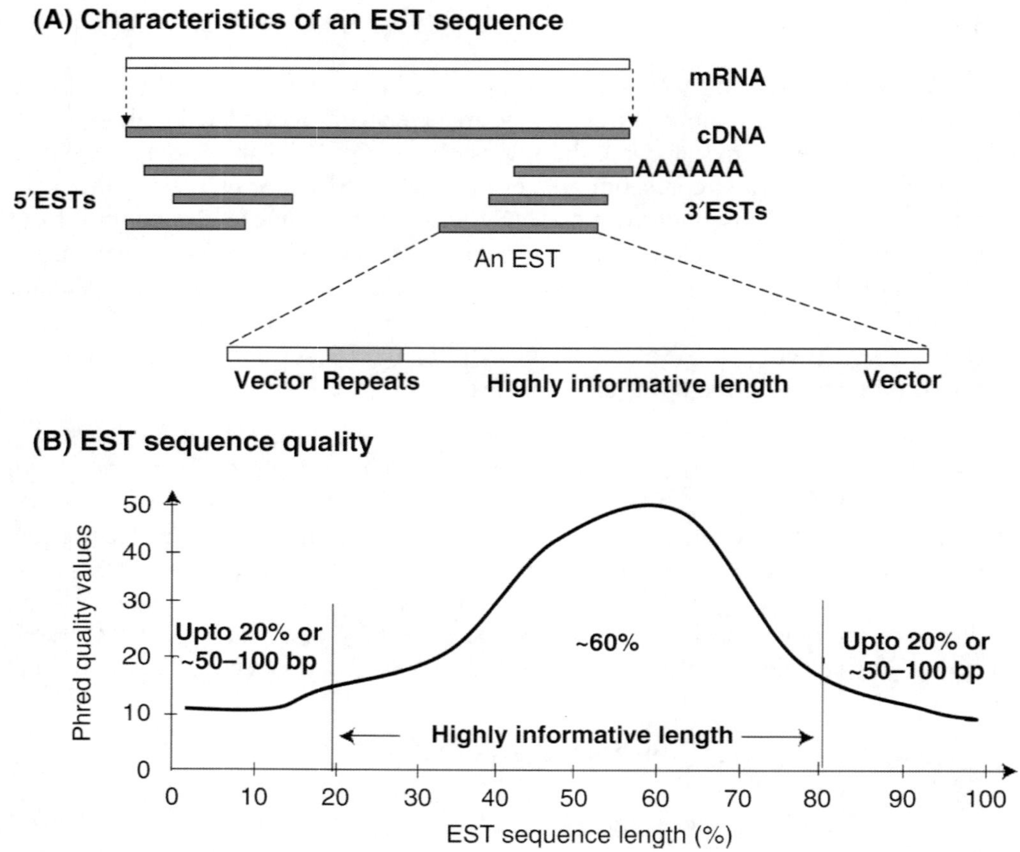

Figure 1: Characteristics of EST sequences. **A**. An EST sequence usually starts and ends with vector-contaminated bases, interspersed with possible repeats or low-complexity regions. **B**. Phred quality scores are plotted as a function of sequence length for a hypothetical EST sequence shown in A.

Redundancy, under-representation and over-representation of selected host transcripts are inherent problems with EST data due to the variable protocols used in their generation. Sequencing artefacts, such as base-calling errors as high as 5% (Aaronson et al. 1996), base stuttering (repeated bases, specifically G and T) and low quality sequences are some of the frequently observed errors in ESTs. There can be possible contaminations from vector, linker, adaptor and chimeric sequences, as also from genomic DNA fragments. In addition, low quality sequence attributes, repeats (simple or tandem), short sequence length and annotation errors can pose problems during downstream analysis. Moreover, natural sequence variations, such as RNA editing and genomic variations, due to SNPs will bring about additional challenges as it is not trivial to distinguish between sequencing artefacts and naturally-occurring substitutions, and insertion/deletion of events in a given EST data set.

ESTs and Untranslated Regions (UTRs)

The 5' and 3' UTRs of eukaryotic mRNA have been experimentally shown to contain sequence elements essential for gene regulation, expression and translation (Mignone et al., 2002). In this context, EST data has proven to be important for mining UTRs as both 5' and 3' ESTs contain significant sections of the UTRs along with protein coding regions. The CORG (COmparative Regulatory Genomics) resource (Dieterich et al., 2005) supports promoter analysis using assembled ESTs, while more than half of the Eukaryotic Promoter Database (Schmid et al., 2006) entries are based on 5' EST sequences. Mach (2002) has developed the PRESTA (PRomoter EST Association) algorithm for promoter verification and identification of the first exon, by mapping EST 5' ends.

Polyadenylation or poly (A) tails found in 3' UTR of the majority of mRNA transcripts are implicated in mRNA metabolism (Mignone et al., 2002). Gene boundaries have been predicted using poly (A) sites from 3' EST clusters (Kan et al., 2001). Differential poly(A) produces mRNAs with specific properties, attributable to post-transcriptional regulation mechanisms. Computational analyses of alternative poly (A) (Beaudoing & Gautheret, 2001; Gautheret et al. 1998; Yan & Marr, 2005) have advanced our understanding of mRNA regulation. Gautheret *et al.* (1998) identified previously unreported poly (A) sites in human mRNAs. Yan and Marr (2005) used 32 ESTs with poly(A) tails and demonstrated that at least 49% of human) 31% of mouse and 28% of rat polyadenylated transcription units show alternative poly (A) sites resulting in new protein isoforms. The presence of poly (A) tails in ESTs can also be used to distinguish untranslated mRNAs from productive transcripts, leading to protein isoforms.

4. EST Data Resources

The largest, freely-available repository of EST data is dbEST (Boguski et al., 1993; Boguski & Schuler, 1995). UniGene from the National Center for Biotechnology Information, USA (NCBI) stores unique genes and represents a non-redundant set of gene-oriented clusters generated from ESTs. Other specialized EST resources created for specific organisms include the The Institute for Genome Research, USA (TIGR) Gene Indices (Lee et al., 2005), the Rat EST project (University of Iowa) and the Cancer Genome Anatomy Project (Table 1).

Table 1 : Resources for EST data

Resource	Contents of EST resource	Organisms	Category*
ApiEST-DB	Raw	Apicomplexan parasites	F, D
dbEST at NCBI	Raw	All	F, D
Diatom EST database	Raw and clusters	Diatoms	F, D
ESTree	Raw and clusters	Peach	F, D
Fungal genomics project	Raw	Fungal	F, D
Honey bee brain EST project	Raw and clusters	Honey Bee	F, D
Nematode ESTs at the Sanger Institute	Raw and clusters	Parasitic nematodes	F, D
NEMBASE- parasitic nematode ESTs	Raw and clusters	Parasitic nematodes	F, D
Parasitic and free-living nematode EST resource	Raw and clusters	Nematodes	F, D
Phytopathogenic Fungi and Oomycete EST database	Plant pathogenic fungi ESTs	Fungi and oomycetes	F, D
Plant Gene Research, Kazusa DNA Research Institute	Raw	Heterogeneous set	F, D
Plant Genome database	Raw and clusters	Plants	F, D
Rat EST data at University of Iowa	Raw	Rat	F, D
Sanger Institute *Xenopus tropicalis* EST project	Raw and clusters	Xenopus	F, D
The TIGR Gene Indices	Raw and gene indices	All	F, D
UniGene database at NCBI	Raw and clusters	All	F, D

*F, free for academic users; D, data available for download.

Limitations of EST Data

Although ESTs are an excellent source of sequence data, these data are not of as high a quality as sequences determined by conventional means. Because EST sequences are generated in a single pass, they have a higher error rate than sequences that are verified by multiple sequencing runs, on the order of 3% (Boguski et al., 1993). In contrast, the standard for the human genome project is an error rate of $<0.01\%$. ESTs may contain substitutions, deletions, or insertions compared with the parent mRNA sequence. The region of an EST between positions 100 and 300 may be the most accurate part of the sequence (Hillier et al., 1996).

Hillier et al. (1996) & have performed a comprehensive analysis of potential EST artifacts. They found that ESTs may contain bacterial, mitochondrial, or vector sequence contamination. Most EST cDNA libraries are oligo(dT) primed, and the 3' EST derives from the 3' untranslated region of the gene. However, Hillier et al. found that 1.5% of oligo(dT)-primed 3' ESTs do not align with the known 3' end of the mRNA. These ESTs either represent nonspecific priming or indicate alternative splicing. cDNA for some libraries is synthesized with random primers, so the location of the 3' EST is unknown. Another potential problem comes from inverted clones in directionally cloned libraries, in which the 5' and 3' EST are mislabeled. cDNA inserts may be inverted because of failures in the directional cloning procedure, or simply because of human error. Hillier et al. found that 6.25% of ESTs that match a known mRNA align in an inverted orientation. Chimeric clones, in which the 5' EST matches one mRNA and the 3' EST another mRNA, may arise either during library construction or sample handling. Hillier et al. found a chimera frequency of 1%, but a separate study estimated the frequency at 11% (Wolfsberg & Landsman, 1997).

5. Overview of EST Sequence Analysis

An individual raw EST has negligible biological information. Analysis using different combinations of computational tools augments this weak signal and when a multitude of ESTs are analysed, the results enable the reconstruction of transcriptome of that organism. While diverse research groups have used different combinations of tools for extraction of data from specific databases followed by analyses (Hotz-Wagenblatt et al., 2003, Mao et al., 2003; Kumar et al., 2004; Parkinson, 2004; Miller et al., 1999; Paquola et al., 2003), a generic protocol of the different steps in the analysis of EST data sets is shown in Figure 2.

Chromatograms or EST sequences extracted from databases are pre-processed (Step 1, Figure 2) into high-quality ESTs wherein they are screened for sequence repeats, contaminants and low-complexity sequences, which are eliminated. Subsequently, high-quality ESTs are grouped into 'clusters' (Step 2, Figure 2) based on sequence similarity. The maximum informative consensus sequence is generated by 'assembling' these clusters, each of which could represent a putative gene. This step serves to elongate the sequence length by culling information from several short EST sequences simultaneously. Database similarity searches are subsequently performed against relevant DNA databases (Step 3, Figure 2) and possible functionality is assigned for each query sequence if significant database matches are found. Additionally, a consensus sequence can be conceptually translated to a putative peptide (step 4, Figure 2) and then compared with protein sequence databases (step 5, Figure 2). Protein centric functional annotation, including domain and motif analysis, can be carried out using protein analysis tools. It must be noted that the entire transcriptome is not translatable into protein products.

Each of these steps is briefly described subsequently, with special emphasis on the software tools available, followed by EST analyses tools for specific applications such as open reading frame (ORF) prediction, gene finding and detection of SNPs and alternative splicing.

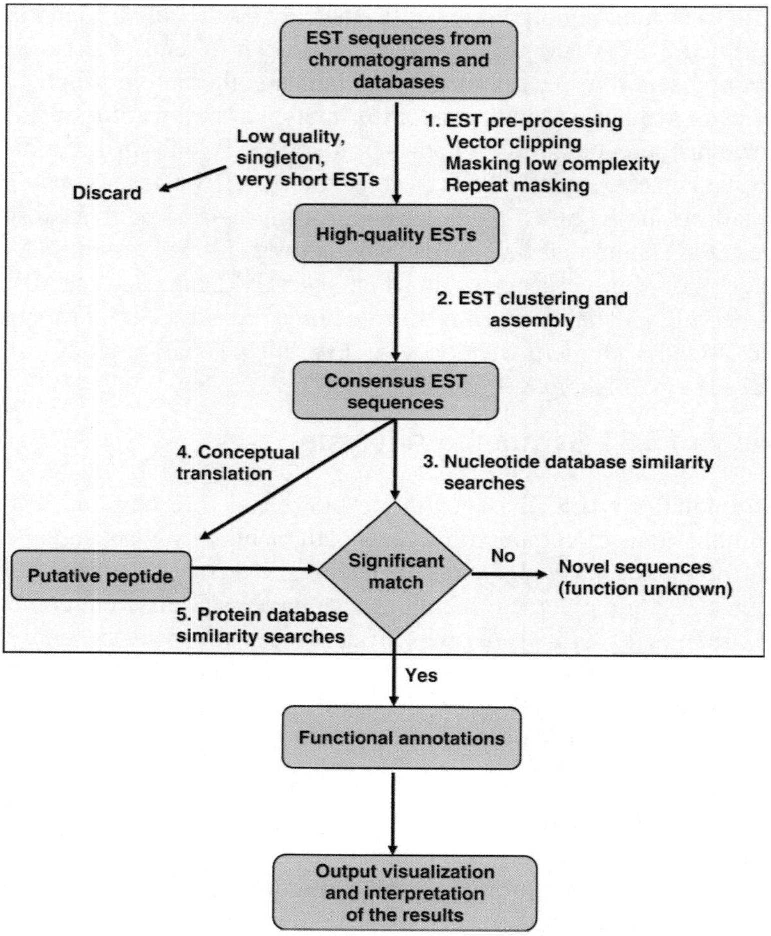

Fig. 2 : Generic steps involved in EST analysis. 1. Raw EST sequences are checked for vector contamination, low complexity and repeat regions, which are excised or masked. Low quality, singleton and very short sequences are also removed. 2. ESTs are then clustered and assembled to generate consensus sequences ('putative transcripts'). 3. DNA database similarity searches are carried out to assign, identify homologues and sign possible function. 4. Putative peptides are obtained by conceptual translation of consensus sequences. 5. Protein database similarity searches are performed to assign putative function(s). The analysis is extended to functional annotation followed by visualization and interpretation of results. The steps enclosed by the grey box alone are implemented in the currently available pipelines.

EST Pre-Processing

Pre-processing reduces the overall noise in EST data to improve the efficacy of subsequent analyses. Vector contamination is prevalent in ESTs, and often a part of the vector is also sequenced along with the EST sequences (Figure 2). These vector fragments have to be clipped before ESTs are clustered. Comparing the ESTs with non-redundant vector databases, such as UniVec and EMVEC with the locally installed tools such as BLAST (Altschul et al., 1997) or Cross_Match (Smith and Green, unpublished work), can identify vector contamination for removal. Any low complexity regions in EST data can be detected and masked using DUST from NCBI or nseg (Wan et al., 2003). Repetitive elements, such as LINEs (Long interspersed elements), SINEs (Short interspersed elements), LTRs (Long terminal repeat) and SSRs (Short simple repeats), can lead to erroneous assembly of sequences. Therefore, they should be 'repeat masked' during the analysis using either RepeatMasker or MaskerAid (Bedell et al., 2000) to screen DNA sequences for low complexity DNA sequences and interspersed repeats.

Poly(A) is not encoded in the genomic sequence and should be trimmed to retain a few adenines (usually 6–10) to get high-quality ESTs for clustering and assembly process. A list of web resources related to EST pre-processing is given in Table 2, grouped according to their functionality.

Table 2 : Resources for EST pre-processing

Name	Description	Category*
UniVec	Vector database	F
EMVec	Vector database	F
Phred/Cross_Match	Base caller/vector trimming and removal	F
Trimest	Poly(A) tail trimmed	F
Trimseq	Ambiguous ends trimmed	F
VecScreen	Vector, linker and adapter identification	F
Vector cleaning	Vector cleaning	C
Vectorstrip	DNA between vector sequences extracted	F
Paracel	EST pre-processing package	C, W
Lucy2	Sequence trimming and visualization	F
Dust	Low-complexity regions masked	F
The TIGR Plant Repeat databases	Plant repeats database	F
MaskerAid	Repeats masked	F
RepeatMasker	Repeats masked	F

*F, free for academic users; D, data available for download., C - Commercial pacakage, W, web interface available

EST Clustering and Assembly

The purpose of EST clustering is to collect overlapping ESTs from the same transcript of a single gene into a unique cluster to reduce redundancy. An EST cluster is a fragmented data, which can be consolidated and indexed using gene sequence information, such that all the expressed data arising from a single gene is grouped into a single index class, and each index class contains information for only that particular gene (Burke et al., 1999). A simple way to cluster ESTs is by measuring the pair-wise sequence similarity between them. Then, these distances are converted into binary values, depending on whether there is a significant match or not, such that the sequence pair can be accepted or rejected from the cluster being assembled. Different steps for EST clustering are described in detail by Ptitsyn and Hide (2005). In their formalism, there are two approaches for EST clustering, 'stringent' and 'loose.' The stringent clustering method is conservative, uses single-pass grouping of ESTs resulting in relatively accurate clusters, but generates shorter-sequence consensuses with low coverage of expressed genes. In contrast, loose clustering is 'liberal' and repeats low quality EST sequence alignments many times to generate less accurate but longer-sequence consensuses. Consequently, there is a better coverage of expressed gene data and alternatively spliced transcripts, but there is a risk of including paralogues in the clusters. stackPACK (Miller et al., 1999) is designed to use loose clustering, while TIGR Gene Indices (Lee et al., 2005) adopt stringent clustering. UniGene clusters lie between the two extremes.

Phrap (Ewing & Green, 1998) and CAP3 (Huang & Madan, 1999) are among the most extensively used programs for sequence clustering and assembly. Quackenbush and co-workers (Liang et al., 2000) have compared Phrap, CAP3 and TIGR Assembler programs with a benchmarking data set of rat ESTs. They assessed the effects of sequencing errors on EST assembly and critically evaluated the tools for EST clustering, assembly and relative accuracy of algorithms. Their results demonstrate that CAP3 consistently out-performed other similar programs, producing high-fidelity consensus sequences and maintaining a high level of sensitivity to gene family members while effectively handling sequencing errors. A newer study by Wang et al. (2004), have recognized two types of errors accruing from CAP3 EST clustering, where ESTs from the same gene do not form a cluster (Type I) and ESTs from distinct genes are wrongly clustered together (Type II). They propose a novel statistical approach for more accurate estimates of the true gene cluster profile. Table 3 provides a list of resources available for EST clustering, assembly and consensus generation.

Table 3 : Programs for EST sequence assembly and consensus generation

Name	Website	Category*
CAP3	http://genome.cs.mtu.edu/cap/cap3.html	F, W
CLOBB	http://zeldia.cap.ed.ac.uk/CLOBB/	F
CLU	http://compbio.pbrc.edu/pti	F
ESTate	http://www.ebi.ac.uk/~guy/estate/	F
ESTs aSSEmbly using Malig	http://alggen.lsi.upc.es/recerca/essem/frame-essem.html	F
megaBLAST	ftp://ftp.ncbi.nih.gov/blast/	F, W
miraEST	http://www.chevreux.org/projects_mira.html	F
Paracel Transcript Assembler	http://www.paracel.com/	C, W
Phrap	http://www.phrap.org/	F
stackPACK	http://www.sanbi.ac.za/Dbases.html#stackpack	F
Xsact and Xtract	http://www.ii.uib.no/~ketil/bioinformatics/	F

*F, free for academic users; C, commercial package; W, web interface available.

The STACK (Sequence Tag Alignment and Consensus Knowledge Base) (Miller et al., 1999) system was created to cluster and assemble ESTs. The major difference between STACK and other resources lies in its initial tissue-specific classification (15 tissue-based categories and one disease category) of EST data, for differential expression analysis. STACK uses a loose, unsupervised clustering strategy to group pre-processed ESTs for a wider gene coverage. The d2_cluster (a non-pairwise alignment algorithm) (Burke et al., 1999), Phrap and CRAW programs are incorporated into STACK to cluster, assemble and analyse EST alignments, respectively. Consensus contigs are then merged based on clone-identification data to obtain the best putative gene representation.

Database Similarity Searches

Once consensus sequences (putative genes) are obtained from assembled ESTs, possible functions can be assigned through downstream annotation, achieved via database similarity searches, employing familiar freely available tools and databases.

Different flavours of BLAST (Altschul et al., 1997) programs from NCBI serve as a universal tool for database similarity searches. BLASTN can be used to search ESTs against nucleotide sequence database and BLASTX to search against protein databases. BLASTX translates a consensus EST sequence (query) into protein products in six reading frames followed by comparisons with protein databases. In addition, one can scan for protein domains by selecting the (CDD) Conserved Domain Database (Marchler-Bauer et al., 2002) and the COG (Cluster of Orthologous Groups) (Tatusov et al., 2003) database using RPS-BLAST (Reverse PSI-BLAST) (Marchler-Bauer et al., 2002). High-throughput EST analysis and annotation involve the generation and interpretation of thousands of BLAST output results. In such cases, BLAST parsers such as MuSeqBox can be used (Xing & Brendel, 2001). Another option is to use SSAHA (Sequence Search and Alignment by Hashing Algorithm) (Ning et al., 2001), a fast and efficient DNA database searching tool. For transcriptome analysis, ESTs are additionally aligned to the genome sequence of the organism itself (if available) or the closest relative, using specialized alignment programs (Table 4) to facilitate genomic mapping and gene discovery. BLAT, GMAP and MGALIGN are considered to be reliable methods for this process (Wu & Watanabe 2005).

Table 4 : Programs for EST/cDNA sequence to genome DNA alignment

Name	Website	Category*
BLAT	http://genome.ucsc.edu/cgi-bin/hgBlat	F, W
est2genome	http://bioweb.pasteur.fr/seqanal/interfaces/est2genome.html	F, W
GMAP	http://www.gene.com/share/gmap/	F
MGAlign	http://origin.bic.nus.edu.sg/mgalign	F, W
SSAHA	http://www.sanger.ac.uk/Software/analysis/SSAHA/	F
Sim4	http://globin.cse.psu.edu/html/docs/sim4.html	F
Splign	http://www.ncbi.nlm.nih.gov/sutils/splign/splign.cgi	F, W

*F, free for academic users; W, web interface available.

Conceptual Translation of ESTs

EST data can be correlated with protein-centric annotations by accurate and robust polypeptide translations, since polypeptides are better templates for identifying domains and motifs, to study protein localization and to assign gene ontologies (GOs).

The first step in translating EST sequences is in identifying the protein-coding regions or ORFs, from consensus EST sequences, to enhance the process of gene discovery and gene boundary predictions. Some tools have been explicitly created for this purpose. For example, OrfPredictor (Min et al., 2005) has been designed specifically to identify protein-coding regions in EST-derived sequences, wherein the program provides six frames of translation and predicts most probable coding regions in all frames. ESTScan (Iseli et al., 1999) and DECODER (Fukunishi & Hayashizaki, 2001) can detect and extract coding regions from low-quality ESTs or partial cDNAs while correcting for frame shift errors, and provide conceptual translations. Table 5 lists key programs related to protein-coding region predictions from EST data. One can also use Prot4EST (Wasmuth & Blaxter, 2004), a pipeline with six tier polypeptide prediction tool, to translate ESTs into polypeptides. Prot4EST effectively incorporates DECODER, ESTScan and BLASTX for more accurate predictions. The putative peptides obtained can be compared with protein databases using BLASTP from the BLAST suite of programs.

Table 5 : ORF prediction and conceptual translation for ESTs

Name	Website	Category*
DECODER	http://www.gsc.riken.go.jp/	F
DIANA-EST	artemis@pcbi.upenn.edu.(upon request from the author)	F
Diogenes	http://analysis.ccgb.umn.edu/diogenes/index.html	F
ESTScan2	http://www.ch.embnet.org/software/ESTScan2.html	F, W
OrfPredictor	https://fungalgenome.concordia.ca/tools/OrfPredictor.html	F, W
TargetIdentifier	https://fungalgenome.concordia.ca/tools/TargetIdentifier.html	F

*F, free for academic users; W, web interface available.

6. Functional Annotations

Once a putative polypeptide is obtained, its function can be predicted by matching against non-redundant protein sequence, motif and family databases using an integrated tool such as Interproscan (Zdobnov & Apweiler, 2001). Protein sequences are better templates for functional annotation, particularly for the construction of multiple sequence alignments, profile and HMM generation, phylogenetic analysis, creation of protein-mass fingerprint libraries for proteomics applications, and domain and motif analysis using Pfam (Bateman et al., 2004) and SMART (Schultz et al., 1998).

7. EST Analysis Pipelines

In any large-scale sequencing project, in which thousands of ESTs are generated daily, it is extremely important to be able to store, organize and annotate using an automated analysis pipeline. Here, a protocol is required to transfer data efficiently between programs without human intervention, based on carefully parameterized threshold criteria. Given the ESTs from a database or from raw sequence chromatograms, such a pipeline would automatically clean, cluster, assemble, generate consensus sequences, conceptually translate and assign putative function based on various DNA and protein similarity searches. Table 6 summarizes the salient features of some of the EST analysis pipelines which comprise Steps 1–5 of Figure 2.

Table 6 : Characteristics of EST pipelines

Name	Pre-processing programs	Clustering and assembly programs	Translation programs	Category*
Edinburgh EST-Pipeline	Phred; Cross_Match	CLOBB; Phrap	DECODER; ESTSCAN	F
EST Analysis Pipeline (ESTAP)	Phred; Cross_Match	D2_cluster; CAP3	BLASTX	F
ESTAnnotator	Phred; RepMask with UniVec data	CAP3	BLASTX	W
ESTIMA	Information not available	BlastClust; CAP3	BLASTX	F
ESTweb	Phred; Cross_Match	None	None	F
Nematode.net	Phred; Consed	Phrap	BLASTX	F
PipeOnline	Phred; Cross_Match	Phrap	BLASTX	F
The TIGR Gene Indices (TGICL)	SeqClean; megaBLAST	CAP3; Paracel Transcript Assembler®	DIANA-EST; EST scan; Framefinder	F

*F, free for academic users; W, web interface available.

The TIGR Gene Indices are a rich resource for the freely-available high-quality gene contigs derived from their in-house EST analysis pipeline, TGCIL (TGI Clustering tools) (Pertea et al., 2003). A high stringency approach to establish sequence similarities during the initial clustering process is employed, that groups closely related genes into distinct consensus sequences. This process also allows the identification and separation of splice variants and then assembles individual clusters to generate longer, virtual transcripts or tentative consensus (TC) sequences, often referenced to mRNA or genomic sequences (Quackenbush et al., 2000). TCs are then annotated with tools for ORF and SNP prediction, long oligonucleotide prediction for microarrays, putative annotation using a controlled vocabulary, GO and enzyme commission number assignments, and mapping to complete or draft genomes and available genetic maps. Presently, TCs have been generated for 77 species and populate resources including EGO and RESOURCERER (Lee et al., 2005).

Two other successful and widely accessed EST 'warehouses' with high-quality data are Unigene and ENSEMBL. We have briefly described the data and software used by these warehouses, since an appreciation of these systems, if it were developed, will form the basis of future EST analysis pipelines.

Unigene uses GenBank mRNA and coding sequence data as 'reference' or 'seed' sequences for supervised cluster generation. Sequence alignment programs identify contaminant (linker, vector, bacterial, mitochondrial and ribosomal) sequences. Repeat and low-complexity regions are suppressed by RepeatMasker and DUST, respectively. ESTs are compared pairwise, using megaBLAST (Altschul et al., 1997) with different stringency levels, and then grouped into clusters. UniGene stores all gene isoforms in a single cluster, but does not generate consensus sequences. Singletons, not belonging to any cluster, are reprocessed at lower stringency and stored separately. With EST data accumulating daily in GenBank, UniGene clusters are updated each week for progressive data management (Boguski & Schuler, 1995).

EST-based gene model validation forms one of the core components of the ENSEMBL automatic gene annotation system (Curwen et al., 2004). The ENSEMBL EST gene build process involves two steps. In the first step, EXONERATE (Eyras et al., 2004) or BLAT is employed for the pairwise alignment of ESTs and unspliced genomic alignments, respectively. Clusters resulting from these alignments are then submitted to the ClusterMerge program (Eyras et al., 2004) which derives a minimal set of non-redundant transcripts compatible with the splicing structure of a set of ESTs. Mapping these clusters onto a genome leads to 'putative genes' called 'ESTgenes' (Curwen et al., 2004, Eyras et al., 2004).

8. Applications of ESTs

ESTs are versatile and have multiple applications. ESTs were first used to construct maps of the human genome (Wilcox et al., 1991), followed by assessment of the gene coverage from EST sequencing alone (Aaronson et al., 1996) and mapping of gene-based site markers (Boguski & Schuler, 1995). With the exponential rise in genomic data from global sequencing projects, EST databases are used for gene structure prediction (Kan et al., 2001), to investigate alternative splicing (Brett et al., 2000; Mironov et al., 1999; Modrek et al., 2001; Kan et al., 2004), to discriminate between genes exhibiting tissue or disease-specific expression (Schmitt et al., 1999) and for the discovery and characterization of candidate SNPs (Buetow et al., 1999; Picoult-Newberg et al., 1999; Useche et al., 2001).

EST-based gene expression protocols have been used in the identification and analysis of coexpressed genes on a large scale (Ewing et al., 1999; Heyer et al., 1999). Ewing et al.(1999), have generated digital gene expression profiles of the rice genome from dbEST classified tissue types and organs. The *in silico* differential display technique allows the identification of genes which are expressed differentially in various tissues (Hawkins et al., 1999) and between normal and diseased individuals (Schmitt et al., 1999). Recently, applying DNA 'bar coding' in EST projects, Qiu et al. (2003), categorized maize cDNA libraries using distinct 6-bp DNA sequences to track the origin of ESTs from specific mRNA pools. ESTs have also become invaluable resources in the area of proteomics for peptide identification and proteome characterization of proteomes, particularly in the absence of complete genome sequence information (Lisacek et al.,

2001; Kim et al., 2003). EST sequencing strategies together with other genomic and proteomics methodologies (transcription profiling and peptide fingerprinting) have been employed for gene and allele identification (Nelson et al., 2005; Mooney et al., 2004). Overall, the usefulness of EST data has extended well beyond its original application in gene finding and in transcriptome analysis, and a comprehensive list of specialized EST application software is provided in Table 7.

Table 7 : EST applications and visualization resources

Name	Description	Application area	Category
ESTgene	Alternative splicing detection	Alternative splicing	F
galaxieEST	Automated phylogenetic analysis	Evolutionary studies	F
GBA server	EST-based digital gene expression profiling	Gene expression profiles	F
ESTminer	Gene and allele identification	Gene structure prediction and alternative splicing	F
ESTminer	Web application and database schema for mining of EST clusters	Candidate gene discovery	F
GeneSeqer	Alternative splicing detection	Gene structure prediction and alternative splicing	F, W
Transcript Assembly program	Predominant and alternative gene structures identified	Gene structure prediction and alternative splicing	F,W
prot4EST	ESTs to protein multiple sequence alignments	Protein alignment	F
miraEST	Human SNP discovery	SNP discovery	F, W
SNP discovery from ESTs	To find SNP candidates in EST assemblies	SNP discovery	F, W
JESAM	EST alignments and clusters	Visualization of clusters	F, W
SpliceNest	Visualization of gene splicing from ESTs	Visualization	F, W

*F, free for academic users; W, web interface available.

Table 8 : Plant specific EST database with their website

Plant EST database	Website
B-EST barley database	http; //:pgrc.ipk-gatersleben.de/est/est/login
Chlamydomonas resource centre	http: //www.biology.duke.edu/chlamy.genome
Kazusa EST database	http: //www.kazusa.or.jp/en/plant/database
NCBI Unigenes	http://www.ncbi.nlm.nih.gov/UniGene/
PlantGDB	http://www.zmdb.iastate.edu/PlantGDB/
Solanaceae genomics network	http://sgn.cornell.edu/
TIGR Plant Gene Indices	http://www.tigr.org/tdb/tgi/plant.shtml
University Minnesota	http://www.ccgb.umn.edu/
Pscroph database	http://www.pscroph.ucdavis.edu
dbEST	http://www.ncbi.nlm.nih.gov/dbEST/index.html
REDB (Rice EST Database)	http://www.ncpgr.cn
Mendel-GFDb and Mendel-ESTS:	http://www.mendel.ac.uk/
US Mirror:	http://genome.cornell.edu/
EGENES	http://www.genome.jp/kegg-bin/create_kegg_
Sputnik	http://www.mips.gsf.de/proj/sputni
ESTree db	http://www.itb.cnr.it/estree
TbestDB	http://www.tbestdb.bcm.umontreal.ca

ESTs and Gene Discovery

ESTs have been widely used for gene discovery (Boguski et al., 1995). Because ESTs outnumber other nucleotide sequences in GenBank, researchers hunting for a novel gene are much more likely to find it in dbEST than in the rest of GenBank. ESTs are not included in the BLAST *nr* database, and sequence similarity searches for ESTs must target the EST database. Gene discovery methods using ESTs include, for example, hunting for new members of gene families in the same species (paralogs), for functionally equivalent genes in other species (orthologs), or even for alternatively spliced forms of known genes. Gene discovery using dbEST is very rapid, requiring only a few minutes for the BLAST search. A TBLASTN search of dbEST with the human protein sequence quickly reveals not only many of these known genes but also an additional likely ortholog in zebrafish. Discovering alternatively spliced transcripts among ESTs is more problematic. For one, it is difficult to determine if the new sequences are due to alternative splicing or to the presence of contaminating genomic DNA sequence in EST libraries (Wolfsberg and Landsman, 1997). An analysis of the TIGR Human Gene Index using a spliced alignment algorithm provided evidence that up to 35% of human genes may undergo alternative splicing and that the majority of these events occur in the 5' untranslated regions (Mironov et al., 1999). The uses of ESTs extend beyond mammals. For example, until recently, the public databases contained little sequence data from *Toxoplasma gondii*, a diseasecausing protozoan parasite of human. A large scale project generated 7,000 5' ESTs, representing _4,000 unique sequences, from *T. gondii* (Ajioka et al., 1998). Comparisons between the ESTs and sequences in public databases identified potential functions for 500 novel *T. gondii* genes. Some ESTs are phylogenetically restricted to *T. gondii* and other members of the Apicomplexa phylum.

The Human Gene Map

ESTs are also being used to create gene maps by the use of sequence-tagged sites (STSs), short stretches of unique sequence identified by polymerase chain reaction (PCR) assays. An international consortium agreed to coordinate a mapping effort for the human genome, using UniGene clusters to represent individual human genes. In the initial effort, the gene-based STSs were mapped relative to two radiation hybrid (RH) maps and one YAC panel; in subsequent work, the STSs have been mapped only to the two RH panels, the Genebridge4 (GB4) and Stanford G3. STSs were generated from the 3' untranslated regions of UniGene clusters. GeneMap '96 reported the mapping of 16,000 gene-based STSs (Schuler et al., 1996), and GeneMap '98 nearly doubled that number to 30,000 (Deloukas et al., 1998). Thus, current maps detail the position of up to one-half of all human protein-coding genes. The gene map is updated as new STSs are mapped. Information about the map location of individual ESTs is provided by UniGene. If an STS exists for an EST in the cluster, the map position of that STS is indicated in the record. This data may confirm what is already known about the location of a mapped human disease gene, but, for the majority of cases involving yet unmapped ESTs, GeneMap '98 provides novel information. Because the STS markers correspond to individual genes, the project also shows the density of genes on each human chromosome.

Gene Prediction in Genomic DNA

Another use of ESTs is to predict or refine computational predictions of the location of genes in genomic DNA. With the appropriate use of sequence alignment parameters, up to 90% of genes annotated on human genomic DNA are also detected by ESTs (Bailey et al., 1998). ESTs can complement other algorithms used for gene prediction because they may do a better job at predicting alternative splicing and 3' untranslated regions. A study is underway to reannotate the *C. elegans* genome using EST sequences. With the use of the acembly program, 126,000 ESTs were aligned to 98 Mb of genomic DNA. The genes predicted by EST clones were compared with those predicted by the *C. elegans* genomic sequencing consortium, which were constructed using GeneFinder with hand editing (The *C. elegans* Sequencing Consortium, 1998). The following points are noteworthy:

1. In about half the cases, the computationally predicted genes were identical to the EST alignments; 25% of the genes were predicted with less accuracy, and the remaining 25% were predicted poorly. In some cases, 5' sequences from ESTs showed that the gene predictions were either too long or too short.

2. Comparisons of the EST sequences to the genomic sequence confirm that the error rate of the worm genome sequence is less than 1 mistake per 10,000 nucleotides. Instances where many ESTs share identical sequence may indicate errors in genomic sequence. Alternatively, these differences could be sequence polymorphisms.
3. About 30% of the ESTs exist in alternatively-spliced forms. Many of these alternative splices are not annotated on the genomic sequence.
4. Computational methods may predict separate genes, whereas EST analysis shows that these segments are actually exons of a single gene. Conversely, the computational method may predict exons in cases that should be separate genes.

ESTs have also been used in genome sequencing projects to make estimates about gene expression along the chromosome. In the complete sequences of *Arabidopsis thaliana* chromosomes 2 (Lin et al., 1999) and 4 (Mayer et al., 1999), about one-third of the computationally predicted genes have an EST match. Histograms plotting the EST distribution along the chromosomes predict that some genes are highly expressed, at least within the tissues from which EST libraries were constructed. On chromosome 4, 75% of the matching ESTs aligned with only 6% of the genes, implying that these genes are transcribed at high rates.

ESTs and Sequence Polymorphisms

Single nucleotide polymorphisms (SNPs) can help to associate sequence variations with heritable phenotypes, facilitate studies in population and evolutionary biology, and aid in positional cloning and physical mapping. On average, SNPs occur every 500–1,000 nucleotides in human DNA. Gene-associated SNPs are found in untranslated regions as well as coding sequences (cSNPs). Because ESTs are sequenced redundantly from libraries prepared from different individuals, they seem an ideal source of polymorphic data. Indeed, a number of recent studies demonstrate that analysis of aligned EST sequences can lead to SNP discovery (Buetow et al., 1999; Garg et al., 1999; Marth et al., 1999; Picoult-Newberg et al., 1999). All rely on alignment of EST sequences, identification of sequence differences, and a method to distinguish real polymorphisms from base calling (sequencing) errors and other artifacts. The public database for SNPs, dbSNP, is maintained at the NCBI (Sherry et al., 1999; Smigielski et al, 2000). dbSNP accepts submissions not only of single nucleotide polymorphisms but of other polymorphisms, such as short deletions and inserts, microsatellites, and polymorphic insertion elements like retrotransposons.

Assessing Levels of Gene Expression Using ESTs

Because ESTs are generated by random sequencing of clones from many different libraries, they appear, at least at first glance, to be a good source of data source for studies of gene expression levels. However, any conclusions about transcript levels must be made very carefully. Many libraries are normalized or generated by subtractive hybridization. Both of these processes change the relative representation of cDNAs. Normalization results in bundant messages being seen less frequently and rare messages more frequently, whereas subtractive hybridization removes entire pools of transcripts from the library. Although libraries made by these processes can provide very general ideas about which genes are expressed at higher levels, detailed analysis is not possible.

NEIBank (Gene Expression in Eye)

NEIBank is an integrated resource for genomics and bioinformatics in vision research. It includes expressed sequence tag (EST) data and sequence-verified cDNA clones for multiple eye tissues of several species (human, mouse, dog, chicken and zebrafish), web-based access to human eye-specific SAGE data through EyeSAGE, and comprehensive, annotated databases of known human eye disease genes and candidate disease gene loci. All expression- and disease-related data are integrated in EyeBrowse, an eye-centric genome browser. NEIBank provides a comprehensive overview of current knowledge of the transcriptional repertoires of eye tissues and their relation to pathology (Wistow et al., 2008).

9. Conclusions

ESTs generation was considered as low-cost alternative to genome sequencing or to complement genome sequencing projects by early characterization of the transcriptome. There are currently numerous databases, web resources and software's are available to store and analyze EST data effectively. Because of these all advances in last decades, ESTs have become a favourite tool to refine the predicted transcripts for those genes, which leads to the prediction of their protein products and ultimately their function. Moreover, the situation in which those ESTs are obtained (tissue, organ, disease state - e.g. cancer) gives information on the conditions in which the corresponding gene is acting. ESTs contain enough information to permit the design of precise probes for DNA microarrays that then can be used to determine the gene expression.

With the rapid convergence of various technologies crossing the 'omics' barrier for holistic solutions to complex biological questions, the analysis and significance of EST data sets will continue to be indispensable in many areas of biotechnology and biomedicine.

References

Aaronson JS, Eckman B, Blevins RA, Borkowski JA, Myerson J, Imran S, Elliston KO.(1996). Toward the development of a gene index to the human genome: an assessment of the nature of high-throughput EST sequence data. *Genome Res* 6:829-45.

Adams MD, Kelley JM, Gocayne JD, Dubnick M, Polymeropoulos MH, Xiao H, Merril CR, Wu A, Olde B, Moreno RF, et al.(1991). Complementary DNA sequencing: expressed sequence tags and human genome project. *Science* 252:1651-6.

Ajioka, J. W., Boothroyd, J. C., Brunk, B. P., Hehl, A., Hillier, L., Manger, I. D., Marra, M.,Overton, G. C., Roos, D. S., Wan, K. L., Waterston, R., and Sibley, L. D. (1998). Gene discovery by EST sequencing in *Toxoplasma gondii* reveals sequences restricted to the Apicomplexa. *Genome Res* 8: 18–28.

Altschul SF, Madden TL, Schäffer AA, Zhang J, Zhang Z, Miller W, Lipman DJ. (1997). Gapped BLAST and PSI-BLAST: a new generation of protein database search programs. *Nucleic Acids Res* 25:3389-402.

Bailey, L. C., Jr., Searls, D. B., and Overton, G. C. (1998). Analysis of EST-driven gene annotation in human genomic sequence. *Genome Res* 8: 362–376.

Barbazuk WB, Bedell JA, Rabinowicz PD. (2005). Reduced representation sequencing: a success in maize and a promise for other plant genomes. *Bioessays* 27:839-48.

Bateman A, Coin L, Durbin R, Finn RD, Hollich V, Griffiths-Jones S, Khanna A, Marshall M, Moxon S, Sonnhammer EL, Studholme DJ, Yeats C, Eddy SR. (2004). The Pfam protein families database. *Nucleic Acids Res* 32:D138-41.

Beaudoing E, Gautheret D. (2001). Identification of alternate polyadenylation sites and analysis of their tissue distribution using EST data. *Genome Res* 11:1520-6.

Bedell JA, Korf I, Gish W. (2000). MaskerAid: a performance enhancement to RepeatMasker. *Bioinformatics* 16:1040-1.

Bennetzen JL. (2002). Mechanisms and rates of genome expansion and contraction in flowering plants. *Genetica* 115:29-36.

Boguski MS, Lowe TM, Tolstoshev CM. (1993). dbEST–database for "expressed sequence tags". *Nat Genet* 4:332-3.

Boguski MS, Schuler GD. (1995). ESTablishing a human transcript map. *Nat Genet 10:369-71.*

Bonaldo MF, Lennon G, Soares MB. (1996). Normalization and subtraction: two approaches to facilitate gene discovery. *Genome Res* 6:791-806.

Brett D, Hanke J, Lehmann G, Haase S, Delbruck S, Krueger S, Reich J, Bork P. (2000). EST comparison indicates 38% of human mRNAs contain possible alternative splice forms. *FEBS Lett* 474:83-6.

Buetow, K. H., Edmonson, M. N., and Cassidy, A. B. (1999). Reliable identification of large numbers of candidate SNPs from public EST data. *Nat Genet* 21: 323–325.

Burke J, Davison D, Hide W. (1999). d2_cluster: a validated method for clustering EST and full-length cDNAsequences. *Genome Res* 9:1135-42.

Camargo AA, Samaia HP, Dias-Neto E, Simão DF, Migotto IA, Briones MR, Costa FF, Nagai MA, Verjovski-Almeida S, Zago MA, Andrade LE, Carrer H, El-Dorry HF, Espreafico EM, Habr-Gama A, Giannella-Neto D, Goldman GH, Gruber A, Hackel C, Kimura ET, Maciel RM, Marie SK, Martins EA, Nobrega MP, Paco-Larson ML, Pardini MI, Pereira GG, Pesquero JB, Rodrigues V, Rogatto SR, da Silva ID, Sogayar MC, Sonati MF, Tajara EH, Valentini SR, Alberto FL, Amaral ME, Aneas I, Arnaldi LA, de Assis AM, Bengtson MH, Bergamo NA, Bombonato V, de Camargo ME, Canevari RA, Carraro DM, Cerutti JM, Correa ML, Correa RF, Costa MC, Curcio C, Hokama PO, Ferreira AJ, Furuzawa GK, Gushiken T, Ho PL, Kimura E, Krieger JE, Leite LC, Majumder P, Marins M, Marques ER, Melo AS, Melo MB, Mestriner CA, Miracca EC, Miranda DC, Nascimento AL, Nobrega FG, Ojopi EP, Pandolfi JR, Pessoa LG, Prevedel AC, Rahal P, Rainho CA, Reis EM, Ribeiro ML, da Ros N, de Sa RG, Sales MM, Sant'anna SC, dos Santos ML, da Silva AM, da Silva NP, Silva WA Jr, da Silveira RA, Sousa JF, Stecconi D, Tsukumo F, Valente V, Soares F, Moreira ES, Nunes DN, Correa RG, Zalcberg H, Carvalho AF, Reis LF, Brentani RR, Simpson AJ, de Souza SJ. (2001) The contribution of 700,000 ORF sequence tags to the definition of the human transcriptome. *Proc Natl Acad Sci USA* 98:12103-8.

Curwen V, Eyras E, Andrews TD, Clarke L, Mongin E, Searle SM, Clamp M. (2004). The Ensembl automatic gene annotation system. *Genome Res* 14:942-50.

da Mota AF, Sonstegard TS, Van Tassell CP, Shade LL, Matukumalli LK, Wood DL, Capuco AV, Brito MA, Connor EE, Martinez ML, Coutinho LL. (2004). Characterization of open reading frame-expressed sequence tags generated from *Bos indicus* and *B. taurus* mammary gland cDNA libraries. *Anim Genet* 35:213-19.

Deloukas, P., Schuler, G. D., Gyapay, G., et al. (1998). A physical map of 30,000 human genes. *Science* 282: 744–746.

Dias Neto E, Correa RG, Verjovski-Almeida S, Briones MR, Nagai MA, da Silva W Jr, Zago MA, Bordin S, Costa FF, Goldman GH, Carvalho AF, Matsukuma A, Baia GS, Simpson DH, Brunstein A, de Oliveira PS, Bucher P, Jongeneel CV, O'Hare MJ, Soares F, Brentani RR, Reis LF, de Souza SJ, Simpson AJ. (2000). Shotgun sequencing of the human transcriptome with ORF expressed sequence tags. *Proc Natl Acad Sci USA* 97:3491-6.

Dieterich C, Grossmann S, Tanzer A, Ropcke S, Arndt PF, Stadler PF, Vingron M. (2005) Comparative promoter region analysis powered by CORG.*BMC Genomics* 21;6(1):24

Dong Q, Kroiss L, Oakley FD, Wang BB, Brendel V. (2005). Comparative EST analyses in plant systems. *Methods Enzymol* 395:400-8.

Ewing B, Green P. (1998) Base-calling of automated sequencer traces using phred. II. Error probabilities. *Genome Res* 8:186-94.

Ewing RM, Ben Kahla A, Poirot O, Lopez F, Audic S, Claverie JM. (1999). Large-scale statistical analyses of rice ESTs reveal correlated patterns of gene expression. *Genome Res* 9:950-9.

Eyras E, Caccamo M, Curwen V, Clamp M. (2004). ESTGenes: alternative splicing from ESTs in Ensembl. *Genome Res* 14:976-87.

Fukunishi Y, Hayashizaki Y. (2001). Amino acid translation program for full-length cDNA sequences with frameshift errors. *Physiol Genomics* 5:81-7.

Garg, K., Green, P., and Nickerson, D. A. (1999). Identification of candidate coding region single nucleotide polymorphisms in 165 human genes using assembled expressed sequence tags. *Genome Res* 9: 1087–1092.

Gautheret D, Poirot O, Lopez F, Audic S, Claverie JM. (1998). Alternate polyadenylation in human mRNAs: a large-scale analysis by EST clustering. *Genome Res* 8:524-30.

Hawkins V, Doll D, Bumgarner R, Smith T, Abajian C, Hood L, Nelson PS. (1999). PEDB: the Prostate Expression Database. *Nucleic Acids Res* 27:204-8.

Heyer LJ, Kruglyak S, Yooseph S. (1999). Exploring expression data: identification and analysis of coexpressed genes. *Genome Res* 9:1106-15.

Hillier L, Lennon G, Becker M, Bonaldo MF, Chiapelli B, Chissoe S, Dietrich N, DuBuque T, Favello A, Gish W, Hawkins M, Hultman M, Kucaba T, Lacy M, Le M, Le N, Mardis E, Moore B, Morris M, Parsons J, Prange C, Rifkin L, Rohlfing T, Schellenberg K, Bento Soares M, Tan F, Thierry-Meg J, Trevaskis E, Underwood K, Wohldman P, Waterston R, Wilson R, Marra M et al. (1996). Generation and analysis of 280,000 human expressed sequence tags. *Genome Res.* 6:807–828.

Hotz-Wagenblatt A, Hankeln T, Ernst P, Glatting KH, Schmidt ER, Suhai S. (2003). ESTAnnotator: A tool for high throughput EST annotation. *Nucleic Acids Res* 31:3716-19.

Huang X, Madan A. (1999). CAP3: a DNA sequence assembly program. *Genome Res* 9:868-77.

Iseli C, Jongeneel CV, Bucher P. (1999). ESTScan: a program for detecting, evaluating, and reconstructing potential coding regions in EST sequences. *Proc Int Conf Intell Syst Mol Biol* 138-48.

Jongeneel CV (2000). Searching the expressed sequence tag (EST) databases: panning for genes. *Brief Bioinform* 1:76-92.

Kan Z, Castle J, Johnson JM, Tsinoremas NF. (2004). Detection of novel splice forms in human and mouse using cross-species approach. *Pac Symp Biocomput* 42-53.

Kan Z, Rouchka EC, Gish WR, States DJ. (2001). Gene structure prediction and alternative splicing analysis using genomically aligned ESTs. *Genome Res* 11:889-900.

Kim SI, Kim JY, Kim EA, Kwon KH, Kim KW, Cho K, Lee JH, Nam MH, Yang DC, Yoo JS, Park YM. (2003). Proteome analysis of hairy root from Panax ginseng C.A. Meyer using peptide fingerprinting, internal sequencing and expressed sequence tag data. *Proteomics* 3:2379-92.

Kirschner MW. (2005). The meaning of systems biology. *Cell* 121:503-4.

Kumar CG, LeDuc R, Gong G, Roinishivili L, Lewin HA, Liu L. (2004). ESTIMA, a tool for EST management in a multi-project environment. *BMC Bioinformatics* 5:176.

Lee BT, Tan TW, Ranganathan S. (2003). MGAlignIt: a web service for the alignment of mRNA/EST and genomic sequences. *Nucleic Acids Res* 31:3533-6.

Lee Y, Tsai J, Sunkara S, Karamycheva S, Pertea G, Sultana R, Antonescu V, Chan A, Cheung F, Quackenbush J. (2005). The TIGR Gene Indices: clustering and assembling EST and known genes and integration with eukaryotic genomes. *Nucleic Acids Res* 33:D71-4.

Liang F, Holt I, Pertea G, Karamycheva S, Salzberg SL, Quackenbush J. (2000). An optimized protocol for analysis of EST sequences. *Nucleic Acids Res* 28:3657-65.

Lin, X., Kaul, S., Rounsley, S., Shea TP, Benito MI, Town CD, Fujii CY, Mason T, Bowman CL, Barnstead M, Feldblyum TV, Buell CR, Ketchum KA, Lee J, Ronning CM, Koo HL, Moffat KS, Cronin LA, Shen M, Pai G, Van Aken S, Umayam L, Tallon LJ, Gill JE, Adams MD, Carrera AJ, Creasy TH, Goodman HM, Somerville CR, Copenhaver GP, Preuss D, Nierman WC, White O, Eisen JA, Salzberg SL, Fraser CM, Venter JC. (1999). Sequence and analysis of chromosome 2 of the plant Arabidopsis thaliana. *Nature* 402: 761–768.

Lisacek FC, Traini MD, Sexton D, Harry JL, Wilkins MR. (2001). Strategy for protein isoform identification from expressed sequence tags and its application to peptide mass fingerprinting. *Proteomics* 1:186-93.

Mach V. (2002). PRESTA: associating promoter sequences with information on gene expression. *Genome Biol* 3. *research0050.*

Maheswari U, Montsant A, Goll J, Krishnasamy S, Rajyashri KR, Patell VM, Bowler C. (2005). The Diatom EST Database. *Nucleic Acids Res* 33:D344-7.

Mao C, Cushman JC, May GD, Weller JW. (2003). ESTAP – an automated system for the analysis of EST data. *Bioinformatics* 19:1720-2.

Marchler-Bauer A, Panchenko AR, Shoemaker BA, Thiessen PA, Geer LY, Bryant SH. (2002). CDD: a database of conserved domain alignments with links to domain three-dimensional structure. *Nucleic Acids Res* 30:281-3.

Marth, G. T., Korf, I., Yandell, M. D., Yeh RT, Gu Z, Zakeri H, Stitziel NO, Hillier L, Kwok PY, Gish WR. (1999). General approach to single-nucleotide polymorphism discovery. *Nat Genet* 23: 452–456.

Mayer K, et al. (1999). Sequence and analysis of chromosome 4 of the plant *Arabidopsis thaliana.* Nature. 402:769–777

Mignone F, Gissi C, Liuni S, Pesole G. (2002). Untranslated regions of mRNAs. *Genome Biol 3:REVIEWS0004.*

Miller RT, Christoffels AG, Gopalakrishnan C, Burke J, Ptitsyn AA, Broveak TR, Hide WA. (1999). A comprehensive approach to clustering of expressed human gene sequence: the sequence tag alignment and consensus knowledge base. *Genome Res* 9:1143-55.

Min XJ, Butler G, Storms R, Tsang A. (2005). OrfPredictor: predicting protein-coding regions in EST-derived sequences. *Nucleic Acids Res* 33:W677-80.

Mironov AA, Fickett JW, Gelfand MS. (1999). Frequent alternative splicing of human genes. *Genome Res* 9:1288-93.

Modrek B, Resch A, Grasso C, Lee C. (2001). Genome-wide detection of alternative splicing in expressed sequences of human genes. *Nucleic Acids Res* 29:2850-9.

Mooney BP, Krishnan HB, Thelen JJ. (2004). High-throughput peptide mass fingerprinting of soybean seed proteins: automated workflow and utility of UniGene expressed sequence tag databases for protein identification. *Phytochemistry* 65*:*1733-44.

Nelson RT, Grant D, Shoemaker RC. (2005). ESTminer: a suite of programs for gene and allele identification. *Bioinformatics* 21*:*691-3.

Ning Z, Cox AJ, Mullikin JC. (2001). SSAHA: a fast search method for large DNA databases. *Genome Res* 11*:*1725-9.

Nunes FM, Valente V, Sousa JF, Cunha MA, Pinheiro DG, Maia RM, Araujo DD, Costa MC, Martins WK, Carvalho AF, Monesi N, Nascimento AM, Peixoto PM, Silva MF, Ramos RG, Reis LF, Dias-Neto E, Souza SJ, Simpson AJ, Zago MA, Soares AE, Bitondi MM, Espreafico EM, Espindola FS, Paco-Larson ML, Simoes ZL, Hartfelder K, Silva WA Jr. (2004). The use of Open Reading frame ESTs (ORESTES) for analysis of the honey bee transcriptome. *BMC Genomics* 5*:*84.

Paquola AC, Nishyiama MY Jr, Reis EM, Verjovski-Almeida S (2003). ESTWeb: bioinformatics services for EST sequencing projects. *Bioinformatics* 19*:*1587-8.

Parkinson J, Anthony A, Wasmuth J, Hedley A, Blaxter M (2004). PartiGene–constructing partial genomes. *Bioinformatics* 20*:*1398-404.

Pertea G, Huang X, Liang F, Antonescu V, Sultana R, Karamycheva S, Lee Y, White J, Cheung F, Parvizi B, Tsai J, Quackenbush J. (2003). TIGR Gene Indices clustering tools (TGICL): a software system for fast clustering of large EST datasets. *Bioinformatics* 19*:*651-2.

Picoult-Newberg, L., Ideker, T. E., Pohl Taylor SL, Donaldson MA, Nickerson DA, Boyce-Jacino M. (1999), Mining SNPs from EST databases. Genome *Res* 9: 167–174.

Ptitsyn A, Hide W. (2005). CLU: a new algorithm for EST clustering. *BMC Bioinformatics 6 Suppl* 2*:*S3.

Qiu F, Guo L, Wen TJ, Liu F, Ashlock DA, Schnable PS. (2003). DNA sequence-based "bar codes" for tracking the origins of expressed sequence tags from a maize cDNA library constructed using multiple mRNA sources. *Plant Physiol1*33*:*475-81.

Quackenbush J, Liang F, Holt I, Pertea G, Upton J. (2000). The TIGR gene indices: reconstruction and representation of expressed gene sequences. *Nucleic Acids Res* 28*:*141-5.

Rudd S. (2003*)*. Expressed sequence tags: alternative or complement to whole genome sequences? *Trends Plant Sci* 8*:*321-9.

Schmid CD, Perier R, Praz V, Bucher P. (2006). EPD in its twentieth year: towards complete promoter coverage of selected model organisms. *Nucleic Acids Res* 34*:*D82-5.

Schmitt AO, Specht T,Beckmann G, Dahl E, Pilarsky CP, Hinzmann B, Rosenthal A. (1999). Exhaustive mining of EST libraries for genes differentially expressed in normal and tumour tissues. *Nucleic Acids Res* 27*:*4251-60.

Schuler, G. D., Boguski, M. S., Stewart, E. A., Stein, L. D., Gyapay, G., Rice, K., White, R. E., Rodriguez-Tome, P., Aggarwal, A., Bajorek, E., Bentolila, S., Birren, B. B., Butler, A., Castle, A. B., Chiannilkulchai, N., Chu, A., Clee, C., Cowles, S., Day, P. J., Dibling, T., Drouot, N., Dunham, I., Duprat, S., East, C., Hudson, T. J., et al. (1996). A gene map of the human genome. *Science* 274: 540–546.

Schultz J, Milpetz F, Bork P, Ponting CP. (1998). SMART, a simple modular architecture research tool: identification of signaling domains. *Proc Natl Acad Sci USA* 95*:* 5857-64.

Sherry, S. T., Ward, M., and Sirotkin, K. (1999). dbSNP-database for single nucleotide polymorphisms and other classes of minor genetic variation. *Genome Res* 9: 677–679.

Smigielski, E. M., Sirotkin, K., Ward, M., and Sherry, S. T. (2000). dbSNP: a database of single nucleotide polymorphisms. *Nucleic Acids Res* 28:352–355.

Tatusov RL, Fedorova ND, Jackson JD. Tatusov RL, Fedorova ND, Jackson JD, Jacobs AR, Kiryutin B, Koonin EV, Krylov DM, Mazumder R, Mekhedov SL, Nikolskaya AN, Rao BS, Smirnov S, Sverdlov AV, Vasudevan S, Wolf YI, Yin JJ, Natale DA. (2003). The COG database: an updated version includes eukaryotes. *BMC Bioinformatics* 4*:*41.

Useche FJ, Gao G, Harafey M, Rafalski A. (2001). High-throughput identification, database storage and analysis of SNPs in EST sequences. *Genome Inform Ser Workshop Genome Inform* 12*:*194-203.

Wan H, Li L, Federhen S, Wootton JC. (2003). Discovering simple regions in biological sequences associated with scoring schemes. *J Comput Biol* 10*:*171-85.

Wang JP, Lindsay BG, Leebens-Mack J, Cui L, Wall K, W. C. Miller, and C. W. dePamphilis (2004). EST clustering error evaluation and correction. *Bioinformatics* 20*:*2973-84.

Wasmuth JD, Blaxter ML. (2004). prot4EST: translating expressed sequence tags from neglected genomes. *BMC Bioinformatics* 5*:*187.

Wilcox AS, Khan AS, Hopkins JA, Sikela JM. (1991). Use of 32 untranslated sequences of human cDNAs for rapid chromosome assignment and conversion to STSs: implications for an expression map of the genome. *Nucleic Acids Res* 19*:*1837-43.

Wistow G, Peterson K, Gao J, Buchoff P, Jaworski C, Bowes-Rickman C, Ebright JN, Hauser MA, Hoover D.(2008). NEIBank: genomics and bioinformatics resources for vision research. Mol Vis. 18; 14:1327-37.

Wolfsberg, T. G., and Landsman, D. (1997). A comparison of expressed sequence tags (ESTs) to human genomic sequences. *Nucleic Acids Res* 25: 1626–1632.

Wu TD, Watanabe CK. (2005). GMAP: a genomic mapping and alignment program for mRNA and EST sequences. *Bioinformatics* 21*:*1859-75.

Wu X, Walker MG, Luo J, Wei L. (2005). GBA server: EST-based digital gene expression profiling. *Nucleic Acids Res* 33*:*W673-6.

Xing L, Brendel V. (2001). Multi-query sequence BLAST output examination with MuSeqBox. *Bioinformatics* 17*:*744-5.

Yan J, Marr TG. (2005). Computational analysis of 32 -ends of ESTs shows four classes of alternative polyadenylation in human, mouse, and rat. *Genome Res* 15*:*369-75.

Zdobnov EM, Apweiler R. (2001). InterProScan – an integration platform for the signature-recognition methods in InterPro. *Bioinformatics* 17*:*847-8.

❑❑❑

CHAPTER – 10

Transgenic Plants Production, Methodologies and its Application in Crop Improvements

Arpita Sharma[1], Indu Ravi[2] and Kambaska Kumar Behera[1]
[1]Department of Bioscience and Biotechnology, Banasthali University, Rajasthan - 304022, India
Email : Kambaska@yahoo.co.in
[2]IGNOU Regional Centre, Jaipur - 302020, Rajasthan, India

Summary

Genetic engineering can be used to introduce specific traits into plants. It may not replace conventional breeding but can add to the efficiency of crop improvement. It is possible due to the fact that plants are totipotent, enabling regeneration of a new plant from an isolated cell. Transformation of dicots is usually carried out using the bacterium, Agrobacterium tumefaciens. Genes are cloned into plant expression vectors that carry the 25-bp repeat sequences at the borders of the onc region, known as the left and right borders (LB and RB) and the virulence genes (vir) of the Ti plasmid. They are introduced into plants with the aid of a des-arcmed Ti plasmid whose virulence gene products allow the genes to be transferred to the plant nucleus where they are integrated into the genome. Monocots are usually transformed by a biolistic process, using a "gene gun". Genes can also be introduced into mitochondria and chloroplasts, present in many copies in a cell, as alternatives to delivery into the nucleus. This has the advantage of increasing the number of copies of the transgene. In both cases callus tissue is regenerated on media containing an antibiotic or herbicide to select for transformants. The exception is transformation of the common wall cress, Arabidopsis thaliana, by floral dip method where tissue culture is not required. For many years the only alternative to A. tumefaciens-mediated transformation was the direct uptake of naked DNA by plant protoplasts, achieved by electroporation or mediated by polyethylene glycol (PEG). This depends on the ability of plants to regenerate from protoplasts, which varies considerably between species. Success has subsequently been obtained with Arabidopsis, rice, maize and forage grasses. Important parameters for successful PEG-mediated protoplast transformation include ion concentrations, the presence of inert carrier DNA, the molecular weight and concentration of PEG and the physical configuration of the nucleic acid. Now the most widely used alternative to transformation by A. tumefaciens is biolistics. Other techniques, which have been reported but not extensively used, include pollen co-cultivation, microinjection of somatic embryos and liposome fusion with protoplasts. At the posttranscriptional level, the transgene RNA could be specifically degraded if tagged by a small complementary RNA molecule. It is often advantageous for plants to express the introduced transgenes in specific tissues or under specific conditions. As a result many genes are cloned downstream of tissue-specific or inducible promoters. High expression of a transgene may be required under certain circumstances. The 35S promoter (camv 35s) of the cauliflower mosaic virus is commonly used in dicots while the maize ubiquitin promoter is the monocot promoter of choice. Targetting genes to organelles such as chloroplasts can also permit the introduction of thousands of copies of foreign genes per plant cell, and generates extraordinarily high levels of foreign protein. Little is understood of the way in which genes are integrated into the plant chromosomes. In many cases multiple inserts occur at one locus. Screenable marker 'reporter genes' have also been developed from bacterial genes coding for easily assayed enzymes, such as chloramphenicol acetyl transferase (CAT), β-galactosidase, β-glucuronidase (GUS), luciferase (LUX), green fluorescent protein (GFP), nopaline synthase and octopine synthase. The utility of any particular marker gene construct as a transformation marker varies depending on the plant species and explant involved. The most popular selectable marker genes used in plant transformation vectors include constructs providing resistance to antibiotics such as kanamycin, chloramphenicol and hygromycin, and genes that allow growth in the presence of herbicides such as phosphinotricin, glyphosate, bialaphos and several other chemicals. For successful selection, the target plant cells must be susceptible to relatively low concentrations of the antibiotic or herbicide in a non-leaky manner. The resulting transgenic plants are able to express new traits that can confer on them an enhanced potential and commercial value. In this regard, new traits such as herbicide resistance, pest and disease resistance are being introduced through genetic engineering. Crop improvement programs based upon GE also require improved in vitro regeneration techniques such as macro and micropropagation and a more thorough understanding of the key metabolic pathways of plants. The potential impact on nearby ecosystems is another greatest concerns expressed about transgenic plants. Most domesticated plants mate with wild relative at some location where they are grown, and gene flow from domesticated crops (irrespective of whether they are transgenic or non-transgenic) can have potentially harmful consequences. In addition biosafety issues related to release of transgenics have to be also considered.

1. Introduction

Conventional plant breeding, combined with improved agricultural practices and modern technology, has contributed to dramatic crop improvements over the past 50 years, and will continue to provide future benefits. However, there are strong pressures for further improvements in crop quality and quantity exerted from population growth, social demands, health requirements, environmental stresses and ecological considerations. When world population is twice its current level, higher levels of production will be needed. Also, if we are to avoid increasing the area of land under cultivation and the erosion of more natural habitats, the current area of cultivated land will need to be more productive. Transgenic crops are not the only way to improve crop productivity, but they are likely to play an increasingly important role in world agriculture in the coming decades.

Notwithstanding the impressive gains in productivity so far, there are limitations to conventional plant breeding technology, either due to the limited gene pool or to the restricted range of organisms between which genes can be transferred due to species barriers. Although these methods have proved to be useful, they depend ultimately upon reliable testing and proper selection, and do not involve the identification and manipulation of genetic targets as defined in molecular terms. Plant biotechnology continues the trend of improving crops with more precise methods, permitting the transfer of a single gene with a known function into existing crop varieties, in contrast to the cross-breeding techniques which transfer thousands of genes of unknown functions into crops. Genetic engineering has been used to complement traditional breeding methods in crop improvement. Transfer of genes from heterologous species provides the means of selectively introducing new traits into crop plants and expanding the gene pool beyond what has been available to traditional breeding systems. Hence, new biotechniques, in addition to conventional plant breeding, are needed to boost the yield of crops that feed the world. The newly acquired ability to transfer genes among organisms without sexual crossing provides breeders with new opportunities to improve the efficiency of production and to increase the utility of agricultural crops. Plants with new traits, such as resistance to herbicides, insect pests, and viruses, have been genetically engineered using genes from unrelated organisms. However, it should be emphasized that the biotechnology is not a substitute or replacement for conventional breeding methods. Rather, it can improve on past, conventional methods. The major differences between conventional breeding and biotechnology lie neither in goals nor processes, but rather in speed, precision, reliability, and scope. Plant biotechnology offers new ideas and techniques applicable to agriculture. It uses the conceptual framework and the technical approaches of plant tissue culture and molecular biology to develop commercial processes and products. Hence, with the rapid development of biotechnology, agriculture has moved from a resource-based to a science-based industry. The technology required for engineering transgenic plants is considerably more sophisticated than the one for producing somatic hybrid plants, and much more, so that required for production of hybrid plants by cross-fertilization. Non-sexual DNA transfer techniques make possible manipulations that are outside the repertoire of breeding or cell fusion techniques. Genes can be accessed from exotic sources of plant, animal, bacterial, even viral, and introduced into a crop. Because the DNA elements that control gene expression can, and often must, be modified for proper function in the new host, it is possible to control timing, tissue specificity, and expression level of transferred genes. Endogenous plant genes may even be reprogrammed through the reintroduction of an engineered gene. There are also important challenges for agriculture from the influence of gene patenting issues, business monopolies and intensifying international market forces.

Conventional plant breeding practices are in many respects less precise and more unpredictable than GM breeding. The insertion of known gene sequences into crops provides the opportunity to analyze the introduced genes with a precision, that is impossible for most genes transferred by traditional breeding. GM crops provide an important opportunity to reduce dependence on chemical inputs into agriculture. They also have the potential to make a valuable contribution to improving care for the environment and wildlife diversity. However, there are very important issues that need to be debated about plant breeding and genetically modified (GM) crops.

1.1. Genetic Engineering of Plants

With the advent of recombinant DNA methods and genetic transformation procedures, it is possible to transfer genes into crop plants from unrelated plants, microbes, and animals. Most of the modifications being carried out, or envisaged, are for disease, pest, or herbicide resistance. Because of these possibilities is now feasible to introduce into crop plants, genes that have previously been inaccessible to the conventional plant breeder or which did not exist in the crop of interest. However, the availability of efficient transformation methods to introduce foreign DNA can be a substantial barrier to the application of recombinant DNA methods in some crop plants. Despite significant advances over the past decade, the development of efficient transformation methods take many years of painstaking research. The major components for the development of transgenic plants are:

(1) The development of reliable tissue culture regeneration systems

(2) Preparation of gene constructs and transformation with suitable vectors

(3) Efficient transformation techniques for the introduction of genes into the crop plants;

(4) Recovery and multiplication of transgenic plants;

(5) Molecular and genetic characterization of transgenic plants for stable and efficient gene expression;

(6) Transfer of genes to elite cultivars by conventional breeding methods if required

(7) Evaluation of transgenic plants for their effectiveness in alleviating the biotic and abiotic stresses without being an environmental biohazard.

Gene transfer (GT) is the transfer of genetic material directly to a living cell or an organism followed by its expression. It occurs widely among prokaryotes. GT has been shown to engage members of the same species, of different species, or even of different domains of life. GT is distinguished from the process of vertical gene transfer, which is the passage of genetic information from parents to offspring as it occurs in reproduction. GT plays a major role in the evolution of genomes: it is considered to be more important in the adaptation of bacteria to new environments than the alteration of gene function through point mutations. It is likely that in early evolution, horizontal gene transfer was pervasive and that it was dominating the evolutionary dynamics. Ancient cells were simpler and more modular in design than are modern cells. Retrospective research, such as the comparison of the genomes of bacteria and plants, provides more and more evidence for the singular and modular transfer of genes involving different mechanisms between lower and higher organisms. It has been suggested that horizontal transfer occurs more frequently for operational genes (those involved in housekeeping) than for informational genes (those involved in transcription, translation, and related processes), because the latter are typically members of large, complex systems thereby making horizontal transfer of such gene products less probable. It has been shown that the genome of *E. coli* strain MG1655 differs from related *E. coli* strains by 67 events, including 37 gene transfers and 30 gene losses. Gene transfer requires a number of steps including uptake of DNA, integration and expression. If the acquired DNA does not provide selective advantage, it is likely to be lost in the population. The probability that a specific gene will be successfully transferred to a new host depends on the specific mechanistic details of the transfer (transformation, transduction, or conjugation), on the relationships of these mechanisms to the types of nucleic acids that are being transferred (single-stranded, double-stranded, linear, circular, etc.), and even on factors such as the distribution of integrases in organisms. Gene transfer as a biosafety issue has been addressed in several studies and a number of potential hazards have received much attention and have been debated intensively in the scientific and popular press. A lot of attention has been spent on the transfer of antibiotic resistance genes from GM plants to soil and plant-related micro-organisms. GT to bacteria in these environments has been shown by marker rescue experiments via homologous recombination. However, GT may also occur under different conditions, for instance transformation of bacteria in the food chain. Free DNA persists in some materials for weeks, and furthermore, some bacteria develop natural/ chemical competence to take up DNA from the environment. In addition, in the gastrointestinal tract of man and husbandry animals, DNA may remain stable for some time, particularly in the colon. However, degradation already begins before the DNA or the material containing the DNA arrives at the critical sites for GT, which are generally believed to be the lower part of the small intestine, caecum, and the colon. And if DNA reaches these sites, the DNA is most probably fractionated in fragments smaller than the length of a complete gene. Thus, breakdown of DNA in the gut, combined with the breakdown of the DNA due to food processing, reduces the risk of dissemination. It is generally accepted that risk assessment is composed of four different steps, namely hazard identification (i.e. the identification of the hazard, the danger it presents, the impact in terms of human health, and the circumstances under which it has an impact), hazard characterisation (i.e. the qualitative and/ or quantitative evaluation of the adverse effects of the hazard on human health), exposure assessment (i.e. qualitative and quantitative evaluation of the degree of consumption or intake of the hazardous agent that is likely to occur) and finally the risk characterisation (i.e. the integration of the three prior steps into an estimate of the adverse effect likely to occur in the target population). This framework can also serve in the risk assessment when the relevance of GT for the safety of food and feed derived from GM plants has to be investigated. Strictly spoken, the agent of interest for hazard identification is the DNA as it is the "chemical" which is delivered in the recipient plants for the construction of the genetically modified organism (GMO). However, DNA in food has never been considered to constitute a health risk. It is rather the eventual subsequent transformation of cells by DNA from a GMO that should be the substance of the hazard identification process. Up to now, micro-organisms are known to be the only transformable organisms under natural conditions (i.e. cells taking up free DNA and incorporating it into the genome) and therefore, the acquisition of new heritable information is of concern only for micro-organisms. Evidently, this type of DNA transfer is not restricted to the DNA, that has been newly inserted into plants and thus, the hazard identification has to be focused in a case by case consideration on the unique properties of the microbial transformants that may result from the newly combined plant DNA. The first GM plant introduced into the market was the Flavr Savr tomato, developed in 1990, the safety assessment of which may serve as an example for a case to demonstrate the principle of risk analysis with GM plants. Among numerous hazards considered, given by Calgene, the safety of the newly introduced npt II gene encoding resistance against the aminoglycoside antibiotics kanamycin and neomycin was of special concern with respect to kanamycin and kanamycin resistance, among which an unacceptable increase in resistant, potentially pathogenic microorganisms and thus the impaired therapeutic use of the antibiotic has been identified as a defined hazard. The closest contact of the GM-plant material and the human associated microflora takes place in the large bowel, were high microbial counts can be detected. For example, in the faeces ca. 40% of the weight consists of bacterial biomass (equivalent to $4x10^{11}$ bacteria per gram), and >1500 species can be detected. Once one bacterium acquires the genetic information for antibiotic resistance and selective pressure is exerted (by therapeutic use of the agent), this information will spread via all mechanisms of GT in this population. It has to be taken into account that kanamycin and neomycin are rather seldomly used in human therapeutics because of their potential side effects. Of importance is also the knowledge of transformation frequencies in a given habitat, such as the content of the human intestine. Evaluation of the likelihood that a consumer suffers from adverse health effects as a result of gene transfer from GM plants is case-dependent, but available data are still very limited. The entire risk assessment process requires the use of scientific data of the highest quality and the application of established scientific procedures. While research and scientific studies continue to provide answers needed to make informed decisions in risk analysis related to hazards in food, uncertainty and unresolved questions still cause concern to decision-makers. These data enable the individuals responsible for risk management to decide on the measures to handle the risk and to communicate adequately with the public in order to provide the private and public sectors with the information necessary for preventing, reducing, or minimising risks to acceptably safe levels through mandatory or voluntary systems of quality and safety management.

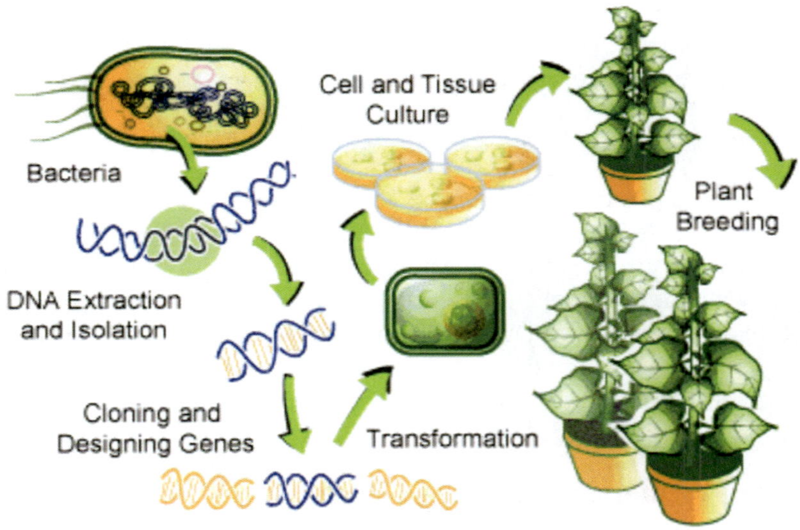

Fig.1 : General representation of genetic transformation in plants (Transgenic Crops)

2. Delivery Systems

A delivery system involves the use of specific technologies for the transfer of isolated DNA into a viable host cell. There are two main classes of delivery systems: biological and physical. We will present a comprehensive overview of the different systems and of the various factors that may affect subsequent gene transfer.

2.1. Biological Delivery Systems

2.1.1. *Agrobacterium* mediated plant transformation

Only two bacterial species, *Agrobacterium tumefaciens* and *Agrobacterium rhizogenes* can at present be used as delivery systems for plant transformation. These bacteria, which normally cause crown gall disease, both have the unique ability to transfer portions of their DNA into the genomes of plants. These strains contain a large megaplasmid (more than 200 kb) which plays a key role in tumour induction, and for this reason it was named the Ti plasmid (after tumour inducing), or Ri (after root-inducing) in the case of *A. rhizogenes*. During bacterial infection of plant tissue the T-DNA (transfer DNA), a mobile segment of Ti or Ri plasmid, is transferred to the plant cell nucleus and integrated into a plant chromosome.

Mechanism

The machinery for transfer of T-DNA by *Agrobacterium* evolved from bacterial conjugation systems, which mobilise plasmids for transport between bacterial cells. Proteins involved in the transfer of T-DNA are closely related, both structurally and functionally, to the proteins in bacterial conjugation systems.

Contact Between Plant Cells and Bacterial Cells

Wounding of a plant, which often occurs in the region of the stem-root interface, allow the entry of bacterium. Bacteria are attracted to the wounded plant in response to signal molecules that are released by the plant cells; they multiply in the wound sap and attach to the walls of plant cells. The specific attachment of *Agrobacterium* to plant cells is a prerequisite for subsequent transfer of DNA. Products of several chromosomal genes (*chv A, chv B, pac A and att*) in *Agrobacterium* and surface receptors of plant cells, which might include both protein and carbohydrates, are believed to be involved in this process.

Activation of Virulence (*vir*) Genes

The T-DNA fragment is flanked by 25-bp direct repeats, which act in cis as a signal for the transfer apparatus. The process of T-DNA transfer is mediated by the co-operative action of proteins encoded by genes of the Ti plasmid virulence region (vir genes) and in the bacterial genome. Following transfer and expression of the T-DNA into the plant nucleus, the plant cells proliferate and synthesize certain types of opines, depending upon the T-DNA type that has been integrated. These opines in turn can serve as a sole C and N source for *Agrobacterium*. The 30 kb vir region is a regulon organised of seven operons that are essential for T-DNA transfer (*virA, virB, virD, and virG*) and for the increasing of transfer efficiency (*virC and virE*). Several genes of the bacterial chromosome have been shown to function in the attachment of *Agrobacterium*, to the plant cell and in bacterial colonisation. The first protein that is involved in the activation of the T-DNA transfer is *VirA*, a transmembrane dimeric sensor protein that detects signal molecules, mainly, small phenolic compounds such as acetosyringone, released from wounded plants. Activated VirA has the capacity to transfer its phosphate to the cytoplasmic VirG that in turn functions as a transcription factor regulating the expression of the other vir genes.

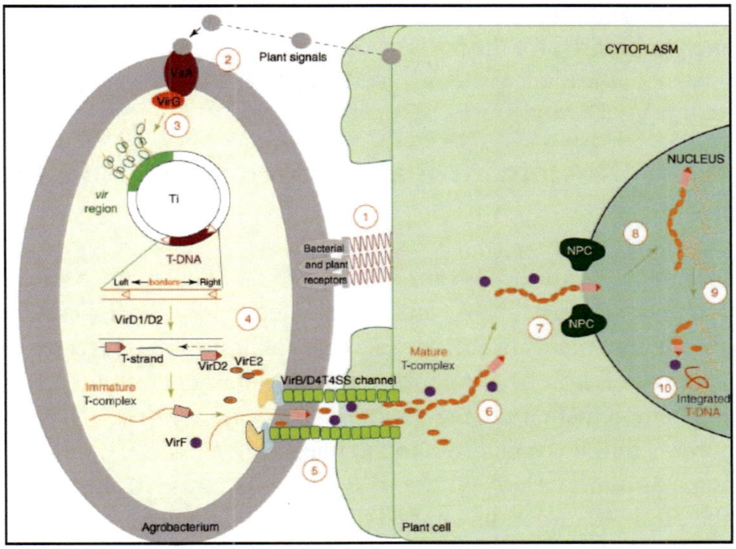

Fig. 2 : A model for the *Agrobacterium*–Mediated genetic transformation: The transformation process comprises 10 major steps and begin with recognition and attachment of the *Agrobacterium* to the host cells (1) and the sensing of specific plant signals by the Agrobactirum Vir A/Vir G two component signal-transduction system (2). Following activation of the vir gene region (3), a mobile copy of the T-DNA is generated by the Vir D1/D2 protein complex (4) and delivered as a Vir D2-DNA complex (immature T-complex), together with several other Vir proteins, into the host-cell cytoplasm (5). Following the association of Vir E2 with the T-strand, the mature T-complex forms, travels through the host-cell cytoplasm (6) and is actively imported into the host- cell nucleus (7). Once inside the nucleus, the T-DNA is recruited to the point of integration (8) stripped of its escorting proteins (9) and integrated into host genome (10).

Generation and Transport of the T-Complex

Both ends of T-DNA are flanked by 25- bp imperfect direct-repeat sequences. VirD2 protein recognises these border repeats and makes nicks between the third fourth bases in the "bottom strand" of the T-DNA, with the help of Vir D1, Vir C1 and possibly, Vir D3. Vir D2 attaches covalently to the 5′ end of the bottom strand of the T-DNA, which is then released from the Ti plasmid by replacement DNA synthesis to form the T- strand, which is the intermediate in the transfer of the T-DNA. Vir E2 proteins then coat the T-strand to make the T-complex. Vir E2 is a protein that binds to single-stranded DNA. Only the 25-bp direct repeats at the ends of the T-DNA are required, in *cis,* for the mobilization of T-DNA. Virtually any DNA sequence flanked by such border repeats can be transferred to plant cell by the virulence systems. These findings have allowed the modification of the *Agrobacterium* system to yield various plant-transformation vectors.

Vir B proteins and Vir D4 are thought to form a channel, which is a type IV secretion system, between the bacterium and plant cell. The T-complex travels through this apparatus into the cytoplasm of the plant cell. Many of these proteins have characteristics of membrane proteins, and some of them are ATPases that probably provide the energy required for the passage of the T-complex. Transfer of the T-complex from the cytoplasm of the plant nucleus is mediated by nuclear localization signals (NLS) that are located in Vir D2 and Vir E2.

Integration of T-DNA into the Plant Chromosome and Expression of T-DNA Genes

Soon after the entry of the T-complex into the nucleus, many of the T-strands are converted to double-stranded DNA, and genes encoded by the T-DNA are expressed without integration into plant chromosomes. This phenomenon is referred to as transient expression. The expression of vir genes results in the generation of single-stranded DNA (ss DNA) complementary to the bottom T-DNA strand. Any DNA placed between the T-DNA borders will be transferred to the plant cell, as ss DNA, and integrated into the plant genome as duplex DNA. T-strand synthesis is initiated at the right border and proceeds in the 50 to 30 direction. An ss T-DNA-protein complex is transferred to the plant nucleus. This DNA-protein complex must pass through three membranes (bacterium, plant cell and nucleus). According to the most widely accepted model, the ssT-DNA-Vir D2 complex is coated by the 69 kDa Vir E2 protein, an ss DNA binding protein. The association occurs co-operatively and prevents the attack of nucleases and, in addition, stretches the ssT-DNA strand reducing the complex diameter to approximately 2 nm. This is thought to make the translocation through membrane channels easier. Vir E2 contains two plant nuclear location signals (NLS) and Vir D2 one. This suggests that both proteins (and probably also Vir F) presumably play important roles once the complex is in the plant cell, mediating uptake of the complex into the nucleus. The final step of T-DNA transfer is its integration into the plant genome. The mechanism effecting T-DNA integration is not yet known. It is, considered that integration occurs by illegitimate recombination processes. According to this model, pairing of a few bases, known as microhomologies, are required for a pre-annealing step between T-DNA strand coupled with Vir D2 and plant DNA. The microhomologies can provide, a minimum specificity required for the recombination process. The Ti system probably inserts only one T-DNA segment per bacterium into the host's nuclear genome, but several bacteria could infect the same host cell, effectively increasing the T-DNA copy number in the genome. Some of the inserts may be located on the same or on different chromosomes. Upon integration, some of them may have undergone rearrangements or may be aberrant.

The T-DNA encodes enzymes that are involved in the biosynthesis of phytohormones, such as auxin and cytokinin, as well as amino acid derivatives known as opines. Over-production of phytohormones eventually results in the neoplastic growth that is visible as a crown gall. Opines are catabolised specifically by the stains of *Agrobacterium* that has induced a particular tumour. The enzymes for the catabolism of opines are encoded in the Ti plasmid.

Vector Systems for *Agrobacterium*-Mediated Transformation

As soon as it has been demonstrated that the formation of a crown gall represented the genetic transformation of plant cells, attempts to develop vector systems were initiated. Various sophisticated plant-transformation vectors, designed on the basis of this naturally occurring gene-transfer mechanism, have been developed. Two problems had to be overcome initially: (1) oncogenes cause tumours and inhibit plant regeneration an (2) Ti plasmids are more than 190-kb long, and they are too big for standard molecular cloning techniques. The first achievement in this field was removal of wild type T-DNA from Ti plasmids to create 'disarmed strains' such as LBA4404. Then two strategies were invented for the introduction of engineered T-DNA into *A. tumefaciens.*

In one case, *E. coli* vectors are used that can be integrated into a disarmed Ti plasmid. Various genes, including selection marker genes and genes of interest, can be placed in the vector and eventually introduced into a disarmed strain. This strategy is referred to as the intermediate vector. The second strategy exploits the facts that the process for transfer of T-DNA is active even if the virulence genes and the T-DNA are located on separate replicons in an *Agrobacterium* cell. This strategy employs small plasmids that can be replicated in both *Agrobacterium* and *E. coli.* Artificial T-DNA is constructed within the plasmids and then introduced into disarmed strains, in a so-called binary vector system.

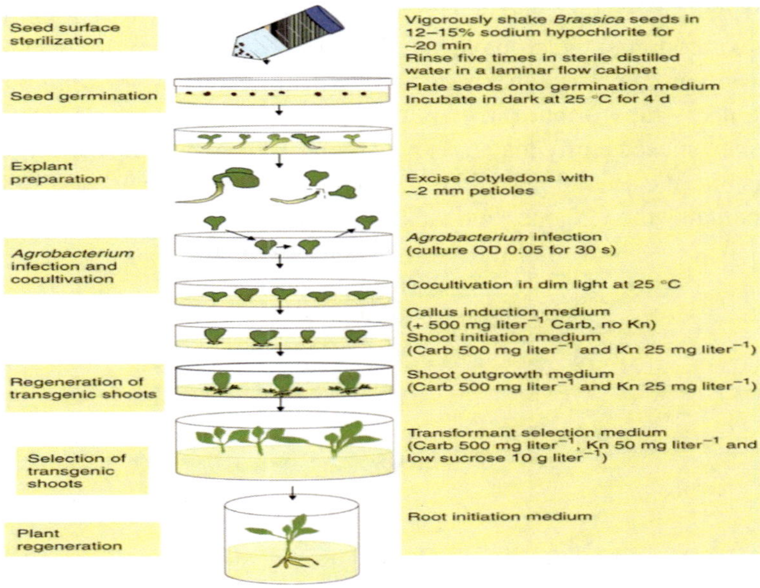

Fig. 3 : A schematic representation of the procedure for *Agrobacterium*-Mediated gene transfer and regeneration (Nature Protocols 3, 181-189; 2008. *Agrobacterium*-mediated transformation of *Brassica napus* and *Brassica oleracea*).

Since 1983, many different strains of *Agrobacterium* have been used as delivery systems. Each strain differs in its ability to infect and transform plants. Thus, certain bacteria have proven to be more effective on some plants and less effective on others. Although the composition of the Ti plasmid may be different between *Agrobacterium* strains, the mechanism of integration of the T-strand is the same. The first method employed to produce a vector to deliver genes of interest to plants by using Ti plasmids relied on cointegration of a plasmid with the gene of interest into the T-region in the Ti plasmid. This method was used to overcome many of the problems associated with the manipulation of large pieces of DNA. However, cointegration vectors have the disadvantage that they contain large areas of DNA that are not required subsequent to the transformation and regeneration of transformed cells. Therefore they are not considered suitable to be used for the production of agronomically important GM crops. As the technology progressed, the long and cumbersome cointegration method was replaced by the binary system, which is now the standard method for *Agrobacterium* vector production. This system was developed when it was discovered that the genes responsible for, integration might act in *trans* independent of the presence of the T-region. The binary vector system consists of two plasmids; one is a broad host vector that replicates in *E. coli* as well as in *A. tumefaciens*, and that contains a bacterial selectable marker, a bacterial origin of replication, and an integration region into which the gene of interest is inserted. The integration region is bordered by the T-region left and right border sequences and contains a multiple cloning site and, for example, an nptII gene for KanR selection in plants. The gene of interest is fused to a strong eukaryotic promoter for expression in the crop plant and is inserted into the multiple cloning site. Although the binary vector can be introduced directly into *Agrobacterium via* electroporation or transformation, it is often first transformed into *E. coli* and transformants are subsequently mated with a strain of *Agrobacterium* that possesses a vir helper plasmid (the second plasmid of the binary system), which is essentially a Ti plasmid that contains the vir genes and lacks the T-region.

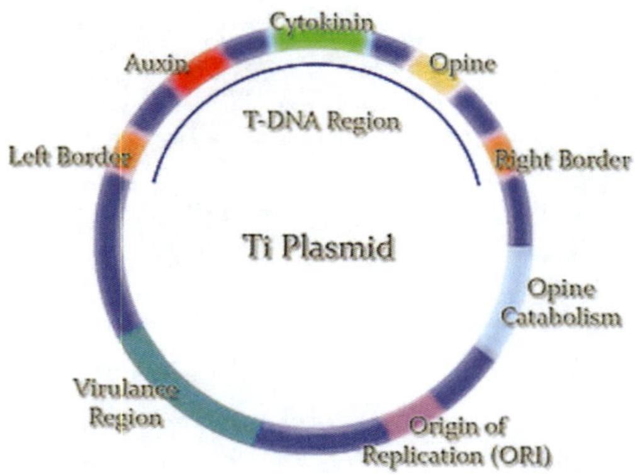

Fig. 4 : The general organization of Ti plasmid

Agrobacterium cells containing the vir helper plasmid and the binary vector are selected and allowed to infect plant cells/protoplasts. The trans-acting vir gene products from the helper plasmid allow the transfer of the binary vector's integration region along the normal *Agrobacterium* integration pathway, as described above. Studies of natural tumours have shown that foreign DNA is confined to the T-DNA region only, but recent work has shown that, when using binary vectors, transformed cells can sometimes contain sequences from outside of the T-DNA, i.e. from the vector backbone. The structure of the small plasmid carrying the vir genes may frequently result in integration of vector backbone sequences. One example is a transgenic potato that has been submitted for marketing approval. There are several advantages to the binary system over the cointegration system, one most important being that it is possible to limit the transferred DNA to its smallest size, which is bordered by the left and right border sequence.

2.1.2. Agrolistics

The agrolistics approach is the combination of biolistic and *Agrobacterium* T-DNA insertion mechanism, minimizing the regions of homology contributing to genetic and/or epigenetic instability. Biolistic transformation is the method of choice for some plant species but many of the integration events are not desirable resulting from these transformation. By combination of *Agrobacterium*-mediated transformation it is possible to target relatively predictable inserts in plants that are not able to transformed using *Agrobacterium*. Transformation by particle bombardment often leads to intregation at one locus of complex array of multiple copies of the introduced genes including plasmid vector, often fragmented and rearranged. These multiple copies are genetically linked; they cannot be segregated during subsequent breeding. Multiple copies of transgenes can interact to inactivate each other and related host genes by epigenetics mechanism variously labelled "co-suppnession" or "gene silencing". In addition, multiple copies may cause genetic instability during homologous recombination. Reduction of the copy number of inserted transgenes and simplification of their arrangement should prove beneficial for maintaining the fidelity and expression of introduced genes to over come these reasons. The foreign genes integration pattern introduced via *Agrobacterim*-mediated transformation is strikingly different from particle bombardment of plant cell. The number of copies of intact and rearranged transgenes resulting from biolistic delivery exceeds, than, the copy number of transgenes introduced by the *Agrobacterium* system into plants. The agrobiolistic approach thus combines the advantages of efficient biolistic delivery with the elegance and precision of the *Agrobacterium* T-DNA insertion mechanism to afford a new, widely applicable technology for producing transgenic crop plants of value to agriculture. Agrolistic transformation allows integration of the gene of interest without the undesired vector sequence, using plant expression cassettes for vir D1 and vir D2 genes co-delivered with a vector containing T-DNA border sequences flanking a gene of interest, resulting in production of transformants without the extraneous vector DNA as a result of T-DNA border cleavage by vir D1 and vir D2 gene products.

2.1.3. Sonication-assisted agrobacterium-mediated transformation (SAAT)

An important modification in *Agrobacterium*-mediated transformation involves subjecting the plant tissue to brief periods of ultrasound in the presence of *Agrobacterium*. SAAT treatment produces a large number of small and uniform wounds throughout the tissue, allowing easy access to the *Agrobacterium*, resulting in improved transformation efficiency in several different plant tissues including immature cotyledons, leaf tissue, suspension cultures, somatic and zygotic embryos. A 100-1400-fold increase in transient β-glucuronidase expression has been demonstrated in different plant species such as soybean, cowpea, white spruce, wheat, and maize, allowing increase in transformation rates in these species which are more recalcitrant to *Agrobacterium*-mediated transformation. The major problem in the development of transformation systems is providing induced *Agrobacterium* with access to cells capable of dedifferentiation followed by regeneration. In species such as tobacco and *Brassica napus*, this combination can be achieved with relative ease; however, it is often difficult to combine transformation competence with totipotency. Tissue culture independent transformation systems have also been demonstrated in various crops such as soybean, *Arabidopsis*, sunflower, and peanut. Over the past decade and half, several tissue culture independent methods for *Agrobacterium*-mediated transformation have been developed. *Arabidopsis* seeds infected with *Agrobacterium* and allowed to grow into mature plants *in vivo* resulted in about 1% transformation frequency. Inoculation of *Agrobacterium* onto wounded sites arising from cutting away inflorescences of *Arabidopsis* yielded transformed seeds from newly emerging inflorescences. Progress has been made on non-tissue culture-based approaches for generating transgenic.

2.1.4. Floral-dip method

In this method the plants are transformed by direct application of *Agrobacterium* to floral tissues, bypassing the tissue culture technique and thereby eliminating the somaclonal variations. Subsequent studies demonstrated the use of female gametophytes of immature flowers as targets of floral-dip transformation in *Arabidopsis*. The floral-dip method requires considerably less time and effort than vacuum infiltration, resulting in greater yields. This technique is an improvement of vacuum infiltration, either by simply dipping of developing floral tissues into a solution of *Agrobacterium* or a floral spray or a combination of floral dipping and vacuum infiltration. This efficient plant transformation procedure relies on the gene transfer by *Agrobacterium* to the female reproductive tissues in the flower. Seeds are collected from the transformed plant and transformants can subsequently be selected for either by antibiotic or herbicide resistance markers present in the transferred DNA.

2.1.5. Vacuum infiltration method

The vacuum infiltration method of transformation has been applied to a number of crops, particularly monocots, to avoid both in vitro culture and regeneration steps during transformation. The cells of a plant when subjected to a vacuum environment establish a more intimate contact with *Agrobacterium*. This method has been reported for Rapeseed, *Brassica*, and *Arabidopsis*. Concentrated solutions of *Agrobacterium* were applied in plant and infiltrated the tissue under a vacuum. It is thought that the negative pressure forces the bacteria into the host cells, thus increasing the susceptibility of the host. Furthermore, the vacuum process stimulates the plants wound response, again increasing the virulence of the bacteria. The method of vacuum infiltration has made it possible to obtain transformed plants without plant tissue culture or regeneration. It is simple and reliable, and transformed plants can be obtained at sufficiently high rates. Improvements related to this technique floral dip method.

2.1.6. Limitations of biological delivery systems

Currently, *Agrobacterium* vectors are used in the successful transformation of many dicots. However, the fundamental limitation of this system is the reduced ability of the system to transfer DNA to cells of monocots. The *Agrobacterium* system requires physically injured plant tissue that the bacteria can enter. Whether the procedure involves the whole plant or plant culture, such as calli, protoplasts, or leaf disc, the process requires the regeneration of a viable plant (as with plant tissue cultures), or the generation of viable transformed gametes. Thus, the *Agrobacterium* system will not function with plants that cannot be regenerated from culture or do not tolerate physical injury. In addition, if surface sterilisation has not been carried out sufficiently well, the Agrobacteria may proliferate further, along with the transformed plant and consequently may be released into the environment where uncontrolled transformation may further occur. Finally, the *Agrobacterium* binary vector system relies completely on the functioning of the Vir proteins, which target only to the nuclear genome and not to organelle genomes. For those purposes, alternative delivery systems, such as particle bombardment must be employed.

2.2 Physical Delivery Systems

2.2.1. Biolistic transformation

In 1987, the microprojectile-mediated transformation was published, which involved shooting of DNA-coated metal particles into plant cells. The "brute force" approach gives the microprojectile technique the unique ability to transform any plant tissue.

Mechanical forces, generated by high velocity particles, drive the foreign DNA past all biological barriers, allowing for genomic integration. Furthermore, this approach has the unique ability to target the transforming DNA to the chromosomal DNA of a cell and also to the chloroplast and mitochondrial DNA. In the initial experiments, DNA was bound to tiny tungsten particles, approximately one micrometer in diameter and accelerated with a gunpowder driven piston, fired at the target cell with a velocity of about 430 meters per second. Host cells in the front line are usually destroyed, but just behind, an area of cells exists where the projectiles penetrate the cell without killing them. Some of the cells that survive the bombardment incorporate the DNA from the projectile into the genome. After successful incorporation, and if the inserted DNA did not knock out any critical gene function, host cells may begin to express the gene product of the foreign DNA. A variation on the microprojectile transformation is an electric discharge particle acceleration device (DPAD). This apparatus uses shock waves to accelerate the particles. The force of the shock wave drives a mylar sheet containing particles and precipitated DNA towards a retaining screen. The mylar sheet is stopped by the screen, but the particles and DNA, now at maximum velocity, continue past the screen and collide with the target tissue. As technology progressed, other improvements, which significantly lowered the cost of these experiments, were developed.

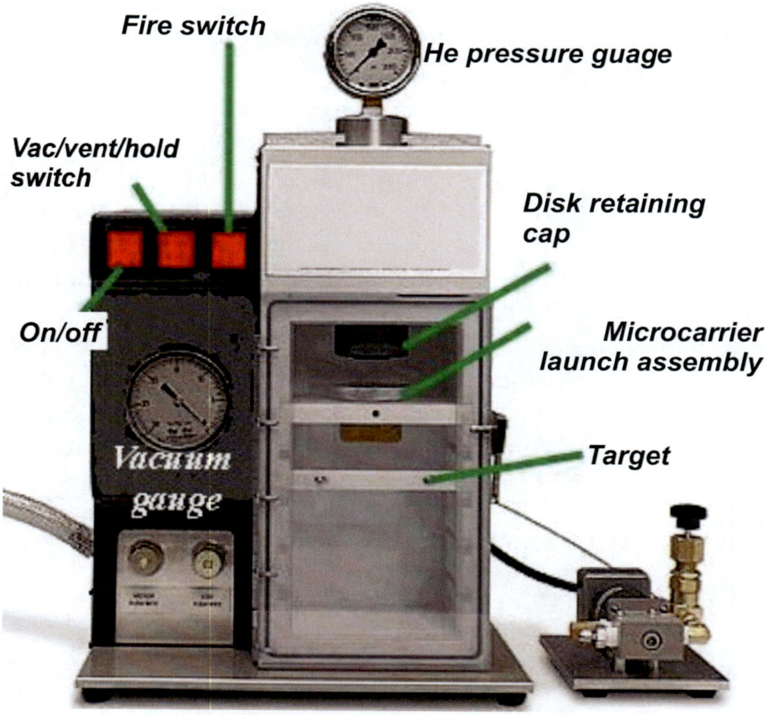

Fig. 5. Biolistic Gene (Gunhttp://www.uoguelph.ca/plant/research/homepages/raizada/Maize.html)

DNA Preparation and Delivery

DNA Construct: Most commonly, plasmids are used in microprojectile treatments. However, the size of the plasmid is limited, as large plasmids will be sheared on impact and gene transfer to the cell is reduced. The plasmids contain a bacterial selectable marker and an E. coli origin of replication for propagation in *E. coli*, a multiple cloning site for insertion of foreign DNA, and a marker gene fused to a promoter active in plants for identification of transformants. The constructs are generated *in vitro*, amplified in *E. coli*, and fused to the delivery particles.

Microprojectiles: Typically, metal particles have been used for microprojectile transformation, as a minimum density (19 g/cm3) is needed to generate enough force to penetrate plant cells. A further requirement is that the size of the particles is proportional to the target cell size. Tungsten and gold particles are the most common, as they are sufficiently dense and chemically inert.

Binding DNA to microprojectiles: To prepare particles for tissue bombardment the DNA must be precipitated onto the particle. The most commonly used precipitation procedure for plasmids and tungsten uses $CaCl_2$ and spermidine. This $CaCl_2$/ spermidine procedure has also been maximised for different tissues and microprojectile transformation systems.

Integration: The particle/plasmid precipitate suspension is applied to the microprojectile acceleration device. The tissue is bombarded by the particles and the plasmid DNA dissociates from the particle upon entering the cell. Once in the nucleus, plastid or mitochondrion, the plasmid is free to undergo recombination to become incorporated into the genome.

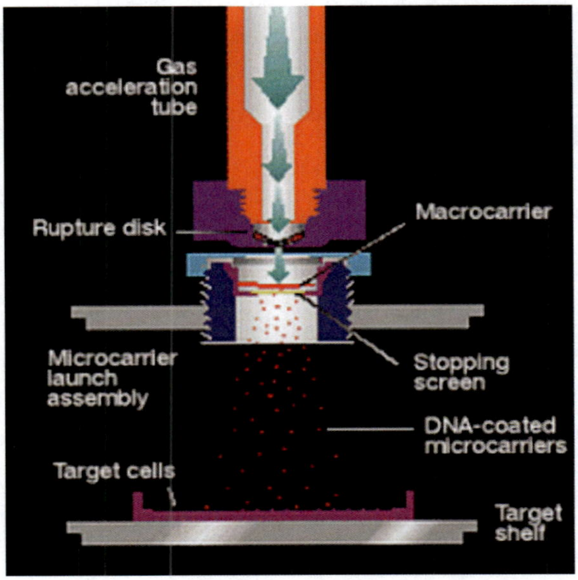

Fig. 6 : A representation of components of particle gene gun device: The microprojectiles are coated with the DNA to be delivered in the target tissue. Target tissue is usually placed below the stopping screen according to explant.

2.2.2. Polyethylene glycol (PEG) mediated genetic transformation

A number of methods are there for introduction of DNA into plant protoplasts, but the most commonly used technique is the polyethylene glycol (PEG-mediated DNA uptake). The PEG-mediated transformation is simple and efficient, to allow processing of many samples simultaneously. Survival and division rates of transformed cell population is high in this process. The method utilizes inexpensive supplies and equipments, and helps to overcome a hurdle of host range limitations of *Agrobacterium*-mediated transformation.

The PEG-mediated DNA transfer can be readily adapted to a wide range of plant species and tissue sources. PEG-mediated direct gene transfer technique was established originally for tobacco protoplasts. This method was used for different system sugarcane, orchardgrass, maize, barley and rice, *Arabidopsis thaliana*, wheat etc. Protoplasts from leaf mesophyll tissue, root and cell suspension cultures can be use in this method. Protoplast is washed with 0.45mM mannitol, and protoplast pellet is resuspend in MaMg solution (Mannitol, $MgCl_2$, $6H_2O$, MES morpholine ethane Sulfonic acid, pH 5.8). Plasmid DNA is added to protoplast suspension shake gently for 4 to 5 min. 40% PEG solution is used for transformation.

Polyethylene glycol (PEG) mediated genetic transformation of plant protoplasts with Ti plasmid was first reported by Krens et al.(1991). This gene transfer method was further developed as a Ti plasmid-independent direct gene transfer approach. Protoplasts isolated from suspension culture cells are a target tissue in the genetic transformation of cereals, particularly in rice, and maize. Old suspension culture cells of 3 to 5 days after subculturing are normally used to isolate protoplasts for transformation, and protoplasts from 3-20 months old suspension cultures after initiating have been successfully used to produce transgenic plants.

Protoplast of every system has proven transformable, though with different efficiency. The recovery of transgenic plants from protoplast is major problem. Plant regeneration from protoplasts is a delicate process and depends upon parameters not under experimental control (e.g., species and genotype-dependent competence for wound response and regeneration). PEG transformation has generally led to relatively low frequencies of transformation (less than 1% of treated cells). Nevertheless, due to the availability of a large number of cells in such systems, a number of transgenic plants can be produced by using effective selection systems.

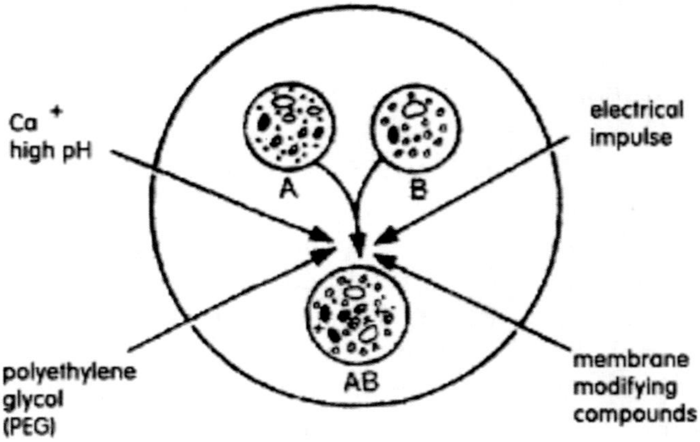

Fig. 7 : Polyethylene glycol (PEG) mediated genetic transformation

The most common method of delivering foreign DNA into plant protoplasts involves treatment with PEG. The mechanism of PEG mediated transformation is not clear, how PEG causes protoplasts to take up DNA? But it is thought that it induces reversible permeabilisation of the cytoplasmic membrane, which allows passage of macromolecules. The yields of transformants (approximately 1–2%) in this system i. e. very low and the regenerated into viable plants from protoplasts (approximately 0.1%) is difficult in many species, making the whole system only about 0.0004% effective.

2.2.3. Lipofection

An advance upon PEG-mediated transformation was the liposome-mediated transformation technique. Foreign DNA is encapsulated in a spherical lipid bilayer termed a liposome. In the presence of PEG, the host protoplast will bind and envelop the liposome through endocytosis. Endocytosis, allows the DNA to recombine and integrate into the host genome. As with other transformation systems, a variety of vectors including viral vectors can be employed in this system. Liposomes are small lipid bags, in which large number of plasmids are enclosed. They can be induced to fuse with protoplasts using devices like PEG, and therefore have been used for gene transfer. The technique offers following advantages: (i) protection of DNA/RNA from nuclease digestion, (ii) low cell toxicity, (iii) stability of nucleic acids due to encapsulation in liposomes, (iv) high degree of reproducibility and (v) applicability to a wide range of cell types. In this technique, DNA enters the protoplasts due to endocytosis of liposomes, involving the following steps.

Fusion of Plasmid-Filled Liposomes with isolated protoplasts as a method of Plasmid uptake. (i) adhesion of liposomes to the protoplast surface, (ii) fusion of liposomes at the site of adhesion and (iii) release of plasmids inside the cell. The technique has been successfully used to deliver DNA into the protoplasts of a number of plant species (e.g., tobacco, petunia, carrot,wheat etc.) and transformation efficiencies reported were in the order of 6%.

2.2.4. Electroporation

An alternative to the method of permeabilising a membrane is electroporation, the exposure of cells to a short-time intense electric field. This technique, which is commonly used in bacterial transformation, is also being applied to plant protoplasts. Transformation by electroporation has been achieved in a variety of species and tissue types, but often species-specific protoplast regeneration problems still necessitate the use of other techniques. Electroporation is a technique that utilizes a high intensity electric pulse to create transient pores in the cell membrane thereby facilitating the uptake of macromolecules like DNA. The survival percentage of protoplasts after electroporation depended largly on the tissue of origin. This method is based on the use of short electrical pulses of high field strength. These pulses increase the permeability of protoplast membrane and facilitate entry of DNA molecules into the cells, provided the DNA is in direct contact with the membrane. In view of this, for delivery of DNA to protoplasts, electroporation is one of the several routine techniques for efficient transformation. However, since regeneration from protoplasts is not always possible, cultured cells or tissue explants are often used. Consequently, it is important to test whether electroporation could transfer genes into walled cells. In most of these cases no proof of transformation was available.

The electroporation pulse is generated by discharging a capacitor across the electrodes in a specially designed electroporation chamber. Either a high voltage (1.5 kV) rectangular wave pulse of short duration or a low voltage (350 V) pulse of long duration is used. Successful transfer of genes was achieved with the protoplasts of tobacco, petunia, maize, rice, wheat and sorghum. In most of these cases CAT gene associated with a suitable promoter sequence was transferred. Transformation frequencies can be further improved by (i) using field strength of 1.25 kV/cm, (ii) adding PEG after adding DNA, (iii) heat shocking protoplasts at 45°C for 5 minutes before adding DNA and (iv) by using linear instead of circular DNA. Electroporation of protoplasts thus considerably improved the gene transfer efficiency as compared to that achieved with PEG mediated approach.

The technique of electroporation, initially utilized for the introduction of foreign genes into protoplasts, was later extended for the transformation of organized tissues. In wheat, the transfer of reporter genes by elecroporating zygotic wheat embryos has been reported. The embryos were arranged with the scutella side facing the negative electrode, so that the negatively charged plasmid DNA will contact the target cells while moving towards the positive electrode. Transient expression of the anthocyanin regulatory gene and the *gus* gene facilitated the identification of transformed scutellum cells. Introduction of *gus* gene was also reported by electroporating organized tissue from slow-growing, embryogenic calli. The factors which played a significant role in affecting the electroporation efficiency were the electroporation voltage, the pulse length, volume of the electroporation buffer, osmoticum of the electroporation buffer, and medium, the osmoticum of the pre-electroporation medium and the pre-electroporation incubation time and temperature. Recently in 2000, the production of fertile transgenic wheat plants by this method from intact immature embryos was reported by at a transformation frequency of 0.4%. Molecular evidence confirmed the stable integration and the inheritance of the *gus* and *bar* transgenes in the T_0 and T_2 plants. The electroporation efficiency was found to be dependent on a combination of factors including pretreatments of the recipient tissue and the culture conditions. It thus appears that tissue electroporation may emerge as a routine transformation method for wheat due to its technical simplicity and low cost associated with this procedure.

2.2.5. Silicon carbide fibres

The size, shape, and chemical composition of silicon carbide fibres make them capable of puncturing the cells without killing them. The technique is quite simple: plasmid DNA (containing a marker gene), silicon carbide fibres, and a suspension of cultured plant cells are added to a tube and mixed vigorously. The hydrodynamic forces drive the fibres and DNA into the cells. The advantages of this system are that it is rapid, inexpensive, and easy to setup and that it can work on a variety of cell types. The successful use of silicon carbide fibers in stable transformation of maize and *Nicotiana* raised the possibility of its use for the introduction of foreign DNA into monocots and dicots as a physical method of transformation. In wheat, the preferential expression of the *gus* gene in the scutellum and the epiblast of mature embryos was observed following vortexing of the explants in a solution containing silicon carbide fibers and plasmid DNA. Later on Silicon carbide fibers have been used for causing wounds in the immature wheat embryos for *Agrobacterium*- mediated genetic transformation. Immature embryos were put in a 5% silicon carbide fiber (SCF) suspension and vortexed for 2 or 3 min followed by co-cultivation with *Agrobacterium tumefaciens* strain LBA4404 SCFs can be used for wounding to improve frequency of transformation by disarmed *Agrobacterium* strains.

2.2.6. Microinjection

This technique uses fine glass needles, which inject the foreign DNA directly into protoplasts or cultured embryonic cells. Although it has a fairly high transformation frequency (20-50%), microinjection is a time consuming process that requires specific equipment and considerable training. Simmonds and coworkers reported the use of microcapillaries for the introduction of plasmid DNA into the germ line precursor cells of apical meristems by microinjection. Despite a positive indication by PCR amplification of DNA isolated from injected apices, this approach has failed to yield any transgenic plants. When cells or protoplasts are used as targets in the technique of microinjection, glass micropipettes each with 0.5-10μm diameter tip are used for transfer of macromolecules into the cytoplasm or the nucleus of a recipient cell or protoplast. The recipient cells are immobilized on a solid support (cover slip or slide, etc.) or artificially' bound to a substrate or held by a pipette under suction (as done in animal systems). Often a specially designed micromanipulator is used for microinjecting the DNA. Although, this technique gives high rate of success, the process is slow, expensive and requires highly skilled and experienced personnel. The main disadvantage of this technique is the production of chimeric plants with only a part of the plant transformed. However, from this plant transformed plants of single cell origin can be subsequently obtained. Utilizing this approach, transgenic chimeras were actually obtained in several crops including oilseed rape (*Brassica napus*).

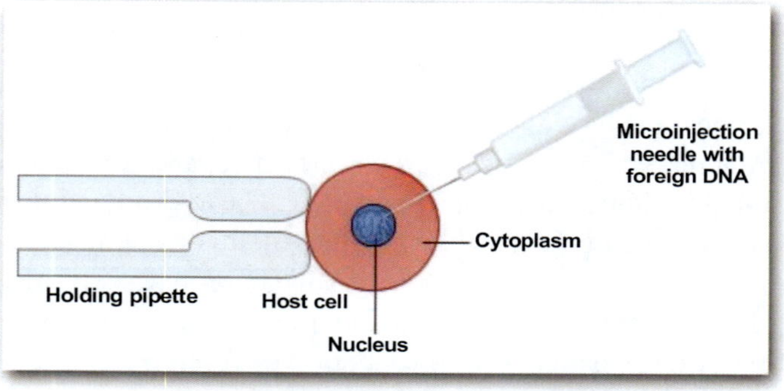

Fig. 8 : Microinjection: Fine glass needles, injecting the foreign DNA directly into protoplasts.

2.2.6. Desiccation

Research repot has shown that dried embryos have the unique ability to take up DNA during rehydration. This approach takes advantage of the natural changes that occur within the embryonic plant cell wall during rehydration. Dried embryos can be mixed with a nutrient solution containing the foreign DNA. The DNA would be taken up as the embryo rehydrates and seedlings can be germinated in the presence of a selection medium to assess the incorporation of the foreign DNA.

2.2.8. Limitations of physical delivery systems

The most obvious drawback of the microprojectile systems is tissue damage and the relatively low yield of stable transformants. Generally, the outer layers of cells, coined the "zone of death", will be destroyed by the impact of the particles, particle fragmentation, or air blasts. The system is also limited in the size of DNA constructs that can be used. As was done with the biological delivery systems, viral vectors have been developed to increase transformation efficiency and marker gene expression. It has been reported that both DNA viruses such as Gemini viruses and ribonucleic acid (RNA) viruses, zucchini yellow mosaic virus (ZYMV) can be successfully integrated in microprojectile systems. As noted before, viral vectors can enhance expression through increased copy number by self-replication and integration. Yeast artificial chromosome (YAC) is also being applied to the microprojectile technique. YAC was specially designed for this process and the results suggest it could be used for map-based cloning of plant genes. Furthermore, YAC could also be used to create genetic libraries in transgenic plants.

3. Transformation Vector

Most commonly used plant transformation vectors have features required for various recombinant DNA manipulations that include multiple unique restriction sites, bacterial origins of replication, and prokaryotic selectable markers for plasmid selection and maintenance in *Escherichia coli* (e.g., antibiotic resistance). In addition, these vectors contain specific selectable marker genes engineered for expression in plants that may be used directly as transformation vectors in physical DNA delivery strategies such as particle bombardment. However, for *Agrobacterium*-mediated gene transfer, these vectors need additional features such as wide host range replication and transfer functions to allow conjugation from *E. coli* to *Agrobacterium* and plasmid maintenance in both bacterial hosts.

3.1. Scorable Markers

Screenable marker 'reporter genes' have also been developed from bacterial genes coding for easily assayed enzymes, such as chloramphenicol acetyl transferase (CAT), β-galactosidase, β-glucuronidase (GUS), luciferase (LUX), green fluorescent protein (GFP) nopaline synthase and octopine synthase. The utility of any particular marker gene construct as a transformation marker varies depending on the plant species and explant involved. Initial steps for genetic transformation involve delivery of a gene cassette into recipient cells followed by analysis of the expression of delivered gene. The results of the above events can be detected by assaying the expression of a reporter gene introduced into plant cell cultures or intact tissues. The reporter genes produce a visible effect, directly or indirectly, due to their activity in the transformed cells. Analysis of reporter gene expression does not require the integration of the transgene into the host genome and is commonly used to test promoter and gene functions. Initial studies on the introduction of foreign DNA to plant have relied on the use of the *E. coli* gene for the enzyme chloramphenicol acetyl transferase (*cat*). However, the detection of transgene by enzymatic and immunochemical methods for measurement of CAT activity and amount of CAT protein, respectively, was tedious; and moreover, presence of inhibitors of CAT activity and endogenous CAT activity hampered its use as a reporter gene. With the availability of a protocol for the rapid assay of β-glucuronidase (*gus*) gene from *E. coli*, this gene has emerged as the most widely used scorable marker in plant transformation. GUS enzyme hydrolyzes β-glucuronide compounds and gives reaction products that can be quantified spectrophotometric or spectrofluorometrically. The *gus* reporter gene system is extremely useful for optimisation of parameters for genetic transformation, due to the availability of a simple histochemical detection procedure. One of the major limitations of *gus* reporter gene system, however, is the destructive nature of its assay.

Fig. 9 : 35S: GUS marker gene

Thus, to study the fate of introduced transgenes in living cells, vital reporter genes encoding for anthocyanin biosynthesis, green fluorescent protein, and firefly luciferase have been used successfully. The R genes of *Zea mays* which stimulate endogenous anthocyanin accumulation in the vacuoles of plant tissues is useful as a scorable marker in mature and differentiated cells. Due to high sensitivity and ease of visualization, the use of R genes in wheat transformation have been reported by different groups working on it. A synthetic, spectrally modified, version of green fluorescent protein (*gfp*) from the jellyfish, *Aqueorea victoria* has also been used as a vital marker in wheat transformation. For the first time researchers had accessibility to a marker gene that was useful in vivo, could be detected in real-time, and required no substrates or cofactors. Subsequently, the gene was expressed in *Escherichia coli* and *Caenorhabditis elegans*. Shortly thereafter, the *gfp* gene was subjected to mutagenesis and several new variants were created ranging from blue to yellow. These events marked the beginning of what has often been called the "Green Revolution" of biotechnology. When a novel protein is introduced into an organism, toxicity is a concern and it has been suggested that GFP is cytotoxic to plant cells. The argument was made that GFP fluorescence caused a photonic disturbance that created free radicals and eventual oxidative damage. However, many researchers failed to observe this toxicity in plants that were clearly expressing high levels of GFP. To address the toxicity issue at the whole plant level, Harper et al. tested tobacco in the field using three different GFP variants (mGFP4, mGFP5-ER, and sGFP). Over two growing seasons, seed yield and biomass were recorded. In this study, there was no cost to yield or biomass; therefore, it was concluded GFP is not toxic to plant cells. But Liu et al. showed GFP toxicity to mammalian cells.

Therefore, a transgenic monitoring system utilizing FPs to tag whole plants would be useful. In the past, ecological genetics focused on phenotypic traits or DNA markers. However, with FPs one would only have to look for a fluorescent signature. With this type of monitoring system, FPs can be genetically linked so that they are inherited along with transgenes in the event of gene flow. This would allow the tracking of transgene movement and their ecological effects, an important component in biotechnology risk assessment. This approach has been demonstrated as an effective tool when either whole canola plants or pollen alone were tagged with GFP. When whole plants were tagged, GFP was inherited along with the *Bt* insect resistance gene after transgenic canola was crossed with its wild relative *Brassica rapa* (field mustard). It has also been observed that GFP and Bt were still present together after several successive back-crosses to *B. rapa*. When pollen was tagged with GFP, it was observed that transgenic pollen captured several meters away could be distinguished from non-transgenic pollen. It would be useful to track transgenic pollen in this manner because the information gathered could then be used in risk assessment. It is also important to note that FP-tagged transgenic plants can be monitored non-destructively, in real time, and perhaps remotely. This type of monitoring could be extended to insects ,domestic animals and aquatic organisms (Millwood et al. 2008).

Fig. 10 : Transgenic fluorescent organisms. A Glofish , expressing GFP, RFP, and GFP + RFP. The top leaf is wild type and all others are expressing varying amounts of the protein].

Luciferase gene from the fire fly, *Photinus pyranus*, has been successfully used in stable transformation. Luciferase and modified versions of *gfp* permit non-destructive analysis of transgene activity, ease to follow the fate of introduced transgenes in individual tissue samples, facilitate rapid assessment and comparison of different transformation procedures, and provide valuable insight into the conditions influencing the efficiency of DNA integration and stable expression.

3.2. Selectable Markers

Most vectors used for the genetic transformation of plants carry 'marker' genes that allow the recognition of transformed cells, by either selection or screening. These genes are dominant, usually of microbial origin, and placed under the control of strong and constitutive, eukaryotic promoters, often of viral origin. The most popular selectable marker genes used in plant transformation vectors include constructs providing resistance to antibiotics such as kanamycin, chloramphenicol, and hygromycin, and genes that allow growth in the presence of herbicides such as phosphinotricin, glyphosate, bialaphos, and several other chemicals. For successful selection, the target plant cells must be susceptible to relatively low concentrations of the antibiotic or herbicide in a non-leaky manner. The selection regimes for transformed cells are based on the expression of a gene termed as the selectable marker producing an enzyme that confers resistance to a cytotoxic substance often an antibiotic or a herbicide. The most commonly used selection marker in plant transformation is the *bar* gene (bialaphos resistance gene) encoding for phosphinothricin acetyl transferase (*pat*). Both *bar* and *pat* genes isolated from different *Streptomyces* species, encode for phosphinothricin acetyl transferase. Amongst the antibiotic resistance markers, the bacterial neomycin phosphotransferase II (*npt*II) gene providing resistance to aminoglycoside antibiotics is commonly used in plant transformation. Herbicide resistance genes offer an alternative to antibiotic-resistant markers. Genes coding for 5-enolpyruvyl-shikimate-3-phosphate synthase (EPSPS), a critical enzyme for aromatic amino acid biosynthesis, and phosphinothricin acetyl transferase (PAT) provide tolerance to glyphosate and glufosinate ammonium herbicides, respectively. The enolpyruvylshikimate-phosphate synthase (*CP4*) gene isolated from *Agrobacterium* strain CP4 and the glyphosate oxidoreductase (*GOX*) also provide tolerance to glyphosate by degrading glyphosate into aminomethyl phosphoric acid. The hygromycin phosphotransferase (*hpt)* gene, widely used in rice transformation, has also been reported as an efficient selectable marker for achieving stable genetic transformation. The efficiency of *hpt* as a selectable marker is comparable to the other widely used selectable markers encoding resistance for antibiotics and herbicides. Mannose-6-phosphate isomerase (MPI), encoded by *manA* from *E. coli,* had been developed as a selectable marker. This marker allows transformants to grow on mannose as a sole carbon source and has a positive mode of action which encourages the growth of the transformed tissues rather than just permitting it. Cyanamide hydratase (*Cah*) gene is also use as a selectable marker for transformation. The *Cah* gene gives the transformed tissues the ability to grow on cyanamide-containing media by converting cyanamide into urea (which can be used as a fertilizer source).

4. Chloroplast Engineering

The concept of chloroplast transformation was conceived in the mid-1980s and this concept was further advanced by inserting transgenes into functional operons and transcriptionally active spacer regions of the chloroplast genome (Daniell et al., 1998). Chloroplast transformation is an environmental friendly approach to plant genetic engineering that minimizes out-crossing of transgenes to related weeds or crops and reduces the potential toxicity of transgenic pollen to non-target insects. Because the plastid genome is highly polyploid, transformation of chloroplasts permits the introduction of thousands of copies of foreign genes per plant cell, and generates extraordinarily high levels of foreign protein. Chloroplast transformation vectors use two targeting sequences that flank the foreign genes and insert them, through homologous recombination, at a precise, predetermined location in the organelle genome. This results in uniform transgene expression among transgenic lines and eliminates the 'position effect' often observed in nuclear transgenic plants. Gene silencing, frequently observed in nuclear transgenic plants, has not been observed in genetically engineered chloroplasts. The ability to express foreign proteins at high levels in chloroplasts and chromoplasts, and to engineer foreign genes without the use of antibiotic resistant genes, make this compartment ideal for the development of edible vaccines.

Moreover, the ability of chloroplasts to form disulfide bonds and to fold human proteins has opened the door to high-level production of biopharmaceuticals in plants. Furthermore, foreign proteins observed to be toxic in the cytosol are non-toxic when accumulated within transgenic chloroplasts.

4.1. Applications of Chloroplast Engineering

Herbicide Resistance

Glyphosate is a potent, broad-spectrum herbicide that is highly effective against grasses and broad-leaf weeds. Glyphosate works by competitive inhibition of an enzyme in the aromatic amino acid biosynthetic pathway, 5-enol-pyruvyl shikimate-3-phosphate synthase (EPSPS). Unfortunately, like most commonly used herbicides, glyphosate does not distinguish crops from weeds, thereby restricting its use. Engineering crop plants for resistance to the herbicide is a standard strategy to overcome the lack of herbicide selectivity. However, this approach raises the concern that if the engineered resistance gene escapes via pollen dispersal, it might result in resistant weeds or might cause genetic pollution among other crops. Engineering foreign genes through chloroplast genomes (which are maternally inherited in most crops) offers a solution to this problem.

Bacterial Operons Via Chloroplast Genomes

Typical plant nuclear mRNAs are monocistronic (i.e. only one polypeptide chain is translated per messenger molecule). This poses a serious drawback when engineering multiple genes. By contrast, most chloroplast genes are co-transcribed as polycistronic RNAs (i.e. they encode multiple proteins that are separately translated from the same mRNA molecule), which are subsequently processed to form translatable transcripts.

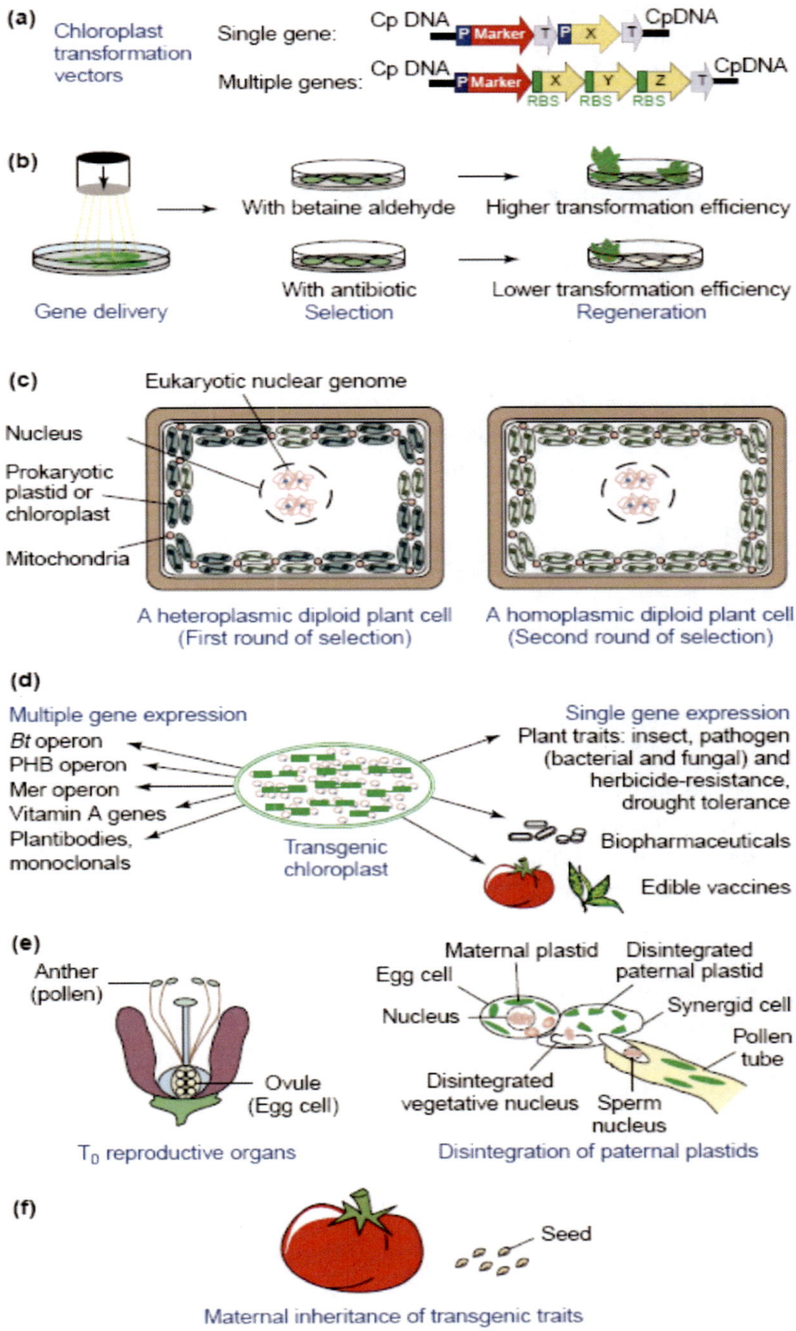

Fig. 11 : Different steps involved in chloroplast genetic engineering (Daniell et al. 2002).

Therefore, introduction of multiple chloroplast transgenes arranged in an operon should allow expression of entire pathways in a single transformation event. Recently, the *Bt cry*2Aa2 operon was used as a model system to test the feasibility of multigene operon expression in engineered chloroplasts. *Cry*2Aa2 is the distal gene of a three-gene operon.

Chloroplast Genome for Pathogen Resistance

Because plant diseases have plagued global crop production, it is highly desirable to engineer plants that are resistant to pathogenic bacteria and fungi. Amphipathic peptides such as magainin have known antimicrobial properties, but their possible uses in agriculture have not been fully explored yet. Recently, a synthetic antimicrobial peptide (MSI-99) was expressed via the chloroplast genome. MSI-99 is an amphipathic α-helical molecule with affinity for the negatively charged phospholipids found in the outer-membrane of bacteria and fungi. Upon contact with these membranes, individual peptides aggregate to form pores, resulting in microbial lysis. Because of the concentration-dependent action of antimicrobial peptides, MSI-99 was expressed via the chloroplast genome to accomplish high-dose release at the point of infection.

Chloroplast Genome for Drought Tolerance

Water stress caused by drought, salinity or freezing is a major limiting factor in plant growth and development. Trehalose is a non-reducing disaccharide of glucose whose synthesis is mediated by the trehalose-6-phosphate (T6P) synthase and trehalose-6-phosphate phosphatase complex in *Saccharomyces cerevisiae*. Because it accumulates under stress conditions such as freezing, heat, salt or drought, there is general consensus that trehalose protects against damage imposed by these stresses. Therefore, engineering high levels of trehalose in plants might confer drought tolerance. Investigations have confirmed that trehalose functions by protecting the integrity of biological membranes rather than regulating water potential. Therefore, this study shows that compartmentalization of trehalose within chloroplasts confers drought tolerance without undesirable phenotypes.

Chloroplast Genome Without Antibiotic Resistance Genes

One disadvantage of chloroplast genetic engineering technology is the use of the antibiotic resistance gene, *aadA*, as a selectable marker. The *aadA* gene product inactivates its selective agents, spectinomycin and streptomycin, by transferring the adenyl moiety of ATP to the antibiotics. Because these antibiotics are commonly used to control bacterial infection in humans and animals, there is concern that their over-use might lead to the development of resistant bacteria. Therefore, several studies have explored strategies for engineering chloroplasts that are free of antibiotic-resistance markers. The spinach betaine aldehyde dehydrogenase (BADH) gene has been developed as a plant-derived selectable marker to transform chloroplast genomes. The selection process involves conversion of toxic betaine aldehyde (BA) by the chloroplast-localized BADH enzyme to non-toxic glycine betaine, which also serves as an osmoprotectant. Because the BADH enzyme is present only in chloroplasts of a few plant species adapted to dry and saline environments, it is suitable as a selectable marker in many crop plants

Chloroplast Genome to Overproduce Biopharmaceuticals

Although pharmaceutical proteins have been synthesized from plant nuclear transgenes, expression levels (particularly of human proteins) are generally disappointingly low. Chloroplasts, with their highly polyploid genomes offer an ideal compartment for overproduction of foreign proteins. An additional significant advantage of using chloroplasts is their potential to process eukaryotic proteins, including folding and formation of disulfide

Chloroplast Genome of Edible Crop Plants

To exploit the chloroplast compartment for the production of orally delivered pharmaceuticals, it is important to extend chloroplast transformation to edible crops. For this reason, the recent development of chloroplast transformation for both potato and tomato is particularly exciting. Western analysis of transgenic potato tissues revealed that the chloroplast transgene (green fluorescent protein, GFP) was expressed to high levels in leaves (5.0% total soluble protein), but accumulated much lower levels in the microtubers (0.05% total soluble protein). This study predicts the feasibility of expressing high-levels of foreign proteins in the plastids of edible plant organs.

Chloroplast Genome to Develop Edible Vaccines

Another major cost of biopharmaceutical production lies in purification; for example, in insulin production, chromatography accounts for 30% of the production cost and 70% of the set-up cost. Therefore, oral delivery of properly folded and fully functional biopharmaceuticals should significantly cut down the production cost. Bioencapsulation of pharmaceutical proteins within plant cells offers protection against digestion in the stomach but allows successful delivery to the target tissues. The a subunits of enterotoxigenic *E. coli* (LTB) and cholera toxin of *Vibrio cholerae* (CTB) are candidate vaccine antigens. Integration of an unmodified CTB gene into the tobacco chloroplast genome results in accumulation of up to 4.1% of total soluble leaf protein as functional CTB oligomers (410-fold higher than the unmodified LTB gene expressed via the nuclear genome).

4.2. Limitations of Chloroplast Genetic Engineering

One barrier to developing plastid transformation for crop plants, including cereals, has been their regeneration from non-green embryonic cells (containing proplastids)

rather than leaf cells (containing chloroplasts). Plastid transformation in rice has been shown using vectors containing a translational fusion of *aadA* with *gfp*. Inspection of streptomycin-resistant rice shoots by confocal microscopy revealed a few proplastids with GFP fluorescence. Although this report only showed heteroplasmy, it represents a step in the right direction towards monocot plastid transformation.

A major limitation in accomplishing this goal might be the low level of marker gene expression in non-green plastids because of low genome copy number and low rates of protein synthesis. Identification of promoters and UTRs active in non green tissues should help to overcome this limitation.

Another limitation is the lack of information on genome sequences for several important crop species to locate intergenic sequences for integration of transgenes. Because of the conservation of the plastid genome sequence across many plant species, the concept of using targeting sequences from one species to transform the plastid genome of another unknown species was developed. Both potato and tomato were transformed using chloroplast vectors with tobacco targeting sequences, and tobacco was transformed with petunia targeting sequences.

Yet another limitation is the challenge of delivering foreign DNA through the double plastid membrane. Among the methods used to transform plant chloroplasts, only particle bombardment has proven to be efficient in transforming different plant species. PEG treatment of protoplasts has been used to stably transform tobacco chloroplast genomes and although laborious, it is inexpensive when compared to particle bombardment. Microinjection is effective for foreign gene expression in chloroplasts, although application of this technology for stable transformation of chloroplast genomes awaits further investigation. Development of new methods of gene delivery systems into chloroplasts should facilitate further progress in this field.

5. Marker-Free Strategies

For commercialization of transgenic plants it would simplify the regulatory process and improve consumer acceptance to remove gene sequences that are not serving a purpose in the final plant variety. For scientific purposes, eliminating the marker genes from the final plant would permit the use of experimental marker genes that have not undergone extensive biosafety evaluations or that may generate negative pleiotropic effects in the plants. Furthermore, it would permit the recycling of useful marker genes for recurrent transformation of transgenic plants if they were eliminated prior to the next round of transformation. Although a number of strategies have been described in the scientific literature for generating marker-free transgenic plants, all are more difficult to implement or are less efficient than procedures which leave the marker genes in the plant. Presently, sufficient data has been accumulated to indicate that co-transformation of non-selected genes with selectable marker genes followed by rounds of segregation will create marker-free plants. However, this process is labor intensive requiring the production of several fold more transgenic plants to isolate the plant of interest and further crossing steps after the initial transformation experiment. Furthermore, the strategy is not suitable for vegetatively-propagated species. For vegetatively-propagated species the use of transposons or homologous recombination to eliminate the marker genes may work but at very low efficiency. The use of transposons to reposition genes into a stable chromosomal location may provide an advantage for certain applications. Currently, the research area of greatest promise is the use of site-specific recombinases under the control of inducible promoters to excise the selectable marker genes and excision machinery once selection has been achieved (Ow, 2001). Concerns exist about pleiotropic effects induced by the action of recombinases on cryptic excision sites in the plant genomes, but the use of inducible promoters may limit the extent of damage. Presently, many of these processes are experimental and insufficient information is available to rate the commercial significance of the technologies.

5.1. Co-Transformation and Segregation of Marker Genes

Co-transformation involves the simultaneous delivery and integration of two or more separate genes. This may result in linkage of the genes at a single locus as often occurs with biolistic-mediated transformation or it may result in independently-segregating, unlinked loci, as often occurs with *Agrobacterium*-mediated transformation. Co-transformation provides unique advantages for the production of transgenic plants. It allows the simultaneous insertion of a large number of genes, independent of gene sequence, into a plant with a limited number of selectable marker genes. For example, in rice, two to thirteen transgenes have been simultaneously inserted using biolistics. The co-transformation frequencies were very high, for example, 85% in the R_0 generation for at least two genes. 17% of R_0 plants contained more than nine different transgenes (Chen et al. 1998). As the co-transformed genes integrated at a single locus they segregated together. Similar results were obtained in soybean (Hadi et al. 1996). The high incidence of linkage using biolistic-mediated transformation would be important for the manipulation of multi-genic traits using cloned genes but would be impractical for the elimination of marker genes from transgenic plants. An advantage of *Agrobacterium*-mediated cotransformation technologies over biolistic transformation is that co-transformed genes often integrate into different loci in the plant genome. Unlinked selectable marker genes can then be segregated away from the genes of interest and allow the production of marker-free transgenic plants (reviewed by Ebinuma et al. 2001). This technology is not useful for plants that reproduce vegetatively as segregation is essential for the separation of the marker genes from the genes of interest.

5.1.1. Co-transformation with separate plasmids in one or two *agrobacterium* strains

Agrobacterium-mediated co-transformation of non-selected genes with selectable marker genes has been demonstrated at relatively high frequencies in a variety of dicot and cereal species. This has been demonstrated in a number of ways. Two separate strains of *A. tumefaciens* (Depicker et al. 1985; McKnight et al. 1987) or *A. rhizogenes* (McKnight et al. 1987) have been shown to co-transform tobacco and/or tomato at frequencies of about 50% or better. The T-DNA insertions were generally unlinked; however, co-transformation of *B. napus* with nopaline strains of *A. tumefaciens* resulted in a higher than expected occurrence of linked insertions indicating that variations in plants and strains could alter linkage relationships (DeBlock and Debrouwer, 1991). The tendency towards multiple T-DNA insertions by nopaline strains may contribute to these observations although the mechanisms involved are unknown. Using a single octopine *A. tumefaciens* strain containing two separate binary vectors, co-transformation frequencies of >50% were obtained in tobacco and *B. napus* for the GUS gene and *npt*II selectable marker gene. Insertions at different loci allowed segregation of the genes and therefore the recovery of marker-free plants. When compared to methods that produce plants where the marker gene is linked to the gene of interest, this method requires about a four-fold greater production of transgenic lines to recover a comparable number of marker-free plants (Daley et al. 1998).

5.1.2. Co-transformation with single plasmids carrying multiple T-DNA regions

An alternative approach for co-transformation proposed by Komari et al. (1996) is the use of octopine strains with binary vectors that carry more than one T-DNA region. They demonstrated that this approach yields higher frequencies of co-transformation than mixtures of *A. tumefaciens* strains carrying separate vectors. In this study, the GUS and *hpt* genes co-transformed tobacco and rice with about 50% frequency at unlinked loci permitting segregation of the GUS gene from the *hpt* selectable marker to create marker-free plants. Although it is believed that the interaction between the bacterial and plant cells is the major factor influencing transformation efficiency (Depicker et al. 1985), it was recently found that the relative size of the co-transforming T-DNA has a major impact. Co-transformation frequencies of 100% were achieved in tobacco when the selected T-DNA was two-fold larger than the non-selected T-DNA. The elevation of co-transformation efficiency to practical levels has been demonstrated (McCormac et al. 2001). In maize, co-transformation with an octopine strain carrying a binary vector with two T-DNAs yielded co-transformation frequencies of 93% for the *bar* and GUS genes in the R0 generation. 64% of the R1 progeny segregated as *bar*-free plants expressing GUS. This contrasted dramatically with the 11.7% co-transformation frequency with mixed *Agrobacterium* strains (Miller et al. 2002). In barley, a similar approach with more compact vectors yielded 66% co-transformation frequencies but only 24% of these segregated as marker-free plants perhaps because nopaline strains were required for barley transformation (Matthews et al. 2001). The studies clearly demonstrate that marker-free plants can be generated at varying efficiencies using

Agrobacterium-mediated co-transformation followed by segregation of the genes in the subsequent sexual generations. This technology is not suitable for all plant species and its efficiency is clearly dependent on a number of variables including the *Agrobacterium* strain used and the plant tissue being transformed.

5.2. Transposon-Mediated Repositioning of Genes

5.2.1. Tranposition-mediated repositioning of the gene of interest

The maize *Ac*/*Ds* transposable element system has been used to create novel T-DNA vectors for separating genes that are linked together on the same T-DNA after insertion into plants. Once integrated into the plant genome, the expression of the *Ac* transposase from within the T-DNA can induce the transposition of the gene of interest from the T-DNA to another chromosomal location. This results in the separation of the gene of interest from the T-DNA and selectable marker gene. The system is functional in a wide range of plants. It only requires the activity of the *Ac* transposase which can be expressed from plant promoters to enhance activity (Honma et al. 1993) and the approximately 200 bp terminal repeat target sequences which must surround the gene to be transposed. Although the creation of marker-free transgenic plants is one outcome, the repositioning of the gene of interest within the genome can also result in favourable position effects that can enhance the expression profile of the gene of interest without creating more transformation events. In tomato, transposition of the GUS marker gene and the generation of *npt*II-free plants were demonstrated for plants with both single and multiple T-DNA insertions. In rice, a related approach was used to create *hpt*-free rice plants that expressed the *Bt* endotoxin coded by the *cry 1B* gene (Cotsaftis et al. 2002). In this study, the *cry1B* gene was placed in the leader sequence of a *gfp* marker gene so that transposition could be monitored by the activation of GFP activity. It was found that excision and reinsertion occurred at very high frequencies (37 and 25%) and plants were recovered with high levels of resistance to striped stem borer (Cotsaftis et al., 2002). The stability of the transposed gene seems to include a tendency to less gene silencing as shown for a transposed *bar* gene in barley (Koprek et al. 2001). This technology relies on crossing plants to segregate the gene of interest from the marker gene and the transposase; therefore, this technology is of limited use in plants that are vegetatively propagated or have a long reproductive cycle. This technology also has limitations for pyramiding multiple genes because introduction of the transposase in subsequent rounds of transformation and marker gene removal may result in the transposition of the first transgene into another chromosomal location.

5.2.2. Tranposition-mediated elimination of the selectable marker gene

An alternative strategy for exploiting the *Ac*/Ds system is to transpose the genes coding for the selectable marker and the transposase from the T-DNA leaving only the gene of interest in the inserted copy of the T-DNA. This research generated the *ipt*-type MAT (multi-auto-transformation) vector system which uses the *ipt* gene as a selectable marker and is designed to remove the *ipt* gene after transformation by using the *Ac* transposable element. This vector system supports recurrent transformation for the pyrimiding of genes into plants. Transgenic tobacco and hybrid aspen were transformed using the *ipt* gene as the selectable marker. The *ipt* gene was interesting in this study as it was used as both a negative and positive selectable marker. In the first positive selection step, transformed tissue proliferated as adventitious shooty material that was abnormal in morphology and could not regenerate due to the overproduction of cytokinin. In the second negative selection step, after several weeks or months in culture, normal shoots appeared (due to the elimination of the *ipt* and transposase genes by transposition) and regenerated into transformed marker-free plants. This occurred at a frequency of about 5%. As the system does not require a sexual reproduction step, it is an alternative for vegetatively propagated germplasm and plants with a long reproductive cycle (Ebinuma et al. 1997).

5.3. Intrachromosomal Homologous Recombination to Remove Selectable Marker Genes

Studies on the use of homologous recombination to eliminate selectable marker genes after insertion are few and presently poorly understood. The 352 bp attachment P (attP) region of bacteriophage is the target for three specific proteins that mediate the integration and excision of the phage within the *E. coli* genome. In tobacco the attP region appears to function without the proteins to effect excision of DNA sequences flanked by the attP repeats (Zubco et al. 2000). Transgenic tobacco shoots transformed with a T-DNA vector in which the gene of interest was separated from the region carrying the marker genes *npt*II, *gfp* and *tms 2* by attP repeats were examined in the presence of naphthalene acetamide (NAM). The *tms 2* gene from *A. tumefaciens* codes for an enzyme that converts NAM to the auxin NAA, which prevents root development and induces callus production. The regeneration of roots under this counter selection strategy was indicative of marker gene elimination by intrachromosomal homologous recombination. This strategy is not always associated with homologous recombination and larger deletions may occur as a result of illegitimate recombination (Zubco et al. 2000).

5.4. Site-Specific Recombinase-Mediated Excision of Marker Genes

Several simple bacterial and fungal recombination systems have been described in which single enzymes (e.g. Cre, FLP, R) acting on specific target sequences (*lox*, *FRT*, *RS*, respectively) have been adapted for use in plants (reviewed by Ow, 2002). Each of the target sites is similar in that short oligonucleotides surrounded by short inverted repeats determine the orientation of the target site. Recombinase-mediated DNA rearrangements can include site-specific excision, integration, inversion and interchromosomal recombination; therefore, the range of applications for this technology is very broad. Rapid progress has been made in the development of these technologies for generating marker-free transgenic plants. The technologies have implications for additional benefits such as the modification of copy number at insertions sites. For example, complex multicopy integration patterns generated by biolistics-mediated transformation of wheat were reduced to single-copies by Cre-mediated recombination of the outermost copies (Srivastava et al. 1999). A concern is that high levels of recombinase expression may result in genome rearrangements at cryptic-target sites in plants. Although such sites have not been described in nuclear genomes of plants, chloroplast cryptic *lox* sites have been described (Hajdukiewicz, 2001). Furthermore, the constitutive overexpression of Cre has been correlated with phenotypic aberrations in plants (Coppoolse et al. 2003). Solutions to this potential problem included the use of inducible promoters (reviewed by Hare and Chua, 2002) or transient expression strategies to limit expression of the recombinase (Vergunst et al. 2000) acting on nuclear genes. Selectable markers have also been successfully removed from plastids using the Cre–*lox* system (Corneille et al. 2001).

5.4.1. Cre–lox

The Cre–*lox* system from bacteriophage P1 was the first of the recombination systems shown to be effective in the generation of marker-free plants. The T-DNA vector carrying the gene of interest was constructed with *lox* sites flanking the *hpt* selectable marker gene and inserted into tobacco. The Cre recombinase was then introduced by a second round of transformation to achieve precise excision of the marker gene (Dale and Ow, 1991). This was subsequently confirmed with other plants and other marker genes. To avoid the introduction of marker genes along with the Cre gene, it was found that transient expression of the Cre-gene construct without selection was sufficient to yield enough Cre recombinase to create a small number of lines (0.25%) that were free of selectable markers and the Cre gene (Gleave et al. 1999). A significant refinement of the strategy was developed using the -estradiol-inducible promoter system in which an artificial transcription factor, XVE was constructed for use in plants with its target promoter (Zuo et al. 2001). In this system, the gene of interest was separated from its promoter by a fragment containing the genes coding for the XVE transcription factor, the *npt*II selectable marker and the Cre recombinase (under the control of the inducible promoter) surrounded by *lox* sites. Transformation of *Arabidopsis* was achieved by selection for kanamycin resistance. Subsequent induction with estradiol resulted in the excision of the complete induction system along with the Cre recombinase and selectable marker genes. The final product was the reconstituted gene of interest, in this case GFP. In *Arabidopsis*, excision occurred in all of the plants with high efficiency in the germline cells (29–66%) using a single transformation (Zuo et al. 2001).

5.4.2. FLP–FRT

The FLP–*FRT* system derived from the *Saccharomyces cerevisiae* 2_ plasmid has also been tested in plants. In tobacco and *Arabidopsis*, plants transformed with the FLP recombinase were crossed with plants transformed with T-DNA in which the GUS coding region is separated from the 35S promoter by a *hpt* gene bracketed by *FRT* sites. This resulted in excision of the *hpt* gene and activation of the GUS gene in all cases (Kilby et al. 1995). Interestingly, the soybean *Gmhsp17.6L* heat shock promoter was used and performed as an inducible promoter in a subset of cells. In transgenic maize callus similar results were obtained and transient expression was shown to result in excision at a frequency of 2–3% (Lyznik et al. 1996).

5.4.3. R–RS

The *R*–RS system from *Zygosaccharomyces rouxii* has been used in the MAT vectors as an alternative to the *Ac* transposase-mediated transposition of the genes as described above (reviewed by Ebinuma and Komamine, 2001). Tobacco plants were transformed with T-DNA vectors in which the *ipt* selectable marker gene and the gene coding for the *R* recombinase were surrounded by RS sites. The *ipt* gene provided the initial selection for morphological abnormalities (i.e. the shooty phenotype). The *A. rhizogenes rol* genes (Table 8) which confer the rooty phenotype have also been used (Ebinuma et al., 1997b; Cui et al. 2001). Co-expression of the *R* recombinase, under the control of the 35S promotor, eventually excised the *ipt* and *R* genes resulting in the development of normal marker-free shoots at very high frequencies (39–70%; Sugita et al. 1999). 67% of marker-free transgenic tobacco plants had more than three T-DNA insertions. This was presumably due to the strong constitutive expression of the *R* gene by the 35S promoter, which resulted in the removal of the *ipt* gene in low-copy-number callus before transgenic shoots could be generated. To control excision events, the 35S promoter controlling the R gene was replaced with the chemically inducible glutathione-*S*-transferase (GST-II-27) promoter from maize. By driving the *R* gene with the GST-II-27 promoter, the frequency of marker-free plants increased to 88%. Furthermore, 86% of these had single T-DNA insertions (Sugita et al. 2000). The GST-II-27 promoter was induced by the herbicide antidote 'Safener R29148' in tissue culture for 2 weeks after transfer of the *ipt*-induced shooty explants to hormone-free solid media. As sexual crossing was not required for the recovery of marker-free plants, the system was tested in hybrid aspen as a model for vegetatively propagated plants. Transgenic marker-free aspen were recovered with 21% efficiency (Matsunaga et al. 2002). A potential criticism of the technology is the dependence on organogenesis whereas most economically important crops are regenerated by embryogenesis. However, in rice the system has performed effectively (25% efficiency) in generating transgenic marker-free plants through organogenesis in a single step without forming *ipt*-shooty intermediates using the 35S-driven *R* gene (Endo et al. 2002).

6. Applications of Transgenic Plants in Crop Improvements

Following the development of modern and traditional ways to plant production improvement, transgenic methods, during the last two decades, are now widely used in both basic and applied studies in plant biology. It is an important tool to help molecular biologists to understand the organization and regulation of eukaryotic genes. Likewise, transgenics fined application in almost every branch of biology such as genetics, biochemistry, physiology, developmental biology, virology etc. In addition, the feasibility of mobilizing and expressing foreign genes into plants has opened up whole new era of genetically crop plants capable of defending against biotic and abiotic stresses and producing better quality products or novel compounds of pharmaceutical and industrial value.

Crop losses from insect pests can be staggering, resulting in devastating financial loss for farmers and starvation in developing countries. Farmers typically use many tons of chemical pesticides annually. Consumers do not wish to eat food that has been treated with pesticides because of potential health hazards, and run-off of agricultural wastes from excessive use of pesticides and fertilizers can poison the water supply and cause harm to the environment. Growing GM foods such as Bt corn can help eliminate the application of chemical pesticides and reduce the cost of bringing a crop to market.

For some crops, it is not cost-effective to remove weeds by physical means such as tilling, so farmers will often spray large quantities of different herbicides (weed-killer) to destroy weeds, a time-consuming and expensive process, which requires care so that the herbicide doesn't harm the crop plant or the environment. Crop plants genetically-engineered to be resistant to one very powerful herbicide could help prevent environmental damage by reducing the amount of herbicides needed. For example, Monsanto has created a strain of soybeans genetically modified to be not affected by their herbicide product Roundup. A farmer grows these soybeans which then only require one application of weed-killer instead of multiple applications, reducing production cost and limiting the dangers of agricultural waste run-off.

Environmental constraints that include abiotic stress factors severely limit crop productivity. Improvement of crop plants with traits that confer tolerance to these stresses was practiced using traditional and modern breeding methods. Molecular breeding and genetic engineering contributed substantially to our understanding of the complexity of stress response. For instance, unexpected frost can destroy sensitive seedlings. An antifreeze gene from cold water fish has been introduced into plants such as tobacco and potato. With this antifreeze gene, these plants are able to tolerate cold temperatures that normally would kill unmodified seedlings.

Malnutrition is common in third world countries where impoverished peoples rely on a single crop such as rice for the main staple of their diet. However, rice does not contain adequate amounts of all necessary nutrients to prevent malnutrition. If rice could be genetically engineered to contain additional vitamins and minerals, nutrient deficiencies could be alleviated. For example, blindness due to vitamin A deficiency is a common problem in third world countries. Researchers at the Swiss Federal Institute of Technology Institute for Plant Sciences have created a strain of "golden" rice containing an unusually high content of beta-carotene (vitamin A).

Medicines and vaccines often are costly to produce and sometimes require special storage conditions not readily available in third world countries. Researchers are working to develop edible vaccines in tomatoes and potatoes. These vaccines will be much easier to ship, store and administer than traditional injectable vaccines. Not all GM plants are grown as crops. Soil and groundwater pollution continues to be a problem in all parts of the world. Plants such as poplar trees have been genetically engineered to clean up heavy metal pollution from contaminated soil.

Further progress in improving crop resistance to stresses will depend on further gene discoveries, leading to a complete understanding of stress protection pathways, efficient transformation technologies applicable to a wider range of species and the involvement of scientists with diverse expertise in evaluation of transgenic material. Ideal gene engineering strategies should:

(1) Be based on knowledge of all major components of desired plant resistances.

(2) Identify major key points of each significant protection pathway.

(3) Design regulation circuits allowing timely switching between these introduced or enhanced pathways; and, importantly.

(4) Ensure that the achieved genetic manipulation does not compromise crop yield or food safety.

Increasing plant productivity is necessary to meet future worldwide food demands and may also be needed to alleviate the dependence on fossil fuels by providing more plant- based alternative energy sources. Photosynthesis, the process through which plants accumulate biomass by using light energy to convert inorganic carbon to carbohydrates, is a major target for improving plant productivity via conventional breeding practices and crop biotechnology (Richards, 2000; Sinclair et al. 2004). Enhancing photosynthesis is a promising approach for increasing plant productivity. Advances in plant transformation technology make it possible to manipulate photosynthesis by overexpressing particular genes for alleviating bottleneck steps, diverting the flux of Calvin cycle intermediates and photo assimilates, or introducing new enzymes and pathways that can positively influence photosynthesis (Genhai et al. 2010). In field conditions, drought is a very common stress which affects plant photosynthesis almost instantly by limiting CO_2 diffusion from the atmosphere into the chloroplasts by reducing stomatal opening. A tight positive relationship between the grain yield of wheat and maize and stomatal conductance has been observed (Evans and Fischer, 1999). Rubisco (ribulose-1,5-bisphosphote carboxylase/ oxygenase) plays a central role in the photosynthetic reaction and is often a rate-limiting enzyme under many physiological conditions. One possibility is to increase the Rubisco activation state under certain conditions. Rubisco must be "activated" in order to fix CO_2 (Lorimer and Miziorko, 1980). The Rubisco activation state is the ratio of catalytically competent sites to total Rubisco sites. The net photosynthetic rate is proportional to Rubisco activation state, but not necessarily to total Rubisco sites (Perchorowicz et al. 1981). Rubisco activation in vivo is controlled by Rubisco activase (Portis, 1992), which is a thermolabile protein (Feller et al. 1998). Inhibition of plant photosynthesis by moderately elevated temperatures appears primarily due to temperature damage to Rubisco activase and perhaps also the specific activity of activase via influencing ATP/ ADP ratio, which results in the loss of Rubisco activation state (Crafts-Brandner and Salvucci, 2000a). Rubisco from crop plants is considerably more thermostable and its catalytic activity increases with increases in temperature beyond 40–45°C. At these temperatures, Rubisco activase activity for most crop plants is significantly reduced or diminished (Feller et al., 1998). Engineering a thermostable Rubisco activase that will stabilize or even increase plant photosynthesis at moderately elevated temperatures (Crafts-Brandner and Salvucci, 2000a, b). Recent advances in plant transformation technology make it possible to manipulate photosynthesis by overexpressing particular genes or introducing new enzymes or pathways that can positively influence photosynthesis (reviewed by Parry et al. 2003; Raines, 2006). It has been reported that overexpression of the Calvin cycle enzymes, fructose-1,6 bisphosphtase or sedoheptulose-1,7-bisphosphatase (SBPase) in tobacco plants, not only increased RuBP concentration, but also Rubisco activation state (Tamoi et al. 2006). A 1.2-fold higher activation state than that of the untransformed wild-type resulted in both photosynthetic rate per leaf area basis and growth of the transgenic plants being significantly increased (Miyagawa et al. 2001). Although it is not clear whether the increased SBPase activity in the chloroplast enhances Rubisco activation state due to elevated RuBP concentrations, the research demonstrates an alternative approach to manipulate plant photosynthesis by other key Calvin cycle enzymes besides Rubisco. Other research showing positive effects on plant photosynthesis by introducing a single enzyme include:

(1) The overexpression of sucrose-phosphate synthase, which influences partitioning of photoassimilates and has resulted in the extended duration of older leaf photosynthesis and increased the biomass of transgenic tobacco and tomato plants (Baxter et al. 2003; Lunn et al. 2003).

(2) The overexpression of a *Escherichia coli* gene, *otsA*, for trehalose synthesis in tobacco that enhanced Rubisco activity and photosynthesis as well as biomass (Pellny et al. 2004).

In attempts to increase the CO_2 concentration at the Rubisco site, overexpression of a C4 cycle enzyme, phosphoenolpyruvare carboxylase, in rice, a C3 plant, as well as a cyanobacterial gene, *ictB*, involved in $HCO3^-$ accumulation in *Arabidopsis* and tobacco showed positive effects on photosynthesis in the transformed plants (Ku et al. 1999;

Lieman- Hurwitz et al. 2003). A successful example for introducing an *E. coli* glycolate catabolic pathway into *Arabidopsis* chloroplasts aimed to alleviate photorespiratory losses has been recently published by Kebeish et al. (2007). In their approach, three *E. coli* enzymes, i.e. glucolate dehydrogenase, glyoxylate carboligase and tartronic semialdehyde reductase, expressed with chloroplast targeting peptides in transgenic *Arabidopsis*, convert glycolate directly into glycerate within the chloroplast. This short-circuited photorespiratory pathway releases CO_2 around Rubisco site which facilitates CO_2 refixation without extra energy input, and reduces NH_3 release that saves energy for NH_3 refixation, resulting in better plant growth and biomass accumulation (Genhai et al. 2010).

During the last two decades, genetic engineering (GE) has been progressing at a steady pace in the forest trees. However, there are questions regarding stability of transgene expression, and transgene escape that need to be addressed in the long-lived forest trees. Although relatively stable transgene expression has been reported for several target traits, including herbicide resistance, insect resistance, and lignin reduction in the vegetative propagules of several forest tree species, there were still unintended unstable events in transgenic plants. Long-term stability of transgene expression involved in these traits and others affecting yield (impacting growth) would be desirable in the vegetative propagules, and also in the generative progeny of the forest trees. Transgene escape through pollen, seed, and vegetative propagules from GE trees to native forest populations, although inevitable, remains an important regulatory issue. However, it may be possible to manage/ minimize the risk of transgene spread via isolation in confined areas, and use of incompatible genotypes of feral tree populations in the vicinity of transgenic forest trees.

7. Current Global Picture of Modern Transgenic Crops

7.1 Ecological Risks

The potential impact on nearby ecosystems is one of the greatest concerns expressed about transgenic plants. Most domesticated plants mate with wild relative a some location where they are grown, and gene flow from domesticated crops (irrespective of whether they transgenic or non-transgenic) can the have potentially harmful consequences. The main concerns are of 1. evolution of increased weediness; 2. increased likihood of extinction of wild-relatives. There are known instances of unwanted weediness of hybrids created by unintended gene flow from domesticated crops to wild-relatives. For instance in California, cultivated rye hybridises with the wild *Secale montanum* to produce a weed, and this has led many Californian farmers to abandon rye as a crop.

Transgenes and other new traits such as mutation to herbicide tolerance present in domesticated crop created by conventional breeding have the potential for significant ecological impact if the plants receiving the trait can increase in frequency and persist in natural populations. This can occur:

1. If transgenic plants "escape" from cultivated to uncultivated areas.
2. If transgenic plants mate with similar wild plants, the transgene could be incorporated into the offspring.
3. If these new transgene plants become weedy or invasive, which could reduce.
4. If the transgenic crop trait confers a selective advantage in natural environments.

Gene flow may affect biodiversity and might affect entire ecosystems. Pollen flow from conventional crop plants to native species also poses gene-flow derived ecological risks, as crop plants are not selected to have optimal selective advantages in natural environments, and farm fields are different to natural ecosystems. Conventional varieties also posses new traits such as pest resistance that have been deliberately transferred into the crop variety from other species.

There are at least three possible avenues of hybridization leading to escape of a transgene:

1. Hybridization with non-transgenic crop plants of the same species and variety.
2. Hybridization with wild plants of the same species.
3. Hybridization with wild plants of closely related species, usually of the same genus.

However, there are a number of factors which must be present for hybrids to be created.

1. The transgenic plants must be close enough to the wild species for the pollen to reach the wild plants.
2. The wild and transgenic plants must flower at the same time.
3. The wild and transgenic plants must be genetically compatible.
4. The hybrid offspring must be viable, and fertile.
5. The hybrid offspring must carry the transgene.

Studies suggest that a possible escape route for transgenic plants will be through hybridization with wild plants of related species.

1. It is known that some crop plants have been found to hybridize with wild counterparts.
2. It is understood, as a basic part of population genetics, that the spread of a transgene in a wild population will be directly related to the fitness effects of the gene in addition to the rate of influx of the gene to the population. Advantageous genes will spread rapidly, neutral genes will spread with genetic drift, and disadvantageous genes will only spread if there is a constant influx.

3. The ecological effects of transgenes are not known, but it is generally accepted that only genes which improve fitness in relation to abiotic factors would give hybrid plants sufficient advantages to become weedy or invasive. Abiotic factors are parts of the ecosystem which are not alive, such as climate, salt and mineral content, and temperature.

Over the past several years it has become possible to isolate genes from any class of living organism and introduce them into most of our crop plants. This provides a wider choice of genes for crop improvement than is available by conventional plant breeding. International studies worldwide on genome organization, genetic mapping, and the isolation of genes are providing many new opportunities to modify crops. Although there is considerable optimism about the future prospects for transgenic crops among the scientific and plant breeding community, public concerns are being expressed in some countries.

7.2. Are Transgenic Crops Safe for Food and the Environment?

Members of the general public are often unaware of the extensive testing and evaluation necessary to produce conventionally bred crops. In addition to agricultural performance, it is often necessary to test for toxicity and other adverse effects of the crop. Conventional plant breeding is very much more sophisticated than that practiced by Mendel in his early pioneering genetics experiments. Many novel plant breeding methods have been developed and refined over this century. These include sexual hybridization with the aid of ovary and embryo culture, mutagenesis, haploidization and polyploidization. Compared with transgenic breeding, conventional breeding is very imprecise in many respects but has proved remarkably successful. This is because plant breeders have learned how to test and evaluate plants, select only those lines that have desirable characters, and reject the rest. Modern molecular methods of genetic modification now make it possible to introduce genes into crops from a wide range of different organisms; therefore, there is international agreement that there should be an additional set of safety assessments. All of the tests from conventional breeding are practiced, plus a wide range of additional safety assessments for genetically modified crops. These tests evaluate actual and potential impact on food, human and animal health, and the environment. With these additional safety tests, and the ability to analyse transgenes with a much greater precision than is usually possible for genes introduced by conventional breeding, I believe there is compelling evidence that genetically modified crops are as safe as conventionally bred crops. A major tactic of Integrated Pest Management (IPM) is to preserve natural enemies associated with crop pests. Tritrophic interactions involving crops, insect pests and their natural enemies must be taken into account when evaluating the environmental impact of transgenic crops. Natural enemies of pest species include generalist predators such as carabid beetles or specific parasites such as parasitoid wasps. Although insect-resistance factors expressed in crops might not have a direct effect on natural enemies of pests, indirect effects are almost inevitable.

For example, prey fed on plant material expressing Bt proteins were compromised in development and consequently nutritional quality, which affected the growth and development of a carabid beetle, with early instars being more sensitive than later instars and adult beetles. The Bt toxin did not accumulate through the different trophic levels in these experiments, but was excreted. The effects of insect-resistant crops on non-target insects should also be evaluated in the right context (i.e. compared with other measures farmers take to control insects). Compared with traditional chemical pesticides, which affect benign insects directly, transgenic crops expressing Bt are more environmentally friendly because of their specificity and absence of direct effects on biological control agents and non-targets. All measures to protect crops against insect pests will reduce the numbers of available prey for predators and parasites, even if there is no direct effect. Many excellent accounts of the economic, environmental and health benefits of insect-resistant transgenic crops have been published. In The USA, the six biotechnology- derived crops planted in 2003 (canola, corn, cotton, papaya, squash and soybean) produced an additional 2.4 million tonnes of food and fibre and increased farm income by US$1.9 billion. These biotechnology- derived crops also reduced the use of pesticides by 21 000 tonnes. Studies have shown that these Bt crops can be successfully deployed in agriculture, which has led to a decrease in pesticide usage, and that they are ecofriendly.

7.3. What Is the Public Concerned About?

A number of the concerns have been expressed and many of these concerns are intensified in people's minds by the activity of pressure groups and the use of pejorative terms such as "Frankenstein food," "gene foods," "mutant crops," and "genetic pollution." There is little doubt that there is a deliberate attempt to demonize the process of transgenic plant breeding by giving the impression that it is innately bad, flawed, undesirable, unnatural and therefore unacceptable. The pressure groups are skilled at manipulating the media, and many appear to be not interested in a careful scientific response to concerns. Fortunately, there are a significant proportion of people in society that are genuinely seeking scientific answers to their concerns. There are important questions raised by transgenic crops, but many are also issues associated with conventional plant breeding and agriculture. Many of these concerns are not scientific but often passionately held and are probably a significant driving force in influencing opinion. The concerns include:

1. **Multinational companies :** There is a perception that powerful multinational companies can have a significant influence on agriculture and food production. There is also the view that traditional farming practices may be undermined, for example, farmer-saved seed.
2. **World trade :** World trade agreements may undermine the consumer's freedom to make choices in what they buy. These choices can be made for ideological reasons but are seen to be part of an individual's democratic right.

3. **Human health :** A series of food scares in the United Kingdom and Europe (e.g., bovine spongiform encephalopathy, *Salmonella* and *Escherichia coli* food poisoning) have sensitized some people to the belief that food is unsafe.

4. **Environment :** There are concerns in some countries that agriculture is responsible for reducing wildlife biodiversity, for example, numbers of birds within certain species. These are issues about the environmental impact of agriculture generally.

References

Baxter, C.J., Foyer, C.H., Turner, J., Rolfe, S.A. and Quick, W.P. (2003). Elevated sucrose–phosphate synthase activity in transgenic tobacco sustains photosynthesis in older leaves and alters development. J. Exp. Bot. 54:1813–1820.

Chen, L., Marmey, P., Taylor, N.J., Brizard, J.P., Espinoza, C., D'Cruz, P., Huet, H., Zhang, S., De Kochko, A., Beachy, R.N. and Fauquet, C.M. (1998). Expression and inheritance of multiple transgenes in rice plants. Nat. Biotechnol. 16:1060–1064.

Coppoolse, E.R., Vroomen, M.J., Roelofs, D., Smit, J., Van Gennip, F., Hersmus, B.J.M., Nijkamp, H.J.J. and Van Kaaren, M.J.J. (2003). Cre recombinase expression can result in phenotypic aberrations in plants. Plant Mol. Biol. 51:263–279.

Corneille, S., Lutz, K., Svab, Z. and Maliga, P. (2001). Efficient elimination of selectable marker genes from the plastid genome by the Cre–*lox* site-specific recombination system. Plant J. 27:171–178.

Cotsaftis, O., Sallaud, C., Breitler, J.C., Meynard, D., Greco, R., Pereira, A. and Guiderdoni, E. (2002). Transposon-mediated generation of T-DNA- and marker-free rice plants expressing a *Bt* endotoxin gene. Mol. Breeding. 10:165–180.

Crafts-Brandner, S.J. and Salvucci, M.E. (2000a). Rubisco activase constrains the photosynthetic potential of leaves at high temperature and CO_2. Proc. Natl. Acad. Sci. USA. 97:13430–13435.

Crafts-Brandner, S.J. and Salvucci, M.E. (2000b). Sensitivity of photosynthesis in a C4 plant, maize, to heat stress. Plant Physiol. 129:1773–1780.

Cui, M., Takayanagi, K., Kamada, H., Nishimura, S. and Handa, T. (2001). Efficient shoot regeneration from hairy roots of *Antirrhinum majus* L. transformed by *rol* type MAT vector system. Plant Cell. Rep. 20:60–66.

Dale, E. and Ow, D. (1991). Gene transfer with subsequent removal of the selection gene from the host genome. Proc. Natl. Acad. Sci. USA. 88:10558–10562.

Daniell, H., Datta, R., Varma, S., Gray, S. and Lee, S.B. (1998). Containment of herbicide resistance through genetic engineering of the chloroplast genome. Nat. Biotechnol. 16:345-348.

Daniell, H., Khan, M.S. and Allison, L. (2002). Milestones in chloroplast genetic engineering: an environmentally friendly era in biotechnology. Review. Trends in Plant Science. 7:84-89.

Depicker, A., Herman, L., Jacobs, A., Schell, J. and Van Montagu, M. (1985). Frequencies of simultaneous transformation with different T-DNAs and their relevance to the *Agrobacterium*/plant cell interaction. Mol. Gen. Genet. 201:477–484.

Ebinuma, H. and Komamine, A. (2001). MAT (multiauto-transformation) vector system. The oncogenes of *Agrobacterium* as positive markers for regeneration and selection of marker-free transgenic plants. In Vitro Cell Dev. Biol. Plant. 37:103–113.

Ebinuma, H., Sugita, K., Matsunaga, E., Endo, S., Yamada, K. and Komamine, A. (2001). Systems for the removal of a selection marker and their combination with a positive marker. Plant Cell. Rep. 20:383–392.

Ebinuma, H., Sugita, K., Matsunaga, E. and Yamakado, M. (1997a). Selection of marker-free transgenic plants using the isopentyl transferase gene. Proc. Natl. Acad. Sci. USA. 94:2117–2121.

Ebinuma, H., Sugita, K., Matsunaga, E., Yamakado, M. and Komamine, A. (1997b). Principle of MAT vector. Plant Biotechnol. 14:133–139.

Endo, S., Sugita, K., Sakai, M., Tanaka, H. and Ebinuma, H. (2002). Single-step transformation for generating marker-free transgenic rice using the *ipt*-type MAT vector system. Plant J. 30:115–122.

Evans, L.T. and Fischer R.A. (1999). Yield potential: definition, measurement, and significance. Crop Sci. 39:1544–1551.

Feller, U., Crafts-Brandner S.J. and Salvucci, M.E. (1998). Moderately high temperatures inhibit ribulose-1,5-bisphosphate carboxylase/oxygenase (Rubisco) activase-mediated activation of Rubisco. Plant Physiol. 116:539–546.

Zhu, G., Kurek, I. and Liu, L. (2010). Engineering Photosynthetic Enzymes Involved in CO_2–Assimilation by Gene Shuffling. The Chloroplast: Basics and Applications. 307–322.

Gleave, A.P., Mitra, D.S., Mudge, S.R. and Morris, B.A.M. (1999). Selectable marker-free transgenic plants without sexual crossing: transient expression of *cre* recombinase and use of a conditional lethal dominant gene. Plant Mol. Biol. 40:223–235.

Hajdukiewicz, P.T. (2001). Multiple pathways for Cre/lox-mediated recombination in plastids. Plant J. 27:161–170.

Hare, P.D. and Chua, N.H. (2002). Excision of selectable marker genes from transgenic plants. Nat. Biotechnol. 20: 575–580.

Honma, M.A., Baker, B.J. and Waddell, C.S. (1993). High-frequency germinal transposition of *DsALS* in *Arabidopsis*. Proc. Natl. Acad. Sci. USA. 90:6242–6246.

Kebeish, R., Niessen, M., Thiruveedhi, K., Bari, R., Hirsh, H.J., Rosenkranz, R., Stabler, N., Schonfield, B., Kreuzaler, F. and Peterhansel, C. (2007). Chloroplastic photorespiratory bypass increases photosynthesis and biomass production in *Arabidopsis thaliana*. Nature Biotech. 25:593–599.

Kilby, N.J., Davies, G.J., Snaith, M.R. and Murray, J.A.H. (1995). FLP recombinase in transgenic plants: constitutive activity in stably transformed tobacco and generation of marked cell clones in *Arabidopsis*. Plant J. 8:637–652.

Komari, T., Hiei, Y., Saito, Y., Murai, N. and Kumashiro, T. (1996). Vectors carrying two separate T-DNAs for co-transformation for higher plants mediated by *Agrobacterium tumefaciens* and segregation of transformants free from selection markers. Plant J. 10:165–174.

Koprek, T., Rangel, S., McElroy, D., Louewerse, J.D., WilliamsCarrier, R.E. and Lemaux, P.G. (2001). Transposonmediated single copy delivery leads to increased transgene expression stability in barley. Plant Physiol. 125:1354–1362.

Ku, M.S.B., Agarie, S., Nomura, M., Fukayama, H., Tsuchida, H., Ono, K., Hirose, S., Toki, S., Miyao, M. and Matsuoka, M. (1999). High-level expression of maize phosphoenolpyruvate carboxylase in transgenic rice plants. Nat Biotech. 17:76–80.

Lieman-Hurwitz, J., Rachmilevitch, S., Mittler, R., Marcus, Y. and Kaplan, A. (2003). Enhanced photosynthesis and growth of transgenic plants that express ictB, a gene involved in HCO3 accumulation in cyanobacteria. Plant Biotech. J. 1:43–50.

Lorimer, G.H. and Miziorko, H.M. (1980). Carbamate formation on the e-amino group of a lysyl residue as the basis for the activation of ribulosebisphosphate carboxylase by CO_2 and Mg^{2+}. Biochemistry 19:5321–5328.

Lunn, J.E., Gillespie, V.J. and Furbank, R.T. (2003). Expression of a cyanobacterial sucrose-phosphate synthase from Synechocystis sp. PCC 6803 in transgenic plants. J Exp. Bot. 54:223–237.

Lyznik, L.A., Rao, K.V. and Hodges, T.K. (1996). FLP-mediated recombination of *FRT* sites in the maize genome. Nucleic Acids Res. 24:3784–3789.

Matsunaga, E., Sugita, K. and Ebinuma, H. (2002). Asexual production of selectable marker-free transgenic woody plants, vegetatively propagated species. Mol. Breeding 10:95–106.

Matthews, P.R., Waterhouse, P.M., Thornton, S., Fieg, S.J., Gubler, F. and Jacobsen, J.V. (2001). Marker gene elimination from transgenic barley, using co-transformation with adjacent 'twin T-DNAs' on a standard *Agrobacterium* transformation vector. Mol. Breeding 7:195–202.

Mc Cormac, A.C., Fowler, M.R., Chen, D.F. and Elliot, M.C. (2001). Efficient co-transformation of *Nicotiana tabacum* by two independent T-DNAs, the effect of T-DNA size and implications for genetic separation. Trans. Res. 10:143–155.

Mc Knight, T.D., Lillis, M.T. and Simpson, R.B. (1987). Segregation of genes transferred to one plant cell from two different *Agrobacterium* strains. Plant Mol. Biol. 8:439–445.

Miller, M., Tagliani, L., Wang, N., Berka, B., Bidney, D. and Zhao, Z.Y. (2002). High efficiency transgene segregation in co-transformed maize plants using an *Agrobacterium tumifaciens* 2 T-DNA binary system. Trans. Res. 11:381–396.

Miyagawa, Y., Tamoi, M. and Shigeoka, S. (2001). Overexpression of a cyanobacterial fructose-1,6-/ sedoheptulose-1,7- bisphosphatase in tobacco enhances photosynthesis and growth. Nat. Biotech. 19:965–969.

Ow, D.W. (2001). The right chemistry for marker gene removal? Nat. Biotechnol. 19:115–116.

Parry, M.A.J., Andralojc, P.J., Mithell, R.A., Madgwick. P.J. and Keys, A.J. (2003). Manipulation of Rubisco: the amount, activity, function and regulation. J. Exp. Bot. 54:1321–1333.

Perchorowicz, J.T., Raynes, D.A. and Jensen, R.G. (1981). Light limitation of photosynthesis and activation of ribulose bisphosphate carboxylase in wheat seedlings. Proc. Natl. Acad. Sci. USA. 78:2985–2989.

Portis, A.R. (1992). Regulation of ribulose 1,5-bisphosphate carboxylase/oxygenase activity. Annu. Rev. Plant Physiol. 43:415–437.

Raines, C.A. (2006). Transgenic approaches to manipulate the environmental responses of the C3 carbon fixation cycle. Plant Cell Environ. 29:331–339.

Richards, R.A. (2000). Selectable traits to increase crop photosynthesis and yield of grain crops. J. Exp. Bot. 51:447–458.

Sinclair, T.R., Purcell, L.C. and Sneller, C.H. (2004). Crop transformations and the challenge to increase yield potential. Trends Plant Sci. 9:70–75.

Srivastava, V., Anderson, O.D. and Ow, D.W. (1999). Single-copy transgenic wheat generated through the resolution of complex integration patterns. Proc. Natl. Acad. Sci. USA. 96:11117–11121.

Sugita, K., Kasahara, T., Matsunaga, E. and Ebinuma, H. (2000). A transformation vector for the production of marker-free transgenic plants containing a single copy transgene at high frequency. Plant J. 22:461–469.

Sugita, K., Matsunaga, E. and Ebinuma, H. (1999). Effective selection system for generating marker.-free transgenic plants independent of sexual crossing. Plant Cell. Rep. 18:941–947.

Tamoi, M., Nagaoka, M., Miyagawa, Y. and Shigeoka, S. (2006). Contribution of fructose-1,6-bisphosphatase and sedoheptulose-1,7-bisphosphatase to the photosynthetic rate and carbon flow in the Calvin cycle in transgenic plants. Plant Cell Physiol. 47:380–390.

Vergunst, A.C., Schrammeijer, B., Den Dulk-Ras, A., De Vlaam, C.M.T., Regensburg-Tuink, T.J.G. and Hooykaas, P.J.J. (2000). VirB/D4-dependent protein translocation from *Agrobacterium* into plant cells. Science. 290:979–982.

Zubco, E., Scutt, C. and Meyer, P. (2000). Intrachromosomal recombination between attP regions as a tool to remove selectable marker genes from tobacco transgenes. Nat. Biotechnol. 18:442–445.

Zuo, J., Niu, Q.W., Moller, S.G. and Chua, N.H. (2001). Chemical-regulated, site-specific DNA excision in transgenic plants. Nat. Biotechnol. 19:157–161.

❑❑❑

CHAPTER – 11

Molecular Farming

Ranjan Gupta[1], Anita Yadav[1] and Neeraj Aggarwal[2]
[1]Department of Biochemistry, Kurukshetra Uniersity, Kurushetra (India)
[2]Department of Microbiology, Kurukshetra University, Kurushetra (India)
Email : gupta-8@hotmail.com

Summary

Molecular farming is the large scale production of pharmaceutically important and commercially valuable proteins in plants and animals. Molecular farming in plants has the potential to provide an inexpensive and convenient system in the production of unlimited quantities of recombinant proteins for use as therapeutics, diagnostics, vaccines, antibodies, plasma proteins, cytokines and growth factors. Plants are also being used for the production of oils, starches, plastics, enzymes for food processing and other uses and spider silk etc. However, quality and safety of the final product and wider effects of molecular farming on health and environment need to be addressed. The benefits of molecular farming should not be outweighed by risks to human health and environment.

1. Introduction

Transgenic plants have been produced for a number of years for genetic improvement of crop plants for providing them with resistance against a variety of biotic and abiotic stresses and improvement in nutritional quality of food crops. An additional goal for the production of transgenic plants has been their use for large scale production of novel pharmaceuticals and biochemicals like interferon, insulin, immuno -globulins etc. which were traditionally produced by cultured mammalian cells, bacteria and fungi. Transgenic crops can also be used for the manufacturing of useful biopolymers like polyhydroxy butyrate which are not produced by normal plants and for improvement in the production of sugars, fatty acids, starches, celluloses, rubber, wax etc.

Molecular farming is an application of the genetically modified plants and animals. It is the production of recombinant pharmaceuticals and other commercially valuable proteins on a large scale outside their natural source.

1.1 Plant Pharming Through Transgenic Plants

Cultivation of transgenic plants for the recovery of pharmaceutical compounds is popularly known as 'pharming'. The first recombinant plant – derived pharmaceutical protein (PDP) was human serum, albumin, produced initially in 1990's in transgenic tobacco and potato plants. After twenty years, plants are now a source of pharmaceutical proteins such as mammalian antibodies (Fischer et al., 1999; Daniell et al., 2001), blood substitutes (Magnuson et al. 1998; Walmsley and Arntzen, 2000; Daniell et al., 2001), cytokines, growth factors, hormones, recombinant enzymes (Twyman et al., 2003), human and veterinary vaccines (Walmsley and Arntzen, 2000; Daniell et al., 2001; Twyman et al., 2003). Furthermore, several PDP products for the treatment of human diseases are approaching commercialization, including gastric lipase for the treatment of cystic fibrosis and antibodies for the prevention of dental caries and the treatment of non-Hodgkin's lymphoma (Christou et al., 2003).

1.2 Animal Pharming Through Transgenic Animals

The creation of transgenic animals for mass production of heterologous proteins and human medicine is known as animal pharming. Through transgenesis attempts were made to use animals as bioreactors to synthesize pharmaceutically important proteins (Table 1). The most successful of this approach has been to explore the properties of milk as renewable resource to produce functional enzyme, antibodies and structural protein using existing dairy manufacturing processes. The strategy has been to use transcriptional promoters of specific mammary genes to direct the expression of soluble transgenic protein. As milk borne proteins do not cross into the blood or lymphatic fluid of the transgenic animal, there is less danger of undesirable side effects being produced in the animals

2. Transgenic Plants and Animals as Bioreactors

Plants provide an inexpensive and convenient system for large-scale production of valuable recombinant proteins. Over the past decade, several efficient plant based expression systems have emerged and many different recombinant proteins have been produced in a range of different species. Plants have many advantages over other production systems, particularly in terms of practicality, economy and safety.

Table 1 : Transgenic Proteins in transgenic animals

Sr.No	Animals Used	Drugs / protein	Uses
1	Sheep	Alpha I-anti trypsin (Carver et al., 1993)	Treatment of emphysema
2.	Sheep & goat	Human tissue plasminogen activator (Ebert et al., 1994)	Treatment of thrombosis
3.	Sheep & pig	Human blood clotting factor VIII and IX	Treatment of hemophilia
4.	Sheep	Fibrinogen	Treatment of wound healing
5.	Goat	Human protein C	Treatment of thromobosis
6.	Goat	Antithrombin 3	-do-
7.	Goat	Glutamic acid decarboxylase	Treatment of type -1 diabetes
8.	Goat	Pro 542	Treatment of HIV
9.	Cow	Lactoferrin infection & infectious arthritis	Treatment of G.I. tract
10.	Cow	Human serum albumin	Maintaining blood flow
11.	Cow	Collagen I and II rheumatoid arthritis	Tissue repair & treatment of
12.	Cow	Fibrinogen	Wound healing
13.	Cow	Factor VIII	Hemophilia treatment
14.	Cow	Alpha-lactalbumin	Against infection
15.	Chicken cow and goat	Monoclonal antibody	For vaccine production

(Sinha and Kumar, 2008)

2.1 Advantages of Plants and Animals as Bioreactors Over Traditional Method

1. Constant monitoring and sampling of cell and bacterial culture is required.
2. Purchase and maintenance of equipments for expansion is more costly.
3. Isolation and purification of proteins is more difficult than obtaining pure form from plants or from animal's milk.

2.2 Advantages of Plants as Bioreactors

From the commercial point of view, transgenic plants offer following benefits

1. Reduced set-up and running costs.
2. No requirement for expensive equipment, media or skilled personnel.
3. A transgenic plant line once produced can be maintained, harvested and processed using existing agricultural infrastructure & unskilled labour.

4. Plants are easy to grow and can generate considerable biomass.
5. Plants do not carry pathogens that might be dangerous to human health.
6. At the level of pharmacologically active proteins, there are no proteins in plants that are similar to human proteins.
7. Plants are able to correctly process and configure both animal and human proteins.
8. Seeds and fruits provide sterile packaging containers for valuable therapeutics and guarantee a certain storage life.

Disadvantages

Though, recombinant mammalian proteins produces in plants are folded, assembled and modified in very similar ways to their native counterparts, and are in most cases functionally equivalent, there are however certain differences in glycan chain structure that could effect the performance and / or safety of recombinant proteins administered to human.

2.3 Advantages of Animals as Bioreactors

1. It is more cost effective because of their inherent capabilities to effect post-translational modification (e.g. glycosylation, single peplide cleavage, sulfuration, phosphorylation, amidation etc.) of complex mammalian proteins maintaining functional integrity in large unlimited amount.
2. Proteins synthesized in milk pose less danger to transgenic animals and are obtained in the pure form.

Disadvantages

1. Setting up of transgenic animal project is extremely expensive
2. Gestation period is long, litter size is large and maintenance cost of animals is high.
3. There may be high mortality rate and other deleterious effects on animals used by researches to create transgenic breed for e.g. transgenic sheep expressing growth hormone may show diabetic like condition.
4. Large number of recipients is required for embryo transfer because of low transgenic rate.

The demand for safe, recombinant pharmaceutical proteins are expanding rapidly, both the amounts of proteins needed and protein complexity are increasing as novel pharmaceutical activities are identified or designed into macromolecules (Rainer and Emans, 2000). Molecular farming in plants will become a pre-eminent method for the production of pharmaceuticals molecules and other valuable proteins in the year to come.

It has been possible to produce carbohydrates, fatty acids, polypeptides, vaccines, industrial enzymes and biodegradable plastics using the molecular farming approach.

Table 2 : The use of plants as bioreactors for the production of lipids, carbohydrates and proteins.

Compound	Origin of gene(s)	Application	Plant species used
Lipids			
Medium chain fatty acids	California bay tree (*Umbellularia californica*) – Thioesterase	Food, detergent, industrial	Oilseed rape
Mono-unsaturated fatty acids	Rat – desaturase	Food	Tobacco
Poly-hydroxy-butyric acid	*Alcaligenes eutrophus*	Biodegradable plastics	*Arabidopsis*, oilseed rape, soybean
Saturated fatty acids –Linolenic acid	*Brassica rapa,* Desaturase (Synechocystis); *Mortierella alpine*	Food, confectioneries For reducing coronary diseases	Oilseed rape Tobacco; canola
Ricinoleic acid	Castor oleate hydroxylase	Nylon paints, vamishes, lubricants	*Arabidopsis*
Carbohydrates			
Amylose-free starch	*Solanum tuberosum* (GBSS)	Food, industrial	Potato
Cyclodextrins	*Klebsiella pneumoniae*- Cyclodextrin glucosyl transferase	Food, pharmaceutical	Potato
Fructans	*Bacillus subtilis*- Fructosyl transferase	Food, Industrial	Tobacco, Potato
Increased amount of Starch	*E. Coli* (glgC16)	Food, Industrial	Potato
Trehalose	*E.Coli*	Food stabilizer	Tobacco
Sucrose	Maize sucrose phosphate synthase (SPS)	Sweetener	Tomato
Proteins			
Alpha – trichasantin	Chinese medicinal plant	Inhibitition of HIV replication	*Nicotiana benthamiana*
Angiotensin tomato converting enzyme inhibitor	Milk	Anti-hypersensitive effect	Tobacco,
Antigens	Bacteria, viruses	Orally administered vaccines	Tobacco, tomato, potato, lettuce
Antigens	Pathogens	Subunit vaccine	Tobacco
Enkephalin	Human	Opiate activity	Oilseed rape, *Arabidopsis*

3. Industrial Molecular Farming

Molecular farming can produce starches, oils, quality proteins, minerals, vitamins, plastics, Industrial enzymes and spider silk

3.1 Production of Specialty Starches for Industry

Starch is one of the components of plant cells. Starch is mainly used as a thickener (sauce) and sweetener (drinks, confectionery) in different foods, in paper, packaging and textile industry or as a raw material in chemical industry (ethanol, fructose and gluconate). For its industrial use, starch needs to be modified according to need. Starch with reduced levels of amylose has been developed in potato using antisense approach for inhibiting the granule bound starch synthase enzyme. Introducing a bacterial gene encoding AGPase into potato could increase amount of starch. Starch precursors have also been rerouted into the biosynthetic pathways of other storage carbohydrates. Tobacco and potato plants, which do not store fructan, have been transformed with fructosyl transferase gene from *Bacillus subtilis* to accumulate fructan. Transfer of mannitol-1-phosphate dehydrogenase gene (mt/D) from *E.coli* to tobacco led to the increased levels of mannitol in transgenic tobacco plants. Similarly, incorporation of myo-inositol o-methyl transferase gene from *Mesembrayanthemum crystallimum* (ice plant) has been incorporated into tobacco to produce pinnitol a myo-inositol derived cyclic sugar alcohol. Trehalose is an additive that improves processed and dried food by making them tastier. The plant biotechnological companies MOGEN (Leiden, The Netherlands) and Calgene (Davis, USA) have reported the synthesis of small quantities of trehalose in transgenic tobacco plants (Chawla, 2002).

Increased sucrose synthesis has been achieved by overexpression of maize sucrose phosphate synthase (SPS), the key enzyme in sucrose synthesis, in transgenic tomato plants.

Genetic engineering of a totally different type of starch was reported in 1991 by scientists from Calgene. They transferred to potatoes a bacterial gene that encodes an enzyme for synthesizing small cyclic starch molecules, rings that have seven or eight glucose molecules. These molecules, called *cyclodextrins,* can form complexes with other small molecules that are enclosed in the center of the ring. In this way, the "guest" molecules acquire new properties such as increased stability or greater solubility. Cyclodextrins are used in the pharmaceutical industry for drug delivery, and to remove unwanted compounds (such as caffeine) from foods. Cyclodextrins are expensive to produce, and their cost remains a problem for large scale use. The production of cyclodextrins in potatoes would undoubtedly decrease the cost of cyclodextrins so that they could now be used more widely by industry.

3.2 Production of Valuable Industrial Oils

Nutritional value of oils in terms of monounsaturated fatty acids could be enhanced in tobacco when a rat desaturase gene was introduced into it. Levels of palmitoleic acid (16:1) and oleic acid (18:1) levels increased. Lauric acid, a 12-carbon fatty acid used to make soaps and detergents has been obtained commercially from a transgenic rape seed plant variety. Transfer of thioesterase gene from California bay tree into *Arabidopsis* lead to the accumulation of medium chain 12-carbon fatty acid as storage lipids for the manufacture of detergents and for specialized nutritional applications.

Constitutive expression of Δ^6-desaturase gene from cyanobacterium *Synechocystis* in transgenic tobacco resulted in the production of –Linolenic acid (GLA) which is important in alleviating hypercholesterolemia and many coronary heart diseases. This gene has also been cloned from filamentous fungi (*Mortierella alpina*) and transferred to canola.

Expression of Oleate hydroxylase gene from castor in *Arabidopsis* resulted in accumulation of ricinoleic acid, which is used in industries in the manufacture of nylon paints, varnishes, resins, lubricants, and cosmetics. An oleate 12-hydroxylase: desaturase gene has also been transferred from *Lesquerella fendleri* to *Arabidopsis* for production of hydroxy-lated fatty acid (Chawla,2002).

3.3 Protein Quality

Nutritional quality of legume storage protein has been improved by expression of the gene responsible for a methionine-rich protein. Transgenic tobacco accumulated Brazil nut *(Bertholletia excelsa)* seed protein, 2S albumin resulting in 30% increase in the methionine content in the transgenic tobacco seed. Similar expression and accumulation of the Brazil nut 2S albumin protein has also been reported in transgenic *Arabidopsis,* rapeseed, soybean, French bean seeds, and potato. As Brazil nut 2S albumin was found to be allergic, sunflower seed albumin, rich in methionine, has been used for develop-ment of transgenic plants. Transgenic potato containing seed storage protein gene *(Ama*1), having well balanced amino acid composition (Raina and Datta, 1992), from Amaranthus could accumulate *Ama1*, which was found to be 0.3% of total soluble protein. More digestible and protein rich transgenic rice was produced by introducing soybean glycinin gene. In addition, normal rice could be replenished with important amino acids, such as lysine. Human milk β-casein gene under the control of an auxin-inducible promoter was expressed in transgenic potato (Chawla, 2002).

3.4 Vitamin and Mineral

Vitamin A content in the diet plays a major role in preventing disorders such as color blindness. Rice, the staple food for billions of people, does not contain any provitamin A in the endosperm. Immature rice endosperm is capable of synthe-sizing the early intermediate geranyl geranyl diphosphate (GGPP), which can be used to produce the uncolored carotene phytoene by expressing the enzyme phytoene synthase in rice endosperm. The synthesis of b-carotene requires three additional plant enzymes: phytoene desaturase and x-carotene desaturase, each catalyzing the introduction of two double bonds, and lycopene b-cyclase, encoded by the *Icy* gene. Ye et al. (2000) used the *Agrobacterium*-mediated transformation method in rice to introduce the entire b-carotene biosynthetic pathway into the rice endosperm. Plant phytoene synthase *(phy)* gene originated from daffodil *(Narcissus pseudonarcissus),* bacterial phytoene desaturase *(crt)* originating from *Erwinia uredovora* placed under the control of endosperm-specific glutelin (Gt) and the constitu-tive CaMV35S promoter respectively. The phytoene synthase cDNA contained a 5'-sequence coding for a functional transit peptide, and the *crtl* gene contained the transit peptide *(tp)* sequence of the pea Rubisco small subunit. These two direct the formation of lycopene in the endospeum plastids, the site of geranyl geranyl diphosphate formation. A third gene, lycopene b-cyclase, from *Narcissus pseudonarcissus,* con-trolled by the rice glutelin promoter, which carried a functional transit peptide allowing plastid import and selectable marker *aphlV,* were introduced. The resulting transgenic rice plants produce sufficient -carotene converted to vitamin A and rice has been named as "Golden rice".

Incorporation of phytoene synthase gene into *Brassica napus* resulted in increased content of carotenoides. This work has been conducted by Monsanto to enhance the carotenoid levels of oilseed crops, with a focus on the accumulation of β-carotene in the seed of canola.

The major micronutrient deficiencies world-wide are iron deficiency, with 24 percent of the world's population (up to 60% in developing countries) or 1.4 billion women suffering from iron-deficiency anemia, and vitamin A deficiency, affecting approximately 400 million children, or 7% of the world population. Iron deficiency is the consequence of (a) a very low amount of iron in the endosperm, (b) a high concentration of phytate (the major cause for inhibition of iron resorption in the intestine) and, (c) lack of high sulfur-containing proteins, which enhance iron resorption.

Consequently, Potrykus and his coworkers at Switzerland aimed to (a) increase iron content with a ferritin transgene from *Phaseolus vulgaris,* (b) reduce phytase in cooked rice with a transgene for heat-stable phytase from *Aspergillus fumigatus,* and (c) increase the resorption-enhancing effect from a transgenic sulfur-rich metallothionin-like protein from *Oryza sativa.* All transgenes are under endosperm-specific regula-tion (Chawla, 2002).

3.5 Biodegradable Plastic

Polyhydroxybutyrate or PHB, a biodegradable, thermoplastic polymer was synthesized in transgenic plants by transfer of phbB gene (acetoacetyl-CoA reductase) and phbC gene (Polyhydroxy butyrate) from the bacterium *Alcaligenes euthophus* into

Arabidopsis thaliana. Addition of a sequence to the gene helps the enzymes to be targeted to plastids protecting rest of the plant cell from the possible harmful effects of PHB accumulation.

3.6 Industrial Enzymes for Food Processing and Other Uses.

Enzymes are used extensively in the food industry, and the commercial production of these enzymes is a $400-million industry. Enzyme suppliers may list as many as 20 different enzymes that can be purchased in bulk. These enzymes are produced by industrial fermentation, using yeasts and other fungi. Many of the enzymes are heat stable and are optimally active between 55°C and 90°C. They are produced by micro-organisms that have been carefully selected (domesticated) over the years. Such strains do not occur in nature. The genes for these enzymes can easily be introduced into plants and expressed with leaf-specific or seed-specific control regions that will cause the enzymes to accumulate in these organs. The enzymes would have to be purified and separated from the other plant proteins and substances, before they could be used in the food industry.

One advantage of using fermentation technology is that very little enzyme purification is needed, because the enzymes are secreted into the fungal culture medium, where they accumulate at very high levels. The use of plant material that contains an enzyme (as opposed to using a purified enzyme) for modifying animal feeds is quite promising. For example, phytase produced by fermentation is added to animal feed to help digest the phytic acid it contains. It would be simple to produce phytase in tobacco leaves or potato tubers, and simply mix these with phytic acid-containing animal feed.

Cellulase is another example of an enzyme that could be produced in transgenic plants and used as an animal feed additive. When added to plant material, cellulase helps digest cellulose, and feedstuffs could be partially pre-digested if they were mixed with cellulase-rich plant material. In this case, the plant parts that contain cellulase would have to be finely dispersed, as a powder or a slurry, to release the cellulase from the tissues.

Recently biotechnologists in San Diego, California, introduced the gene for bovine lysozyme into tobacco plants and showed that the leaves contained high levels of this enzyme. This enzyme is extremely useful for ridding seeds of the pathogenic bacteria that may adhere to them. Whether the enzyme will need to be purified from a tobacco leaf extract, or can be applied to the seeds as a powder made from dried pulverized leaves remains to be investigated.

Laccase, developed by researchers using transgenic plants can be used in paper and textile production or water treatment plants instead of toxic processes.

3.7 Spider Silk

Spider silk has been produced using transgenic plants whose tensile strength is comparable to that of the synthetic superfibre, Kevlar, and can be stretched to more than twice its length before breaking. It is useful for a variety of industrial and medical purposes, including organ transplants.

4. Medical Molecular Farming

Genetic engineering of plants to produce desired pharmaceuticals has revolutionized the use of therapeutically valuable proteins in a variety of clinical treatments.

4.1 Production of Edible Vaccines

Genes for peptide epitopes of pathogens can be used for the production of vaccines against a variety of human and animal diseases. This concept of vaccine production in transgenic plants was introduced in 1992. In the year 1995, a biopharmaceutical company ProDigene from Texas (USA) conducted clinical trials, where pigs, fed with an edible maize vaccine, were found protected against the transmissible gastroenteritis virus (TGEV). The company had also taken patent on this and other edible vaccines produced in transgenic crops. Such vaccines, whenever successful, will provide a technology for production and delivery of inexpensive vaccines, especially in the developing world. This technology will also be cheaper than the expensive recombinant cell culture-based expression systems.

A **'mucosal immune response'** is necessary particularly for combating infections due to bacteria / viruses invading epithelial membrane. Mucosal immune system involves production of secretory IgA (S-IgA) at mucosal surfaces such as gut and respiratory epithelia. Such a response is more easily achieved by oral vaccine rather than by parenteral (administered elsewhere than in alimentary canal) antigen delivery system. Particulate antigens rather than subunit or soluble antigens have been shown to be effective oral immunogens. The subunit vaccines whenever used require larger amounts (mg versus mg) of antigens and will be expensive if produced in recombinant cell culture expression system, which will need fermentation technology. In contrast to this, transgenic plants can be made to express the antigens in particulate form in the edible tissue, which can then be used as food, thus providing an inexpensive oral vaccine production and delivery system.

4.1.1 Factors affecting choice of transgenic plants for vaccines

While using transgenic plants for vaccines, several factors needs to be taken into consideration : (i) it Should stimulate a mucosal immune response and therefore diarrheal diseases were the first target for oral vaccine through transgenic plants; (ii) antigen should assemble into ordered structures such as virus-like particles (VLP), which are resistant to digestion and reach the gut associated lymphoid tissue (GALT), being perceived there as foreign antigen for production of antibodies; (iii) plant system should be such that it may be used as food/feed, so that although tobacco was initially used, but it is not suitable. Potato and banana were subsequently used/suggested, with an expression system, where antigen is produced or stored in the tuber in potato and in the fruit of banana. Alfalfa, coarse grain crops and beans can also be used for animal vaccines.

4.1.2 Transgenic tobacco, potato and banana as edible vaccines

Several examples of successful production of prospective edible vaccines through transgenic plants are listed in Table 3. It may be seen that in majority of cases, transgenic tobacco plants were used, although other plants like potato, cowpea and blackeyed bean were also used. In some of these cases, the oral immunogenicity was tested in mice and positive results were obtained and recovery of antibodies was possible. It is hoped that eventually transgenic banana will be the fruit of choice to be used as edible vaccine, because banana is grown extensively throughout the world and can be consumed raw without any modification or processing of the edible part. The diseases that will be targeted for production of edible banana vaccines may initially include measles, polio, diphtheria, yellow fever, viral diarrhea and hepatitis B, which will be successfully produced and used in the developing countries.

Table 3 : The production of vaccines in transgenic plants

Application	Plant	Protein
Hepatitis B	Tobacco	Recombinant HbsAg
	Tobacco	Murine hepatitis epitope
Dental caries	Tobacco	*Streptococcus mutans* surface protein SpaA
Autoimmune diabetes	Potato	*Vibrio cholerae toxin* B subunit-human insulin fusion
	Potato	Glutamic acid decarboxylase
Cholera and *E. coli* diarrhea	Tobacco / potato	*E. coli* heat-labile enterotoxin LT-B
Cholera (Oral vaccine)	Potato	*V. cholerae* toxin Ctox A and Ctox B subunits
Mucosal vaccines not requiring adjuvants	Cowpea	D2 peptide of fibronectin - binding protein B of *Staphylococcus aureus*
Diarrhea due to Norwalk virus	Tobacco / potato	Coat protein of Norwalk virus
Rabies	Tobacco / spinach	Rabies virus glucoprotein
HIV	Tobacco / bean	HIV epitope (gp 120)
Rhinovirus	Blackeyed bean	Human rhinovirus epitope (HR 14)
Foot & mouth	Blackeyed bean	Foot & mouth virus epitope (VP 1)
Resistance of mink, Dog and cat against virus	Blackeyed bean	Mink enteritis virus (VP2)
Malaria	Tobacco	Malarial B-cell epitope
Influenza	Tobacco	Hemagglutinin
Cancer	Tobacco	c-Myc

4.1.3 Transgenics as edible vaccines for hepatitis B

An important target for edible vaccine is hepatitis B, a virus causing a chronic liver disease. Serum derived antigens were initially used for immunization against this disease, but due to a variety of safety concerns shown against this vaccine, recombinant vaccines were developed in yeast. These recombinant vaccines also failed due to high expense and lack of facilities at several places for their production. Therefore, in 1990s and in the year 2000, transgenic tobacco and potato were produced, which expressed hepatitis B surface antigen (HbsAg). In the year 2002 transgenic carrot plants were also produced at Giessen University in Germany which have the potential as an editle vaccine against hepatitis B. In future, HbsAg may also be produced in banana, which can be eaten raw. Thus, it is hoped that the transgenic plants, without any treatment of edible parts that are ingested, will provide for edible vaccines against hepatitis B for developing countries. It has already been shown that transgenic potato expressing HbsAg, when orally administered to mice, can elicit humoral immune responses.

4.1.4 An alternative to transgenic plants for vaccine production

An alternative to the production of transgenic plants for vaccine production is to produce a chimeric viral coat protein gene and introduce it into a plant for transient expression. This strategy will require inoculation of host plants with viral vectors for their transient expression, which can not be tissue specific as in case of transgenic plants. This approach has, however been successfully utilized for the following antigen capsule fusion proteins : (i) malaria epitope with TMV capsid proteins; (ii) zona-pallucida with TMV capsid protein and (iii) rhinovirus-14 and HIV-1 with cow pea mosaic virus (CPMV) capsid protein (Gupta, 2008).

4.2 Antibody Production in Plants

In the decade since the expression and assembly of immunoglobulin (Ig) heavy and light chains into functional antibodies was first shown in transgenic tobacco, plants have proven to be versatile production systems for many forms of antibodies. These include full-sized IgG and IgA, chimeric IgG and IgA secretory IgG and IgA, single-chain Fv fragments (scFv), Fab fragments and heavy-chain variable domains. Recently, this list has been extended to include bispecific antibodies, which are made by the genetic fusion of two different scFvs via a flexible peptide linker. Plants have great potential as a virtually unlimited source of inexpensive monoclonal antibodies (dubbed plantibodies) for human and animal therapeutics.

Most antibodies expressed to date have been in tobacco, although recently potatoes, soybean, alfalfa, rice and wheat have also been used successfully. The major advantage of using green tissue (tobacco, alfalfa, and soybean) is sheer productivity. Both alfalfa and tobacco can support several crops (cuttings) per year, with potential annual biomass yields of 25 tonne ha^{-1} and > 100 tonne ha^{-1}, respectively. By contrast, the maximum yields of wheat, rice and corn seed and 3 tonne ha^{-1}, 6 tonne ha^{-1} and 12 tonne ha^{-1}, respectively. Other advantages of tobacco include its relative ease of genetic manipulation, production of large numbers of seeds (upto a million per plant) and an impending need to explore alternate uses for this hazardous crop. However, seeds are likely to have fewer phenolic compounds and a less complex mixture of proteins and lipids than green leaves, which might be an advantage in purification.

Another advantage of seeds or tubers is their ability to be stored for long periods. Levels of scFv in rice seeds did not show a significant decline after storage at room temperature for six months. Potato tubers in cold storage for 18 months lost only 50% of functional antibody. For short periods of time (five to seven days), dried tobacco and alfalfa leaves can also be stored with little loss of scFv or IgG antibody. Purification of antibody from stored plant material has the advantages that the processing facility need not be near the field and can be used continually all year, rather than for just a few large batches.

To date, only four antibodies have been made in plants that are potentially useful as human therapeutics. Only one of these has been tested in humans: a chimeric secretory IgG-IgA antibody against a surface antigen of *Streptococcus mutans,* the primary causal agent of tooth decay. This tobacco-produced antibody was applied topically to teeth and found to be as effective as an IgG produced in a murine hybridoma at preventing recolonization by *S. mutans.* The second antibody, a humanized anti-herpes-simplex virus (HSV) antibody made in soybean, was effective in the prevention of vaginal HSV-2 transmission in a mouse model. Its activity was indistinguishable both *in vitro* and *in vivo from* the monoclonal antibody produced in cell culture. A third antibody, against carcinoembryonic antigen (CEA), has recently been expressed in rice and wheat. CEA, a cell-surface glycoprotein, is one of the best-characterized tumor-associated antigens. Antibodies against CEA are used for *in* vivo tumor imaging, as well as in antibody-based cancer therapy. Levels of scFv in seeds did not show a significant decline after storage at room temperature for six months. This same antibody has been expressed in a rice cell culture.

The fourth antibody is an example of both a novel use of plant-produced antibodies and an alternative production system. A plant virus vector has been used to produce a tumor-specific vaccine transiently in tobacco for the treatment of lymphoma. The antibody genes for expression of an scFv were derived from a mouse B-cell lymphoma. The plant-produced scFv was used to immunize mice, which generated anti-idiotypic antibodies (antibodies against the binding portion of the antibody). These mice were protected against infection by the lymphoma that produced the original antibody. Other groups have used modified plant viral vectors to produce therapeutically useful antibodies in plants, including an antibody against the colorectal-cancer-associated antigen GA733-2. Although these vectors might find limited usefulness if the rapid production of an antibody is necessary (perhaps in greenhouse production), their acceptability to regulatory agencies (e.g. the US Food and Drug Administration, Dept of Agriculture and Environmental Protection Agency) has not been tested.

There are no plantibodies yet in commercial production, therefore estimates of cost are difficult to find and involve many assumptions. The costs of producing an IgG from alfalfa grown in a 250m^2 greenhouse are estimated to be USS500-600 g^{-1}, compared with US$5000 g^{-1} for the hybridoma-produced antibody. Planet Biotechnology (Mountain View, CA, USA) has compared the cost per gram of purified IgA made by cell culture, transgenic goats, grain (7.5 tonne ha^{-1}) and green biomass (120.0 tonne ha^{-1}). Expression levels will have a significant impact on the costs but, at the best expression level reported (500 μg^{-1} leaf for a secretory IgA), the final cost should be well below US$50 g^{-1}. This significantly undercuts the costs of cell culture (US$1000 g^{-1}) or transgenic animal production systems (US$ 100 g^{-1}). The biggest component of cost with plantibodies will be purification. However, expression in seeds of rice and wheat opens up the possibility of oral administration of some therapeutic antibodies without the need for expensive purification.

Some of the properties of Igs depend on their glycosylation (e.g. binding to monocyte Fc receptors). There is one conserved Af-glycosylation site in the CH_2 domain of IgG. The plantibody N-glycans were more structurally diverse, with 40% being of the high-mannose type. The other 60% of the plantibody oligosaccharides had b-(1,2)-xylose and a-(1,3)-fucose linked to the $Man_3GlcNAc_2$ core. These linkages are typical of plants but are not found in mammalian N-glycans. The plantibody also lacked sialic acid, which represented ~10% of the sugar content of the mouse monoclonal antibody. These differences in glycan structure appear to have no effect on antigen binding or affinity *in vitro* and might not be significant *in vivo* either. An IgG produced in alfalfa had a serum half-life in Balb/c mice that was indistinguishable from that of the hybridoma-produced antibody. However, there is some concern about the potential immunogenicity and allergenicity of plantibodies used as human therapeutics. For mucosal applications, this is not likely to present problems for most people because plant glycoproteins are ubiquitous in the human diet. There has been no evidence of allergic reaction or of a human anti-mouse antibody (HAMA) response in 60 patients receiving topical oral application of a secretory IgA specific to *S. mutans* (Daniell et al., 2001).

Table 4 : Production of antibodies in transgenic plants

	Antibody aimed	Plant	Protein
I	**Immunoglobulins**		
	Antibody for dental caries	Tobacco	Hybrid slgA G for S. mutans
	Full length IgGI	Tobacco	IgG for S mutans
II	**Single-chain Fv**		
	Strong in tubers	Potato	Phytochrome binding scFv
	For Hodgkin's lymphoma	Tobacco	sc Fv of IgG of mouse B-cells
	Fv for tumour associated antigen	Cereals	scFv T84.66 against carcino-embryonic antigen

4.3 Production of Pharmaceuticals and Proteins

Cultured mammalian cells, bacteria and fungi provided excellent production systems in the past for the production of biopharmaceuticals. However, with cultured mammalian cell systems, the yields are low and culturing of cells makes production expensive. The cultured cells are also sensitive to shear forces occurring during industrial-scale culture and also to the variation in temperature, pH, dissolved oxygen and certain metabolites. This makes it necessary to control culture conditions thus adding to the cost. However, the bacterial and fungal systems are more robust, but they are not ideal for synthesis of many mammalian proteins, due to differences in metabolic pathways, protein processing, codon usage and the formation of inclusion bodies.

Transgenic plants have been considered as a better production system for the production of many pharmaceuticals to be used by humans and livestock, despite the fact that differences also occur between mammalian and plant systems with respect to codon usage and post- translational processing. In view of this, many biotechnology companies are developing transgenic plants to be used as factories for the production of biopharmaceuticals. For this purpose expression systems for high expression of introduced genes are being patented and transgenic plants produced using these expression systems are being already field tested. Clinical trials are also in progress using biopharma-ceuticals derived from these transgenic plants.

Like edible vaccines, the biopharmaceuticals derived from transgenic plants can be stored and distributed as seeds, tubers and fruits, which can be either directly used for oral ingestion or for extraction of the biopharma-ceutical. This will make the immunization programmes in the developing countries cheaper and easy to administer, since the delivery by direct ingestion of modified plant product eliminates the need for product purification, which is an expensive process in pharma-ceutical industry. Other advantages of using transgenic plants as production systems for the production of biopharmaceuticals include (i) comparatively higher yields at a relatively low cost (according to some estimates, the production of a recombinant protein in transgenic plants can be 10 to 50-fold cheaper than producing the same protein by *E. coli);* (ii) reduced health risk, which is common in conventional pharmaceuticals (iii) relatively little capital investment, due to the existing infrastructure.

The biopharmaceuticals that have been shown to express in transgenic plants include erythropoietin, insulin, enkephalins, a-interferon, human serum albumin, and two of the most expensive drugs, glucocerebrosidase and granulocyte-macrophage colony-stimulating factor.

Table 5 : Production of biopharmaceuticals in transgenic plants

	Application	Plant	Protein
I	**Anticoagulants**		
	Protein C Pathway	Tobacco	Hybrid slgA G for S. mutans
	Indirect thrombin inhibitors	Tobacco, oilseed rape, Ethiopian mustard	Human hirudin variant 2
II	**Recombinant hormones / proteins**		
	Neutropenia	Tobacco	Human granulocyte-macrophage
	Anemia	Tobacco	Colony-stimulating factor
	Antihyperanalgesic by opiate activity	Thale cress, oilseed	Human enkephalins
	Wound repai / control of cell proliferation	Tobacco	Human epidermal growth factor
	Hepatitis C and B treatment	Rice, turnip	Human interferon-
	Liver cirrhosis	Potato, tobacco	Human serum albumin
	Blood substitute	Tobacco	Human hemoglobin
	Collagen	Tobacco	Human homotrimeric collagen I
III	**Protein / peptide inhibitors**		
	Cystic fibrosis, liver disease, and hemorrhage	Rice	Human -I antitrypsin
	Trypsin inhibitor for transplantation surgery	Maize	Human aprotinin
	Hypertension	Tobacco / tomato	Angiotensin-I converting enzyme
	HIV therapies	Nicotiana bethamiana	-trichosanthin from TMV-UI subgenomic coat protein
IV	**Recombinant enzymes**		
	Gaucher's disease	Tobacco	Glucoccrebrosidase
V	**Neutraceuticals**		
	Provitamin A deficiency	Rice	Daffodil phytoene synthase

Many more therapeutic proteins will become available in future through investments currently being made in genomics research. The biotechnology company 'Applied Phytologics' (API) in California (USA) produced transgenic rice producing a-1-antitrypsin, commonly used for treatment of cystic fibrosis liver disease, and hemorrhages. Trials of a-1-antitrypsin extracted from malted grain of transgenic rice started in 1998 and it is hoped that regulatory approval for its commercial use should become available by 2004.

4.3.1 Production of glucocerebrosidase for 'Gaucher's disease'

As mentioned above, one of the most expensive biopharmaceutical that has been shown to express in transgenic plants is glucocerebrosidase, which is an important lysosomal hydrolase enzyme, and whose deficiency causes 'Gaucher's disease' in humans. The drug for treatment of this disease is generally prepared from the enzyme purified from human placentas, and 10-12 tonnes of placentas per year arc required for production of glucocerebrosidase needed for the treatment of a patient. This makes it the most expensive medicine in the world, in recent years, in order to reduce the cost, the drug has been produced in mammalian cell cultures, but it still remains one of the 'most expensive' drugs. In the year 1999, the production of this drug in transgenic tobacco was patented by scientists at Virginia Tech and State University, and Crop Tech (both in Blacksburg, Virginia, USA). It is hoped that commercial production of this and other lysosomal enzymes in transgenic plants will be possible in future.

4.3.2 Production of hirudin (an anticoagulant) for 'thrombosis'

Hirudin is another important drug that is used as anticoagulant for treatment of thrombosis. It was originally isolated from the leech *(Hirudo medicinaiis),* but is now mostly produced by recombinant bacteria and yeast. Transgenic plants (oilseed rape, tobacco and Ethiopian mustard) were initially produced, which although expressed hirudin, but purification of hirudin from the seed proved rather expensive. It has been shown that hirudin in transgenic plants can be produced as a fusion protein with oleosin, which is transported to oil bodies thus making the purification cheaper and easier (oleosin-hirudin fusion proteins are extracted with oil bodies and can be easily separated by flotation centrifugation). Between the two genes (for hirudin and oleosin), the fusion gene also contained a sequence that encoded an endopeptidase recognition site, thus enabling easy cleavage of the fusion protein releasing hirudin from oleosin-hirudin fusion protein. This strategy also ensures that hirudin becomes active only after extraction, thus reducing environmental risks, if any. Transgenic oilseed rape expressing hirudin is now commercially grown by SemBiosis in Calgary, Canada (Gupta, 2008).

5. Biosafety Concerns

Use of plants in molecular farming has several advantages over the traditional production systems such as absence of endotoxins, human pathogens and oncogenic DNA sequences. However, there are biosafety concerns relating the effects of molecular farming on health and environment. The potential biosafety risks include: Spread of transgene in the environment, recombinant protein accumulation in the ecosystem, contamination of food and feed chains with transgenes and their products, product quality and safety.

5.1 Transgene Pollution and Biodiversity

Transgene pollution is the spread of transgenes beyond the intended genetically-modified species by natural gene flow mechanisms. The transgenes might spread from transgenic to non-transgenic populations of the same crop species growing nearby, from transgenic crops to related wild species, or even from transgenic plants to microorganisms and animals. Transgene pollution is undesirable because the impact of novel genes on the performance and fitness of other species is unpredictable, and in the worst case scenario could severely disrupt natural and agricultural ecosystems, even leading to extinction in some species. The spread of transgenes to "natural" crops i.e. crops that were never intended to be modified, is also undesirable because this provides an early opportunity for foreign DNA sequences and their products to enter animal and human food chains. The standard method for producing a transgenic plant line is to introduce the primary transgene along with a selectable marker, which allows the propagation of transformed plant material at the expense of non-transformed material. The use of selectable markers is perhaps one of the major issues in biosafety, since traditional markers which exploit herbicide or antibiotic resistance as selectable traits each represent significant environmental or health threats. It is also a standard practice to transform plants with plasmid vectors containing the expression cassette. This results in the integration of vector backbone sequences along with the functional transgenes. Not only are such sequences superfluous to requirements, but they also have numerous undesirable effects in transgenic plants, acting as triggers for *de novo* methylation and promoting extensive rearrangement of the foreign DNA sequences prior to integration. They may also carry additional functional DNA sequences such as selectable markers, promoters and origins of replication used in bacteria, which could become active in environmental organisms polluted by gene flow. It is possible, although quite laborious, to screen plant cells for the incorporation of a primary transgene using the polymerase chain reaction, without relying on any type of maker. However, most "Marker-Free" transformation strategies involve removal of selectable markers after transformation has been achieved. An alternative approach is to use an innocuous scorable marker gene i.e. a gene that would have no conceivable negative effects in wild populations. Examples of such markers include growth regulators (e.g. ipt or CKII) and metabolic markers (e.g. manA or BADH) under inducible control. For indispensible primary transgenes, the only way to avoid transgene pollution is by containment, which aims to prevent seed and pollen dispersal, prevent the survival of dispersed seeds and pollen or prevent gene flow from viable pollen. The containment may be physical and artificial (i.e. trangenic plants can be maintained in greenhouses or the flowers/fruits may be concealed in plastic bags in the field) or it may be based on isolation (a sufficient distance between the transgenic crop and any potential recipients of the pollen will minimize the chances of outcrossing). Isolation zones around transgenic crops can be barren, but a more suitable alternative for insect-pollinated crops is to provide a zone of non-insect pollinated plants which would discourage the insects from leaving the transgenic zone. Barrier crops, i.e. a border of non-GM plants of the same species as the transgenic crop, are also useful as these can absorb much of the pollen released by transgenic plants and can then be destroyed after flowering (Commandeur et al.,2003).

5.2 Protein Pollution

The recombinant proteins produced by transgenic plants constitute another form of pollution and represent a further threat to the environment. One immediate concern is the possible negative effect of recombinant proteins on non-target organisms, particularly insects and microorganisms that interact directly with the plant and herbivores that may eat transgenic plant material laced with biopharmaceutical molecules. Such proteins might have direct toxicity effects, or they might accumulate in the food chain and therefore affect animals that do not interact with the transgenic plants at all.

Like transgene pollution, protein pollution can be addressed to a certain extent by physical environment, since this restricts the impact of protein toxicity to a much localized environment and limits knock on effects to non-target organisms. However, further barriers to protein toxicity can be put in a place by controlling transgene expression, and therefore limiting the availability of the protein to interacting organisms. The exposure of non-target organisms to recombinant proteins can be minimized by restricting expression to particular tissues. For example, a number of promoters have been identified that restrict gene expression to seeds or fruit. This would prevent the consumption of the protein by insects and other animals feeding on green plant tissue and would likewise avoid other forms of contact, such as the exposure of pollinating insects to recombinant proteins expressed in pollen grains. By avoiding transgene expression in roots, leaching of the recombinant protein into the soil (and consequent disruption of the rhizophere) would also be prevented. If restricted expression strategies were used in combination with effective management (e.g. specific harvesting times) then vegetative transgenic material could safely decay in the environment with no risk of protein toxicity. An alternative strategy is to bring the transgene under inducible control, such that the recombinant protein would be expressed only when the plant was exposed to a certain chemical inducer (Zuo and Chua, 2000).

In addition to the control of gene expression, the protein can also be targeted to a specific intracellular compartment. This would not necessarily protect herbivores from exposure to the protein, but it might limit adventitious contact. For example, by adding a KDEL tetrapeptide sequence to the C-terminus of the recombinant protein, there is efficient retrieval from the Golgi apparatus to the endoplasmic reticulum (ER) (Munro and Pelham, 1987). This helps to prevent proteins being secreted to the apoplast, phloem and xylem, where contact with the plant's environment, including microbes and insect pests, becomes more likely (Commandeur et al., 2003).

5.3 Contamination of the Food Chain during Processing

There are a number of ways in which transgenic plant material, or products derived therefrom, could enter the food chain of humans or domestic animals. These include transgene spread to food and feed crops, contact between transgenic plants and non-target organisms, adventitious herbivory of transgenic plants and leaching of recombinant proteins into the environment through poor waste management. Another major source of contamination is the unintentional mixing of transgenic and non-transgenic crops during harvesting, transport, refining and processing, which has resulted in some highly-publicized incidents including the discovery of recombinant DNA in Linda McCartney organic food products and the discovery of unregistered StarLink maize in maize products. A more pertinent example in molecular farming is the ProdiGene incident in which stray maize plants expressing pharmaceutical proteins were found growing among a soyabean crop. The accidental mixing of transgenic and non-transgenic harvest products is more likely when those products appear visually identical. Therefore an important step towards identity preservation is the use of non-commercial crop varieties, visually striking varieties (e.g. white tomatoes) or non-food/feed crops that could possibly be introduced into food or feed processing by misidentification (Commandeur et al., 2003).

5.4 Product Safety

The impact of plant produced proteins on human and animal health is of paramount importance in molecular farming. Though, recombinant proteins produced in plants are in most regards identical to their native counterparts, still there are two potential areas of concern.

1. The purified protein may be contaminated with toxic substances from the plant or applied to the plant, e.g. plant-derived metabolites, allergens, field chemicals (e.g. herbicides, pesticides fungicides), fertilizers, dung and manure.
2. The product itself, due to intrinsic properties, may be harmful.

One aspect in which plant-derived recombinant proteins can differ significantly from their native counterparts is glycosylation. While most of the protein post-translational modification steps in plants and animals are conserved, there are minor differences in the structures of glycan chains, such as the absence of terminal sialic acid residues and the presence of plant-specific residues α-1,3-fucose and β-1,2-xylose in complex glycans. There is a concern that such differences could render plant-derived proteins inactive, make them harmful or make them immunogenic. Glycan chains play a number of roles including structural (helping proteins to fold correctly), chemical (making proteins more soluble), biochemical (making proteins more stable by protecting them for proteases) and functional (allowing proteins to be recognized by other molecules, including cell surface receptors). Even minor differences can have profound effects on the protein's stability, biodistribution and biological activity. For example, the normal glycoform of the hormone erythropoietin has a half life of 6-8 hours but a varient lacking the terminal sialic acid residue has a half life of less than 5 minutes. Studies have shown that recombinant antibodies with plant-specific complex glycans do not provoke an immune response in mice. It is also possible to modify the glycan structure and make it more similar to the pattern normally found in humans by co-introducing transgenes encoding human glycosyltransferase (Commandeur et al., 2003).

6. Conclusions

Molecular farming provides an opportunity for the economical and large-scale production of pharmaceuticals, industrial enzymes and technical proteins that are currently produced at great expense and in small quantities. Plants, in molecular farming, have several advantages for serving as an expression system for pharmaceutically important, commercially valuable proteins. At the current rate of progress, molecular farming may become the premier expression system for many pharmaceuticals proteins in the next decade. This is partly because of the utility of plants, with their higher eukaryotic protein synthesis pathway, to make mammalian proteins. Importantly, the economical advantages brought by using plants, which can be grown with such a minimal investment in their cultivation, makes them attractive as an industrial scale expression system. It is anticipated that plants will be the premier expression system for diagnostic and therapeutic proteins. Plant expression systems have the potential to make them as abundant tomorrow as prescription drugs are today.

It is a must to ensure that these benefits are not outweighed by risks to human health and the environment by transgene pollution, the accumulation of recombinant proteins and the contamination of food and feed crops. These risks can be addressed through rigorous comparisons with equivalent products isolated from their natural sources or produced in other systems, and comprehensive testing and safety monitoring from the field through to the extraction and isolation of the pure pharmaceutical product.

References

Chawla, H.S.(2002).Introduction to Plant Biotechnology, Second Edition, Oxford & IBH publishing Co. Pvt. Ltd. New Delhi.

Christou,P; Ma, JK and Drake, PM.(2003). "Genetic modification": The production of recombinant pharmaceutical proteins in plants". Nature Reviews Genetics. http://www.nature.com/nrg/journal/v4/n10/abs/nrg 1177. html. Retrieved 2008-09-13.

Commandeur, U; Twyman, RM and Fischer, R.(2003). The biosafety of molecular farming in plants. Ag Biotech. 5:1-9.

Daniell, H; Streatfield, SJ and Wycof, K.(2001). Medical molecular farming: Production of antibodies, biopharmaceuticals and edible vaccines in plants. Trends in plant science, 6(5): 219-229.

Fischer, R and Emans, N. (2000). Molecular farming of pharmaceutical proteins. Transgenic Research, 9: 279-299.

Fischer, R; Drossard, J; Commandeur, U; Schillberg, S and Emans, N. (1999). Towards molecular farming in the future: Moving from diagnostic protein and antibody production in microbes to plants. Biotechnol Appl Biochem, 30:101-108.

Giddings,G. (2001). Transgenic plants as protein factories. Current Opinion in Biotechnology, 12 (5): 450 – 454.

Gupta,P.K. (2008).Biotechnology and Genomics. Rastogi Publications.

Magnuson, NS; Linzmaier, PM; Reeves, R; An, G; Hay Glass, K and Lee JM.(1998). Secretion of biologically active human interleukin-2 and interleukin-4 from genetically modified tobacco cells in suspension culture. Protein Expr Purif, 13:45-52.

Munro,S and Pelham,H.R. (1987). A C-terminal signal prevents secretion of luminal ER proteins. Cell, 48: 899- 907.

Scheller, J and Conrad, U. (2005). Current Opinion in Biotechnology, 8 (2): 188 – 196.

Sijmons, Peter J. (1997) "Production of Correctly Processed Human Serum Albumin in transgenic plants". Mogen International NV, Einsteinweg 97,2333CBLeiden,TheNetherlands,http://www.nature.com/nbt/journal/v8/n3/abs/nbt 0390-217. html. Retrieved 2008-09-13.

Sinha, BK; and Kumar, R.(2008). Principles of Annual Cell Culture. First edition. International Book distributing Co.

Twyman, RM; Stoger, E; Schillberg, S; Christou, P and Fischer, R.(2003). Molecular farming in plants: host systems and Expression technology. TRENDS in Biotechnology, 21(12):570-578.

Walmsley, A and Arntzen, C. (2000). Plants for delivery of edible vaccines. Current Opinion in Biotechnology, 11:126-129.

Zuo, J and Chua,N.H. (2000).Chemical inducible systems for regulated expression of plant genes. Current Opinion in Biotechnology, 11: 146-151.

❑❑❑

CHAPTER – 12

Scope of Light Trap as IPM Tool for Sustainable Cultivation of Rice

Amit Kumar Sharma, Swati Barche and P.K. Mishra
Department of Entomology, JNKVV, College of Agriculture, Tikamgarh 482004 (M.P.)
Email - amitsharma2375@gmail.com

Summary

Although the concept of IPM, with emphasis on minimum use of pesticides: is well established and widely practiced in India, pest population still continue to be unabated, often reaching to alarming levels .Development of non chemical methods of pest control as component of IPM tactics is a priority area of research today. Development of multiple resistance in larval stages against recommended insecticides ,spoiling of crop ecosystem with toxic chemicals and consequences of health hazards is the best alternative to control the pest in adult stage before they reproduce. Use of light trap is one such potential methods of control for the species known to be positively phototropic in adult stage. Rice is one of the most important cereal crop grown in different parts of world , serving as staple diet for millions of people . The losses due to insect pests in rice are very high and the pesticidal use in this crop is second highest after cotton in world .The positive feature of using light trap as IPM tool for survey ,monitoring , detection and control tool is that most of the insect pest species of rice are phototropic in nature and collected in sufficient numbers in light trap .Therefore use of light trap for imparting eco-integrity by minimizing indiscriminate use of chemicals ,reduction of pest management cost and enhancement of farm income can be very useful for achieving the ultimate goal of sustainability in rice cultivation.

Introduction

The term green revolution was coined by Dr. Williom Gaud of US Department of Agriculture in 1968 to describe the revolutionary progress taking place in wheat and rice field of South Asia ,in terms of increase yield per hectare. The pathway of green revolution involved dwarf and semi dwarf varieties ,the luxuriant growth of high yielding varieties attracted a variety of pests and therefore ,chemical pesticides needed to applied periodically .The overuse and misuse of pesticides created many ecological problems .Dr M.S. Swaminathan coined the term "Evergreen Revolution" in 1990 to emphasize the need of enhancing productivity perpetuity without associated ecological harms(Swaminathan,1996,2006). In view of the second generation pest problems created by green revolution technology ,several pest management tactics were evolved and refined for the sustainable agriculture.

Many of the insect pests mostly nocturnal and few diurnal species are positively phototropic and are attracted towards light. The light trap works on this principle of phototropic response and automatically capture adult insects flying during the night hours. Although the exact mechanism that lead to an insect capture is far from clear (Southwood, 1966) but it seems that there is an area, some times called radius of light trap or "catchments area", within which insects come into influence of light during their random flight. The detailed study of Verheijen (1960) have shown that light trap catches insects because the light illumination of trap relative to the surrounding, interferes with normal photic orientation, this results in the insect moving towards the light source. Hence, anything that reduces this contrast e.g. the illumination from surrounding houses or street lamps or moon light will have the effect on reducing the catch.

The idea of building automatic trap to capture nocturnal insects is very old as conceived by the entomologist long ago. The use of a gas filled container to function as a fumigation chamber with an electric trap probably dates back to about 1880. This was the year when first carbon filament electric lamp became available and when it was put to use by some entomologist for his insect collection (Hollingsworth, 1961). The first authors to investigate the usefulness of incandescent lamps were Parrot (1929) and Peterson and Haeussner (1928) [quoted by Mezokhin-Porshnyakov 1969]. In India the use of light trap for insect control purposes was known to the entomologists much earlier than this period.

Importance of IPM

Basic tactics of integrated pest management (IPM) where used to defend crop plants against ravages of pests long before, the term was accepted by the scientific community in 1972 .Till now more then 100 definition of integrated control , pest management and integrated pest management have been proposed ,however the most appropriate definition is the one adopted by the FAO panel of experts (FAO,1967)i.e. integrated pest control is the pest management system that in the context of associated environment and population dynamics of the pest species ,utilizes all suitable techniques and methods as compatible manner as possible and maintains pest population levels below those causing economic injury.

The thrust of IPM programmes have been to reduce the pesticide load in agro ecosystem .The first major IPM project lunched in USA, commonly called Huffaker project (1972-78) involved six crops ,viz .alfaalfa, citrus, cotton, pines, pome , stone fruit and soybean resulted in to 70-80 per cent reduction in use of environmentally polluting insecticides in 10 years .In Asia most outstanding example is that of FAO sponsored farmers participatory National IPM programme in Indonesia, which has led to decrease in pesticide input and increase in rice out put .In India ,IPM programmes have been developed for major agricultural crops .A landmark example of bio-intensive IPM (1998-2001)is of rained cotton of Astha village ,district Nanded (Maharashta). The cost of plant protection was reduced by 94.3 percent and seed yield increased by 56.9 percent (Dhaliwal,2008).

The contribution of IPM are critical to meeting the economic, environmental and social mandates in assuring sustainability of agro ecosystem (Dhaliwal and Arora ,2006).The future road map of pest management passes through blend of traditional ecological principles and frontier technologies such as Ecologically based integrated pest management (EBIPM).

Importance of Light Trap as IPM Tool

Light traps as pest control tool, were extensively used in India prior to 1940 against cutworms in Bihar (Dutt, 1919), hairy caterpillars in Gujrat (Javeri, 1921) and Punjab (Hussain, 1930) and against paddy stem borer in South India (Ayyar and Ananthanaryan, 1937). However, after 1940, with the advent of insecticides such as DDT, BHC and organophosphate and their instant success in insect control, use of this practice was forgotten until recent years, around seventies, when importance of non chemical methods of pest control was revived again in IPM system.

The indiscriminate and consistent use of insecticides to achieve 100% control of pest for over two decades, badly disturbed the stable ecosystem and gave rise to several problems such as environmental pollution, hazards to human being and beneficial fauna & flora, insecticidal resistance, sudden outbreak of insect pest and resurgence of new insect pest species. Entomologists soon realized these problems and diverted their attention to develop non chemical methods of insect pest control based on sound ecological principles.

In recent years use of light trap has occupied an important place in entomological studies all over the world for survey, detection and control of insect pest population. Extensive work has been carried out on various aspects of light trap studies at J.N.K.V.V., Jabalpur under the leadership of Dr.S.M.Vaishampayan for more than three decades leading to development of technology of use of light trap as IPM tool, covering studies on trap design, light sources, interpretation of trap catch data using correction factors to monitor the pest activity and effective suppression of pest population. Very promising results were obtained against pod borer (*Heliothis armigera*), cutworm (*Agrotis ipsilon*) and hairy caterpillar (*Amsacta moorei*).

The use of light trap in insect pest management studies have more advantages with few limitations. The advantages are :- (i) They operate continuously through out the year irrespective of weather condition. (ii) They will attract insects irrespective of the physical condition of field i.e. water lodged condition or the bad weather (iii) During rainy season insecticides are generally less effective in insect control while light trap operates well during rains also. Inseticide application is also difficult in this season (iv) Their use can be integrated with other approaches of insect pest control to reduce the use of chemical insecticides, which is the basic principle of IPM. Chemicals free produce is good for health also (v) Cost of operation is low and one time purchase of trap can work year after year (vi) A group of famers can be benifitted by one instrument, because adults of target pest are flying insects (vii) Use of light trap helps in forcasting of insect pest out break situation in certain species, giving enough time to evade the damage done to the crop.

There are however, few limitations including: (i) Problem of availability of electric power & electric connection in fields in remote areas (villages) and the availability of substitute of energy power sources like Battery operated and Solar energy operated systems at affordable cost (ii) It's success is limited against photopositive insect species only (iii) Prediction or forecast of pest activity based on trap catches is a skilled job which requires the help of technical personnel like the entomologists.

Case Studies

Rice is one of the most important cereal crop grown in different parts of the world serving as staple diet of millions of people. It is an important cereal food crop of India with wide physiological adaptability and is grown in all the states of India. Major rice growing states are Andhra Pradesh, Tamil Nadu, Panjab, Odisha, U.P. and Madhya Pradesh etc. Though India has largest area comprising 27 per cent of the total area under rice in the world, its production is hardly $1/5^{th}$ of the world produce. Among the several causes of low production of paddy in India, one is atributed to large infestation of insect pests and diseases including use of high yielding dwarf cultivars associated with high doses of fetilizers. Intensification of agriculture has enabled the emergence of various pests in paddy crop, of these are some major and some are minor. In last 30 years indiscriminate use of inecticides, to achieve 100 per cent control of insect pest in such crops, has created many problems as mentioned earlier. In addition to these problems pesticidal toxicity causes mortality in paddy-cum-fish culture also. Therefore, evaluation of an alternative pest management technique leading to reduction in use of pesticides is important in this crop.

There are few reports of work down on light trap against insect pests of paddy crop in different states including Chhattisgarh (Bhatnagar et al. 1999). Tamil Nadu (Ramakrishnan et al., 1994), Karnataka (Shrinivasa et al. 1990), West Bengal (Sarkar and Gayen, 1992) and Uttar Pradesh (Rai et al. 1990).

Ganathilagaraj and Kumar (1997) gave a list of 537 pests of rice in India, which includes 385 species of insects. Kalode (1976) reported out 350 species of insects, only 20 can be considerd of major significance.

The losses due to various insect pest in paddy crop are very high. Insect pest-damage the crop at different stages of crop and in different ways. Abraham and Khosla (1967) reported expected crop loss of 11 to 17 per cent by insect pests in paddy. Besides the presence of insect pests in rice ecosystem, the presence of complex of parasitoides & predators also play an important role as biocontrol agents in the management of insect pest population in paddy crop ultimately helping in minimizing the yield losses.

Seasonal activity of various pests of paddy through light trap catches in Jabalpur district(M.P.), as reported by Sharma et al. (2004), includes *Sogatella furcifera* H., *Cnaphalocrocis medinalis* Gen., *Mythimna separata* W. and Grass hopper (complex) as major pests and *Nephotettix virescens* Dist., *Leptocorisa varicornis* T and *Dicladispa armigera* O., *Tryporyza incertulas* W., *Nilaparvata lugens* S. and *Melanitis ismene* C. as minor pests. All these pests are phototropic in nature and thus provide great scope of using light trap as IPM tool in paddy.

Sharma and Vaishampayan (2009)evaluated light trap as direct control tool against major pest species of paddy in Jabalpur M.P.,With exception of Army Worm (*Mythimna separata*) the results have consistently proved the utility of light trap as direct control tool against remaining three major species namely (*Sogatella furcifera*),Leaf Folder (*Cnaphalocrocis medinalis*) and Grass hopper (complex) (G*astrimargus transversus* and *Trilophidia cristata*). Effective coverage area was between 100 to 200 m distance from trap covering around 5 to 6 hactare of crop area per trap.

A total of 64 species of insects were collected in paddy ecosystem by using light trap.These insects belongs to 8 orders and 35 families ,Lepidoptera was the largest order followed by Hemiptera (14 species),Coleoptera (12 species), Orthoptera (4 species).Hymenoptera ,Odonata,Isoptera and Dictioptera are the order of minor significance. Based on economic importence the collection was represented by 45 species of insect pests among them 14 species belongs to rice pests (Sharma et al., 2006).

Besides insect pests natural enimies such as parasitoides and predators are also collected in light trap.Seventeen predatory species were collected in light trap through out the kharif season in paddy ecosystem (Sharma et al. 2006).Atwal et al. (1969); De bach (1974); Larcohelle (1974); Ismael (1974); Patil et al. (1982); Khan (1983) and Huddleston and Gauld (1988) also reported the activity of parasitic and predatory species through light trap catches.Mazokhin and Parshnikov (1969) reported that light trap is far more convenient then indiscriminate use of insectisides ,because majority of parasitic & predatory species and honey bees are diurnal and are not attracted by light traps and spared .

Application in Sustainable Agriculture

Development of non-chemical ,ecofriendly methods of pest control as a component of IPM tactics is a priority area of research in sustainable agriculture today .With the reported development of multiple resistance in larval stages against recomonded insecticides ,spoiling of crop ecosystems with toxic chemicals and consecuent helth hazards, the best alternative is to control the pest in adult stage before they reproduce . Use of light trap is one of such potential methods of control against the species known to be positively phototropic in adult stage (Fig. 2).

Fig. 1 : New Design of Light Trap

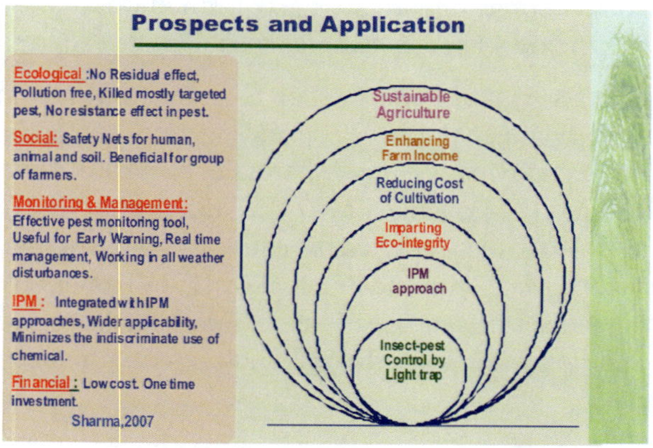

Fig. 2 : Prospects and Applications

Applications of light trap in sustainable agriculturecan be made through monitoring of adult activity and supressiopn of adult population as direct control tool in paddy and many other important crops.

As Monitoring Tool of Adult Activity

Light traps have widely used in past for survey and monitoring of seasonal activity in several crops (Otman,1964;Kovitvadhi and Cantelo, 1966; Brown et al. (1969) and Betts, 1976). Naganagoud et al. (1999) from Tamil Nadu, reported seasonal incidence of *Nilaparvata lugens*, *Sogatella furcifera* and *Cnaphalocrocis medinalis* in rice crop through light trap catches. Harinkhare et al. (1998) observed peak activity of *Cnaphalocrocis medinalis* in September followed by October, at Waraseoni in Balaghat district (M.P.). Yellow stem borer, leaf folder and case worm were observed at low level in light trap catches during first three months of rice cropping season (Bhatnagar et al. 1999).Similarly Yoshimeki(1967)studied seasonal abundance of rice stem borer moths using standard light trap in about 6340 hactares of rice crop in Japan.

Study on seasonal activity of insect pest species collected in light trap will be very useful for prediction, development of the forecasting module and management strategy against major phototropic insect pests well in advance.

As Direct Control Tool

1. Philosophy is to kill the noxious insects at the time and place when they are most concentrated, immobile and accessible. Light trap concentrates the adult population present in the surrounding environment within a trap zone, arrest their movement by trapping (make them immobile) and are killed easily using a drop of inscticide.
2. Concept is to control pest population in its nucleus stage prior to oviposition, when it is concentrated and confiend within the ovaries of female adults.

Light trap caught sufficient numbers of *Scirpophaga incertulas, Nilaparvata lugens, Sogatella furcifera* and *Orseolia oryzae* to hold their population level below economic threshholds in many cases and at the 3sites ,pestisides use was reduced by 60.00,50.00 and 57.63 percent respectivly in Karnataka (Shrivstava et al. 1992).

Conclusion

Light trap works as pest menagnment tool indirectly through monitoring the adult activity,which decids the need of pesticide application ,it's timing if necessory and directly by mass trapping of adults (male and female)and disrupting reproduction and oviposition. Mass trapping minimizing infestation in field significantly without any toxic hazards.

The concept of "Use of light trap as a component of adult oriented strategy of pest management "(Vaishampayan, 2002)is to detect the potential danger of poulation exsplosion well in advance by monitoring the peak periods of activity of adults with their reproductive potential, and decide the proper time of application of insecticids synchronizing with these peaks. Forecasting of pest population level in field shall be more accurate and reliable if adult activity is also taken into account .

Therefore in view of information available on utility of light trap as Integrated Pest Management tool for the sustainability of rice cultivation , use of light trap should be promoted in pest management for imparting ecointegrity,reduction of cultivation cost , enhensment of farm income and ultimatly achiving the goal of sustainable agriculture.

Aknowladgement

I would like to express my deep regards to Dr. S.M. Vaishampayan Retd. Prof. and Head Deptt. of Entomology JNKVV,Jabalpur (M.P.)for his valuble guidance. I shall always be indebted to him for the ease with which he laid bare before me to draw from his life time experiences as an Entomologist.

Table 1 : Impact of light trap operation in reduction in pest population in Paddy.

	Insect species observed	Pest population level [Average of two locations (mean of 20 samples), % in round figure]				Per cent reduction over non trap a rea (control) [round figure]		
			Trap area					
		500 m away from trap	200 m away from trap	100 m away from trap	50 m away from trap	200 m away from trap	100 m away from trap	50 m away from trap
		43.74	*14.44 33% survival*	*15.83 36% survival*	*22.66 52% survival*	*67%*	*64%*	*48%*
2.	*Cnaphalocrocis medinalis*	*29.49*	*15.08 51% survival*	*8.49 29% survival*	*15.91 54% survival*	*49%*	*71%*	*46%*
3.	*Mythimna separata*	*22.80*	*14.00 61% survival*	*7.90 34% survival*	*15.10 66% survival*	*39%*	*65%*	*34%*
4.	*Grass hoppers (Complex) (Trilophidia cristata) (Gastrimargus transversus)*	*16.62*	*5.93 36% survival*	*5.37 32% survival*	*8.99 54% survival*	*64%*	*68%*	*46%*

(Sharma and Vaishampayan 2009)

References

Atwal, A.S., Caceres, O. and Monzon, A. (1969). Testing of home made light traps for monitoring adults of white grubs (*Phyllophaga* spp.) in Esteli, Nicaragua. Revista Nicaraguense de Entomologia, No. 48:1-13.

Abraham, T.P. and Khosla, R.K. (1967). Assessement of losses due to incidence of pests and diseases on rice crop. *J. Indian Soc. Agril. Stat.* 19 (1):69-82.

Ayyar, T.V.R. and Anantanarayan, K.P. (1937). The stem borer pest of rice schoenobius incertulas in South India. Agril. and Livestock in India. 7:171-179.

Betts E. (1976). Forecasting infestations of tropical migrant pest :the desert locust and army worm .pp.113-114 in rainey ,R.C.(Ed.) Insect Flight -287 PP. Oxford.

Bhatnagar, A., Saxena, R.R. and Bhatnagar, A. (1999). Environmental correlation of population build up of rice insect pests through light trap catches. *Oryza;* 36 (3): 241-245.

Brown,E.S.,Betts,E.and Rainey,R.C.(1969). Seasonal changes in distribution of Africa army worm ,*Spodoptera exempta* (Wik)(Lep.:Noctuidae) with special reference to Eastern Africa *Bull.Ent. Res.* 58, 661-721.

Gunathilagaraj, K. and Kumar, M.G. (1997). Rice insect out breaks : an analysis. *Madras Agric. J.* 84(6) : 298-311.

De bach, Paul (1974). Biological control by Natural enemies. Cambridge University Press, p. 323.

Dubey, O.P. (1970). Effect of date of planting and fertilizer doses on the incidence of major pests of paddy with particular reference to *Sogatella furcifera*. Horv. Unpulished M.Sc. Thesis JNKVV, Jabalpur.

Dutt, H.L. (1919). The methods of control of *Agrotis ipsilon* in Bihar. *Rep. Proc. 3rd Ent. Mtgs. Pusa* II : 622-625, Calcutta.

Dhaliwal G.S. and Arora R. (2006). Integrated Pest Management :Concepts and Approaches. Kalyani Publishers,New Delhi.

Dhaliwal G.S. (2008). Pest management in global context .Key Note Address ,2nd Congress on Insect Science ,Feb.17-19 2008.

FAO.(1967). Report of first session of the FAO panel of experts on integrated pest control. Food and Agriculture Organization ,Rome.

Harinkhare, J.P., Kandalkar, V.S., Bhomick, A.K. (1998). Seasonal abundance and association of light trap catches with field incidence of rice leaf folder (*Cnaphalocrocis medinalis* G.) *Oryza*. 35(1) : 91-92.

Hollingsworth, J.P. (1961). Relation of wavelength to insect response. *USDA ARS* (Ser.) 20(10) : 9-25.

Huddleston, T. and Gauld, I. (1988). Parasitic wasps (Ichneumonoidea) in British light traps. *Emtomologist*. 107(2) : 134-154.

Hussain, M.A. (1930). Studies on *Amsacta* sp. of maize in Hoshiarpur District. *Ent. Rep. Dept. Agril. Punjab*. 29(1) : 53-63.

Ismael, I.I. (1974). The effect of certain weather factors on the activity and population density of tiger beetle *Cicindela melancholica* F. (Coleoptera, Cicindelidae). *Bull. Soc. Ent. Egypt*. 58 : 345-348.

Javeri, T.N. (1921). Notes on "Kutra" (Hairy caterpillars) and their, controlling measures. *Res. Proc. 4th Ent. Mtgs.* Pusa : 98-100.

Kalode, M.B. (1976). Insect pest control in high yielding rice varieties, *Seeds and Farmers*. 11(7) : 7-16.

Khan, R.M. (1983). Studies on the common predatory and parasitic species of insects collected on light trap at Jabalpur. Unpublished M.Sc. Ag. thesis, JNKVV, Jabalpur, pp. 49.

Koitvadhi,K. and Cantelo,W.W. (1966). The use of light traps to measure relative population densities of the rice gall midge, *Pachydiplosis oryzae* and other insect pests of rice. *Int. Rice Comm. News Letter*15(2):16-19.

Larchohelle, A. (1974). Ground beetles Flying under an electric light (Coleoptera : Carabidae). *Proc. Ent. Sco.* Wash. 74(4) : 424-427.

Mezokhin-Porshnyakov, G.A. (1969). Design and operation of insect traps with source of ultraviolet(In Russian). *Ent. Obozr*. 37(2) : 464-471.

Naghagound, A., Patil, B.V., Sreenivas, A.G. (1999). Studies on light trap catches of major pests of rice in Tiungabhadra project area. *Karnataka J. Agril. Sci. 12* (1-4):191-192.

Otman, M.E.R. (1964). Orcheard insects Surveys with black light traps *J. econ. Entomol.* 57(1):6-8.

Patil, B.V., Devaiah, M.C. and Thontadarya, T.S. (1982). Studies on attraction of predatory insects to mercury bulb light trap. *Ind. J. Eco*. 9(1) : 108-112.

Rai, A.B., Singh, J. and Rai, L. (1990). Rice bug (*Leptocorisa varicornis* F.) appearance on light trap in eastern Uttar Pradesh, India. *Oryza*. 27(1) : 66-72.

Ramakrishana, A., Sundram, A. and Uthamaramy, S. (1994). Path coefficient analysis of abiotic factors affecting light trap catches of green leaf hopper *Nephotettix virescens* (Distant) on rice. *J. Ins. Sci.* 7(2) : 208-209.

Salem, M.M., Al-Gamal, M.M., El-Sebaey, H.A. and Negm, F.H. (1999). Survey and population studies on catches of Hemiptera by a light trap at Al-Arish city(North Sinai). *Egyptian J. Agril. Res*. 77(1) : 187-193.

Sarkar, T.K. and Gayen, P. (1992). Population dynamics of paddy stem borer(*Scirpophaga incertulas* walk, (Lepidoptera : Pyralidae) with respect to some weather factors. *Ann. Agril. Res*. 13(3) : 247-252.

Shrivastava, V.K., Diwakar, M.C. and Pawar, A.D. (1992). Light trap and rice pest management. *Plant Protc. Bull*. Faridabad. 44(3) : 39-41.

Southwood, T.R.E. (1966). Ecological Methods. University printing house. p. 223.

Srinivasa, N., Viraktamath, C.A. and Thonatadarya, T.S. (1990). Influence of moon phase and weather on the light trap catches of insect pests of rice. *Oryza,* 27(2) : 183-190.

Swaminathan, M.S. (1996). Sustainable Agriculture :Towards an Evergreen Revolution .Konark Publication New Delhi.

Swaminathan ,M.S. 2006.An evergreen revolution .*Crop Sci.* 46:2293-2303 .

Sharma, A.K., Vaishampayan, S. and Vaishampayan, S.M.(2006). Documentation of texonomic distribution of insect pest species of paddy ecosystem. JNKVV Journal

Sharma, A.K., Vaishampayan, S.M. (2009). Evaluation of light trap as eco-friendly control tool in the management of insect pest species of paddy. *JNKVV Journal*.

Sharma, A. K. (2007). Evaluation of light trap as eco-friendly control tool in the management of insect pest species of paddy crop. Paper presented in the Bhartiya Vigyan Sammelan 2007-Silver Jubilee all India young

Sharma, A.K. (2004). Scope of use of light trap as IPM tool in paddy eco system. Unpublished Ph.D Thesis.R.D.V.V.Jabalpur.

Vaishampayan, S.M. (2002). Use of light trap as a component of Adult oriented strategy of pest management. In resource management in plant protection. *Pant Protc. Anore. India Hydrabad publication* pp. 139-144.

Verheijen, F.J. (1960). The mechanism of the trapping effect of artificial light sources up on animals. *Arch. Neerland Zool*. 13 : 1 -107.

Yoshimeki, M. (1967). Summery of the forecasting programme for rice stem borer in Japan. The major insect pests of rice plant. Baltimore, Maryland. The John Hopkins Press, pp. 729.

❑❑❑

CHAPTER – 13

Agricultural By-products and their Biotechnological Approach

Parkash Kumar Sarangi and Hara Prasad Sahoo
Department of Botany and Biotechnology, Ravenshaw University, Cuttack - 753003, India
Email : sarangi77@yahoo.co.in

Summary

The huge accumulation of agro-industrial waste-materials generated by the milling, brewing and various agriculture and food based industries which in total amount to be about 500 million tones per year in India. Most of these by-products contain three major structural polymers such as cellulose, hemicellulose and lignin. These are also the main source of feed for domestic animals. A high proportion of this waste material is carbohydrate and phenolic compounds. Accumulation of this kind of biomass not only results in the deterioration of the environment, but also loss of potentially important materials. Biological degradation, now-a-days, has become an increasingly popular alternative for the treatment of agricultural, industrial, organic as well as toxic wastes. Cellulose and hemicellulose being the major constituents of these materials can be referred to as valuable resources for a number of reasons, mainly due to the fact that they can be bio-converted easily into valuable products. These wastes can be processed to yield a number of value added products, such as biofuels like biohydrogen, bioethanol along with a variety of chemicals. Various treatment methods such as physical, chemical, biological are employed for the production of value-added compounds. Various phenolics like ferulic acid and vanillin are world's most prized flavor compound mostly obtained through microbial treatment of agro-wastes.

1. Introduction

The yearly accumulation of agro-industrial waste-materials generated by the milling, brewing and various food based industries, has led to the consideration of extracting high value residues to offset the cost of treating and disposing of the residues. Most of these by-products are currently used as animal feed, but attention towards obtaining of high-value compounds from these wastes is being emphasized by various academic as well as industrial researchers. Biological degradation, for both economic and ecological reasons, has become an increasingly popular alternative for the treatment of agricultural, industrial, organic as well as toxic wastes. Agricultural by-products which contain three major structural polymers, cellulose, hemicellulose and lignin, are the main sources of feed for ruminants. However, their quality as feed, especially in terms of digestibility, depends not only on the amount and structure of these polymers but also on the amount of readily available substances such as sugars and amino acids. A high proportion of this waste material is carbohydrate and phenolic in nature. Specific sugar residues can be released from the cell-wall by the action of carbohydrases. On the other hand phenolics moieties are linked to polysaccharides by ester-bond forming structural cross-linking bridges between various polymers. Some phenolic compounds, such as ferulic and p-coumaric acids, are also present in cell wall. Many species of microorganisms are able to degrade agricultural by-products thus releasing a vast amount of carbon which otherwise would be locked away in plant secondary metabolites such as lignin. Recently, the increasing interest in natural products led to develop different value-added phenolic compound production via biotechnological processes involving microorganisms. An important attribute of microbial biocatalysts in bioreactors is the ability to synthesize products in a consistent and predictable manner. Application of microbial biotransformation tools is a promising alternative to obtain natural products. In view of the above facts, this chapter focuses on the biotechnological approaches to develop technologies for efficient use of this lignocellulosic material and to utilize agro waste materials as cheap sources of phytochemicals, antioxidants, biofuels and value-added phenolic compounds.

2. Crop Residues

Agro-waste materials like crop residues are primarily used as fuel in many countries around the world. Crop residues are more widely burnt than animal waste and forest litter. Crop residues are the non-edible plant parts that are left in the field after harvest. These also include remains that are generated from crop-packing plants or that are discarded during crop processing into the generic category of crop residue (Ernest and Buffington, 1981).

On the surface of our planet, around 200 billion tons per year of organic matter are produced through the photosynthetic process (Zhang 2008). However, the majority of this organic matter is not directly edible by humans and animals so creating a source of environmental problem. Moreover, today's society, in which there is a great demand for appropriate nutritional standards, is characterized by rising costs and often decreasing availability of raw materials together with much concern about environmental pollution (Laufenberg et al. 2003). Therefore, there is an extensive emphasis on recovery, recycling and upgrading of wastes. This is predominantly valid for the agro-food industry, which furnishes large volumes of solid wastes, residues and by-products, produced either in the primary agro-forestry sector or by secondary processing industries, posing serious and continuously increasing environmental pollution problems (Boucque and Fiems 1988; Koopmans and Koppejan 1997; Lal 2005). It is also to be noted that only crop residues production is estimated to be about 4 billion tons per year, 75% originating from cereals (Lal 2008). Crop residues also vary widely in properties and decomposition rates. Rather than direct measurement, estimates of crop residue production are made on the basis of data on the area and production of different crops, and research information on the straw/grain ratio (Eq. 1).

Residue production= grain production × straw / grain ratio ———— Eq.1

The straw/grain ratios commonly used for different crops are shown in Table 1.

Table 1 : Dry weight ratio of straw to grain for different crops

Crop	Range of straw/grain ratio	
	Gupta et al. (1979)	Lal (1995)
Barley (*Hondeum vulgare* L.)	1.5	1.5
Corn (*Zea may* L.)	1.0	1.0
Cotton (*Gossypium hirsutum* L.)	1.0	1.5
Legumes	1.0	1.0
Oats (*Avena sativa* L.)	2.0	1.0
Potato (*Solanum tuberosum*)	—	0.25
Rapeseed (Brassica campestris)	—	1.5
Rice (*Oryza sativa* L.)	1.5	1.5
Rye (*Secale cereale* L.)	1.5	1.5
Sorghum (*Sorghum bicolor* L.)	1.0	1.5
Soybeans (*Glycine max* L.)	1.5	1.0
Sugarbeet (*Beta vulgaris* L.)	—	0.25
Sugarcane (*Saccharum officinarum* L.)	—	0.25
Wheat (*Triticum aestivum* L.)	1.7	1.5

According to the survey by Lal. (2005) regarding the data on estimates of residue production in the world, the annual residue production is estimated at 2.8 billion Mg of cereals, 305 million Mg of legumes, 108 million Mg of oil crops, 373 million Mg of sugar crops and 170 million Mg of tubers. The total crop residue production in the world is estimated at 3.8 billion Mg, of which 74% are of cereals, 8% of legumes, 3% of oil crops, 10% of sugar crops and 5% of tubers. The most useable crop residue, however, is that of cereals.

3. Biotechnology for Crop Residues

Agro-residues such us cereals straw, corn cobs, cotton stalks, various grasses and reed stems, maize and sorghum stover, vine prunings, sugarcane and tequila bagasse, coconut and banana residues, corn husks, coffee pulp and coffee husk, cottonseed and sunflower seed hulls, peanut shells, rice husks, sunflower seed hulls, waste paper, wood sawdust and chips, are some examples of residues and by-products. These also include all agricultural wastes such as straw, stem, stalk, leaves, husk, shell, peel, lint, seed/ stones, pulp, stubble, etc. which come from cereals (rice, wheat, maize or corn, sorghum, barley, millet), cotton, groundnut, jute, legumes (tomato, bean, soya) coffee, cacao, olive, tea, fruits (banana, mango, coco, cashew) and palm oil. These are widely used as household fuel in developing countries. Agro-residues can be recovered and upgraded to higher value and useful products by chemical or biological processes (Wang 1999; Fan et al. 2000a; Pandey et al. 2000b, c; Webb et al. 2004). In fact, the chemical properties of such lignocellulosic agricultural residues make them a substrate of huge biotechnological value. They can be converted into various different value-added products those may be used as biofertilizer or biopesticide, enzymes, organic acids, ethanol, flavours, biologically active secondary metabolites and also for bioremediation of hazardous compounds, biological detoxification of agro-industrial residues, biopulping etc. (Pandey et al. 2000a ; Bennet et al. 2002 ; S´anchez et al. 2002 ; Tengerdy and Szakacs 2003 ; Howard et al. 2003 ; Kim and Dale 2004 ; Nigam et al. 2004 ; Zervakis et al.2005 ; Manpreet et al. 2005 ; Krishna, 2005). The energy content of residue varies among crop species. For example, the energy content is 3015 kcal/kg for rice straw and 3738 kcal/kg for hay (Stout, 1990). The approximate fuel value per Mg of crop residue is equal to 2 barrels of diesel, 18.6×10^9 J or 3×10^6 kcal. Based on these approximations, estimates of the energy value of total crop residue produced in the world are shown in Table 2. The global energy value of crop residue is 7516×10^6 barrels of diesel or 69.9×10^{18} J of energy.

Energy production is among numerous and competing alternative uses of crop residue (Fig. 1). The share of global biomass energy consumption varies widely, with 47% in Asia, 25% in Africa, 19% in Latin America, 5% in North America, 3% in Europe and 1% in Oceania (Lansink et al. 2002). Goldenberg (2003) estimated that out of the total energy use of 3.9×10^9 toe (total oil equivalent) by developing countries, 22% came from the biomass. In contrast, out of the total energy use of 6.7×10^9 toe worldwide, only 4% came from biomass. Biomass may be defined as "all organic material of vegetal or animal origin, which is produced in natural or managed ecosystems (agriculture, aquaculture, forestry), all or not industrially transformed". Bioenergy which is the renewable energy released from biomass. Bioenergy is definitely projected to contribute significantly in the mid to long term as far as the energy crisis is concerned. According to the International Energy Agency (IEA), bioenergy will provide a valuable system for mitigating 50% of our world energy needs in the 21st century. Agriculture is a source of energy by producing huge biomass. Biotechnological technologies are now applied for the development of genetic engineering plants specifically of industrial or energy crops (Van Beilen 2008).The production of bioenergy and chemicals from agricultural waste and agro-industrial residues may somewhat change the concept more towards the production of biofuel and chemical rather than food use (Morris 2006; Zhang 2008).

Table 2 : Energy value of crop residues produced around the world

Parameters	Values
Total crop residue (106 Mg/year)	3758
Oil equivalent (106 barrels)	7560
Energy equivalent:	
1: Exajoules (10^{18} J)	69.9
2: Quads	60.0
3: 10^{15} kcal	11.3

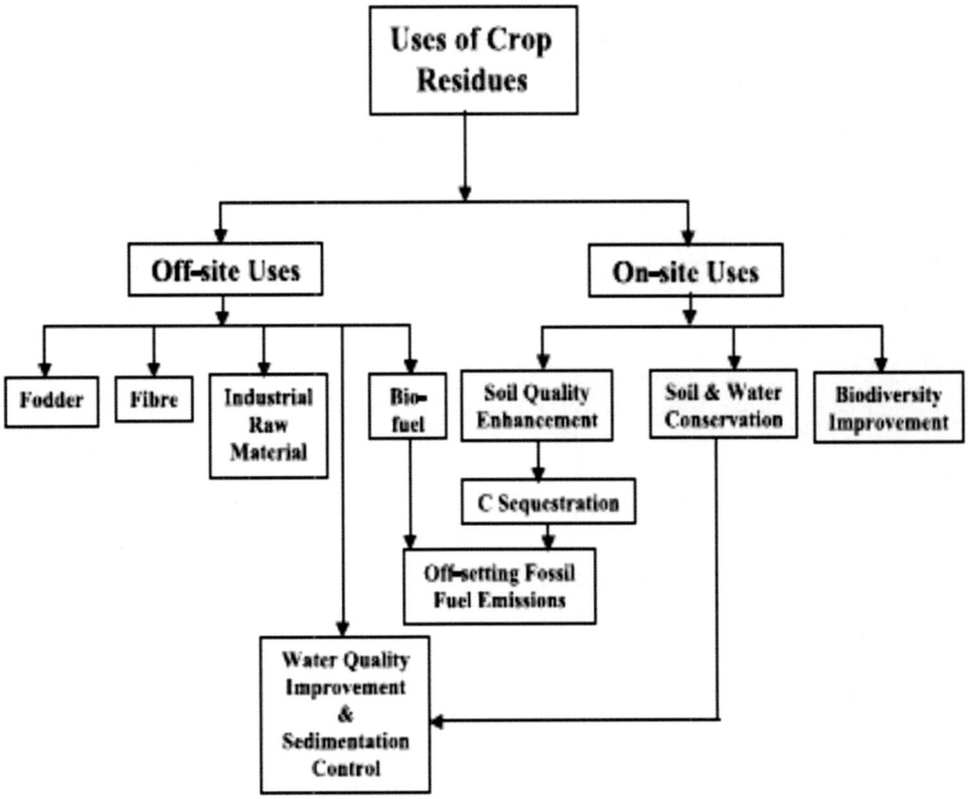

Fig. 1 : Alternative and challenging uses of crop residues (Lal 2005).

Agricultural crops such as corn, wheat, rice and other cereals, sugar cane and beet, potato, tapioca, etc. are now already processed in the starch and sugar refineries into relatively pure carbohydrate like starch, sugars etc. those are used as primary substrates for the food industries. These are also providing good substrates for most industrial fermentation processes and for some chemical processes (Dahod 1999; Kamm and Kamm 2004). Wide variety of valuable chemical products, including biofuels such as bioethanol, and organic solvents such as butanol are produced using fermentation processes from agro-feedstocks (Demain, 2000, 2007; Kunz, 2008; Soetaert and Vandamme 2005, 2008; Wall et al. 2008).

3.1 Bio-Fuels

There are two classes of Bio-fuels such as gaseous and liquid biofuels. Development biomethane can be obtained by purification of the conventional biogas. Biohydrogen is a relatively new type of gaseous biofuel, which is produced by anaerobic fermentation of agro-industrial wastes by the action of methanogenic, acidogenic and hydrogenic bacteria. On the other hand, liquid biofuels are either bioethanol or biodiesel. While bioethanol has recently gained greater importance for mitigation of energy crisis worldwide whereas, biodiesel occupied the centre stage as a potential substitute for petroleum diesel in the last two decades.

3.1.1 Biogas

The conventional biogas is produced using anaerobic digestion of organic wastes including manures by mixed microbial cultures. It is composed mostly of methane (typically 55%–70% by volume) and carbon dioxide (typically 30%–45%) and may also include smaller amounts of hydrogen sulfide (typically 50–2000 ppm), water vapor (saturated), oxygen, and various trace hydrocarbons (Amigun et al. 2008; Kashyap et al. 2003; Siso 1996). Biomethane is extremely similar to natural gas which contains 90% methane only differs that it comes from renewable sources (Krich et al. 2005). Biogas can also used as vehicle fuel. It has been estimated that 40–50% in vehicle maintenance costs achieved due to operation of biogas, (Parawira 2004). Fig. 2 shows formation of simple sugars, amino acids, fatty acids and alcohols from complex polymers like carbohydrates, proteins and fats respectively. The simple molecules are converted into CH_4 and CO_2 on acetogenesis and methanogenesis.

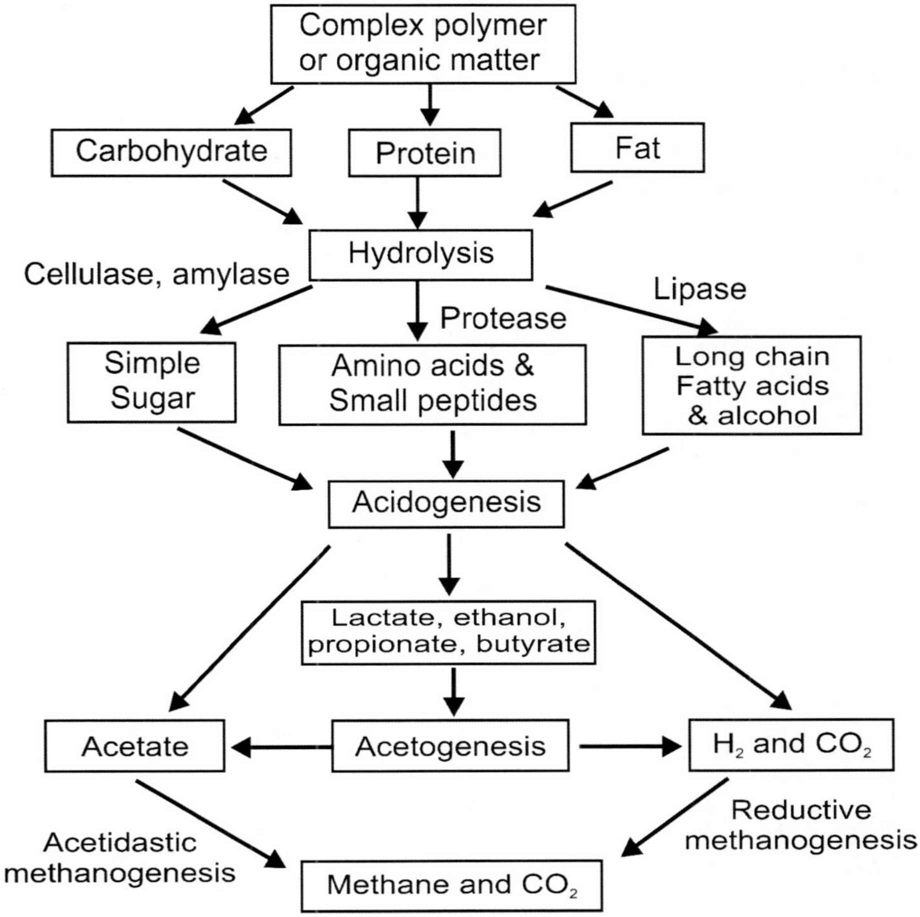

Fig. 2 : Methane and carbon dioxide formation from complex polymers and organic matter (Chattopadhyay et al. 2009).

3.1.2 Biohydrogen

Hydrogen is a very high energy (122 kJ/g) yielding fuel in comparison to methane or ethanol. Photoautotrophically growing bacteria and some algae, utilize light as primary energy source to split water into hydrogen and oxygen by the enzyme hydrogenase. Several forms of organic waste streams, ranging from solid wastes like rice straw and waste water from different sources have been successfully used for hydrogen production. Most experiments have shown considerable hydrogen production with the limited number of thermophilic strains used. It has been examined that CSTR (Continuously stirred tank reactor) gives the best performance with a yield of 0.30 L H_2/g of carbohydrate and production rate of 4.50 mmol H_2/L reactor.

3.1.3 Bioethanol

Bioethanol is a biofuel used as a petrol substitute. It is produced by simple fermentation processes utilizing agricultural carbohydrate feedstock and yeasts as main biocatalysts. A variety of common sugar feedstocks including sugarcane stalks, sugar beet tubers and sweet sorghum are used as the substrates for production of bioethanol. The fermentation process is mediated by two enzymes invertase and zymase, produced by the yeast cells. The overall process is given Fig.3 (Zanichelli et al. 2007).

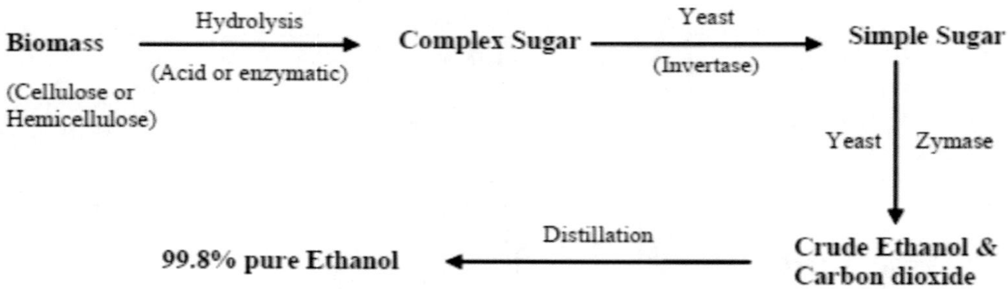

Fig. 3 : Flow chat shows the steps of ethanol production.

Ethanol is produced from different renewable and cheap agricultural products. Not only it solves the energy crisis but also reduces the green house gas emissions like CO_X, NO_X and SO_X thus eliminates smog from the environment. Bioethanol produced from agro- industrial residues can also be used for the transesterification of vegetable oils to produce mono ethyl esters of fatty acids as biodiesel.

4.0 Agro-Industrial Residues

Agro-industrial residues are derived from the processing of a particular crop or animal product usually by an agricultural firm. Materials like molasses, bagasse, oilseed cakes and maize milling by-products and brewer's wastes are also included under these categories. Different Agro-industrial residues are as follows.

1. Milling industry by-products: bran; waste flour; wastes resulting from grain cleaning processes; wheat, corn and rye germs.
2. By-products of the sugar industry: beet pulp, molasses.
3. By-products of the starch industry: potato pulp, potato cell juice.
4. By-products of the fermentation industry include grain, molasses.
5. A portion of this waste material is also generated by the coir industry using coconut mesocarp.

Numerous by-products generated by agro-industries, including corn-cobs, sugarcane bagasse, bran and straw from wheat and rice, sugar beet pulp and coffee pulp, have been used as cheap substrates for biological degradation to obtain value added products by an environment-friendly bioengineering process. This type of biological degradation using enzymes, for both economic and ecological reasons has become an increasingly popular alternative for the treatment of agricultural, industrial, and toxic wastes. A high proportion of these waste materials are carbohydrates and phenolic in nature. Specific sugar residues can be released from these waste materials by the action of enzymes. A number of non-pathogenic micro organisms have been used as producers of these enzymes. The micro organisms are 'Generally Recognized As Safe (GRAS)'. Of the several enzymes used industrially, about a half comes from fungi and one third from bacteria. The processes of microbial degradation of biological waste materials have many industrial and medicinal applications. Research works in this field are primarily focused on discovering new enzymes with novel processes.

Microbial feruloyl esterases are a group of such enzymes that act on the plant cell wall polymers and thus represent the key tools for degradation and modification of plant cell walls components. The enzymes have a large number of potential applications due to their role in many biotechnological processes and in various industries such as chemicals, animal feed, textile and laundry, pulp and paper, food and agriculture and pharmaceutical industries. The prospect of broad applications of feruloyl esterases has fueled much interest in this field and an increasing number of feruloyl esterases have been discovered in several organisms in recent years. The field is however, at an early stage and many problems are there to commercialize it. Till date there is no fully purified feruloyl esterase for commercial use. There is lack of sufficient enzyme stocks too. Efforts have to be made with the help of new technology to achieve cheap production of feruloyl esterases in heterologous hosts and their modification by chemical means or protein engineering to obtain more robust and active purified feruloyl esterase enzyme. The enzyme has potential applications in obtaining ferulic acid from agro- industrial waste materials such as those produced by milling, brewing, and sugar producing industries.

5. Ferulic Acid Production

Ferulic acid is a most abundant ubiquitous hydroxycinnamic acid derived from phenolic compounds. It is distributed widely throughout the plant kingdom, in cereals, legumes, pulses, vegetables fruits and spices (Mathew & Abraham 2004; Mori et al. 1999; Lee 2005). It is found conjugated with esterified or etherified polysaccharides and lignin in the cell wall (Sun et al. 2002; MacAdam and Grabber 2002). It is a renewable resource for the biocatalytic or chemical conversion to other useful aromatic chemicals from agricultural by-products in nature (Ernest 1992). The enzyme FAE (Ferulic acid esterase) is used in biocatalytic degradation of sugar-phenolic acid linkages in the cell wall. Various microorganisms have been reported to biotransform these ferulic acid into value-added products.

Comprehensive studies have been conducted on the nature and biochemistry of ferulic acid obtained from agro-industrial residues and their bioconversion into value-added vanillate derivatives. Biological treatment of different agro-industrial residues using various microorganisms is well studied by many people around the world. Studies on the classification, microbial production, biophysicochemical properties, and biotechnological applications of the enzyme FAE have also been made by several researchers. Benoit et al. presented a schematic overview of industrial applications of feruloyl esterases and ferulic acid. Ferulic acid (Fig..4) is a phenolic compound universally found in plant tissues and relatively abundant in cell walls. These are found covalently linked to polysaccharides by ester bonts and to components of lignin by ester or ether bonds. Ferulic acid represents upto 1.5 % dry weight of cell wall materials in the members of Poaceae. This influences the mechanical properties like extensibility and biodegradability of the cell wall (Fry 1980). It acts as antioxidant and plays an important role in plant defence response. It is found beneficial for health due to its chemo-preventive activities against diseases. It is considered as an abundant class of bioresources for its manifold beneficial activities including antimicrobial and anticancer activity. Several in vivo and in vitro studies have been conducted in humans, animals and cell culture (Priefert et al. 2001) to establish the function of ferulic acid. Some important functions of ferulic acid are : (1) inhibition or prevention of cancers of the breast, colon, lung, stomach, and tongue (2) prevention of brain damage by Alzheimer's proteins (3) inhibiting prostate growth (4) strengthening of bone (5) prevention of diabetes-induced free radical formation (6) expansion of pancreatic islets (7) reduction of elevated lipid, triglyceride and blood glucose level (8) protection of skin from aging effects of UV light (9) stimulation of the immune system and (10) stimulation of retinal cell growth in degenerative retinal diseases.

Ferulic acid, *p*-coumaric acid, Cafeic acid and Sinapic acid are common phenolic compounds belonging to an important and most abundant bio-molecule formed sequentially from cinnamic acid via phenyl propanoid pathway (Fig.4). Ferulic acid shows two isomers: cis (a yellow oily liquid) and trans (crystalline). Its nomenclature is made after *Ferula asafoetida*, a species belonging to family Apiaceae from which this active compound was isolated for the first time in 1866.

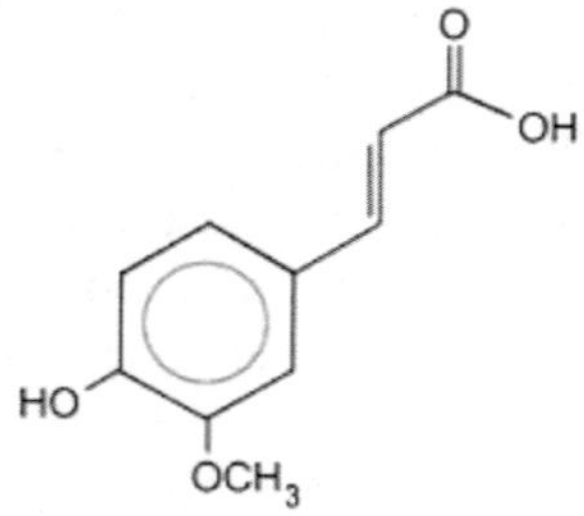

Fig. 4 : Structure of ferulic acid

There are three possible pathways for the extraction of ferulic acid from natural resources:

(1) From low molecular weight conjugates,

(2) From plant cell walls; and

(3) From tissue culture or microbial fermentation.

Ferulic acid can also be chemically synthesized by condensation reaction of vanillin with malonic acid catalyzed by piperidine (Adams et al. 1952). However, this requires long time to complete the reaction. Latter two methods were developed to break the cross link and release ferulic acid from plant cell wall materials. One of the methods is an enzymatic method using FAEs. Alkaline hydrolysis is another way to release ferulic acid from polysaccharides (Oosterveld et al. 2000). Sugarcane bagasse and wheat bran are suitable natural resources for the extraction of ferulic acid by alkaline hydrolysis.

5.1 Vanillin

Natural vanillin is used as the world's most highly prized natural flavour in foods, beverages, perfumes and pharmaceuticals (Clark 1990). A growing interest is seen world wide in the flavour industry to produce natural vanillin by using biotransformation technology, utilizing either plant cells or microbes, which ensures a stable supply, quality and price (Funk and Brodelius 1994). As compared to the price of cured vanilla pods, the price of chemically synthesized "nature-identical" vanilla flavour is very low. The price of natural vanillin extracted from vanilla pods usually varies between \$1200 per kg and \$4000 per kg, as compared to the price of synthetic vanilla, which is less than \$15 per kg. The limited availability of *Vanilla* pods causes high price of natural vanillin. This is mainly due to climate dependent variations in harvest yields along with labour-intensive cultivation, harvesting and curing of *Vanilla* pods. The worldwide demand for natural *Vanilla* beans cannot be met by the available supply from limited growing areas in countries such as Mexico, Madagascar, Tahiti and Indonesia. On account of the limited supply and high price of natural vanillin and the predominance of chemically synthesized vanillin, there is an encouragement to explore and develop biological sources of "natural vanillin" and vanilla-type flavouring that could be marketed as an alternative to the chemically synthesized substances.

Biotransformation is a broad and growing field of biotechnology and involves the use of enzymes or microorganisms to perform chemical reactions in which the starting substances and products are of comparable chemical complexity (Dordick et al. 1998). A large number of natural products are obtained via microbial biotransformation. At present biotransformation is widely used for the preparation of valuable chemicals for the pharmaceutical, agricultural, flavour and fragrance, nutritional and chemical industries (Beilin and Li 2002; Cheetham 1993).

The hydroxycinnamic acids are released from plant wastes as breakdown products from lignin (Peng et al. 2003). Vanillin can be produced by chemical breakdown of lignin. Microbes involved in lignin degradation possess enzymes for natural vanillin formation. Organisms capable of degrading ferulic acid have been of particular interest, since the phenolic substituents in these compounds are the same in those of vanillin and vanillic acid (Fig. 5).

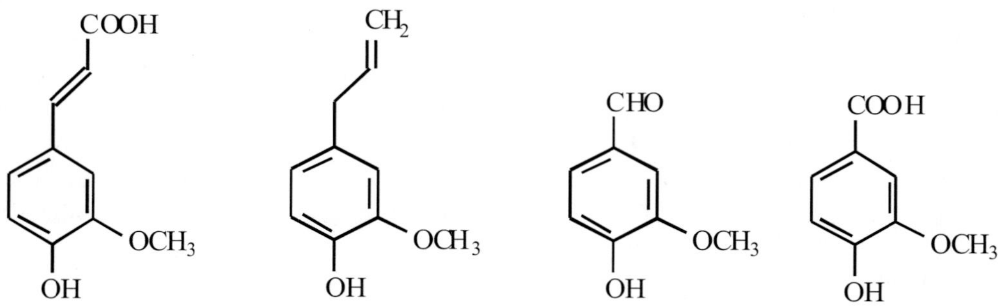

Fig. 5 : Identical phenolic substitutes of potential substrates and products of natural vanillin.

Among various phenylpropanoid compounds, the hydroxycinnamic acids are quite amenable for microbial degradation. Microbes transform hydroxycinnamic acids (e.g. ferulic acid, *p*-coumaric acid, sinapic acid) to their corresponding hydroxybenzoates which are important components of natural flavours and fragrances. A large number of microbes such as bacteria, fungi and yeast were explored for the laboratory-scale production of vanilla flavour metabolites from various phenylpropanoid precursors such as ferulic acid, eugenol, isoeugenol, vanillyl alcohol, vanillylamine etc. (Rao and Ravishankar 2000). Of these, the two most promising substances as indicated earlier are eugenol, a major component of clove oil and ferulic acid, a very common constituent of plant cell wall wastes such as, wheat and oat bran, sugarbeet pulp and sugarcane bagasse.

Because of abundant availability of ferulic acid in nature, there is considerable interest in utilizing this as a cheap source of natural substrate for microbial transformation, aimed at producing value-added phenolic products such as vanilla flavour metabolites (Tripathi et al. 2002). A number of microorganisms have shown to be capable of degrading ferulic acid into vanillate derivatives (Walton et al. 2003). The catabolic routes of ferulic acid degradation differ from organism to organism. Four major pathways of ferulic acid degradation can be distinguished with respect to the phenylpropanoid side chain-shortening mechanism. These are, (1) nonoxidative decarboxylation, (2) side chain reduction, (3) coenzyme-A-independent deacetylation and (4) coenzyme-A-dependent deacetylation.

5.1.1 Biotechnological production of vanillin

5.1.1.1 Plant tissue/cell culture-mediated production

Works on plant tissues / cell culture mediated production of vanillin have been carried out by many researchers. Westcott et al.1994 developed a process for producing natural vanillin from ferulic acid as precursor with aerial roots and charcoal as a product reservoir. Biotransformation of ferulic acid to vanillin has been reported in immobilized cell cultures of *Capsicum frutescens* treated with the phenylpropanoid-pathway-intermediates-ferulic acid and vanillylamine (Johnson et al. 1996). Rao and Ravishankar (1996, 1998, 1999) observed vanillin flavour metabolism using precursors such as isoeugenol, coniferyl aldehyde. However, the production of vanilla flavour metabolism through plant tissue/ cell culture technology is not commercially viable due to high costs associated with the maintenance of plant tissue cultures.

5.1.1.2 Production of natural vanillin using microbial biodegradation technologies

Application of microbial biotransformation tools is a promising alternative to obtain natural products. According to the US and European legislatures, compounds obtained from living cells, or their components, including enzymes were defined as 'natural flavour'. A microbial transformation is the conversion of one substance (substrate) to another (product) by a microorganism. It requires a substrate whose structure is closely related to the anticipated product (Yamada and Shimizu 1988). Microbes transform hydroxycinnamic acids (e.g. ferulic acid) to their corresponding hydroxybenzoates, which are important components of natural flavours and fragrances. In principle, elimination of the phenylpropionic C_2–side chain from ferulic acid (C_6-C_3 compound) can generate vanillate derivatives (C_6-C_1 compounds). Due to the abundance of these natural aromatic products, much scientific interest has been focused on the ability of microorganisms to metabolize phenolic compounds such as hydroxycinnamic acids, in order to discover novel systems for the formation of phenolic flavouring and aromas. The concept of natural biodegradation has led to the evolution of biotransformation technologies, which are important tools for renewing natural resources by conversion into commercially valuable products. The biodegradation of ferulic acid has also been investigated in many bacteria. The bacterial species have been shown to form vanillin as an intermediate in ferulic acid metabolism. In *Pseudomonas acidovorans* Toms and Wood (1970) have reported vanillin formation as a result of ferulic acid biodegradation. Accumulation of vanillin (9.8 mg/l) was also obtained in the culture media containing 5.0 mM ferulic acid on 7th day of incubation (Sarangi and Sahoo, 2010). Our previous results show that maximum accumulation of vanillin (10.3 mg/l) was obtained in the culture media containing 5.0 mM ferulic acid on 7th day of incubation using *Streptomyces* isolate S 10 (Sarangi et al. 2009).

6. Conclusion

Crop residues are a significant resource, with different challenging uses. Due to rapidly amplification of crude oil prices there is the increasing the demand towards the biofuel technologies leading for utilization of agro-industrial residues. The major gaseous biofuels, namely, biomethane and biohydrogen and the major liquid biofuels, namely, bioethanol and biodiesel have evolved as potential alternative to the fossil fuel resources. Bioethanol and biodiesel are gaining importance as alternative fuels to petrol and diesel respectively. Bioethanol, which is conventionally produced from cane molasses by yeast fermentation, can also be produced from various agro-industrial residues and plant wastes. Agro-industrial wastes constitute interesting source of phytochemicals that can be readily extracted by use of microbial enzymes. There is essential to guarantee the product purity using quality control analyses. Biological activity of these phytochemical extracts needs to be demonstrated by *in vivo* studies and clinical assays. The financial side and environmental consequences of challenging uses of crop residue must be assessed objectively with a holistic approach and long-term perspective.

References

Adams, R. and Bockstahler, E. (1952). Preparation and reaction of hydroxycinnamic acids and esters. J Amer Chem Soc 74: 5346-5348.

Amigun, B., Sigamoney, R. and von Blottnitz, H. (2008). Commercialisation of biofuel industry in Africa: A review. Renewable Sustainable Energy Rev 12: 690–711.

Beilen, J. B. V. and Li, Z. (2002). Enzymes Technology: an overview. Curr Opin Microbiol 13: 334-338.

Bennet, J. W., Wunch, K. G. and Faison, B. D. (2002). Use of fungi Biodegradation. In: Hurst CC (ed) Environmental Microbiology. 2nd edition, ASM Press Washington DC.

Boucque, C. V. and Fiems, L. O. (1988). Vegetable by-products of agro-industrial origin. Livest Prod Sci 19: 97–135.

Benoit, I., Danchin, E. G. J., Bleichrodt, R. J. and De Vries, R. P. (2008). Biotechnological applications and potential of fungal feruloyl esterases based on prevalence, classification and biochemical diversity. Biotechnol Lett. 30: 387–396.

Chattopadhyay, S. (2009). Biotechnology for Agro-Industrial Residues Utilisation. Springer Science and Business Media B.V.

Cheetham, P. (1993). The use of biotransformations for the production of flavours and fragrances. Trends Biotechnol 11: 478-488.

Clark, G. S. (1990). Vanillin. Perfumer Flavorist 15: 45-54.

Dahod, S. K. (1999). Raw materials selection and medium development for industrial fermentation processes, pp. 213–220 In "Manual of Industrial Microbiology and Biotechnology" (2nd ed); (Demain, A.L. and Davies, J.E., eds.) ASM Press, Washington DC.

Demain, A. L. (2000). Small bugs, big business: the economic power of the microbe. Biotechnol Adv 18: 499–514.

Demain, A. L. (2007). The business of biotechnology. Ind Biotechnol 3: 269–283.

Dordick, J. S., Khmelnitsky, Y. L. and Sergeeva, M. V. (1998). The evolution of biotransformation technologies. Curr Opin Microbiol 1: 311-318.

Ernest, R. K. and Buffington, L.E. (1981). Crop residue.In: McClure TA, Lipinsky ES, editors. Handbook of bisolar research materials. Boca Raton, FL7 CRC Press.

Ernst, R. K. (1992). Antioxidant potential of ferulic acid. Free Radic Biol Med 13: 435-448.

Fan, L., Pandey, A and Mohan, R. (2000a). Use of various coffee industry residues for the cultivation of Pleurotus ostreatus in solid state fermentation. Acta Biotechnol 20(1): 41–52.

Fry, S. C. (1980). Feruloylated disaccharides of D-galactose and L-arabinose from spinach polysaccharide. Biochem J 203: 493-504.

Funk, C. and Brodelius, P. E. (1994). Vanilla planifolia Andrews: in vitro biosynthesis of vanillin and other phenylpropanoid derivatives. In: Biotechnology in Agriculture and Forestry. Medicinal and aromatic plants VI, Vol 26, (Ed Bajaj Y P S) Springer, Berlin. pp. 377-402.

Gupta, S. C., Onstad, C. A. and Larson, W. E. (1979). Predicting the effects of tillage and crop residue management on soil erosion. J Soil Water Conserv 34:77 – 9.

Howard, R. L., Abotsi, E. and Jansen van Rensburg, E. L. (2003). Lignocellulose biotechnology: issues of bioconversion and enzyme production. African J Biotechnol 2(12): 602–619.

Johnson, T. S., Ravishankar, G. A. and Venkataraman, L. V. (1996). Biotransformation of ferulic acid and vanillylamine to capsaicin and vanillin in immobilized cell cultures of Capsicum frutescens. Plant Cell Tiss Org Cult 44: 117-121.

Kashyap, D. R., Dadhich, K. S., and Sharma, S. K. (2003). Biomethanation under psychrophilic conditions: a review. Bioresour Technol 87: 147–153.

Laufenberg, G., Kunz B. and Nystroem, M. (2003). Transformation of vegetable waste into value added products. Bioresource Technology 87: 167–198.

Lee, Y. S. (2005). Role of NADPH oxidase-mediated generation of reactive oxygen species in the mechanism of apoptosis induced by phenolic acids in HepG2 human hepatoma cells. Arch Pharm Res 28: 1183-1189.

Mathew, S. and Abraham, T. E. (2004). Ferulic acid: An antioxidant found naturally in plant cell walls and feruloyl esterases involved in its release and their applications. Crit Rev Biotechnol 24: 59-83.

Mori, H., Kawabata, K., Yoshimi, N., Tanaka, T., Murakami, T., Okada, T. and Murai, H. (1999). Chemopreventive effects of ferulic acid on oral and rice germ on large bowel carcinogenesis. Anticancer Res19: 3775-3778.

MacAdam, J. W and Grabber, J. H. (2002). Relationship of growth cessation with the formation of diferulate cross-links and p-coumaroylated lignins in tall fescue leaf blades. Planta 215: 783-793.

Kamm, B. and Kamm, M. (2004). Principles of biorefineries. Appl Microbiol Biotechnol 64: 137–145.

Kim, S. and Dale, B. E. (2004). Global potential bioethanol production from wasted crops and crop residues. Biomass Bioenerg 26: 361–375.

Koopmans, A. and Koppejan, J. (1997). Agricultural and forest residues – Generation, utilization and availability. Regional Consultation on Modern Applications of Biomass Energy, Kuala Lumpur, Malaysia.

Krich, K., Augenstein, D. and Batmale, J. P. (2005). Biomethane from Dairy Waste: A Sourcebook for the Production and Use of Renewable Natural Gas in California.

Krishna, C. (2005). Solid-state fermentation systems-an overview. Critic Rev Biotechnol 25: 1–30.

Kunz, M. (2008). Bio-ethanol: Experiences from running plants, optimization and prospects. Biocat Biotransf 26: 128–132.

Lal, R. (2005). World crop residues production and implications of its use as a biofuel. Environ Int 31 (4): 575–584.

Lal, R. (2008). Crop residues as soil amendments and feedstock for bioethanol production. Waste Manag 28 (4): 747–758.

Manpreet, S., Sawraj, S. and Sachin, D. (2005). Influence of process parameters on the production of metabolites in solid-state fermentation. Mal J Microbiol 1(2): 1–9.

Morris, D. (2006). The next economy: From dead to living carbon. J Sci Food Agric 86: 1743–1746.

Nigam, P., Robinson, T. and Singh, D. (2004). Solid-state fermentation: An overview. In: Arora D (ed) Handbook of Fungal Biotechnology, Mycology 20, CRC Press, London.

Oosterveld, A., Beldman, G. and Voragen, A. G. (2000). Oxidative cross-linking of pectic polysaccharides from sugar beet pulp. Carbohydrat Res 328: 199-207.

Parawira, W. (2004). Doctoral Dissertation, Biotechnology Department, Lund University, Sweden.

Pandey, A., Soccol, C. R. and Mitchell, D. (2000a). New developments in solid state fermentation: I – Bioprocesses and products. Process Biochem 35: 1153–1169.

Pandey, A., Soccol, C. R. and Nigam, P. (2000b). Biotechnological potential of agro-industrial residues: I – Sugarcane bagasse. Bioresour Technol 74: 69–80.

Peng, X., Misawa, N. and Harayama, S. (2003). Isolation and characterization of thermophillic bacilli degrading cinnamic, 4-coumaric, and ferulic acids. Appl Environmen Microbiol 69: 1417-1427.

Priefert, H., Rabenhorst, J. and Steinbüchel, A. (2001). Biotechnological production of vanillin. Appl Microbiol Biotechnol 56: 296-314.

Ramachandra Rao, S. and Ravishankar, G. A. (1996). Production of vanillin and capsaicin in immobilized cell cultures of Capsicum frutescens. Poster presented at Int Symp on Principles Regulating Biosynthesis and Storage of Secondary Products. Halle Abstract 38.

Ramachandra Rao, S. and Ravishankar, G. A. (1998). Biotransformation of phenylpropanoids to vanilla flavour metabolites in immobilized cell cultures of Capsicum frutescens. Paper presented at 4th Int Food Convention, Mysore, and Abstract D-30.

Ramachandra Rao, S. and Ravishankar, G. A. (1999). Biotransformation of isoeugenol to vanilla flavour metabolites and capsaicin in suspended and immobilized cell cultures of Capsicum frutescens: study of the influence of beta-cyclodextrin and fungal elicitor. Process Biochem 35: 341-348.

Ramachandra Rao, S. and Ravishankar, G. A. (2000). Vanilla flavour: production by conventional and biotechnological routes. J Sci Food Agric 80: 289-304.

Sarangi, P. K., Sahoo, H. P., Ghosh, S. and Mitra, A. P. (2009). Biotransformation of ferulic acid into vanillin by a Streptomyces isolate S10. Asian J Microbiol Biotechnol Environ Sci 2: 273-278.

Sarangi, P. K. and Sahoo, H. P. (2010). Biotransformation of ferulic acid into value added-products. Asian J Microbiol Biotechnol Environ Sci 12(3): 37-39.

S´anchez, A., Ysunza, F. and Beltr´an-Garc´ýa, M. J. (2002). Biodegradation of Viticulture Wastes by Pleurotus: a source of microbial and human food and its potential use in animal feeding. J Agric Food Chem 50: 2537–2542.

Siso, M. I. G. (1996). Biotechnological utilization of cheese whey: review. Bioresour Technol 57:1–11.

Soetaert, W. and Vandamme, E. J. (2005). Biofuel production from agricultural crops. In: Biofuels for fuel cells: Renewable energy from biomass fermentation, pp. 37–50 In "Series Integrated Environmental Technology" (Lens, P., Westerman, P., Haberbauer, M., and Moreno, A., eds.) IWA Publ., UK.

Stout, B. A. (1990). Handbook of energy for world agric. New York7 Elsevier Science; 1990. p. 504.

Sun, R., Sun, X. F., Wang, S. Q., Zhu, W. and Wang, X. Y. (2002). Ester and ether linkages between hydroxycinnamic acids and lignin from wheat, rice, rye and barley straws, maize stems and fast-growing poplar wood. Ind Crops Prod 15: 179-188.

Tengerdy, R. P. and Szakacs, G. (2003). Bioconversion of lignocellulose in solid substrate fermentation. Biochem Eng J 13: 169–179.

Tripathi, U., Rao, S. R. and Ravisankar, G. A. (2002). Biotransformation of phenylpropanoid compounds to vanilla flavor metabolites in cultures of Haematococcus pluvialis. Process Biochem 38 (3) 419-426.

Tomes, A. and Wood, J. M. (1970). The degradation of trans-ferulic acid by Pseudomonas acidovorans. Biochemistry 9: 337-343.

Wall, J. D., Harwood, C. S. and Demain, A. L. (2008). "Bioenergy". ASM-Press, Washington DC.

Walton, N. J., Mayer M. J. and Narbad, A. (2003). Vanillin. Phytochem 63: 505-515.

Wang, H. H. (1999). Development and/or reclamation of bioresources with solid state fermentation. Proc Nat Sci Counc 23(2): 45–61.

Webb, C., Koutinas, A. A. and Wang, R. (2004). Developing a sustainable bioprocessing strategy based on a generic feedstock. Adv Biochem Eng Biotechnol 86: 195–268.

Westcott, R. J., Cheetham, P. S. J.and Barraclough, A. J. (1994). Use of organized viable Vanilla plant aerial roots for the production of vanillin. Phytochem 35: 135-138.

Yamada, H. and Shimizu, S. (1988). Microbial and enzymatic processes for the production of biologically and chemically useful compounds. Angewandte Chemie International Edition 27: 622-42.

Zanichelli, D., Carloni, F. and Hasanaj, E. (2007). Production of Ethanol by an Integrated valorization of olive oil byproducts. Envron Sci Pollut Res 14: 5–6.

Zervakis, G., Papadopoulou, K. and Ehaliotis, C. (2005). Use of composts deriving from Mediterranean agro-industrial wastes in vegetable crops: effects on disease suppression and plant growth. In: de Kreij C, Warmenhoven M (eds) Proceedings of the International Symposium on the Use of Composted Organic Wastes in Horticulture, Wageningen.

Zhangm, Y. H. P. (2008). Reviving the carbohydrate economy via multi-product lignocellulose biorefineries. J Ind Microbiol Biotechnol 35: 367–375.

❑❑❑

CHAPTER – 14

Lignocellulosic Bio-ethanol Production A Second Generation Biofuel

Nibedita Sarkar[1] and Kaustav Aikat[1]
[1]Department of Biotechnology, National Institute of Technolgy, Durgapur - 713209, India
Email : kaustav.aikat@bt.nitdgp.ac.in

Summary

Over the past several decades, the exponential population and social civilization expansion have been accompanied by rapid industrialization. As such, there has also been a proportionate expansion in the global energy demand. Till date, crude oil has been the only source to meet the increasing energy demand. Nowadays ethanol is the most acceptable product in the fuel market. The importance of ethanol is increasing due to a number of reasons such as global warming and climate change, limited reserves of fossil fuel and reduction on imports of transport fuel. Lignocellulosic biomass is an abundant renewable source for commercial alternative fuel. Maximum bioethanol production depends on the whole process and optimum process parameters itself. Bioethanol from lignocellulosics is a type of second generation biofuel. Many researchers have been focused on bioethanol production from renewable sources. This chapter highlights the processing steps for second generation bioethanol production from lignocellulosics.

1. Introduction

Crude oil accounts for up to 35% of the world's primary energy supply(Scragg, 2009). Oil primarily finds application in the production of transport fuels petrol and diesel. However, the environmental problem associated with the usage of these synthetic fuels is well established (Balat, 2008). There is an urgent need for clean fuels. Under such circumstances, if biofuels are to replace the liquid fuels produced from oil, the scale of the replacement needs to be appreciated. Oil consumption is increasing at an alarming rate whereas sources of fossil fuel are limited. The development of energy sources is not proportionate with the high rate of consumption. In order to run internal combustion engines and jet engines liquid transport fuels are generally required. In this perspective, biofuels have gained considerable interest from a global viewpoint.

Bioethanol or simply ethanol is a type of biofuel which can replace petrol. Depending on the production type and substrate it is divided into first and second generation bioethanol. First generation bioethanol is produced from sugar crops which have a food value whereas second generation bioethanol is produced from lignocellulosic substances which are abundant, renewable and inexpensive. First generation bioethanol is well established. This chapter deals with second generation bioethanol production. Ethanol can be described as the fuel of the future. It was already proposed in the late 1800s by car manufacturers, so ethanol and ethanol – petrol blends are old and renowned fuels for the internal combustion engine (Scragg, 2009). In addition, the issue of global warming has generated increased interest in the development of methods to reduce green house gas emissions. Green house gas emissions are the biggest threat of the present day (Lombardi, 2003).

CO_2 is the most important greenhouse gas (GHG). But other gases are also considered as green house gas. Present world is threatened by continued gas emission by fossil fuels and global warming. (Demirbas et al. 2004). The combustion of fossil fuels is responsible for 73% of the total CO_2 production (Wildenborg and Lokhorst, 2005).

Several efforts have been taken to control such emissions. (Lombardi, 2003). Some well developed technologies are considered that includes : (i) energy consumption minimization, (ii) efficient energy conversion or utilization technique, (iii) switching over to lower carbon content fuels, (iv) enhancing natural sinks for CO_2, and (v) capturing and storing CO_2 etc. Lombardi, 2003 has proposed several technologies.

i. Semiclosed gas turbined cycle where exhaust gases are recirculated to the compressor inlet, allowing previous gas to cool for substantial recovery of water vapour as condensate.

ii. Integrated coal gasification combined cycle where CO_2 is chemically absorbed from the syngas. In this technique maximum amount of hydrogen is recovered from syngas. The hydrogen is used to produce maximum amount of water during the period of combustion. This technique is able to remove carbonyl sulphide , hydrogen sulphide, carbon monoxide and carbon dioxide.

iii. O_2 / CO_2 cycle is an efficient cycle for removing carbon dioxide.

iv. Life cycle assessment is another cycle to reduce CO_2 emission. Basically power plants are built on the basis of this assessment where energy and power output is considered as the final product.

Reducing the use of fossil fuels would considerably reduce the amount of CO_2 released, as well as reduce the levels of pollutants (Demirbas, 2005). Renewable energy sources are gaining wide spread attention due to less CO_2 emission and reduction of global warming. Bioethanol has been recognized as a potential alternative to petroleum-derived transportation fuels for the reduction of net contribution of green house gas emissions to the atmosphere. (Govindswami and Vane, 2007).

More importantly, green house gas emissions from transport fuel will be substantially minimized in the near future. Actions towards this aim have been initiated. According to the European Commission plans, 20% of conventional fossil fuels will be replaced with alternative fuels in the transport sector by 2020. The US President, in his state of the union address, set the goal to replace more than 75% of imported oil with alternative fuels by the year 2025 (Herrera, 2006).

1.1 Fuel Properties of Bioethanol

Properties of bioethanol are similar or superior to gasoline which includes higher octane number, broader flammability limits, higher flame speeds and higher heats of vaporization than petrol. These properties allow for a higher compression ratio, shorter burn time and leaner burn engine, which ultimately leads to theoretical efficiency advantage over petrol in an internal combustion engine (Balat, 2007). However, ethanol has certain disadvantages which include corrosive nature to steel tanks as it mixes with water, lower energy density than petrol, low flame luminosity, lower vapor pressure and toxicity to ecosystems (Maclean and Lave, 2003).

Several countries worldwide, including Brazil, United States, Canada, Japan, India, China and Europe, are concentrated in developing their internal biofuel markets (Musatto et al. 2010). Recently they have established plans for use of these biofuels. Ethanol production worldwide has strongly increased since the oil crises in 1970. Its market grew from less than a billion liters in 1975 to more than 39 billion liters in 2006, and is expected to reach billion liters in 2015 (Musatto et al. 2010). Ethanol is expected to be one of the dominating renewable biofuels in the transport sector within the coming 20 years. The 'Aggressive Biofuel growth' scenario aims to replace 15 % gasoline production with biofuels by 2015, and 20% by 2020 throughout most of the world (Chen and Qiu, 2010) In China, the Chinese government law of renewable energy published in January 1, 2006, indicated that biofuels from renewable sources would account for 10% by 2020 (Chen, 2009).

At present, the dominating cost element in fermentative production of fuel ethanol is the substrate, i.e. starch, though some fermentable starch and sucrose are easily obtainable from agricultural by-products like wheat and corn straw and cane bagasse, respectively. Generally sugarcane and starch are used as the feedstock for bioethanol production in Brazil and USA respectively. To reduce the production cost and increase the use of ethanol as a fuel additive lower-value substrates such as lignocellulose offer a great potential towards expansion of fermentation process. Lignocellulosic biomass in nature is by far the most abundant raw material from hardwood, softwood, grasses, and agricultural residues. The biological process for converting the lignocellulose to fuel ethanol involves:

(a) Delignification of lignocellulose to liberate cellulose and hemicellulose from their complex with lignin.

(b) Saccharification of the carbohydrate polymers to produce free sugars.

(c) Fermentation of resulted oligosaccharides to produce ethanol.

2. Lignocellulose

Lignocellulosic substance is a complex of hemicellulose, cellulose and lignin. Sources of lignocellulosics include wood, agricultural residues, water plants, grasses, and other plant substances. Although lignocellulosics are used as biomass for bioethanol production it can specifically be called photomass as these are results of photosynthesis. Lignocellulose is a substance with consistent, uniform, continuous, predictable, and reproducible properties. Lignocellulosics have been used in bioethanol production because they are economical, low in processing energy, renewable, and it contains large amount of cellulose and hemicellulose. The basic structure of all lignocellulosic biomass consists of three basic polymers: cellulose $(C_6H_{10}O_5)x$, hemicelluloses such as xylan and lignin (Arin and Demirbas, 2004).

Cellulose is a linear polymer of D-glucopyranose sugar units (actually the dimer, cellobiose) linked by β,1,4 – glycosidic bonds. The average cellulose chain includes about 9,000 to 10,000 polymerized units. Basically two types of cellulose can be seen in the cell wall. Approximately 65 percent of the cellulose is highly oriented, crystalline, and not accessible to water or other organic solvent (Stamm, 1964). The remaining cellulose associated with hemicellulose and lignin, is only partially accessible to water and other solvents. Generally cellulose is not in direct contact with the lignin in the cell wall. The hemicelluloses are a group of polysaccharide polymers which includes mainly pentose sugars such as D-xylose and L-arabinose and the hexose sugars such as D-glucose, D-galactose, D-mannose, and 4-O-methylglucuronic acid. The hemicelluloses are not crystalline, these are highly branched with a much lower degree of polymerization than cellulose. Depending on the source, hemicelluloses vary in structure and sugar composition.

Lignins are composed of nine carbon units derived from substituted cinnamyl alcohol; that is, coumaryl, coniferyl, and syringyl alcohols. Lignins are highly branched, not crystalline like cellulose. Lignins are associated with the hemicelluloses and play a role in the natural decay resistance of the lignocellulosic substance. As it is highly hydrophobic, lignin is the major obstacle for cellulose and hemicellulose hydrolysis. Apart from these three polymers lignocellulose consists of water, organic soluble extractives, and inorganic materials. The inorganic portion (ash content) of lignocellulosics can vary from a few percent to over 15 percent, with respect to total dry weight depending on the source.

The strongest component of the lignocellulosic resource is the cellulose polymer. The fiber is thus the smallest unit that can be used to produce high-yield of monomers from lignocellulosic composites.(Rowell, 1990).

Table 1 : The chemical component of various fiber are shown

Type of fiber	Cellulose (%)	Lignin (%)	Pentosan (%)	Ash (%)	Reference
Rice straw	32-47	5-24	19-27	12.4	Saha, 2003, Peijii, 1997
Wheat straw	35-45	8-15	20-30	9.6	Peijii, 1997
Corn straw	42.6	8.2	21.3	4.3	Peijii, 1997
Bagasse	50	25	25	2.4	Pandey, 2000
Barley	31-34	14-15	24-29	5-7	Rowell,1990
Oat	31-37	16-19	27-38	6-8	Rowell, 1990
Bamboo	26-43	21-31	15-26	1.7-5	Rowell, 1990

Hydroxyl and other oxygen-containing groups inside the cell wall polymer attract moisture through hydrogen bonding that results lignocellulosic dimension change with changing moisture content (Stamm,1964). Though hemicellulose is mainly responsible for moisture sorption, accessible cellulose, non crystalline cellulose, lignin, and surface of crystalline cellulose also play major roles.

Cellulolytic microorganisms recognize carbohydrate polymers of lignocellulose and hydrolyze those polymers to monomers by several enzymes. Biodegradation of the high molecular weight cellulose weakens the lignocellulosic cell wall because crystalline cellulose is primarily responsible for the strength of the lignocellulosics (Rowell, 1988). Strength is lost as the cellulose polymer undergoes degradation through oxidation, hydrolysis, and dehydration reactions. The same types of reactions take place in the presence of acids and bases.

3. Lignocellulose Pretreatment

The main aim of pre-treatment technique is to alter or remove structural and compositional impediments to hydrolysis in order to improve the rate of enzyme hydrolysis and increase yields of fermentable sugars from cellulose or hemicellulose (Moiser et al. 2005). Pre-treatment is one of the critical steps for cellulose conversion processes. After pre-treatment the structure of cellulosic biomass alters and becomes more accessible to the enzymes that convert the carbohydrate polymers into fermentable sugars. A successful pre-treatment must meet the following requirements.

i. High yields for multiple crops
ii. Highly digestible pretreated solid
iii. Minimization of sugars degradation
iv. Proper detoxification
v. Non-production of solid-waste residues
vi. Solid state process for lower moisture content
vii. High sugar yield
viii. Fermentation compatibility
ix. Total delignification
x. Minimum heat and energy requirements
xi. Milling or grinding the raw material to small particle sizes before pretreatment are energy-intensive and costly technologies. Operation in reasonable size and moderate cost reactors are desirable.

3.1 Factors Affecting Pretreatments are

1. Cellulose crystallinity
2. Cellulose degree of polymerization (DP, number of glycosyl residues per cellulose chain)
3. Substrate available surface area
4. Lignin barrier
5. Hemicellulose content
6. Feedstock particle size
7. Porosity
8. Cell wall thickness

4. Hydrolysis

Pretreatment is followed by saccharification step. There are two types of processes to hydrolyze the feed stocks into monomeric sugar constituents required for fermentation. Most commonly used hydrolysis methods are acid (dilute and concentrated) and enzymatic hydrolysis. In case of enzymatic hydrolysis, the lignin-hemicellulose net work has to be loosened for the better amenability of cellulases to residual carbohydrate fraction to improve sugar recovery. Dilute acid treatment is employed for the degradation of hemicellulose leaving lignin and cellulose network in the substrate. Generally sulfuric acid and hydrochloric acid are used at high (40% - 70%) and low (0.5%- 2%) concentration in acid hydrolysis.

Table 2 : Different pretreatment methods are shown

Method	Process	Mechanism of changes on biomass	References
Physical pretreatments	Ball milling Knife milling Hammer milling Roller milling Vibro-energy milling Hydrothermal Pyrolysis Gamma ray irradiation Microwave irradiation	Increase in accessible surface area; De-crystallizes cellulose; Removes hemicellulose; Removes lignin; Alters lignin structure. Mosier et al (2005)	Buaban et al. 2010 Nigam, 2001 Choi and Mathews, 1997 Choi and Mathews, 1997 Sun and Cheng, 2002 Garrote et al.1999 Mutt , 2009 Mtui, 2009
Physico-chemical and chemical pretreatments	Explosion • Steam • Ammonia • CO_2 • SO_2	Delignification; Decrease cellulose crystallinity and its degree of polymerization; partial and complete hydrolysis of hemicelluloses	Moniruzzaman, 1995 Sendelius, 2005; Sun and Cheng, 2002; Hamelinck et al. 2005; Mtui, 2009
	Liquid hot water process		Hamelinck, 2005; Mtui,2009
	Alkali • Sodium hydroxide • Ammonia • Lime • Urea		Sun and Cheng, 2002; Hamelinck et al. 2005; Mtui, 2009
	Acid • Sulfuric acid • Hydrochloric acid • Phosphoric acid • Sulfur dioxide		Takahashi et al. 2000; Sun and Cheng, 2002; Hamelinck et al. 2005;
	Oxidising Agents • Hydrogen peroxide • Wet oxidation		Sun and Cheng, 2002 Klinke et al. 2002
	Cellulose solvents • Cadoxen		Mosier et al. 2005
	Organosolv process		Sun and Cheng, 2002; Mtui, 2009
Biological pretreatments	Actinomycetes Fungi	Delignification and reduction in degree of polymerization	Mtui, 2009 Sun and Cheng, 2002

4.1 Acid Hydrolysis

In acid hydrolysis lignocellulosic materials are exposed to an acid for certain period at specific temperature and monomers are generated from celluloses and hemicelluloses. Sulfuric acid is mostly used, though hydrochloric acid also is considered for acid hydrolysis. There are two types of acid hydrolysis process commonly used - dilute and concentrated acid hydrolysis. High temperature and pressure is required for dilute acid process. At this condition tightly bound cellulose and hemicellulose is loosened and converted to fermentable reducing sugars. The reaction time of dilute acid process is seconds or minutes. Concentrated acid hydrolysis is different from dilute process, as concentrated acid hydrolysis involves mild temperature, high acid concentration and minimum pressure. Reaction times in concentrated acid process are typically much longer than in dilute acid process.

4.1.1 Dilute acid hydrolysis

Dilute acid hydrolysis is the oldest technology for lignocellulosic biomass conversion to bioethanol (Graf, 2000). As hemicelluloses are less crystalline than cellulose, these are readily polymerized in low temperature conditions (less than 200°C). Hemicelluloses are mainly converted to pentose such as xylose and arabinose and other sugars (Chandel et al. 2007). Most dilute acid processes are limited to a sugar recovery efficiency of around 50% w/w. (Badger, 2002). Percolation reactors have been used in most of the wood sugar processes. Strong acids can reduce the crystalline region but they degrade glucose (Lee, 2005). Dilute acid process is also used as pretreatment of lignocellulose prior to enzymatic hydrolysis. The main drawback of dilute acid process is degradation of monomers and generation of toxic substances like HMF, furfural, acetic acid at high temperature. These toxic substances have inhibitory effect to the fermentation step.

Two step dilute acid hydrolysis is more advantageous compared to one step because of less energy consumption. High rate of cellulose and hemicellulose conversion is obtained in two stage dilute acid hydrolysis process. In two step dilute acid hydrolysis, the first stage is performed at low temperature where maximum hemicellulose is hydrolyzed and for cellulose hydrolysis high temperature is applied at second step (Chandel, 2007). The first stage is conducted under mild process conditions to recover the 5-carbon sugars while the second stage is conducted under harsher conditions to recover the 6-carbon sugars.

4.1.2 Concentrated acid hydrolysis

In this method, the substrate is primarily treated with concentrated acid followed by diluted with water to dissolve and finally hydrolyzed the mixture into sugar constituents. This process provides complete and rapid conversion of cellulose to glucose and hemicellulose to xylose with a little degradation. The process is advantageous than dilute acid process as high yield of monomers are generated. This process is operated at low temperature and high acid concentration (30-70%). Concentrated acid hydrolysis takes longer time to complete. The sugar degradation will be minimized by low temperature and pressure. The hydrolyzed material is then washed to recover the sugars. In the next step, the cellulosic fraction has to be depolymerized. Sugar steams from hydrolysis step are used for subsequent ethanol production (Chandel et al. 2007). The main disadvantage of concentrated acid hydrolysis is acid recovery which is an energy demanding process.

4.2 Enzymatic Hydrolysis

Enzymatic hydrolysis is a saccharification technique which is operated at mild condition (Ghose and Bisaria, 1979).Pretreated lignocellulosics are hydrolyzed by several microbial enzymes such as cellulase and hemicellulase. Cellulase and hemicellulase consists of endoglucanase, exoglucanase, cellobiohydrolase, â-glucosidase, xylanase and mannase. These enzymes act on cellulose or hemicellulose. Regular efforts have been made for the isolation of efficient cellulolytic bacterial and fungal strain (Aro et al. 2005) There is a group of microorganisms (*Clostridium*, *Cellulomonas, Tricho-derma*, *Penicillium*, *Neurospora*, *Fusarium*, *Aspergillus* etc.) showing a high cellulolytic and hemicellulolytic activity. It is essential to pretreat the lignocellulosic substances for lignin removal and to make cellulose and hemicellulose available for enzyme action.

Different enzymes have different mechanisms. Endo-1,4- glucanase acts on low crystalline regions of the cellulose fiber and creates free chains. Exoglucanase removes the cellobiose unit from the free chain ends thus further degrades the sugar chain. β–glucosidase completes the hydrolysis process by catalyzing the hydrolysis of cellobiose to glucose (Heikinheimo, 2002).

Though few bacteria and actinomycetes are able to produce cellulase, fungi are largely attributed for cellulase enzymes. *Trichoderma reesei* is popular for commercial cellulase production. Several factors that affect enzymatic hydrolysis are substrate concentration, enzyme loading, surfactant addition, temperature and pH. Surfactant prevents enzyme from unproductive binding with lignin, thus improves enzyme productivity. To reduce the cost of enzymatic hydrolysis cellulase recycling is a better option. The rate of enzymatic hydrolysis of the cellulosic materials always decreases rather quickly. Generally, enzymatic lignocellulose degradation is characterized into two phases; rapid initial phase followed by slow secondary phase that may last until all substrate is consumed. This has been explained most often by the rapid hydrolysis of the readily accessible fraction of cellulose, strong product inhibition, and slow inactivation of absorbed enzyme molecules (Nutt, 2006).

5. Fermentation

Following hydrolysis, reducing sugars thus released are fermented to bioethanol. There are three different types of fermentation techniques:

i. Separate hydrolysis and fermentation (SHF)
ii. Simultaneous saccharification and fermentation (SSF)
iii. Consolidated bioprocessing (CBP)

5.1 Separate Hydrolysis and Fermentation (SHF)

Enzymatic hydrolysis performed separately from fermentation step is known as SHF (Chandel et al. 2007). Saccharified lignocellulosics are allowed to pass through a vessel for fermentation. In separate enzymatic hydrolysis and fermentation first the pretreated lignocellulosic substances are enzymatically hydrolyzed at optimum condition in a vessel. After hydrolysis, monomer solution is fermented by ethanol producing microorganism at optimum temperature. SHF is advantageous because two separate optimum conditions for saccharification and fermentation can be generated. Cellulase generally acts at 45°C-50°C depending on the cellulolytic strain (Wingren et al. 2003) where as most ethanol producing microorganism acts at 30°C to 37°C. The main drawback of SHF is inhibition of cellulase activity by accumulation of glucose and cellobiose. Another drawback is microbial contamination as hydrolysis process is long (Hahn-Hagerdal et al. 2006).

5.2 Simultaneous Saccharification and Fermentation (SSF)

The most important process improvement made for the enzymatic hydrolysis of biomass is the introduction of simultaneous saccharification and fermentation (SSF). In SSF enzymatic hydrolysis of pretreated lignocelluloses and fermentation is operated at one step. The process has been improved to include the co-fermentation of multiple sugar substrates (Wingren et al. 2003). In this process pretreated lignocellulose is saccharified by hydrolyzing enzymes and produced monomer mixture is immediately consumed by fermenting microorganism present in the same vessel. In SSF technique, both C5 and C6 sugar fermentation is carried out by wild and recombinant yeast strain which has the ability to utilize both C5 and C6 sugars. SSF is advantageous than SHF and enzyme loading is low (Chandel, 2007). Presence of both cellulolytic microorganism and ethanol producing microorganism in the same vessel as well as presence of ethanol, reduce the risk of contamination. In SSF designing yeast preparation is carefully considered. Adaptation of yeast to the inhibitors present in the medium is an important factor for consideration in the design of SSF process. Low glucose concentration is maintained in SSF which combines enzymatic hydrolysis with ethanol fermentation. The accumulation of ethanol in the fermenter does not inhibit cellulase action as much as high concentration of glucose; so, SSF is a good strategy for increasing the overall rate of cellulose to ethanol conversion. However maintaining optimum condition such as pH and temperature for enzymatic hydolysis and fermentation is very difficult. So the optimum condition for both step should be as close as possible. It is the main drawback of SSF technique because optimum temperature for cellulase is generally 45°C but ethanol producing microorganism has an optimum temperature between 30°C and 35°C. Due to its low rate of cellulose hydrolysis SSF is not feasible to meet all the challenges at industrial level and most microorganisms employed for ethanol fermentation can not utilize all sugars derived after hydrolysis. The problem is overcome by using cellulolytic enzyme cocktail which is more stable in a wide range of pH and temperature. Also the fermenting microorganisms (yeasts or bacteria) should be able to ferment a wide range of C5 and C6 sugars.

5.3 Consolidated Bioprocessing (CBP)

Bioethanol production is implemented by consolidated bioconversion process. The process is also known as direct microbial conversion (DMC). In consolidated bioconversion technique a single microorganism is used for both enzymatic hydrolysis and fermentation as the microorganism has the ability to produce required enzymes for enzymatic hydrolysis as well as ethanol production. DMC is advantageous because it is not only dedicated to the production of cellulase enzyme. Cellulase enzyme production is costly. Hence DMC increases the cost effectiveness of the whole process as energy demand is less and total processing cost is also low. However, DMC is not considered the leading process alternative. As there is no robust organism available containing cellulases or other cell wall degrading enzymes in conjunction with ethanol with a high yield, DMC can not be considered as alternative process. *Fusarium oxysporum* have the potential of bioconverting cellulose and hemicellulose directly to ethanol through the enzymatic hydrolysis of lignocellulose and simultaneously fermentation of saccharified lignocellulose (Xiros and Christakopoulos, 2009).

6. Conclusion and Future Prospect

Global depletion of energy supply due to continuing overutilization is being a major problem of the present and future world. Though first generation bioethanol production is largely established, it is least attractive as the used feedstocks are mainly sugar crops which have food value also. Second generation bioethanol from lignocellulose has gained a widespread interest. The feedstock is abundant, renewable and less costly. Appropriate pretreatment can improve the sugar recovery from lignocelluloses after saccharification. Among various fermentation technique, SSF is more efficient though it has disadvantage as enzymatic hydrolysis and fermentation require different optimum conditions. Recombinant yeast strains can be used to convert both C5 and C6 sugars. An integrated and dedicated approach can help in realizing large scale commercial production of lignocellulosic bioethanol and can contribute toward a cleaner and more energy efficient world.

References

Arin, G. and Demirbas, A. (2004). Mathematical modeling the relations of pyrolytic products from lignocellulosic materials. Energy Sources.,26 : 1023- 32.

Aro, N., Pakula, T., Penttiillae, M. (2005). Transcriptional regulation of plant cell wall degradation by filamentous fungi. FEMS Microbiol. Rev., 29 : 719-739.

Badger P.C. (2002). Ethanol from cellulose : a general review. In : Janick J, Whipkey A, editors. Trends in new crops and new uses. Alexandria, VA : ASHS Press., p.17-21.

Balat, M. (2007). Global biofuel processing and production trends. Energy Explor Exploit., 25 : 195-218.

Buaban, B., Inoue, H., Yano, S., Tanapongpipat, S., Ruanglek, V., Champreda, V., Pichyangkura, R., Rengpipat, S., Eurwilaichitr, L. (2010). Bioethanol production from ball milled bagasse using an on-site produced fungal enzyme cocktail and xylose-fermenting *Pichia stipitis*. Journal of Bioscience and Bioengineering., 110,1:18–25.

Chandel, A.K., Chan, E.C., Rudravaram, R., Narasu, M.L., Rao, L.V.and Ravindra P.(2007). Economics and Environmental Impact of Bioethanol Production Technologies: An Appraisal. Biotechnol. Mole. Biol. Rev.,2: 14-32.

Chen, H. Z. (2009). Process engineering in plant based products. New York : Nova Science Publishers.

Chen, H., Qiu, W. (2010). Key technologies for bioethanol production from lignocellulose. Biotechnology advances., 28, 5:556-562.

Choi, C.H., Mathews, A.P.(1996). Two-step acid hydrolysis process kinetics in the Saccharification of Low-Grade Biomass: 1. Experimental Studies on the formation and degradation of sugars. Bioresource Technology., 58 : 101-106.

Demirbas, F., Bozbas, K., Balat, M. (2004). Carbon dioxide emission trends and environmental problems in Turkey. Energy Explorer Exploit., 22 : 335-65.

Demirbas, A. (2005). Estimating of structural compositions of wood and non- wood biomass samples. Energy Sources ., 27 : 761-7.

Garrote, G., Domýnguez, H., Parajo, J.C. (1999). Mild auto hydrolysis: an environmentally friendly technology for xylooligosaccharide production from wood. J Chem Technol Biotechnol.,74 : 1101-1109 .

Ghose, T.K., Bisaria, V.S. (1979). Studies on mechanism of enzymatic hydrolysis of cellulosic substances. Biotechnol Bioeng., 21 : 131-146.

Govindswami, S. and Vane, L.M. (2007). Kinetics of growth and ethanol production on different carbon substrates using genetically engineered xylose fermenting yeast. Bioresource Technol ., 98 : 677-85.

Hahn-Hagerdal, B., Galbe, M., Gorwa-Grauslund, M.F., Liden, G., Zacchi, G.(2006). Bioethanol- the fuel of tomorrow from the residues of today. Trends Bioethanol., 24 : 549-56.

Hamelinck, N.C., Hooijdonk, G. and Faaij, P.C. (2005a). Ethanol from lignocellulosic biomass: techno-economic performance in short-, middle- and long-term. Biomass and Bioenergy ., 28 : 384-410.

Heikinheimo, L. (2002). Trichoderma reesei cellulases in processing of cotton. Espoo 2002, VTT Publications 483, VTT Technical Research Centre of Finland.

Herrera, S. (2006). Bonkers about biofuels. Nat. Biotechnol., 24 : 755-760.

Klinke, H.B., Ahring, B.K., Schmidt, A.S. and Thomsen, A.B. (2002). Characterization of degradation products from alkaline wet oxidation of wheat straw. Bioresource Technology., 82 : 15-26.

Lee, Y.J. (2005). Oxidation of sugarcane bagasse using a combination of hypochlorite and peroxide. Master's Thesis, Department of Food Science, Graduate Faculty of Louisiana State University and Agricultural and Mechanical College.

Lombardi, L. (2003). Life cycle assessment comparison of technical solution for CO2 emissions reduction in power generation. Energy Convers Manage., 44 : 93-108.

Maclean, H.L. and Lave, L.B. (2003). Evaluating automobile /propulsion system technologies. Prog Ener Combust Sci., 29 : 1-69.

Moiser N., Wyman, C., Dale, B., Elander, R., Holtzapple, Y.Y.L.M. and Ladisch, M. (2005). Features of promising technologies for pretreatment of lignocellulosic biomass. Bioresources Technol., 96 : 673-686.

Moniruzzaman, M. (1995). Alcohol fermentation of enzymatic hydrolysate of exploded rice straw by *Pichia stipitis*. World Journal of Microbiology & Biotechnology.,11 : 646-646

Mtui, Y.S.G. (2009). Recent advances in pretreatment of lignocellulosic wastes and production of value added products. African Journal of Biotechnology., 8,8 : 1398-1415.

Mussatto, S.I., Dragone, G., M.R. Guimarães, P., A. Silva, J.P., Carneiro, L.M., Roberto, I.C., Vicente, A., Domingues, L. and Teixeira, J.A. (2010). Technological trends, global market, and challenges of bio-ethanol production. Biotechnology Advances., 28: 817–830

Nigam, J.N. (2001). Ethanol production from wheat straw hemicellulose hydrolysate by *Pichia stipitis.* Journal of Biotechnology., 87 : 17–27

Nutt, A. (2006). Hydrolysis and oxidative mechanisms involved in cellulose degradation. Acta Universitatis Dissertations from the faculty of Science and Technology 185, Upsala, May 31, 51pp

Pandey, A., Soccol, C.R., Nigam, P. and Soccol, V.T. (2000). Biotechnological potential of agro-industrial residues. I : sugarcane bagasse. Bioresource Technology., 74 : 69-80.

Peiji, G., Yinbo, Q., Xin, Z., Mingtian, Z., and Yongcheng, D. (1997). Screening microbial strain for improving the nutritional value of wheat and corn straws as animal feed. Enzyme and Microbial technology., 20 : 581-584.

Rowell, R.M., Esenther, G.R., Youngquist, J.A., Nicholas, D.D., Nilsson, T., Imamura, Y., Kerner-Gang, W., Trong, L. and Deon, G. (1988). In Proc. IUFRO wood protection subject group, Honey Harbor, Ontario, Canada, Canadian Forestry Service, pp. 238-266.

Saha, B.C. and Cotta, A.M. (2006). Ethanol production from alkaline peroxide pretreated enzymatically saccharified wheat straw. Biotechnol Prog., 22 : 449-453.

Scragg A.H. (2009). Biofuels Production, Application and Development. CAB International, Wallingford, Oxfordshire, UK .

Sendelius, J. (2005). Steam Pretreatment Optimization for Sugarcane Bagasse in Bioethanol Production. Master of Science thesis. Lund University.

Stamm, A.J. (1964). Wood and Cellulose Science. The Ronald Press Co: New York.

Sun, Y. and Cheng, J. (2002). Hydrolysis of lignocellulosic materials for ethanol production: a review. Bioresource Technology., 83 : 1-11.

Takahashi, C.M., Lima, K.G.C., Takahashi, D.F. and Alterthum., F. (2000). Fermentation of sugar cane bagasse hemicellulosic hydrolysate and sugar mixtures to ethanol by recombinant Escherichia coli KO11. World Journal of Microbiology & Biotechnology., 16 : 829-834

Wildenborg, T. and Lokhorst, A. (2005). Introduction on CO2 geological storage classification of storage options. Oil Gas Sci Technol Rev IFP. 60 : 513-5.

Wingren, A., Galbe, M. and Zacchi, G. (2003). Technoeconomic evaluation of producing ethanol from softwood : Comparison of SSF and SHF and identification of bottlenecks. Biotechnol. Prog., 19, 4 : 1109-1117.

Xiros, C. (2009). Christakopoulos, P., Enhanced ethanol production from brewer's spent grain by a *Fusarium oxysporium* consolidated system. Biotechnology for Biofuels., 2 : 4.

❑❑❑

CHAPTER – 15

Bioenergy Ethics and Prospective

Kambaska Kumar Behera and Nidhi Srivastava
Department of Bioscience and Biotechnology, Banasthali University, Banasthali Vidyapith, Rajasthan-304022,India
Email : kambaska@gmail.com.

Summary

Bioenergy is the subject of increasing attention around the globe and represents an important issue in the present context. The rise of commodities prices, the negative impact on food security, and climate change represent different challenges to be overcome before the full potential of bioenergy can be realized. Bioenergy issues related to agriculture needs special attention for its development. Bioenergy development is always linked with rural economy. Farmers, having the choice to convert their food crops to fuels crops, naturally expect a high return from their farmland, thereby generating a scenario where food production gets depleted. The optimal production of bioenergy can be generated by using the wasteland present in different area. This way, food security for the future generation can be maintained along with optimal growth of bioenergy.

Introduction

Nowadays, bioenergy has become one of the most dynamic and rapidly changing sectors of the global energy economy (UN-Energy 2007). It is considered as a clean and renewable source of energy that could dramatically improve the environment, reduce dependency on fossil fuels and revitalize socioeconomic conditions of rural communities by creating new jobs (Gokcol 2009). At present, biomass is by far the largest global contributor of renewable energy (RE) and represents about 10% of global annual energy consumption, mostly used for cooking and heating (IEA 2010a). It accounts for almost 80% of the total primary energy supply in many developing countries, whereas the figure is usually less than 5% in the industrialized countries (Keam and McCormick

2008). In general, three main categories of bioenergy resources are used globally such as forestry biomass, agricultural biomass, and wastes biomass (Resch et al.2008). Despite its much benefit over fossil fuels, some recent controversies surrounding bioenergy have attracted wider international attention. Biofuels were held responsible for increasing food prices and decreasing food availability in many developing countries (FAO 2008; Mitchell 2008;UNCATD 2007). In addition, one can find much discussion related to the impacts of biofuel production on water scarcity(Gerbens-Leenes 2009) accelerating deforestation, and emissions of greenhouse gases in the tropical countries(Fargione et al.2008; Laurence 2007; Searchinger et al.2008).The development of the modern bioenergy sector varies greatly across countries. Currently, only a few countries have well-developed forest- and agriculture-based bioenergy sectors such as Finland, Sweden, Germany, Austria, Brazil, and the USA with their supportive policies and infrastructure.However, the development of the modern bioenergy sector is in the preliminary stage in many other countries and varies greatly among those countries. Majority of studies in the past have analyzed different aspects of bioenergy production and their impacts on global economy, environment, and society (Bush 2008; Hill et al.2006; Landis et al.2008; Lapola et al.2010;Wegner and Kelly,2008).However, none of them focused on young citizens' awareness of bioenergy from an international perspective. Young citizens are the future decision makers deciding on all aspects of society including energy issues and their positive attitudes toward REs in general and bioenergy in particular have great policy relevance. Bioenergy is a new and challenging area of development that includes several socioeconomic and environmental dimensions demanding special consideration from both energy and educational policy makers to create awareness of it among young people (Halder et al.2010).

Bioenergy in Indian Context

In the past decade, policy makers in central as well as some state governments and international development agencies have shown considerable interest in bioenergy (Francis et al. 2005). Instability of oil prices and in oil producing regions, surging energy demand in developing countries, and greater awareness about climate change threats due to fossil fuel usage have primarily contributed to this renewed interest in bioenergy. Fortunately, alongside this increasing interest, also marked by the development of more efficient and cost-effective bioenergy technologies (Cushion et al. 2010).India is currently experiencing a surge in energy demand for their growing population for sustainable development. Growing import dependence and greater consensus at policy level about the need and utility of tapping into national bio-resources for energy supply has created impetus for greater research initiatives, private investment and promotion of bioenergy technologies which are :-

1. Baseline situation in India in terms of import dependence for transportation energy, deficit in power generation, and rural energy poverty in terms of dependence on biomass-based traditional cooking and unreliable power situation.

2. Scale of current actual biomass usage (by state) as against projected potential for
   a. traditional cooking and heating
   b. biomass based power generation/cogeneration
   c. biogas
   d. biofuels (first generation ethanol and bio-diesel)
3. Current impact of biomass energy usage.
4. Analysis of policies and stakeholders who are part of the bioenergy sector and who can/should be roped into it.
5. Description of the Village Energy Security Program (VESP) project which is being undertaken by Government of India.

India's Energy Profile

With a population of about 1.2 billion growing at about 1.5% annually, India's economy has been recording growth of over 9% p.a. in 2006-2008 and over 6% even at the time of global recession. In concurrence with Gross Domestic Product (GDP) growth, India's energy demand is also rising rapidly. India's energy demand is expected to be more than three to four times the current level in another 25 years (Balachandra and Ravindranath 2009).Per capita energy consumption in India has risen by 42.1% in the last two decades (1990-2008), compared to the world average of 9.5%. Yet, in comparison to world average Total Primary Energy Supply (TPES of 1.83 toe/capita), India still lags behind at only 0.54 toe/capita in 2008.Coal dominates the Indian commercial energy basket followed by oil and gas.

Bioenergy Usage in Different Parts of India

In the total energy mix of India, biomass fuels play a vital role especially in its rural areas, as it constitutes the major energy source to majority of households in the rural and peri urban India apart from some rural enterprises. Ravindranath and Balachandran (2009) have reported that non-commercial energy sources, predominantly fuel wood, chips and dung cakes, dry leaves, dry straw of pulses and cereals which contribute around 30% of the total primary energy consumed in the country. It has been reported that 46% of households using firewood and chips in rural India obtain these fuels at zero cash outlay; about 21.14% of households depend on home grown stock; and 23.7% make cash purchase. In comparison, two-thirds of urban households using firewood need to purchase the same. Yet, bioenergy does not figure in most energy analyses as they are confined to "non-commercial" energy. Given that most of the biomass used in households is not transacted on the market, bioenergy data is "inadequate and not up-to-date (FAO, 2007). What is clear though, based on studies by TERI, and is that biomass delivers nearly 90% of energy used in rural households and about 40% of energy used in urban households use. FAO (2007) reported that in 2005, the contribution of bioenergy was 29.4% of India's TPES. An analysis of India's Energy Balance (IEA 2008) also substantiates the role of biomass-based energy in India's energy basket. Table 2 below provides the breakup by energy carrier type for India's Total Final energy Consumption (TFC) in 2007, indicating the prominent part played by biomass-based energy in India's energy basket.

Bio-Energy in India

Historically, traditional biomass has been a major source of household energy in India. About 78 percent of rural households use biomass for their energy needs, while in urban areas only 20 percent use biomass for their energy requirements (PISCES 2010). Currently, about 40 percent of the total energy supply in India is from non-commercial energy sources such as wood and cow dung. Rural households in India predominantly use wood and cow dung as fuel for cooking and water heating due to lack of reliable supply of electricity or cooking gas. Modern biomass energy is derived from organic material and can be used in a variety of conversion processes to yield power, heat/ steam, and fuel. In India, the renewable's use is focused on waste materials such as municipal, agricultural, or forest residues. Biomass is generally divided into three categories: biogas, solid biomass, and liquid biofuels. Biogas has been mostly used for small, rural, and off-grid applications, and the majority of gasifiers in India are providing for individual households. About 4 million family-size biogas plants were installed in India and Indian Government's Ministry of New and Renewable Energy (MNRE) estimates that the annual biogas generation potential is about 17,340 million cube meters (m3), which could support the installation of up to about 12 million family-size biogas plants. Larger scale biogas facilities use wastewater generated from beverage, meat processing, pulp and paper, food packaging, and other industrial sectors which together produce about 91 MW equivalent of electricity. Another 73 medium biogas plants with aggregated capacity of 461 kW have been installed under Distributed/Grid Power Generation Program of the Ministry. Indian Government adopted a National Policy on Biofuels in December 2009(GOIMNRE 2009). Indian government's biofuel strategy is currently focused on using non-food feedstock for the production of biofuels, mainly sugar molasses and non-edible oils. The Indian approach to biofuels, in particular, is somewhat different to the current international approaches which could lead to conflict with food security. It is based solelyon non-food feed stocks to be raised on degraded or wastelands that are not suited to agriculture, thusavoiding a possible conflict of fuel vs. food security. As a part of this policy, the government extends the minimum support price (MSP) to non-edible oil seeds for production of ethanol. The commercial production of biodiesel is very limited, and what is produced is mostly sold to the unorganized rural sector. The existing biodiesel producers in India are using non-edible oilseeds, non-edible oil waste, animal fat, and used cooking oil as feedstock.

Table 1 : Contribution of Energy (by type) in India's energy Usage

Sl.No.	Energy Type	Total Primary Energy supply
1	Coal and peat	42.1%
2	Crude oil	27.3%
3	Gas	5.7%
4	Hydro	1.6%
5	Combustible and renewable and wastes	26.3%

Source: EIA, 2010

Table 2 : Sector-wise Usage of energy in India

Sl.No.	Sector	Usages Pattern
1	Residential	78.7%
2	Industrial	17.4%
3	Others	3.9%

Source: EIA, 2010

There are mainly three energy sources: fossil, renewable, and fissile. The fossil energy sources are petroleum, coal, natural gas, bitumens, oil shales, and tar sands; the renewable energy sources covered are biomass, solar, wind, geothermal, and hydropower; and the fissile energy sources are uranium and thorium. Oil is a finite resource. Various studies put the date of the global peak in oil production between 1996 and 2035. In developed countries, there is a growing trend towards employing modern technologies and efficient bioenergy conversion using a range of bio-fuels, which are becoming cost wise competitive with fossil fuels (Puhan et al. 2005).Energy production from fossil fuels results in high greenhouse gas emissions. Global utilization of fossil fuels or energy needs is rapidly resulting in critical environmental problems throughout the world. The availability of fossil fuels, renewable energy sources, and nuclear sources strongly affects current and future energy supply systems. Declining availability of fossil fuels may cause increases in fuel prices and declining security of energy supply in the current energy system. Limited availability of fossil fuels and renewable energy sources on a local scale may affect the sustainability of future energy systems.The global and regional availability has been estimated for wind energy; solar energy, especially photovoltaic (PV) cell; and biomass; together with a description of factors influencing this availability.

On the long term, countries with surplus biomass potential could develop into exporters of bioenergy. Renewable energy sources or renewables contributed 2% of the world's energy consumption in 1998, including 7 exajoules from modern biomass and 2 exajoules for all other renewables (UNDP 2000). The renewables are clean or inexhaustible and primary energy resources. Renewable technologies like water and wind power probably would not have provided the same fast increase in industrial productivity as fossil fuels did (Edinger and Kaul 2000). The share of renewable energy sources is expected to increase very significantly (to 30–80% in 2100). Hydropower and traditional biomass are already important factors in the world's energy mix, contributing about 18% of the total world energy requirements, whereas the renewables contribute only about 2% of the present world primary energy use. Biomass, wind, and geothermal energy are commercially competitive and are making relatively fast progress.

Biomass Crop Yield

India's total land area is 328.7 Mha out of which 42.5 Mha is not available for cultivation. The existing land usage in India is provided in the below table.

Table 3 : Potential land areas in India with biomass potential (based on 2004 field data) Land details

Sl.No.	Land details	Area(kha)	%of Indians total land area
1	Forest	69.8	22.9%
2	Net sown area	141.9	46.5%
3	Uncultivated land excluding fallow land	26.9	8.8%
4	Fallow land	24.2	7.9%
5	Not available for cultivation	42.5	13.9%
6	Total	328.7	100%

Table 4 : Potential of biomass yield by source

Sl.No.	Type of resource	Area(kha)	Biomass generation (kt/yr)	Biomass surplus (kt/yr)
1	Agri-residue	16423	95512	43162
2	Forestland	64570	89119	59678
3	Wasteland	54253	66355	44369
4	Total	135246	250986	147210

Source: Biomass Atlas of India, Version 2.0

Impact of Bioenergy Technologies

It is important to examine inter-linkage and balance between key social, economic and ecological sustainability concerns related to small and large applications of modern biomass energy technologies (BETs) in the context of rising concerns regarding sustainable development in the energy sector (Demirbas 2005).

Promotion of Energy Security

Exposure of the Indian economy to increasingly unstable international energy markets is high. In some instances, recent advances in BETs are expected to provide locally produced bioenergy for local agriculture, industrial and household usage at less than the cost of fossil fuels (UN 2007). Instead of directing scarce resources to foreign countries to pay for oil, it makes business sense to invest in local agriculture and manufacturing sectors with additional benefits of strengthening local economies and rise in livelihood opportunities. In terms of plant capacity, the potential of biomass gasification projects could reach 31 GW that can generate more than 67 TWh electricity annually (Purohit 2009) which would directly contribute to energy self-sufficiency.

Poverty Reduction

Livelihood benefits

As in any development project, the essence of sustainability of bioenergy projects lies in how the community benefits from the project activity. The primary driving force for acceptance of such project activity from the community point of view will most probably be employment or job creation, contribution to regional economy and income improvement. Other "big issues" such as carbon emissions, environment protection, security of energy supply on a national level are an "added bonus" (Domac et al. 2005).

Bioenergy-related employment opportunities include direct employment, comprising jobs involved in fuel or crop production, in the construction, operation and maintenance of conversion plants and in the transport of biomass; and indirect employment, comprising jobs generated within the economy as a result of expenditures related to biofuel cycles (Faaij 1997).

Bioenergy is possibly the most labour intensive energy source and there is little doubt that bioenergy development will bring about significant job creation in unskilled and semi-skilled labour in India depending on the scale of production (large scale plantations, or medium and small scale operations) and on the degree of mechanisation – new employment opportunities arise for unskilled workers (FAO 2007). On average, labour intensive biofuels would generate about 100 times more workers per joule of energy content produced in comparison to capital-intensive fossil fuel industry (UNDP 2010). However, actual direct and indirect employment opportunities for biomass gasification and biogas vary considerably due to local factors such as physical infrastructure, density of plants, feedstock type, soil quality, etc. For example, TERI experience suggests that installation of a 2 m^3 biogas plant requires 10 skilled and 40 semi-skilled person days. However, the job generation for servicing and maintenance will vary depending on the number of biogas plants installed in adjoining areas. Say, if 10 biogas plants are installed, 10 skilled person days will be generated for periodic visits to the installed plants every week. However, the presence of 100 plants would lead to a full time direct employment for 1 skilled worker deputed for service and maintenance. In the case of biomass gasifier technology, there is potential of regular employment generation. One skilled and one semi-skilled person are required for daily maintenance and operation for a 20 kWe biomass gasification system. Job creation for fabrication of one plant is difficult to calculate. Staff on permanent company payroll deal with fabrication and person days/unit of plant would depend on the number of orders executed by the company in a month. The potential for generating employment opportunities in modern bioenergy applications among developing countries is a topic worthy of serious study and a country and technology specific study should be commissioned to understand the direct, indirect and induced benefits from selected case studies.

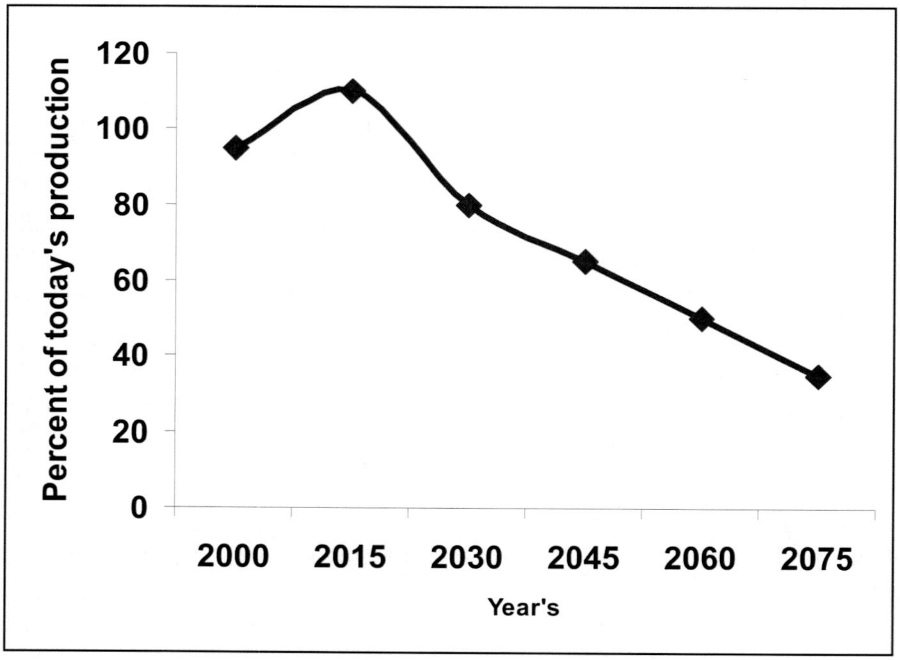

Fig. 1 : Global oil production scenarios based on today's production.

Ethical concerns are being raised regarding bioenergy production. However, the ethical issues often do not stand out very clearly. This may generate ethical problems mainly for two reasons. First, bioenergy production may compete, directly or indirectly, with food production; and as consequence the food security of poor people may be adversely affected (social aspects arguments). Secondly, the production of bioenergy may directly or indirectly lead to deforestation and other changes of land use that have a negative effect on greenhouse gas emissions (environmental arguments).

Bioenergy Ethical Debate

Increasingly, biofuels are being subjected to ethical analysis that focuses directly or indirectly on the land-use issue. A number of commentators have highlighted important issues here. For example, Thompson (2008) places biofuels in a broader agricultural ethics perspective; Mudge (2008) asks how we are to assess whether biofuels are environmentally or ethically sound; Gomiero et al. (2009) discuss biofuels in relation to the appropriation of ecosystem services; Landeweerd et al. (2009) reflect on the distribution of responsibilities; Mol (2010) focuses on the role of environmental authorities; Sodano (2009) examines the human rights and gender aspects of biofuels; and the Nuffield Council on Bioethics (2011) examine whether there is a duty to produce biofuels if this can be done in an environmentally and socially benign way through sustainability benchmarking. Building on this literature, this chapter will present an analysis of the ethical issues that focuses on land use and draws a distinction between the social and environmental aspects of land use. It should be noted that the term "aspect" instead of "impact" is used here, as the latter may imply an exclusively consequentialist approach to the issues, whereas in fact, non-consequentiality arguments will be examined as well. The social aspects of the discussion are essentially "people and community" issues: here we examine questions about livelihood and access to food, about who benefits, who may bear the cost in relation to choice of technology, and so forth. The principal environmental aspects to be considered are: reduction of GHG (green house gas) emissions, biodiversity degradation, and loss of natural habitat.

Social Aspects

The arguments to be considered in this section focus on the way biofuel production may compete with, or displace, food crops, and lead to food scarcity. These arguments tend to concentrate on the commercial establishment of first-generation biofuels i.e., fuels derived from basic feedstock (such as seed or grain) that could in theory enter the animal or human food chain. It has been alleged that biofuel production of this sort is unwise, and some have claimed outright that it is morally unacceptable: "10 years from now the rapid expansion of biofuel production may look foolish, or worse unethical, if it leads to environmental degradation, high food prices, and increases the number of undernourished people"(Cassman and Liska 2007).An argument of this kind can be found in the 2007 address made by Jean Zeigler, the UN Special Rapporteur on the Right to Food (2000-2008): "It is a crime against humanity to divert arable land to the production of crops which are then burned for fuel." In a similar vein, referring to liquid biofuels, it has been argued that if "you start to fuel cars with crops you are instantly putting the world's one billion starving people in competition with the world's one billion motorists."

Runge and Senauer (2007) conducted a food and fuel balance calculation; they state that "filling the 25-gallon tank of an SUV with pure ethanol requires over 450 pounds of corn which contains enough calories to feed one person for a year." Statements and calculations such as these (Rice 2010) have sparked fierce debate, and the points made by Zeigler and others have been countered. Matthews (2008), for example, points to the need for economic development in many countries, arguing that "the real crime against humanity is to block Africa's potential, by blocking its exports and stunting its growth." In essence it is claimed that developing countries should be given the economic opportunity to "grow not only food but also cash crops, and that the cash crop in greatest demand right now is biofuels" (Matthews 2008). However, according to the FAO (2008), it is exactly this driver that has contributed to higher food prices, posing an immediate threat to food security via a dramatic impact on the prices poorer food consumers have to pay in urban and rural areas. This issue is complex, as exemplified by the studies from the World Bank (Cushion et al. 2010) and The International Food Policy Research Institute IFPRI (Boue et al. 2010). Thus there is notable disagreement about the exact reasons for steep increases in worldwide food prices, such as that seen in the food price spike between autumn 2007 and summer 2008, where a number of commentators identified the use of corn for bioethanol (Bach 2009) as a causal, and perhaps significant, contributory factor. So in terms of food security, biofuels have been framed both as contributing directly to worldwide hunger and as a mechanism to alleviate poverty through rural economic empowerment, thereby reducing hunger. From a consequentiality perspective, the questions we need to ask about the relationship between food and fuel production, and whether this is developing in distinctively new ways, are these: How much land is available? If land is not used to grow fuels, will it be used to grow food?

The main critical argument seems to rely on assumptions that are all highly controversial: that the most important limiting factor in efforts to address global malnutrition which according to the World Health Organization is the biggest contributor to child deaths, representing one third of all child deaths (UNICEF et al. 2010) is the availability of agricultural land; that abstaining from growing bioenergy crops will ensure land is available for increased food production; and that this in turn will lead to the alleviation of food poverty (Pimentel et al. 2009).

The conversion of land to biofuel production within Europe and North America may, as an indirect LUC, provide markets within which farmers outside Europe can produce staple crops at better prices on the global market. It has been argued that in the short term rising food prices threaten the food security of many people in developing countries. However, over the medium to long-term, if farmers are given the resources to respond, a reversal of the long-term trend of depressed agricultural prices could allow them to increase production and revitalize many countries' agriculture sectors (FAO 2008b). Similarly, the cultivation of biofuels as cash crops in these countries may lead to increased investment in infrastructure that is needed to support thriving or emergent agricultural markets (FAO 2008b). This observation stems from the paradox that land for crop production is both considered a scarce global resource and often under-utilized. For example, land is left fallow within the EU because of CAP set-aside policies; outside the EU it is left fallow because it is not economically viable. In a number of developing countries the quality of the land or a lack of agricultural investment are significant limiting factors. However, with greater resources and more investment in infrastructure, soil quality could be improved, after which yields would increase. The criticism of biofuels illustrated mainly by the complexity of food poverty. It is also, in effect, a criticism of most cash crops, not just first-generation biofuels. Moreover, when it comes to second-generation biofuels that are not yet commercially established the consequences for global food security are by no means clear-cut.

These biofuels may not present a simple and transparent solution to the complex issues raised by first-generation biofuels. It was once claimed that there would be no competition between second-generation fuels and food crops (Antizar- Ladislao and Turrion-Gomez 2008). Again, Doornsbosch and Steenblik (2007) have argued that "second-generation technologies could, in theory, make it possible to avoid competing land-use claims by growing biomass feedstocks on marginal and degraded land and using residual biomass materials." However the social implications of using marginal land (e.g., associated with willow production) or agricultural co-products (e.g., straw) to produce second-generation biofuels have been questioned. Franco et al. (2010) observe that individuals and communities often depend on land categorized (but not by them) as "marginal." Such land might be relied upon in various ways: it can be used in subsistence agriculture or as a source of community fuel (e.g., firewood). Some have also questioned, in principle, the economic viability of an agricultural commodity that is produced on marginal land when it is competing against a similar commodity produced on prime land. Much food produced on marginal land within the EU cannot compete with food produced on prime land outside the EU, unless subsidized. The suspicion that the same competitive advantage will apply to biofuels seems reasonable. It has been shown that Jatropha planted on marginal land produces uneconomic, marginal yield (Nuffield Council on Bioethics 2011). This raises important questions about the production of second-generation biofuels on marginal land in Europe. One is whether this production system will be unable to compete with cheaper biofuels produced on prime land outside the EU. Another is: If marginal land is to be used economically to produce biofuels, will this only be feasible, and acceptable, if a policy stipulating that only marginal land can be used is in place? Such a policy would seem unworkable in the current global commodity market. In addition,how would policy-makers formally define "marginal" land?Would the adoption of a strict definition affect other important uses of the supposedly"marginal"sites?.

It has also been claimed that second-generation biofuels could provide more energy output per unit of land than their first-generation predecessors, so that the former involve a more efficient use of land (whether marginal or prime). This implies that significant social advantages will arise from these biofuels in terms of the energy security benefits; however, this needs further investigation. The FAO (2008: 34) states that if second-generation fuel production were to use 25% of agricultural land it would be able to replace 14% of transport fuels. Discussions of second-generation biofuels often assume implicitly that such fuels will be able to overcome the problems faced by first-generation biofuels if they are more energy efficient and "sustainable." However, the utilization of 25% of currently available agricultural land appears to be a very high price to pay for 14% of our current transport fuel needs. Given that biofuel production will undoubtedly incur costs connected with competition for resources and environmental damage, irrespective of the technology used, a key question might be: What are the comparative benefits of biofuel production in terms of economic development and energy security? The consequences of second-generation biofuels, both positive and negative, need to be evaluated not only in comparison with first-generation equivalents, but in relation to other land uses across the board. From a deontological perspective, the first claim that biofuels compete with food crops could require an assessment of whether biofuels in fact constitute the most appropriate or right use of land. This requires us to explore further questions: What is land for? What principles should we apply when seeking to determine the levels of food energy produced on land and transport energy and electricity? Estimations of efficiency of this kind are likely to depend on the outcome of a full Life Cycle Assessment, and this assessment should be complemented by some form of social impact assessment. Holistic assessment of this sort may reveal that other forms of bioenergy are more efficient than biofuels, whether first-generation or second-generation. For example, the cultivation of crops such as Miscanthus or willow in short rotation coppice for non-liquid fuel production may turn out to be a good use of the land. Such cultivation can be efficient at least, as long as the crops are used effectively in stationary facilities such as combined heat and power plants (sometimes referred to as "CHPs") rather than used to produce biofuels (Yuan et al. 2008).

Miscanthus and willow, which are regarded as high-yielding, yet low in put, perennial crops with low, or zero, fertilizer requirements may deliver a number of additional environmental benefits (Campbell et al. 2009; Rawlings 2007). Given this alternative approach, it could be argued that it is ethically problematic to use land to produce biofuels for transport if we have a global scarcity of viable land, particularly if the fundamental targeted purpose of biofuel production is to increase energy supply. One cannot properly address the question whether biofuels lead to food scarcity or displacement without considering possible inefficiencies and inequalities of current food production systems. Livestock production is widely felt to be a very inefficient way to produce food because it uses grains as animal feed that can be directly consumed by humans (Steinfield et al. 2006). Hence depending on how we value land across several land use options i.e., for meat production, staple food production, or energy production it might well be suggested lower levels of animal production are a better option than reducing, or altogether avoiding, the cultivation of crops for first-generation biofuel production. In other words, it is important to place the bioenergy debate in a wider agricultural context.

Our reasons for valuing food energy (and other forms of energy) need to be made explicit if we are to structure land use priorities around specific values. Indeed it is helpful to differentiate two kinds of arguments: those supporting the use of land for bioenergy with the primary intention of providing benefits to communities in developing countries, on the one hand, and those supporting policies that do not have this explicit intention, but that point to this as an additional beneficial consequence, on the other. The main aim of developing bioenergy could be to provide a greater volume of safe, reliable energy to communities in developing countries (UNEP 2009). European biofuels policy could have the beneficial, secondary consequence of strengthening agricultural sectors in developing countries, as claimed above. Although raised prices might stimulate supply, significant challenges to equitable market access may remain for farmers in developing regions wanting to produce food and/or biofuel crops. Again, bioenergy production needs to be assessed in terms of market opportunities, specifically to determine whether the farmers involved will obtain fair returns, or whether, instead, they will face unique challenges when attempting to access an equitable market. A study by German et al. (2011) focusing on biofuel production in six developing countries found that the potential environmental and rural-development benefits have largely not materialized. This study can be used to support the idea that the beneficial outcomes of bioenergy production depend not just on the form of land use, but also on the policies that affect trade and equitable market access. Hence the market values implicit within the prevailing economic "arrangements" must be addressed in tandem with those applied through bioenergy policy.

If the stimulation of agricultural markets is not the intended consequence of bioenergy policy, and if it is claimed merely that this may come about despite certain obstacles, the skepticism of some commentators about the likelihood of an equitable market emerging is understandable. It could therefore be argued, in deontological terms, that bioenergy policy should not be justified on the basis of this possible, secondary positive consequence. These observations serve to highlight the fact that there is a genuine ethical debate to be had about the relationship between the production of bioenergy and global food security. In taking a position on this issue, one needs to be aware of a number of its controversial empirical aspects: principally, the consequences for food security in the short-term and the long-term, and the land-use effects of different forms of bioenergy. In addition there are various ways of characterizing the ethical concerns involved, regarding both the balancing of effects on different communities over time and the nature of the intentions connected with different bioenergy policies.

Environmental Aspects

When biofuels are based on co-products (sometimes referred to as "waste products"), or are produced through the cultivation of marginal land, or very few, detrimental land-use practices are involved and negative environmental impacts are reduced. Thus it is often argued that these forms of bioenergy whose prospects have been contemplated for more than 30 years (Pimentel et al. 1981) would require minimal resources in terms of water and land, reduce GHG emissions, and avoid degradation of habitats and biodiversity (Nuffield Council on Bioethics 2011). Lignocellulosic biofuels, i.e., second-generation biofuels that use indigestible co-products from food or fodder crops or other biomass sources such as Miscanthus, have been promoted as environment-friendly, and hence ethically attractive, bioenergy solutions. This assumption, or set of assumptions, has been challenged, however. Some observers allege that the production of biomass for second-generation biofuels in fact presents a number of threats to environmental conservation and biodiversity, and therefore is not sustainable. Specifically, it is claimed that the use of biomass from crop and forest residues could disrupt local carbon cycles and the fertility of agricultural soil; and that applications of artificial fertilizer to compensate for the disruption may well result in eutrophication and the loss, in effect, of any potential savings in GHG emissions(Doornsbosch and Steenblik 2007).

Assessments of the benefits of second-generation biofuels must take into account the impacts of indirect LUC. Yet, as stated above, this aspect of assessment is controversial (Searchinger et al. 2008). It would be possible in principle, presumably, to manage the sequelae of indirect LUC by ensuring that imported biomass is not sourced from converted areas that were previously forest, or naturally carbon-rich in some other way, or such as to support high levels of biodiversity. If feedstocks could be sourced from abandoned or underutilized arable land (Croezen et al. 2010), and if yields could be increased in way which does not require high levels of fertilizer input.

Thus the second-generation biofuel production is claimed to deliver greater savings in GHG emissions than first-generation biofuel production (Delshad et al. 2010), these savings could be absorbed, or lost, in an indirect LUC, if production of the necessary biomass simply results in the displacement of cropland on to land that presently acts as a carbon sink, such as forest and pasture (Melillo et al. 2009). It is no easy matter evaluating the empirical evidence used to determine the sustainability credentials of second-generation biofuels, and establishing both the direct and indirect environmental impacts over a longer time horizon is particularly difficult. The currently dominant Life Cycle Assessment approach to the development of policy instruments like environmental certification and labeling may adequately address only the vertical dimension of environmental impact: that is, the direct effects of production of a particular product (UNEP 2009). It has been claimed that current approaches are less able to cope with the horizontal dimension: indirect and cumulative environmental impacts such as eutrophication and GHG emissions from indirect land-use change (UNEP 2009). From the consequentialist perspective, the two questions introduced earlier become pertinent once again in connection with environmental aspects: What is “marginal” (or, as it is sometimes stated, “abandoned” land)? What are “byproducts” or “waste” material?

On the first of these questions, even from an agrarian perspective, of course there is always an alternative use for a piece of land. Moreover, if the land in question is converted to production, the net carbon sequestering through natural re-growth and the wider impacts on biodiversity will need to be calculated. And, as has been pointed out by Bindraban et al. (2009), the exploitation of this land will often require high inputs of nutrients and water, which in turn has a production cost and impacts upon GHG savings. On the second question, though second-generation biofuels use co-products that are sometimes accurately characterized as surplus to current agricultural need, rather than referred to as waste, indirect land use change (LUC) may be problematic if, for example, it is not included in emissions accounting. According to Howarth and Bringezu (2008) "Indirect land use will be responsible for substantially more carbon loss (up to twice as much) than direct land use; however, because of predicted increases in fertilizer use, nitrous oxide emissions will be more important than carbon losses themselves in terms of warming potential." If this view is substantiated by empirical evidence, it will be hard to resist the conclusion that second-generation biofuels could lead to higher GHG emissions than their fossil fuel counterparts and as such result in substantially greater carbon loss (up to twice as much) than direct land use. In addition, the availability of some "co-products," like straw and other coproducts of crop production,may not be as unlimited as originally hoped.

Use of residual co-produced material such as straw as a feedstock on a large scale may, in practice, be unfeasible. Assumed to be an "available resource," such co-products may turn out to be very limited feedstocks: it is important to acknowledge that straw is still part of the crop, and that therefore there is a natural limit to how much is available. The quantity of a co-product like straw available depends on factors such as crop rotation, field management practice, and climate (Pahkala and Kontturi 2009). More importantly, straw is useful as a farm resource in number of ways for example, it is used as bedding for livestock, and it can be used (when dug back) to improve soil fertility, control erosion, and maintain soil carbon (Gomiero et al. 2009). In short, the casual description of some co-products as "waste material" conceals various "uses," greater exploration of value is needed. It can be seen, then, that the specific "details" of production, and the differences between different kinds of production, affect assessments of bioenergy and land use. In turn, if we apply a consequentialist approach, this affects the assessment of acceptability. For example, diverting land currently used to produce the animal feed used in highly (energy) inefficient meat production systems to first-generation biofuel production may be a more acceptable and sustainable practice (under the assumption of a drop in the consumption of the relevant kinds of meat). From a deontological perspective, a key task in the discussion of the potential environmental benefits of producing biofuels from co-products, is that of reframing the (predominantly consequentialist) argument. Questions to be addressed here include: Are the co-products used to mitigate climate change or for other reasons? Is land being treated merely as a means or as an end itself? Clearly, one of the most important goals of biofuel and bioenergy policy is to mitigate climate change: the intention behind the policies is to limit global warming. One of the reasons that biofuels have faced such harsh criticism from environmental NGOs is the view that there is a gross mismatch between intention and reality in this respect.

At first glance, it may seem inappropriate to examine the environmental dimension of land use in Kantian terms, since at the micro-level the relevant decisions are made, not so much with the intention of bringing about, or avoiding, particular environmental outcomes, but rather with market conditions and resource constraints in mind: that is, people do not intend to release carbon dioxide into the atmosphere, or to destroy ecosystems, when they cut down a forest to grow palm oil for biodiesel, but they do so because it is profitable and/or is deemed to support development. Indirect LUC is also very problematic because it is by definition an unintended consequence of biofuel production. As in the case of the social dimensions of biofuels production, this raises questions about the prioritization of the intention to mitigate climate change by using biofuels and shows that a realistic appraisal of the work needed to align intention and reality is required. It is possible to examine the market conditions and resource constraints under which biofuels are produced and traded within a Kantian framework. One might conclude that, in the biofuels commodity market, land is viewed as private property, and that this is tantamount to treating land merely as a means and not as an end in itself (O'Neill 2000). Some take a rather different view of land and land use. Jasanoff (2010) states that greater awareness of current environmental problems is leading to deeper reflection on questions about whom and what has standing in our ethical reasoning.

She points out that India and Ecuador have included "rights of the environment" within their constitutions. Similarly, it was recently reported that Bolivia has granted nature equal rights to those of humans in a Law of Mother Earth. Does this constitute treating land as an end, not a means? Certainly, it raises questions about the interpretation of nature's "ends": nature is not one homogeneous entity and it cannot be said to have plans or guiding principles. The argument under consideration here also raises questions about the way we deal with situations where people's rights conflict with notions of the ethical standing of nature. Debate in environmental ethics has often returned to the contentious issue of whether the environment should be valued for its own sake or instrumentally, because of the pleasure or benefits it provides for us (Jamieson 2008). In the case of carbon sequestration, this could be seen as something of a false dichotomy, as the existence of trees benefits us. It shows that our relationship with the land is not always a one-way relationship of exploitation: humans actively use passive land as a resource, and depend on its activities. If the rights of nature are interpreted as the right to exist without interference from humans, one might be tempted to discern here a moral and legal imperative to bring to a halt the deforestation caused by biofuels, though this in turn may cause many problems of interpretation and conflict-management. Other win–win outcomes for people and the environment have been predicted for biofuels: as well as being useful to us, they are "natural" products that sequester carbon as they grow. However, industrial agricultural methods and land grabbing have created disillusionment with this claim, as discussed above. This suggests that when we treat natural resources like land as means to our own ends, as we are obliged to do in producing biofuels, we must investigate the conditions in which these means are used. How should we change our model of resource use when we switch from fossil fuels (regarded by many as essentially a "free" gift from nature) to biofuels (a costly technology to produce involving many difficult resource tradeoffs)? Would this involve changing the end use of the resource or the nature of the markets through which it is sold? Moreover, when we address the question of how to value trade-offs between the non-environmental benefits of bioenergy production (income and energy) and environmental benefits from the same piece of land, we need according to the line of moral thinking in play here to consider the relative prioritization of the intentions behind bioenergy policy when they conflict, as many maintain they do. We must ask: What does a given policy implicitly assume to be the most important use of land?

Conclusions

The biofuel debate has moved beyond discussion of energy security, less reliance on fossil fuels and the production of more environmentally responsible energy sources. An important observation is that talk about a debate is misleading, because the assessment of bioenergy compresses many debates that are highly complex and challenging to unpack in a structured and accessible way. Secondly, it is important to understand that these debates are not only about empirical uncertainties but also involve the clarification and relative weighting of potentially conflicting values. These values are very much up for discussion. One of the most important issues in the bioenergy debate turns on questions about land-use: What direct and indirect consequences are at stake here? What is the right, or the best, use of land? How should utilization priorities be set? This paper has explored the conceptualization, or framing, of biofuel production in light of these questions in order to throw a more nuanced light on the basis of judgments about the acceptability of bioenergy, and specifically biofuels. The burden of the paper has been to bring a range of questions to the foreground: What underlying assumptions or perspectives are being applied (who matters, where)? What impacts or effects are being considered? Should things other than impacts matter in judgments of acceptability? How are the risks being framed? What kinds of benefit are being taken into account? How should different concerns be balanced? Attempts to analyze this complex ethical issue will rely, more or less implicitly, on a conceptualization that involves various assumptions. Some of these assumptions will concern the data to be included in each level of assessment (e.g., social or environmental assessments), and which aspects are to be emphasized or prioritized. Others, perhaps less obviously, will be evidence of a broadly consequentiality or deontological ethical perspective. And of course the various framings will result in actors, including policymakers, potentially taking different positions on the ethical acceptability of bioenergy and biofuels in terms of the land-use issue. One practical message of the analysis presented in this paper is, we suggest, that deeper ethical engagement in the debate about the future role of bioenergy is needed. Such engagement can only complement and fruitfully extend the current technical and economic approaches.

References

Antizar-Ladislao, B., & Turrion-Gomez, J. L. (2008). Second-generation biofuels and local bioenergy systems. Biofuels, Bioproducts & Biorefining 2: 455–469.

Arthur P. J. (2010), The Future of Transparency: Power, Pitfallsand Promises, in: Global Environmental Politics 10(3):132-143.

Bach, F. B. (2009). Biofuels: Hunger, subsidies and lack of effect on CO_2 emission. In M. Gjerris C. Gamborg J. E. Olesen&J. Wolf (Eds.), Earth on fire (pp. 169–174) Copenhagen: The Institute of Food and Resource Economics, The Faculty of Life Sciences, University of Copenhagen. Available at: \http://curis.ku.dk/portal-life/files/14971824/Earth_on_Fire_forbedret.pdfhttp://curis.ku.dk/portallife/files/14971824/Earth_on_Fire_forbedret.pdf. Accessed 2 March 2011.

Balachandra, P and Ravindranath, N.H (2009), Energy efficiency in India: Assessing the policy regimes and their impacts, Energy Policy, Elsevier Science.

Bindraban, P. S., Bulte, E. H., and Conijn, S. G. (2009).Can large-scale biofuels production be sustainable by 2020? Agricultural Systems 101: 197–199.

Boue, A., Dimaranan, B. V.and Valin, H. (2010). Modeling the global trade and environmental impacts of biofuel policies. IFPRI discussion paper 01018.The International Food Policy Research Institute (IFPRI).Available at http: // www. ifpri.org/publication/modeling-global-trade-and environmental impacts bio fuel- policies. Accessed 7 August 2011.

Bush SR (2008) The social science of sustainable bio-energyproduction in Southeast Asia. Biofuels Bioprod. Bioref 2(2):126–132.

Campbell, J. E., Lobell, D. B.and Field, C. B. (2009). Greater transportation energy and GHG offsets from bioelectricity than ethanol. Science 324:1055–1057.

Cassman, K. G., & Liska, A. J. (2007). Food and fuel for all: Realistic or foolish? Biofuels, Bioproducts and Biorefining1:18–23.

Connor, D. J. and Hernandez, C. G. (2009). Crops for biofuels: Current status and prospects for the future. In R.W. Howarth & S. Bringezu (Eds.), Biofuels: Environmental consequences and interactions with changing land use (pp. 65–80).Available at http://cip.cornell.edu/biofuels/. Accessed 2 March 2011.

Croezen, H. J., Bergsma, G. C., Otten, M. B. J.and van Valkengoed, M. P. J. (2010). Biofuels: Indirect land use change and climate impact. Delflt: CE Delft.

Cushion, E., Whiteman, A.and Dieterle, G. (2010). Bioenergy development: Issues and impacts for poverty and natural resource management. Washington, DC: The International Bank for Reconstruction and Development/The World Bank. Available at http://www-wds.worldbank.org/. Accessed 7 August 2011.

Delshad, A.B., Raymond, L., Sawicki, V.and Wegener, D. T. (2010). Public attitudes towards political and technological options for biofuels.Energy Policy 38: 3415-3425.

Demirbas A. (2005) Potential applications of renewable energysources, biomass combustion problems in boiler powersystems and combustion related environmental issues. Progr.Energy Combus. Sci. 31(2): 171-192.

Domac, J., Richards K. and Risovic, S. (2005). "Socio-economic drivers in implementingbioenergy projects." Biomass and Bioenergy 28: 97-106.

Doornsbosch R and Steenblik R (2007). Biofuels: is the cure worse than the disease? Roundtable on Sustainable Development. OECD: Paris

Doornsbosch, R.and Steenblik, R. (2007).Biofuels: Is the cure worse than the disease? OECD Roundtable on Sustainable Development. Available at http: // media. ft. com/cms/fb8b5078-5fdb-11dc-b0fe-0000 779fd2ac. pdf.Accessed 2 March 2011.

Dwegener and J.R. Kelly (2008), Social Psychological Dimentions of Bioenergy Development and Public Acceptance,BioEnergy Research 1(2): 107-117.

Edinger,R.and Kaul S.(2000) .Human's kind detour towards sustainability.past, present and future of renewable energies and electric power generation. Renew Sustain Energy Rev. 4:295-313.

European Commission. (2010). Report from the commission on indirect land-use change related to biofuels and bioliquids. Available at: http: // ec. Europa / energy / renewables/biofuels/doc/land-usechange/ com_2010_811_report.en.pdfhhtp.

Faaij, A.P.C.(1997). Energy from biomassand waste. Utrecht, the Netherlands,University of Utrecht.

FAO.(2008a) The state of food and agriculture 2008. Biofuels: prospects, risks and opportunities. Food andAgriculture Organization,Rome, Italy

FAO.(2008b).Bioenergy, food security and sustainability towards an international framework. FAO. Available at: http: //www. fao. Org / fileadmin/user_upload /foodclimate/HLCdocs/HLC08-inf-3-E.pdf. Accessed 2 March 2011.

Fargione, J., Hill, J., Tilman, D., Polasky, S.and Hawthorne, P. (2008).Land clearing and biofuel carbon debt. Science 319:1235–1238.

Franco, J., Levidow, L., Fig, D., Goldfrab, L., Honicke, M., and Mendonca, M. L. (2010).Assumptions in the European union biofuels policy: Frictions with experiences in Germany, Brazil and Mozambique. Journal of Peasant Studies, 37(4), 661–698.

Franke, G.R.and Nadler, S.S.(2008).Culture, economic development,and national ethical attitudes. J Bus Res 61(3):254–264.

Gamborg, C., Madsen, K. H. and Sandae, P. (2009). Keeping warm in an ethical way is it acceptable to use food crops as fuel? In K. Millar, P. Hobson-West, & B. Nerlich (Eds.), Ethical futures: Bioscience and food horizons. Nottingham: Wageningen Academic Publishers106-110.

Gerbens-Leenes W, Hoekstra AY, van der Meer TH (2009) The water footprint of bioenergy. PNAS 106(25):10219–10223.

German, L., Schoneveld, G., Skutch, M., Andriani, R., Obidzinski, K., Pacheco, P., Komarudin, H., Andrianto, A., Lima, M. and Dayang Norwana, A. A. B. (2011). The local social and environmental impacts of biofuel feedstock expansion: A synthesis of case studies from Asia, Africa and Latin America.online] availableat:http://www.cifor.cgiar.org/publications/pdf_files/infobrief/3295- infobrief. pdf. Accessed June 2011.

Gokcol, C., Dursun, B., Alboyaci, B. and Sunan, E. (2009). Importance ofbiomass energy as alternative to other sources in Turkey. EnergyPolicy 37(2):424–431.

Gomiero, T., Paoletti, M.G., and Pimentel, D.(2009).Biofuels:Efficiency, ethics, and limits to human appropriation of ecosystem services. J.Agricul.and Envir.Ethics. doi: 10.1007/s10806-009-9218-x.

Government of India Ministry of New and Renewable Energy (2009), National Policy on Biofuels.

Halder, P., Pietarinen, J, Nuutinen, S.and Pelkonen, P.(2010).Youngcitizens' knowledge and perceptions of bioenergy and futurepolicy implications. Energy Policy 38(6):3058–3066.

Hill, J., Nelson, E., Tilman, D., Polasky, S.and Tiffany,D.(2006)Environmental, economic, and energetic costs and benefits ofbiodiesel and ethanol biofuels. PNAS 103(30):11206–11210.

Howarth, R.W. and Bringezu, S. (2008). Biofuels: Environmental consequences and interactions with changing land use. Available at http://cip.cornell.edu/biofuels/. Accessed 2 March 2011.

IEA.(2007).Energy policies of IEA countries, Finland 2007.OECD/IEA, Paris, pp 7–1146. IEA (2010a) Bioenergy annual report 2010. http: // www. ieabioenergy. com/DocSet.aspx?id=6506&ret=lib. Accessed 15 April 2010.

IEA.(2008).From 1st to 2nd-generation biofuels technologies. http: // www. iea. Org / papers/2008/2nd_ Biofuel_Gen.pdf Accessed 20 March 2011.

Jamieson, D. (2008). Ethics and the environment: An introduction. Cambridge: Cambridge University Press. Jasanoff, S.(2010).A new climate for society. Theory, Culture and Society 27(2–3):233–253.

Jamieson, D. (2008). Ethics and the environment: An introduction. Cambridge: Cambridge UniversityPress.

Jasanoff, S. (2010). A new climate for society. Theory, Culture and Society, 27(2–3): 233–253.

Keam, S. and McCormick, N.(2008) Implementing sustainablebioenergy production a compilation of tools and approaches. IUCN, Gland, Switzerland, pp 1–32.

Landeweerd, L.,Osseweijer, P., and Kinderlerer, J.(2009).Distributing responsibility in the debate about sustainable biofuels. Science Engineering Ethics 15: 531–543.

Landis, D.A., Gardiner, M.M., van der Werf, W.and Swinton, S.M.(2008) Increasing corn for biofuel production reduces biocontrol services in agricultural landscapes. PNAS 105(51):20552–20557.

Lapola, D.M., Schaldach, R.and Alcamo, J.(2010) Indirect land-usechanges can overcome carbon savings from biofuels in Brazil. PNAS 107(8):3388–3393.

Laurence, W.F.(2007).Switch to corn promotes Amazon deforestation. Science 318:1721.

Levidow, L. and Paul, H.(2008).Land-use bioenergy and agro-biotechnology. Berlin: WBGU.

Mathews,J.A.(2008).Opinion: Is growing biofuel crops a crime against humanity? Biofuels, Bioproducts and Biorefining 2: 97–99.

McCormick, K.and Kaberger, T.(2007).Key barriers for bioenergy in Europe: economic conditions, know-how and institutional capacity, and supply chain co-ordination. Biomass Bioenergy 31(7):443–452.

Melillo, J. M., Reilly, J. M., Kicklighter, D. W., Gurgel, A. C., Cronin, T. C., Paltsev, S., et al. (2009). Indirect emissions from biofuels: How important? Science, 326, 1397–1399.

Mitchell D (2008) A note on rising food prices. Policy ResearchWorking Paper 4682. Development Prospects Group, the World Bank, 1–21.

Mol,A.P.J.(2007).Boundless biofuels?Between environmental sustainability and vulnerability. Sociologica Ruralis 47(4): 297–315.

Mudge, S. M. (2008). Is the use of biofuels environmentally sound or ethical? Journal of Environmental Monitoring10: 701–702.

Nuffield Council on Bioethics. (2011). Biofuels: Ethical issues. London: Nuffield Council on Bioethics. O'Neill, O. (2000). A simplified account of Kantian ethics. In J. E. White (Ed.),Contemporary Moral Problems. Belmont CA: Wadsworth.

O'Neill, O. (2000). A simplified account of Kantian ethics. In J. E. White (Ed.), Contemporary MoralProblems. Belmont CA: Wadsworth.

Pahkala, K.and Kontturi, M.(2009).Potential of straw as bio-energy raw material in northern European countries. NJF Report, 5(3):56.

Palmer,J.(2010).Stopping the unstoppable? A discursive-institutionalist analysis of renewable transport fuel policy. Environment and Planning C: Government and Policy 28(6):992–1010.

Pimentel, D., Marklein, A., Toth, M. A., Karpoff, M. N., Paul, G. S., & McCormack, R. (2009).Food versus biofuels: Environmental and economic costs. Human Ecology 37:1–12.

Pimentel, D., Moran, M. A., Fast, S., Weber, G., Bukantis, R. and Balliett, L. (1981). Biomass energy from crop and forest residues. Science 212: 1110–1115.

Policy Innovation Systems for Clean Energy Security (PISCES). (2010).Status report on use of fuel wood in India.

Puhan, S.,Vedaraman, N., Rambrahaman, B. V., and Nagarajan, G.(2005).Mahua (*Madhuca indica*)seed oil: A source of renewable energy in India. J. Sci. Ind. Res. 64:890–896.

Purohit, P.(2009). Economic potential of biomass gasification projects under clean development mechanism in India. Journal of Cleaner Production 17:181–193.

Rawlings, A. (2007). An expanding liquid biofuel market: Investigating the likely impacts on welfare and environment. Environmental Ethics 1:26–37.

Resch, G., Held, A., Faber, T., Panzer, C.and Toro, F.(2008). Potentialsand prospects for renewable energies at global scale. EnergyPolicy 36(11):4048–4056

Rice, T. (2010). Meals per gallon. The impact of industrial biofuels on people and global hunger. London: ActionAid UK. Available athttp:// www. Europolitics .info /pdf /gratuit_en/266253-en.pdf. Accessed 19 March 2011.

Runge, F. and Senauer, B. (2007). How biofuels could starve the poor. Foreign Affairs 41–53.

Searchinger T, Heimlich R, Houghton RH et al (2008) Use of UScroplands for biofuels increases greenhouse gases throughemissions from land use change. Science 319:1238–1240.

Searchinger, T., Heimlich, R., Houghton, R. A., Dong, F., Elobeid, A. and Fabiosa, J., et al. (2008). Use of US croplands for biofuels increases greenhouse gases through emissions from land-use change. Science 319:1238–1240.

Sodano, V. (2009). Food security, human rights and gender equality. In K. Millar, P. Hobson-West, & B. Nerlich (Eds.), Ethical futures: Bioscience and food horizons. Nottingham: Wageningen Academic Publishers.

Steinfeld, H., Gerber, P., Wassenaar, T., Castel, V., Rosales, M. and de Haan, C. (2006). Livestock's long shadow—environmental issues and options. FAO report. http :// www.fao.org/docrep/010/ a0701e/ a0701e00.HTM. Accessed 7 June 2011.

TERI.(2010).Biomass energy in India. A background paper prepared for the International Institute for Environment and Development (IIED) for an international ESPA workshop on biomass energy, 19-21 October 2010, Parliament House Hotel, Edinburgh. TERI, New Dehli, India

Thompson,P.A.(2008).The agricultural ethics of biofuels: A first look. Journal of Agricultural and Environmental Ethics 21(2):183–198.

Tilman, T., Socolow, R., Foley, J. A., Hill, J., Larson, E. and Lynd, L. (2009). Beneficial biofuels–the food, energy, and environment trilemma. Science, 325, 270–271.

UNEP. (2009). Towards sustainable production and use and of resources: Assessing biofuels. [online] available at: http :// www.biomassenergycentre. org.uk/pls/portal/docs/PAGE/RESOURCES/ REF_LIB_RES/PUBLICATIONS/UKBIOMASSSTRATEGY.PDF. Accessed February 2011.

UNCATD (United Nations Conference on Trade and Development).(2007). The Biofuels Controversy. http://www.unctad.org/en/docs/ditcted200712_en.pdf Accessed 24 April 2010.

UN-Energy.(2007).Sustainable bioenergy: a framework fordecision makers. http://esa.un.org/un-energy/ pdf/susdev.Biofuels.FAO.pdf Accessed 18 April 2010106.

UNICEF, WHO, UNESCO, UNFPA, UNDP, UNAIDS, WFP, The World Bank. (2010). Facts for life. (4th ed.). New York: United Nations Children's Fund. Availble at http://www.factsforlifeglobal.org/ resources/factsforlife-en-full.pdf. Accessed 8 August 2011.

Wegner DT, Kelly JR. (2008) Social psychological dimensions ofbioenergy development and public acceptance. Bioenerg Res 1(2):107–117.

Wenzel, H. (2007). Bio-ethanol: Is the world on the wrong track? Analysis of energy issues. In H. Gani & K. Dam-Johansen (Eds.), ECCE6 Book of Abstracts, 1. Lyngby: Department of Chemical Engineering, Technical University of Denmark.

Yuan, J. S., Tiller, K. H., Al-Ahmad, H., Stewart, N. R. and Stewart, C. N. (2008).Plants to power: Bioenergy to fuel the future. Trends in Plant Science, 13(8): 421–429.

❑❑❑

CHAPTER – 16

Biotechnology and Intellectual Property Rights

Kambaska Kumar Behera, Sanskriti Gautam, Shaili Yadav and Shailja Dubey
Department of Bioscience and Biotechnology Banasthali University,
Rajasthan-304022, India
Email : kambaska@yahoo.co.in

Summary

Idealistic theories of property are projected to offer general and explicit statements of the justification for deciding legal and moral questions about the status property claims. Debates over property rights in biotechnology were occasioned by specific legislative proposals such as the US Animal Patent Act of 1986, and by filing of patent applications for DNA sequences and processes in the early 1990s. While these debates make sporadic petition to philosophical theories of property, moral claims were entwined with questions about filing requirements, tests for worth, and the rules for licensing and defending patents. The comparative scarcity of discussion on intellectual property at that time dictated the general approach of the chapter: Review basic philosophical approaches to property rights, and speculate on how one might use these approaches in constructing an argument pertinent to agricultural biotechnology. This chapter deals with different approaches to intellectual property in the field of Plant Sciences especially the Biotechnology. It incorporate many different types of intellectual property and a variety of incommensurable values play a role in the justification of intellectual property.

Introduction

What is Intellectual Property?

Property is simply a collection of legal rights to control something (Honore, 1977; Dreyfuss, 1989). The legal system defines the types of matter that may be treated as property as well as the scope of rights with respect to those matters. For example, a proprietor of a house has legal rights to possess, use, modify, destroy, transfer, sell, or rent the house. Various laws and regulations may prohibit the homeowner from painting his house in various colours, operating a business from his home, or keeping a horse in his backyard. Since properties are collections of legal rights, the property regime is a social institution that serves particular purposes in society (Feinman, 2000). We are pertinent to think of objects, such as, land, as the properties, but these objects only have their status as property in relation to people in society. If no people were around, land would not be someone's property; it would just be the land. We mark borders and create deeds to represent the social and legal status of a person's rights and duties with respect to the land. Thus, the concept of property is a three-place relationship between an object, an individual (i.e. a person, group of people, or corporation), and society.

Improvement and development of new ideas is primary to the progress of evidence based clinical practice. Intellectual property rights are basic to the protection of bio-information and biotechnology but may hamper progress and dispersal of information. At the same time as the application of patents and copyrights are common place in companies profiting from information technology, their application in the field of biotechnology is less understood and publicized.

Intellectual property (IP) is a term referring to a number of distinct types of creations of the mind for which property rights are documented—and the corresponding fields of law. Under intellectual property law, owners are granted certain exclusive rights to a variety of indefinable assets, such as musical, literary, and artistic works; discoveries and inventions; and words, phrases, symbols, and designs. Common types of intellectual property include copyrights, trademarks, patents, industrial design rights and trade secrets in some jurisdictions (Murashige 2002).

Although many of the legal principles governing intellectual property have evolved over centuries, it was not until the 19th century that the term intellectual property began to be used, and not until the late 20th century that it became usual in the United States. The British Statute of Anne 1710 and the Statute of Monopolies 1623 are now seen as the origin of copyright and patent law respectively (Cimoli, 2002).

Modern usage of the term intellectual property goes back at least as far as 1888 with the founding in Bern of the Swiss Federal Office for Intellectual Property. When the administrative secretariats recognized by the Paris Convention (1883) and the Berne Convention (1886) merged in 1893, they also located in Berne, and also adopted the term intellectual property in their new combined title, the United International Bureaux for the Protection of Intellectual Property. The organization subsequently relocated to Geneva in 1960, and was succeeded in 1967 with the establishment of the World Intellectual Property Organization (WIPO) by treaty as an agency of the United Nations. It was only at this point that the term really began to be used in the United States (which had not been a party to the Berne Convention), and it did not enter popular usage until passage of the Bayh-Dole Act in 1980.

Historical Background

"The history of patents does not begin with inventions, but to a certain extent with royal grants by Queen Elizabeth I (1558-1603) for command constitutional rights. Approximately 200 years after the end of Elizabeth's preeminence, however, a patent represents a legal [right] obtained by an inventor providing for exclusive control over the production and sale of his mechanical or scientific invention the evolution of patents from royal privilege to common-law doctrine."

In 1818, the French liberal theorist, Benjamin Constant, argued against the recently introduced idea of "property which has been called intellectual." The term intellectual property can be found used in an October 1845 Massachusetts Circuit Court ruling in the patent case Davoll et al. v. Brown., in which Justice Charles L. Woodbury wrote that "only in this way can we protect intellectual property, the labors of the mind, productions and interests are as much a man's own...as the wheat he cultivates, or the flocks he rears." (1 Woodb. & M. 53, 3 West.L.J. 151, 7 F.Cas. 197, No. 3662, 2 Robb.Pat.Cas. 303, Merw.Pat.Inv. 414). The statement that "discoveries are property" goes back earlier. Section 1 of the French law of 1791 stated, "All new discoveries are the property of the author; to assure the inventor the property and temporary enjoyment of his discovery, there shall be delivered to him a patent for five, ten or fifteen years." In Europe, French author A. Nion mentioned propriété intellectuelle in his Droits civils des auteurs, artistes et inventeurs, published in 1846.

The concept's origins can potentially be traced back further. Jewish law includes several considerations whose effects are similar to those of modern intellectual property laws, though the notion of intellectual creations as property does not seem to exist – notably the principle of Hasagat Ge'vul (unfair encroachment) was used to justify limited-term publisher (but not author) copyright in the 16th century. The Talmud contains the prohibitions against certain mental crimes (further elaborated in the Shulchan Aruch), which some have interpreted as prohibiting theft of ideas, though the doctrine is principally concerned with fraud and deception, not property (Correa, 2001).

Objectives of Intellectual Property

Copyright and its importance

Intellectual property (IP) protects ideas. There are four main types of **Intellectual property** (IP):

1. Patents protect the technical aspect of the product
2. Trademarks protect signs of the trader
3. Copyright protects creative material such as music and literature
4. Design registration protects the visual appearance of the product.

Copyright law varies between nations. In the UK it originated from the Statute of Anne 1709, before the passing of the Copyright Act in 1911. The Copyright, Designs and Patents Act of 1988 was statute until October 2003 when the EU issued a directive to acknowledge the impact of electronic information and the internet on IP. Copyright law acknowledges the creator of a work as the author with the right to govern how the work is disseminated and prevents distortion of the work by third parties. This applies to literary works (such as scientific papers) as well as artistic creations and even computer programs. Copyright is automatic in the UK on creation of an original work. It will protect the work itself but not the idea behind it. Thus the idea of hip resurfacing would not be protected but the Birmingham Hip Resurfacing components would be. Commissioning or freelancing does not influence ownership which is still exclusive to the original author. For literary works such as scientific papers, computer programs, commercial documents and song lyrics the Copyright, Designs and Patents Act of 1988 states that the duration of copyright is 70 years from the time the work was written or first published in the public domain. For typographical arrangements of published editions such as magazines, periodicals and journals, copyright duration is 25 years from the end of the calendar year of publication. A scientific paper prior to publication can be registered for copyright that will last for 70 years. On publication, the author transfers copyright to the journal which then holds the copyright for 25 years. Thus the author has the right to be identified as the author of the work and to object to critical treatment. It is an offence to copy, adaptor show the work in public without the consent of the owner. However, the work can be copied for the purposes of research or educational purposes without infringing on the copyright. On submission of a manuscript for publication, most publishers require a transfer of copyright agreement. This allows the publisher to register a single copyright product (if the paper is attributed to multiple authors) and protects the author, editors and publishers from copyright violation due to unlawful duplication once the manuscript has been published as part of a journal. Authors do not have to sign a transfer of copyright as a condition of publication and may retain copyright of their manuscript but would have to sign a formal statement licensing publication exclusively in the relevant journal. Most publishers would also expect authors to acknowledge that their work is original and has not previously been published to ensure no copyright infringement prior to publication (Elliott, 2007).

Authors still retain the proprietary right of their work once the transfer of copyright has been signed over to a publisher since 'moral rights' are conferred to authors by the UK Copyright Act of 1988. Authors are therefore credited whenever the publishing house uses the work.

Authors can also reproduce their own papers provided the publisher is acknowledged and a citation provided to the original publication. Most publishers would require notification if the paper is to be used in a volume in which the author acts as editor or author. Policies regarding electronic postings on websites vary between publishers with some allowing pre-print postings on the web with a link to the publication and acknowledgement. Most publishers will allow broadcasting of the published article by the author provided it is for educational purposes and not open access to the general public (Coriat and Orsi, 2001).

At the same time as copyright is an automatic process on conception of a body of work in the UK, a copyright notice will formalize this process. The copyright notice takes the form of a copyright symbol ©, the year the work is published or written and the name of the owner, e.g. Copyright © CB Hing 2007. Additional information such as 'all rights reserved' or 'any unauthorized broadcasting, public performance, copyrighting or re-recording will constitute an infringement of copyright' can also be added but are not a necessity to establish copyright. In order to support the claim of copyright, the body of work can be registered with the UK Copyright Service in order to strengthen the claim of ownership in case of dispute. Supporting evidence for a claim of copyright includes all earlier drafts of the work, watermarks or electronic file comments as well as 'footprints' (calculations, mistakes etc.) (Crespi, 2000).

Joint works should include an agreement between co-authors regarding ownership of copyright. The principal author may claim copyright and take the work with them if they leave. In the event of one author leaving with joint ownership of copyright, the remaining authors must agree what will happen to the copyright (Murphy, 2002).

If the work is produced during employment, the employer will own the work. However, if the work is freelance or commissioned, rights will belong to the author of the work. Once copyright has expired, the author loses intellectual property rights to his work and it is freely available in the public domain. The work thus becomes public property and no other individual can claim copyright. An interesting exception in the UK is the copyright to 'Peter Pan'. Lord Callaghan amended the 1988 Copyright Designs and Patents Act to allow the copyright for 'Peter Pan' indefinite status such that any royalties are passed on to Great Ormond Street Hospital. In October 2003, changes were made to the UK copyright law in response to a European Union directive and the impact of electronic information sharing (Coriat and Orsi, 2001). This has meant that copying for commercial purposes where income will be generated is no longer allowed without the consent of the owner of copyright. However, copying for educational purposes, such as libraries, allows limited amounts of copying without the consent of the author. Thus copying research papers for scientific education would be allowed but copying papers to give to private patients would not as this potentially generates income and is classified as a commercial purpose (Azoulay et al., 2007).

Intellectual property rights transcend countries and the Berne Convention for the Protection of Literary and Artistic Works was introduced in 1886 as an agreement between countries to uphold this. The World Intellectual Property Organization (WIPO) administers the latest version of the convention, the Paris Act of 1971. There are approximately 147 members of the World Trade Organization that have agreed to the act, which protects the intellectual property of individuals' resident to the member countries. Outside of those member countries, protection is afforded by a reciprocal agreement incentive. Consequently the author has rights to protect work from reproduction, adaptation and translation (http://www.iprcommission.org).

Financial Incentive

These exclusive rights allow owners of intellectual property to benefit from the property they have created, providing a financial incentive for the creation of an investment in intellectual property, and, in case of patents, pay associated research and development costs.

Economic Growth

The existence of IP laws is credited with significant contributions toward economic growth. Economists estimate that two-thirds of the value of large businesses in the U.S. can be traced to intangible assets. "IP-intensive industries" are estimated to generate 72 percent more value added (price minus material cost) per employee than "non-IP-intensive industries".

A combined research project of the WIPO and the United Nations University measuring the impact of IP systems on six Asian countries found "a positive correlation between the strengthening of the IP system and subsequent economic growth." Other models, such as the Nash equilibrium, would not expect that this correlation necessarily means causation: The Nash equilibrium model predicts that patent holders will prefer to operate in countries with stronger IP laws. In some of the cases, as was shown for Taiwan after the 1986 reform, the economic growth that comes with a stronger IP system might be due to an increase in stock capital from direct foreign investment (Mooney, 2000).

IPR in the Field of Plant Sciences

Plant breeders' rights (PBR), also known as **plant variety rights** (PVR), are rights granted to the breeder of a new variety of plant that give them exclusive control over the propagating material (including seed, cuttings, divisions, tissue culture) and harvested material (cut flowers, fruit, foliage) of a new variety for a number of years (Pushpagandhan, 1996).With these rights, the breeder can choose to become the exclusive marketer of the variety, or to authorize the variety to others. In order to qualify for these exclusive rights by plant breeders' rights, a variety must be new, distinct, uniform and stable. A variety is new if it has not been commercialized for more than one year in the country of protection. A variety is different if it differs from all other known varieties by one or more important botanical characteristics, such as color, height, maturity, etc. A variety is homogeneous if the plant characteristics are consistent from plant to plant within the variety. A variety is stable if the plant characteristics are genetically fixed and therefore remain the same from generation to generation, or after a cycle of reproduction in the case of hybrid varieties. The breeder must also give the variety an acceptable "denomination," which becomes its generic name and must be used by anyone who markets the variety (Warner, 2001).

In general, plant variety rights are granted by national offices, after examination. Seed is submitted to the plant variety office, which grow it for one or more seasons, to ensure that it is distinct, stable, and uniform. If these tests are passed, exclusive rights are granted for a period of 20 years (or 25 years, for trees and vines). Annual renewal fees are required to maintain the rights (Altieri, 2003).Breeders can bring suit to implement their rights and can recover damages for violation. Plant breeders' rights contain exemptions from violation that are not recognized under patent law. Commonly, there is an exemption for farm-saved seed. Farmers may store the production in their own bins for their own use as seed, but this does not necessarily extend to brown-bag sales of seed. Further sales for propagation purposes are not allowed without the written approval of the breeder. There is also a breeders' exemption (research exemption in the 1991 Act) that allows breeders to use protected varieties as sources of initial variation to create new varieties of plants (1978 Act), or for other experimental purposes (1991 Act). There is also a provision for compulsory licensing to assure public access to protected varieties if the national interest requires it and the breeder is unable to meet the demand (Visser et al., 2002).

The United States of America passed the Plant Patent Act in 1930 (US) at the urging of such notable figures as Thomas Edison and Luther Burbank's widow. Plant patents provided a special form of patent protection, which relaxed certain requirements of the utility patent law as applied to asexually reproduced varieties of plants. In 1957, the French Government held a conference in Paris concerned with the protection of new varieties. This led to the creation of the Union Internationale pour la Protection des Obtentions Végétales (UPOV) and adoption of the first text of the International Convention for the Protection of New Varieties of Plants (UPOV Convention) in 1961. The purpose of the Convention was to ensure that the member states party to the Convention acknowledges the achievements of breeders of new plant varieties by making available to them an exclusive property right, on the basis of a set of uniform and clearly defined principles.

The Convention was revised in Geneva in 1972, 1978 and 1991. Both the 1978 and the 1991 Acts set out a minimum scope of protection and offer member States the possibility of taking national situation into account in their legislation. Under the 1978 Act, the minimum scope of the plant breeder's right requires that the holder's prior authorization is essential for the production for purposes of commercial marketing, the offering for sale and the marketing of propagating material of the protected variety. The 1991 Act contains more detailed provisions defining the acts relating to propagating material in relation to which the holder's authorization is required. The breeder's authorization is also required in relation to any of the specified acts done with harvested material of the variety, unless the breeder has had reasonable opportunity to exercise his right in relation to the propagating material, or if not doing so could constitute an "Omega Threat" situation. Under that provision, for example, a flower breeder who protects his variety in the Netherlands could block importation of cut flowers of that variety into the Netherlands from Egypt, which does not grant plant breeders' rights, because he had no opportunity to exercise any rights in Egypt. Member countries also have the option to require the breeder's authorization with respect to the specified acts as applied to products directly obtained from the harvested material (such as flour or oil from grain, or juice from fruit), unless the breeder has had reasonable opportunity to exercise his right in relation to the harvested material (Schrell et al., 2007).

The UPOV Convention also establishes a multilateral system of national treatment, under which citizens of any member state are treated as citizens of all member states for the purpose of obtaining plant breeders rights. It also sets up a multilateral priority filing system, under which an application for protection filed in one member state establishes a filing date for applications filed in all other member states within one year of that original filing date. This allows a breeder to file in any one member country within the one-year period required to preserve the novelty of his variety, and the novelty of the variety will still be recognized when he files in other member countries within one year of his original filing date. However if the applicant does not wish to make use of priority filing he or she has four years in which to apply in all other member states, excepting the USA, for all species except tree and vine species in which case he or she has six years to make application. See article 10 1 (b) of Council Regulation EC No 2100/94 of 27 July 2004 on the website www.cpvo.eu. The trigger to start the four or six year period is not actually the date on which the first filing is made but the date on which the variety was first commercialized (www.wipo.int).

The UPOV Convention is not self-executing. Each member state must adopt legislation consistent with the requirements of the convention and submit that legislation to the UPOV Secretariat for review and approval by the UPOV Council, which consists of all the UPOV member states acting in committee. In compliance with these treaty obligations, the United Kingdom enacted the Plant Variety and Seeds Act 1964. Similar legislation was passed in the Netherlands, Denmark, Germany, and New Zealand. In 1970 the United States followed the lead of seventeen Western European nations and passed the Plant Variety Protection Act 1970 (US). This legislation provided protection to developers of novel, sexually reproduced plants. However, the United States originally acceded to the UPOV Convention on the basis of the Plant Patent Act and did not bring the PVP Act into compliance with UPOV requirements until 1984 when the Commissioner of Plant Variety Protection promulgated rules to do so. Since the 1980s, the US Patent Office has granted patents on plants, including plant varieties: this provides a second way of protecting plant varieties in the USA. Australia passed the Plant Variety Protection Act 1987 (Cth) and the Plant Breeders Rights Act 1994 (Cth). Australian patent law also permits the patenting of plant varieties. In total, 65 countries have signed the UPOV

Convention and adopted plant breeders' rights legislation consistent with the requirements of the convention.

The WTO's Agreement on Trade-Related Aspects of Intellectual Property Rights (TRIPs) requires member states to provide protection for plant varieties either by patents or by an effective sui generis (stand alone) system, or a combination of the two. Most countries meet this requirement through UPOV Convention-compliant legislation. India has adopted a plant breeders' rights law that has been rejected by the UPOV Council as not meeting the requirements of the treaty.

IPR and Biotechnology

Biotechnology constitutes a new promise of great profits as, in the course of genetic engineering, new products or products with new properties can be made. Thus, the attainment of specific biological resources that possess properties of commercial concern becomes precious. This fact, along with the introduction of patents on biological material, lead pharmaceutical, food, and seed companies in an unprecedented chase around the world for the collection of living organisms that may be of some use, ranging from soil micro-organisms to animals, including human genes. However, the biological material is often removed in great discretion from the country in question, without even its national government's consent (Fowler, 1995).

Intellectual property is a form of knowledge that societies have decided can be assigned specific property rights (Commission on Intellectual Property Rights, 2002, p. 11). Intellectual property rights (IPR) are the rights awarded to individuals or organizations chiefly over creative works: inventions, literary and artistic works, symbols used in commerce. They include patents, trademarks and copyrights, geographical indications, and trade secrets. IPRs are the rights to make, use, and sell a new product or technology that are granted for a specific period of time (e.g., 20 years), solely to the inventor or the corporation that files a claim on the inventor's behalf. "Intellectual property rights are either based on originality (Iañez Pareja, 2001).

The basic economic argument justifying the existence of IPRs is that, unless an industry can ensure the capitalization on its investments in technological development, it will not invest in it in the first place. That would make technological breakthroughs unachievable. Patents and other IPRs address this fundamental market failure: if a product takes considerable effort and research but can be easily copied, it is unlikely that there will be sufficient financial incentive to devote resources to invention (Commission on Intellectual Property Rights, 2002, p. 14). By conferring temporary market exclusivities, patents allow producers to recoup the costs of investment and reap a profit (Katz, 2005).

The first patent on a living organism was granted in the US in June 1980, with the granting of a patent to Ananda Chakrabarty. The organism in question was a bacterium, a *Pseudomonas* strain, engineered to contain added plasmid DNA, which made it more effective in dispersing oil spills than the naturally occurring precursor strain. This decision constituted the basis for extending patent application to higher life forms. As Crespi (2000) points out, the decision was "a watershed in the development of patent law, as a precedent for extension to cell lines, such as hybridoma that produce monoclonal antibodies, and genetically modified plants and animals."

As stated by Ehrlich and Ehrlich, 1981; EFB, 2001, some countries are richer than others in terms of biodiversity. This can be calculated through the number of existing species of plants and animals. Although it is often difficult to distinguish the exact ecological role of a single species in the ecosystem, it is fair to accept that the absence of close relatives when a distinctive species is concerned will lead to far less overlap in its functional role. The direct instrumental value of biodiversity is demonstrated in the fact that human societies derive many essential goods from natural ecosystems such as pharmaceuticals, food items, and building material. These numerous natural ecosystems, composed of a diverse biological base, perform fundamental life-support services, without which life as we know it would cease to exist. Several authors highlight the existence of direct and indirect economic benefits associated with species preservation and point out that species should be preserved because of their beauty, symbolic value, or intrinsic value.

Technological breakthroughs including biotechnology and its applications can increase considerably the instrumental value of biodiversity, as new products based on existing natural resources can now be made. Most of the mega-biodiversity countries are developing countries, which could profit substantially in financial terms from their wealth in biodiversity. It also appears that, in many cases, those extraordinary properties of biological resources that now get commercial interest were already known to indigenous communities and used for centuries. This knowledge, generated, refined, and passed from generation to generation, is often described as traditional knowledge. Following a parallel track, formal plant breeding programs have utilized varieties conserved and developed by farmers, in order to develop improved varieties of higher productivity, or with other desirable characteristics (Crespi, 2000).

It is rather difficult to deliver a precise definition of traditional knowledge. The World Intellectual Property Organization (WIPO) refers to it as "tradition-based literary, artistic or scientific works; performances, inventions, scientific discoveries, designs, marks, names and symbols, undisclosed information and all other tradition-based innovations and creations resulting from intellectual activity in the industrial, scientific, literary or artistic fields" (WIPO, 2001). This definition delivers the agony that all forms of traditional knowledge should be included. In addition, traditional knowledge can be codified, that is, formalized in some way (e.g., textile designs), but in many cases, depending upon geographic origin among other factors, it is non-codified (e.g., "tribal" or "indigenous" medicine) (Correa, 2000).

The question that rises is whether traditional knowledge is an intangible component of the resource itself and thus, should be rewarded. In order to try to answer this question, the example of plant genetic resources will be employed. Research and development on plant genetic resources is a dynamic procedure that involves a variety of agents ranging from small farmers to seed companies. "On-farm innovation by farmers has happened continuously since settled agriculture began" (Dbar, 2002). Small farmers undertake experiential research enhancing the value of plant genetic resources through the selection of the best-adapted farmers' varieties, as well as through the invention of new production techniques. On the other hand, research institutions use plant genetic resources to "undertake basic and applied research, including agro-biotechnology, and to enhance existing varieties and the availability of gene pools" (Correa, 2003).

However, while research institutions obtain materials and information from small farmers generally on a non-market basis, they later function solely within the market. When an improved variety enters the commercial channels, the seed company can protect it under plant breeders' rights (PBRs) or patents, and benefit from it. Nevertheless, small farmers are not compensated for the added value they have created: seed companies' researchers are not charged a price for the samples they obtain, neither is there any compensation or sharing of benefits with the farmers. The contribution made by farmers in conserving, improving, and making available plant genetic resources is seeking, therefore, its recognition in the international regulatory and policy framework, particularly through the concept of "Farmers' Rights." This concept was recently incorporated in the International Treaty on Plant Genetic Resources for Food and Agriculture (ITPGR), adopted under the auspices of FAO's Commission on Genetic Resources for Food and Agriculture (CGRFA).

In the case of commercialization of products incorporating traditional knowledge, indigenous and local communities suffer a double loss. On one hand, they cannot share the benefits arising from the new product, as this right stays with the holder of the patent. On the other hand, applications of modern biotechnology often eliminate the need for commercial growing of the original plant, depriving thus the local population from another source of income. The communities can only take advantage of the scientific breakthrough – in which they have also contributed – if they pay the premium that is demanded by the company through the price of the final product. As India states in its WTO paper on biopiracy, "traditional knowledge saves time and money for modern biotechnology industry by providing leads for development of useful products" (India, 1999). Hence, traditional knowledge is to be rewarded because it enters and influences directly the economic function. The reward constitutes, in economic terms, part of the profit associated with the development of the product. It is therefore self-evident that indigenous and local communities in the developing countries should share the benefits arising from the commercialization of the final product. Not only do they supply the necessary genetic resources, but also, in a great number of cases, they have themselves identified and used the relevant properties for many centuries (Lichtenberg, 2000).

International Regulation

Paris Convention for the Protection of Industrial Property

It was signed in 1883 and has been the object of various revisions. It applies to patents, trademarks, geographical indications, industrial designs and models, and unfair competition.

Patent Cooperation Treaty (PCT)

It facilitates the application for and registration of patents abroad for the signatory countries.

International Convention for the Protection of New Varieties of Plants (UPOV)

It was signed in 1961 and was amended in 1978 and 1991. It establishes the criteria for protecting plant varieties by means of "plant breeders' rights." The 1991amendments stipulate strengthening the rights of plant breeders over the multiplication, commercialization, exportation, and importation of the material to be propagated, including improvements to the potential protection of all genres and species of plants. These provisions were incorporated into the European Union's Plant Varieties Law in 1995 (Marris, 2005).

UPOV also introduced the concept of "essentially derived varieties" to allow plant breeders to control the use of random mutations. Understood as such are varieties "it is predominantly derived from the initial variety, or from a variety that is itself predominantly derived from the initial variety, while retaining the expression of the essential characteristics that result from the genotype or combination of genotypes of the initial variety" (UNCTAD–ICTSD, 2005). Thus the rights of plant breeders who would otherwise lose the value of their intellectual property in a short time are protected (as would often occur when a plant breeder made "cosmetic" changes to a previous variety without paying for it).

Other UPOV stipulations recognize the right of farmers to keep seeds for the next sowing cycle without the need to ask permission and that "small farmers" are exempt from payment of royalties and may keep particular varieties for 7 years.

Despite these flexibilities, the 1991 UPOV revision brought the regime of plant breeders' rights closer to that of patents. The Convention extends the coverage of intellectual property rights to imports, exports, and harvested crops; broadens the duration of plant breeders' rights, aligning them with that of patents; restricts free access to protected varieties (although it maintains it for purposes of reproduction of new varieties); and extends plant breeders' rights to all varieties considered "essentially derived" from the protected variety, limiting the use of new varieties that are developed. Although the new norms allow using protected varieties for research purposes, any improvement obtained should show significant changes in the phenotype. Otherwise, the variety is not considered "new" and will continue to be the property of the first plant breeder. Furthermore, the right to use formerly permitted, such as the accumulation and reproduction of varieties protected in "gene banks" aimed at preserving genetic diversity, is restricted. Finally, farmers' privilege of keeping seeds for later sowing seasons is eliminated.

Governing Bodies and Contradictions

Over the past few years, the international community has attempted to regulate the conservation and sustainable use of biological diversity, as well as the issue of access to biological materials, their control and ownership.

Many bodies engage in discussions on biodiversity and intellectual property, the objectives of which are sometimes overlapping or even contradictory. Given the ongoing struggle of many companies and research institutes for the acquisition of genetic resources, the issue of the sharing of benefits arising from their use and commercialization is receiving major attention. The call contained in the Plan of Implementation agreed in the Johannesburg World Summit on Sustainable Development for launching negotiations on an international regime on benefit sharing in the framework of the UN Convention on Biological Diversity (CBD) can only be seen in that regard (Platt, 2001).

The principles and practice of access to biological resources and fair and equitable sharing of the benefits arising from their use are being debated in two major international discussions: the CBD and the CGRFA. Benefit sharing is one of the three main objectives of the CBD, the Parties to which have recently agreed on a set of voluntary guidelines, called the Bonn Guidelines on access to genetic resources and benefit sharing. These guidelines pave the way for an internationally agreed, multilateral system on access and benefit sharing. On the same side, the recent completion of the lengthy negotiations for the revision of the non-binding International Undertaking on Plant Genetic Resources in harmony with the CBD, under the auspices of the FAO, has resulted in the innovative and legally binding International Treaty on Plant Genetic Resources for Food and Agriculture, "which covers 64 food crops accounting for about 85% of global human nutrition" (RAFI, 2002). The Treaty establishes a multilateral system for facilitated access to a specified list of plant genetic resources for food and agriculture, balanced by benefit sharing in the areas of information exchange, technology transfer, capacity building, and commercial development.

While the CBD calls for the protection of indigenous knowledge, the WTO TRIPS agreement (Trade-related aspects of Intellectual Property Rights) obligates its members to adopt patents or a sui generis system for plant varieties. Conservation and privatization are seen by many as contradictory goals, and the TRIPS agreement is accused for restricting the availability of genetic resources and causing risks for food security and sustainable livelihoods of local communities by jeopardizing farmers' rights and displacing their varieties. In addition, Western-style intellectual property regimes do not fit the characteristics of traditional knowledge, and can therefore hardly be used for its protection.Although until recently, WIPO was considered to be the most important international body dealing with intellectual property issues, the creation of WTO and the adoption of the TRIPS agreement changed the scenery.

The TRIPS agreement brought intellectual property to center stage in multilateral trade negotiations. Of particular relevance to biological diversity, Article 27.3b states that plants and animals, as well as essentially biological processes may be excluded from patentability. Since "all biotechnological processes (including microbiological) are actually biological processes" (Mae-Wan Ho and Traavik, 1999), their exclusion from patentability under 27.3b remains open. As Tansey points out, words like "plants, animals, microorganisms, non-biological, effective, microbiological, sui generis systems are defined differently in different international and national legislation" (Tansey, 1999). So, they are, in some degree, open to interpretation. In addressing that problem, there have been suggestions of creating a single body to deal with gene biotechnology in its very sense instead of addressing the application in particular areas (Saunders, 1999).

As stated above, WTO Members must offer protection for plant varieties either by patents and/or by an effective sui generis system, meaning a system of rights that is unique for a specific item or technology.Because of the difficulty in reaching consensus during the negotiations, the Members agreed that the text would be reviewed. Article 27.2 then allows the exclusion from patenting of inventions contrary to public order or morality: inventions dangerous to human, animal or plant life or health, or seriously prejudicial to the environment. It is still open to whether the public morality exclusion of TRIPS can be used to reject patents on life forms or controversial new technologies such as the genetic use restriction technologies.

Patents

Patents in the field of biotechnology can be applied to drugs and chemicals, genes and proteins, micro-organisms, plants and animals. An early example of the historical application of patents in biotechnology is the US patent awarded in 1873 to Louis Pasteur for his work on a 'yeast, free from organic germs of disease, as an article of manufacture'(Pasteur, 1873). Patents are designed to grant the inventor an exclusive monopoly on his design for a limited period of time. In return the inventor discloses information regarding the invention into the public domain to allow dissemination of knowledge and use of the invention within the framework of the patent. Once the patent expires, the invention can be made, sold or used with no limits (Tansey, 1999).

Application and Granting of Patents

Background

In 1873 the United States Patents and Trademarks Office (USPTO) granted patent No. 141,172 to Louis Pasteur for yeast free of disease germs, considering it a manufacture. It was subsequently decided not to continue along that line, with some exceptions (a breed of bacteria in 1977). The criterion that prevailed was considering that living organisms were not patentable because they were regarded as products of nature, or because they were not subject to sufficient written description, as demanded by the patents system. This excluded bacteria and fungi that produced antibiotics, for which reason prior to 1980 most patents were granted to processes, mainly those that used bacteria to treat wastewater or produce chemical substances, antibiotics, etc. (Somberg, 2005).

In 1930 the United States Congress approved the Plant Patent Act, which permitted protection for asexually reproduced plants with the exclusive right to propagate the plant for 17 years (Dunwell, 2010). To date the USPTO has granted around 6,000 patents for plant holders, mainly fruit trees, flowers, ornamental trees, grapes, and other horticultural species. In 1961 the International Convention for the Protection of New Varieties of Plants (UPOV) was signed in Paris. It has been submitted to three revisions, the last one in 1991. In 1970 the United States introduced an adapted version of the UPOV System of Plant Breeders' Rights into its legislation to protect new sexually reproduced varieties of plants.

In Europe, after the signing of UPOV, several countries recognized protection titles for new plant varieties. Nevertheless, in 1973 the European Patent Convention in Munich excluded the patenting of plant varieties and the essentially biological procedures for their production (Correa, 2001). In 1977 the first patent on a breed of bacteria was granted in the United States, but it was not until 1980 that the first major changes were made to intellectual property regulations.

In 1972 microbiologist Chakrabarty filed a patent application for 36 claims related to the invention of a genetically designed bacterium of the Pseudomonas species, capable of degrading the multiple components of crude petroleum. The application was rejected, but Chakrabarty appealed, and on June 16, 1980, the United States Supreme Court decided in a historic ruling that the said bacterium was a "manufacture" or "composition of matter" which met the criteria of novelty (inexistent as such in nature and not obvious for science at the time) was derived from an inventive step (it had been produced in a laboratory by transfer of plasmids) and met the requirement of usefulness (its purpose was to use it in oil spill clean-up work).

The ruling included the following phrase: patents can be granted "to anything under the sun that is made by man." Thus the objection to patents on living beings for the simple fact of being living was eliminated. The Budapest Treaty (1977, in force since 1980) establishes as a requirement for patent applications on microorganisms the deposit of cultures in well-known collections to that effect.

This decision marked an important change on the subject and led to modifications of the regulatory framework for Intellectual Property Rights in the United States. The perception of deterioration of the United States' comparative advantages in the technology-intensive industry brought about significant changes in forms of protection of intellectual property rights (Hunt, 1999). The loss of competitiveness was attributed to the fact that the system was geared to basic research, was weak, and rather inefficient in producing results for firms, whereas the results of research were easily appropriable by competitors due to lack of protection (Coriat and Orsi, 2001).

Regulatory Levels and Frameworks

Patents can be applied for and obtained at national, regional, and international level. For the first level there are the specialized national offices, for the second the regional authority (e.g., the European Patent Office), and for the third the International Patent Cooperation Treaty (PCT) administered by WIPO, under which valid patents can be registered in all the signatory countries. Patents do not grant the same rights in all countries, for each one has its own law. For the same reason, authorizations may have different coverage, so that comparisons with information from different countries and systems entail difficulties.

USPTO, the European Patent Office (EPO), and the Japan Patent Office (JPO) are the ones that receive the most applications and authorize the largest number of patents. The agents that apply for and obtain patents are usually classified into companies, individuals and government, nationals, and foreigners. It is also possible for patents registered abroad to be applied for by the head office in some cases and by subsidiaries in others.

Patenting Trends

According to available data, applications for patents grew from 2.3 million in 1994 to more than 8 million in 2001 and more than 12 million in 2004 throughout the world. Applications under the Patent Cooperation Treaty show a similar growth rate, from just over 1.1 million to 5 million during the period.

As for the fields in which patents are applied for, there are some differences between USPTO and EPO. According to the international classification, in USPTO 17% of applications fall under the category "Human Needs," which include foods, food production, and transgenics. In EPO the most important category is chemicals and metallurgy, although not much more than others. A research project with USPTO information for the past 10 years shows that the United States is the most active country, since between 1980 and 2000 the number of patents authorized more than doubled, and between 2000 and 2004 these grew 60%. Although the innovating agents that patent the most are individuals and national firms, as well as governments, foreigners are the most dynamic.

Japan, Germany, the United Kingdom, France, and Canada are the most active countries within the United States, the country that concentrates the most foreign patents. One outstanding fact is the high concentration of patenting activities in few countries. Japan is the country that patents the most in the United States, for it holds 452,737 patents, more than four times the number of the United Kingdom (101,330) and almost 20 times more than the Chinese province of Taiwan (24,646).

Latin America and the Caribbean are in a very marginal position. Mexico is in first place with position 24th in the world and 1,907 patents. Fairly far behind is Brazil (place 28th with 1,263 patents), Argentina (32nd place with 904 patents), and Venezuela (36th with 557 patents). At the corporate level, IBM of the United States occupies first place, followed by Canon and Toshiba of Japan and Samsung of the Republic of Korea.

The most dynamic fields for patenting in the United States are biotechnology and molecular microbiology, pharmaceuticals and compounds, electronics, and optic systems. In the field of biotechnology, transgenics and associated processes, genes, and gene sequences are outstanding.

Main Patenting Agents

Biotechnology	Agricultural biotechnology
Dow Chemical	Monsanto Group (5.9% of total patents)
Basf	Dupont Pioneer Group (0.2%)
Ciba Geigy	Singenta (13.8%)
Monsanto	Aventis (15.7%)
United States Government	Universities under contract with the above-mentioned groups and on their own account (56%)

Source: USPTO databases.

In transgenics and advanced biotechnology applications on plants, research with USPTO data indicated that so far, 4,609 patents have been obtained, of which 902 correspond to higher cultivable plants, seed plants, or parts of plants (Angiosperms and gymnosperms); 501 patents for transgenic soy; 423 for pathogen-resistant transgenics; and 336 for transgenics with male sterility (Table1).

Table 1 : United States: patents obtained on transgenics and new plant varieties

Category	No. of patents
Total	4,609
Higher cultivable plants, seed plants, or parts of plants (angiosperms and gymnosperms)	902
Maize	783
Soya bean	501
Pathogen-resistant transgenics	423
Transgenics with male sterility	336
Insect-resistant transgenics	316
Brassica (Canola)	187
Herbicide-resistant transgenics	176
Tobacco	173
Rice	117
Potatoes	103
Wheat	95
Cotton	89
Herbicide-resistant maize	82
Sunflower	76
Beans	27
Marrow, Zucchini	22
Cucumber	21
Others	180

Source: USPTO databases.

With regard to genes and gene sequences, the ranking is as follows:

a) United States Government

b) University of California

c) Smith Kline Beecham Incyte Pharmaceuticals Inc.

The Challenges

Latin America and the Caribbean as well as many other developing countries possess an enormous biodiversity and its native inhabitants have for centuries developed knowledge and learning on how to utilize that enormous wealth in a sustainable manner. One hectare of Latin America's tropical forest possesses more biodiversity than the entire European continent. Thanks to the work of the original peoples of the region, humanity has foods such as maize and potatoes, among many others. The region also has an important base of scientists and technicians in agricultural innovation, but it benefits little from their knowledge. Conservative estimates indicate that in Latin America and the Caribbean more than 150 million people live in the rural area, and a significant part of them in conditions of poverty and indigence.

The advantages of patent registration include:

1. Dissemination of information into the scientific community
2. Protection of financial profitability of an invention
3. Signal a firm's innovative capacity and attraction of capital.
4. Facilitate emerging markets from exchange of new technologies, e.g. university licensing to new venture start-ups

In spirit a patent affords the inventor a transitory exclusion right allowing development of an invention to a marketable product. Thus infringement of patent occurs if the same means is used in the same application. In the UK a patent takes on average three to four years to grant and if renewed each year will last for up to twenty years. An application for a patent must describe:

1. The background of the invention
2. Correctly identify the inventor(s).
3. The essential features of the invention
4. The preferred but non-essential features of the invention
5. How the invention will be used
6. The claims of the invention

The patent is usually written by a lawyer specializing in patent law and filed with the patent offices of a specified country. If the patent application successfully fulfils the legal criteria, a patent is granted. In Europe, the EU Directive on the Legal Protection of Biotechnological Inventions 98/44/EC was intended to harmonize patent law in biotechnology and genetic engineering across Europe but has instead created some confusion and uncertainty with some European member states implementing the directive while others have not. Current analysis aims to determine the variation in patents granted between the national and European patent offices in order to delineate the boundaries between ethics and legal statutes with regard to research particularly on human embryos.

More specifically in the UK, the House of Lords in 2005 made a landmark decision on the infringement of a European patent for erythropoietin. Infringement analysis involves determining the extent of the patent and more specifically what would a person skilled in the art have understood the patent to mean. Thus the level of knowledge of the inventor in the field of the patent must be sufficient that he understands the claims of the patent at the time of submission.

Protocol questions are used to aid analysis of patent infringement by a new variant:

1. Does the new variant have a material effect on the way the original invention works
2. Would the fact that the new variant had no effect on the original patent been obvious at the time of granting the original patent
3. Would experts in the field have understood that the patentee intended strict patent compliance?

A decision on when to file a patent is therefore crucial to the protection of intellectual property rights and economic potential. In Europe intellectual property rights are protected on a first to file rather than a first to invent basis. Patents filed early protect the inventor from competitors but are at risk of not providing sufficient robust experimental data to support the patent. Delaying patent submission until sufficient data has been collected risks competitors making use of the invention. In order for a patent to be filed, sufficient technical data must be included to allow reproducibility of the invention. If insufficient data is presented, the patent will be rejected as late filed experimental evidence will not be accepted. Specifically in Europe, inventions relating to therapeutic treatment must include data from experiments to show burden of proof of therapeutic efficacy for the invention.

As the economic importance of patenting has become more recognized in the scientific community, the number of patent applications has steadily increased. In the USA, trends has emerged where income from patents is greater in institutions with a larger amount of sponsored research, was funded by national institutes, were male, had a PhD or were experienced faculty members. Patents may be economically highly lucrative but unpredictable as development of an invention can prove a costly, high risk strategy. Thus in order to achieve economic rewards an institution must have a patent in a key area to draw industry interest or have a high level of sponsored research to increase the chances of earning income from patents and subsequent licensing agreements. Patents should therefore not be relied upon to generate sufficient income to reinvest into subsidizing further research. A further concern is that companies will favour investment in lucrative inventions to the detriment of diseases affecting primarily developing countries thus creating further disparity between the rich and the poor.

The IT industry and biotechnology industry have different agendas regarding patent infringement. IT companies want patent reform to prevent the larger companies buying up patents with the aim of charging other smaller companies for patent infringement or for licensing agreements. However, in biotechnology where a small change to a prior invention (e.g. inhaled insulin) can have far reaching economic consequences, patents are strongly contested as they can attract substantial investment from venture capitalists. In the USA anticipated reforms with 'first-to-file' provision would bring it into line with

Europe and Japan with the hope that eventual unification of patent offices worldwide could one day be a reality.

New advances in biotechnology raise the issue of patents on forms of life, genetic sequences and disease states. Is human biological material obtained from an identifiable source to be considered as personal data? If it is, then the use of that human material or genetic sequence to obtain a patent is considered as processing of personal data under European directives and thus subject to data protection legislation. The implication of patents on issues of morality and ethics in biotechnology also needs to be addressed. In Europe the Munich Convention states that European patents shall not be granted if the invention is contrary to law or morality. In the UK this dates back to the Statute of Monopolies of 1624 when it was recognized that as patents were granted in the name of the Crown, protecting immoral inventions could reflect unfavourably on the Crown.25 However, patents are issued by staff not trained in issues of morality which can change as society evolves with time. The scope of application of an invention is not defined by the patent which can also raise later issues on morality.

There is a major controversy regarding the possibility of patenting living organisms. This goes back many years, but the scenario has undoubtedly been complicated by the development of genetic engineering and the possibility of industrially developing products based on the use of modified living organisms. World Trade Organization (WTO), through the TRIPS Agreement, has put forward criteria on the possibility of granting intellectual protection to biotechnological innovations. According to Article 27.3(b) of TRIPS, plants and animals other than microorganisms are patentable subject-matter, as are essentially biological processes for the production of plants or animals other than non-biological and microbiological processes. However, TRIPS also stipulates that Members shall provide for the protection of plant varieties either by patents or by an effective sui generis system or by any combination thereof and that such provisions shall be reviewed 4 years after the date of entry into force of the WTO Agreement.

Article 27.3(b) is one of the most controversial in the TRIPS Agreement since on the one hand it describes patentable subject-matter, and on the other hand, it obliges Members to protect microorganisms and certain biological processes. The above reflects the strong conflict of interests between developed countries desiring to obtain protection for their biotechnological innovations, the differences between various countries on the scope of protection, and the concern of the developing countries about patents on life forms (UNCTAD-ICTSD, 2005).

Added complexity arises as research into genomic material reveals that biological information operates on multiple interconnected hierarchical levels with upstream research influencing downstream economic investment and technological development. Whilst current patent laws assume a relatively simplistic linear relationship between innovation and development, system biology consists of multiple complex levels of knowledge.

Granting broad patents for the system may thus prevent opportunistic development of medical products in a multidisciplinary setting.

The major challenges for the region in terms of intellectual property and innovation fall within this general picture. The consolidation of the new regulatory frameworks of TRIPS can increase the gap between developing and developed countries, unless the former take the necessary measures to reinforce their scientific development and protect their heritage and their creations simultaneously. The linkage of these agreements with the free trade agreements points up the need for the countries to articulate their efforts to co-ordinate both aspects, particularly biodiversity. The region has an important amount of scientific advances without intellectual protection. Many of them have been generated in institutes and public universities with the involvement of private parties. In view of the investments made and their capabilities, it is a question of subject-matters that can and should be addressed from a public-policy perspective.

Innovative efforts in the field of scientific and technological development require economic scales that the region itself can provide, for instance in agriculture and in the protection and utilization of biodiversity, where a critical mass of qualified scientists is available, whose coordination could be the object of multinational efforts.

In this regard, public policy should consider elements such as the following:

1. Creation of a regulatory framework that favors care and sustainable use of biodiversity and recognition and protection of the traditional learning and knowledge of indigenous peoples and peasants. This involves making national laws compatible with agreements between countries on biodiversity, the fight against biopiracy, and norms on bio-prospecting contracts.
2. National and regional innovation policies that make possible the sustainable use of the region's capabilities, protection of locally generated knowledge, adoption of new technologies with regard to biodiversity and the environment, and the entry of new national innovators. In keeping with the above, establish regulatory measures to prevent monopolistic or almost monopolistic control of knowledge.
3. In line with the above, establish measures to neutralize or at least counter the control of complete productive chains via genetic coding of seeds and the exclusion of small farmers.

Discussion

The focal point of research on translational projects has raised the profile of intellectual property in biotechnology. Whilst patents, copyrights and trademarks aim to protect intellectual property they should not impede scientific progress favouring larger institutions with the resources to invest in and defend them. Consciousness of commercial implications with worthwhile investment by the pharmaceutical and implant industries has also resulted in an increase in patent applications amongst the scientific community. Restriction of generic drugs due to patent protection of brand names has created a divide between the rich and poor. As surgery turns increasingly to pharmaceutical interventions and tissue engineering, the divide can only increase. At their heart, patents should still allow dissemination of knowledge within all of the scientific community and not disadvantage poorer countries (Altieri, 2003). There are criticism on the term as Richard Stallman argues that, although the term intellectual property is in wide use, it should be discarded altogether, because it "systematically distorts and confuses these issues, and its use was and is promoted by those who gain from this confusion." He claims that the term "operates as a catch-all to lump together disparate laws originated separately, evolved differently, cover different activities, have different rules, and raise different public policy issues" and that it confuses these monopolies with ownership of limited physical things. Stallman advocates referring to copyrights, patents and trademarks in the singular and warns against abstracting disparate laws into a collective term. There are also the limitation of it, some critics of intellectual property, such as those in the free culture movement, point at intellectual monopolies as harming health, preventing progress, and benefiting concentrated interests to the detriment of the masses, and argue that the public interest is harmed by ever expansive monopolies in the form of copyright extensions, software patents and business method patents.

There is also criticism because strict intellectual property rights can inhibit the flow of innovations to poor nations. Developing countries have benefitted from the spread of developed country technologies, such as the internet, mobile phone, vaccines, and high-yielding grains. Many intellectual property rights, such as patent laws, arguably go too far in protecting those who produce innovations at the expense of those who use them. The Commitment to Development Index measures donor government policies and ranks them on the "friendliness" of their intellectual property rights to the developing world. Some libertarian critics of intellectual property have argued that allowing property rights in ideas and information creates artificial scarcity and infringes on the right to own tangible property (Resnik, 2003).

Imagine the time when men lived in caves. One bright guy let's call him Galt-Magnon decides to build a log cabin on an open field, near his crops. To be sure, this is a good idea, and others notice it. They naturally imitate Galt-Magnon, and they start building their own cabins. But the first man to invent a house, according to IP advocates, would have a right to prevent others from building houses on their own land, with their own logs, or to charge them a fee if they do build houses. It is plain that the innovator in these examples becomes a partial owner of the tangible property (e.g., land and logs) of others, due not to first occupation and use of that property (for it is already owned), but due to his coming up with an idea. Clearly, this rule flies in the face of the first-user homesteading rule, arbitrarily and groundlessly overriding the very homesteading rule that is at the foundation of all property rights (King and Eyzaguirre, 1999).

Other criticism of intellectual property law concerns the tendency of the protections of intellectual property to expand, both in duration and in scope. The trend has been toward longer copyright protection (raising fears that it may someday be eternal). In addition, the developers and controllers of items of intellectual property have sought to bring more items under the protection. Patents have been granted for living organisms, and colors have been trademarked. Because they are systems of government-granted monopolies copyrights, patents, and trademarks are called intellectual monopoly privileges, (IMP) a topic on which several academics, including Birgitte Andersen and Thomas Alured Faunce have written. In 2005 the RSA launched the Adelphi Charter, aimed at creating an international policy statement to frame how governments should make balanced intellectual property law.

Conclusions

Western outlook and valuations have largely twisted the current global system governing intellectual property. This is perfectly portrayed in the international concord that establishes enforceable minimum standards on intellectual property for the first time. As biotechnology is a new, profitable technology, easily copied, patents appear to be the only viable form of assuring control and ownership of biotechnological innovations using life forms or material found in nature (Stiglitz, 2006). In Indian context, Ganguli (1998) all the way through paid attention on patent where a detail Indian scenario has been discussed. Patents were not needed for biological materials at the time when there were more efficient control mechanisms, namely under colonial rule and control of production. They became necessary with the rise of political independence in the developing world, the decline of plantation economies, and the growing importance of plant breeding. As Fowler states, "the mode of control has simply shifted from physical to legal" (Fowler, 1995).

As the conflict intensifies over questions of ownership and control of biological materials, IPRs are at the center of controversies. In a great number of cases, the interests of the commercial users of the system dominate in the evolution of intellectual property rights' policy. Thus "policy tends to be determined more by the interests of a few, than by an impartial conception of the greater public good" (Commission on Intellectual Property Rights, 2002). As long as there is an obvious bias in favor of the Western, developed world a - yet again - new form of colonialism cannot be considered out of the question. Looking at the history of the last 500 years of trade, conquest, slavery, colonialism, neo-colonialism, and globalizing arrogance, all led initially by Europeans, one gets even more tempted to hold this view. Shiva is definitely not standing alone when writing, "it seems that the Western powers are still driven by the colonization impulse to discover, conquer, own and possess everything, every society, every culture" (Shiva, 1997). As the developed world enjoys the benefits of investments made on the basis of biological material obtained in the South, the countries of origin of genetic resources can no longer be expected to freely donate them for the enrichment of others and the improvement of northern crops. The developing world has started to develop its own control mechanisms: the CBD, recognizing national sovereignty over a country's genetic resources, can be a useful tool in that respect (Amanda and Pablo, 1999; Webber, 2007).

In that regard, "patent systems which recognize the contributions of some but not of others are based on injustice and will be unstable, if not unsustainable" (Fowler, 1995). Steps are needed to protect and compensate for forms of innovation currently not covered by the traditional IPR systems. Humanity's ability to conserve genetic resources, have access to them, and ensure food security, ultimately depends on the full implementation of the principle of prior informed consent, the development of equitable benefit sharing systems, and the recognition of farmers' rights, elements that highlight the links between justice, environmental conservation, and sustainable development. To achieve these goals, the patent system needs either total re-conceptualization, or restructuring through benefit-sharing systems to address and redress some of its inequalities.

References

Altieri, M. A.(2003) The Myths of Agricultural Biotechnology: Some Ethical Questions, Discussion paper (University of California, Berkeley, 1998), http://www.cnr.berkeley. edu/˜agroeco3/the_myths.html (retrieved on 19/2/2003).

Amanda B. King and Pablo B.(1999) Eyzaguirre Intellectual property rights and agricultural biodiversity: Literature addressing the suitability of IPR for the protection of indigenous resources Agriculture and Human Values 16: 41-49.

ANAPQUI, RAFI, the Genetics Forum, No Patents on Life, Global 2000, Steve Emmot, Council for Responsible Genetics.(1997). Submission to the International Peoples' Tribunal on Human Rights and the Environment, June 22-23. Available:http://www.rafi.ca.quinoa/tribunal.htm

Azoulay P, Michigan R and Sampat BN.(2007) The anatomy of medical school patenting. The New England Journal of Medicine; 357:20:2049-56.

Cimoli, M. (2002) "Public policy and technological innovation in Latin America: An evolutionary perspective", presented at the seminar "Políticas de Innovación", Brazil, Universidad Federal de Río de Janeiro.

Copyright Act of 1911 (c. 46) Chapter 46_1&2_Geo5

Coriat, B. and F. Orsi (2001) The Installation in the US of a New Regime of Intellectual Property Rights; Origins, Contents, Problems, París 13 University, June.

Correa, C. M. (2000). Integrating public health concerns into patent legislation in developing countries. Geneva:South Centre.

Correa, C. M. (2002). Protection of traditional systems of medicine, patenting and promotion of medicinal plants. In R. R. Chaudhury, & U. M. Rafei (Eds.), Traditional medicine in Asia (pp. 227-245). New Delhi:WHO South- East Asia Regional Office.

Correa, C. M. (2003) Options for the Implementation of Farmers' Rights at the National Level, Working paper (South Centre, 2000), http://www.southcentre.org/publications/ farmersrights/ toc.htm#TopOfPage (retrieved on 2/3/2003).

Correa, C.M. (2001). Traditional knowledge and intellectual property: Issues and options surrounding the protection of traditional knowledge. Geneva: The Quaker United Nations Office.

Crespi, R. S. (2000) "An Analysis of Moral Issues Affecting Patenting Inventions in the Life Sciences: A European Perspective," Science and Engineering Ethics 6(2), 157-180.

Dbar, B. (2002). Sui generis Systems for Plant Variety Protection, Discussion paper (Quaker United Nations Office, Geneva, 2002).

Dreyfuss, R. (1989). 'General Overview of the Intellectual Property System', in V. Weil and J. Snapper (1989)

Dunwell, J.M. (2010) Transgenic Crop Plants, (eds.), C. Kole et al.,DOI 10.1007/978-3-642-04812-8_12, # Springer-Verlag Berlin Heidelberg.

ECLAC (Economic Commission for Latin America and the Caribbean). (2005). "Science and technology policies in open economies: The case of Latin America and the Caribbean".

EFB (European Federation of Biotechnology), Bioethics and Public Perceptions on Biotechnology, Advanced Workshop (Oxford, 2001).

Ehrlich, P. R. and A. H. Ehrlich (1981). Extinction: The Causes and Consequences of the Disappearance of Species (Ballantine, New York).

Elliott, G. (2007). Basics of US patents and the patent system. The AAPS Journal; 9:3:E317-E324.

Feinman, J. (2000). Law 101 (Oxford University Press,New York).

Fowler, C. (1995). "Biotechnology, Patents and the Third World," in V. Shiva and I. Moser (eds.), Biopolitics: a Feminist and Ecological Reader on Biotechnology (Zed Books/Third World Network, London/ Penang).

Ganguli P. (1998). Gearing up for Patents ± The Indian Scenario. Hyderabad: Universities Press (India).

Honore, A. (1977). 'Ownership', in P. Smith (ed.), The Nature and Process of Law (Oxford University Press, New York), pp. 370-375.

Hunt, R. (1999). What Happened in the 1980's, Branscomb, Kodama and Florida.

Iañez Pareja, E. (2001). Patentes y Biotecnología, Departamento de Microbiología e Instituto de IICA (Inter-American Institute for Cooperation on Agriculture), "Situación global de.

India. (1999). Discussion Paper on Review of Article 27.3(b) submitted to the TRIPs Council in Geneva. Intelectual en la América del Sur, Brasilia, D. F.

Katz, J. (2005).”Ciencia, Tecnología e Innovación, la Cooperación Regional y las Políticas Nacionales”, regional scientific conference in Cuba, UNESCO, Havana, 30 November.

Lichtenberg, E. (2000). “Costs of regulating genetically modified ... plants”, National Research Council, Genetically Modified Pest-Protected Plants: Science and Regulation, Washington, DC, National Academy Press, University of Maryland.

Mae-Wan Ho and Traavik T. (1999). Why We Should Reject Biotech Patents from TRIPS. ScientificBriefing on TRIPS Article 27.3(b) (1999), http://www.i-sis.org.uk/trips99.php (retrieved

Marris, E. (2005). Patent reform prompts intellectual tug-ofwar. Nature 437:1230-31.

Mooney, P. (2000). "The ETC Century", Development Dialogue 1999, 1-2, y RAFI 2000. Paris Convention for the Protection of Industrial Property (1883).

Murashige, K. (2002). "Patents and research-an uneasy alliance," Academic Medicine 77(Part 2), 1329-1338.

Murphy, A. (2002). Biopiracy, an Emerging Human Rights Issue (2000), http://www.law.ecel.uwa.edu.au/ elawjournal/Volume%201/Articles%20Vol_1/biopiracy.pdf (retrieved on 24/11/2002).

Pasteur, L. (1873) US Patent 141072.

Platt, L. (2001) "Securitization of biotechnology research patents", Convergence: The Biotechnology Industry Report, Millennium Edition, Ernst & Young.

Pushpagandhan P. (1996). In: Swaminathan MS, editors. Agrobiodiversity and Farmers Rights. Konark Publishers.

Resnik, D.B. (2003). A Pluralistic Account of Intellectual Property Journal of Business Ethics 46: 319-335.

Rural Advancement Foundation International, Biopiracy+10 (Rafi Communique, 2002).

Rural Advancement Foundation International, The Patenting of Human Genetic Material (Rafi Communique, 1994), http://www.rafi.org/communique/fltxt/19941.html.

Saunders, P. (1999). Ethical Aspects of Biotechnology and Genetically Modified Organisms, http://www.i-sis.org.uk/response.php (retrieved on 16/12/2002).

Schrell, A. et al. (2007). Biotechnology patenting policy in the European Union - as exempli_ ed by the development in Germany. Adv Biochem Engin/Biotechnol;107:13-39.

Shiva, V. (1997). Biopiracy: The Plunder of Nature and Knowledge (South End Press, Boston MA).

Somberg, J. (2005). Intellectual Property. American Journal of Therapeutics;12:285.

Statute of Anne 1709, Vol 3, Article 1, Section 8, Clause 8, Document 2. The Founder's Constitution. Published University of Chicago 1987.

Stiglitz, J.E. (2006). Editorials. Scrooge and intellectual property rights. BMJ;333:1279-80.

Tansey, G. (1999). Trade, Intellectual Property, Food and Biodiversity, Discussion paper (Quaker United Nations Office, Geneva).

TRIPS .(2006).Agreement on Trade-Related Aspects of Intellectual Property Rights.

UK Copyright Law.(2006). www.coyrightservice. co.uk.

UNCTAD and ICTSD.(2005). Resource Book on TRIPS and Development, Project on IPRs and Sustainable Development, Cambridge University Press, United States.

UNDP (United Nations Development Programme) (2000), Informe de Desarrollo Humano y Globalización.

UPOV (International Union for the Protection of New Plant Varieties) (www.upov.int).

US Food and Drug Adminstration. www.fda.gov

Visser B., D. Eaton D., N. Louwaars N., I. Van der Meer I, J. Beekwilder, and F. Van Tongeren (2002) Potential Impacts of Genetic Use Restriction Technologies (GURTS) on Agrobiodiversity and Agricultural Production Systems. FAO Background Study Paper No. 17.

Warner, K. D. (2001). "Are Life Patents Ethical? Conflict between Catholic Social Teaching and Agricultural Biotechnology's Patent Regime," Journal of Agricultural and Environmental Ethics 14, 301-319.

Webber, P.M. (2007). The patenting of novel biotech products. I Drugs;10:7:459-62.

WIPO (World Intellectual Property Organization). (2006). "Arreglo de Madrid" [on line] (www.wipo.int).

WIPO (World Intellectual Property Organization). (2006). "Tratado de Cooperación sobre Patentes" [on line] (http://www.wipo.int).

WIPO. (2001). Matters Concerning Intellectual Property and Genetic Resources, Traditional Knowledge and Folklore, Report (Geneva, 2001).

❑❑❑

CHAPTER – 17

Animal Tissue Culture Principle andProcedure Limitation

Renu Bist and Anupama Sharma
Department of Bioscience and Biotechnology, Banasthali University, Rajasthan-304022,India
Email : renu_bisht22@yahoo.co.in

Summary

Animal tissue culture is a biological technique in which a piece of tissue is taken from an animal specimen and cells from this tissue are grown and maintained in an in vitro or artificial environment (Harrison, 1907). Aim of cell / tissue culture is actually to make cells as happy in the in vitro conditions, as they are in their natural environment. The objective of this chapter is to highlight those principles and procedures that have particular relevance to the use of cell culture in tissue engineering.

1. Introduction

Tissue culture is used in numerous biological research applications and is a critical technique in biotechnology which may lead to further discoveries in human or animal health, genetics, disease diagnosis and its prevention and treatment. Animal tissue culture helps to learn more about cell and organ structure and their functions, diagnose viruses, learn the effects and toxicity of drugs, develop treatments to organ failure and much more. Products developed from cell culture include vaccines, hormones, enzymes and anticancer drugs. Animal tissue collection can help diagnosis of an illness, before it spreads.

2. History

Animal cell culture became a common laboratory technique in the mid 1900s, when the 19th-century English physiologist Sydney Ringer developed salt solutions containing the chlorides of sodium, potassium, calcium and magnesium suitable for maintaining the beating of an isolated animal heart outside of the body (Bodnar et al. 1998). The concept of maintaining live cell lines separated from their original tissue source was discovered in the 19th century (Witkowski, 1983) In 1885 Wilhelm Roux removed a portion of the medullary plate of an embryonic chicken and maintained it in a warm saline solution for several days, establishing the principle of tissue culture (Baker, 1987). Ross Granville Harrison, working at Johns Hopkins Medical School and then at Yale University, published results of his experiments from 1907–1910, establishing the methodology of tissue culture (Schiff and Judith, 2006).

Cell culture techniques were advanced significantly in the 1940s and 1950s to support research in virology. Growing viruses in cell cultures allowed preparation of purified viruses for the manufacture of vaccines. The injectable polio vaccine developed by Jonas Salk was one of the first products mass-produced using cell culture techniques. This vaccine was made possible by the cell culture research of John Franklin Enders, Thomas Huckle Weller, and Frederick Chapman Robbins, who were awarded a Nobel Prize for their discovery of a method of growing the virus in monkey kidney cell cultures (Alberts et al. 2008).

3. Salient Features of Animal Cell Culture

a) Animal cells can grow in simple glass or plastic containers in nutritive media but they grow only to limited generations.

b) Animal cells exhibit contact inhibition. In culture the cancer cells apparently differ from the normal cells. Due to uncontrolled growth and more rounded shape, they loose contact inhibition and pile over each other.

c) There is a difference in the in vitro and in vivo growth pattern of cells. For example:

   (i) Absence of cell-cell interaction and cell matrix interaction,

   (ii) Lack of three-dimensional architectural appearance, and

   (iii) Changed hormonal and nutritional environment. They provide a way of adherence to glass or plastic container in which they grow, cell proliferation and shape of cell results in alterations.

d) Maintenance of growth of cells under laboratory conditions in suitable culture medium is known as Primary cell culture.

e) Cells are dissociated from tissues by mechanical means and by enzymatic digestion using proteolytic enzymes.

f) Cells can grow as adherent cells (anchorage dependent) or as suspension cultures (anchorage independent).

g) The primary culture is sub cultured in fresh media to establish Secondary cultures.

h) The various types of cell lines are categorized into two types as Finite cell line and Continuous cell line.

Finite cell lines are those cell lines which have a limited life span and grow through a limited number of cell generations. The cells normally divide 20 to 100 times (i.e. is 20-100 population doublings) before extinction. Cell lines transformed under in vitro conditions give rise to continuous cell lines. The continuous cell lines are transformed, immortal and tumorigenic.

i) The physical environment includes the optimum pH, temperature, osmolality and gaseous environment, supporting surface and protecting the cells from chemical, physical, and mechanical stresses.

j) Nutrient media is the mixture of inorganic salts and other nutrients capable of sustaining cell survival in vitro

k) Serum is essential for animal cell culture and contains growth factors which promote cell proliferation. It is obtained as exuded liquid from blood undergoing coagulation and filtered using Millipore filters.

l) Cryopreservation is storing of cells at very low temperature (-180°C to -196 °C) using liquid nitrogen. DMSO is a cryopreservative molecule which prevents damage to cells.

m) In order to maintain the aseptic conditions in a cell culture, a LAF hood is used. Based on the nature of cells and organism the tissue culture hoods are grouped into three types: Class I, Class II, and Class III.

n) CO_2 incubators are used and designed to mimic the environmental conditions of the living cells.

o) An inverted microscope is used for visualizing cell cultures *in situ*.

p) For most animal cell cultures low speed centrifuges are needed.

q) Neuronal cells constitute the nervous system. In culture the neuronal cells cannot divide and grow.

r) The cells that form connective tissue (skin) is called fibroblast. The fibroblast can divide and grow in culture to some generations after which they die. All normal animal cells are mortal.

s) Organ culture- The culture of native tissue that retains most of the in vivo histological features is regarded as organ culture.

t) Histotypic culture- The culturing of the cells for their reaggregation to form a tissue-like structure represents histotypic culture.

u) Organotypic culture- This culture technique involves the recombination of different cell types to form a more defined tissue or an organ.

4. Requirements for Animal Cell Culture

Among the essential requirements for animal cell culture are special incubators to maintain the levels of oxygen, carbon dioxide, temperature, humidity as present in the animal's body. The synthetic media contains vitamins, amino acids and fetal calf serum. The parameters which are essential for successful animal cell culture are as follows:

4.1. Temperature

In most of the mammalian cell cultures, the temperature is maintained at 37ºC in the incubators as the body temperature of *Homo sapiens* is 37ºC.

4.2. Culture Media

The culture media is prepared in such a way that it provides-

4.2.1. The optimum conditions of factors like pH, osmotic pressure, etc.

4.2.2. Nutrients

It should contain chemical constituents which the cells or tissues are incapable of synthesizing. Generally the media is the mixture of inorganic salts and other nutrients capable of sustaining cells in culture such as amino acids, fatty acids, sugars, ions, trace elements, vitamins, cofactors, and ions. Glucose is added as energy source- the concentration of glucose varies according to the requirement. Phenol Red is added as a pH indicator of the medium. There are two types of media used for culture of animal cells and tissues- the natural media and the synthesized media.

4.2.3. Natural media

The natural media are the natural sources of nutrient sufficient for growth and proliferation of animal cells and tissues. The Natural Media used to promote cell growth fall in three categories.

a) Coagulant, such as plasma clots. It is now commercially available in the form of liquid plasma kept in silicon ampoules or lyophilized plasma. Plasma can also be prepared in the laboratory taking out blood from male fowl and adding heparin to prevent blood coagulation.

b) Biological fluids such as serum. Serum is one of the very important components of animal cell culture which is the source of various amino acids, hormones, lipids, vitamins, polyamines, and salts containing ions such as calcium, ferrous, ferric, potassium etc. It also contains the growth factors which promotes cell proliferation, cell attachment and adhesion factors. Serum is obtained from human adult blood, placental, cord blood, horse blood, and calf blood. The other forms of biological fluids used are coconut water, amniotic fluid, pleural fluid, insect haemolymph serum, culture filtrate, aqueous humour, from eyes etc.

Advantages of serum in culture medium are: -

(i) Serum binds and neutralizes toxins.

(ii) Serum contains a complete set of essential growth factors, hormones, attachment and spreading factors, binding and transport proteins.

(iii) It contains the protease inhibitors.

(iv) It increases the buffering capacity.

(v) It provides trace elements.

Disadvantages of serum in culture medium are:

(i) It is not chemically defined and therefore its composition varies a lot.

(ii) It is sometimes source of contamination by viruses, mycoplasma and prions etc.

(iii) It increases the difficulties and cost of down stream processing.

(iv) It is the most expensive component of the culture medium.

4.2.4. Synthetic media

Syntheic media are prepared artificially by adding several organic and inorganic nutrients, vitamins, salts, serum proteins, carbohydrates and cofactors etc. Different types of synthetic media can be prepared for a variety of cells and tissues to be cultured. Synthetic media are of two types- Serum containing media (media containing serum) and serum- free media (media with out serum). Examples of some media are: minimal essential medium (MEM), RPMI 1640 medium, CMRL 1066 and F12 etc.

4.3. Tissue Extracts for Example Embryo Extracts

Extracts from tissues such as embryo, liver, spleen, leukocytes, tumor, bone marrow are also used for culture of animal cells.

4.4. pH

Most media maintain the pH between 7 and 7.4. A pH below 6.8 inhibits cell growth. The optimum pH is essential to maintain the proper ion balance, optimal functioning of cellular enzymes and binding of hormones and growth factors to cell surface receptors in the cell cultures. The regulation of pH is done using a variety of buffering systems. Most media use a bicarbonate-CO_2 system as its major component.

4.5. Osmolality

A change in osmolality can affect cell growth and function. Salt, Glucose and Amino acids in the growth media determine the osmolality of the medium. All commercial media are formulated in such a way that their final osmolality is around 300 m/Osm.

5. Types of Cell Cultures

5.1. Primary Cell Culture

The maintenance of growth of cells dissociated from the parental tissue (such as kidney, liver) using the mechanical or enzymatic methods, in culture medium using suitable glass or plastic containers are called Primary Cell Culture (Nicosia and Ottinetti, 1990, Fig.1).

The primary cell culture could be of two types depending upon the kind of cells in culture.

a) Anchorage Dependent /Adherent cells- Cells shown to require attachment for growth are set to be Anchorage Dependent cells. The Adherent cells are usually derived from tissues of organs such as kidney where they are immobile and embedded in connective tissue. They grow adhering to the cell culture (Pretlow and Pretlow 1989). b) Suspension Culture/Anchorage Independent cells - Cells which do not require attachment for growth or do not attach to the surface of the culture vessels are anchorage independent cells/suspension cells. All suspension cultures are derived from cells of the blood system because these cells are also suspended in plasma in vitro e.g. lymphocytes.

5.2. Secondary Cell Cultures

When a primary culture is sub-cultured, it becomes known as secondary culture or cell line. Subculture (or passage) refers to the transfer of cells from one culture vessel to another culture vessel.

Sub culturing- Sub culturing or splitting cells is required to periodically provide fresh nutrients and growing space for continuously growing cell lines. The process involves removing the growth media, washing the plate, disassociating the adhered cells, usually enzymatically. Such cultures may be called secondary cultures (Maas-Szabowski, 2002).

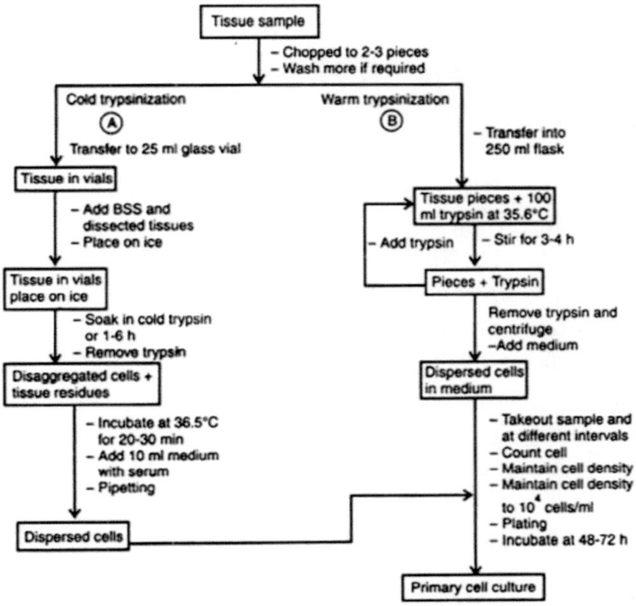

Fig.1 : Establishment of Primary cell culture.

Types of Tissue Culture

The term "tissue culture" is often used interchangeably with "cell culture," the only difference being that cell cultures are often derived from a primary tissue culture source, which originated directly from tissue. "Primary" cell lines have a limited lifespan whereas "immortal" or established cell lines can proliferate indefinitely, table 1.

Table 1 : Types of Cultures (Organ culture, explant culture and cell culture) showing different features.

Category	Organ Culture	Explant Culture	Cell Culture
Source	Embryonic organs, adult tissue fragments	Tissue Fragments	Disaggregated tissue, primary culture, propagated cell line
Effort	High	Moderate	Low
Characterization	Easy, Histology	Cytology and markers	Biochemical, molecular, immunological and cytological assays
Histology	Informative	Difficult	Not applicable
Biochemical differentiation	Possible	Heterogenous	Lost, but may be reinduced
Quantitation	Difficult	Difficult	Easy

6. Cell Line

A Cell Line or Cell Strain may be finite or continuous depending upon whether it has limited culture life span or it is immortal in culture. On the basis of the life span of culture, the cell lines are categorized into two types:

a) Finite cell Lines - The cell lines which have a limited life span and go through a limited number of cell generations (usually 20-80 population doublings) are known as Finite cell lines. These cell lines exhibit the property of contact inhibition, density limitation and anchorage dependence. The growth rate is slow and doubling time is around 24-96 hours. Senescence is determined by a number of intrinsic factors regulating cell cycle, such as Rb and p53 (Munger and Howley, 2002), and is accompanied by shortening of the telomeres on the chromosomes (Wright and Shay, 2002). Once the telomeres reach a critical minimum length, the cell can no longer divide. Telomere length is maintained by telomerase.

b) Continuous Cell Lines - Cell lines transformed under laboratory conditions or in vitro culture conditions give rise to continuous cell lines. The cell lines show the property of ploidy (aneupliody or heteroploidy), absence of contact inhibition and anchorage dependence. They grow in monolayer or suspension form. The growth rate is rapid and doubling time is 12-24 hours.

c) Monolayer cultures - When the bottom of the culture vessel is covered with a continuous layer of cells, usually one cell in thickness, they are referred to as monolayer cultures.

d) Suspension cultures - Majority of continuous cell lines grow as monolayers. Some of the cells which are non-adhesive e.g. cells of leukemia or certain cells which can be mechanically kept in suspension, can be propagated in suspension. There are certain advantages in propagation of cells by suspension culture method.

The advantages are: (a) The process of propagation is much faster.,(b) The frequent replacement of the medium is not required.,(c) Suspension cultures have a short lag period,(d) Treatment with trypsin is not required, (e) Homogenous suspension of cells is obtained, (f) The maintenance of suspension cultures is easy and bulk production of the cells is easily achieved.(g) Scale-up is also very convenient.

6.1. The Cell Lines are Known by

a) A code e.g. NHB for Normal Human Brain.

b) A cell line number- This is applicable when several cell lines are derived from the same cell culture source e.g. NHB1, NHB2.

c) Number of population doublings, the cell line has already undergone e.g. NHB2/2 means two doublings, Fig. 2.

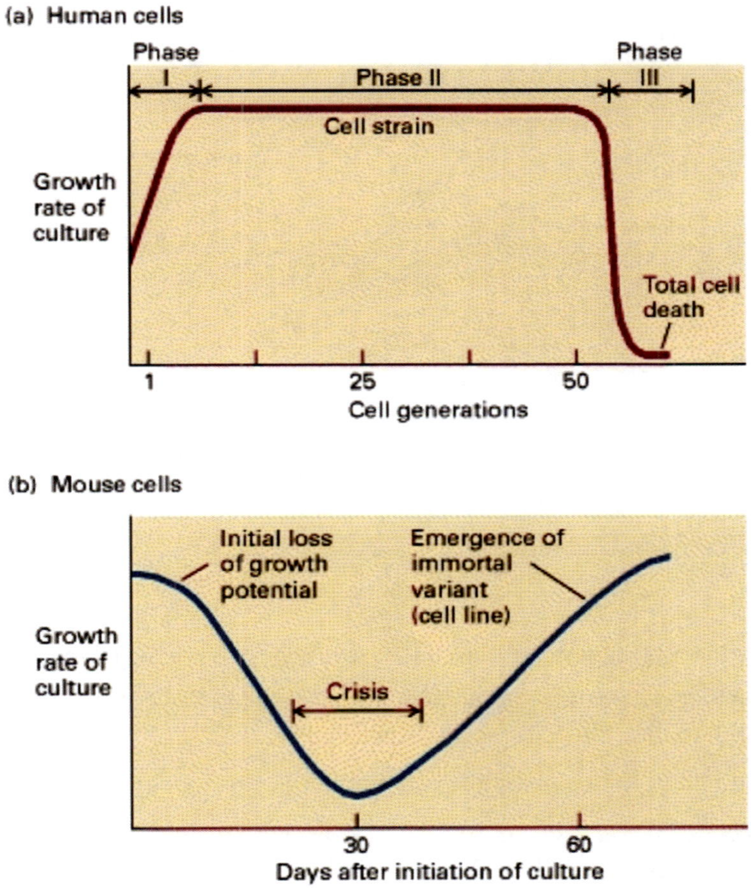

Fig. 2 : Stages in the establishment of a cell culture. (Lodish et al. 2000)

6.2. Characterization of Cell lines (Westermark and Wasteson, 1975 Davis, 1994)

The cell lines are characterized by their

a) Growth rate

b) Karyotyping.

a) Growth Rate - A growth curve of a particular cell line is established taking into consideration the population doubling time, a lag time, and a saturation density of a particular cell line. A growth curve consists of:

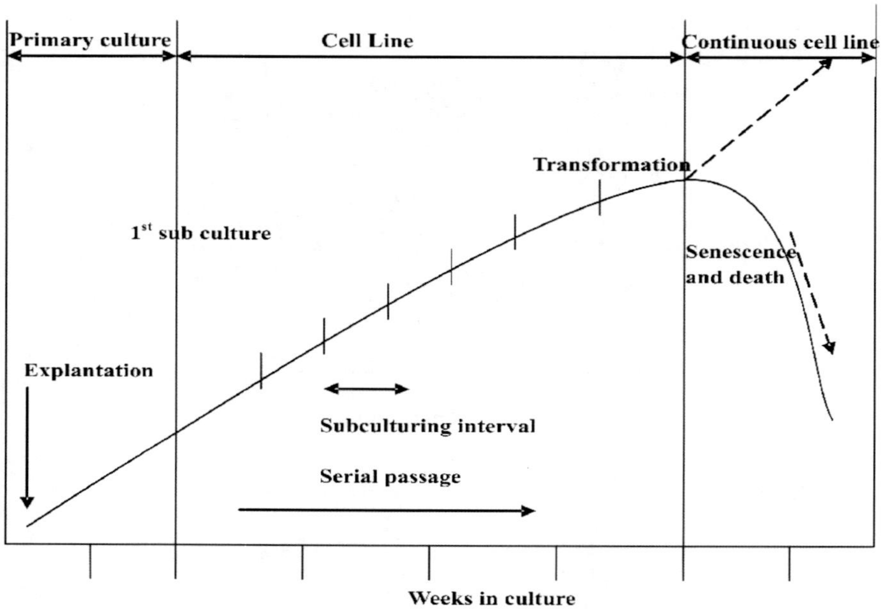

Fig. 3 : Salient features of cell culture with evolution of a cell line (Freshney, 2005)

1) Lag Phase: The time the cell population takes to recover from such sub culture, attach to the culture vessel and spread.

2) Log Phase: In this phase the cell number begins to increase exponentially.

3) Plateau Phase : During this phase, the growth rate slows or stops due to exhaustion of growth medium or confluence, Fig. 3.

b) Karyotyping - Karyotyping is important as it determines the species of origin and determine the extent of gross chromosomal changes in the line. The cell lines with abnormal karyotype are also used if they continue to perform normal function. Karyotype is affected by the growth conditions used, the way in which the cells are sub cultured and whether or not the cells are frozen.

c) There are certain terms that are associated with the cell lines.

These are as follows: (i) Split ratio- The divisor of the dilution ratio of a cell culture at subculture. (ii) Passage number- It is the number of times that the culture has been cultured. (iii) Generation number- It refers to the number of doublings that a cell population has undergone.

7. Animal Tissue Culture Techniques

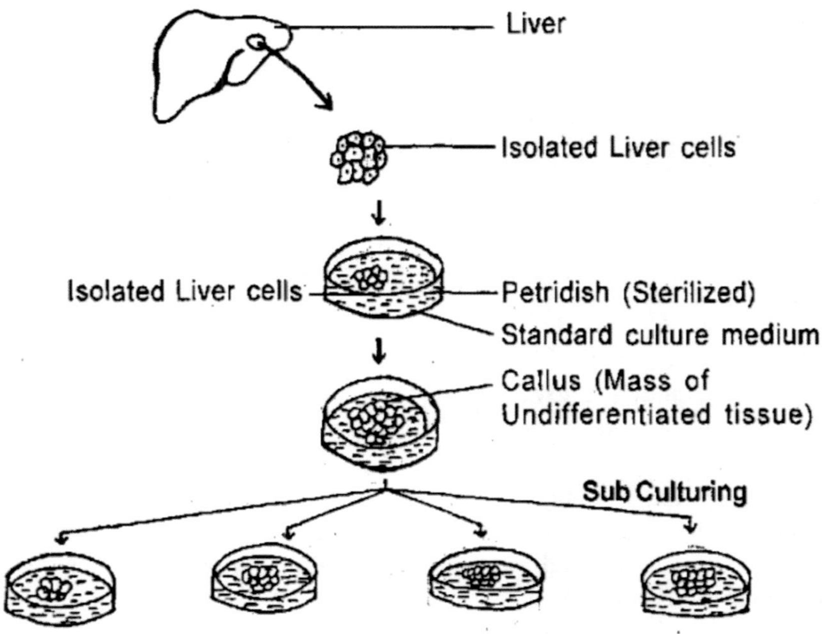

Fig. 4 : Method of Animal cell Culture

7.1. First Step: Isolation/Collection of Cells

Cells can be isolated from tissues for *ex vivo* culture in several ways. Cells can be easily purified from blood; however only the white cells are capable of growth in culture. Mononuclear cells can be released from soft tissues by enzymatic digestion with enzymes such as collagenase, trypsin, or pronase, which break down the extracellular matrix. Alternatively, pieces of tissue can be placed in growth media, and the cells that grow out are available for culture. This method is known as explant culture (Caputo 1996).

Cells that are cultured directly from a subject are known as primary cells. With the exception of some derived from tumors, most primary cell cultures have limited lifespan. After a certain number of population doublings (called the Hayflick limit) cells undergo the process of senescence and stop dividing, while generally retaining viability.

An established or immortalized cell line has acquired the ability to proliferate indefinitely either through random mutation or deliberate modification, such as artificial expression of the telomerase gene. There are numerous well established cell lines representative of particular cell types, Fig. 4.

7.1.1. Blood specimens

Most animal tissue cultures are taken through the technique of blood drawing. This method is completed by inserting a hollow needle into the vein of the animal and attaching a vacuum tube to the needle, drawing the blood out of the vein. This method is slightly painful and can cause an animal to spook or draw back, but it is overall harmless and minimally invasive. The tissue or blood that is drawn into the vials can then be tested for viruses and virus antibodies. Bacteria can also be diagnosed through blood work and various illnesses can be diagnosed in part with the changes in blood gas and content levels.

7.1.2. Biopsy

Biopsies allow the medical and scientific teams working with the animal to collect a larger sample of tissue for study. This tissue can come from any portion of the animal. The most common method of obtaining a biopsy sample is the punch method. This method removes a small amount of skin or other tissue by creating a small hole in the tissue, collecting what was there previously. This method can be painful, and most animals that undergo a punch-style biopsy are sedated to prevent trauma. A needle biopsy can also be used, which the tissue is drawn into a hollow needle and placed in a sterile tube. This method is frequently used if the area being sampled is small or if the tissue is difficult to reach with a biopsy punch. Both biopsy styles can be painful and invasive.

7.1.3. Swabbing

Swabbing is a common method of non-invasively collecting animal tissue cultures. This technique involves wetting a swab with a sterile solution and wiping the mouth or skin of the animal with the swab. That swab should then be allowed to dry in a sterile environment and can then be tested for DNA or genome testing. Oral or skin infections can also be detected using this technique of tissue culture collection.

7.2. Maintaining Cells in Culture

Cells are grown and maintained at an appropriate temperature and gas mixture (typically, 37°C, 5% CO_2 for mammalian cells) in a cell incubator. Culture conditions vary widely for each cell type and variation of conditions for a particular cell type can result in different phenotypes being expressed (Freshney, 2005).

Aside from temperature and gas mixture, the most commonly varied factor in culture systems is the growth medium. Recipes for growth media can vary in pH, glucose concentration, growth factors, and the presence of other nutrients. The growth factors used to supplement media are often derived from animal blood, such as calf serum. One complication of these blood-derived ingredients is the potential for contamination of the culture with viruses or prions, particularly in biotechnology medical applications. Current practice is to minimize or eliminate the use of these ingredients wherever possible and use chemically defined media, but this cannot always be accomplished. Alternative strategies involve sourcing the animal blood from countries with minimum BSE/TSE risk such as Australia and New Zealand, and using purified nutrient concentrates derived from serum in place of whole animal serum for cell culture.

Plating density (number of cells per volume of culture medium) plays a critical role for some cell types. For example, a lower plating density makes granulosa cells exhibit estrogen production, while a higher plating density makes them appear as progesterone producing theca lutein cells. (Portela et al. 2010).

Cells can be grown in suspension or adherent cultures. Some cells naturally live in suspension, without being attached to a surface, such as cells that exist in the bloodstream. There are also cell lines that have been modified to be able to survive in suspension cultures so that they can be grown to a higher density than adherent conditions would allow. Adherent cells require a surface, such as tissue culture plastic or micro carrier, which may be coated with extracellular matrix components to increase adhesion properties and provide other signals needed for growth and differentiation. Most cells derived from solid tissues are adherent. Another type of adherent culture is *organotypic culture* which involves growing cells in a three-dimensional environment as opposed to two-dimensional culture dishes. This 3D culture system is biochemically and physiologically more similar to *in vivo* tissue, but is technically challenging to maintain because of many factors (e.g. diffusion) (Atala, 2004).

8. Medicines from Animal Cell Culture

Medicines from Animal Cell Culture focuses on the use of animal cell culture, which has been used to produce human and veterinary vaccines, interferons, monoclonal antibodies and genetically engineered products such as tPA and erythropoietin. It also addresses the recent dramatic expansion in cell-based therapies, including the use of live cells for tissue regeneration and the culture of stem cells.

9. Applications of Animal Cell Culture

The animal cell cultures are used for a diverse range in research and development. These areas mainly include:

a) The production of antiviral vaccines, which requires the standardization of cell lines for the multiplication and assay of viruses.

b) Cancer research, which requires the study of uncontrolled cell division in cultures.

c) Cell fusion techniques.

d) Genetic manipulation, which is easy to carry out in cells or organ cultures.

e) Production of monoclonal antibodies requires cell lines in culture, table 2.

f) Production of pharmaceutical drugs using cell lines.

g) Chromosome analysis of cells derived from womb.

h) Study of the effects of toxins and pollutants using cell lines.

i) Use of artificial skin.

j) Study the function of the nerve cells.

Table 2 : Some animal cell lines and the products obtained from them

Cell line	Product
Human tumour	Angiogenic factor
Human leucocytes	Interferon
Mouse fibroblasts	Interferon
Human Kidney	Urokinase
Transformed human kidney cell line, TCL-598	Single chain urokinase-type plasminogen activator (scu-PA)
Human kidney cell (293)	Human protein (HPC)
Dog kidney	Canine distemper vaccine
Cow kidney	Foot and Mouth disease (FMD) vaccine
Chick embryo fluid	Vaccines for influenza, measles and mumps
Duck embryo fluid	Vaccines for rabies and rubella
Chinese hamster ovary (CHO) cells	1. Tissue-type plasminogen activator (t-PA)
	2. B-and gamma interferons
	3. Factor VIII

10. Advantages and Limitations of the use of Animal Tissue Culture (Freshney, 2002):

10.1. Advantages

1. The behaviour of cells in culture is easily interpreted and regulated properly.
2. Controlled physiochemical environment (pH, temperature, osmotic pressure, O_2, CO_2, etc.)
3. Controlled and defined physiological conditions (constitution of medium, etc.)
4. Homogeneity of cell types (achieved through serial passages)
5. Economical, since smaller quantities of reagents are needed than *in vivo.*

6. Legal moral and ethical questions of animal experimentation are avoided.

10.2. Disadvantages

1. Expertise is needed, so that behaviour of cells in culture can be interpreted and regulated.
2. Ten times more expensive for same quantity of animal tissue; therefore reasons for its use should be compelling.
3. Unstable aneuploid chromosome constitution.

References

Alberts, B., Johnson, A., Lewis, J., Raff, M., Roberts, K., Walter, P. (2008). Molecular biology of the Cell. Newyork: Garland Science. Classic textbook of cell and molecular biology.

Atala, A. (2004). Tissue engineering and regenerative medicine: concepts for clinical application. Rejuvenation Res. 7:15-31.

Baker, F. (1987). Diffusion of a new methodology. Goldberg A.M. (ed.): *In vitro* toxicology approaches to validation (Alternative methods in toxicology. Vol. 5). Mary Ann Liebert, Inc., New York.

Bodnar, A.G., Ouellette, M., Frolkis, M., Holt, S.E., Chiu, C.-P., Morin, G.B., Harley, C.B., Shay, J.W., Lichsteiner, S., Wright, W.E. (1998). Extension of life-span by introduction of telomerase into normal human cells. Science 279: 349–352.

Caputo, J.L. (1996). Safety Procedures. In Freshney, R.I., Freshney, M.G., eds., Culture of Immortalized Cells. New York, Wiley-Liss, pp. 25–51.

Davis, J.M. (1994). Basic cell culture: A practical approach. IRL press at oxford. University Press. Oxford.

Freshney, R.I. (1999) Culture of animal cells: A multimedia guide. New york.

Freshney, R.I. (2002) Cell line provenance. Cytotechnology 39: 3–15.

Freshney, R.I. (2005) Culture of Animal Cells, a Manual of Basic Technique, 5th Ed. Hoboken NJ, John Wiley & Sons.

Harrison, R. (1907) Observations on the living developing nerve fiber. Anat. Rec. 1:116-128; Proc. Soc. Exp. Med, N.Y. 140-143.

Lodish, H., Berk, A., Zipursky, S.L. (2000). Molecular Cell Biology. 4th edition. New York: W. H. Freeman.

Maas-Szabowski, N., Stark, H.-J., Fusenig, N.E. (2002) Cell interaction and differentiation. In Freshney. R.I. and Freshney. M.G., eds, Culture of Epithelial Cells, 2nd Ed. New York, John Wiley & Sons, pp. 31–63.

Munger, K., Howley, P.M. (2002) Human papillomavirus immortalization and transformation functions. Virus Res. 89(2): 213–228.

Nicosia, R.F.,and Ottinetti, A. (1990). *In vitro*. 26: 119.

Portela, V.M., Zamberlam, G., Price, C.A. (2010). "Cell plating density alters the ratio of estrogenic to progestagenic enzyme gene expression in cultured granulosa cells". Fertil. Steril. 93 (6): 2050–2055.

Pretlow, T.G., and Pretlow, T.P. (1989). Cell separation by gradient centrifugation methods. Methods Enzymol. 171: 462–482.

Schiff., Judith, Ann. (2006). An unsung hero of medical research. Yale Alumni Magazine, 04-19.

Westermark, B., Wasteson, A. (1975). The response of cultured human normal glial cells to growth factors. Adv. Metab. Disord. 8: 85–100.

Witkowski, J. A. (1983). Experimental pathology and the origins of tissue culture: Leo Loeb's contribution. Medical History, 1983, 27: 269-288.

Wright, W.E., Shay, J.W. (2002) Historical claims and current interpretations of replicative aging. Nat. Biotechnol. 20: 682–688.

❑❑❑

CHAPTER – 18

Cell-Cell Communication

Zeeshan Mushtaq and Vimlesh Kumar*
Indian Institute of Science Education and Research-Bhopal ITI Building, Govindpura,
Bhopal-23, India
**Email: vimlesh@iiserbhopal.ac.in*

Summary

Multicellular organisms are made up of diverse set of cells. Cells respond to the changes in the particular microenvironment by means of cellular communication. This cell-cell communication is extremely important for the development, maintenance and survival of the organism. This communication is equally important for the unicellular organisms. Cells communicate to one another and respond to the environmental changes using diverse chemical languages, as discussed in this chapter. However, in many cases like multicellular organisms, cell communication is primarily mediated by specific receptor(s) expressed on the cell membranes. These receptors bind to the extracellular signaling molecules (called ligands) and activate specific signaling pathways to bring about particular cellular responses. These different receptors, ligands and signaling pathways are briefly discussed here.

1. Principle of Cell Communication

Signals that pass between cells are far simpler than the sorts of messages that humans ordinarily exchange. A typical communication between cells is mediated mainly by extracellular signaling molecules. While some of these signaling molecules operate over long distances-signaling to cells far away; others act only on immediate neighbors. Most cells in multicellular organisms both emit and receive signals. Reception of the signals depends on receptor proteins, usually (but not always) located at the cell surface, which bind the signaling molecule. The binding of signaling molecule activates specific receptor, which in turn activates one or more intracellular signaling pathways. These relay chains of molecules—mainly intracellular signaling proteins process the signal inside the receiving cell and distribute it to the appropriate intracellular targets. These targets are generally effector proteins, which are altered when the signaling pathway is activated and implement the appropriate change of cell behavior. Depending on the signal and the nature and state of the receiving cell, these effectors can be gene regulatory proteins, ion channels, components of a metabolic pathway, or parts of the cytoskeleton—among other things.

The signal transduction begins when the receptor protein on the target cell receives an incoming extracellular signal and converts it to the intracellular signals. This in turn regulates cellular behavior in many ways.

2. Cellular Communication in Unicellular Organisms

Unicellular organisms like bacteria and yeasts respond to physical and chemical changes in their environment and can also communicate and influence others behavior. This is accomplished by translating signals that they receive from environment. Cellular communications help microorganisms in tracking down nutrients, avoiding predators and differentiating between light and dark.

2.1. Cellular Communication in Yeast

The yeast, *Saccharomyces cerevisiae* communicate with one another for the preparation of mating. The ready to mate haploid cells secrete a peptide mating factor which binds to specific receptors of opposite mating type and prevent them from further proliferation. After binding to these factors the normal spherical yeast cells extend protrusions towards the source of factor. This in turn results in the fusion of two haploid cells of opposite mating type producing a diploid cell (**Fig. 1**). Many proteins have been identified in yeast mutants that are required for signaling processes. The signaling network of yeast mainly includes cell surface receptor proteins, GTP binding proteins and protein kinases.

2.2. Cellular Communication in Bacteria

The wide arrays of chemical signals that either diffuse through extracellular environment or remain cell-associated help bacteria to communicate and coordinate their activities (Bonnie, 2002). The information provided by these molecules is essential for inter and intracellular coordination of the cells. While Quorum sensing is one of the fundamental coordinating mechanisms, various other modes of communications are also found in bacteria. Here we briefly discuss two of them.

2.2a Physical Interaction in Bacteria

Bacteria in nature are often in mixed communities. It is clear that interactions among multiple bacterial taxa are quite complex, and even the presence of two bacterial species can significantly complicate matters. Bacteria inhabiting complex environments communicate with neighboring cells through physical interactions. Fruiting body formation or biofilm formation are the kind of physical interactions which allow bacteria to functionally coordinate in complex environments. For example, *E. faecalis* a gram-positive bacterium that resides within biofilm by oligopeptide. This signaling is mediated through the FsrAB two component systems. It has been found that FsrA is controlled by gel-A gene, encoding a zinc mettalo-protease which is required to determine the length of the oligopeptide chains formed by *E. faecalis*, promoting degradation of misfolded surface proteins and biofilm formation (Karen and Clay, 2005).

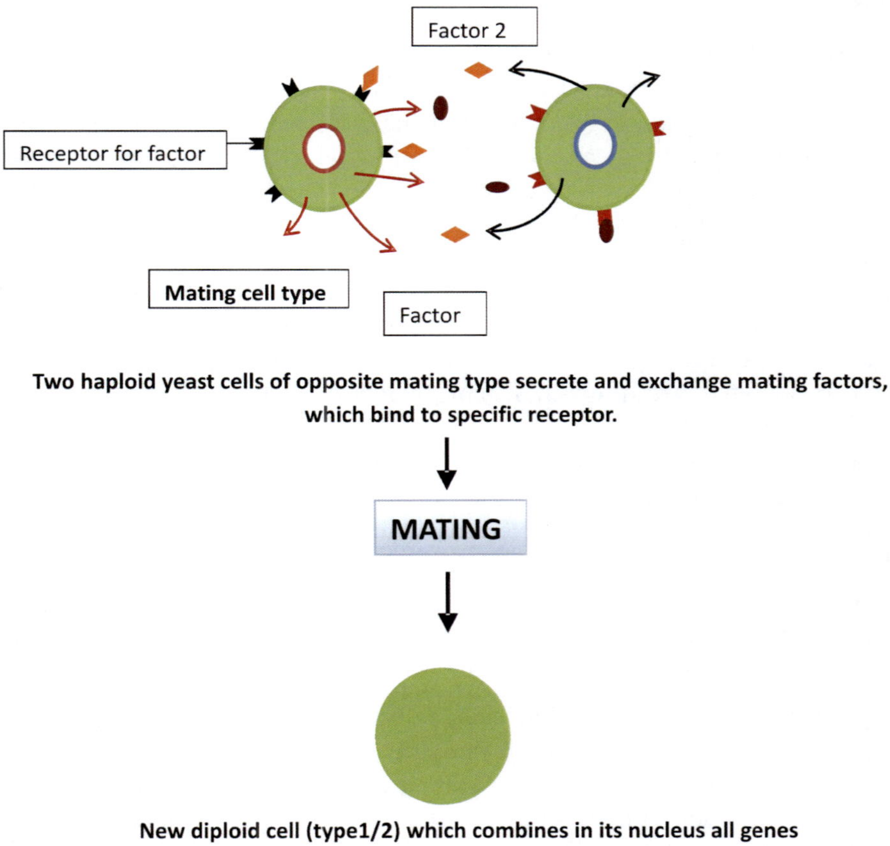

Fig. 1 : Response of Yeast Cell to Mating Factor

Myxococcus xanthus- a soil bacterium cites another example of physical interaction. The fruiting bodies in *M. xanthus* are developed through series of cell reversals from simpler cell aggregates. These reversals involve several different diffusible and cell-contact dependent signals. Motility plays central role in this process and is target for signaling.

2.2b Host Microbe Communication

In addition to communication among different bacterial species, bacteria also communicate with different eukaryotic hosts. For example, *Pseudomonas aeruginosa* produces 3-oxo-C12-HSL which can induce cytokine production in mammalian cells. It also stimulates the mucus production in cystic fibrosis patients. Similarly, zoospores of the marine alga *Ulva intestinalis* respond to the presence of AHL-producing derivatives of *Vibrio anguillarum* and synthetic AHLs, enhancing their colonization of surfaces. Plants also show extensive response to bacterial signals. For example, *Agrobacterium tumefaciens* signals are responded by plants during biofilm formation on plant roots.

2.2c Quorum Sensing

Bacteria have the capacity to coordinately regulate specific sets of genes by sensing and communicating amongst themselves utilizing a variety of signals. They exude diverse range of compounds into the extracellular environment and also secrete molecules from their cell surfaces. This chemical signaling or communication in most of the bacteria involves producing, releasing, detecting, and responding to the higher concentration of self produced small hormone-like molecules termed autoinducers. Some bacterial signals however work as simple autoinduction circuits, perhaps using the stability of signal as an indirect measure of relevant environmental measures, such as pH. These systems are described by a single term **quorum sensing** which regulates the gene expression in response to cell population density (Christopher and Bonnie, 2005). The stimulatory concentration of an extracellular autoinducer requires a minimum bacterial population density known as quorum. Most quorum-sensing-controlled processes are unproductive when undertaken by an individual bacterium acting alone but become beneficial when carried out by a simultaneous action of a group of cells. Quorum sensing regulates biofilm formation, mating, sporulation and bioluminescence. In both Gram-positive and Gram-negative bacteria, it regulates wide array of physiological activities like antibiotic production, conjugation, competence, motility, virulence and symbiosis.

In general two basic types of mechanisms by which bacteria sense and respond to intercellular signals in environment are found (**Fig. 2**). However, *vibrio harveyi*- a gram-negative soil bacterium is an exception having multiple signaling pathways. Vibrio harveyi recognizes at least three signals i.e (i) The AHL 3-hydroxy-butanoyl HSL, (ii) AI-2 and (iii) The CAI-1 signal (Henke and Bonnie, 2004).

In the first type, usually followed by Gram-negative bacteria the signal is perceived through bacterial membrane either by diffusion or by specific transport. Gram-negative bacteria in general use acylated homoserine lactones (AHLs) synthesized cytoplasmically by Lux-I type enzyme as autoinducers. These AHLs diffuse through membrane and accumulate both intra- and extra cellular in proportion to cell density. The target for the AHL is usually a member of the LuxR family of transcriptional activators. A LuxR type protein and several of its homologs, including TraR and LasR, directly bind to their cognate AHLs after the stimulatory concentration of AHL is achieved. These LuxR-AHL complexes then bind to the promoters of quorum sensing-regulated target genes and activate transcription.

In second type, followed by Gram-positive bacteria the signals are actively exported out of the cell and are recognized extracytoplasmically by specific sensor proteins. Gram positive bacteria use short, processed oligopeptides as autoinducers. These oligopeptide signals are recognized by transmembrane sensor kinases which control the phosphorylation cascades of cognate response regulators to alter the gene transcription. Specificity exists because each sensor protein is highly selective for a given peptide signal. Some peptide autoinducers act exclusively from the outside, while others elicit a specific set of gene expression changes from the outside and are also transported back into the cell wall where they trigger a different set of behavioral changes.

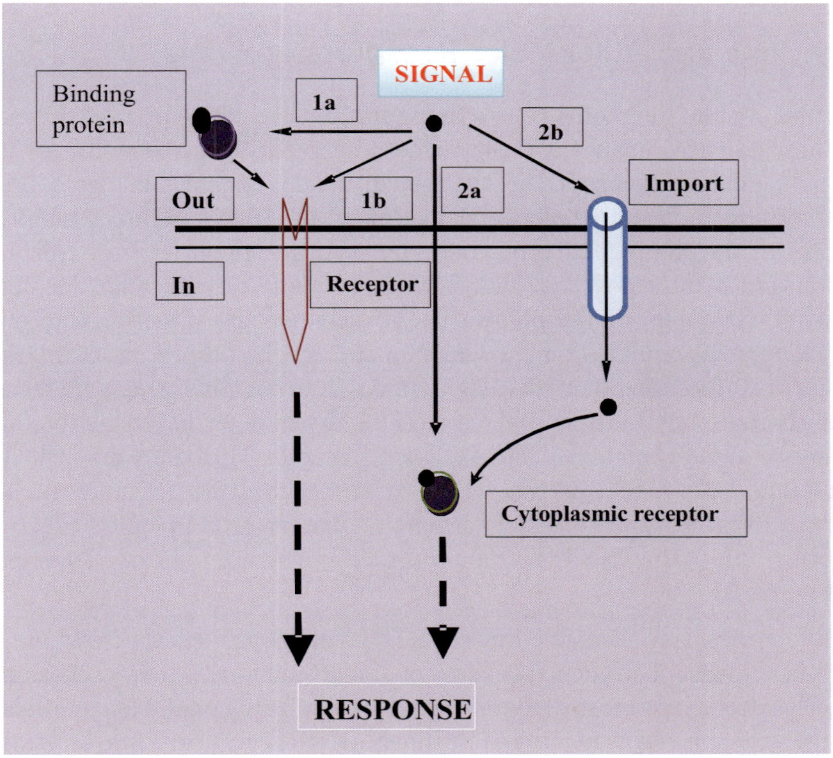

Fig. 2 : Signal Perception Pathways **1a**) Perception through an external intermediate protein that presents signal to the transmembrane receptor **1b**) Direct interaction of signal with transmembrane receptor. **2a)** Signal diffuses passively and binds to an intracellular receptor. **2b)** Signal molecule is internalized through specific import mechanism and binds to an intracellular receptor.

3. Cell-Cell Communication in Plants

Plants and animals have been evolving independently for more than billion years, the last common ancestor being a single-celled eucaryote that most likely lived on its own. Because these kingdoms diverged so long ago when it was still "every cell for itself each has evolved its own molecular solutions to multicellular functioning. Thus, the mechanisms for cell–cell communication in plants and animals evolved separately and would be expected to be quite different. At the same time, however, plants and animals started with a common set of eucaryotic genes including some used by single-cell organisms to communicate among themselves and so their signaling systems should show some similarities. Thus, whereas both plants and animals use nitric oxide, cyclic GMP, Ca2+, and Rho family GTPases for signaling, there are no homologs of the nuclear receptor family, Ras, JAK, STAT, TGFb, Notch, Wnt, or Hedgehog encoded by the completely sequenced genome of *Arabidopsis thaliana*, the small flowering plant. Similarly, plants do not seem to use cyclic AMP for intracellular signaling.

3.1. Ethylene and Other Major Receptors of Plants

Different **plant hormones** (**Growth regulators**) like ethylene, abscisic acid, auxin, gibberellins and cytokinins act as chemical messengers to control the molecular, chemical and physiological events underlying the growth and development in plants. Ethylene-a gaseous hormone is one of the best studied plant signaling systems, which regulates wide array of developmental processes viz. seed germination, fruit ripening, leaf abscission and plant senescence etc. Various ethylene receptors are located in the endoplasmic reticulum of the plants. These receptors are dimeric, with a copper-containing ethylene binding domain which functions as histidine kinases, like bacterial proteins. The Raf-kinase homolog CTR1 is thought to negatively regulate the ethylene response through MAP-kinase pathway. CTR1 depends on its serine/threonine and association of its N-terminal domain to ethylene receptor. The empty ethylene receptors are active and keep CTR1 active, which in turn stimulates the ubiquitination and degradation in proteasomes of nuclear gene regulatory protein called EIN3 (Larissa and Alonso, 2006). This EIN3 is required for the transcription of ethylene responsive genes (**Figure. 3**).

Instead of receptor tyrosine kinase (RTK) - the largest class of enzyme coupled receptors in animals; plants depend mainly on transmembrane receptor serine/threonine kinases. These receptors mostly have a tandem array of extracellular leucine-rich repeats and are therefore known as **leucine-rich repeat (LRR) receptor kinases.** Arabidopsis genome encodes about 175 LRR receptor kinases. Out of which clavata1/clavata2 (clv1/clv2) receptor complex is best characterized. Brassinosteriods are a class of steroids synthesized by plants and were first identified in family *Brassicaceae*. These regulate the growth and differentiation of plants throughout their life cycle by binding to Bril cell receptor kinase.

Auxins and Photoproteins are the other examples of major plant signaling molecules. Auxin- a plant hormone, which is generally indole-3-acetic acid, helps plants to grow towards light, and grow upward rather than branch out. It binds to specific nuclear receptor proteins. Photoproteins like phytochromes, phototropin and cryptochromes sense light by means of a covalently attached light absorbing chromophore.

Phytochromes respond differently and reversibly to red and far-red light, their kinase activity is activated by red light whereas the far-red inactivates it. **Phototropin** is associated with plasma membrane and is partially responsible for phototropism. **Cryptochromes** are generally flavoproteins, structurally related to blue-light-sensitive enzymes called photolyases, which are involved in the repair of ultraviolet-induced DNA damage.

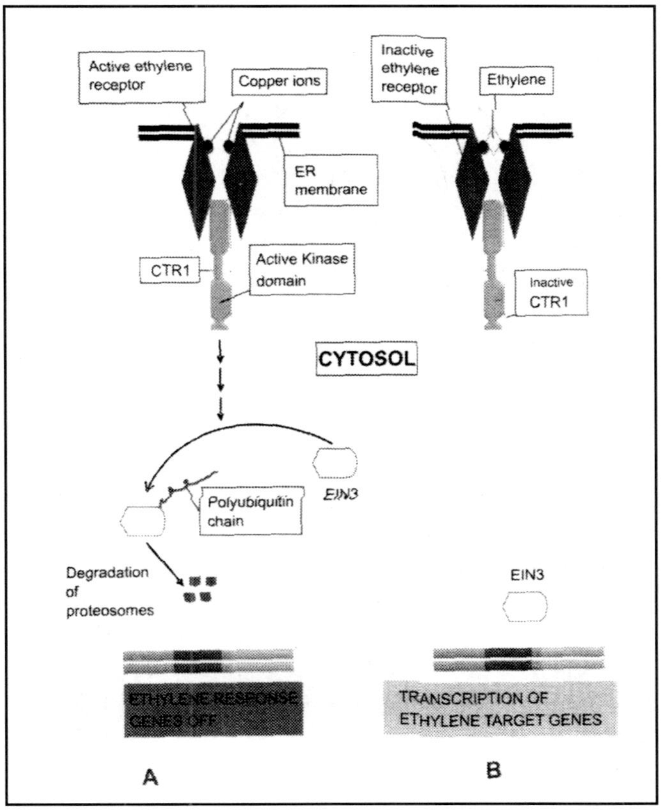

Fig. 3 : Ethylene signaling pathway: A) In the absence of ethylene CTR1 is active, causing ubiquitylation and destruction of EIN3 which in turn results in inactivation of ethylene responsive genes. B) The binding of ethylene inactivates the receptors and disrupts the interaction between the receptors and CTR1. The EIN3 protein is not degraded and can therefore activate the transcription of ethylene responsive genes.

4. Cell-Cell Communication in Animals

Signaling in multicellular organisms is much more complicated than that of unicellular organisms. Cells in multicellular organisms must interpret and respond to diverse array of signals they receive from other cells in order to coordinate their behavior. This wide array of signals include amino acids, small peptides, proteins, nucleotides, retinoids, steroids and even dissolved gases like carbon monoxide and nitric oxide. For example during development, the embryo cells exchange signals to determine physiology and overall behavior of an animal.

Animal cells usually communicate by secreting chemical signals directed to target cells that may be immediately adjacent to the signaling cell or far away from it. On the basis of distance covered by the signal molecule, the signaling can be divided broadly into two types:

4.1. Local Signaling

Local signaling also known as **paracrine signaling** can be defined as the signaling where the signal molecules act as local mediators, and affect only cells in the local environment of the signaling cell (**Fig. 4**). Animal growth factors, for example, are compounds that stimulate nearby target cells to grow and multiply. Molecules regulating inflammation at the site of infection is also an example of paracrine signaling. The nature of signals and target cells are usually of different type in local signaling. However, there are some cells which produce signals and respond to these signals themselves, such a kind of signaling is referred as **autocrine signaling.** For example, cancer cells use this strategy to stimulate their own survival and proliferation. For paracrine signals to act only locally, the secreted molecules must not be allowed to diffuse too far; for this reason they are often rapidly taken up by neighboring target cells, destroyed by extracellular enzymes or immobilized by the extracellular matrix.

Another specialized type of local signaling occurs between nerve cells. The axon of a neuron terminates at specialized functional sites known as chemical synapses. The neurons are stimulated by electric impulse and in turn release a pulse of an extracellular chemical signal known as **neurotransmitter.** They are a diverse group of small hydrophilic molecules which carry signals between neurons or from neurons to other types of target cells. Due to their hydrophilic nature they are unable to cross the plasma membrane of the target cells and thus act by binding to cell surface receptors. About 60 neurotransmitters have been found which fall under four general classes:

i) **Amino acids:** Glutamate and Aspartate are well known excitatory transmitters, while gamma-aminobutyric acid (GABA), glycine and taurine are inhibitory neurotransmitters.

ii) **Biogenic amines :** Serotonin, histamine, and the catecholamines - dopamine and norepinephrine.

iii) **Choline:** of which acetylcholine is amongst the most important neurotransmitter.

iv) **Neuropeptides:** these are formed by longer chains of amino acids (like a small protein molecule). Over 50 of them are known to occur in the brain, and many of them have been implicated in the modulation or transmission of neural information.

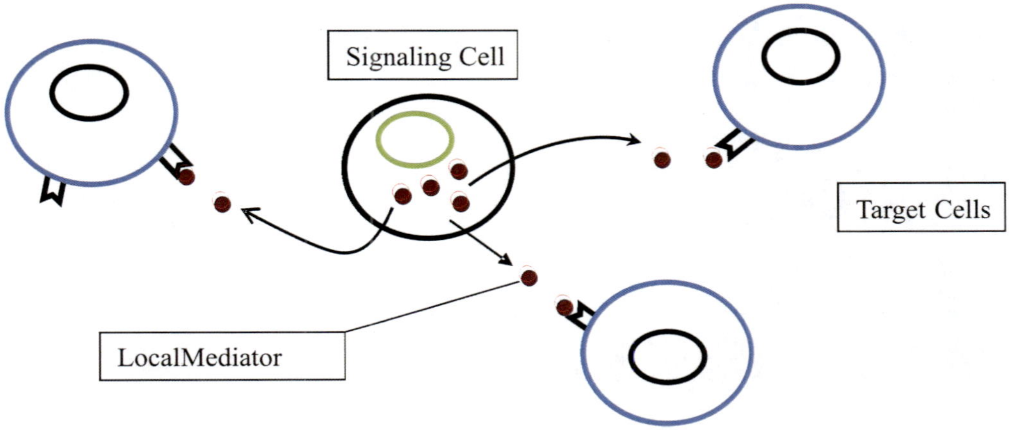

Fig. 4 : Paracrine signaling.

4.2. Hormonal Signaling

A quite different strategy for signaling used by multicellular organisms involves broadcasting the signal throughout the body by secreting it into the blood stream (in an animal) (**Fig. 5**) or the sap (in a plant). Signal molecules used in this way are called as hormones, and in animals, the cells that produce hormones are called as endocrine cells e.g. thyroid gland, ovaries and testes. Most hormones initiate a cellular response by initially combining with either a specific intracellular or cell membrane associated receptor protein. Hormonal signaling, however involves the following steps:

(i) A particular endocrine cell synthesizes a particular hormone (s).

(ii) This hormone is first stored and then secreted in blood stream.

(iii) Blood transports the hormone to target cell(s).

(iv) Recognition of the hormone by an associated cell membrane or intracellular receptor protein.

(v) The received hormonal signal is then amplified by the target cell via signal transduction process which leads to cellular response.

(vi) Degradation of the hormone.

A cell may have different receptors which recognize same hormone and activate different pathways, or a cell may have different receptors which recognize different hormones and activate the same signal transduction pathway. However an important consideration that decides at which level signal pathway activates is the concentration of hormone –receptor complex formed, which is in turn effectively determined by following factors:

1. The number of available particular hormone molecules for complex formation.
2. The number of receptor molecules available for complex formation.
3. The affinity of interaction between hormone and receptor.

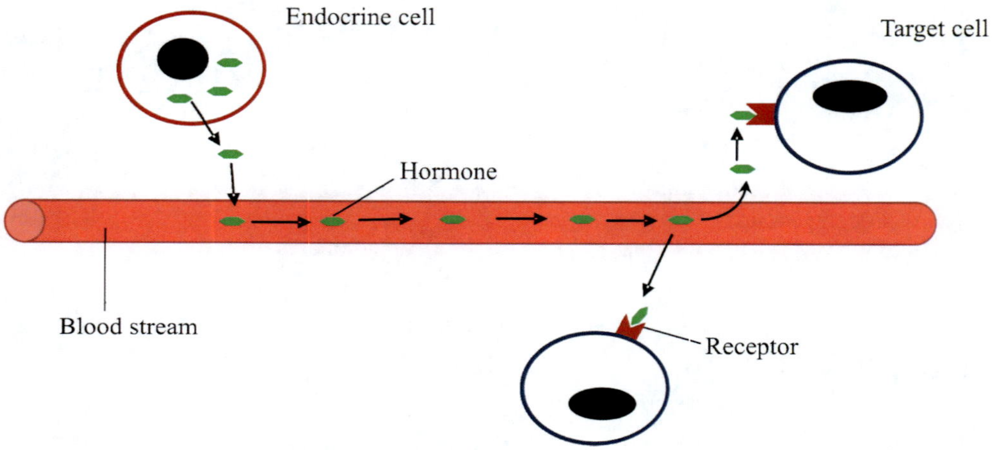

Fig. 5 : Hormonal Signaling

5. Membrane Proteins and Signaling Cascades

Cells in multicellular organisms are exposed to hundreds of different signals which may be soluble, embedded to the extracellular matrix or to the surface of neighboring cells. These signal molecules can be stimulatory or inhibitory and can act in wide array of combinations that can influence cell in almost any aspect. The cell must respond to this muddle of signals selectively, according to the cell's specialized function. The response of cell depends on whether or not it possesses a receptor for that signal. Without the appropriate receptor, a cell will be deaf to the signal and will not react. The receptor or target molecule which is usually activated by only one kind of signal performs the primary signal transduction. After receiving an external signal it produces an internal signal response and starts the intracellular signal transduction processes. The intracellular transduction process then transfers, activates or even generates intracellular signaling molecule until a metabolic enzyme is kicked into action, a gene is switched on, or a cytoskeleton is tweaked in to new configuration.

5.1. Membrane Receptors Involved in Signaling

Cell surface (membrane or transmembrane) receptors are specialized integral membrane proteins that span the plasma membrane and take part in communication between cell and extracellular signaling molecules, as discussed above. These cell-surface receptors act as signal transducers by converting an extracellular ligand-binding event into intracellular signals that alter the behavior of the target cell. Most of these receptors are composed of two or more protein subunits which operate collectively and may dissociate when ligand binds to it. A major in understanding cell-cell signaling is unraveling the mechanisms by which cell surface receptors transmit the signals initiated by ligand binding. Based on the structure and the nature of intracellular signal they produce after binding to an extracellular signal molecule, membrane receptors can be broadly divided into three large classes: ion-channel-linked receptors, G-protein-linked receptors, or enzyme-linked receptors. The number of different types of receptors in these three classes is even greater than the number of extracellular signals that act on them, because for many extracellular signal molecules there is more than one type of receptor e.g. acetylcholine acts on skeletal muscle cells via an ion-channel–linked receptor, whereas in heart muscle cells it acts through a G-protein–linked receptor.

In addition to membrane receptors, there are proteins which serve both as intracellular receptors and as intracellular effectors for the signal. These receptors (not discussed here) like **nitric oxide**, **carbon monoxide** and **steroid hormone** receptors mainly fall under **nuclear receptor super family.**

5.1a. G-protein-linked Receptors (GPCRs)

The largest class of cell surface receptors act by indirectly regulating the activity of a separate plasma-membrane bound target protein, which is generally either an enzyme or an ion channel. They mediate responses to an enormous diversity of extracellular signal molecules, including hormones, local mediators, and neurotransmitters. These receptors act via intermediary action of guanine nucleotide-binding proteins called **G-Proteins.** Although a wide array of signal molecules binds to these proteins, the G-protein coupled receptors are structurally and functionally related proteins. Each receptor is made of a single polypeptide á-helical chain that spans back and forth across the lipid bilayer seven times. This superfamily of seven-pass transmembrane receptor proteins includes rhodopsin (a light-activated protein in the vertebrate eye), the olfactory (smell) receptors in the vertebrate nose, and the receptors in yeasts that recognize the secreted mating factors (William and Hamm, 2008).

The binding of extracellular signal molecules to the extracellular domain of seven-pass transmembrane receptors induces a conformational change that enables cytosolic domain of receptors to activate G-protein attached to the cytoplasmic face of plasma membrane. The activated G protein then dissociates from the receptor and carries the signal to an intracellular target.

Over a thousand of G proteins, each specific for a particular set of GPCRs and for a particular set of target proteins in the plasma membrane have been identified. However, all of these G proteins have a similar general structure and operate in a similar fashion. They are composed of three subunits designated α, β, and γ – two of which are tethered to the plasma membrane by short lipid chains. In order to distinguish them from other guanine nucleotide-binding proteins like Ras proteins, they are often called as heterotrimeric G proteins. The α subunit binds guanine nucleotides, which regulate G protein activity and make it act like guanine nucleotide exchange factor (GEF). In the unstimulated state, the α subunit has GDP bound to it, in a complex with β and γ. When a hormone binds to receptor, it induces conformational change in the receptor, such that the á subunit of G protein is induced to release GDP and exchange it for a molecule of GTP (**Fig. 6**). This activation of G protein makes α subunit to dissociate from β and γ, which remain together and function as βγ complex. Both the active GTP-bound α subunit and βγ complex can roam indepently along the plasma membrane and interact directly with their targets to elicit an intracellular response. The GTPase activity of α subunit is terminated by hydrolysis of bound GTP, and the inactive α subunit then reassociates with βγ complex. The longer the α subunit or βγ complex is bound to target protein, the stronger and more prolonged signal will be. However the time for which G protein remains active depends on how quickly the α subunit hydrolysis its bound GTP. The GTPase activity of α subunit is greatly enhanced by binding to target or to a specific **regulator of G protein signaling (RGS)**, which acts as α subunit-specific **GTPase activating protein (GAP)**. So the time for which G protein remains active is very less and the reunion of α subunit and βγ complex occurs within seconds after the G protein has been activated.

5.1b Enzyme-Coupled Surface Linked Receptors

In contrast to G-coupled receptors, the cytosolic domain of enzyme-coupled receptors has an intrinsic enzymatic activity or are directly linked to other intracellular proteins that act as intracellular enzymes. Like GPCRs these transmembrane receptors display their ligand binding domain on the outer surface of plasma membrane. The subunits of these receptors usually have only one transmembrane segment. These proteins act as receptors mostly for growth factors, however they also mediate direct, rapid reconfigurations of the cytoskeleton, controlling the way a cell moves and changes its shape. There are six principal classes of enzyme-coupled receptors:

(i) Receptor protein tyrosine kinases: are receptors which phosphorylate their substrate proteins on tyrosine residue. These are the largest class of enzyme-linked receptors and will be discussed here in detail.

(ii) Tyrosine-kinase-associated receptors: These receptors directly recruit cytoplasmic tyrosine kinases, as they don’t have intrinsic enzymatic activity.

(iii) Histidine-kinase-associated receptors: In these receptors the kinase phophorylates itself on histidine and then immediately transfers the phosphoryl group to a second intracellular signaling protein.

(iv) Receptor serine/threonine kinases: are protein kinases that phosphorylate serine or threonine, rather than tyrosine, residue on their substrate proteins.

(v) Receptor guanyl cyclases: catalyze the production of cyclic GMP in the cytosol, which acts as a small intracellular mediator.

(vi) Receptor tyrosine phosphatases: remove phosphate groups from phosphotyrosine residues of specific intracellular signaling molecules, thus acting to counterbalance the effects of protein-tyrosine kinases.

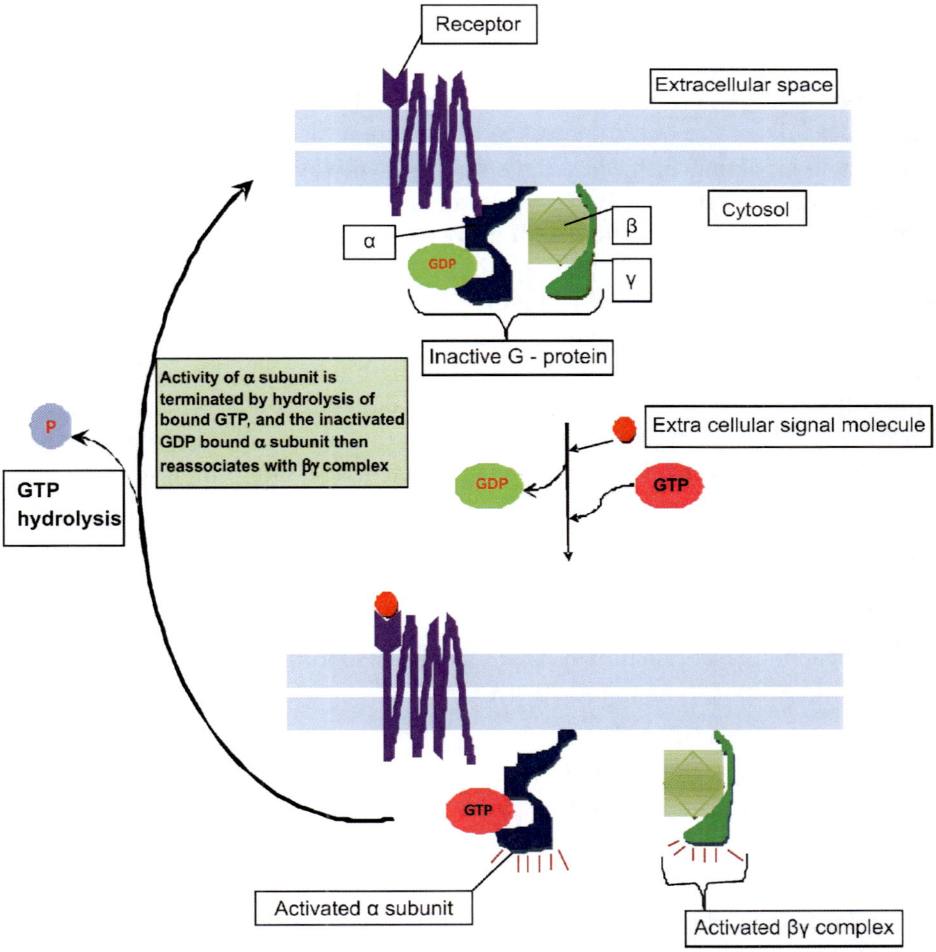

Fig. 6 : Activation & deactivation of G-protein (see text for details)

Receptor tyrosine kinase family includes the receptors for most polypeptide growth factors, so protein-tyrosine kinase phosphorylation has been particularly well studied. This signaling mechanism is involved in the control of animal cell growth and differentiation. Epidermal growth factor (EGF), platelet-derived growth factor (PDGF), fibroblast growth factors (FGFs), hepatocyte growth factor (HGF), insulin, insulin like growth factor-1(IGF1), vascular endothelial growth factor (VEGF), macrophage-colony-stimulating factor (MCSF), and the neurotrophins are some of the extracellular signal proteins which act through RTKs. Unlike GPRCs most of the RTKs span the lipid bilayer as a single α helix, however insulin receptors and some related receptors are dimers consisting of two polypeptide chains (Geoffrey and Hausman, 2009).

The first step in signaling from most receptor protein tyrosine kinases is ligand-induced receptor dimerization. Ligand induced dimerization activates the kinase function of intracellular receptor tails, such that the dimerized polypeptide chains cross-phosphorylate one another (**Fig. 7**). Such a process of phosphorylation is refered to as transautophosphorylation. The phosphorylation on RTKs occur only on specific tyrosine residues located on the cytosolic tail of the receptor. This transautophosphorylation contributes in signaling from these receptors in two ways. First, phosphorylation of tyrosine residues within the catalytical domain increases the protein kinase activity of the enzyme. Second, phosphorylation of tyrosine residues outside the catalytical domain creates specific, high-affinity docking sites for intracellular signaling proteins that transmit intracellular signals downstream of the activated receptors. Once bound to the activated RTK, a signaling protein may itself become phosphorylated on tyrosines and thereby be activated. In many cases, however, the binding alone may be sufficient to activate the docked signaling protein, by either inducing a conformational change in the protein or simply bringing it near the protein that is next in the signaling pathway. The binding of downatream signaling molecules with receptor protein-tyrosine kinases is mediated by protein domains that bind to specific phosphotyrosine-containing peptides. These can be either **SH2 domains** (for Src homolgy region 2) or less commonly **PTB domains** (for phosphotyrosine binding). The recruitment of different collections of intracellular signaling proteins is done by different enzymes like:

(i) Phospholipase: function in same way as phospholipase C to activate the inositol phospholipid signaling pathway.

(ii) Phosphoinositide 3-kinase or phosphatidylinositol 3-kinase (PI 3-kinase): phosphorylates inositol phospholipids in the plasma membrane that then become docking sites for other intracellular signaling proteins.

In order to inactivate the receptor, cells contain enzymes called protein tyrosine phosphatases, which remove the phosphates that were added in response to the extracellular signal. However, some activated receptors are disposed of in a more brutal way: they are pulled into the interior of the cell by endocytosis and then destroyed by digestion in lysosomes.

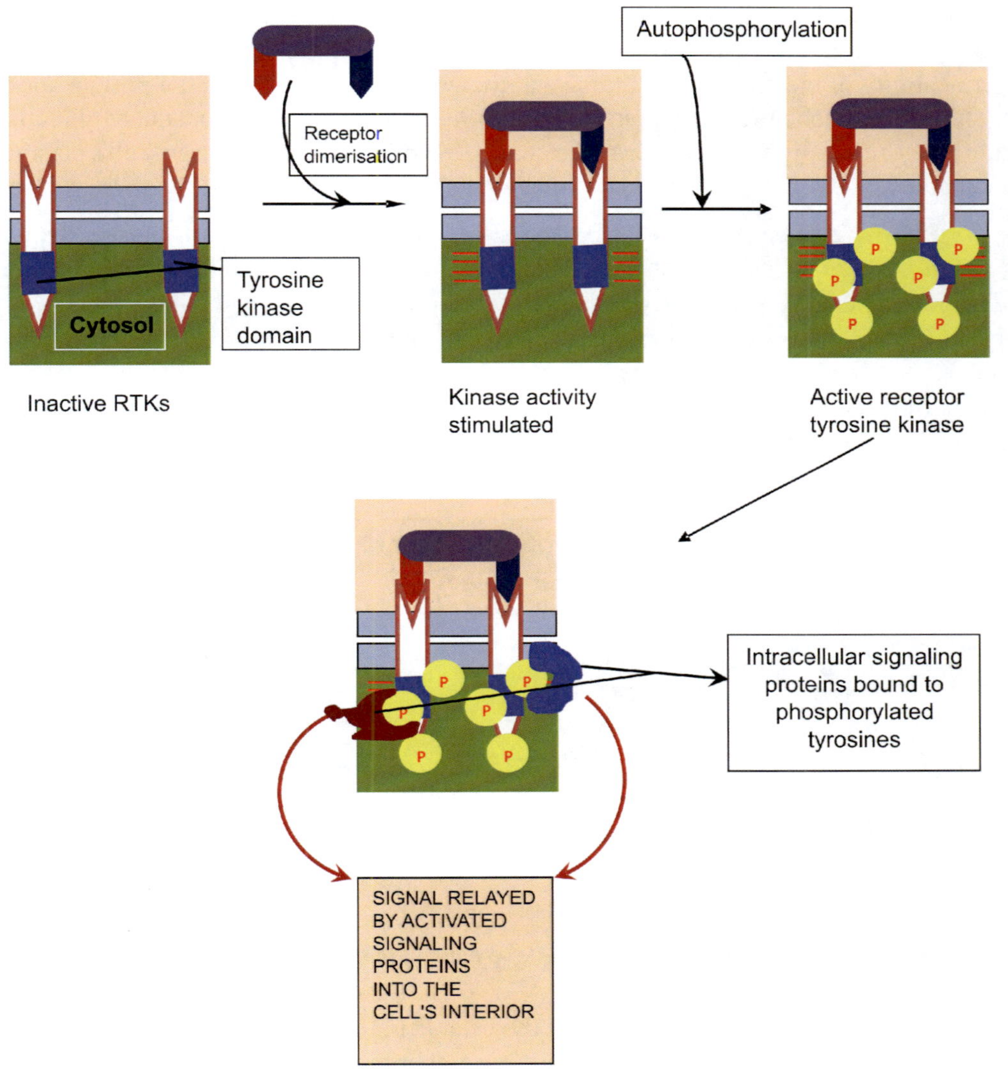

Fig. 7 : Dimerization, autophosphorylation and association of downstream signaling molecules with receptor protein-tyrosine kinases.

5.1c Ion Channel Linked Receptors

Ion channel linked receptors, also known as ligand- or transmitter- gated ion channels or ionotropic receptors are ion (both cation and anion) channels themselves and constitute a large family of multipass transmembrane proteins. These receptors function in the simplest and most direct way and are involved in the rapid synaptic signaling between nerve cells and other electrically excitable target cells. This type of signaling is mediated by a small number of neurotransmitters that transiently open or close an ion channel formed by the protein to which they bind, briefly changing the ion permeability of the plasma membrane and thereby the excitability of the postsynaptic target cell. The chemical signal in the form of neurotransmitter is transduced directly into an electrical signal, in the form of a change in voltage across the target cell's plasma membrane.

There are three properties of ion channels which are central to their function. First in the signal transduction event in a neuron when a neurotransmitter binds, this type of receptor alters its conformation so as to open or close a channel for the flow of specific types of ions of the appropriate size and charge-such as Na^+, K^+, $Ca2^+$, or Cl^- across the membrane. Second, transport through channels is extremely rapid. More than a million ions per second flow through open channels. Third, the opening of ion channels is regulated by gates that transiently open in response to specific stimuli e.g. neurotransmitter (Fig. 8).

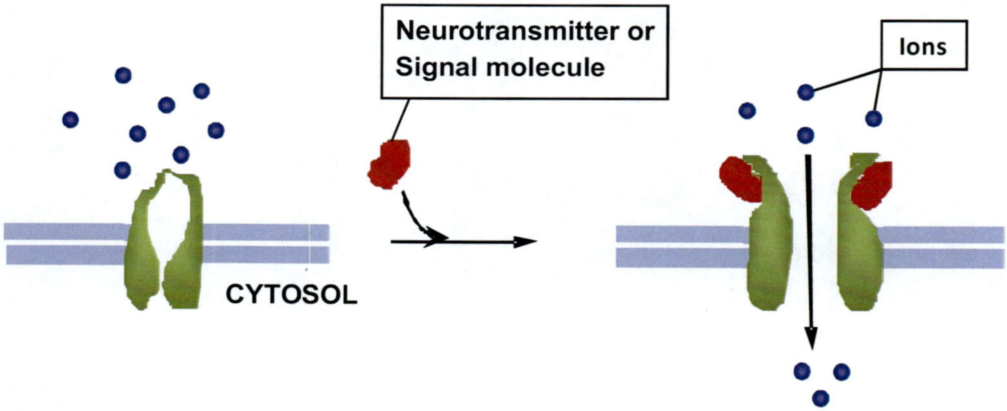

Fig. 8 : Ion channel coupled receptors: Due to binding of signal molecules to receptors, the ion channels open and the ions flow in.

The flow of ions is dependent on the establishment of ion gradients across the membrane. These ion pumps or receptors use energy derived from ATP to actively transport ions across the plasma membrane. As a result, the ionic composition of the cytoplasm is substantially different from that of extracellular fluids. For example, Na^+ is actively pumped out of cells while K^+ is pumped in. The flow of positively charged ions into the cell depolarizes the portion of the membrane nearest the channel. Because this situation is favorable to the subsequent generation of an action potential, ligand gated channel receptors that are permeable to positive ions are called excitatory. Other ligand-gated channels are permeable to negatively charged ions. An increase of negative charge within the cell makes it more difficult to excite the cell and induce an action potential. Such channels accordingly are called inhibitory.

Acetylcholine receptor is kind of cation-channel linked receptor which was first isolated from the electrical organ of Torpedo rays. These receptors consist of 5 subunits i.e. α (2 units), β, γ, and δ arranged as a cylinder in the membrane. This receptor can exist in three conformations. In its original unoccupied or closed conformation, the channel pore is thought to be blocked by the side chains of hydrophobic amino acids. After two molecules of acetylcholine bind simultaneously to the binding sites on α subunits, the conformation of the receptor is altered, such that these hydrophobic side chains shift out of the channel, opening gate that allows the passage of positively charged ions, including Na^+ and K^+. However, the channel remains impermeable to negatively charged ions, such as Cl^-, because it is lined by negatively charged amino acids. The occupied-open state of the receptor can only last for a very short period of time and then the gate is closed again, forming the occupied-closed state. The two molecules of acetylcholine quickly dissociate from the receptor and the receptor returns to its unoccupied-closed state and is ready for next transduction cycle again.

5.2. Signaling Pathways

Many cell surface receptors stimulate intracellular target enzymes, which may be directly linked or indirectly coupled to receptors. These intracellular target enzymes function as downstream functioning molecules to propagate and amplify the signal initiated by ligand binding. In most cases, the message is passed from one intracellular signaling molecule to another, each activating or generating the next signaling molecule and thus signal is transmitted from cell surface to a variety of intracellular targets. This process is called as intracellular signal transduction. Following are some important functions of signaling pathways:

(a) They transform, or transduce, the signal into a molecular form suitable for passing the signal along or stimulating a response.

(b) They relay the signal from the point in the cell at which it is received to the point at which the response is produced.

(c) In many cases, signaling cascades also amplify the signal received, making it stronger, so that a few extracellular signal molecules are enough to evoke a large intracellular response.

(d) The signaling cascades can also distribute the signal so as to influence several processes in parallel: at any step in the pathway, the signal can diverge and be relayed to a number of different intracellular targets, creating branches in the information flow diagram and evoking a complex response.

(e) Each step in this signaling cascade is open to modulation by other factors, including external signals, so that the effects of the signal can be tailored to the conditions prevailing inside or outside the cell.

Different signaling pathways like cAMP, cGMP, CAM kinase, PI 3-kinase, mTOR, MAP kinase, JAK/STAT, NF-KB, Hedgehog, Wnt and Notch pathways have been identified and we contemplate some of these pathways here.

5.2.1. Cyclic AMP Pathway

Cyclic AMP (cAMP) acts as a small intracellular second messenger in hormonal signaling (the first messenger being hormone itself) in all prokaryotic and animal cells that have been studied. Cyclic AMP is formed from ATP by a plasma-membrane bound enzyme called adenylyl cyclase and is continuously degraded to 5'-AMP by cAMP phosphodiesterase. Many extracellular signal molecules work by increasing cAMP concentration, and they do so by increasing the activity of adenylyl cyclase against a steady background of phosphodiesterase activity. Adenylyl cyclase is a large multipass transmembrane protein with its catalytic domain on the cytosolic side of the plasma membrane. There are at least eight isoforms in mammals, most of which are regulated by both G proteins and Ca^{2+}. GPCRs that act by increasing cyclic AMP are coupled to a stimulatory G protein (Gs), which activates adenylyl cyclase and thereby increases cyclic AMP concentration. Another G protein, called inhibitory G protein (Gi), inhibits adenylyl cyclase, but it acts mainly by directly regulating ion channels. Many cell responses like glycogen breakdown, fat breakdown, increase in heart rate and force of contraction by adrenaline, and cortisol secretion by ACTH are mediated by cyclic AMP. Glycogen breakdown is discussed here in more detail.

Glycogen breakdown and most other effects of cAMP in animal cells are mediated by the action of cAMP-dependent protein kinase or protein kinase A (PKA) enzyme. This enzyme is normally held inactive as a tetramer complex containing two regulatory and two catalytic subunits. The binding of cyclic AMP to the regulatory subunits forces a conformational change that leads to the dissociation of regulatory subunits from catalytic subunits. The free catalytic subunits are enzymatically active and are able to phosphorylate specific serine or threonine residues on target proteins. The regulatory subunits which are bound together and to a component of cytoskeleton or a membrane of an organelle by special A-kinase anchoring proteins (AKAPs) are required for localizing the kinase inside the cell. PKA phophorylates two key target enzymes i.e. phosphorylase kinase (a protein kinase) and glycogen phosphorylase in glycogen metabolism. PKA phosphorylates and activates phosphorylase kinase first, which in turn phosphorylates and activates glycogen phosphorylase, which catalyzes the breakdown of glycogen to glucose-1-phosphate (Fig. 9). PKA also phosphorylates the enzyme glycogen synthase, which catalyzes the synthesis of glycogen. In this case, however, phosphorylation inhibits enzyme activity.

In cells that secrete the peptide hormone somatostatin, increases in cAMP activate the transcription of specific target genes (genes that encode this hormone) that contain a regulatory sequence called the cAMP response element or CRE. In this case, the signal is carried from the cytoplasm to the nucleus by the catalytic subunit of PKA. Within nucleus, PKA phosphorylates a transcription factor called CREB (CRE-binding protein), leading to the recruitment of coactivators and transcription of cAMP-inducible genes.

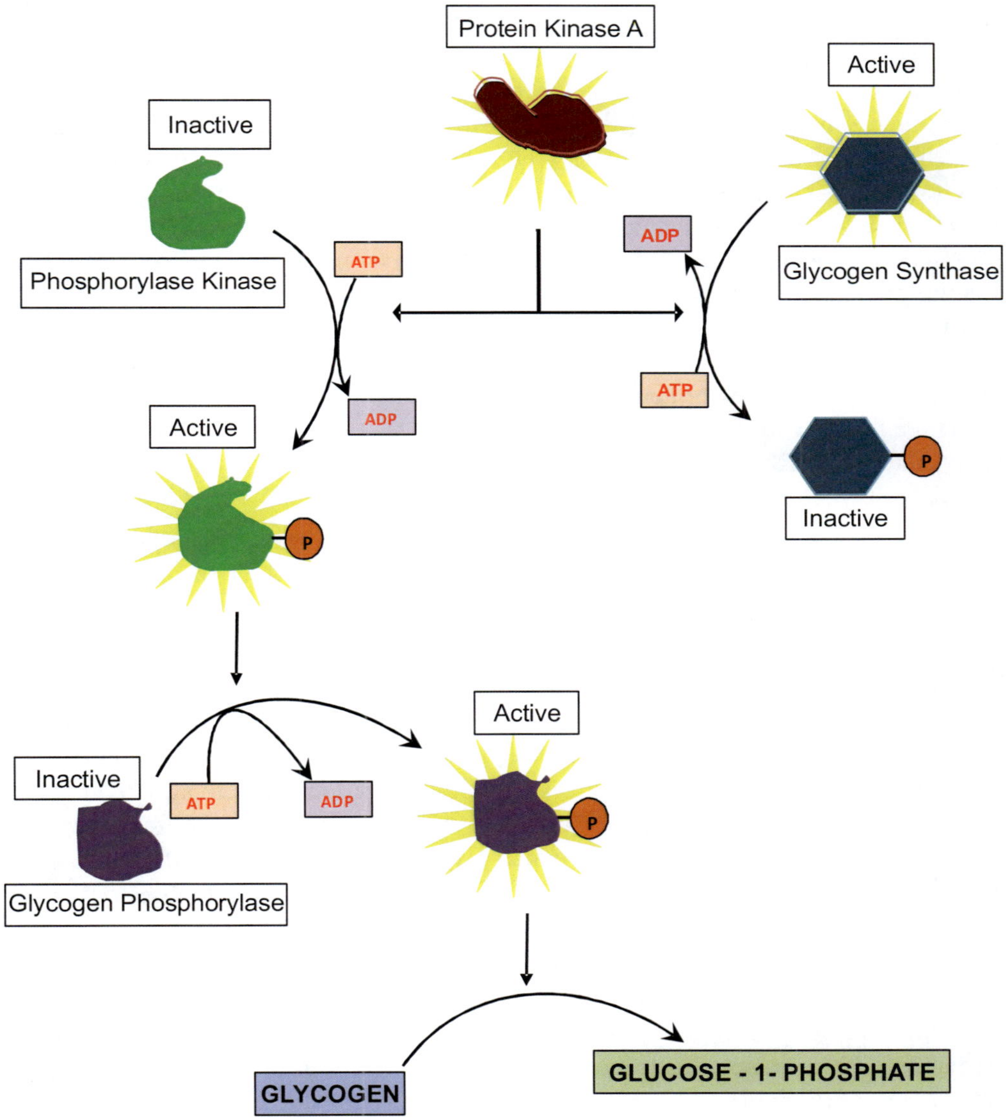

Fig. 9 : Regulation of Glycogen Metabolism by Protein Kinase A

5.2.2. MAP kinase cascade

MAP (mitogen-activated protein) kinase pathways are highly conserved in evolution and play central roles in signal transduction in all eukaryotic cells ranging from yeasts to humans. These relay systems carry signal from plasma membrane to the nucleus. MAP Kinases- a family of protein tyrosine/threonine kinases are the central elements in the pathway, that are activated in response to a variety of growth factors and other signaling molecules. In fact, the name MAP kinase pathway is in honor of the final enzyme i.e. MAP kinase in the cascade. The MAP kinases were initially characterized in mammalian cells and belong to the ERK (extracellular signal-regulated kinase) family. ERK activation plays vital role in signaling cell proliferation induced by growth factors that act through either PTKs or GPCRs (Nicolas et al. 2006). The MAP kinase cascade is activated by a small intracellular protein known as Ras, which is bound to cytoplasmic face of plasma membrane by a lipid tail. The Ras proteins are guanine nucleotide-binding proteins which resemble α subunit of G protein and function as a switch between active GTP-bound and inactive GDP-bound form. In contrast to G protein α subunits, however, Ras functions as a monomer and are thus often called as monomeric GTP-binding proteins. The activity of Ras is mediated by guanine nucleotide exchange factors (GEFs) that stimulate the release of bound GDP and its exchange for GTP. Activity of Ras-GTP complex is then terminated by GTP hydrolysis, which is stimulated by the interaction of Ras-GTP with GTPase-activating proteins (GAPs).

The three components i.e. MAP kinase (ERK/MAPK), MAP kinase kinase (MEK/MAPKK) and MAP kinase kinase kinase (Raf/MAPKKK) of MAP kinase pathway together form a functional signaling module. The ERK activation is mediated by two upstream protein kinases i.e. MEK and Raf. The Ras-GTP binding protein is activated by growth factor receptors which in turn phosphorylates and activates Raf protein serine/threonine kinase. The activated Raf phosphorylates and activates a second protein kinase i.e. MEK, which in turn phosphorylates and thus switches on final enzyme in the cascade i.e. ERK (Fig. 10). MEK is a dual specificity protein kinase that activates components of the MAP kinase cascade by phosphorylation of both threonine and tyrosine residues separated by one amino acid. The activated MAP kinase relays signal downstream by phosphorylating certain gene regulatory proteins and other protein kinases on serines and threonines, altering their ability to control gene transcription and thereby causing a change in the pattern of gene expression. ERK MAP kinase e.g. enters the nucleus and phosphorylates one or more components of gene regulatory complex. Three-component MAP kinase signaling modules operate in all eucaryotic cells, with different modules mediating different responses in the same cell. This specificity of MAP kinase signaling is maintained atleast in part by the organization of the components of each MAP kinase cascade as complexes that are associated with scaffold proteins. For example, the KSR scaffold protein organizes ERK and its upstream activators Raf and MEK into signaling cassette. As a result of the specific association of these protein kinases on the KSR scaffold, activation of Raf by Ras leads to specific and efficient activation of MEK, which in turn activates ERK.

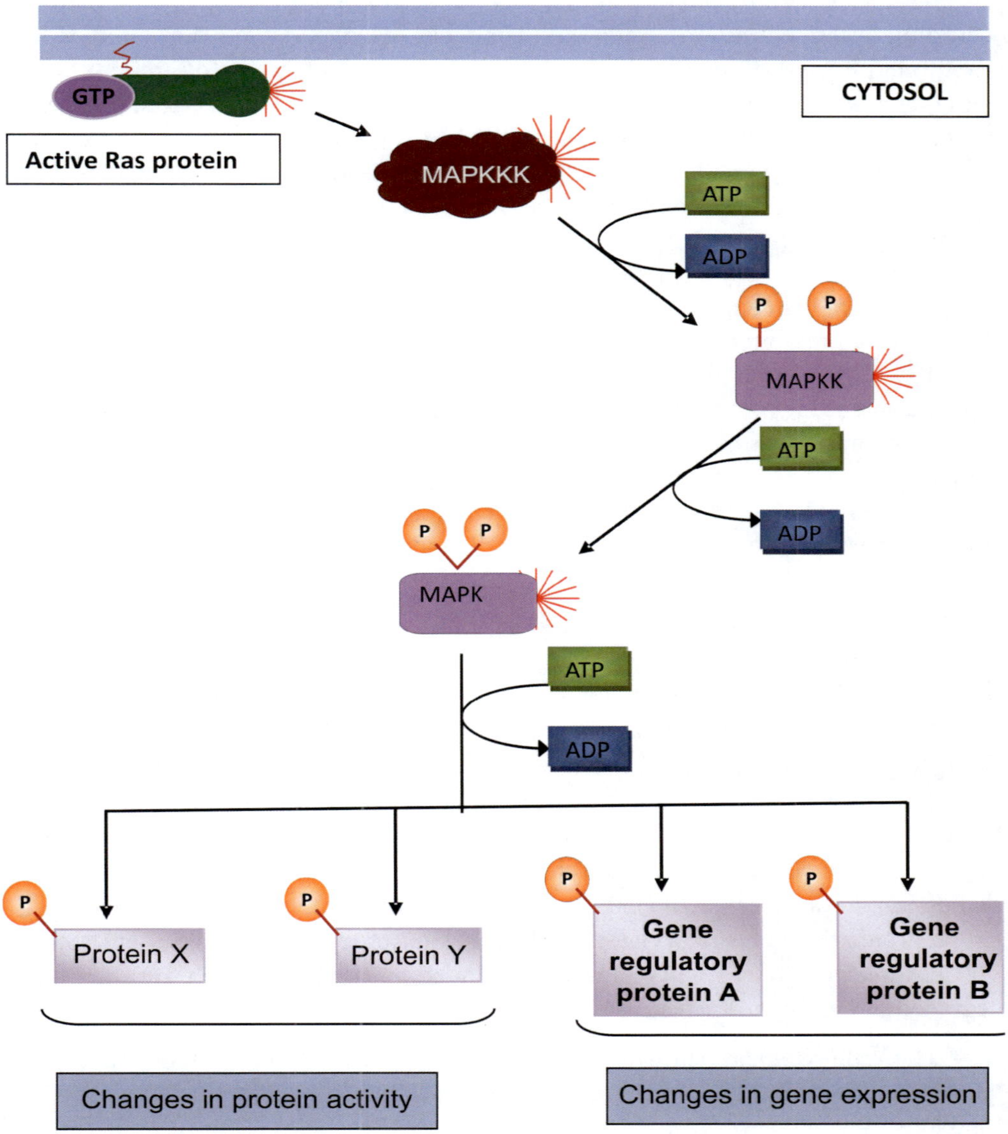

Fig. 10 : Activation of MAP-kinase phosphorylation cascade.

5.2.3. The hedgehog and wnt pathways

The Hedgehog and Wnt pathways are closely related signaling systems that play key roles in determining cell fate during embryonic development. Though these pathways were first described in Drosophila but members of Hedgehog and Wnt families have been found to control a wide range of events during development of embryo in both vertebrates and invertebrates. The hedgehog gene was identified by Nusslein-Volhard et al., in 1980 during genetic screens aimed to provide an understanding of body segmentation in Drosophila. Loss of the secreted Hedgehog signaling protein was found to cause Drosophila embryos to develop as spiny balls reminiscent of Hedgehogs. The Hedgehog trafficking, distribution and signaling is controlled by hedgehog genes that encode secreted proteins, which are modified by the addition of lipids. In Drosophila the receptor for hedgehog ligand is a 12-transmembrane protein called as Patched (Ptc). Ptc acts as an inhibitor of a second 7-transmembrane protein known as Smoothened (Smo), which is related to the Frizzled family of Wnt receptors and to other 7-transmembrane G protein-coupled receptors. The mechanism by which Ptc inhibits Smo is yet to be understood. Downstream of Smo is a multi-protein complex known as the Hedgehog signaling complex (HSC), which comprises the transcription factor Cubitus interruptus (Ci) in Drosophila and Gli in mammals, the serine/threonine kinase Fused (Fu), the kinesin-like molecule Costal 2 (Cos2) and Supressor of fused (Sufu) (Catriona and Nusse, 2004; Cohen, 2003).

The binding of ligand (Hedgehog) inhibits Patched, leading to the activation of Smo, which then activates Hedgehog signaling complex. The HSC in turn activates full length transcription factor i.e. full length Ci or Gli, which in turn exhibits transcriptional activity of Hedgehog target genes. However in the absence of Hedgehog, Ptc represses Smo preventing the activation of Hedgehog signaling pathway. The full length form of Ci or Gli is prevented from nuclear translocation through interactions with Sufu and Cos2. A portion of full length Ci is proteolytically cleaved to produce a repressor form of Ci or Gli, which enters the nucleus leading to the inhibition of Hedgehog target gene expression. In the absence of Hedgehog, in turn absence of smoothened signaling, the Ci/Gli proteins are phosphorylated by Protein kinase A, Casein kinase-1 and GSK-3. The phosphorylation of Ci/Gli leads to their ubiquitination, which are then either degraded or cleaved to fragments that act as repressors (Fig. 11).

The Wnt/Wingless pathway was first discovered in Drosophila and mouse, and is one of the most interesting signal transductions, in which key components have multiple functions. Wnt pathway diversifies into at least three branches: (1) the -catenin pathway (canonical Wnt pathway), which activates target genes in the nucleus; (2) the planar cell polarity pathway, which involves jun N-terminal kinase (JNK) and causes cytoskeletal rearrangements; and (3) the Wnt/Ca2+ pathway (Joerg and Behrens, 2002). The Wnt proteins are glycoproteins which belong to the family of secreted growth factors. However these secreted glycoproteins are modified by adding palmitate on its conserved cysteine. Wnt proteins bind to a complex of receptors i.e. Frizzled (Fz) seven-transmembrane-span receptors, which may be coupled to heterotrimeric G proteins, Low-density lipoprotein (LDL) receptor related protein (LRP) at the cell surface. These receptors transduce a signal to several intracellular proteins that include Dishevelled (Dsh), glycogen synthase kinase-3β (GSK-3), Axin, Adenomatous Polyposis Coli (APC), and the transcriptional regulator, β-catenin.

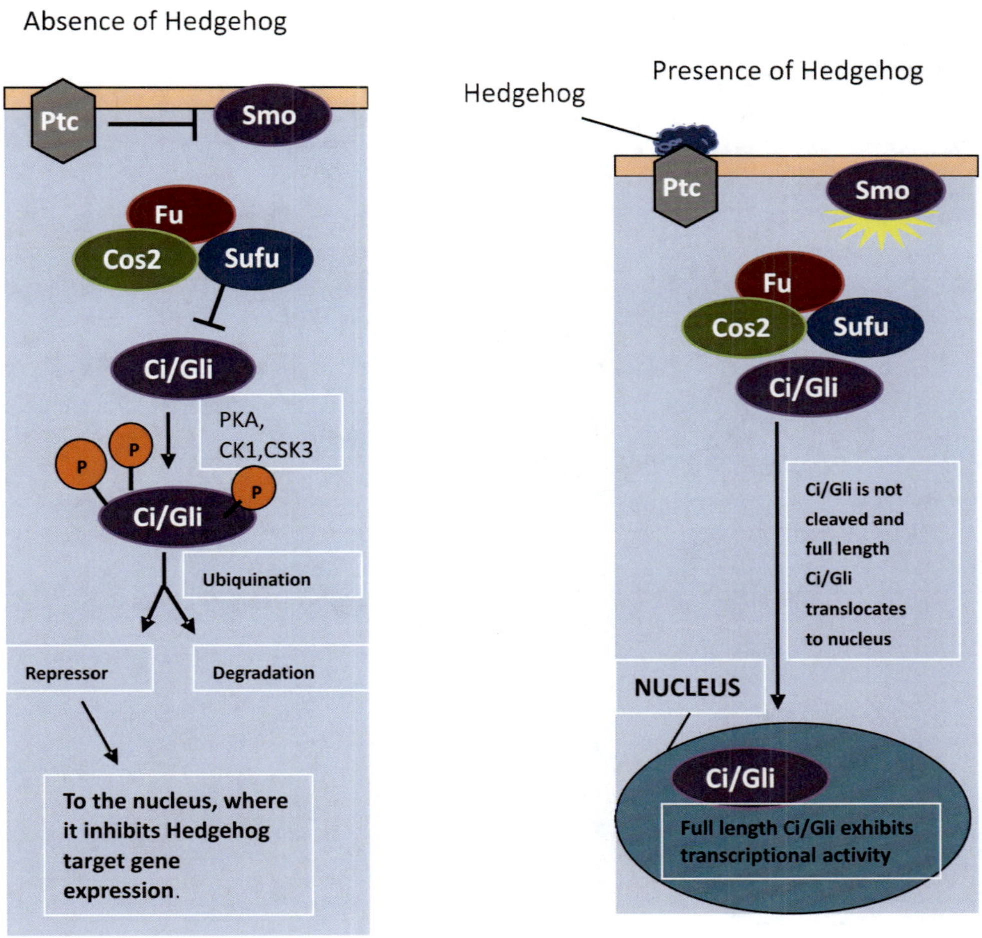

Fig. 11 : Hedgehog Signaling Pathways (See text for details)

Although the main signaling molecule of the pathway is β-catenin, APC is a critical component in the formation of multiprotein complex with axin/axin2, casein kinase I and glycogen synthase kinase 3-β (GSK3-β). Intracellularly, Wnt signalling leads to stabilization of cytosolic β-catenin. In the absence of Wnt signaling, β-catenin is phosphorylated by casein kinase-1 and GSK-3 in a complex with the proteins axin and APC, which in turn results in its ubiquitination and degradation (Fig. 12). However, the binding of Wnt leads to the phosphorylation of Dishevelled, which then promotes phosphorylation of LRP. This provides binding sites for axin, resulting in the association of axin with the receptor and disrupting the complex of axin, casein-kinase and GSK-3 with β-catenin. In response to wnt signaling, or under the circumstances of mutated APC, β-catenin is stabilized, accumulates in the cytoplasm and enters the nucleus, where it finds a partner, a member of the DNA binding protein family, the lymphoid enhancer factor-T cell factor (LEF/TCF).

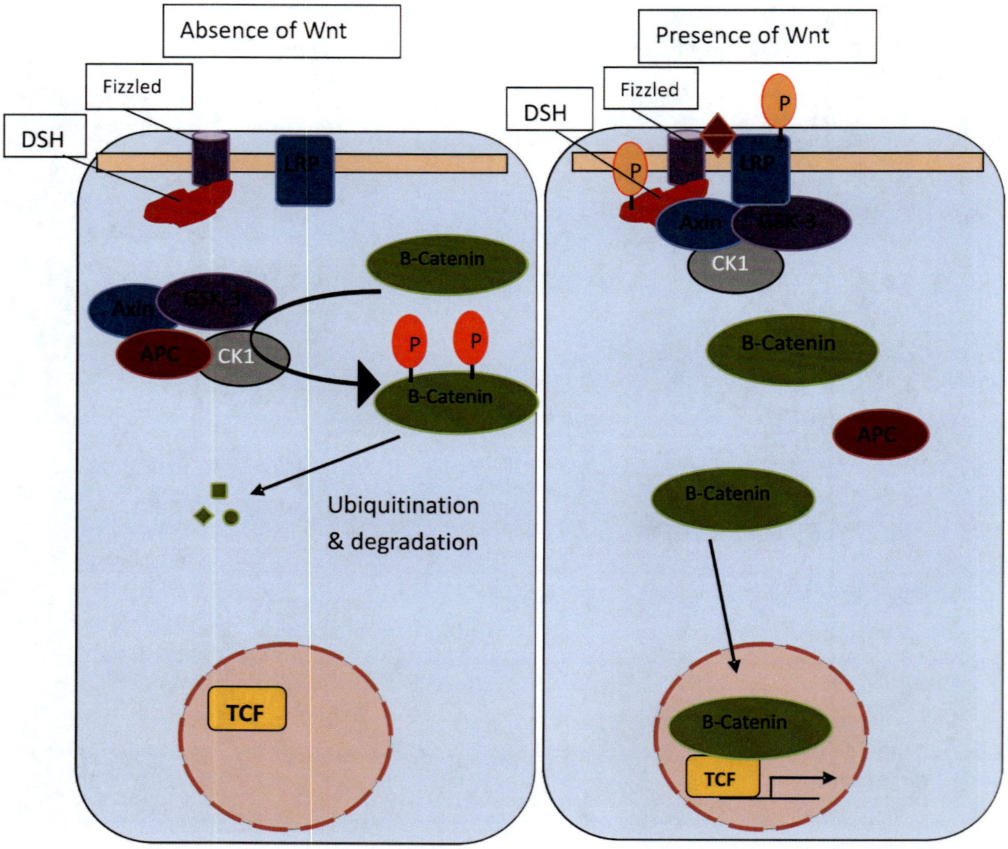

Fig. 12 : The Wnt pathway; Left panel: in cells not exposed to Wnt, β-Catenin is degraded. **Right panel:** Wnt proteins bind to Fizzled/LRP complex and in turn inhibit degradation of β-Catenin.

5.2.4. CAM kinase and PI 3- kinase pathway

Phosphatidylinositol 4, 5-bisphosphate (PIP2) is a minor component of the plasma membrane that is hydrolysed by phospholipase C to two distinct second messengers, diacylglycerol and inositol 1, 4, 5- triphosphate (IP3). These second messengers activate different signaling pathways. Diacylglycerol, for example, activates protein-serine/threonine kinases belonging to Protein kinase C family which play important roles in the control of cell growth and differentiation (Jeffrey et al. 2006). Calmodulin protein that mediates the effects of Ca2+ activates the CaM kinase family, which phosphorylates a number of different proteins, including metabolic enzymes, ion channels and transcription factors.

PIP_2 is phosphorylated to phosphatidylinositol 3, 4, 5-triphosphate (PIP3) by an enzyme phosphatidylinositide-3-kinase. The PIP3 produced in turn activates a protein-serine/threonine kinase known as Akt. This activated Akt in turn phosphorylates a number of target proteins, including proteins that are direct regulators of cell survival.

5.2.5. JAK/STAT pathway

The key elements in the JAK/STAT pathway are STAT proteins (signal transducers and activators of transcription). JAK family and STAT proteins are essential for signaling by many cytokines and growth factors following their binding to specific receptors on the cell surface. JAK signaling is involved in a number of biologic processes, including the formation and development of blood cells and the regulation of immune function.

6. Diseases Due To Deregulation of Signaling Pathways

The need for intercellular signaling is never greater than during animal development. Each cell in the embryo has to be guided along one developmental pathway or another according to its history, its position, and the character of its neighbors. At each step in the pathway, it must exchange signals with its neighbors to coordinate its behavior with theirs, ensuring the correct number and pattern of different cell types in each tissue and organ. These different signals, as discussed above are regulated by feedback loops that control the extent and duration of signaling activity. These pathways are regulated by other mechanisms also. The deregulation of these signaling pathways result in different diseases, and some of these are discussed here.

1. Myeloproliferative disorders (MPDs): The hyperactivation and dysregulation of JAK-STAT signaling pathway has been found to be the central cause for MPDs. People who suffer from MPDs report high levels of fatigue, anemia, pruritus, night sweats, bone pain, sweight loss, and debilitating splenomegaly. People with MPDs also are at risk for thrombotic events, transformation to acute leukemia, and premature death (Murray 2007).

2. The JAK dependent cytokines like interleukin-6, -12, -23 which signal through JAK signaling pathway have been implicated in autoimmune and inflammatory diseases like:

   Psoriasis- a chronic autoimmune condition, Rheumatoid Arthritis (RA) - cytokines like TNF-α, IL-6, IL-15 and GM-CSF have been found to play central role in RA.

3. Fibrodysplasia ossificans progressiva (FOP): Bone morphogenetic protein 4 (BMP 4) signaling pathway is regulated by different BMP proteins-members of the transforming growth factor (TGF)-β superfamily of proteins. BMPs interact with specific BMP receptors (BMPRs) that phosphorylate downstream effector molecules. The dysregulation of BMP-4 signaling pathways accounts for FOP. It is a rare and most severe genetic disorder of heterotopic osteogenesis in humans and results in the postnatal formation of an ectopic skeleton. In FOP for a congenital loss-of-function phenotype causing malformed great toes aswell as a postnatal gain-of-function phenotype resulting in progressive heterotopic ossification in an endochondral pattern (Frederick , 2006).

4. The mTOR which plays a central role in the regulation of cell growth receives input from multiple signaling pathways. The high levels of dysregulated mTOR are associated with several hamartoma syndromes, including tuberous sclerosis complex, the PTEN-related hamartoma syndromes and Peutz-Jeghers syndrome. These disorders are all caused by mutations in tumor-suppressor genes that negatively regulate mTOR.

Conclusion

Cell signaling is a complex system of communication which is highly important for the survival of living organisms. Microorganisms like bacteria, yeasts do not live solitary reclusive life; rather communicate by producing different type of chemicals. Eukaryotic cells interact with their environment by interpreting extracellular signals via proteins that span their plasma membrane called receptors. While plants mainly use enzyme-linked receptors, animals mainly rely on GPRC's. Plants use different growth regulators like auxin, ethylene etc for their development. Animals use three largest families of cell-surface receptors to transduce extracellular signals in different ways. These extracellular signaling mechanisms are usually programmed during development of living organism. Deregulation of many of these signaling pathways result in diseases. Studying the molecular mechanisms underlying cellular signaling will result in better understanding of the molecular players involved in various signaling pathways and may help us in designing strategies to treat diseases effectively.

References

Bonnie L. Bassler. (2002). Small talk: Cell-to-Cell Communication in Bacteria- Minireview. Cell press 109:421-424.

Bruce Alberts et al.(2007). Molecular Biology of the cell- 5th edition. Garland Science text books.

Bruce D. Gelb & Marco Tartaglia .(2006). Noonan syndrome and related disorders: dysregulated RAS-mitogen activated protein kinase signal transduction- Human Molecular Genetics.

Catriona Y. Logan and Roel Nusse. (2004). The Wnt signaling pathway in development and disease. Annu. Rev. Cell Dev. Biol 20: 781-810.

Chris Bakal et al.(2007). Quantitative Morphological Signatures Define Local SignalingNetworks Regulating Cell Morphology. DOI: 10.1126/science.1140324 Science 316,1753.

Christopher M. Waters and Bonnie L. Bassler. (2005). Quorum Sensing: Cell-to-Cell Communication in Bacteria. Cell and Developmental Biology 21: 319-346.

Cohen M. M Jr. (2003). The Hedgehog signaling network. Am. J. Med. Genet. 123A: 5-28.

Fiona M Watt et al.(2008). Epidermal Notch signaling: differentiation, cancer and adhesion. Cell.

Frederick s. Kaplan. (2006). Dysregulation of the BMP-4 Signaling Pathway in Fibrodysplasia Ossificans Progressiva. Ann. N.Y. Acad. Sci. 1068:54-65.

Geoffrey M. Cooper, Robert E. Hausman. (2009). The Cell-A molecular approach-5th edition.

Harvey Lodish et al.(2008) . Molecular Cell Biology, 6th edition.

Henke, J. M., and B. L. Bassler. (2004). J. Bacteriol. 186:6902-6914

Hooper J. E. & Scott M. P .(2005) .Communicating with Hedgehogs. Nat. Rev. Mol. Cell Biol. 6:306-317.

Jeffrey A. Engelman et al. (2006). The evolution of phosphatidylinositol 3-kinases as regulators of growth and metabolism. Nature 7: 606-619.

Joerg Huelsken & Juergen Behrens.(2002). The Wnt signaling Pathway. Journal of Cell Science 115:3977-3978.

Karen L. Visick and Clay Fuqua. (2005). Decoding Microbial Chatter: Cell-Cell Communication in Bacteria. Journal of bacteriology 187.16: 5507-5519.

Ken Inoki et al.(2005).Dysregulation of the TSC-mTOR pathway in human disease. Nature Genetics 37:19 -24.

Larissa M. Benavente and Jose M. Alonso. (2006). Molecular mechanisms of ethylene signaling in Arabidopsis. Mol. BioSyst 2:165-173.

Marco Tartaglia & Bruce D. Gelb. Disorders of dysregulated signal traffic through the RAS-MAPK pathway: phenotypic spectrum and molecular Mechanisms. Ann. N.Y. Acad.Sci. ISSN 0077-8923.20.

Marshall C.J. (1995). Specficity of receptor tyrosine kinase signaling: Transient versus Sustained extracellular signal-Regulated kinase activation. Cell 80.

Melissa B. Miller & Bonnie Melissa B. Miller & Bonnie L. Bassler.(2001).Quorum Sensing in Bacteria. Annual Review of Microbiology 55: 165-199.

Michael J. Berridge.(2005).Unlocking the secrets of cell signaling. Annu. Rev. Physiol.

Monica A. Schwartz & Hiten D. Madhani .(2004).Principles of map kinase signalling specificity in saccharomyces cerevisiae. Annu. Rev. Genet 38:725-48.

Murray PJ.(2007).The JAK-STAT pathway: input and output integration. J Immunol.178:2623-2629.

Nicolas Dard & Matthias Peter.(2006).Scaffold proteins in MAP kinase signaling: more than simple passive activating platforms. BioEssays 28:146-156.

Nives Pecina-Slaus .(20101).Wnt signal transduction pathway and apoptosis: a review.Cancer Cell International 10:22.

Robert B. McGrath & Joseph R. Ecker .(1998).Ethylene signaling in Arabidopsis: Events from the membrane to the nucleus. Plant Physiol. Biochem 36(1-2) 103-113.

Vasili Hauryliuk et al.(2008).Cofactor Dependent Conformational Switching of GTPases. Biophysical Journal 95:1704-1715.

William M. Oldham and Heidi E. Hamm. (2008). Heterotrimeric G protein activation by G-protein-coupled receptors. Nature 9:60-71.

❑❑❑

CHAPTER – 19

Biotechnological Application of Stem Cellsfor Haemoglobinopathies

Nibarana Satapathy, Y. Archana and* Bisnu Prasad Dash
Department of Biosciences and Biotechnology, Fakir Mohan University, Nuapadhi, Balasore-756020, Odisha
Email : bisnu_fmu@yahoo.co.in

Summary

Stem cells are undifferentiated cells which have the potential to divide and multiply indefinitely and can be differentiated to any type of cell of the body of an organism. Due to its peculiar characteristics stem cells are being experimentally used to cure certain dreaded diseases of the humans including different haemoglobinopathies. The stem cells isolated from bone marrow have been cultured, modified and differentiated to get various blood cells like erythrocytes, leukocytes, and thrombocytes etc. Recently haematopoietic stem cells have been transplanted to the sickle cell and beta thalassaemia patients successfully. Gene therapy using the embryonic and cord blood stem cells have been experimentally tried successfully to cure transgenic sickle as well as thalassaemic mice. Induced Pluripotent stem cells (iPS) generated from autologous skin has also been used successfully for treatment of transgenic sickle cell anaemia mouse.

Introduction

Genetic diseases have not been considered yet as important global health problems. 'Haemoglobinopathies' are the commonest genetic disorders. It is predominantly monogenic in nature and are also among the serious public health problems of the world (Weatherall and Clegg 1999). The term 'Haemoglobinopathies' is given to the inherited disorders of structure and synthesis of globin part of haemoglobin molecule and falls into several overlapping groups. It has been estimated that approximately 7%

of the world population are carriers of such disorders and 3,00,000–4,00,000 babies with severe forms of these diseases are born each year (WHO 1994). Although these conditions occur at their highest frequency in tropical regions, due to population migrations, they are now encountered in many countries of the globe (Fig-1).

Haemoglobinopathies have no universal classification. By convention, haemoglobinopathies are classified according to the qualitative nature (structural) of the resultant haemoglobin (i.e., sickle cell disease) and the quantitative (synthesis) amount of haemoglobin produced (i.e., thalassaemia). Haemoglobinopathies were traditionally detected on the basis of ionic charge differences imparted by amino acid substitutions however certain important variants are electrophoretically silent because the amino acid substitution does not alter the net charge. Quantitation of haemoglobin variants by high performance liquid chromatography (HPLC) and DNA (molecular) based studies can provide valuable information regarding the presence of haemoglobin variant in question.

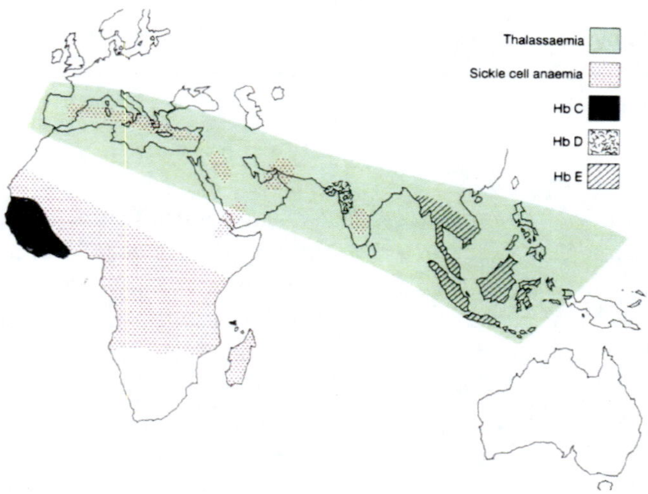

Fig.1 : Global distribution of Haemoglobin Disorders

A comprehensive knowledge on the etiology, patho-physiology, inheritance, clinical presentation and molecular variations of different haemoglobinopathies has been gained through scientific studies. But there is no permanent solution or cure for such health problems yet. The ultimate goal is to replace the defective beta globin (Gene therapy) haemoglobin gene by the normal haemoglobin gene. The discovery of stem cells and their biological potential holds promise in this regards.

Sickle Cell Disease

Sickle cell disease results from a single nucleotide substitution in which valine amino acid replaces glutamic amino acid at the sixth position of the β-globin chain of haemoglobin (Pauling, 1949; Ingram, 1957). This change causes a propensity towards polymerization of haemoglobin. The clinical manifestations of sickle cell disease arise from the tendency of sickle haemoglobin (also known as haemoglobin S [HbS] or $\alpha_2\beta^s_2$) to polymerize at reduced oxygen tensions and deform red cells into the characteristic, rigid sickle cell shape. Such inflexible red cells cannot negotiate the microcirculation efficiently; this results in a chronic haemolytic anaemia and intermittent vaso-occlusion, which affects a wide variety of organs. The clinical course is highly variable, but the complications are often life-threatening.

Red blood cells containing high levels of sickle haemoglobin contribute to the pathophysiological development of sickle cell anaemia in three major ways. First, these cells lose deformability when deoxygenated, leading to vascular obstruction and ischemia. This is a critical factor underlying painful crises, the acute chest syndrome, functional asplenia and acute stroke. Second, membrane damage shortens the life span of the red Blood cell, causing chronic intravascular and extravascular haemolysis. Intravascular haemolysis contributes to decreased availability of nitric oxide, increased vascular tone, and pulmonary-artery hypertension. Third, damaged red cells have abnormal surfaces that result in increased adherence to and damage of vascular endothelium, a process that enhances acute vaso-occlusion and also provokes the proliferation of lesion involving white cells, platelets, smooth-muscle cells, cytokines, growth factors, and coagulation proteins. Such lesions underlie large-vessel stroke and possibly pulmonary-artery hypertension (Platt et al. 1994).

Thalassaemia

The thalassaemias are a heterogeneous group of inherited anaemia caused by mutations affecting the synthesis of haemoglobin. The hallmark of thalassaemia syndromes is decreased (or absent) synthesis of one or more globin chains. Álpha (α) and beta (β) thalassaemias refer to deficits in α and β globin production, respectively. The α and β thalassaemias include clinical syndromes of varying severity. Because the structural gene for α-globin is duplicated on chromosome 16, each diploid cell contains four copies of the α-globin gene. The four α-thalassaemia syndromes: silent carrier, α-thalassaemia trait, HbH disease, and Hydrops foetalis reflect the inheritance of molecular defects affecting the output of 1, 2, 3, or 4 of the α-globin genes, respectively (Weatherall 2001).

The β-thalassaemias also include four clinical syndromes of increasing severity: silent carrier, thalassaemia trait, thalassaemia intermedia, and thalassaemia major. In contrast to the α-thalassamias, the four classes of β-thalassaemia are not correlated with the number of functioning genes. Because a single functional β-globin gene resides on each chromosome 11, a diploid cell normally has two β-globin genes. Thalassaemia is clinically heterogeneous because various genetic lesions variably impair globin-chain synthesis. However, genotypic variability at known loci is often insufficient to explain the disparate phenotypes of individual patients with the same genotype. However, the known genetic factors are insufficient to account for the marked variability, and other genetic modifiers may exist (Kong et al 2004).

Carriers (heterozygotes) of β-thalassaemia are clinically normal and largely unaware of their genetic condition. Inheritance of two copies of β-thalassaemia genes, one from each parent, usually results in life threatening anaemia and transfusion dependence for survival. Intermediate clinical forms of the disease exist as thalassaemia intermedia, which is characterised by moderately severe anaemia with the occasional need for blood transfusion. The mainstay of treatment for the severe forms (thalassaemia major) is blood transfusion and aggressive management of complications including life long iron chelation therapy. The complications of iron overload, together with the sequel of the anaemia and ineffective erythropoiesis, and the chelation therapy itself, are major causes of morbidity and mortality in the transfusion dependent β-thalassaemias (Weatherall and Clegg, 2001).

Stem Cell

Stem cells are unspecialized cells that develop into the specialized cells that make up the different types of tissue in the human body. Stem cells occur in many different somatic tissues and are important participants in their physiology. Populations of cells that derived from stem cells are organized in a hierarchical fashion, with the stem cell residing at the apex of the developmental pathway. Stem cells have three distinctive properties: self-renewal (i.e., at cell division, one or both daughter cells retain the same biologic properties as the parent cell), the capability to develop into multiple lineages, and the potential to proliferate extensively. The combination of these three properties makes stem cells unique.

Self-regeneration is the ability of stem cells to divide and produce more stem cells. During early development, the cell division is symmetrical i.e. each cell divides to gives rise to two daughter cells each with the same potential. Later in development, the cell divides asymmetrically with one of the daughter cells as produced also a stem cell and the other a more differentiated cell. A totipotent stem cell (e.g. fertilized egg) can develop into all cell types including the embryonic membranes. A pluripotent stem cell can develop into cells from all three germinal layers (cells from the inner cell mass). Other cells can be oligopotent, bipotent or unipotent depending on their ability to develop into few, two or one other cell type(s) (Pardal et al. 2003).

Stem cells can be classified into four broad types based on their origin, viz. stem cells from embryos; stem cells from the foetus; stem cells from the umbilical cord; and stem cells from the adult. Each of these types can be grouped into subtypes. Some belive that adult and foetal stem cells evolved from embryonic stem cells and the few stem cells observed in adult organs are the remnants of original embryonic stem cells that gave up in the race to differentiate into developing organs or remained in cell niches in the organs which are called upon for repair during tissue injury (Anderson et al. 2001).

Application of Stem Cell for Curing Sickle Cell Disease

Recently, the potential use of stem cells for regenerative medicine and for the treatment of genetic disease has rarely been out of the news. Discussion has focused mainly on the use of human embryonic stem cells, which have the capacity to generate in all cell types. However, initial hopes for stem-cell therapy have been somewhat dampened by both technical and ethical problems. Recent studies have therefore created a great deal of excitement.

Haematopoietic Stem Cell Transplantation

Fifty years of intensive investigation has had a disappointingly limited effect on the clinical management of the disease. The only cure for sickle cell disease is bone marrow (haematopoietic stem cell) transplantation, which can legitimately be called as adult stem-cell therapy. To date, approximately 250 patients with sickle cell disease have been treated with haematopoietic stem-cell transplantation involving stem cells donated by an HLA-identical sibling. This approach is currently reserved for those with severe disease; the rate of disease-free survival is 85%. Allogeneic HSCT has been used as one of the curative therapy for Sickle cell disease. HSCT is a suitable treatment option which has been applied to cure sickle cell disease when an HLA-matched sibling donor is available. Children and young adult sickle cell disease patient who have severe complications (e.g. stroke, recurrent ACS, refractory pain) and have an HLA-matched donor are the best candidates for transplantation. It has been estimated that about 38% of persons with SCD meet these requirements (Hoppe and Walters 2001; Horan et al. 2005; Iannone 2005; Bernaudin et al 2007; Shenoy 2007; Yesilipek 2007).

Induced Pluripotent Stem cells

Hanna et al (2007) have taken the next important step and have shown, by correcting a mouse model of sickle cell disease, how induced pluripotent stem cells might eventually be used to cure a human disease. They have shown that fully differentiated somatic cells (such as skin fibroblasts) can be reprogrammed to make cells similar to embryonic stem cells called induced pluripotent stem cells. The first approach to attempt this reprogramming was ingenious but complicated. It involved removing the nucleus from a somatic cell and using it to replace the nucleus of a fertilized egg (so called somatic-cell nuclear transfer). Factors remaining in the egg could then reprogramme the somatic nucleus as the reconstituted egg developed into an early human embryo. Epiblast cells from this embryo could then be used to establish a "customized" embryonic stem cell that would be immunologically compatible with the donor of the somatic cell. Perhaps, not surprisingly, the technical difficulties and ethical complexities of this approach were always likely to render it impractical (Yu et al. 2007). Important breakthroughs came from the laboratories of Takahashi and Yamanaka (2006) and, Thomson et al. (1998*)* have shown that mouse and human fibroblasts can be reprogrammed into cells similar to embryonic stem cells by introducing combinations of transcription factors (such as Oct4, Sox2, Klf4, Nanog, Lin28, and c-Myc) that are normally expressed in embryonic stem cells.

An experiment using a humanized mouse model of sickle cell anaemia, Hanna et al. (2007) have shown that such mice can be rescued after transplantation with the use of haematopoietic progenitors derived in vitro from autologous induced pluripotent stem cells (Fig. 2). First, induced pluripotent stem cells were made from a mouse with sickle cell disease, and the sickle mutation (β^s) was corrected in the induced pluripotent stem cells by means of conventional homologous recombination. Next, the repaired induced pluripotent stem cells were differentiated into cells similar to haematopoietic stem cells and transfected with HoxB4, which has previously been shown to confer engraftment potential on haematopoietic cells derived from embryonic stem cells. Finally, these transfected cells were transplanted back into three isogenic mice with sickle cell disease. Stable engraftment resulted in the blood containing approximately 70% of cells derived from corrected induced pluripotent stem cells and 30% endogenous haematopoietic cells. Furthermore, the red blood cell abnormalities, haemolytic anaemia, and associated pathologic features were corrected. With these important caveats, we have good reason to be optimistic. This is a key step forward in the treatment of genetic disease and in regenerative medicine.

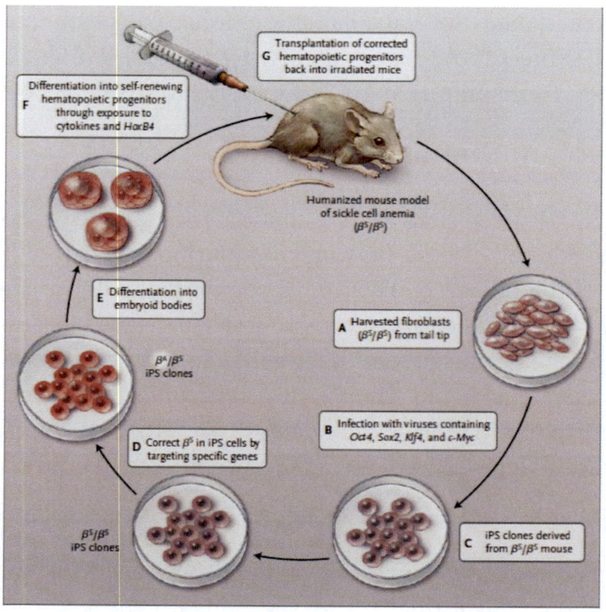

Fig. 2 : Curing Sickle Cell Anaemia in a Humanized Mouse Model, Hanna et al. 2007)

Stem Cell Gene Therapy

Isolating the adult stem cell from the sickle cell patient's bone marrow and silencing the sickle gene and introducing the normal β-globin gene can correct the sickle cell problem. The isolation of the stem cell from the sickle knockout mice and replacement of the sickle cell β-globin by the normal gene into them has been successfully achieved by several authors recently. The corrected stem cells have been grown, programmed and implanted into the bone marrow of the affected mice to have a permanent cure of the disease. Recently, lentiviral vectors have been used for β globin gene transfer to bone marrow stem cells of the sickle knockout mice (Pawliuk et al. 2001; Levasseur et al. 2003; Sadelain et al. 2005; Samakoglu et al. 2006; and Wu et al. 2006).

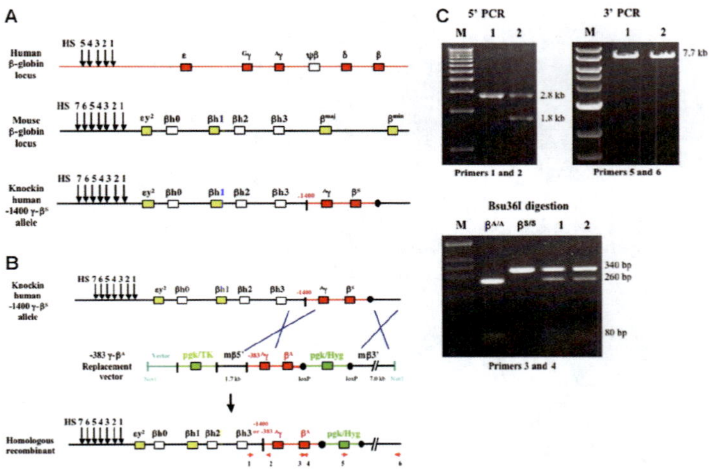

Fig. 3.a. Replacement of the β^S-globin gene with a β^A-globin gene in knock-in sickle ES cells, Wu et al. 2006).

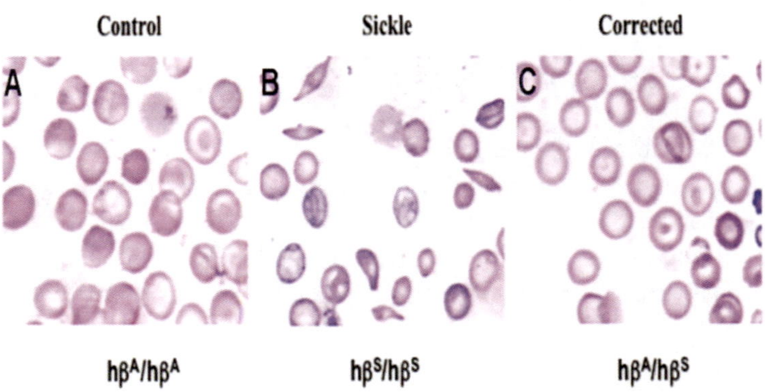

Fig. 3.b. Correction of abnormal RBC morphology in –383 γB^A–corrected mice, Wu et al. 2006)

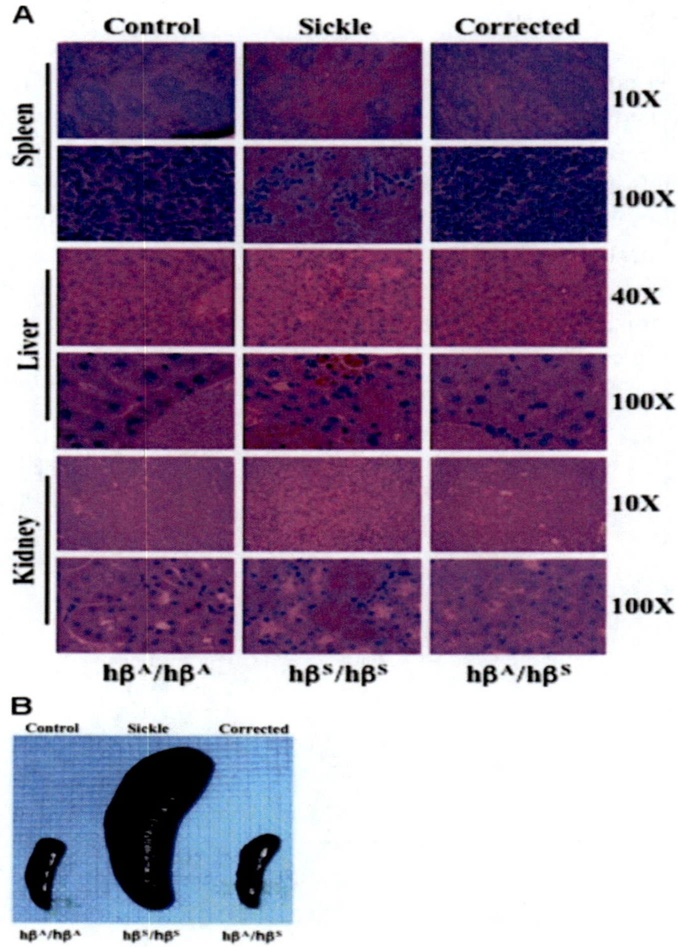

Fig.3.c. Improvement spleen, liver, and kidney pathology in –383 γ-B^A– corrected mice, Wu et al. 2006).

Application of Stem Cell for Curing Thalassaemia

Although haematopoietic stem-cell transplantation is the only available curative approach for thalassaemia, it has been limited by the high cost and the scarcity of HLA-matched, related donors. The steady-state bone marrow can be used as a safe and efficient source of haematopoietic stem cell for gene therapy of β-thalassemia (Frittoli et al. 2011). The past several years have brought progress in the realms of conditioning regimens, donor identification and selection, and the development of alternative sources of haematopoietic stem cells. Allogeneic HSCT has been considered as a potentially curative therapy for selected individuals with thalassaemia major who have an appropriate donor. HSCT has been associated with a non-negligible risk of both transplant-related mortality and morbidity. The outcome of allogeneic HSCT using an HLA-identical family donor is largely dependent on the age and clinical severity of the patient. In children who do not have liver disease and have received regular iron chelation therapy prior to transplantation, their probability of survival with transfusion independence is over 90%. For individuals with good-risk disease with an HLA-compatible sibling donor, the probability of disease free survival is 85–90%. The major cause for transplantation failure has been attributed to the graft versus host disease (GVHD). Recently the use of umbilical cord blood (CBT) has been reported to result in lower rates of GVHD (Hongeng 2006;Jaing 2005; La Nasa 2005; Locatelli and De Stephano 2005; Resnick 2007). Although the clinical application of haematopoietic cell transplantation has relied on bone marrow collected from related and unrelated donors as the primary source of donor haematopoietic cells, umbilical cord blood (UCB) is an alternative source of haematopoietic cells and represents a suitable allogeneic donor pool in the case where bone marrow donor is not available (Kanathezhath and Walters, 2010).

Gene therapy for β-thalassemia requires stable transfer of a β-globin gene into haematopoietic stem cells (HSCs) and high and regulated haemoglobin expression in the erythroblastic progeny. Miccio et al. (2008) have developed an erythroid-specific lentiviral vector driving the expression of the human β-globin gene from a minimal promoter/enhancer element containing two hypersensitive sites from the β-globin locus control region. Transplantation of transduced haematopoietic stem cells into thalassaemic mice leads to stable and long-term correction of anemia with all red blood cells expressing the transgene. A frequency of 30–50% of transduced HSCs, harboring an average vector copy number of 1/cell, was sufficient to fully correct the thalassaemic phenotype. In the mouse model of Cooley's anemia transplantation of transduced cells rescues lethality, leading to either a normal or a thalassemia intermedia phenotype (Fig-4).

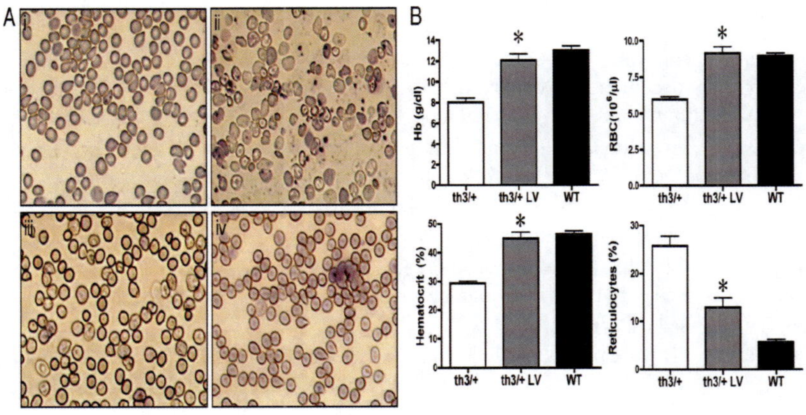

Fig. 4 : Correction of haematological parameters in mice transplanted with transduced haematopoietic stem cells, Miccio et al. 2008)

Conclusion

Stem cell based gene therapy for haemoglobinopathies in transgenic mice have been achieved experimentally but the uses for these disease in human still to be worked out. It requires knowledge about successful isolation of Potential Stem cells from affected patients. These promising results with some modification and refining will certainly help the human society in the near future to get rid of these deadly health problems. Though stem cell based gene therapy for haemoglobinopathies in transgenic mice have been achieved experimentally the cure for these diseases in human needs a thorough knowledge about the successful isolation of potential stem cells from the affected patients and transformation of these stem cells into normal stem cells before being implanted to the patients.

References

Anderson DJ, Gage FH, and Weissman IL (2001) Can stem cells cross lineage boundaries? *Nature.* 7: 393–395.

Bernaudin F, Socie G, Kuentz M, Chevret S, Duval M, Bertrand Y, Vannier JP, Yakouben K, Thuret I, Bordigoni P, Fischer A, Lutz P, Stephan JL, Dhedin N, Plouvier E, Margueritte G, Bories D, Verlhac S, Esperou H, Coic L, Vernant JP, Gluckman E (2007) Long-term results of related myeloablative stem-cell transplantation to cure sickle cell disease. *Blood.*110: 2749-2756.

Frittoli MC, Biral E, Cappelli B, Zambelli M, Roncarolo MG, Ferrari G, Ciceri F, Marktel S (2011) Bone marrow as a source of haematopoietic stem cells for human gene therapy of β-thalassemia. *Hum Gene Ther*. 22:507-513.

Hanna J, Wernig M, Markoulaki S, Sun CW, Meissner A, Cassady JP, Beard C, Brambrink T, Wu LC, Townes TM, Jaenisch R (2007) Treatment of sickle cell anemia mouse model with iPS cells generated from autologous skin. *Science.*318:1920-1923.

Hongeng S, Pakakasama S, Chuansumrit A, Sirachainan N, Kitpoka P, Udomsubpayakul U, Ungkanont A, Jootar S(2006) Outcomes of transplantation with related- and unrelated-donor stem cells in children with severe thalassemia. *Biol Blood Marrow Transplant.* 12:683-687.

Hoppe CC and Walters MC (2001) Bone marrow transplantation in sickle cell anemia. *Curr Opin Oncol*. 13:85-90.

Horan JT, Liesveld JL, Fenton P, Blumberg N, Walters MC (2005) Hematopoietic stem cell transplantation for multiply transfused patients with sickle cell disease and thalassaemia after low dose total body irradiation, fludarabine and rabbit anti-thymocyte globulin. *Bone Marrow transplant.*35:171-177.

Ingram VM (1957) Gene mutations in human haemoglobin: the chemical difference between normal and sickle cell haemoglobin. *Nature*. 180:326-328.

Iannone R, Ohene-Frempong K, Fuchs EJ, Casella JF, and Chen AR (2005) Bone marrow transplantation for sickle cell anemia: progress and prospects. *Pediatr Blood Cancer.* 44:436-440.

Jaing TH, Hung IJ, Yang CP, Chen SH, Sun CF, Chow R (2005) Rapid and complete donor chimerism after unrelated mismatched cord blood transplantation in children with beta-thalassemia major. *Biol Blood Marrow Transplant.*11:349-353.

Kong Y, Zhou S, Kihm AJ, Katein AM, Yu X, Gell DA, Mackay JP, Adachi K, Foster-Brown L, Louden CS, Gow AJ, Weiss MJ (2004) Loss of alpha-hemoglobin-stabilizing protein impairs erythropoiesis and exacerbates β-thalassemia. *J Clin Invest*. 114:1457-1466.

Kanathezhath B and Walters MC (2010) Umbilical cord blood transplantation for thalassemia major. *Hematol Oncol Clin North Am*. 24:1165-1177.

La Nasa G, Caocci G, Argiolu F, Giardini C, Locatelli F, Vacca A, et al.(2005) Unrelated donor stem cell transplantation in adult patients with thalassemia. *Bone Marrow Transplant*. 36:971-975.

Levasseur DN, Ryan TM, Pawlik Kevin M, and Townes TM (2003) Correction of a mouse model of sickle cell disease: lentiviral / antisickling β-globin gene transduction of unmobilized, purified hematopoietic stem cells. *Blood*. 102: 4312-4319.

Locatelli F and De Stephano P (2005) Innovative approaches to hematopoietic stem cell transplantation for patients with thalassemia. *Haematologica*. 90:1592-1594.

Miccio A, Cesari R, Lotti F, Rossi C, Sanvito F, Ponzoni M, Routledge SJ, Chow CM, Pinto FO, and Roberts I (2008) Cord blood stem cell transplantation for haemoglobinopathies. *Br J Haematol.* 141:309-324.

Pardal R, Clarke MF, and Morrison SJ (2003) Applying the principles of stem-cell biology to cancer. *Nat Rev Cancer*. 3: 895-902.

Pauling L, Itano HA, Singer SJ, and Wells IC (1949) Sickle cell anaemia, a molecular disease. *Science.* 110:543-548.

Platt OS, Brambilla DJ, Rosse WF, et al. (1994) Mortality in sickle cell disease: life expectancy and risk factors for early death. *N Engl J Med*.330:1639-1644.

Pawliuk R, Westerman KA, Fabry ME, Payen E, Tighe R, Bouhassira EE, Acharya SA, Ellis J, London IM, Eaves CJ, Humphries RK, Beuzard Y, Nagel RL, Leboulch P (2001) Correction of sickle cell disease in transgenic mouse models by gene therapy. *Science.* 294: 2368-2371.

Resnick IB, Aker M, Tsirigotis P, Shapira MY, Abdul-Hai A, Bitan M, et al. (2007) Allogeneic stem cell transplantation from matched related and unrelated donors in thalassemia major patients using a reduced toxicity fludarabine-based regimen. *Bone Marrow Transplant*. 40:957-964.

Sadelain M, Lisowski L, Samakoglu S, Rivella S, May C, Riviere I(2005) Progress toward the genetic treatment of the beta-thalassemias. *Ann N Y Acad Sci.* 1054:78-91.

Samakoglu S, Lisowski L, Budak-Alpdogan T, Usachenko Y, Acuto S, Di Marzo R, Maggio A, Zhu P, Tisdale JF, Rivière I, and Sadelain M(2006) A genetic strategy to treat sickle cell anemia by co-regulating globin transgene expression and RNA interference. *Nat Biotechnol*. 24: 89-94.

Shenoy S. (2007) Has stem cell transplantation come of age in the treatment of sickle cell disease? *Bone Marrow Transplant*. 40:813-821.

Takahashi K and Yamanaka S (2006) Induction of pluripotent stem cells from mouse embryonic and adult fibroblast cultures by defined factors. *Cell.* 126:663-676.

Thomson JA, Joseph IE, Sander SS, Michelle AW, Jennifer JS, Marshall VS, and Jones JM (1998) Embryonic stem cell lines derived from human blastocysts. *Science*. 282: 1145–1147.

Weatherall DJ and Clegg JB (1999) Genetic disorders of Hemoglobin, *Seminars in Hematology*, 36: 24-37.

Weatherall DJ (2001) Phenotype-genotype relationships in monogenic disease: lessons from the thalassaemias. *Nat Rev Genet*. 2:245-255.

Weatherall DJ and Clegg JB (2001) The Thalassaemia Syndromes. Blackwell Science, Oxford. U.K.

WHO (1994) Educational materials on prenatal diagnosis for sickle cell disorder, hereditary diseases programme. Geneva.

Wu Li-Chen, Sun Chiao-Wang, Ryan TM, Pawlik KM, Ren J, and Townes TM (2006) Correction of sickle cell disease by homologous recombination in embryonic stem cells. *Blood*.108: 1183-1188.

Yesilipek MA (2007). Stem cell transplantation in haemoglobinopathies. *Hemoglobin.* 31:251-256.

Yu J, Vodyanik MA, Smuga-Otto K, Antosiewicz-Bourget J, Frane JL, Tian S, Nie J, Jonsdottir GA, Ruotti V, Stewart R, Slukvin II, Thomson JA (2007) Induced pluripotent stem cell lines derived from human somatic cells. *Science*. 318:1917-1920.

❑❑❑

CHAPTER – 20

Organochlorine Pesticide Toxicity and itsImpact on Animals

Renu Bist
Department of Bioscience and Biotechnology, Banasthali University, Rajasthan - 304022
Email : renu_bisht22@yahooo.co.in

Summary

The indiscriminate use of pesticides and their persistence in the environment enhances the residue levels in meat, dairy products, eggs and food crops, consequently adversely affecting the human health (Akhtar, 1985). In India millions of tons of pesticides are applied to the agricultural fields to increase the productivity for compensation of the nutritional requirements of increasing population from the same land resources. It was the advent of green revolution that really spurred the growth of pesticide industry. Several reports published reveal that a large amount of organochlorine pesticides is present in vegetables, fruits and soft drinks besides other pesticides compared to World Health Organization (WHO) standards (Martin and Duggan, 1968; Corneliussen, 1969; Ulmann, 1972; Hildebrandt et al., 1986). Organochlorine pesticide persistence, biomagnification and toxicity adversely affect the ecological systems (Khanna et al., 2002). There are a large number of different types of pesticides which differ in their mechanisms of action. Some pesticides are very acutely toxic in humans, while others have low acute toxicity but may cause long term health effects in chronically exposed individuals. Each pesticide class has to be considered individually when predicting effects. Therefore, considering organochlorine pesticides induced disruption of body functions and possibly the involvement of free radicals is cardinal to present study.

1. Introduction

Pesticides are those chemicals which are used to kill pests. Their wide structural variety and divergent chemical properties lead to a broad range of applications. They include:

Insecticides: These target harmful or destructive insects.

Herbicides: These target weeds.

Fungicides: These target fungi.

Rodenticides: These target rodents.

Acaricides: These target acarose (mites).

An organochloride, organochlorine, chlorocarbon, chlorinated hydrocarbon, or chlorinated compound is an organic compound containing at least one covalently bonded chlorine atom. The toxicity of these agents varies according to their molecular size, volatility and effects on the CNS. In general, they cause either CNS depression or stimulation, depending upon the agent and dose (Gribble, 1998). Organochlorine pesticides act primarily by altering the movement of ions across the nerve cell membranes, thus changing the ability of the nerve to fire.

Chlorinated hydrocarbon (organochlorine) insecticides, solvents, and fumigants are widely used around the world. This class comprises a variety of compounds containing carbon, hydrogen, and chlorine. These compounds can be highly toxic, and some agents, such as DDT, which have been banned in the United States because of their slow degradation and subsequent bioaccumulation.

2. Classification

2.1. When considering the toxicity of pesticides, it is important to remember that these chemicals were developed to kill organisms, and thus they are inherently acutely toxic. To protect the public, WHO developed a hazard classification system which was used to label all pesticide containers to warn users of the acute hazards associated with each product. This hazard system was based on the LD50 for the pesticide in rats under either oral or dermal exposure conditions (see table 1):

Table 1 : The WHO recommended classification of pesticides by hazard:

		LD_{50} FOR THE RAT (mg/kg body weight)			
		ORAL		DERMAL	
CLASS		SOLIDS	LIQUIDS	SOLIDS	LIQUIDS
Ia	Extremely hazardous	≤5	≤20	≤10	≤40
Ib	Highly hazardous	5-50	20-200	10-100	40-400
II	Moderately hazardous	50–500	200–2000	100–1000	400–4000
III	Slightly hazardous	>500	>2000	>1000	>4000
III+	Unlikely to present hazard in normal use	>2000	>3000	—	—

Source : From Copplestone, 1988,

2.2. Depending on structure, these organochlorine pesticides can be separated into five groups, as follows:

1. Dichlorodiphenyltrichloroethane (DDT) and analogues (eg, dicofol, methoxychlor)
2. Hexachlorocyclohexane (ie, benzene hexachloride) and its isomers (eg, α, β, γ, δ, ε).
3. Cyclodienes (eg, chlordane, heptachlor, aldrin, dieldrin, endrin, endosulfan, isobenzan)
4. Chlordecone, kelevan, and mirex
5. Toxaphene

Table 2 : Structural classification of organochlorine pesticides

Dichlorodiphenylethanes	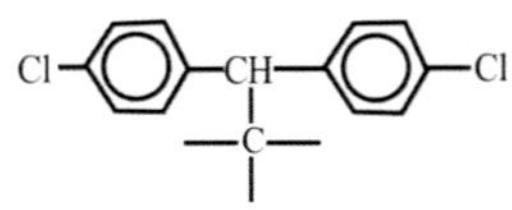	DDT.DDD Dicofol Perthane Methoxychlor Methlochlor
Cyclodienes	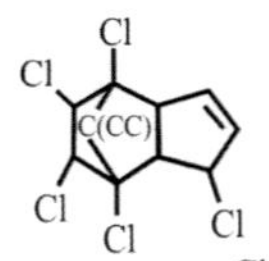	Aldrin, Dieldrin Heptachlor Chlordane Endosulfan
Chlorinated Benzenes Cyclohexanes	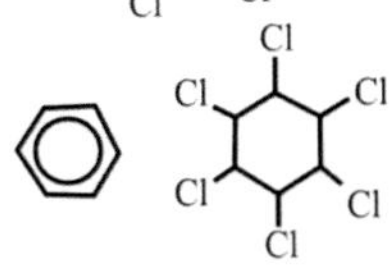	HCB. HCH Lindane (a-BHC)

Table 2 shows three major classes of organochlorine insecticides. All of these are organic compounds with chlorine (Cl) atoms attached to the ring structures. The Cl atoms prevent the organic compounds from being rapidly degraded in the environment, thus these pesticides "persistent" and are active for long periods of time after application.

3. Physical Properties

Chloride substituents modify the physical properties of organic compounds in several ways. They are typically denser than water due to the presence of high atomic weight of chlorine. Chloride substituents induce stronger intermolecular interactions than hydrogen substituents. The effect is illustrated by trends in boiling points: methane (-161.6°C), methyl chloride (-24.2 °C), dichloromethane (40°C), chloroform (61.2°C), and carbon tetrachloride (76.72°C). The increased intermolecular interactions are attributed due to the effects of vander waals bond and polarity.

4. Natural occurrence

Many organochlorine compounds have been isolated from natural sources ranging from bacteria to humans (Engvild, 1986; Gribble, 1999). Chlorinated organic compounds are found in nearly every class of biomolecules including alkaloids, terpenes, amino acids, flavonoids, steroids, and fatty acids(Gribble, 1998; Gribble, 1999). Organochlorides, including dioxins, are produced in the high temperature environment of forest fires, and dioxins have been found in the preserved ashes of lightning-ignited fires that predate synthetic dioxins. In addition, a variety of simple chlorinated hydrocarbons including dichloromethane, chloroform, and carbon tetrachloride have been isolated from marine algae (Gribble, 1996). A majority of the chloromethane in the environment is produced naturally by biological decomposition, forest fires, and volcanoes. The natural organochloride epibatidine, an alkaloid isolated from tree frogs, has potent analgesic effects and has stimulated research into new pain medication.

5. Preparation

Organochlorine pesticides are prepared in one of the following ways:

5.1. From Chlorine

Alkanes and arylalkanes may be chlorinated under free radical conditions, with UV light. However, the extent of chlorination is difficult to control. Aryl chlorides may be prepared by the Friedel-Crafts halogenation, using chlorine and a Lewis acid catalyst. The haloform reaction, using chlorine and sodium hydroxide, is also able to generate alkyl halides from methyl ketones, and related compounds. Chloroform was formerly produced thus.

Chlorine adds to the multiple bonds on alkenes and alkynes as well, giving di- or tetra-chloro compounds.

5.2. Reaction with hydrogen chloride

Alkenes react with hydrogen chloride (HCl) to give alkyl chlorides. For example, the industrial production of chloroethane proceeds by the reaction of ethylene with HCl:

$$H_2C{=}CH_2 + HCl \rightarrow CH_3CH_2Cl$$

Secondary and tertiary alcohols react with the Lucas reagent (zinc chloride in concentrated hydrochloric acid) to give the corresponding alkyl halide; this reaction a method for classifying alcohols:

$$R\text{–}OH + HCl \xrightarrow[\Delta]{Zncl_2} R\text{–}Cl + H_2O$$

5.3. Other Chlorinating Agents

In the laboratory, alkyl chlorides are most easily prepared by reacting alcohols with thionyl chloride ($SOCl_2$), phosphorus trichloride (PCl_3), or phosphorus pentachloride (PCl_5):

$$ROH + SOCl_2 \rightarrow RCl + SO_2 + HCl$$

$$3\ ROH + PCl_3 \rightarrow 3\ RCl + H_3PO_3$$

$$ROH + PCl_5 \rightarrow RCl + POCl_3$$

In the laboratory, thionyl chloride is especially convenient, because the byproducts are gaseous.

Alternatively, the Appel reaction:

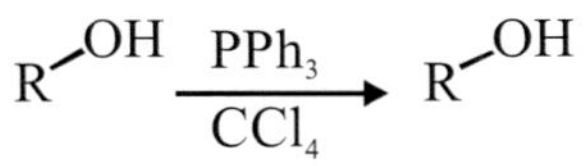

Alkyl chlorides are versatile building blocks in organic chemistry. While alkyl bromides and iodides are more reactive, alkyl chlorides tend to be less expensive and more readily available. Alkyl chlorides readily undergo attack by nucleophiles.

Heating alkyl halides with sodium hydroxide or water gives alcohols. Reaction with alkoxides or aroxides give ethers in the Williamson ether synthesis; reaction with thiols give thioethers. Alkyl chlorides readily react with amines to give substituted amines. Alkyl chlorides are substituted by softer halides such as the iodide in the Finkelstein reaction. Reaction with other pseudohalides such as azide, cyanide, and thiocyanate are possible as well. In the presence of a strong base, alkyl chlorides undergo dehydrohalogenation to give alkenes or alkynes.

Alkyl chlorides react with magnesium to give Grignard reagents, transforming an electrophilic compound into a nucleophilic compound. The Wurtz reaction reductively couples two alkyl halides to couple with sodium.

6. Applications

6.1. Vinyl chloride

The largest application of organochlorine chemistry is the production of vinyl chloride. The annual production in 1985 was around 13 billion kilograms, almost all of which was converted into polyvinylchloride (PVC).

6.2. Chloromethanes

Most low molecular weight chlorinated hydrocarbons such as chloroform, dichloromethane, dichloroethene, and trichloroethane are useful solvents. These solvents tend to be relatively non-polar; they are therefore immiscible with water and effective in cleaning applications such as degreasing and dry cleaning. Several billion kilograms of chlorinated methanes are produced annually, mainly by chlorination of methane:

$$CH_4 + x\ Cl_2 \rightarrow CH_{4-x}Cl_x + x\ HCl$$

The most important is dichloromethane, which is mainly used as a solvent. Chloromethane is a precursor to chlorosilanes and silicones. Historically significant, but smaller in scale is chloroform, mainly a precursor to chlorodifluoromethane ($CHClF_2$) and tetrafluoroethene which is used in the manufacture of Teflon (Magnus et al., 1997).

6.3. Pesticides

Many pesticides contain chlorine. Notable examples include DDT, dicofol, heptachlor, endosulfan, chlordane, aldrin, dieldrin, endrin, mirex, and pentachlorophenol. These can be either hydrophilic or hydrophobic depending on their molecular structure. Many of these agents have been banned in various countries, e.g. mirex, aldrin.

6.4. Insulators

Polychlorinated biphenyls (PCBs) were once commonly used electrical insulators and heat transfer agents. Their use has generally been phased out due to health concerns. PCBs were replaced by polybrominated diphenyl ethers (PBDEs), which bring similar toxicity and bioaccumulation concerns.

7. Toxicity Mechanism

Organochlorine pesticides have significant toxicity to plants or animals, including humans. They exhibit both acute and chronic toxicity. For example, Dioxins are produced when organic matter is burned in the presence of chlorine show persistence and are toxic to various organ systems of body. Some insecticides such as DDT are persistent organic pollutants which pose dangers when they are released into the environment. DDT was widely used to control insects in the mid 20th century, also accumulates in food chains, and causes reproductive problems (as eggshell thinning) in certain bird species.

Table 3 : Categories of organochlorine pesticides and their toxic potential:

	Acute Oral Toxicity
DDT and analogs	Low to moderate
Methoxychlor	Low
Benzene hexachloride	
Gamma-hexachlorobenzene (Lindane)	Moderate
Cyclodienes and related compounds	
Aldrin	High
Chlordance	Moderate
Chlordencone (Kepone)	Moderate
Dieldrin	High
Endosulfan (Thiodan)	High
Endrin	High
Heptachlor	Moderate
Isobenzan	High
Mirex (Dechlorane)	Low
Toxaphene and related compounds	
Toxaphene	Moderate

When chlorinated solvents, such as carbon tetrachloride, are not disposed of properly, they accumulate in groundwater. Some highly reactive organochlorides such as phosgene have even been used as chemical warfare agents.

However, the presence of chlorine in an organic compound does not ensure toxicity. Many organochlorides are safe enough for consumption in foods and medicines. For example, peas and broad beans contain the natural chlorinated plant hormone 4-chloroindole-3-acetic acid (4-Cl-IAA) (Pless et al., 1984; Magnus et al., 1997) and the sweetener sucralose (Splenda) is widely used in diet products. Among total 2004, at least 165 organochlorides are approved worldwide for use as pharmaceutical drugs, including the natural antibiotic vancomycin, the antihistamine loratadine (Claritin), the antidepressant sertraline (Zoloft), the anti-epileptic lamotrigine (Lamictal), and the inhalation anesthetic isoflurane (Magnus et al., 1997). Rachel Carson brought the issue of DDT pesticide toxicity to public awareness with her 1962 book *Silent Spring*. While many countries have phased out the use of some types of organochlorides such as the US ban on DDT, persistent DDT, PCBs, and other organochloride residues continue to be found in humans and mammals across the planet many years after production and use have been limited. In Arctic areas, particularly high levels are found in marine mammals. These chemicals concentrate in mammals, and are even found in human breast milk. Males typically have far higher levels, as females reduce their concentration by transfer to their offspring through breast feeding.

The basic mechanism of action for most pesticides is an alteration in the transfer of a signal along a nerve fiber and across the synapse from one nerve to another or from a nerve to a muscle fiber. The transfer of a signal along a nerve occurs by changes in the electrical potential across the nerve cell membrane which is created by the movement of ions in and out of the cell. At the terminal end of a nerve, the signal is transferred across the synapse to the next nerve cell by the release of neurotransmitters such as acetylcholine (Ach). Different classes of pesticides inhibit this process in different ways, but the end result is an alteration in normal nerve signal propagation. Organochlorine pesticides act primarily by altering the movement of ions across the nerve cell membranes, thus changing the ability of the nerve to fire.

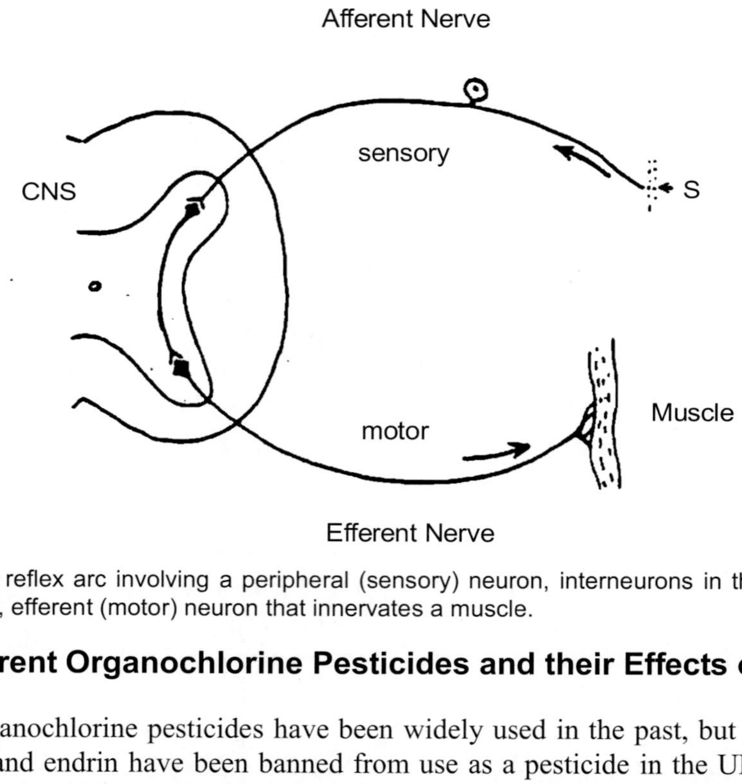

Fig. 1 : A reflex arc involving a peripheral (sensory) neuron, interneurons in the CNS and a peripheral, efferent (motor) neuron that innervates a muscle.

8. Different Organochlorine Pesticides and their Effects on Health

Organochlorine pesticides have been widely used in the past, but aldrin, DDT, dieldrin and endrin have been banned from use as a pesticide in the UK (PSD/HSE 1998). However, the continued persistence and wide distribution of these substances in the marine environment remain a concern. They are List I and Red List (except isodrin) substances. Two other organochlorine pesticides remain in use: lindane and endosulfan. Majority of organochlorine pesticides are persistent in environment and their toxic activities are due to their nature and the intermediates which are formed during their metabolism. All of these organochlorine pesticides exhibit their effects by production of free radicals thereby causing lipid peroxidation and alteration in membrane fluidity.

Lindane is one such organochlorine pesticide that was reported to induce lipid peroxidation in mice brain (Bist and Bhatt, 2009; Bist et al., 2010). Since, organochlorine pesticides show their effects by accelerating rate of free radicals generation therefore it is obvious that they affect each and every organ-system in the body. There is a big list of organochlorine compounds which are toxic in nature but current chapter highlights those organochlorine pesticides whose exposure is easy to humans and which show devastating effects in tissues. Few such organochlorine compounds are mentioned as following:

8.1. Aldrin

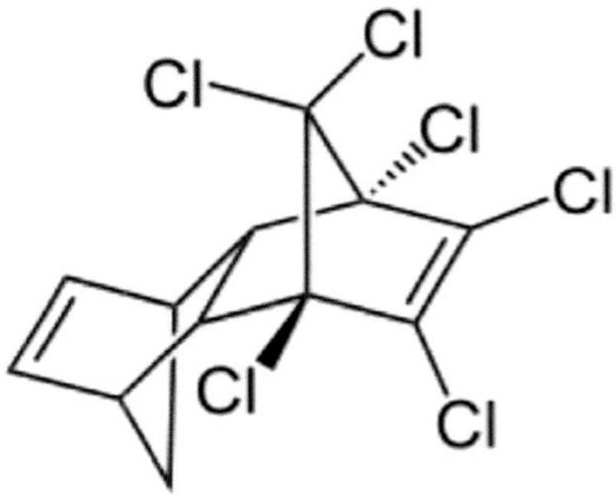

Fig. 2 : Chemical structure of Aldrin

Aldrin is an organochlorine insecticide that was widely used until the 1970s. Later on, it was banned in most countries. It is a colourless solid. The technical compounds are light tan to brown solids or powders. Before the ban, it was heavily used as a pesticide to treat seed and soil. Aldrin is a persistent organic pollutant (Metcalf, 2002).

Aldrin is readily converted to dieldrin in the environment, so these two closely related compounds are considered together by regulatory bodies. Their toxicities do not differ significantly. Aldrin has been used as a soil insecticide to control root worms, beetles, and termites. Dieldrin has been used in agriculture for soil and seed treatment and in public health to control disease vectors such as mosquitoes and tsetse flies.

Mode of exposure: The most likely way that aldrin and dieldrin have entered the human body in the past is by eating food grown in treated soil, by eating products from animals previously exposed to these insecticides (fish, poultry or beef), or by drinking water or milk containing the compounds.

Formation: Aldrin is produced by combining hexachlorocyclopentadiene with norbornadiene in a Diels-Alder reaction to give the product.

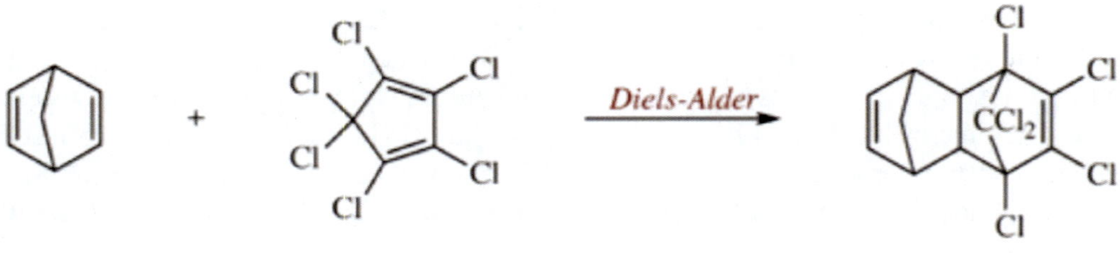

Health effects: Brief exposures at high levels (short-term exposure) of aldrin and dieldrin cause headache, dizziness, irritability, loss of appetite, nausea, muscle twitching, convulsions, and loss of consciousness; death may occur at extremely high exposures or doses. All symptoms disappear with time after removal from a nonlethal exposure. The use of protective clothing and respirators is necessary under conditions when high exposures may occur.

8.2. Dieldrin

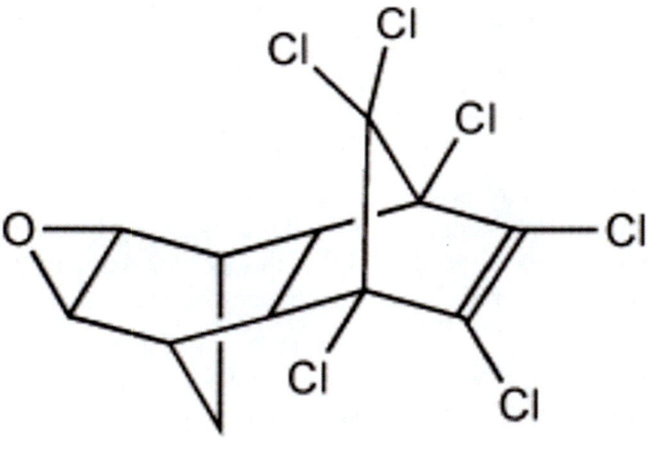

Fig 3: Chemical structure of Aldrin

Dieldrin has a veterinary use as a sheep dip and has been used in treatment of wood and mothproofing of woolen products. Most uses for aldrin and dieldrin were banned in 1975; at present these compounds are no longer produced in or imported into the United States.

Although dieldrin is persistent in soil, environmental background levels are known to be decreasing slowly. Residual contamination may be present at waste sites from the disposal of used stocks. Where this is appreciable, a potential exists for exposure to cleanup workers. Air and water appear to be sources of minor importance to the general population with regard to aldrin and dieldrin exposure. In the past, food products grown in soil treated with aldrin or dieldrin have probably been the primary source of dieldrin residues in fatty tissues of the general population; however, since 1970, dietary intake has shown a significant decrease.

Mode of exposure: Persons could be occupationally exposed to aldrin and dieldrin from inhalation and by absorption through the skin.

Health effects: Studies with animals fed dieldrin have shown that the liver can be damaged and the ability of the immune system to protect against infections can be suppressed. Oral doses of aldrin and dieldrin have caused liver cancer in mice but not in rats. Although there is an inadequate evidence to judge whether aldrin/dieldrin are carcinogenic in humans, the Environmental Protection Agency (EPA) considers aldrin and dieldrin probable carcinogens based on sufficient evidence in animals.

8.3. DDT

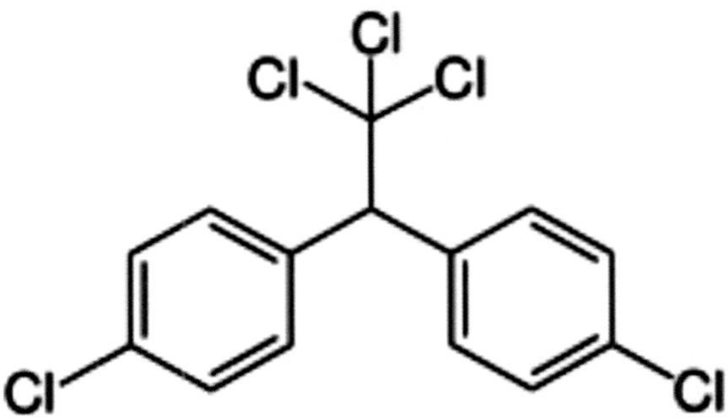

Fig. 4 : Chemical formula of DDT

DDT (dichlorodiphenyltrichloroethane) is a chemical with a long, unique, and controversial history. First synthesized in 1874, DDT's insecticidal properties were not discovered until 1939, and it was used with great success in the second half of World War II to control malaria and typhus among civilians and troops. The Swiss chemist Paul Hermann Müller was awarded the Nobel Prize in Physiology or Medicine in 1948 "for his discovery of the high efficiency of DDT as a contact poison against several arthropods." After the war, DDT was made available for use as an agricultural insecticide, and soon its production and use skyrocketed. The book 'Silent Spring' written by Rachel Carson in 1962, catalogued the environmental impacts of the indiscriminate spraying of DDT in the US and questioned the logic of releasing large amounts of chemicals into the environment without fully understanding their effects on ecology or human health. This book suggested that DDT and other pesticides may cause cancer and that their agricultural use was a threat to wildlife, particularly birds. After that DDT was banned in the US in 1972 (Lear and Linda, 1997) but its limited use in disease vector control continues to this day and remains controversial.

Along with the passage of the Endangered Species Act, the US ban on DDT is cited by scientists as a major factor in the comeback of the bald eagle, the national bird of the United States, from near-extinction in the contiguous US (Stokstad, 2007). At present it is a well known chemical which causes toxicity in the body. Once entering the food chain, it magnifies at each and every trophic level.

8.4. Endrin

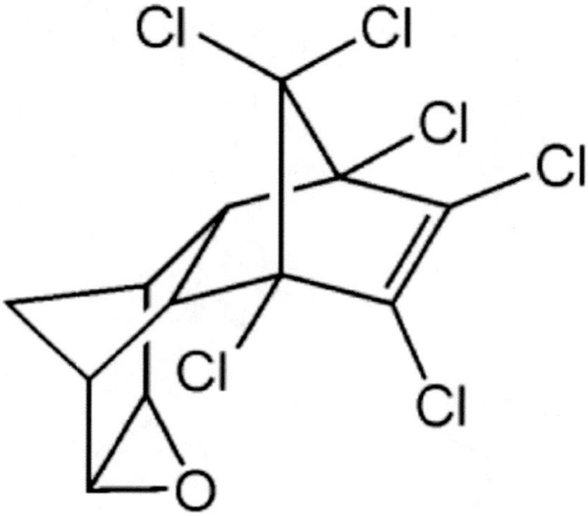

Fig. 5 : Chemical structure of Endrin

Endrin is an organochloride that was primarily used as an insecticide. It is a colourless odorless solid, although commercial samples are often off-white. It is also a rodenticide. This compound became infamous as persistent organic pollutant and for this reason is banned in many countries (Metcalf, 2002).

Production and Uses

Endrin is produced via a multistep route from hexachlorocyclopentadiene.[2] Diels-Alder addition of acetylene gives an isomer of hexachloronorbornadiene. This compound undergoes a second Diels-Alder addition, this time with cyclopentadiene. The resulting polycyclic derivative is epoxidized to give endrin. Endrin is a stereoisomer of dieldrin.

The majority of endrin (about 80%) was consumed as a spray to control insect pests of cotton. It was also used on rice, to some extent on sugar cane, in a limited way on grain crops and sugar beets, and in Australia on tobacco and cole crops. It was occasionally used in orchards as a control of rodents, where it is sprayed on the ground under the trees in autumn or spring, often as a solution in mineral oil. As a seed treatment, it was used for cotton seed in the United States, and for beans seeds in Australia.

Environmental Problems

The use of endrin is banned in many countries. Like related organochlorine pesticides, it is lipophilic. Thus, it tends to accumulate in fatty tissues of organisms living in water. Some estimates indicate its half-life in soil is over 10 years. In comparison with dieldrin, endrin is less persistent in the environment.

Safety

Endrin is toxic with an LD_{50} is 17.8 and 7.5 mg/kg (oral, rat). Acute endrin poisoning in humans affects primarily the nervous system. Food contaminated with endrin caused several clusters of poisonings worldwide, especially affecting children. Orally ingested endrin is eliminated mostly in feces.[3] It is very toxic to aquatic organisms, namely fish, aquatic invertebrates, and phytoplankton. The U.S. EPA has set a freshwater acute criterion of 0.086ug/L and a chronic criterion of 0.036ug/L. In saltwater, the numbers are acute 0.037 and chronic 0.0023ug/L. Human health contaminate criterion for water plus organism is 0.059ug/L. Drinking water limits or maximum contaminant level (mcl) is set to 2 ppb.

8.5. Lindane

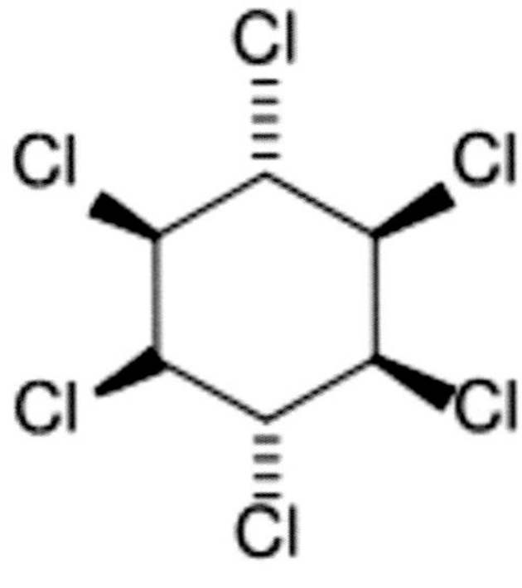

Fig. 6: Chemical structure of lindane

γ-Hexachlorocyclohexane (γ-HCH) also known as lindane is a white crystalline organic solid that turns into a vapor when released into the air. Once released, it looks colorless but has a musty odor (Reigart and Roberts, 1999). It is a member of the organochlorinated pesticides used extensively in agriculture worldwide during the 1970s. The United States Environmental Protection Agency (EPA) has declared lindane to be a persistent, bioaccumulative and toxic chemical.

The use of this pesticide has been restricted in most industrialized countries including US and California; however, it continues to be used in some regions of Europe and Asia for insect control, especially in programs for the prevention of malaria (Popp et al., 2000). A notice was announced by EPA's issuance of final orders cancelling the registrations of all pesticide products containing the pesticide lindane (U.S. EPA Federal Register, 2006).

It is used as an insecticide and as a therapeutic scabicide, pediculicide and ectoparasiticide for humans and animals (Budavari et al., 1989). The pure g-isomer of HCH (of at least 99.5% purity) was named lindane after the Belgian chemist Van Der Lindane, who discovered it in 1912. Its insecticidal properties were first demonstrated by Bender in 1933 (Ulmann, 1972).

Table 4 : Physical and Chemical Properties of lindane

Property	**Value**	**References**
Molecular mass	290.828 g/mol	Windholz *et al.*, 1983
Color	White	Ulmann, 1972
Physical state	Crystalline solid	Ulmann, 1972
Odor	Faint musty to odorless	Ulmann, 1972
Melting point	112.3-113°C	IARC, 1979
Boiling point	323.4°C	IARC, 1979
Vapor pressure	0.03 mm Hg (20°C) 9.4×10^{-6} Hg (20°C)	IARC, 1979 Windholz *et al.*, 1983
Solubility in water	10 mg/L	IARC, 1979
Solubility in other solvents	>5g/100g in acetone, benzene, ethanol and ethyl acetate	
Density	1.85 g/ml	IARC, 1979
Relative gas density	10	
Half life	18 hours	
Protein binding	91%	
EC 50	20.34 mg/ml	

EC: Effective concentration; **IARC:** International Agency for Research on Cancer

Modes of Exposure

There are several ways of γ-HCH exposure such as by eating contaminated food, meat or dairy product (milk), and breathing and by contact at work place. The exposure through contaminated water or by breathing γ-HCH released from waste sites or landfills is another means of exposure. The use of soaps, lotions or shampoo containing γ-HCH that helps to treat and control head and body lice, and scabies also enhances the absorption through skin. Nursing mother that have been exposed to γ-HCH can pass it on to babies in breast milk. This environmentally persistent xenobiotic enters animal tissues through food chain, respiration or dermal contact and is deposited in the membrane bilayer of all cell types (Zhu et al., 1986). Lindane and its metabolites persist in the environment, consequently affecting the human being and wildlife.

Health Effects

Lindane affects almost each and every organ system in the body.

A) Acute Toxicity

Lindane is a central nervous system stimulant. Symptoms of acute exposure in humans can include mental and motor retardation, central nervous system excitation, clonic (intermittent) and tonic (continuous) convulsions, respiratory failure, pulmonary edema and dermatitis.

Other symptoms in humans are more behavioral in nature such as loss of balance and somersaulting (Smith, 1991), grinding of the teeth, and hyperirritability. It was reported that lindane exerts hazardous effects at the neuroendocrine level in birds (Pešut-Mikinèiè et al., 1987). Highly acute dose of lindane is reported to increase lipid peroxidation and decrease in membrane bound enzymes such as acetylcholinesterase (AChE) and adenosine triphosphatase (ATPase) (Bist and Bhatt, 2010; Bist et al., 2010).

Most acute effects have been due to accidental or intentional ingestion, although inhalation occurred (especially among children) when it was used in vaporizers (U.S. National Library of Medicine). The lowest oral dose which may be lethal for a child is estimated to be 180 mg/kg. Lotions (10%) applied for scabies have resulted in severe intoxication in some children and infants.

B) Chronic Toxicity

Female rats exposed to lindane were found to influence estrous cycle whereas the seminiferous tubules become atrophied in male rats (Smith, 1991). Lindane at the concentration of 60 mg/kg bw increased the fetal mortality and decreased the fetal weights (Frohberg and Bauer, 1972). Female rats that received lindane daily during pregnancy had higher post implantation embryonal mortality than controls.

It can be passed from the mother to the developing fetus (U. S. National library of medicine, 1992). It appears that lindane will not cause developmental effects at low levels of exposure and causes reproductive effects at levels approaching the acute toxicity doses.

Lindane was mutagenic in an experiment of dominant lethal test carried out on Drosophila melanogaster (Sinha and Sinha, 1983). It induced chromosomal breakage in the lymphocytes, *in vitro* (Tzoneva-Maneva et al., 1971). Rocchi et al. (1980) reported that in lymphocytes lindane disturbs the schedule of DNA synthesis and inhibits DNA repair after damage by ultraviolet rays.

Cytotoxicity of lindane was found to be more compared to DDT and chlordane pesticides. Moreover, it is a powerful mouse skin tumor promoter (Tsushimoto *et al.,* 1983).

Increased liver weights were recorded in the mice receiving 600 mg/kg diet lindane (Goto et al., 1972). In 1987, the International Agency for Research on Cancer (IARC) classified the lindane as a possible human carcinogen and included it in the category 2B (IARC, 1998).

A dose dependent accumulation of lindane was observed in the brains of male Wistar rats (Tusell et al., 1988). In rats lindane lowered the threshold limit for kindled seizures (Joy et al., 1982, 1983). It is reported to inhibit GABA mediated neurotransmission (Woodward et al., 1992; Pomes et al., 1994)

Environmental fate: Lindane is highly persistent in most soils, with a field half life of approximately 15 months (Wauchope et al., 1992). It is a contaminant in water in the great lakes at very low concentrations as well. It is very stable in both fresh and salt water environments. Lindane is highly persistent and produces many intermediates which may also cause toxicity. Fig. 2 shows the proposed metabolism of lindane.

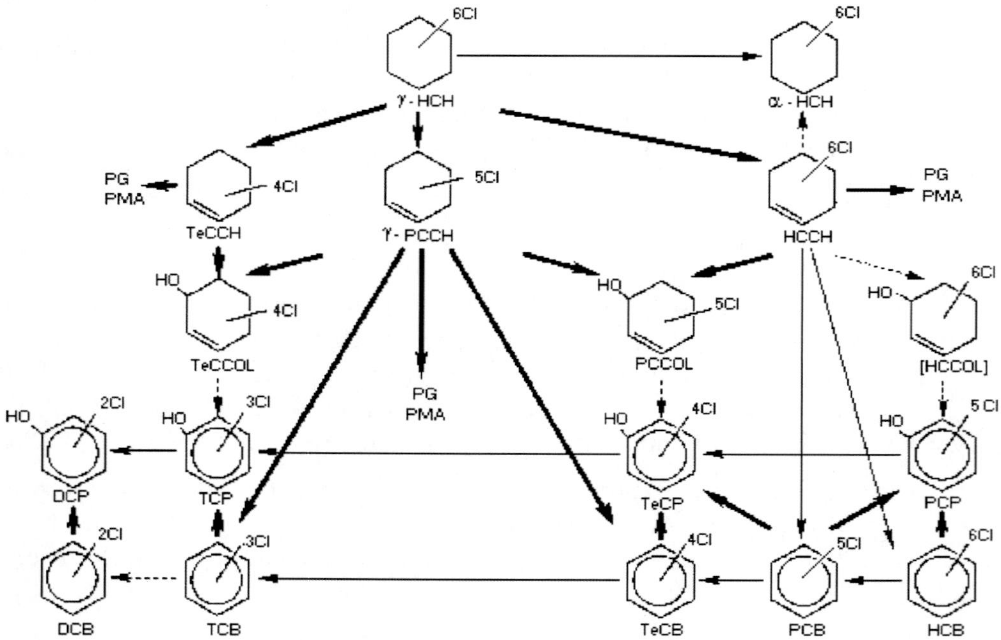

Fig. 7 : Metabolic pathways of γ-HCH metabolism (From Engst et al., 1978)

[→ Confirmed major pathways → Confirmed subordinate pathways,→ Hypothetical but unconfirmed pathways]

[Note: DCB, dichlorobenzene; DCP, dichlorophenol; HCB, hexachlorobenzene; HCCH, hexachlorocyclohexene; HCCOL, hexachlorocyclohexenol; alpha-HCH, alpha-1,2,3,4,5,6-hexachlorocyclohexane; γ-HCH, γ-1,2,3,4,5,6 hexachlorocyclohexane; PCB, pentachlorobenzene; γ-PCCH, γ-2,3,4,5,6-pentachlorocyclohexene; PCCOL, 2,3,4,5,6-pentachlorocyclohexene-(2)-ol-(1); PCP, pentachlorophenol; TCB, trichlorobenzene; TCP, tricholorophenol; TeCCH, 3,4,5,6-tetrachlorocyclohexene; TeCCOL, tetrachlorocyclohexenol; TeCB, tetrachlorobenzene; TeCP, tetrachlorophenol; PG, glucuronide conjugate; PMA , sulfate conjugate]

8.6. Chlordane

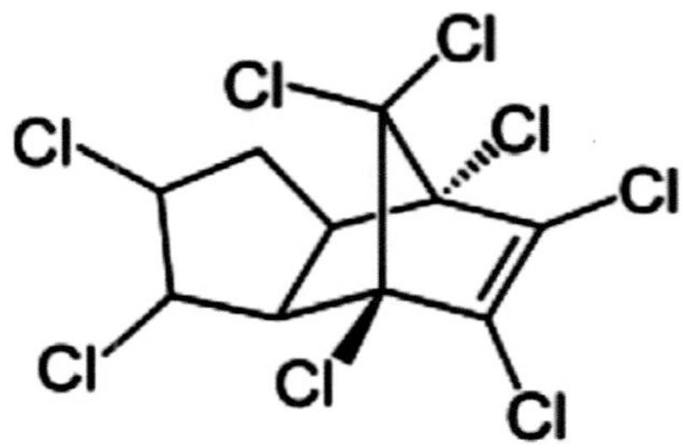

Fig. 8 : Chemical structure of chlordane

Chlordane is a white solid which was sold in the U.S. upto 1983. After that it was used as an insecticide for crops like corn and citrus and on lawns and domestic gardens.

Production and Uses

Chlordane is a cyclodiene pesticide, meaning that it is derived from hexachlorocyclopentadiene. Hexachlorocyclopentadiene forms an adduct with cyclopentadiene, and chlorination of this adduct give two isomers, α and β. The mixture is called chlordane. The β isomer is more bioactive.

United States Environmental Protection Agency (EPA) banned all uses of chlordane in 1983 except termite control. The EPA banned all uses of chlordane in 1988. The EPA recommends that children should not drink water with more than 60 parts of chlordane per billion parts of drinking water (60 ppb) for longer than 1 day.

Mode of Exposure and Excretion

Pathways of exposure to Chlordane include ingestion of crops grown in Chlordane contaminated soil, ingestion of high fat foods such as meat, fish, and dairy, as Chlordane builds up in fatty tissue, as well as through inhalation of air near Chlordane treated homes and landfills. Chlordane is excreted slowly through feces and urine elimination, as well as through breast milk in nursing mothers and is able to cross placenta and become absorbed by developing fetus' in pregnant women.

Health Effects

Exposure to chlordane metabolites may be associated with testicular cancer. The incidence of seminoma in men with the highest blood levels of *cis*-nonachlor was almost double that of men with the lowest levels. Prostate cancer has been associated with *trans*-nonachlor levels, a component of chlordane. Japanese workers who used chlordane over a long period of time had minor changes in liver function.

Heptachlor and chlordane are some of the most potent carcinogens tested in animal models. No human epidemiological study has been conducted to determine the relationship between levels of chlordane/heptachlor in indoor air and rates of cancer in inhabitants. However, studies have linked chlordane/heptachlor in human tissues with cancers of the breast, prostate, brain, and cancer of blood cells like leukemia and lymphoma. Breathing chlordane in indoor air is the main route of exposure for these levels in human tissues. Currently, USEPA has defined a concentration of 24 nanogram per cubic meter of air (ng/M^3) for chlordane compounds over a 20-year exposure period as the concentration that will increase the probability of cancer by 1 in 1,000,000 persons. This probability of developing cancer increases to 10 in 1,000,000 persons with an exposure of 100 ng/M3 and 100 in 1,000,000 with an exposure of 1000 ng/M3.

Other health effects of chlordane compounds (migraines, respiratory infections, diabetes, anxiety, depression, and activated immune system) may affect more people than cancer. The Agency for Toxic Substances and Disease Registry (ATSDR) has defined a concentration of chlordane compounds of 20 ng/M^3 as the Minimal Risk Level (MRLs). ATSDR defines Minimal Risk Level as an estimate of daily human exposure to a dose of a chemical that is likely to be without an appreciable risk of adverse non-cancerous effects over a specific duration of exposure. Chlordane bioaccumulates in animals. It is highly toxic to fish, with an LD_{50} of 0.022–0.095 mg/kg (oral).

8.7. Heptachlor

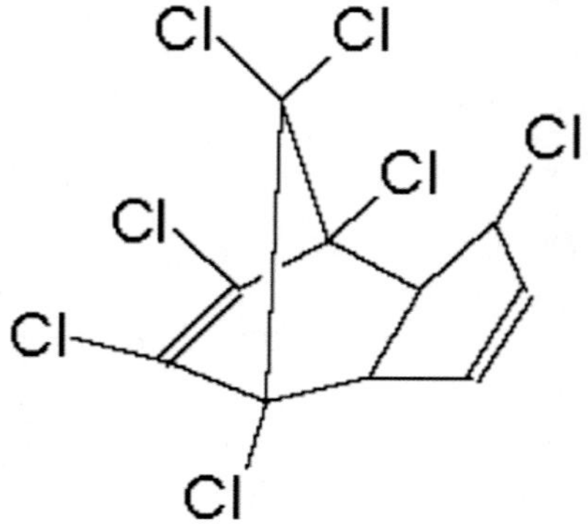

Fig. 9 : Chemical structure of heptachlor

Heptachlor is an organochlorine compound that was used as an insecticide. It was usually sold as a white or tan powder. Due to its highly stable structure, heptachlor can persist in the environment for decades. The US EPA has limited the sale of heptachlor products to the specific application of fire ant control in underground transformers.

Synthesis

Heptachlor is produced via the Diels-Alder reaction of hexachlorocyclopentadiene and cyclopentadiene. The resulting adduct is brominated followed by treatment with hydrogen chloride in nitromethane in the presence of aluminum trichloride or with iodine monochloride (Plimmer, 2003).

Compared to chlordane, it is about 3 - 5 times more active as an insecticide, but more inert chemically, being resistant to water and caustic alkalies (Plimmer, 2003).

Metabolism

Soil microorganisms transform heptachlor by epoxidation, hydrolysis, and reduction. When the compound was incubated with a mixed culture of organisms, chlordane is formed, which was further metabolized to chlordene epoxide. Other metabolites include 1-hydroxychlordene, 1-hydroxy-2,3-epoxychlordene, and heptachlor epoxide. Soil microorganisms hydrolyze heptachlor to give ketochlordene. Rats metabolize heptachlor to the epoxide 1-exo-1-hydroxyheptachlor epoxide and 1,2-dihydrooxydihydrochlordene. When heptachlor epoxide was incubated with microsomal preparations form liver of pigs and from houseflies, the products found were diol and 1-hydroxy-2,3-epoxychlordene (Plimmer, 2003).

Toxicity of Heptachlor and Related Derivatives

The range of oral rat LD_{50} values are 40 mg/kg to 162 mg/kg. Daily oral doses of heptachlor at 50 and 100 mg/kg were found to be lethal to rats after 10 days. For heptachlor epoxide, the oral LD_{50} values ranging from 46.5 to 60 mg/kg. With rat oral of LD_{50}47mg/kg, heptachlor epoxide is more toxic. A product of hydrogenation of heptachlor, â-dihydroheptachlor, has high insecticidal activity and low mammalian toxicity, rat oral LD_{50}>5,000mg/kg (Plimmer, 2003).

Humans are exposed to heptachlor through drinking water and foods, including breast milk (ATSDR, 2007). Heptachlor epoxide is derived from a pesticide that was banned in the U.S. in the 1980s. It is still found in soil and water supplies and can turn up in food and be passed along in breast milk. High levels of it seemed to increase type 2 diabetes risk to about 7 percent (Harmon, 2010).

The International Agency for Research on Cancer (IARC) and the EPA have classified the compound as a possible human carcinogen. Animals exposed to Heptachlor epoxide during gestation and infancy are found to have changes in nervous system and immune function. Higher doses of Heptachlor when exposed to newborn animals caused decrease in body weight and death.

The U.S. EPA MCL for drinking water is 0.0004 mg/L (Heptachlor) and 0.0002 mg/L (Heptachlor epoxide). The U.S. FDA limit on food crops is 0.01 ppm, in milk is 0.1 ppm, and on edible seafoods is 0.3 ppm. The Occupational Safety and Health Administration (OSHA) has limit of 0.5 mg/m^3 (cubic meter of workplace air) for 8-hour shifts and 40-hour work weeks (ATSDR, 2007).

The compound was designated as a Class 2B or "possible human carcinogen" by The International Agency for Research on Cancer (IARC). ATSDR report in 1993 found no studies with respect to death in humans after oral exposure to heptachlor or heptachlor epoxide.

8.8. Isodrin

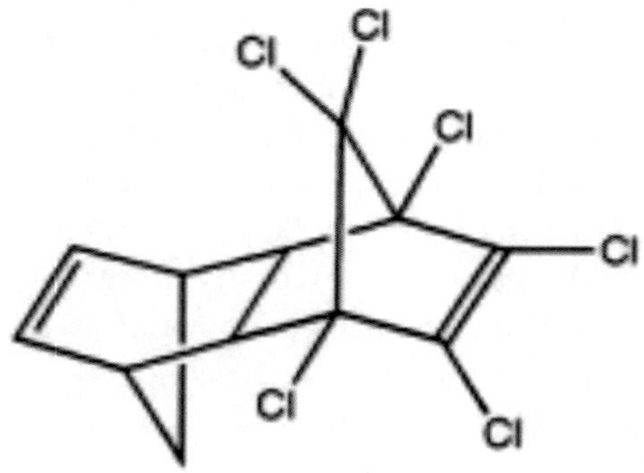

Fig.10: Chemical structure of heptachlor

Together with the other "drins" (aldrin, dieldrin and endrin) isodrin was developed in the 1940s to provide an alternative to DDT. European drin production peaked in 1967 and ceased in 1990. It used to be commonly used as an insecticide in agriculture and to control malaria spreading mosquito's. Since isodrin is no longer manufactured or used in the European Union or the United States, its leakage to the marine environment from these countries is not expected to be significant. In water mostly isodrin adsorbs to suspended particles, organic matter, soils and sediments. It has a low tendency to evaporate and isn't easily degraded in the atmosphere, causing it to be transported over great distances. It is also stable in water and soil. It can take between 0,5 and 6 years to half the isodrin concentration of a soil. Isodrin may bioaccumulate in aquatic organisms and has a tendency to biomagnify through food chains.

Isodrin is a very toxic chemical for fresh water fishes. Fishes die when exposed to concentrations of isodrin above 6 µg/l. Zooplanktons and crustaceans start dying when exposed to concentrations above 1 mg/l.

8.9. Endosulfan

Endosulfan is an acaricide which became a highly controversial agrichemical due to its acute toxicity, potential for bioaccumulation, and role as an endocrine disruptor. Because of its threats to human health and the environment, a global ban on the manufacture and use of endosulfan was negotiated under the Stockholm Convention in April 2011. The ban will take effect in mid 2012, with certain uses exempted for 5 additional years. More than 80 countries, including the European Union, Australia and New Zealand, several West African nations, the United States, Brazil and Canada had already banned it or announced phase outs by the time the Stockholm Convention ban was agreed upon. It is still used extensively in India, China, and few other countries. It is produced by Makhteshim Agan and several manufacturers in India and China.

Uses

Endosulfan has been used in agriculture around the world to control insect pests including whiteflys, aphids, leafhoppers, Colorado potato beetles and cabbage worms. Due to its unique mode of action, it is useful in resistance management; however, as it is non-specific, it can negatively impact populations of beneficial insects. It is, however, considered to be moderately toxic to honey bees.

Production

The World Health Organization estimated world wide annual production to be about 9,000 metric tonnes (t) in the early 1980s. From 1980–89, worldwide consumption averaged 10,500 t per year, and for the 1990s use increased to 12,800 t per year.

Endosulfan is a derivative of hexachlorocyclopentadiene and is chemically similar to aldrin, chlordane, and heptachlor. Specifically, it is produced by the Diels-Alder reaction of hexachlorocyclopentadiene with *cis*-butene-1,4-diol and subsequent reaction of the adduct with thionyl chloride. Technical endosulfan is a 7:3 mixture of stereoisomers, designated α and β. α- and β-endosulfan are conformational isomers arising from the pyramidal stereochemistry of sulfur. α-Endosulfan is the more thermodynamically stable of the two, thus β-endosulfan irreversibly converts to the β form, although the conversion is slow.

Health Effects

Endosulfan is one of the most toxic pesticides on the market today, responsible for many fatal pesticide poisoning incidents around the world. Endosulfan is also a xenoestrogen a synthetic substance that imitates or enhances the effect of estrogensand it can act as an endocrine disruptor, causing reproductive and developmental damage in both animals and humans. Whether endosulfan can cause cancer is debated. With regard to consumers intake of endosulfan from residues on food, the Food and Agriculture Organization of United Nations has concluded that long-term exposure from food is unlikely to present a public health concern, but short term exposure can exceed acute reference doses.

Toxicity

Endosulfan is acutely neurotoxic to both insects and mammals, including humans. The US EPA classifies it as Category I: "Highly Acutely Toxic" based on a LD_{50} value of 30 mg/kg for female rats, while the World Health Organization classifies it as Class II "Moderately Hazardous" based on a rat LD_{50} of 80 mg/kg. It is a GABA-gated chloride channel antagonist, and a Ca^{2+}, Mg^{2+} ATPase inhibitor. Both of these enzymes are involved in the transfer of nerve impulses. Symptoms of acute poisoning include hyperactivity, tremors, convulsions, lack of coordination, staggering, difficulty breathing, nausea and vomiting, diarrhea, and in severe cases, unconsciousness. Doses as low as 35 mg/kg have been documented to cause death in humans, and many cases of sub-lethal poisoning have resulted in permanent brain damage. Farm workers with chronic endosulfan exposure are at risk of rashes and skin irritation.

EPA's acute reference dose for dietary exposure to endosulfan is 0.015 mg/kg for adults and 0.0015 mg/kg for children. For chronic dietary expsoure, the EPA references doses are 0.006 mg/kg/day and 0.0006 mg/kg/day for adults and children, respectively (EPA, 2002).

Endocrine Disruption

Theo Colborn, lists endosulfan as a known endocrine disruptor, and both the EPA and the Agency for Toxic Substances and Disease Registry consider endosulfan to be a potential endocrine disruptor. Numerous *in vitro* studies have documented its potential to disrupt hormones and animal studies have demonstrated its reproductive and developmental toxicity, especially among males. A number of studies have documented that it acts as an anti-androgen in animals. Environmentally relevant doses of endosulfan equal to the EPA's safe dose of 0.006 mg/kg/day have been found to affect gene expression in female rats similarly to the effects of estrogen. It is not known whether endosulfan is a human teratogen (an agent that causes birth defects), though it has significant teratogenic effects in laboratory rats. An assessment concluded that endocrine disruption occurs only at endosulfan doses that cause neurotoxicity.

Reproductive and Developmental Effects

Endosulfan can also affect human development. Birth defects of the male reproductive system including cryptorchidism Re caused by endosulfan. Endosulfan exposure in male children may delay sexual maturity and interfere with sex hormone synthesis (Saiyed et al., 2003). A study conducted by California Department of Public Health (2007) and found that women who lived near farm fields sprayed with endosulfan and the related organochloride pesticide dicofol during the first eight weeks of pregnancy are several times more likely to give birth to children with autism. This is the first study to look for an association between endosulfan and autism, and additional study is needed to confirm the connection. A report concluded that epidemiology and rodent studies that suggest male reproductive and autism effects are open to other interpretations, and also that developmental or reproductive toxicity occurs only at endosulfan doses that cause neurotoxicity.

Cancer

In vitro assays have shown that endosulfan can promote proliferation of human breast cancer cells (Grunfeld et al., 2004). Evidence of carcinogenicity in animals is mixed.

9. Regulation of Organochlorine Pesticides

In the U.S., pesticides are regulated by the Environmental Protection Agency (EPA) under the Federal Insecticide, Fungicide, and Rodenticide Act (FIFRA). States may also regulate pesticides under FIFRA, but only in a more restrictive manner than the EPA. The UN and World Health Organization (WHO) may also influence use of pesticides through treaties and declarations. The Stockholm Convention on Persistent organic Pollutants, a UN treaty, has established global bans on several organochlorine pesticides including DDT, hexachlorobenzene, pentachlorobenzene, chlordane, dieldrin, endrin, heptachlor, mirex, toxaphene, hexachlorocyclohexane (alpha-HCH, beta-HCH, and gamma-HCH (lindane)), and chlordecone. The European Union (EU) recently proposed that endosulfan should be included under provisions of the Stockholm Convention as well. Although the U.S. banned the use of lindane for agricultural uses, the Food and Drug Administration still allows use of lindane in pharmaceutical products for the treatment of head lice and scabies.

References

ATSDR (2007). Heptachlor and Heptachlor Epoxide - Fact Sheet -1

Akhtar, M.H. (1985). The fate of insecticides in economic animals. In: Hutson DH, Roberts TR (eds) John Wiley and Sons Ltd. Progress in pesticide Biochemistry and Toxicology. 4: 219.

Bist, R., Bhatt, D.K. (2009). The evaluation of effect of alpha-lipoic acid and vitamin E on the lipid peroxidation, gamma-amino butyric acid and serotonin level in the brain of mice (*Mus musculus*) acutely intoxicated with lindane. Journal of Neurological Sciences. 276: 99-102.

Bist, R., Bhatt, D.K. (2010). Augmentation of cholinesterases and ATPase activities in the cerebellum and pons-medulla oblongata, by a combination of antioxidants (resveratrol, ascorbic acid, alpha-lipoic acid and vitamin E), in acutely lindane intoxicated mice. Journal of Neurological Sciences. 296 (1): 83-87.

Bist, R., Misra, S., Bhatt, D.K. (2010). Inhibition of lindane-induced toxicity using alpha-lipoic acid and vitamin E in the brain of *Mus musculus*. Protoplasma. 242 (1-4): 49-53.

Budavari, S., O'Neil, M.J., Smith, A. (1989). The Merck Index. *Merck and Co.*, Inc., Rahway, NJ. 866-867.

Corneliussen, P.E. (1969). Residues in food and feed. Pesticide residues in total diet samples (IV). Pestic Monit J. 2: 140-152.

Engst, R., Macholz, R.M., Kujawa, M. (1978). Confirmations of the degradation scheme of gamma-hexachlorocyclohexane. Nahrung. 22: 29-32.

Engvild, K.C. (1986). Chlorine-Containing Natural Compounds in Higher Plants. Phytochemistry. 25 (4): 7891–791.

Frohberg, H., Bauer, A. (1972). Lindane: Testing for mutagenic effect. Dominant lethality test on male mice. Darmstadt, E. Merck

Goto, M., Hattod, M., Miyagawa, T. (1972). Contribution to ecological chemistry. II. Hepatoma formation in mice after administration of high doses of HCH isomers. Chemosphere. 6: 279-282.

Gribble, G.W. (1996). Naturally occurring organohalogen compounds - A comprehensive survey. Progress in the Chemistry of Organic Natural Products. 68(10): 1–423.

Gribble, G.W. (1998). Naturally Occurring Organohalogen Compounds. Acc. Chem. Res. 31 (3): 141–152.

Gribble G.W. (1999). The diversity of naturally occurring organobromine compounds. Chemical Society Reviews. 28 (5): 335.

Harmon, K. (2010). New Mass-Screening Method Finds Additional Environmental Risks for Diabetes. Scientific American.

Hildebrandt, G., Jeep, H., Hurka, H., Stuke, T., Heitmann, M., Boiselle, C. (1986). Breast milk as a biological indicator: study on the organic halogen burden in breast milk and foodstuffs. Berlin, Schmidt E. Verlag. (Report 5/86 of the Federal Office for the Environment)

IARC (1979). IARC Monographs on the evaluation of the carcinogenic risk of chemicals to humans. Some halogenated hydrocarbons. International Agency for Research on Cancer. Lyon. 195.

IARC (1998). IARC Cancer databases and other resources.

Joy, R.M., Stark, L.G., Albertson, T.E. (1982). Proconsulvant effects of lindane: enhancement of amygdaloid kindling in the rat. Neurobehav Toxicol Teratol. 4: 347-354.

Joy, R.M., Stark, L.G., Albertson, T.E. (1983). Proconvulsant action of lindane compared at two different kindling sites in the rat-amygdala and hippocampus. Neurobehav Toxicol Teratol. 5: 461-465.

Khanna, R.N., Das, M., Anand, M. (2002). Influence of phenobarbitol and CCl_4 on the modulation of tissue retention profile of hexachlorocyclohexane in rats. Biomedical and Environmental Sciences. 15: 119-129.

Lear, L. (1997). Rachel Carson: Witness for Nature. New York: Henry Hoyten.

Magnus, V., Ozga, J.A, Reinecke, D.M., Pierson, G.L. Larue, T.A. Cohen, Jerry D; Brenner, Mark, L (1997). "4-chloroindole-3-acetic and indole-3-acetic acids in Pisum sativum. Phytochemistry 46 (4): 675–681.

Martin, R.J., Duggan. RE (1968). Pesticide residues in total diet samples (III). Pestic Monit J. 1: 1-20.Robert L. Metcalf "Insect Control" in Ullmann's Encyclopedia of Industrial Chemistry" Wiley-VCH, Wienheim;

Pešut-Mikinèiè, M., Šimiæ, B., Crnojeviæ, T., Kniewald, J. (1987). Lindane effects on testosterone metabolism in neuroendocrine organs of Cockerel and female. Turkey Bull Environ Contam Toxicol. 38: 911-916.

Pless, T., Boettger, M., Hedden, P., Graebe, J., (1984). Occurrence of 4-Cl-indoleacetic acid in broad beans and correlation of its levels with seed development. Plant Physiology. 74 (2): 320–3.

Plimmer, J. R., (ed.). (2003). The Encyclopedia of Agrochemicals, (Vol. 3). New York: John Wiley & Sons.

Popp. P., Bruggemann, L., Keil, P., Thuss, U., Weiss, H. (2000). Chlorobenzenes and hexachlorocyclohexanes (HCHs) in the atmosphere of Bitterfeld and Leipzig (Germany). Chemosphere. 41: 849-855.

Pomés, A., Rodríguez-Farré, E., Suñol, C. (1994). Disruption of GABA-dependent chloride flux by cyclodienes and hexachlorocyclohexanes in primary cultures of cortical neurons. J Pharmacol Exp Ther. 271: 1616-1623.

Reigart, R.J., Roberts, J.R. (1999). Medical University of South Carolina. Recognition and Management of Pesticide Poisonings. 5th edition. 25.

Rocchi, P., Perocco, P., Alberghini, W., Fini, A., Prodi, G. (1980). Effect of pesticides on scheduled and unscheduled DNA synthesis of rat thymocytes and human lymphocytes. Archives of Toxicology. 45 (2): 101-108

Sinha, R.R.P., Sinha, S.P. (1983). Induction of dominant lethal mutations in *Drosophila melanogaster* by three pesticides: Gammexane, Sevin, and Folidol. Comp Physiol Ecol. 8 (2): 87-89.

Smith, A.G. (1991). Chlorinated Hydrocarbon Insecticides. In Handbook of Pesticide Toxicology. Classes of Pesticides. Wayland J. Hayes Jr. and Edward R. Laws, Jr. editors. Academic Press, Inc., NY. 3: 3-6.

Stokstad, E. (2007). Species conservation. Can the bald eagle still soar after it is delisted?. Science. 316 (5832): 1689–90.

Tsushimoto, G., Chang, C.C., Trosko, J.E., Matsumura, F. (1983). Cytotoxic, mutagenic, and cell-cell communication inhibitory properties of DDT, lindane, and chlordane on Chinese hamster cells *in vitro.* Arch Environ Contam Toxicol. 12: 721-730.

Tussel, J.M., Sunol, C., Gelpi, E., Rodriguez-Farre, E. (1988). Effect of lindane at repeated low doses. Toxicology. 49: 375-379.

Tzoneva-Maneva, M.T., Kaloyaneva, F., Georgieva, V. (1971). Influence of diazinon and lindane on the mitotic activity and the caryotype of human lymphocytes, cultivated *in vitro.* Bibl Haematol. 38 (1): 344-347.

U.S. Environmental Protection Agency, Federal Register (2006). 71: 74905-74907.

U. S. National Library of Medicine (1992). Hazardous Substances Databank. TOXNET, Medlars Management Section, Bethesda, MD.

Ulmann, E. (1972). Lindane Monograph of an Insectcide. Verlag K. Schillinger, Freiberg im Breisgau, Germany.

Wauchope, R.D., Buttler, T.M., Hornsky, A.G., Augutijn-Beckers, P.W., Burt, J.P. (1992). The SCS/ARS/CES pesticide properties database for environmental decision making. Rev Environ Contam Toxicol. 123: 1-155.

Windholz, M., Budavari, S., Blumetti, R.F., Otterbein, E.S. (1983). The Merck Index. Tenth edition. Merck and Co., Inc., Rahway, NJ. 789.

Woodward, R.M., Polenzani, L., Miledi, R. (1992). Effects of hexachlorocyclohexanes on ã-aminobutyric acid receptors expressed in *Xenopus* oocytes by RNA from mammalian brain and retina. Mol Pharmacol. 41: 1107-1115

Zhu, J., Feng, Z., Chen, J. (1986). Studies on the disturbution and fate of [gamma-^{3}H] hexachlorocyclohexane in rats. Pestic Biochem Physiol. 25: 414-419.

❑❑❑

CHAPTER – 21

Nanotechnology – Application in Biomedical Sciences

Punyasloke Bhadury
Department of Biological Sciences, Indian Institute of Science Education and Research - Kolkata, Mohanpur–741252, Nadia, West Bengal
Email : pbhadury@iiserkol.ac.in

Summary

Nanotechnology, the science of nanoscale structures has been applied in various fields including in biomedical sciences. Nanotechnology is being increasingly used to create materials and devices so as to interact with the body at sub-cellular scales with a high degree of specificity. Nanotechnology is used in the fight against cancer, cardiac and neurodegenerative disorders, infection and other diseases. This chapter provides an overview of some of the applications of nanotechnology in the field of biomedical sciences particularly with a focus on tissue engineering and medical diagnostics.

Background

Nanotechnology which deals with understanding and control of matter at dimension of roughly 100 nm and below, has a crosssectoral application and an interdisciplinary orientation. At this scale, the physical, chemical and biological properties of materials differ from the properties of individual atoms and molecules or bulk matter, which enable novel applications. Nanotechnology research and development is directed towards understanding and creating improved materials, devices and systems that exploit these properties as they are discovered and characterized. There are many applications of nanotechnology with in the area of medicine, chemistry and environment, energy, agriculture, information and communication, heavy industry and consumer goods.

In the field of biomedical science there has been diverse application of nanotechnology including the areas of drug delivery and gene therapy and increasingly being explored in emerging areas of molecular imaging and disease diagnostics (eg. Roy et al. 2003; Muller and Keck 2004). In this chapter some of the advancement made in areas of oxygen measurement in cellular systems, tissue engineering, medical diagnostics and in single nucleotide polymorphism analysis based on nanotechnological applications are discussed with examples.

Oxygen Measurement in Cellular systems

Oxygen is a key component for many physiological and pathological processes in living cells (López-Barneo et al 2001; Acker and Acker 2004). A variety of diseases including tumor growth, diabetic retinopathy and rheumatoid arthritis are strongly linked with tissue hypoxia (e.g. Carmeliet et al 1998; Hocker et al 1998; Ikeda 2005). Hence measurement of oxygen level in living cells and tissues is an absolute necessity, particularly from the viewpoint of medical sciences.

Although the measurement of oxygen level is a challenging problem, nevertheless efforts have been mostly focused around development of optical sensors for oxygen (e.g. Rosenzweig and Kopelman 1995; McDonagh et al. 1998). Oxygen sensing by phosphorescence quenching has been implemented for real-time measurements as well as high-resolution oxygen mapping in tissues (Vinogradov et al; 1996; Xu et al. 2001; Koo et al. 2004). There is currently considerable interest in the development of nanoparticle-based optical oxygen sensors (Koo et al. 2004). Typically the sensor consists of phosphorescent dyes encapsulated inside a polymer or silica nanoparticle, which serves to isolate the dyes from the cellular environment (Xu et al. 2001). Additionally, nanoparticles exhibit higher brightness and better photostability than molecular dyes, owing to large numbers of chromophores per particle as well as the protective matrix. It has been also shown that ð-conjugated polymer nanoparticles (CPdots) exhibit extraordinarily high fluorescence brightness under both one-photon and two-photon excitation (Wu et al 2007, 2008). Dye-doped conjugated polymer nanoparticles were found to sensitize bright emission from fluorescent dye dopants by efficient energy transfer from the polymer to the dye (Wu et al. 2006).

There are also reports of development of optical nanosensors for dissolved oxygen measurement in which organically modified silicate (ormosil) nanoparticles act as matrix. The ormosil nanoparticles are prepared through a sol-gel-based process, which includes the formation of core particles with phenyltrimethoxysilane as a precursor followed by the formation of a coating layer with methyltrimethoxysilane as a precursor (Koo et al 2004). The highly permeable structure and the hydrophobic nature of the ormosil nanoparticles, as well as their small size, result in an excellent overall quenching response to dissolved oxygen and a linear response over the whole range, from 0 -100% oxygen-saturated water. The sensor has a higher sensitivity and broader linearity as well as longer excitation and emission wavelengths resulting in reduced background noise during cellular measurement.

Recently a novel nanoparticle architecture for oxygen sensing has been described which consist of π-conjugated polymer molecules doped with an oxygen-sensitive phosphorescent dye. Upon light excitation, the polymer efficiently transfers energy to the phosphorescent dye, which results in bright phosphorescence that is highly sensitive to the concentration of dissolved oxygen. The key features of the nanoparticle sensors, including their small size, extraordinary brightness, and ratiometric emission, together with the demonstration of single-particle sensing and cellular uptake, indicate their excellent potential for quantitative imaging of local molecular oxygen concentration in living cells and tissues (Wu et al. 2009).

Tissue Engineering

Tissue engineering is an evolving interdisciplinary field which integrates biology, engineering, materials science, and medicine with focuses on the development of biological substitutes to restore, replace, maintain, or enhance tissue and organ function. Over the last few decades, continued progress in tissue engineering research has resulted in creation of implantable tissues and some of which are being used in humans or currently under clinical trials. Majority of the tissue engineering strategies rely on the principle that under appropriate bioreactor conditions, cells seeded or recruited into three dimensional (3D) biocompatible scaffolds are able to reassemble into functional structures resembling native tissues. It is generally accepted that to recapitulate proper tissue functionality scaffolds should also establish a tissue specific microenvironment to maintain and regulate cell behavior and function.

Within tissues, cells are surrounded by extracellular matrix (ECM) which is characterized by a natural web of hierarchically organized nanofibers. This integral nanoarchitecture is important because it provides cell support and directs cell behavior via cell-ECM interactions. Recent developments in nanoscience have resulted in designing and fabrication of biomimetic microenvironment at the nanoscale level and thus providing an analog to native ECM. Several studies have already shown the significance of promoting unique microenvironment in ECM so as to foster tissue organization (Ott et al 2008; Gui et al 2009; Grayson et al 2010).

The significance of the overall fibrillar and porous nanoscale topography of the ECM in promoting essential cellular processes has led tissue engineers to replace macroporous scaffolds with biomimetic materials with nanoscale features. For example, to engineer the ECM fibrous mesh, researchers fabricated scaffolds composed of fibres with appropriate diameters that formed a highly interconnected, porous architecture allowing high mass transfer to and waste removal from the developing tissue (Lee et al 2009). In the past decade, there has been major interest in exploring the potency of biomimetic fibres in tissue engineering, and several techniques originally developed for different purposes have been used for this application.

Electrospinning and molecular self assembly are common nanofabrication techniques used to create 3D scaffolds composed of interwoven fibres (Barnes et al

2007, Zhang et al. 2009). Electrospinning is a simple method that uses an electric field to control the deposition of polymer fibres on a target substrate. The fibres are continuous, with high porosity and high spatial interconnectivity. Moreover, the direction of the fibres can be selectively controlled by spinning speed and polymer concentration to fit the needs of various engineered tissues. The fibre bioactivity can also be designed to include delivery systems that control the release of cues for tissue assembly such as growth factors, peptides, enzymes, drugs, DNA and RNA (Cao et al 2010). These scaffolds, which enable sequential release of multiple growth factors or cytokines, may have important applications in tissue engineering, such as promoting vasculogenesis or influencing the fate of stem cells (Yilgor et al 2009).

In vivo, in addition to the biochemical properties of the ECM, cells encounter and interact with many topographical features ranging from folded protein to banded collagen through a phenomenon known as contact guidance. A mimetic approach to study this effect and to create the structure and length scale (5 nm to micrometre scale) of the native ECM is by nanopatterning of different geometries such as nanogroove, nanopost and nanopit arrays (Bettinger et al 2009). For example, epithelial cells elongate and align along patterns of grooves and ridges with feature dimensions as small as 70 nm, whereas on smooth substrates, cells are mostly rounded (Teixeira et al. 2003). Similarly the control of human mesenchymal stem cell (MSC) differentiation using nanoscale symmetry and disorder was demonstrated by Dalby et al. (2007). They showed that highly ordered topographies produce low to negligible cellular adhesion and osteoblastic differentiation, whereas nanoscale topographic disorder stimulated MSCs to produce bone mineral.

Nanodevices can be also used to trigger desired processes in engineered tissue for tissue development. For example, electrodes designed to stimulate neural tissue can be improved by coating them with nanostructures such as carbon nanotubes or other inorganic crystals (Keefer et al. 2008). Individual carbon nanotubes have been shown to form spontaneous junctions with lipid bilayers with cleft spacings that are tight compared with the size of a single protein (Zhou et al. 2007). Such tight interfaces can improve the quality of neural stimulation and recording by reducing the impedance between the device and cell membrane (Kotov et al. 2009).

Nanostructures can be also used to record the electronic signals that are transmitted through cells such as neurons and cardiomyocytes. In a recent study it has been shown that arrays of nanowire devices when interfaced with patterned, cultured neurons can measure extracellular signal fluctuations locally at individual axons, dendrites and soma (Patolsky et al. 2006).

Nanotechnologies clearly have had an impact on tissue engineering and still have great potential to advance therapeutic methods based on tissue engineering. The synthesis of new nanostructures and their incorporation into existing macro- and microtechnologies have led to improvements in the ability to provide a true biomimetic microenvironment to the developing tissue. In recent years, several reports have indicated possible negative effects of carbon nanotubes on cells and their potential to provoke oxidative stress, inflammation, genetic damage and long-term pathological effects. Therefore the biocompatibility and biodegradation of inorganic nanomaterials need to be thoroughly investigated before they can be safely applied in clinical trials. Another challenge that must be addressed is the complexity of creating 3D porous scaffolds with nanotopographies. It is also crucial discover the key factors promoting the assemblies of different tissues and to create specific microenvironments. Developing nanotechnological tools for controlling and guiding cells to desired locations in 3D matrices will be useful for engineering complex multicellular constructs such as epithelial and vascularized tissues.

Medical Diagnostics

Nanoparticles are increasingly used in the field of medical diagnostics. Numerous studies have shown the applicability of nanoparticle based diagnostics for better detection of diseases. For example Bhalgat et al (1998) have shown that streptavidin coated fluorescent polystyrene nanospheres were twenty five times more sensitive than conjugate streptavidin-fluorescin and able to detect epidermal growth factor receptor (EGFR) on A431 cells (human epidermoid carcinoma cells). A group led by Santra et al (2001) developed novel luminescent nanoparticles composed of inorganic luminescent dye, Tris(2,2'-bipyridyl) dichlororuthenium (II) hexahydrate, doped inside a silica network for staining human leukemia cells. These dye doped silica (DDS) nanoparticles have been synthesized using a water-in-oil microemulsion technique in which controlled hydrolysis of the tetraethyl orthosilicate lead to the formation of monodispersed nanoparticles. These nanoparticles can be used as biomarkers since they are much smaller than cellular dimension and highly photostable when compared to most commonly used organic dyes. The DDS nanoparticles can be modified biochemically so as to attach membrane-anchoring groups and has been successfully applied to stain human leukemia cells (Santra et al. 2001). Based on a very similar methodology, He et al (2004) have also developed a method for the recognition of HepG liver cancer cells based on organic dye-doped fluorescent silica nanoparticles and showed that the nanoparticles can identify the target cells selectively and efficiently.

Ambrosi et al (2009) developed an optical enzyme-linked immunosorbent assay (ELISA) for the analysis of CA15-3 antigen, an important biomarker present in blood samples and useful for the follow-up of the medical treatment of breast cancer. In their study gold nanoparticles (AuNPs) were used as carriers of the signaling antibody anti-CA15-3"HRP (horseradish peroxidase) in order to achieve an amplification of the optical signal. In the range between 0-60 U/mL, the assay adopting AuNPs as an enhancer resulted in higher sensitivity and shorter assay time when compared to classical ELISA procedures. In a separate study Georganopolou et al (2005) used the ultrasensitive bio-barcode assay to measure the concentration of amyloid-β-derived diffusible ligands (ADDLs), a potential soluble pathogenic Alzheimer's disease (AD) marker, in the cerebrospinal fluid (CSF) of 30 individuals. The bio-barcode assay involves the homogeneous isolation of specific antigens by means of a sandwich process involving oligonucleotide-modified Au nanoparticles (NPs; with biobarcodes) and magnetic microparticles (MMPs), both functionalized with Abs specific to the antigen of interest. The increased sensitivity advantage derives mainly from the very effective sequestration of antigen and the amplification process that occurs as a result of the large number of barcode DNA strands released for each antigen recognition and binding event. By using this approach and adequate Abs, one can routinely detect as low as 30 copies of a target protein, often in a complex sample such as plasma.

Other studies have shown the applicability of gold nanoparticle probes coupled with dynamic light scattering in a homogenous assay for detection of prostate cancer biomarker (eg. Liu et al. 2008). Mani et al (2009) have used an ultrasensitive immunosensor based on gold nanoparticle film electrodes for detection of cancer biomarkers in serum.

In future, nanoparticles can be engineered with specific binding affinities and could be resuspended into the collecting body fluids, or may be injected directly into the circulation. The nanoparticles, together with the bound molecules, could be directly captured on engineered filters and directly questioned by ultra high-resolution mass spectrometry.

Application in Single Nucleotide Polymorphism Analysis (SNP)

The decoding of the human genome (Sachidanandam 2001, Venter et al 2001) and the resulting greater than 3 million single nucleotide polymorphisms (SNPs) have led to numerous opportunities to study the effects of genetic variation on complex phenotypes. Moreover, the ability to associate disease states and pharmacological responses with individual SNPs and increasingly with haplotypes (Gabriel et al 2002) and related linkage disequilibrium mapping studies (Wall and Pritchard 2003; Zondervan and Cardon 2004) offer the possibility of improved therapeutic and prophylactic measures. To realize this promise at the level of the individual, highly sensitive, robust and inexpensive SNP detection methodologies are needed. Numerous strategies have been developed for SNP discrimination with various degrees of multiplexing capability (Syvanen 2001; Kirk et al. 2002).

Technological developments have also taken place in the field of nano-biotechnology so as to facilitate SNP analysis. Ihara et al (2004) prepared DNA-modified nanospheres by anchoring amino-terminated oligodeoxynucleotides (ODNs) with carboxylates onto a colored polystyrene sphere surface through amido bonds. About 220 ODN molecules were immobilized onto a nanosphere 40 nm in diameter. Preliminary studies using the microspheres with 1 ìm diameter reveal that the specificity of hybridization was retained after modification. Three kinds of differently colored (RGB, red/green/blue) nanospheres bearing unique ODNs on their surface were prepared for detecting SNP in *p53* gene (Ihara et al 2004). Cao et al (2005) synthesized Ag/Au core-shell nanoparticles and based on their optical signatures applied the same for the detection and simultaneous validation of single nucleotide polymorphisms in a DNA target.

SNP analysis using nanoparticles is being increasingly used in infectious disease diagnostics including in tuberculosis (TB). Tuberculosis is one of the leading causes of infection in humans, causing high morbility and mortality all over the world. The rate of new cases of multidrug resistant tuberculosis (MDRTB) continues to increase, and since these infections are very difficult to manage, they constitute a serious health problem. The use of gold nanoparticles derivatized with thiol-modified oligonucleotides (Au-nanoprobes) has led to new approaches in molecular diagnostics. Based on the differential non-cross-linking aggregation of Au-nanoprobes, colorimetric method has been developed for the detection of specific sequences and to apply this approach to pathogen identification and single base mutations/single nucleotide polymorphisms (SNP) discrimination. Au-nanoprobes have been developed for the specific identification of SNPs within the beta subunit of the RNA polymerase (*rpo*B locus), responsible for resistance to rifampicin in over 95% of rifampicin resistant *Mycobacterium tuberculosis* strains (Veigas et al 2010).

Way Forward

The application of nanotechnology can significantly improve the area of biomedical science as evident from some of the examples discussed in this chapter. In future, nanoscale biosensors could be incorporated inside scaffolds to follow the development of engineered tissues after transplantation, and according to biochemical composition or tissue behaviour, triggering the release of compensating cues. It is also envisaged that medical nanodevices could be effectively used as implants or perhaps injected into the blood stream of a patient for health monitoring or repairing systems that may have deviated from the normal pattern. In future it is also expected that smart nanoparticulate systems will recruit stem cells to desired sites in the body and instruct the formation of tissues *in vivo*, on demand, by triggering the release of chemical attractants.

References

Acker T and Acker H (2004) Cellular oxygen sensing need in CNS function - physiological and pathological implications. J Exp Biol 207: 3171-3188

Ambrosi A, Air F and Merkoi A (2009) Enhanced gold nanoparticle based ELISA for a breast cancer biomarker. Anal Chem 82: 1151-1156

Barnes CP, Sell SA, Boland ED, Simpson DG and Bowlin GL (2007) Nanofiber technology: Designing the next generation of tissue engineering scaffolds. Adv. Drug Deliv. Rev. 59: 1413–1433

Bettinger CJ, Langer R and Borenstein JT (2009) Engineering substrate topography at the micro- and nanoscale to control cell function. Angew. Chem. Int. Ed. 48: 5406–5415

Bhalgat MK, Haugland RP, Pollack JS and Swan S (1998) Green and Red fluorescent nanospheres for the detection of cell surface receptors by flow cytometry. J. Immunol. Methods, 219: 57-68

Borisov SM and Klimant I (2007) Ultrabright oxygen optodes based on cyclometalated iridium(III) coumarin complexes. *Anal Chem.* 79: 7501-7509

Briñas RP, Troxler T, Hochstrasser RM and Vinogradov SA (2005) Phosphorescent oxygen sensor with dendritic protection and two-photon absorbing antenna J. Am. Chem. Soc. 127:11851-11862

Cao YC, Jin R, Thaxton CS and Mirkin CA (2005) A two-color-change, nanoparticle-based method for DNA detection. Talanta 67: 449-455

Cao HQ, Jiang X, Chai C and Chew SY (2010) RNA interference by nanofiber based siRNA delivery system. J. Cont. Release 144: 203–212

Carmeliet P, Dor Y, Herbert JM, Fukumura D, Brusselmans K, Dewerchin M, Neeman M, Bono F, Abramovitch R, Maxwell P, Koch CJ, Ratcliffe P, Moons L, Jain RK, Collen D and Keshet E (1998) Role of HIF-1alpha in hypoxia-mediated apoptosis, cell proliferation and tumour angiogenesis. Nature 394:485-490

Dalby MJ et al (2007) The control of human mesenchymal cell differentiation using nanoscale symmetry and disorder. Nature Mater. 6: 997–1003

Gabriel SB, Schaffner SF, Nguyen H, Moore JM, Roy J, Blumenstiel B, Higgins J, DeFelice M, Lochner A, Faggart M, et al (2002) The structure of haplotype blocks in the human genome Science 296: 2225–2229

Georganopoulou DG, Chang L, Nam JM, Thazton CS, Mufson EJ, Klein WL and Mirkin CA (2005) Nanoparticle-based detection in cerebral spinal fluid of a soluble pathogenic biomarker for Alzheimer's disease. Proc Natl Acad Sci USA 102:2273-2276

Grayson WL et al (2010) Engineering anatomically shaped human bone grafts. Proc. Natl Acad. Sci. USA 107: 3299–3304

Gui L, Muto A, Chan SA, Breuer CK and Niklason LE (2009) Development of decellularized human umbilical arteries as small-diameter vascular grafts. Tissue Eng. A 15: 2665–2676

He X, Duan J, Wang K, Tan W, Lin X and He C (2004) A novel fluorescent label based on organic dye-doped silica nanoparticles for HepG liver cancer cell recognition. J Nanosci Nanotechnol. 4:585-589

Höckel M, Schlenger K, Hockel S, Aral B, Schaffer U and Vaupel P (1998) Tumor hypoxia in pelvic recurrences of cervical cancer. Int J Cancer 79: 365-369.

Ihara T, Tanaka S, Chikaura Y and Jyo A (2004) Preparation of DNA-modified nanoparticles and preliminary study for colorimetric SNP analysis using their selective aggregations. Nucleic Acids Res 32:e105

Ikeda E (2005) Cellular response to tissue hypoxia and its involvement in disease progression. Pathol. Int. 55: 603-610.

Keefer EW, Botterman BR, Romero M I, Rossi AF and Gross GW (2008) Carbon nanotube coating improves neuronal recordings. Nature Nanotech 3:434–439.

Kirk BW, Feinsod M, Favis R, Kliman RM and Barany F (2002) *Single nucleotide polymorphism seeking long term association with complex disease* Nucleic Acids Res. 30: 3295–*3311.*

Koo YEL, Cao YF, Kopelman R, Koo SM, Brasuel M and Philbert MA (2004) Real-time measurements of dissolved oxygen inside live cells by organically modified silicate fluorescent nanosensors. Anal Chem 76: 2498-2505.

Kotov NA et al (2009) Nanomaterials for neural interfaces. Adv. Mater. 21: 3970–4004.

Lee S et al (2009) Preparation of macroporous carbon nanofibres with macroscopic openings in the surfaces and their applications. Nanotechnology20:445702.

Liotta LA, Ferrari M and Petricoin E (2003) Clinical proteomics: written in blood. Nature 425: 885-989.

Liu X, Dai Q, Austin L, Coutts J, Knowles G, Zou J, Chen H and Huo Q (2008) A one-step homogeneous immunoassay for cancer biomarker detection using gold nanoparticle probes coupled with dynamic light scattering. J Am Chem Soc 130: 2780-2782.

López-Barneo J, Pardal R and Ortega-Sáenz P (2001) Cellular mechanisms of oxygen sensing. Annu. Rev. Physiol. 63:259–287.

Mani V, Chikkaveeraiah BV, Patel V, Gutkind JS and Rusling JF (2009) Ultrasensitive immunosensor for cancer biomarker proteins using gold nanoparticle film electrodes and multienzyme-particle amplification. ACS Nano 3: 585-594.

McDonagh C, MacCraith BD and McEvoy AK (1998) Tailoring of sol-gel films for optical sensing of oxygen in gas and aqueous phase. *Anal Chem* 70:45-50.

Mills A and Lepre A (1997) Controlling the response characteristics of luminescent porphyrin plastic film sensors for oxygen. Anal Chem. 69:4653-4659.

Mullis K, Faloona F, Scharf S, Saiki R, Horn G and Erlich H (1986) Specific enzymatic amplification of DNA *in vitro*: the polymerase chain reaction. Cold Spring Harbor Symp. Quant. Biol. 51: 263-273.

Muller RH and Keck CM (2004) Challenges and solutions for the delivery of biotech drugs-a review of drug nanocrystal technology and lipid nanoparticles. J Biotechnol 113:151-170

Ott HC et al (2008) Perfusion-decellularized matrix: Using nature's platform to engineer a bioartificial heart. Nature Med. 14: 213–221.

Patolsky F et al (2006) Detection, stimulation, and inhibition of neuronal signals with high-density nanowire transistor arrays. Science 313:1100–1104.

Roy I, Mitra S, Maitra A and Mozumdar S (2003) Calcium phosphate nanoparticles as novel non-viral vectors for targeted gene delivery. Int J Pharmaceutics 250:25-33

Rosenzweig Z and Kopelman R (1995) Development of a Submicrometer Optical-Fiber Oxygen Sensor. *Anal Chem* 67:2650-2654.

Sadeghi SJ, Tsotsou GE, Fairhead M, Meharenna YT and Gilardi G (2001) Rational design of P450 enzymes for biotechnology In: De Cuyper M and Bulte J (Eds.), Focus on Biotechnology.Physics and Chemistry Basis of Biotechnology", Kluwer Academic Publisher, Dordrecht, 71-104.

Sachidanandam R (2001) A map of human genome sequence variation containing 1.42 million single nucleotide polymorphisms. Nature 409: 928–*933.*

Santra S, Wang K, Tapec R and Tan W (2001) Development of novel dye-doped silica nanoparticles for biomarker application. J Biomed Opt 6:160-166

*Syvanen, A.-C. (*2001) *Accessing genetic variation: genotyping single nucleotide polymorphisms* Nature Rev. Genet. 2930–*942.*

Teixeira AI, Abrams GA, Bertics PJ, Murphy CJ and Nealey PF (2003) Epithelial contact guidance on well-defined micro- and nanostructured substrates. J. Cell. Sci. 116:1881–1892.

Venter JC, Adams MD, Myers EW, Li PW, Mural RJ, Sutton GG, Smith HO, Yandell M, Evans CA, Holt RA et al. (2001) The sequence of the human genome. Science 29: 1304-*1351*

Vinogradov SA, Lo LW, Jenkins WT, Evans SM, Koch C and Wilson DF (1996) Non-invasive imaging of the distribution of oxygen in tissue in vivo using near infrared phosphors. *Biophys J.* 70: 1609-1617.

Wu C, Szymanski C, Cain Z and McNeill J (2007) Conjugated polymer dots for multiphoton fluorescence imaging. *J Am Chem Soc.* 129:12904-12905.

Wu C, Bull B, Szymanski C, Christensen K and McNeill J (2008) Multicolor conjugated polymer dots for single molecule fluorescence detection in living cells. *ACS Nano.* 2:2415-2423.

Wu C, Peng H, Jiang Y and McNeill J (2006) Energy Transfer Mediated Fluorescence from Blended Conjugated Polymer Nanoparticles. *J Phys Chem B.* 110:14148-14154.

Wu C, Bull B, Christensen K, McNeill J (2009) Radiometric single-nanoparticle oxygen sensors for biological imaging. Angew Chem Int Ed Engl 48: 2741-2745.

*Wall, JD and Pritchard JK (*2003) *Haplotype blocks and linkage disequilibrium in the human genome* Nature Rev. Genet. 4587*–597.*

Xu H, Aylott JW, Kopelman R, Miller TJ and Philbert MA (2001) A real-time ratiometric method for the determination of molecular oxygen inside living cells using sol-gel-based spherical optical nanosensors with applications to rat C6 glioma. *Anal Chem* 73:4124-4133.

Yilgor P, Tuzlakoglu K, Reis RL, Hasirci N and Hasirci V (2009) Incorporation of a sequential BMP-2/ BMP-7 delivery system into chitosan-based scaffolds for bone tissue engineering. Biomaterials 30:3551–3559.

Zondervan KT and Cardon LR (2004) The complex interplay among factors that influence allelic association Nature Rev. Genet. 589–100.

Zhang G, Chen J, Payne SJ, Kooi SE, Demas JN and Fraser CL (2007) Multi-emissive difluoroboron dibenzoylmethane polylactide exhibiting intense fluorescence and oxygen-sensitive room-temperature phosphorescence. *J Am Chem Soc.* 129:8942-8943

Zhou M, Lucas DA and Chan K (2004) Investigation into the human serum interactome. Electrophoresis 25:1289-1298.

Zhang S (2009) Fabrication of novel biomaterials through molecular self-assembly. Nature Biotechnol. 21:1171–1178.

Zhou XJ, Moran-Mirabal JM, Craighead HG and McEuen PL (2007) Supported lipid bilayer/carbon nanotube hybrids. Nature Nanotech 2:185–190.

❑❑❑

CHAPTER – 22

Protein Purification An Overview

Biswanath Bhunia[1] Kambaska Kumar Behera[3] Nibedita Mahata[1] and Apurba Dey[2]
[1]Department of Biotechnology, Bengal College of Engineering & Technology Durgapur-713212
[2]Department of Biotechnology, NIT Durgapur-Durgapur-713209, West Bengal
[3]Department of Bioscience and Biotechnology, Banasthali University, Rajasthan-304022
Email : Kambaska@yahoo.co.in

Summary

Researchers face a series of similar-yet unique problems to isolate proteins. To solve these they must have dip knowledge to select an appropriate technique. The mechanism, to study the way in which a living system works, is necessary to dismantle the machine and to isolate the component parts so that they may be studied, separately and in their interaction with other parts. The knowledge that is gained in this way may be put to practical use, for example, in the design of medicines, diagnostics, pesticides, or industrial processes. In every case a pure protein is desirable as impurities may either be misleading, dangerous or unproductive, respectively. Protein isolation is, therefore, a very common, almost central, procedure in biotechnology. The purpose of this chapter is thus to fill the beginners with techniques relevant to protein isolation, hopefully to improve their knowledge.

Introduction

A major portion of most biochemical investigations involves the purification of materials under consideration, because these substances must be relatively free of contaminants, if they are to be properly characterized. This has often been a formidable task, because a typical cell contains thousands of different substances, many of which closely resemble other cellular constituents in their physical and chemical properties. Furthermore, the material of interest may be unstable and often exist in minute amounts. The development of techniques and methods for separation and purification of biological macromole such as proteins has been an important prerequisite for many of the advancements made in bioscience and biotechnology over the past three decades. Improvements in materials and utilization of microprocessor-based instruments have made protein separations more predictable, and controllable, although too many are still more an art than a science.

Protein purification has come of age with the rapid growth of high-resolution chromatographic techniques [e.g. HPLC, fast protein liquid chromatography (FPLC)] on an analytical, preparative and industrial scale where several hundred liters of raw materials can be purified in a reproducible & cost-effective manner. Also, new electrophoretic techniques and systems for fast analysis of protein composition and purity have contributed to increased efficiency of the purification process. Another field of increasing importance is micro-preparative chromatography, a consequence of modern methods for amino acid and sequence analysis requiring sub-microgram samples only.

However, not all problems in protein purification can be solved by acquisition of sophisticated laboratory equipment. Difficulties still remain in finding optimum conditions for each individual stages of protein purification, as well as choosing suitable methods for monitoring protein concentration and biological activity. It is this unpredictable nature of proteins coupled with the uncertainties and difficulties of protein purification that makes the subject more attractive to the investigators / researchers.

Extraction Methods

Some biological materials, like snake venoms, blood serum, enzyme, milk, urine etc. are nearly clear protein solution by virtue of being extracellular proteins. So, these comparatively stable mixtures of protein are suitable for direct application to chromatography columns after centrifugation or filtration. However, intracellular protein presents either in cytosol or sub-cellular compartments of cell require suitable extraction techniques.

Major problems in preparing a crude protein extract are general denaturation, oxidation proteolysis, contamination with pyrogens, nucleic acids, bacteria, viruses. These can be limited by proper choice of extraction medium, by reducing preparation time and by using more gentler technique, low temperature, optimum concentration of bacteriostatic agents, reducing agents (Mercaptoethanol, DTT) and protease inhibitors (like EDTA, or PMSF etc.).

Table 1 : Different purification strategy

Characteristic	Fractionation Procedure
1. Solubility	1. Salting in
	2. Salting out (Precipitation)
	i) Ammonium Sulfate Precipitation
	ii) Isoelectric Precipitation
	iii) Precipitation by selective denaturation
	iv) Organic solvent precipitation
	v) Precipitation by organic polymers and other materials
2. Ionic change	1. Ion exchange chromatography
	2. Electrophoresis
	3. Isoelectric focusing
3. Polarity	1. Paper chromatography
	2. Reverse-phase chromatography
	3. Hydrophobic interaction chromatography
4. Molecular size	1. Dialysis and ultra-filtration
	2. Gel electrophoresis
	3. Gel filtration chromatography
	4. Ultracentrifugation
5. Binding Specificity	1. Affinity chromatography
	2. Affinity precipitation

Overview of Fractionation Techniques

The fact that proteins are well defined substances was not widely accepted until after 1926, when James Sumner first crystallised an enzyme, jack bean urease. Proteins are purified by fractionation procedures. In a series of independent steps, the various physicochemical properties of the protein of interest are utilised to separate it progressively from other substances. The idea here is not necessarily to minimize the loss of the desired protein, but to eliminate selectively, the other components of the mixture. The following characteristics are utilised in various separation procedures.

In earlier studies, complex protein mixtures were fractionated mainly by extraction/ precipitation methods. These methods are still used as preliminary steps for crude fractionation or to obtain more concentrated sample solutions. Preparative electrophoretic and chromatographic techniques developed during 1950s and 1960s, made rational purification protocols possible and laid the foundation for further advancements.

Precipitation Techniques

The solubility of protein in aqueous solution is the function of ionic strength (i.e. salt concentration and its ionic charge), pH and temperature. Precipitation of a protein in an extract may be achieved by adding salts, organic solvents, or organic polymers or by varying the pH or temperature of the solution or by denaturation of the target protein.

(i) Salting Out (Ammonium Sulfate Precipitation)

Salt fractionation is one of the most widely used techniques in enzyme purification, where solubility of proteins decreases at high ionic strength or electrolyte concentration of the solution. Precipitation occurs by neutralization of surface charges by the salt, by reducing the chemical activity of the protein and by diminishing the effective concentration of available water.

The property of a particular salt as a precipitating agent is described in the Hofmeister series.

Anions: $PO_4^{3-}, SO_4^{2-}, CH_3, COO^-, Cl^-, Br^-, NO_3^-, ClO_3^-, I^-, SCN$

Cations: NH_4^+, K^+, Na^+, guanidine $C(NH_2)_3^+$

←

Increasing precipitation (Salting out)

→

Increasing chaotropic effect (Salting in)

The anti-chaotropic salts increase the hydrophobic effect in the solution and promote protein aggregation by association of hydrophobic surfaces. The salting out of a protein can be described by the equation

$$LogS = B - Kc$$

Where,

S = the solubility of the protein in g/L of solution.

B = intercept constant

K = salting out constant

c = salt concentration in mol/L

The value of B depends on the salt used, the pH, the temperature and the protein itself. K depends on the salt used and the protein.

Among a number of different neutral or slightly acidic salts, ammonium sulfate [$(NH_4)_2SO_4$] is the most useful for differential precipitation of proteins. The major advantages are :-

(1) At saturation, it has sufficiently high molarity, that causes the precipitation of most proteins;

(2) Low heat of solution, so that the heat generated is easily dissipated;

(3) Even its saturated solution (4.04 M at 20°C) has a density (1.235 g/cm^3) that is not so large that it interferes with the sedimentation of most precipitated proteins by centrifugation;

(4) Concentrated solutions will prevent or limit most bacterial growth;

(5) Its solution will protect most protein from denaturation (i.e. anti-chaotropic effect).

The most productive protocol is to add increasing proportions of ammonium sulfate to the extract with intermittent centrifugation steps. These stepwise precipitates are often referred to as ammonium sulfate "cuts".

(ii) Salting in

The solubility of protein at low ionic strength (i.e., 0 to 0.5 M) generally increases with increasing salt concentration. This phenomenon, where additional electrolyte counter-ions more effectively shield the protein molecules' multiple ionic charges, thereby increasing the protein's solubility is called "salting in".

These chaotropic salts, actually bind and interact directly with the proteins, thus having a destabilizing effect. The "salting-in' range can be exploited in purification, in two distinct ways:-

1. Aggregation may occur by dilution or dialysis and if the required protein is present in the aggregate, improves the probability of purification.
2. Isoelectric precipitation can be used by exploiting the variation of solubility with pH, without changing the ionic strength.

Here, if the required protein is present entirely in the precipitate and remains fully active, then this step is likely to be very effective.

(iii) Organic Solvent Precipitation

Organic solvents like acetone, ethanol, methanol and butanol etc. promote the precipitation of proteins due to decrease in water activity in the solution as water is replaced by organic solvent. The principle causes of aggregation are likely to be electrostatic and dipolar forces, similar to those occurring in the salting-in range in the absence of organic solvent. Hydrophobic attractions are less involved because of solubilizing influence of organic solvents on these areas. The following five variables should be considered:

(a) Concentration of organic solvent,

(b) Protein concentration,

(c) pH

(d) Ionic strength

(e) Temperature.

Low temperature is often necessary to avoid protein denaturation. The addition of an organic solvent decreases the freezing point of the solution and temperatures below 0°C can be used.

The solvent used must be completely water-miscible, un-reacting with proteins, have a good precipitating effect and safe both in terms of flammability and noxious vapours. For the last two reasons, dioxan, 2-methoxy ethanol and other ethers should be avoided. This technique has been useful especially in the fractionation of serum proteins.

(iv) Precipitation with Organic Polymers and Other Materials

Organic polymers, either neutral or charged can cause aggregation of proteins without denaturation. They perform in the similar way to that of organic solvents. The most widely used polymer is polyethylene glycol (PEG), with an average molecular weight of either 6000 or 20,000. It is nontoxic, uncharged can be handled more easily and inexpensive. Rather low concentrations (< 25%) are required to precipitate most proteins. The only disadvantage is that concentrated solutions of PEG are highly viscous. So it can be difficult to remove from protein solutions. Other precipitations include caprylic acid salts, polyacrylic acid salts and rivanol etc. These have been tried in plasma protein fractionation.

(v) Isoelectric Precipitation

The isoelectric point (pI) of a protein is defined as the pH at which the net charge on the protein is zero, thus having lowest solubility in aqueous solutions. The pH of protein mixture is adjusted to the pI of the target protein so as to selectively minimize its solubility. In practice, this technique is combined with salting out so that the protein being purified is usually salted out near its pI. This technique has been successfully used in the purification of insulin and calmodulin.

(vi) Precipitation by Selective Denaturation

The basic principle behind this technique is selective denaturation and precipitation of the unwanted proteins from impure preparation leaving the desired protein in the supernatant. This is possibly only, when the target protein can withstand quite extreme conditions in terms of (a) temperature (b) pH and (c) organic solvent.

Chromatographic Techniques

Separation by chromatography depends on the differential affinity of proteins between a stationary phase (the chromatographic medium or the adsorbent) and a mobile phase (the buffer solution). Normally the stationary phase is packed into a vertical column of plastic, glass or stainless steel, whereas the buffer is pumped through the column. An alternative is to stir the protein solution with the adsorbent and then pour the slurry onto an appropriate filter and make the washings and de-sorption on the filter. As a cell contains huge number of different proteins, many of which closely resemble the target protein in terms of its properties, one has to combine several chromatographic methods to achieve complete purification of a protein from a crude biological extract. Generally, adsorption or partition chromatography is not used for protein purification.

(1) Ion Exchange Chromatography (IEC)

Principle: In the process of ion-exchange chromatography (IEC), ions that electrostatically bound to an insoluble and chemically inert matrix are reversibly replaced by ions in solution. In IEC, the stationary phase carries ionizable functional groups, fixed by chemical bonding.

IEC is named on the basis of the sign of these displaceable charges. Thus, in anion-exchange matrix the fixed charges are positive and cation-exchange matrix the fixed charges are negative.

STEP-I(Binding & retention)

$$R^+ A^- + P^- \rightleftharpoons R^+ P^- + A^-$$

R^+ = fixed charge (anion exchanger)

A^- = counterion

P^- = charged protein molecule.

STEP-II (Elution)

$$R^+P^- + S^- \rightleftharpoons R^+ S^- + P^-$$

S^- = New counterion

IEC involves two separate events. These are :-

(a) Binding of the protein to the fixed charges by electrostatic interaction by replacing the non-fixed counterion (A^-) [for anion exchange chromatography]

(b) Elution, in turn, involves displacement of the protein from immobilized matrix by a new counterion with greater affinity for the fixed charges than the protein, which then becomes the new, non fixed ion.

Types of Ion Exchangers

Wide used solid-phase packing includes cellulose, dextrans, agarose and polystyrene. The exchange groups used include DEAE (diethyl-amino-ethyl), a weak base or QA (quaternary amino) groups, a strong base, that will have net positive charge below their pKs, will therefore bind and exchange anions. On the other hand, CM (carboxy-methyl), a weak acid or sulfonic acid functional groups, all are strong acids with negative charge above their pKs will bind and exchange cations only. DEAE-cellulose and CM-cellulose are the most popular choice for anion-exchange and cation-exchange chromatography respectively.

Starting Conditions for IEC

Buffer: Maintenance of pH in ion-exchange chromatography is of paramount importance. Since the ionic interactions are so dependent on pH and buffers themselves are ionic and may be taking part in the ion-exchange process. Choice of suitable buffer depends on the following criteria:-

1. Buffering ions should not themselves interact with the adsorbent, that is, the charged form(s) of buffer should be of the same sign as the substituent on the adsorbent.
2. In absence of any reason to do otherwise, only simple anions (e.g. Cl^-, *acetate*) and simple cations (e.g. K^+, Na^+, Mg^{2+}, $Htris^+$ for pH < 7) should be present for anion & cation exchange chromatography respectively.
3. To maximize the buffering power at the least possible ionic strength to ensure adsorption of a weakly binding protein are :-
   (a) The pKa of the buffer should be not more than 0.5 unit and preferably within 0.3 unit of the pH being used.
   (b) One of the buffering species should be uncharged, and so, not contribute to the ionic strength.

For anion exchanger at pH 8.1, a Tris-Cl buffer obeys all the three rules because, the buffering species are H $Tris^+$ (non-interactive) and Tris (neutral), the pK_a of Tris is 8.1 and the counterion is Cl^-.

Nevertheless, many examples are there in the literature, where these rules have been successfully broken. For example, phosphate buffer has been used successfully with DEAE-cellulose.

pH: The initial pH of the experiment must be determined empirically within the pH range of 5.0 to 9.0, unless the activity being purified cannot withstand changes in pH. Theoretically, for anion exchange chromatography, the experimental pH should be more than pI (isoelectric pH of protein (target) and less than pK_a of the ion-exchange group.

i.e., pI < experimental pH < pK_a of ion-exchange group.

Generally, the initial pH should be 0.5 pH unit higher than the point at which the activity begins to be bound by the ion-exchanger. But, if the pH is too high, it will require very high salt concentration to elute the proteins, which is not desirable.

For cation exchange chromatography:

pK_a (ion-exchange < experimental pH < pI (target protein)

Here, activity (protein) will be bound at lower pH (< pI).

Ionic Strength

Generally, starting ionic strength of the experiment should be determined empirically at optimum buffer and predetermined pH. It is not necessary to load at zero salt, because there can be enormous purification potential gained by allowing contaminating proteins to "flow through" the resin. Initial ionic strength should be at least 0.1 M salt below the point of elution of target protein or activity.

Capacity

The adsorptive capacities of ion-exchangers for proteins can be very high, but it depends on molecular size of protein. Small proteins bind in amounts in excess of 100 mg/cm^3, but very large proteins (> 10^6 MW) may not penetrate the exchanger and bind only to the surfaces of the beads, resulting in very low capacities. So, it is preferable to determine the capacity empirically for the protein mixture under consideration. It should be kept in mind that capacity involve the total protein applied, not just the activity in question.

Preparation and Loading Sample

Loading of IEC columns involve applying the sample onto the packing all at a time. Any volume can be applied as long as the total amount of protein does not exceed the binding capacity of the packing. To minimize the proteolytic degradation of the sample, during IEC, working temperature should be kept at 4°C and optimum concentration of protease inhibitor(s) should be used in the buffers.

Elution Procedure

Theoretically, two general methods for eluting proteins are available :-

1. To change the buffer pH to a value, where binding is weakened - lower pH for anion exchanger or higher pH for cation exchanger.
2. To increase the ionic strength, thereby weakening the electrostatic interaction between protein and adsorbent.

Practically method (1) is very difficult to implement, unless there is very high buffering capacity, because, sudden large pH changes happen as proteins become eluted, so there is little separation of individual components. Method (2) or salt gradients are the commonest means of eluting proteins from ion exchangers. Usually, either *KCl* or *NaCl* is used to generate the gradient at a constant pH. A linear gradient is most widely used by a simple mixing device consisting of two chambers joined by a tube, where the salt concentration of eluent solution varies linearly with the volume of solution passed. Here, the solute concentration, C (in the solution being withdrawn from the mixing chamber) is expressed as :-

$$C = C_2 - (C_2 - C_1)f$$

Where, C_1 = solution of initial concentration in the mixing chamber

C_2 = concentration in the reservoir chamber

f = remaining fraction of combined volumes of the solutions initially present in both reservoirs

However, gradients of different shapes can be generated by using two or more chambers of different cross-sectional areas or programmed mixing devices. Elution of unknown proteins can be done by stepwise salt elution, where bound proteins are eluted by sequential use of buffered salt solution of increasing molarity.

Graphical Presentation of Data

The gradual change in protein concentration in the eluent from IEC columns can be conveniently determined by monitoring the eluent at 280 nm using a suitable UV-visible spectrophotometer. This is because most proteins contain aromatic residues (such as tyrosine & tryptophan) which have an absorption maxima in this region. In absence of such residues (e.g. collagen) it is possible to monitor the effluent at 210-220 nm, the absorption region of the peptide linkage.

Gel Filtration Chromatography

Gel filtration chromatography, which is also called size exclusion and molecular sieve chromatography is a method for separation of molecules on the basis of their size and shape. The stationary phase in this technique consists of beads of a hydrated, sponge-like material containing pores that span a relatively narrow size range of molecular dimensions. If an aqueous solution containing molecules of various sizes is passed through a column containing such molecular sieves, the molecules that are too large to pass through the pores are excluded from the solvent volume inside the gel beads. These larger molecules therefore transverse the column around the beads rapidly and smaller eluent volume will achieved. The molecules that are small enough to enter the pores of the beads will emerge from the column later.

Exclusion Limit: The molecular mass of smallest molecule unable to penetrate the pores of a given gel is said to the gel's exclusion limit. This will also partially depend on the molecular shape, because elongated molecules, as a consequence of their higher radius of hydration, are less likely to penetrate a given gel pore than spherical molecules of same molecular volume.

Void volume (V_O): It is defined as the volume of solvent space surrounding the beads. So if V_X is the volume occupied by the gel beads and V_t is the total bed volume then,

$$V = V_0 + V_x$$

Generally, V_o is typically ~35% of V_t.

Elution Volume (V_e): The volume of solvent required to elute the solute from the column after it has first contacted the gel.

K_{av}: A protein will elute at a reproducible and characteristic position from a particular resin with a specific buffer. This property is usually reported as the K_{av}, which is :-

$$K_{av} = \frac{V_e - V_0}{V_t - V_0}$$

Similar to R_f value of Thin-Layer-Chromatography. A protein that elutes in the void volume, will have a $K_{av} = 0$, whereas, a protein that elutes at V_t will have a $K_{av} = 1$. A gel filtration resin is most effective when the K_{av} of the protein to be purified is approximately 0.5. A related parameter is V_o/V_e ratio, called the relative elution volume, a quantity that is independent of the size of the particular column used.

Determination of V_o (Void Volume)

The void volume of a column will be the elution volume of a solute, whose molecular mass is larger than the exclusion limit of the gel. Blue dextran, a polysaccharide, to which the blue dye Cibacron blue is covalently attached, is used for this purpose. Its molecular weight is around 2 million. So it completely excluded from the pores.

Gel Filtration Resins

The gel filtration resins commonly composed of the following three polymers.

(i) Dextran (a polysaccharide, produced by the bacterium *Leuconostoc mesenteroides*)

(ii) Agarose (a linear polymer of alternating D-galactose and 3,6-anhydro-L-galactose from red algae).

(iii) Polyacrylamide

The properties of some commonly used gels are listed below :-

Name	Code	Fractionation Range (KD)	Type
Sephadex	**G-10**	**0.05 0.7**	**Dextran**
	G-25	1 - 5	Dextran
	G-50	1-30	Dextran
	G-100	4-150	Dextran
	G-200	5-600	Dextran
Biogel	**P-10**	**1.5 - 20**	**Polyacrylamide**
	P-30	2.4 - 40	Polyacryl
	P-100	5-100	Polyacryl
	P-300	60-400	Polyacryl
Sepharose	**6B**	**10-4000**	**Agarose**
	4B	60-20,000	Agarose
	2B	70 - 40,000	Agarose
Bio-gel	**A-5**	**10 - 5000**	**Agarose**
	A-50	100 - 50,000	Agarose
	A-150	**1000 - 150,000**	**Agarose**

Column Profile

For maximal resolution with gel filtration it is better to use a long, thin column rather than short, fat column of equivalent packed bed volumes. Ideally, height of column should be 20-40 times of the diameter of the column. Though, thin, long column will make the procedure less sensitive to imperfections in the packing of resin, it can lower the flow-rate below acceptable limit.

Sample Profile

Sample volume should be less than 5% of the total bed volume. The protein concentration in the sample can be as high as 50 mg/ml. The sample is diluted by a factor of 2 to 3 following gel filtration.

Elution Profile

Molecules with molecular masses ranging below the exclusion limit of a gel will elute from the gel in order of their molecular masses, with the largest eluting first. This is because the pore size in any gel vary over a limited range, so that larger molecules have less of the gel's interior volume available to them than do smaller molecules.

Flow Rate

Because beads used in the gel filtration are porous, they can be sensitive to moderate pressure. So, chromatography must be performed according to the recommended flow rates provided by the manufactures. The unit for flow rate is cm/hour, which is identical to $ml/cm^2/hour$, i.e., ml of buffer per cm^2 of cross-sectional area of the column per hour.

Graphical Presentation of Data

Absorbence of the each eluted fraction is measured at 280 nm with respect to elution buffer as blank. Generally, effective fractionation volume is about 50% of the total column volume. Resolution of the experiment depends on the bead size, homogenity of beads and ratio of sample size & column volume. Gel filtration chromatography can be used to estimate Molecular Masses, because, linear relationship exists between relative elution volume (V_e/V_o) of a protein and logarithm of its molecular mass (log FW) over a considerable molecular mass range. A standard calibration curve (log FW vs. V_e/V_o) can be made by using Gel filtration. Calibration Kit (contain mixture of std. proteins of varying molecular mass), the molecular mass of unknown substrate can be estimated from this plot. This precision of this technique is limited by the accuracy of the underlying assumption that the known and unknown macromolecules have identical shapes.

Hydrophobic Interaction Chromatography (HIC)

Hydrophobic interaction between protein and immobilized stationary phase forms the basis of this chromatography. In HIC, the stationary phase is only lightly substituted with hydrophobic ligands, usually octyl or phenyl residues. The resulting hydrophobic interactions in HIC will therefore be relatively weak, thus maintaining the native protein structures. The two most common matrices in use are 4% and 6% cross-linked agarose and the synthetic co-polymer materials.

The protein-binding capacity of the matrix will increase with increasing degree of substitution of the immobilised ligand, before it reaches a plateau. But the strength of binding will continue to increase in proportion to the degree of substitution. Thus, proteins bound to highly substituted matrices will be difficult to elute. The eluents in HIC, whose gradients must progressively reduce these weak hydrophobic interactions are aqueous buffers, for example, decreasing salt concentrations (on, HI is strengthen by increased ionic strength), increasing concentrations of detergents or water-miscible alcohols or increasing pH. Thus, HIC separates native proteins according to their degree of surface hydrophobicity.

Affinity Chromatography

A distinct characteristic of many protein, is their ability to bind specific molecules tightly but non covalently. This property can be used to purify such proteins by affinity chromatography. In this technique, a molecule, known as a ligand, which specifically binds to the protein of interest, is covalently attached to an inert and porous matrix. When an impure protein solution is passed through this chromatographic material, the desired protein binds to the immobilized ligand, whereas other substances are washed through the column with the buffer. The target protein can then be recovered in highly purified form by changing the elution conditions such as :-

1. Eluting with a solution of a compound that has higher affinity for the protein-binding site than the bound ligand.
2. Altering the solution conditions by changing the pH, ionic strength, and/or temperature, so that protein-ligand complex becomes unstable.

All these changes result in release of protein from the chromatographic matrix. The chromatographic matrix in affinity chromatography must be chemically inert, have high porosity and have large number of functional groups capable of forming covalent linkages to ligands. Agarose, having numerous free hydroxyl groups, meet all these criteria. The ligand, having a primary amino group that is not essential for its binding to the protein of interest, can be covalently linked to agarose in a two-step process.

1. Agarose is reacted with cyanogen-bromide to form an "activated" but stable intermediate.
2. Ligand reacts with activated agarose to form covalently bound product.

Sometimes, if cyanogen-bromide-coupled ligands cause steric interference during protein binding, then "epoxy-activated" agarose, in which a spacer group links the resin to a reactive epoxy group, can be used for forming ligand-matrix complex.

The ligand used in the affinity chromatography must have an affinity for the protein high enough to immobilise it on the agarose gel, but not so high as to prevent its subsequent release. If the ligand is a substrate for an enzyme being isolated, the chromatography condition must be such that the enzyme does not function catalytically or ligand will be destroyed. The great advantage of affinity chromatography is its ability to exploit the desired proteins unique biochemical properties rather than the small differences in physico-chemical properties between proteins that other chromatographic methods must utilize. It has been used to isolate such substances as enzymes, antibodies, transport proteins, hormone receptors, membranes and even whole cells.

Agarose -$OH + N \equiv C - Br$ (cyanogen bromide)

OH^{-} ↓ Step-I

Agarose -$O - C - Br$ (Activated agarose)

$\|$

NH

Step-II $RNH_2 + H_2O$

O

$\|$

Agarose -$O - C - NHR + NH_3 + HBr$

Fig. Covalent linking of ligand to agarose.

Manipulation of Protein Solutions

Protein purification is a multistep process. Very often it is observed that the solution of protein that is resulted at the end of a particular step is not in the right state for the next fractionation procedure or for long-term storage or for measurement of activity. It might be in the wrong buffer, at the wrong pH, have too much salt, be too dilute or too turbid or have desorption agents from an affinity chromatography step, that must be removed before the use of the protein.

All the above factors necessitate the use of following techniques to obtain the protein solution in desired state.

(a) Dialysis

Dialysis is the process of separating molecules according to size through the use of semi-permeable membranes containing pores of less than macromolecular dimensions. These pores allow small molecules, such as those of solvents, salts and small metabolites, to diffuse across the membrane but block the passage of larger molecules.

Cellophane (cellulose acetate) is the most commonly used dialysis material, although several other substances as cellulose and collodion are similarly employed. These are available in a wide variety of molecular weight cutoff values (the size of the smallest particle that cannot penetrate the membrane) that range from 0.5 to 500 kD.

Use

1. Dialysis is the classic way of routinely changing buffering species and buffer composition of a protein solution. The dialysis bag consisting the protein solution is placed in a larger stirred vessel containing the desired buffer, which is changed several times to reach equilibrium between in and outside of the dialysis bag.
2. Dialysis can be used to remove traces of salts (ammonium sulfate) that remain in protein solution following ammonium sulfate precipitation of the crude extract.
3. Finally, it can also be employed to concentrate a protein solution by packing a solution filled dialysis bag in polymeric desiccant, such as polyethylene glycol [$HO\text{-}CH_2(CH_2\text{-}O\text{-}CH_2)CH_2OH$] The concentration is affected as water diffuses across the membrane to be absorbed by the polymer.

The disadvantages of dialysis are -

1. Slow process
2. One needs to change the dialysate repeatedly
3. Chances of proteolytic degradation will be there, so, it is advisable to do it at 4°C

Ultrafiltration

Ultrafiltration is a process of forcing a macromolecular solution through a semi-permeable membranes disk either under pressure or by centrifugation. Solvent and small solutes will pass through the membrane, which is collected as 'ultrafiltrate', leaving behind a more concentrated macromolecular solution, is called as 'retentate'. Since ultrafiltration membrane of different pore sizes is available, it can be used to separate different sized macromolecules like proteins or for changing the buffer composition of protein solutions. Although, it is mainly used as a rapid method to concentrate dilute protein solutions without expecting significant purification.

Lyophilization

Most biological processes occur in water solution, so the best way to stop it is to remove the water from solution. Freeze-drying or Lyophilization is the method of choice for biologically active proteins. In this method, the protein solution is frozen below the eutectic point of the solution (<40°C), which is then placed under high vacuum until sublimation of the water molecules are complete. It is generally a safe and gentle process, where the biological activity can be fully recovered upon addition of water. This is one of the commonest ways of preserving any proteins including enzymes for commercial use that can be reconstituted simply by adding water.

Centrifugation and Microfiltration

These two techniques are routinely used for clarification of turbid protein solution to make it suitable for chromatographic applications. These will remove particulate and most non-protein materials from protein solution. Differential centrifugation can be used to obtain an extract containing the desired sub-cellular component rich with the target protein. Refrigerated high-speed centrifuges covering operating speeds from 20,000 to 75,000 rpm generating about 40,000 X g to about 500,000 X g are routinely used in a small scale to clarify a cell homogenate. On the other hand, clarification of cell extracts on a large scale is carried out by membrane ultrafiltration, using a membrane with a pore-size of about 0.2 mm.

Electrophoresis

Electrophoresis is defined as the migration of ions or charged molecules in an electric field, which is widely used for analytical and preparative separation of biological molecules, like proteins, nucleic acids, charged sugars, nucleotides, amino acids and peptides.

Principle: A protein molecule in solution at a pH other than its isoelectric pH will have a net average charge. According to laws of electrostatics,

$$F_{electric} = qE$$

Where, q is the charge of an ion in an electric field of strength E. The resulting electrophoresis migration of the ion through the solution is opposed by a frictional force, $F_{friction,}$ So,

$$F_{electric} = vf$$

Where, v is the velocity of the ion and *f* is its frictional co-efficient, which is a measure of the drag that the solution exerts on the moving ion and is dependent on the size, shape and state of the solvation of the ion as well as on the viscosity of the solution. In an constant electric field,

$$qE = vf$$

So, each ion moves with a constant characteristic velocity. An ion's electrophoretic mobility, *m*, is defined as :-

$$\mu = \frac{v}{E} = \frac{q}{f}$$

The above equation really applies only to ions at indefinite dilution in a non conducting solvent. In aqueous solutions, poly-electrolytes such as proteins are surrounded by counterions, which impose an additional electric field making the equation, at best, a poor approximation of reality. Electrophoresis of macromolecules is normally carried out by applying a thin layer of sample to the solution stabilized by a porous matrix. At the end of the separation, the different species present in the sample are detected as bands at different positions in the matrix. A matrix, which acts as a stabilizing medium, is required to nullify the electrophoretic heat induced diffusive and convective mixing of bands. The matrix can be a paper, cellulose acetate or gels made of poly-acrylamide, agarose or starch.

Gel Electrophoresis

Polyacrylamide and agarose gels are the most common stabilizing media used in research laboratories. The gels are usually formed as thin, flat slabs or sheets or as cylinders in tubes. Polyacrylamide is the most common matrix for separating proteins.

Polyacrylamide Gel Electrophoresis (PAGE)

Gels are formed by a free-radical induced polymerization of acrylamide and cross-linker *N,N'*-methylene-bis-acrylamide (bis) in the buffer of choice. The polymerization reaction is initiated either by chemical peroxide or by a photochemical method. The most common method uses ammonium persulfate as the initiator peroxide, whose chemical decomposition supplies the sulfate radical ($S_2O_8^{2-} \rightarrow 2SO_4^-$) for polymerization. In photochemical polymerization, riboflavin in presence of traces of O_2 and long-wave UV light act as the initiator. In either case, *N,N,N',N'*-tetra-methyl-ethylene-di-amine (TEMED), a free radical stabilizer is used as catalyst. Photodecomposition of riboflavin is used, when the ionic strength in the gel must be very low since only a minute amount of riboflavin is required. It is also used if the protein studied is sensitive to ammonium persulfate or the byproduct of peroxide initiated polymerization.

Rapid polymerization of acrylamide can generate too much heat, causing convection inconsistencies in the gel structure and occasionally breaking glass plates. To prevent excessive heating, the concentration of initiator-catalyst reagents should be adjusted so that complete polymerization requires 20-60 minutes. The most commonly used poly-acrylamide concentration are in the range of 3 to 15% with the amount of TEMED usually fixed at 5% of total acrylamide present.

There are two types of buffer systems in electrophoresis, continuous and discontinuous. A continuous system has only a single separating gel and uses the same buffer in the tanks and the gel.

In discontinuous system, a method first developed by *Ornstein* (1964) and *Davis* (1964), and modified by *Laemmli* (1970) a non-restrictive large pore gel, called a stacking gel is layered on the top of the separating (running) gel. This two-gel system requires several different buffers, namely -

1. Stacking gel buffer (pH 6.8)
2. Resolving gel buffer (pH 8.8)
3. Sample buffer (pH 6.8)
4. Tank buffer (pH 8.3) or upper reservoir buffer (contains weak acid, glycine).

When the current is switched on, the buffer ions from the upper reservoir migrate into the stacking gel as the stacking gel buffer ions migrate ahead of them. As this occurs, the upper reservoir buffer ions encounter a pH that is much lower than their pK. They therefore assume uncharged (or in case of glycine, zwitterionic) form and become electrophoretically immobile. This causes a deficiency of charge carriers, and results in high electrical resistance (R) in this region. So, according to Ohm's law ($I = E/R$), to maintain a constant (I) current throughout the electrical circuit, there is a highly localized increase in electric field E. In response to this increased field, the macromolecular anions migrate rapidly until they reach the region containing the stacking gel buffer ions (leading ions), where they show down because at this point there is no ion deficiency. This effects causes the macromolecular ions to the approach the resolving gel as stacks of very narrow (~0.01 mm thick) bands or disks and lie between the migrating ions of the upper reservoir (trailing ions) and leading ions. As the macromolecular ions enter the resolving or running gel, they slow down as a result of gel filtration effects. This permits the trailing ions to overtake the macromolecular bands as because of running gel' has higher pH, they assume their fully charged form. So from resolving gel onwards, electrophoretic separation proceeds normally. However, the compactness of the macromolecular bands entering the running gel greatly increases the resolution of the macromolecular separation, when compared with continuous slab gel electrophoresis.

Polyacrylamide gel electrophoresis (PAGE) can be of two types :-

(1) Native PAGE :- where proteins are separated according their intrinsic charge, size and shape.

(2) SDS-PAGE (sodium dodecyl sulfate PAGE) :- Where molecular separation takes place not by intrinsic charge of the proteins but by their molecular weight.

SDS is an anionic detergent that denatures proteins by wrapping around the polypeptide backbone, thus importing a net negative charge to the polypeptide in proportion to its length. The polypeptides become rods of negative charges with equal "charge densities" or charge per unit length. The relative mobilities of proteins on such gels vary linearly with the logarithm of their molecular masses. In practice, molecular mass of unknown proteins are routinely to an accuracy of 5-10% using the plot, which is drawn after electrophoresing several standard "marker" proteins of known molecular masses.

SDS-PAGE yields the molecular mass of proteins subunit rather than intact protein, unless the subunits are disulfide linked, as SDS treatment disrupts the non-covalent interactions between subunits of protein. However, marcaptoethanol treatment will reductively cleave these disulfide bonds.

Gel Staining Techniques

Bands resulting from gel electrophoresis can be located by a variety of techniques. They are: -

1. Staining
2. Radioactive counting or
3. Immunoblotting.

1. Staining: It can be classified under following headings.
   (a) Organic staining: Most widely used staining technique which employs cromassie brilliant blue in an acidic, alcoholic (methanol) solution. (R-250 or G-250).

The dye molecules form complex with gel-bound proteins through electrostatic interaction with the amino ($-NH_2$) group of protein in acidic medium as well as by vander-Waals forces. This reversible binding is facilitated by the presence of lysine, arginine (basic amino acids) residues, in the polypeptide chain. Protein bands containing as little as 0.1 mg protein can be detected, following de-staining procedure, with an acidic solution.

(b) Fluorescent Staining

Fluorescamine is a non fluorescent molecule that reacts with primary amines, such as, lysine of the polypeptides to yield an addition product that is known to fluoresce under UV radiation.

(c) Metal based staining

Gel bands containing nanogram amount of protein can be visualized with metal-based staining, such as -

(i) Silver staining: This offers 100 fold increases in sensitivity than the coomassie-staining.

(ii) Copper staining: This is a telativelty simple and a more sensitive technique as compared to coomassie.

2. Autoradiography

Proteins can be detected through the UV-absorption of the gel along its length. The level of radioactivity can be determined using a X-ray film (autoradiography) or a scintillation counter. The last method yields quantitatively more accurate results than autoradiography.

3. Immunobloting: (Western blotting)

This technique employs an antibody to the protein of interest that will specifically detect the protein on a gel in presence of many other proteins.

Steps in a Purification Strategy

The design of purification scheme will be different depending on the material at hand and the purpose. If the starting material is very precious, one should favour high yield over speed and convenience. The choice of purification protocol will depend on certain factors:-

(1) Characteristics of the target protein like its approximate molecular weight, pI, degree of hydrophobicity, presence of carbohydrate (glycoprotein) or free sulphahydryl group and so on.

(2) Stability of the target protein at varying temperature, pH, concentration of organic solvents, oxygen, heavy metals and mechanical shear.

(3) Susceptibility of the protein towards proteolytic degradation.

(4) Properties of the most important impurities present in the crude extract.

The purification scheme consists of following three stages.

1. The preliminary or initial fractionation stage (capture stage).
2. The intermediate purification stage.
3. The final polishing stage.

1. Capture Stage

The crude sample is generally turbid from the presence of cellular debris, particulates, pigments, lipophilic mateials or viscous from the presence of DNA and will typically represent the largest volume to be processed during purification procedure. The purpose of the initial stage is to obtain a protein solution suitable for chromatography and electrophoresis by using clarification through centrifugation, microfiltration, ultrafiltration and coarse fractionation by ammonium sulfate precipitation followed by dialysis.

Contamination of crude extract with lipids and nucleic acids can be removed by sequential treatment with organic solvent (dichloromethane, acetone etc) and precipitants like streptomycin sulfate, protamine sulfate etc, respectively. Another way to eliminate nucleic acid is to add nucleases that will cleave nucleic acids into smaller pieces, thereby reducing the viscosity.

2. Intermediate Purification Stage

At this point, only small amount of other types of biomolecules (e.g. lipids, & nucleic acids) should be present in the extract. It is advisable to design the sequence of chromatographic steps in such a way that buffer changes and concentration steps are avoided. It is most convenient to start with the methods that allow the application of large volumes of samples and that have highest capacities and low material cost. The obvious candidates are ion-exchange chromatography and hydrophobic interaction chromatography. As the latter often requires the addition of salt for adequate protein binding, it is preferably applied after salt precipitation or after salt displacement from ion-exchange chromatography, thereby excluding the need for desalting step. The peaks eluted from an ion-exchanger can, regardless of the ionic strength, can be applied to a gel filtration column (Sephadex G-10), which functions as a desalting column.

3. Polishing Stage

The goal of the final polishing stage is to remove structural and functional variants of the target protein as well as any trace contaminants that share very similar physicochemical attributes with the target protein. All high-resolution chromatographic techniques, like gel filtration, chromatofocusing, reversed-phase chromatography (RPC) and affinity chromatography are useful at this stage. RPC is very effective for resolving structural variants of a target protein, but generally leads to loss of activity of the target protein due to presence of organic solvents. Affinity chromatography is a biospecific method thus gives a homogeneous product with respect to biological activity. But it often requires special eluent to recover the target protein, which may have to be removed before the purified target protein can be used. Gel filtration does not generally require any special eluent and is often the best technique for polishing especially, when the protein is used for subsequent physicochemical characterization (like molecular weight determination).

Crude preparations of alkaline proteases are generally used for commercial use. However, the purification of alkaline proteases is important from the perspective of developing a better understanding of the functioning of the enzyme (Takagi, 1993; Tsai et al., 1988).

STANDARD PROTEIN PURIFICATION SCHEME :

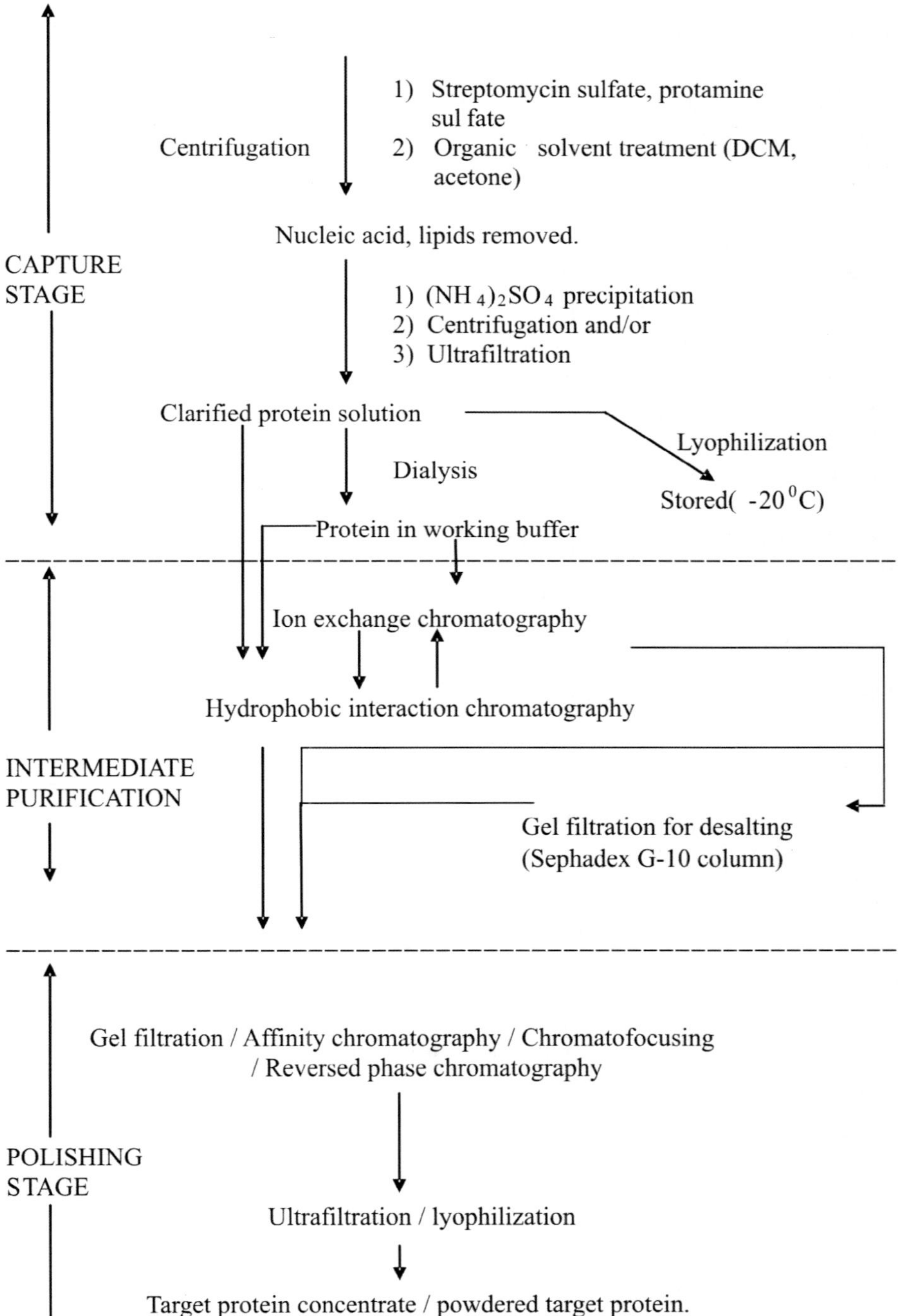

Purification of alkaline proteases : an example

Recovery

After successful fermentation, broth was kept in low temperature (to below 4°C) to prevent microbial contamination as well as to maintain enzyme activity and stability. The removal of the cells, solids, and colloids from the fermentation broth is the primary step in enzyme downstream processing. The vacuum rotary drum filters and cold centrifuges are commonly used. To prevent the losses in enzyme activity caused by imperfect clarification or to prevent the clogging of filters, it is necessary to perform some chemical pretreatment of the fermentation broth before commencing separation (Aunstrup, 1980; Mukhopadhyay et al., 1990). Changes in pH may also be suitable for better separation of solids (Tsai et al., 1983). Furthermore, the fermentation broth solids are often colloidal in nature and are difficult to remove directly. In this case, addition of coagulating or flocculating agents becomes vital (Boyer & Byng, 1996). Flocculating agents are generally employed to effect the formation of larger flocs or agglomerates, which in turn accelerate the solid–liquid separation. Cell flocculation (Bautista et al. 1986) can be improved by neutralization of the charges on the microbial cell surfaces, which includes changes in pH and the addition of a range of compounds that alter the ionic environment. The flocculating agents commonly used are inorganic salts, mineral, hydrocolloids, and organic polyelectrolytes. In some cases, it becomes necessary to add a bioprocessing filter aid, such as diatomaceous earth, before filtration (Boyer & Byng, 1996; Tsuchida et al. 1986).

Isolation and purification

When isolating enzymes on industrial scale for commercial purposes, the prime consideration has been the cost of production in relation to the value of the end product.

Precipitation

Precipitation is the most commonly used method for the isolation and recovery of proteins from crude biological mixtures (Bell et al. 1983). It also performs both purification and concentration steps. It is generally affected by the addition of reagents such as salt or an organic solvent, which lowers the solubility of the desired proteins in an aqueous solution. The precipitation by ammonium sulphate has been widely used for many years. Ammonium sulphate found wide utility only in acidic and neutral pH values and developed ammonia under alkaline conditions (Aunstrup, 1980). Hence, the use of sodium sulphate or an organic solvent formed the preferred choice. Despite better precipitating qualities of sodium sulphate over ammonium sulphate, the poor solubility of the salt at low temperatures restricted its use for this purpose (Shih et al. 1992). Many reports revealed the use of acetone at different volume concentrations: 5 volumes (Horikoshi, 1971), 3 volumes (Kim et al. 1996; Tsujibo et al. 1990), and 2.5 volumes (Kumar et al. 1997), as a primary precipitation agent for the recovery of alkaline proteases. Precipitation was also reported by various workers with acetone at different concentrations: 80% (v/v) (Durham, 1987; Kwon et al., 1994), 66% (v/v) (Yamagata et al., 1995); or 44, 66, and 83% (v/v) (El-Shanshoury et al. 1995), followed by centrifugation and/or drying. Precipitation of enzymes can also be achieved by the use of water-soluble, neutral polymers such as polyethylene glycol (Larcher et al. 1996).

Concentration

Because the amount of enzyme present in the cell-free filtrate is usually low, the removal of water is a primary objective. So, membrane separation processes have been widely used for downstream processing (Strathmann, 1990). Ultrafiltration (UF) is one such membrane process that has been largely used for the recovery of enzymes (Bohdziewicz, 1996; Bohdziewicz, 1994; Bohdziewicz & Bodzek, 1994) and formed a preferred alternative to evaporation. This pressure-driven separation process is inexpensive, results in little loss of enzyme activity, and offers both purification and concentration (Sullivan et al., 1984), as well as diafiltration, for salt removal or for changing the salt composition (Boyer & Byng, 1996; Manachini et al., 1988; Peek et al., 1992). However, a disadvantage underlying this process is the fouling or membrane clogging due to the precipitates formed by the final product. This clogging can usually be alleviated or overcome by treatment with detergents, proteases, or acids and alkalies. Han and associates (Han et al., 1995) used a temperature-sensitive hydrogel ultrafiltration for concentrating an alkaline protease. This hydrogel comprised poly (*N-isopropyl-acrylamide*), which changed its volume reversibly by the changes in temperature. The separation efficiency of the enzyme was dependent on the temperature and was 84% at temperatures of 158^0C and 208^0C. However, at temperatures above 258^0C, a decrease in the separation efficiency was observed.

References

Aunstrup, K. (1980). Proteinases. in: Economic Microbiology: Microbial Enzymes and Bioconversions, (Ed.) A.H. Rose, Vol. 5, Academic Press. New York, pp. 50–114.

Bautista, J., Chico, E., Machado, A. (1986). Cell removal from fermentation broth by flocculation-sedimentation. Biotechnol Lett 8, 315–318.

Bell, D.J., Hoare, M., Dunnill, P. (1983). The formation of protein precipitates and their centrifugal recovery. Adv Biochem Eng Biotechnol 26, 1–72.

Bohdziewicz, J. (1996). Ultrafiltration of technical amylolytic enzymes. Process Biochem, 31, 185–191.

Bohdziewicz, J. (1994). Ultrafiltration of technical proteolytic enzymes. Process Biochem, 29, 109-118.

Bohdziewicz, J., Bodzek, M. (1994). Ultrafiltration preparation of pectinolytic enzymes from citric acid fermentation. Process Biochem, 29, 99-107.

Boyer, E.W., Byng, G.S. 1996. Bacillus proteolyticus species which produce an alkaline protease, US Patent No: 5518917.

Durham, D.R. (1987). Utility of subtilisin GX as a detergent additive. J Appl Bacteriol 63, 381-386.

El-Shanshoury, A.R., El-Sayed, M.A., Sammour, R.H., El-Shouny, W.A. (1995). Purification and partial characterization of two extracellular alkaline proteases from Streptomyces corchorusii ST36. Can J Microbiol, 41, 99-104.

Han, J., Park, C.H., Ruan, R. (1995). Concentrating alkaline protease, subtilisin, using a temperature-sensitive hydrogel. Biotechnol Lett, 17, 851-852.

Horikoshi, K. (1971). Production of alkaline enzymes by alkalophilic microorganisms. Part I. Alkaline protease produced by Bacillus No. 221. Agric Biol Chem, 35, 1407–1414.

Kim, W., Choi, K., Kim, Y., Park, H., Choi, J., Lee, Y., Oh, H., Kwon, I., Lee, S. (1996). Purification and characterization of a fibrinolytic enzyme produced from Bacillus sp. strain CK 11-4 screened from Chungkook-Jang. Appl Environ Microbiol, 62(7), 2482-8.

Kumar, C.G., Tiwari, M.P., Jany, K.D. (1997). Purification and characterization of two alkaline proteases from an alkalophilic Bacillus sp. Zeitschrift für Ernährungs-Wissenschaft, 36, 48-48.

Kwon, Y.T., Kim, J.O., Moon, S.Y., Lee, H.H., Rho, H.M. (1994). Extracellular alkaline protease from alkalophilic Vibrio metschnikovii strain RH530. Biotechnol Lett 16, 413-418.

Larcher, G., Cimon, B., Symoens, F., Tronchin, G., Chabasse, D., Bouchara, J.P. (1996). A 33 kDa serine proteinase from Scedosporium apiospermum. Biochem J, 315 (Pt 1), 119-26.

Manachini, P.L., Fortina, M.G., Parini, C. (1988). Thermostable alkaline protease produced by Bacillus thermoruber—a new species of Bacillus. Appl Microbiol Biotechnol 28, 409-413.

Mukhopadhyay, A., Chakrabarti, S.K., Bajpai, P.K. (1990). Treatment and clarification of fermented broth in bacterial enzyme production. Biotechnol Techniques, 4, 121–126.

Peek, K., Daniel, R.M., Monk, C., Parker, L., Coolbear, T. (1992). Purification and characterization of a thermostable proteinase isolated from Thermus sp. strain Rt41A. Eur J Biochem, 207(3), 1035-44.

Shih, Y.C., Prausnitz, J.M., Blanch, H.W. (1992). Some characteristics of protein precipitation by salts. Biotechnol Bioeng, 40(10), 1155-64.

Strathmann, H. (1990). The use of membranes in downstream processing. Food Biotechnol, 4, 253-272.

Sullivan, T.G.O., Epstein, A.C., Korchin, S.R., Beaton, N.C. (1984). Applications of ultrafiltration in biotechnology. Chem Eng Prog, 80, 68-75.

Takagi, H. (1993). Protein engineering on subtilisin. Int J Biochem, 25(3), 307-12.

Tsai, Y.C., Juang, R.Y., Lin, S.F., Chen, S.W., Yamasaki, M., Tamura, G. (1988). Production and further characterization of an alkaline elastase produced by alkalophilic bacillus strain ya-B. Appl Environ Microbiol, 54(12), 3156-61.

Tsai, Y.C., Yamasaki, M., Yamamoto-Suzuki, Y., Tamura, G. (1983). A new alkaline elastase of an alkalophilic bacillus. Biochem Int, 7(5), 577-83.

Tsuchida, O., Yagima, Y., Ishizuka, T., Arai, T., Yamada, J., Takeuchi, M., Ichishima, E. (1986). An alkaline proteinase of an alkalophilic Bacillus sp. Curr Microbiol, 14, 7-12.

Tsujibo, H., Miyamoto, K., Hasegawa, T., Inamori, Y. (1990). Purification and characterization of two types of alkaline serine proteases produced by an alkalophilic actinomycete. J Appl Bacteriol, 69(4), 520-9.

Yamagata, Y., Ishiki, K., Ichishima, E. (1995). Subtilisin Sendai from alkalophilic Bacillus sp.: Molecular and enzymatic properties of the enzyme and molecular cloning and characterization of the gene, apr S. Enzyme Microb Technol, 17, 653-663.

❑❑❑

CHAPTER – 23

Cancer Biology and RNAi

Reema Gabrani
Department of Biotechnology, Jaypee Institute of Information Technology, Noida
Email: rmgabrani@gmail.com

Summary

RNA interference has revolutionized the concepts in regulation of gene expression. RNA interference is the mechanism that controls sequence specific gene silencing by double stranded RNA. These non-coding RNAs are gene specific regulators critical in the transformation of normal cells to a malignant stage. Alterations in the non-coding RNAs have been linked to specific cancers and can serve as diagnostic markers. The ability of exogenously introduced siRNAs to specifically bind its complementary mRNA and down regulate its expression has various applications. It has markedly influenced the way the role of certain genes is investigated in tumorigenesis. The siRNA technology has been effectively functional in identification of new components and blocking the expression of critical mediators in tumorigenesis.

1. Introduction

Regulation of gene expression is a complex, multi-layered process that controls the correct and the timely expression of proteins to maintain cell architecture and regulate its interaction with the environment. The profile of expressed gene products vary extensively within a cellular life-span and only a fraction of proteins are expressed at a given time. Regulation of the gene expression takes place at both levels i.e. transcription of DNA into RNA and translation of RNA into protein. Transcription is governed by various molecular signals that direct the binding of regulatory proteins to DNA. These factors can alter the accessibility of nucleosomes to transcription factors and/or modulate the initiation, elongation and rate of transcription. Once transcribed, mRNA is regulated through alternative splicing which at times can lead to different protein products. Another form of regulation is by small RNA molecules which can destroy its homologous RNA, inhibit translation and modify the structure and function of heterochromatin. This phenomenon of repression known as RNA interference (RNAi) is considered as one of the most important discoveries of RNA mediated gene regulation in recent times. RNAi has been identified as an endogenous process for regulation of genes during development. Moreover it is an important defense mechanism to silence the foreign genetic elements. It is highly conserved and similar in plants, fungi and mammals (Fire et al. 1998). Non-coding RNAs as gene-specific regulators are critical in the transformation of normal cells to a malignant stage. The role of non-coding RNAs in mediating the genetic and epigenetic changes in the cancer cells is being increasingly established. These ncRNAs can act as tumor suppressor genes or oncogenes in human cancers. Rearrangements, deletions, amplification and mutations of ncRNAs in cancer suggest their role in cancer pathogenesis. Moreover, exogenously generated small interfering RNA (siRNA) has been successfully demonstrated to selectively bind its homologous target gene and down-regulate its expression. This opens the new avenues for identification of new components and therapies for tumorigenesis and metastasis.

2. RNA Interference

Small non-coding RNAs (ncRNAs)(19–31-nucleotides) play a crucial role in RNA interference as regulators of gene expression in many eukaryotes, including plants, animals, and fungi (Kutter and Svoboda, 2008). These ncRNAs are an integral part of regulation as evident from their conserved sequence and expression patterns as well as their role in regulation of gene expression during development and disease. Detailed studies of these small RNAs and their role in sequence specific silencing has highlighted heterochromatin silencing, co-suppression of transgenes, repression of transposon expression mainly in the germline as other mechanisms also mediated by RNAi. These small noncoding, regulatory RNAs are grouped under small interfering RNAs (siRNAs), microRNAs (miRNAs), and Piwi-interacting RNAs (piRNAs).

2.1. RNAi: Brief History

In early 1990s, RNAi type of phenomenon was discovered in three types of species almost at the same time. In plants, Napoli and colleagues (1990) tried to over-express the chalcone synthase gene of anthocyanin biosynthesis responsible for the deep violet coloration in order to generate violet petunias. It resulted in white petunias instead of expected violet color. The phenomenon was named as ‘co-suppression’. Another related mechanism of suppression of endogenous RNA was discovered in *Neurospora crassa* (Romano and Macino, 1992) by homologous RNA and named as ‘quelling’. In the nematode worm *Caenorhabditis elegans*, the similar phenomenon of ‘RNA interference’ was observed by Guo and Kemphues (1995) and to their surprise the down regulation of mRNA levels were not only observed with antisense strand but also with sense strand. In 1998, Andy Fire and co-researchers carried out the studies in *C. elegans* to explain the above phenomenon. They observed silencing effects on endogenous RNA using both sense or antisense single strand preparations and also double strand RNA. The dsRNA targeting the same mRNA was much more effective than purified ssRNA and they reasoned that silencing effects of sense strand observed by Guo and Kemphues was due to contamination of antisense strand forming the double stranded RNA and thus resulting in inhibition. RNAi was thus observed as a sequence specific post-transcriptional gene silencing (PTGS) mechanism.

2.2. siRNA

Endogenous-small-interfering (endo-si) RNA pathways were first discovered in plants in response to viral infection and since shown in *C. elegans* and only recently have been discovered in *Drosophila melanogaster* and mouse oocytes (Tam et al. 2008). EndosiRNAs are part of the genomic structures such as inverted repeats, sites of bidirectional transcription and antisense transcripts from various loci, including retrotransposons and pseudogenes.

The first step in RNAi mediated gene silencing is the generation of small RNAs. Double stranded RNA is processed by ribonuclease dicer to generate siRNA. The dicer belongs to third class of dsRNA ribonuclease III family. It contains two RNase III motifs and additional helicase and PAZ motifs. One strand of siRNA duplex (the guide strand), is loaded on the RNA-induced silencing complex (RISC), after the cleavage by Dicer. This guide strand is complementary to the mRNA which needs to be cleaved. The other strand known as passenger strand is ejected. RISC is a protein–RNA effector nuclease complex and consists of an Argonaute protein, Dicer and a dsRNA-binding protein. It uses Argonaute protein (slicer) and the guide strand to cleave the specific mRNA. Cleaved mRNA gets degraded resulting in the inhibition of translation. RISC complex leaves the cleaved mRNA and can carry out its next round of target cleavage (Fig. 1).

Argonaute proteins have two characteristic domains namely PAZ and Piwi. Piwi domain is solely present in Argonaute proteins whereas PAZ domain is also common to dicer proteins and may have a role in assembly of RISC complex.

The Argonaute RNA-binding protein family based on sequence homology can further be divided into several subsets: the Argonaute subfamily, the Piwi subfamily, and the *C. elegans*-specific subfamily. The core effector for miRISC or siRISC belongs to the Argonaute subfamily: Ago1–4 in humans; whereas the core effector for piRISC is PIWI subfamily of Argonaute proteins. In humans, all four proteins Ago1-4 can associate with miRNA whereas in Drosophila, Ago1 and Ago2 have distinct roles: Ago1 is part of miRISC whereas Ago2 is part of siRISC (Chu and Rana, 2009).

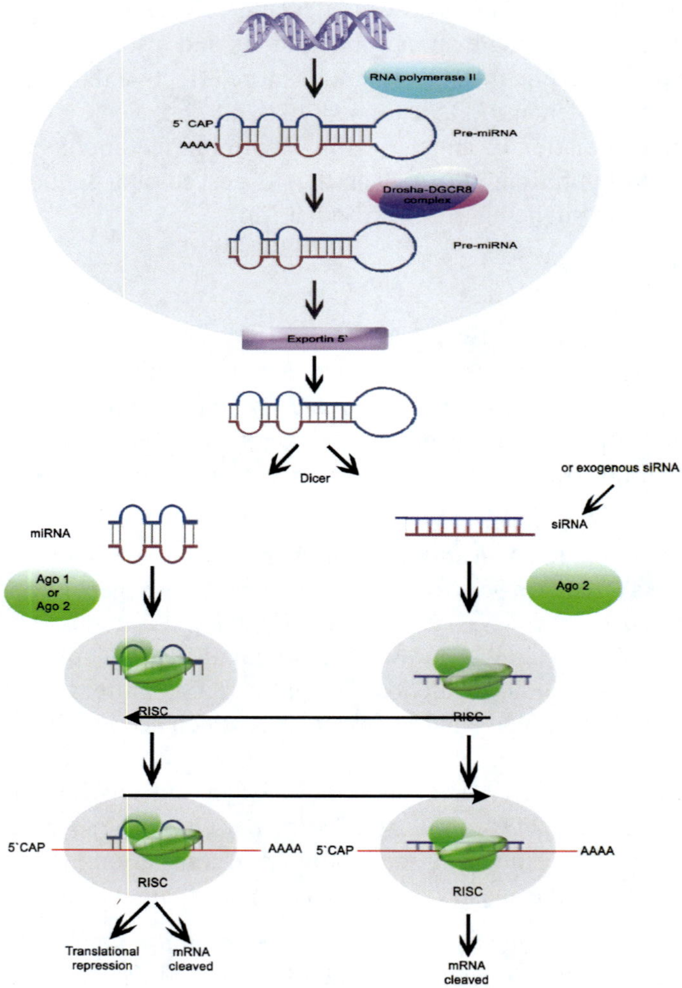

Fig. 1 : Biogenesis of miRNA and siRNA.

2.3. miRNA

miRNAs are evolutionary conserved in many eukaryotes including fruitflies, yeast, plants and humans. They play an important role in development, hematopoiesis, proliferation, apoptosis and regulating chromatin structure. miRNAs are transcribed in the nucleus from RNA polymerase II promoter into long primary transcripts with 5' m7G caps and 3' poly-A tails. Most of these miRNAs are grouped as clusters such that multiple miRNAs are included in one primary miRNA (pri-miRNA) in the nucleus. MiRNAs as part of pri-miRNAs form the hairpin loop structure and are recognized by RNase III enzyme drosha along with its dsRNA binding partner DGCR8 (DiGeorge syndrome critical region gene-8) and this complex is also known as the Drosha–Pasha complex in *Drosophila* and *C. elegans*. The Drosha–DGCR8 complex cleaves pri-miRNAs into the precursor miRNAs (pre-miRNAs) with characteristic 5' phosphate end and a 2 nucleotide 3' overhang. Pre-miRNAs are transported by the nuclear export factor, exportin-5 with the help of another protein Ran bound to GTP. Further processing of pre-miRNAs take place by cytoplasmic RNAse III dicer and results in 18-24 nt duplex. Fully processed miRNAs duplex thus formed is unwound and associate with miRISC complex having core Argonaute proteins. Out of these two strands, only guide strand of miRNA remains part of the miRISC complex and passenger strand (miRNA*) is removed. The choice of the strand depends on the thermodynamic stability of the ends of miRNA/miRNA* duplex. The strand with less stable 5' end is loaded into the miRISC. This single stranded guide miRNA direct the miRISC complex to target mRNAs to mediate translational silencing (Fig. 1). In plants, miRNAs are perfect complementary to their target RNAs and thus cleave mRNA whereas in others like C. elegans, Drosophila and mammals, miRNAs are partially homologous and lead to translational silencing of the target (Chu and Rana, 2009).

2.4. piRNA

piRNAs are short 24-30 nucleotide long single stranded RNAs which arise from a Dicer-independent pathway. They associate with the PIWI subfamily of Argonaute proteins. They have been implicated in germline development from *Drosophila* to mammals (Chu and Rana, 2009). The first piRNAs was identified in *Drosophila*, where they mainly originate from retrotransposons and other repetitive sequence elements and are required for retro-transposon silencing (Aravin et al. 2003). Piwil2, one member of Piwi family, is expressed in most tumors, and is classified as cancer-testis antigens (Lu et al. 2010). Many attempts have been made to identify piRNA associated with cancer and small data set in cancer cell lines have been identified but further detailed characterization needs to be carried out.

3. Carcinogenesis

Cancer is caused by genetic and epigenetic changes in genome leading to deregulation of pathways critical for maintenance of cell growth and differentiation. Molecular changes in the tumor suppressor genes and oncogenes as well as epigenetic changes such as abruptions in DNA methylation, histone modification and chromatin modeling have been increasingly linked to human cancer.

An oncogene is a gene that when mutated or altered contributes to converting a normal cell into a cancer cell. The term oncogene is derived from the Greek word "oncos," meaning tumor. The activation to oncogene results in loss of regulation and may result in unrestrained cell growth. The cellular oncogenes in their normal form are called proto-oncogenes and code for a variety of normal enzymes, growth factors and receptors that transmit signals to a cell nucleus and stimulate growth and proliferation.

The proto-oncogene gets transformed into oncogene by: (a) change in the structure of gene leading to mutated or altered product having an aberrant function; (b) change in the regulation of gene expression resulting in the production of increased amount of normal growth-promoting protein (Table 1) (Chumbalkar et al. 2011; Kiuru-Kuhlefelt et al. 2000; Kong-Beltran et al. 2006; Pierotti et al. 2003).

These changes bring about the critical "hallmarks" or processes necessary for the transformation of a normal cell to a cancer cell as described by Hanahan and Weinberg (2000). These include self-sufficiency in growth signals; insensitivity to antigrowth signals; evasion of programmed death (apoptosis); endless replication potential; tissue invasion and metastasis; and sustained angiogenesis.

Table 1 : Mechanism of conversion of proto-oncogene to oncogene.

Type of change	Oncogene	Disease	Function of oncogene
Point mutation	*H-ras*	Colon, lung cancer	GTPase
	c-fms	acute myeloid leukemia, hepatocellular carcinoma	CSF-1 receptor
Deletion	*met*	Lung cancer	hepatocyte growth factor receptor
Deletion	*ÄEGFR*	glioblastoma	Epidermal growth factor-receptor
Insertion	*int2*	Kaposi's sarcoma (insertion of mouse mammary tumor virus)	Type of FGF family
Gene amplification	*myc*	Small cell lung cancer, breast cancer, esophageal cancer	Transcription factor
Gene amplification	*mdm2*	sarcoma	Inhibits p53 activity
Chromosomal translocation (gene fusion)	*bcr-abl*	chronic myelogenous leukemia (CML).	ABL-tyrosine kinase activated by BCR
Chromosomal translocation (with IgH locus)	*bcl-2*	Follicular lymphoma	Anti-apoptotic factor

3.1. Cell Cycle Deregulation in Tumorigenesis

A cell cycle mainly comprises of duplication of cellular contents followed by its division. Cell cycle is responsible for the regulation of growth, repair and maintenance of organisms. The cell cycle is divided into four stages: G1 (gap phase 1)- responds to stimuli (growth factors), checks the integrity of DNA and prepares the cell for DNA replication; S (synthesis) – replication of DNA; G2 (gap phase 2) – checks the integrity of replicated DNA and prepares the cell for mitosis; M (mitosis) – cell division. One important characteristic of the cell cycle is that DNA in chromosomes are accurately replicated and precisely distributed to the daughter cells. Another important parameter is that growth of cell is coordinated with cell division.

Regulation of cell cycle is very critical as any defect in the cell cycle can lead to uncontrolled cell division leading to tumor formation. Major regulatory proteins in controlling the cell cycle are cyclins and cyclin-dependent protein kinases (CDKs) and they are sequentially expressed to progress through successive phases. CDKs are ubiquitously expressed in the cell whereas cyclins are synthesized cyclically. When growth factor or stimulus binds to its receptor on a cell, it activates Ras pathway or focal adhesion kinase (FAK) signal transduction pathway resulting in the expression of cyclin D. Cyclin D forms a complex with CDK4 which phosphorylates and inactivates Rb (retinoblastoma) protein. Rb is a tumor suppressor protein and is found deleted in patients with retinoblastoma. Rb binds E2F transcription factor in its dephosphorylated form and blocks the transcription of proteins required for transition to next cell cycle stage. Cyclin D-CDK4 complex releases E2F from Rb negative control and initiates the first step of cell division. Activation of E2F results in the induction of cyclin E and cyclin A expression which in turn regulates the proteins required for DNA synthesis in S phase i.e. DNA polymerase, helicase, topoisomerase, histones and nucleotide synthesis machinery. Cyclin B complexes with CDK1 during G2 phase and prepare the cell for mitosis. CyclinB-CDK1 complex is further required for condensation of DNA during mitosis phase and separation of chromatids during anaphase (Table 2). Further more, these Cyclin-CDK complexes are regulated by CIP/KIP inhibitors and CDKs are inhibited by members of the INK4 family. These inhibitors are described as tumor suppressors and their mutation or deletion can result in onset of cancer (Gomberts, 2002).

Table 2 : Cyclins and CDKs.

Cyclin	CDK	Cell-cycle Stage
Cyclin D	CDK4/CDK6	G1
Cyclin E	CDK2	G1/S
Cyclin A	CDK2	S
Cyclin B	CDK1	M

The changes in the regulatory pathways involved in cell cycle control can lead to tumor formation. The transition from G1 to S controlled by (CDK)-cyclin D/ retinoblastoma (pRb)/E2F cascade is most commonly deregulated in cancer such as malignant sarcomas, gliomas, and breast carcinomas (Graf et al. 2009).

Cell-cycle is further controlled by checkpoints at different stages of cell cycle which checks the integrity of DNA. Detection of any defects at any stage of cell cycle can trigger the cell-cycle arrest and activation of DNA repair mechanism. Tumor suppressor gene *p53* plays a pivotal role in regulation of cell cycle and apoptosis especially in maintenance of genomic DNA integrity. Any damage to DNA triggers the induction of p53 which play an important role in DNA repair processes. Mutations in the *p53* tumor suppressor gene have been observed in about 50% of all human tumors including lung, breast, colon, prostate, non melanoma skin making it the most frequent target for genetic alterations in cancer.

Levels of p53 are kept in check by the product of an oncogene, Mdm2 protein in a unique manner. Mdm2 protein can bind p53 and inactivate it whereas p53 binds specifically to the *mdm2* gene and stimulates its transcription. Inactivation of p53 or over-expression of Mdm2 can lead to constitutive inhibition of p53 and thereby promote cancer. Mdm2 accumulation is also found in numerous cancers, such as sarcoma or breast carcinomas (Oren, 1999).

Apoptosis or programmed cell death is highly conserved in evolution. Tumor cells evade apoptosis and thus sustain uncontrolled division. Apoptosis can be extrinsic being mediated by death receptors or intrinsic being controlled by mitochondria. Intrinsic pathway is regulated by Bcl-2 (B cell lymphoma-2) family of proteins which are divided into pro-apoptotic proteins (Bax, Bid, Bad) and anti-apoptotic proteins (Bcl-2, Bcl-xL). The activation of pro-apoptotic proteins results in removal of the inhibitory control of Bcl-2 on apoptosis. Bax and Bad help to change the permeability of mitochondrial membrane resulting in the release of cytochrome C which along with cytoplasmic Apaf-1 (Apoptotic protease activating factor 1) and dATP forms the Apoptosome complex. This complex activates the executioner caspases 3 and 7 which cleave the DNA, degrade proteins, destroy cellular compartments and repair machinery. Changes in the levels of anti-apoptotic and pro-apoptotic proteins can lead to formation of tumor. Bcl-2 plays an important role in generation of B cell lymphomas. The tumor arises due to chromosomal translocation of Bcl-2 near hyperactive IgH locus resulting in its over-expression. This inhibits apoptosis, increases the life-span of B cells and results in neoplasia.

Another family of proteins, inhibitors of apoptosis (IAPs) which suppress apoptosis include NAIP, XIAP, cIAP-1, cIAP-2, ILP-2, Livin and Survivin. X-linked inhibitor of apoptosis protein (XIAP) is often expressed in high amounts in various cancer cells. XIAP can directly inhibit caspases-3, -7, and -9 and has also been shown to interfere with the action of various pro-apoptotic agents, including transforming growth factor beta (TGF-β) tumor necrosis factor alpha (TNF-α) and chemotherapeutic drugs. XIAP has also been shown to promote poly-ubiquitination and proteasomal degradation of PTEN (phosphatase and tensin homolog deleted on chromosome ten) (Van Themsche et al. 2010). PTEN is a phosphatase which acts on phosphatidylinositol 3,4,5-triphosphates (PI(3,4,5)P3) resulting from phosphatidylinositol 3-kinase (PI3K) activation and thus directly opposes PI3K. PI3K is activated through tyrosine kinase and G protein coupled receptors resulting in activation of Akt, or protein kinase B, an important intracellular signaling molecule that affects various biochemical pathways involved in cellular proliferation and angiogenesis. Akt deregulation has also been linked to various neoplasms (Nelson et al. 2007).

3.2. Tumor Angiogenesis

In 1971, Judah M. Folkman proposed a hypothesis that tumor growth is angiogenesis-dependent and tumor mass will not grow beyond few millimeter diameters without new capillary blood vessels. Angiogenesis is in indeed a necessity for both tumor expansion as well as metastasis. These newly formed blood vessels help to provide the nutrients, oxygen, and growth factors critical for the survival and growth of tumor cells. Moreover, they help the tumor cells to invade and grow in normal tissues. The formation of a tumor vasculature is called 'angiogenic switch' and occurs at various stages of the tumor progression. Tumor cells as well as other cells within the tumor microenvironment are responsible for vascularization by the release of angiogenic, growth, and survival factors. Various factors like vascular endothelial growth factor (VEGF) and basic fibroblast growth factor (bFGF), angiogenin, platelet derived growth factor (PDGF) epidermal growth factor (EGF) TGF-α and TNF-α have shown to be critical for angiogenesis. Some of these secreted factors like VEGF and bFGF bind the receptors on nearby vascular endothelial cells and promote cell proliferation, migration, and invasion into the tumor. These activated endothelial cells secrete matrix metalloproteinases (MMPs), into the extracellular space and help to break down the extracellular matrix that fills the spaces between cells. Thus they make the space for the migration of endothelial cells permitting the growth of new capillary vessels.

Tumor cells during their incessant growth can develop areas of hypoxia as its growth surpasses the blood supply. The sufficient oxygen does not reach all the regions of tumor. Tumor cells respond to this hypoxia by elevating the levels of transcription factor, hypoxia inducible factor (HIF). Apart from hypoxia, HIF also gets upregulated by growth factors and cytokines such as (TNF-α), interleukin-1β (IL-1β), EGF and insulin-like growth factor-1 (IGF-1). The tumors induce angiogenesis, switch from aerobic metabolism to anaerobic glycolysis, and express a variety of stress proteins which regulate cell death or survival. Moreover, it also promotes tumor progression and makes the tumor resistant to radiation and chemotherapy (Koh et al. 2010).

3.3. Tumor Immunity

The immune system in the host has been developed to generate a remarkable level of diversity against the potential limitless array of antigens but equally fascinating is its ability to protect the host against autoimmunity i.e. immune cells targeting self. One of the major factors in development of tumorigenesis is the role of the immune system. Tumor cells employ different strategies to evade detection and elimination by the cells involved in immunity.

There are two main pathways to target antigen or tumor-associated antigen (TAA) in case of cancer. Antigens need to be presented with major histocompatibility complex (MHC) molecules to be recognized by immune system. MHC molecules are cell membrane proteins and there are two main types of MHC molecules: class I and class II MHC molecules. Class I MHC molecules are expressed by nearly all nucleated cells of vertebrate species whereas Class II MHC molecules are expressed only by antigen-presenting cells (APCs). APCs include macrophages, B lymphocytes, and dendritic cells.

In the MHC class I-restricted pathway for antigen presentation, the intracellular antigen is processed by proteasome and combined with MHC class I molecules in the endoplasmic reticulum, then eventually presented on the surface of the cell. The antigen associated with MHC class I molecules are recognized by the $CD8^+$ restricted cytotoxic T cells. Cytotoxic T cells can attack and kill these self altered cells (Goldsby et al. 2000).

In comparison, in the MHC class II-restricted pathway, extracellular antigen is taken up into APCs by endocytosis or phagocytosis into endosome or phagolysosome, respectively, where the antigen is degraded and part of antigen is loaded onto MHC class II molecules and expressed on the surface of APCs. $CD4^+$ helper T cells recognize the antigen –MHC class II molecule complex on APCs. These helper T cells secrete various growth factors known as cytokines to activate B cells, cytotoxic T cells, macrophages, and various other cells of the immune system. The presentation of antigen to APCs as well as T helper and T cytotoxic cells are regulated by various immunosuppressive pathways. There is a third population of T cells known as Regulatory T cells which suppress the immune response. APCs can also express the molecules involved in suppressing antigen presentation or T cell activation and function.

The tumor microenvironment can negatively affect professional APCs, dendritic cells (DCs) maturation and function (Rabinovich et al. 2007). Some of these strategies involve increased population of regulatory T cells, decreased expression of co-stimulatory or MHC class I molecules. Both MHC class I and co-stimulatory molecules associate with antigen and are required for their recognition by $CD8^+$ cytotoxic T cells and subsequent killing of the tumor cell. Tumor cells secrete cytokines such as IL-10, (VEGF), macrophage colony stimulating factor, and IL-6 to disrupt the function of APCs. Moreover DCs have also been shown to express molecules such as A20, and suppressor of cytokine signaling 1 (SOCS1). A20 is involved in the attenuation of T cell-mediated responses and SOCS1 is a broadly immunosuppressive protein. SOCS1 curtails T cell expansion and suppresses antigen presentation to T cells (Mao and Wu , 2010).

Certain cancers, like neuroblastoma and Hodgkin Reed–Sternberg tumors, secrete abundant amounts of the immunosuppressive cytokine TGF-β. It results in growth arrest, and induction of tolerance in T cells. It also forms a major route in evasion of immune response by tumors.

Small interference technology can be applied to target the immunosuppressive pathways that attenuate the host immune response to maximize the function of helper and cytotoxic T cells (Mao and Wu, 2010).

4. RNAi in Tumorigenesis

The human genome consists of only 2% of protein coding sequences and is not sufficient to explain the complexity of the human system. Research is currently being focused on identifying the role of these 98% sequences that do not code for any protein. Many of these non-coding sequences are transcribed into long and short RNAs which play a critical role in regulation during development, exhibit cell specific expression and are associated with certain diseases. Various alterations in both small and long noncoding RNAs identified are implicated in cancer and are being discussed.

4.1. Long Noncoding RNA

Multiple studies have shown that long noncoding RNAs are biologically significant and play an important role in development and can function at molecular level to regulate the expression of proteins (Wilusz et al. 2009; Malecová and Morris 2010). The first lc RNA associated with tumors was identified as H19 gene (Hao et al. 1993). H19 locus is epigenetically controlled by methylation. It is expressed in various cancers including breast, ovarian and liver cancer. It enables the tumor cells to survive under the stress conditions of limiting nutrition and low oxygen levels. Since H19 many ncRNAs have been associated with cancer and it has also been reported that large ncRNAs can generate one or more (miRNAs). For example, the ncRNA BIC gene, which has been implicated in growth control and oncogenesis in cancer cells generates two miRNA, one of which is miR155, which is over expressed in Burkitt and B cell lymphomas (Metzler et al. 2004).

The human *HOX* loci were analyzed for non coding RNA transcripts by conducting the comprehensive transcriptional activity. Homeobox genes specify the anterior-posterior axis during development and are preferential targets for DNA methylation in cancer cell lines (Pfeifer and Rauch, 2009). 231 ncRNA transcripts were identified by the researchers and the majority of these (74%) were transcribed in an antisense orientation from the intergenic regions as compared with the *HOX* genes (Rinn et al. 2007). In the same study, large ncRNA in the *HOX* loci, *HOTAIR* (*Hox* antisense intergenic RNA), was identified. *HOTAIR* is being increasingly expressed in primary breast tumours and epithelial cancer cells and its expression level in primary tumours is a powerful predictor of eventual metastasis and death (Gupta et al. 2010).

Abnormal expression of another lcRNA, MALAT-1 (Metastasis Associated in Lung Adenocarcinoma Transcript 1) gene is implicated in many cancers, including breast, lung and liver cancers and recently its role in normal cells has been elucidated in regulating pre-mRNA splicing which is a critical step in the protein production. (Bernard et al. 2010). It has been characterized as a highly reliable prognostic parameter for (NSCLC) patient survival, and has been used for identification of early-stage NSCLC patients who are at high risk to develop metastases (Ji et al. 2003).

A long ncRNA in 8q24, prostate cancer non-coding RNA 1 (PRNCR1) expression is upregulated in some of the prostate cancer cells as well as precursor lesion prostatic intraepithelial neoplasia and is involved in prostate carcinogenesis possibly through androgen receptor activity. Thus it elucidates the role of genetic factors in prostate carcinogenesis. (Chung et al. 2011). Thus various studies indicate that lcRNA associated with cancer can play an important role as diagnostic marker or can be targeted for specific therapy (Abdelrahim et al. 2006).

4.2. miRNA in Cancer

Small ncRNA, miRNA plays a critical role in the cancer pathogenesis. MiRNAs are frequently deleted, amplified or mutated in cancer and they have been linked to various stages of cancer development and progression. First endogenous small RNA to be identified was lin-4 miRNA in *C. elegans* (Lee et al. 1993). Subsequently many functional small RNAs have been identified in diverse organisms using molecular cloning and computational approaches. miRNAs can be further classified as tumor suppressor miRNAs and oncolytic miRNAs.

4.2.1. Tumor suppressor miRNAs

Tumor suppressor miRNAs mean that loss of function of miRNA is linked to cancer. These miRNAs regulate the genes responsible for cell differentiation and apoptosis. Many miRNAs have been identified as tumor suppressors (Table 3) (Cho 2007; Fornari et al. 2009; Friedman and Jones, 2009; Hwang and Mendell, 2006; Inomata et al. 2009; Lin et al. 2008; Zhang, 2008). miRNA *let-7* is one of the foremost members of the miRNA family which is conserved in *C. elegans*, *Drosophila* and vertebrates. *Let-7* negatively regulates an important oncogene, High mobility group AT-hook 2 (*Hmga2*). HMGA2 is a member of nonhistone chromatin proteins and acts as architectural transcription factor. It is expressed at low levels in adult tissues. Defects in *let-7* and thus HMGA2 over-expression have been linked to human neoplasias such as non-small-cell carcinoma of the lung (Lee and Dutta, 2007).

In addition to HMGA2, loss of *let-7* is also related to RAS deregulation which is a key event in lung cancer and has been associated with pathogenesis. Deregulation of RAS- MAPK (mitogen activated protein kinase) signal transduction pathway results in chromatin remodeling and continuous expression of transcription factors leading to uncontrolled growth.

Table 3 : Selected miRNAs as tumor suppressors.

miRNA	Target gene	Associated cancer
Let-7 family	HMGA2, RAS	Breast cancer, lung cancer
miR-15a/16-1cluster	BCL2,DDND1,WNT3A	B-cell chronic lymphocytic lymphomas, prostate cancer
miR-34 family	CCNE2,CDK4, MET	Colorectal cancer
miR-127	BCL6	Burkitt lymphoma
miR-124a	CDK6	Acute lymphoblastic leukemia
miR-223	NFI-A	Acute myeloid leukemia
miR-203	ABL1, BCRABL1	Leukemias andLymphomas
miR-1	FoxP1, HDAC4, MET	Hepatocellular carcinoma
miR-29 family	YYIDNMT3A/B	Lung cancer
miR-128	BMI1, E2F3a	Human glioblastoma

Other miRNAs as tumor suppressors, miR-15a and miR-16-1 are located as cluster at 13q14. Their downregulation and deletion have been reported in chronic lymphocytic leukemia (CLL). They act as tumor suppressors probably by down regulating anti-apoptotic protein Bcl-2. Bcl-2 family is a complex network of proteins responsible for proper development and functioning of multicellular organisms. Bcl-2 is antiapoptotic protein which prevents the intrinsic apoptotic pathway and finally the cell death. Levels of miR-15a and miR-16-1 are inversely correlated with Bcl-2 protein and loss of these miRNAs in CLL may lead to elevated Bcl-2 protein expression and thus evade programmed cell death.

Epigenetic silencing of miRNAs has been reported which leads to the permanent upregulation of the miRNA target genes. miR-127 and mir-136 are located near CpG island and probably have a tumor suppressor function. Hypermethylation of CpG islands can lead to gene silencing. The proto-oncogene BCL6 (B-cell lymphoma-6) is translationally downregulated by miR-127 and inactivation of miR-127 can lead to over-expression of BCL-6 ensuing neoplasia. Chromatin-modifying drugs can direct demethylation or acetylation by inhibiting histone methylase or deacetylase and can activate expression of miRNAs that may act as tumor suppressors (Friedman and Jones 2009).

4.2.2. Oncolytic miRNAs

Amplification and over-expression of oncolytic microRNAs promote tumor development or inhibit apoptosis (Table 4) (Cho, 2007; Friedman and Jones, 2009; Hwang and Mendell, 2006; Inomata et al. 2009; Zhang, 2008). They negatively inhibit tumor suppressor genes and/or genes that control cell differentiation or apoptosis. Many miRNA genes have been identified that are over-expressed in various cancers.

Table 4 : Selected miRNAs as oncogenes.

miRNA	Target gene	Associated cancer
miR-17~92 cluster	HIF-1α, Bim, CDKN1A/p21, E2F	B cell lymphomas, Lung cancer, Human Hepatocellular carcinoma
miR-221/222	p27(CDKN1B), p57(CDKN1C)	Thyroid cancer, Glioblastomas, Hepatocellular carcinoma
BIC/miR-155	TP53INP1	Lymphomas, Breast cancer
miR-21	BCL2, PTEN	Glioblastomas, Breast cancer
miR-19	PTEN, PDCD4, TM1	Human hepatocellular cancer, Non-small cell lung cancer (NSCLC), Breast cancer

The first oncogenic miRNA to be recognized was miR-155, which is associated with B-cell lymphomas and is generated from the B-cell integration cluster (*BIC*) non-coding RNA. MicroRNA-155 targets the mRNA of tumor suppressor protein, TP53INP1 (tumor protein 53-induced nuclear protein 1) that mediates apoptosis and cell cycle arrest. TP53INP1 is induced under different stress conditions and its overexpression leads to cell cycle arrest (G1 phase) and p53-dependent or independent apoptosis.

Oncogenic miRNA17-92 cluster comprises of 7 miRNAs and is located on human chromosome 13q31. This region gets amplified in lymphomas and lung cancer. These miRNAs are regulated by proto-oncogene c-Myc at the transcriptional level. MicroRNAs in this cluster as well as miR-20 in turn regulate the translational efficiency of pro-apoptotic transcription factor E2F1. E2F1 is constitutively nuclear and exhibits dual control in cell cycle regulation, sometimes as a tumor-suppressor and others as an oncogene. It is active during the G1 to S transition, and thus its target genes, which include regulatory elements of the cell cycle, cyclin E, and essential components of DNA replication machinery. In response to DNA damage, E2F1 is upregulated and may be involved in apoptosis or DNA repair. It has been shown that mi-RNAs of *mir-17* cluster interact in a cooperative manner and over-expression of these miRNAs and not individual miRNA have oncogenic properties. It has recently been shown that E2F and Myc are inhibited at the posttranscriptional step by members of a cluster of microRNAs (miRs) called miR-17-92. In return, E2F and Myc induce the transcription of miR-17-92, thus forming a negative feedback loop in the interaction network. (Aguda et al. 2008).

To study the role of miRNAs in pancreatic cancer, differential expression of 95 miRNAs was analyzed by real time RT-PCR. Eight miRNAs were significantly upregulated in most pancreatic cancer tissues and cell lines, including miR-196a, miR-190, miR-186, miR-221, miR-222, miR-200b, miR-15b, and miR-95. The incidence of upregulation of these eight genes between normal control subjects and tumor cells or tissues ranged from 70–100% (Zhang et al. 2009).

Another microRNA miR-21 has experimentally shown to be anti-apoptotic and has been upregulated in various cancers including HCC (Hepatocellular carcinoma) (Meng et al. 2007) and NSCLC (Zhang et al. 2010). MicroRNA-21 high levels have been associated with mutated form of PTEN which is implicated in various cancers.

4.3. siRNA Design

Introduction of artificial double stranded RNA in invertebrates led to inhibition of gene expression whereas in vertebrates, initial studies were unsuccessful. The introduction of longer dsRNA always led to cell death. It was explained later that in mammalian cells, longer dsRNAs activate NF kappa B pathway which lead to the induction of Type I Interferons. IFNs response can activate RNase L which degrades RNA. Moreover, IFNs and dsRNA activate serine threonine dependent protein kinase which results in systemic shut down of protein synthesis rather targeting the specific protein. These cellular responses have probably been evolved in mammals in response to viral infection (Galun, 2005). After various attempts, Tuschl and colleagues (2003) were able to design siRNAs smaller than 30 bp and were capable of blocking the expression of specific gene by degrading the homologous mRNA.

These siRNAs need to be very carefully designed. Various online tools are available to help design siRNAs (Dharmacon, Applied Biosystems, Clontech, Invitrogen) Appropriate delivery method depending on the system should be chosen so that siRNAs can reach its target. siRNAs can be designed according to the guidelines given by Tuschl which are commonly known as the MPI (Max-Planck-Institute) principles. Length of siRNA should be 21 nt preceded by an AA or NA with and GC content between 30% to 70%. Regions within 50-100 bp of start and stop codons as well as introns should be avoided (Tuschl et al. 2003). It implies that siRNA should be designed based on coding region and be free of translational or regulatory proteins.

4.4. Delivery Methods and Challenges

RNA interference is one of the most powerful techniques in silencing the particular gene and its ability to block the overproduction of a particular protein has led to its application in therapeutics. The major obstacle faced by small interfering RNA as therapeutic agent is their effective delivery at the site of action. There has been a surge to develop the effective delivery vehicle as it is recognized as major bottleneck in RNAi based therapy. The standardization and optimization of the delivery vehicle will considerably reduce the effective dose to be administered to get the desired response. It will also reduce the cost of the production and off-target side effects of siRNA.

Naked siRNAs have been used both *in vitro* and *in vivo* for delivery (Tseng et al. 2009). siRNA is taken up by the liver making it the most effective target site. It is also targeted to lungs, spleen, pancreas and kidneys though to a lesser extent. The multi level challenges in delivery lie at maintaining stability of siRNA via systemic route and also effective uptake at the target site. The half life of naked siRNAs in circulatory system is approximately 5-6 min. and a very high dose is required for effective targeting. The administered dsRNA can get degraded by plasma components and need to escape the renal excretion and reticuloendothelial system (RES) to stay intact and be taken up only by the target sites. Once taken inside the cell through endosome mediated route, siRNA should be able to escape as intact molecules and only favorable pharmacokinetic properties can mediate RNA interference with minimum side effects and cytotoxicity.

Various strategies have been adopted for the delivery of siRNA, siRNAs are complexed to cationic lipids or polymer, chemically conjugated/modified or linked along with protein/peptide to improve the efficacy to reach its target. One of the strategies involves packaging of negatively charged RNA inside the self -assembled micelle composed of polyethylene glycol (PEG). These polyplex micelles composed of RNA and polyethylene glycol (PEG) have gained popularity because of their nanometric size, reduced non-specific interaction and biocompatibility.

New classes of nano-carrier have also been developed which consists of (i) Semiconductor quantum dot-amphipol complex which helps in efficient delivery and real time imaging of siRNA (Qi and Gao, 2008); (ii) Non-ionic, nano-transforming, degradable polymersomes which allows endolysosomal escape and siRNA delivery into the cytosol. This method has shown promising results in mouse model of muscular dystrophy (Kim et al. 2009); (iii) Gold nanoparticles modified with PEG, conjugated to siRNA and further coated with poly-β-amino ester (PBAE) to facilitate intracellular delivery (Lee et al. 2009). The main problems in the Liposome/cationic-lipid approach are physical size, bioavailability, cytotoxicity and induction of interferon or tumor necrosis factor TNF related genes.

Proteins/peptides can also be used for efficient uptake of siRNA by the cell. Peptide transduction domain (PTD) or cell penetrating peptides (CPP) is a cationic peptide (10-30 amino-acids) which masks the negative charge of RNA and helps in non-covalent packaging of siRNA molecules into stable nanoparticles. PTDs interact with cell membrane components and mediate their own internalization through macropinocytosis. PTDs can escape the endosomes and thus delivers siRNA to the intracellular environment by retrograde transport where siRNA can mediate the blockage of its target mRNA (Meade and Dowdy, 2008). Some well studied PTDs are human immunodeficiency virus (HIV-1) transactivator TAT peptide, polyarginine, transportan, homeodomain transcription factors such as Antennapedia, the herpes simplex virus (HSV) type 1 protein VP22, penetratin, MPG peptide (27 amino acid) composed of basic domain of nuclear localization signal from SV40 large T antigen and a hydrophobic domain from HIV-1 gp41. The main bottleneck in PTD based delivery is the restricted entry inside the cell and limited release of PTD coupled siRNA from endocytic vesicles. It also results in aggregation and cytotoxicity.

Recently PTDs have been modified with double stranded RNA-binding domain (DRBD). DRBD binds with high avidity to RNA and results in an efficient uptake inside the cell (Eguchi et al. 2009). siRNA is non-covalently linked to DRBD from RNA-dependent eIF-2 alpha protein kinase which is fused to TAT peptide. This method targets variety of cells including primary cell types which are difficult to transfect and without cytotoxicity. RNA interference has been shown in the entire cell population. This technology provides improved pharmacokinetics, distribution and bioavailability over other methods but further *in vivo* studies are required to prove its efficacy.

4.5. RNAi as an Investigative Tool

The advent in sequencing technologies has resulted in accumulated RNA databases, amongst them function of several RNAs need to be further characterized. Loss of functional studies by the traditional methods such as homologous recombination and antisense methods are time consuming and at times cumbersome. RNAi provides a system in which dsRNA is specifically designed to target homologous mRNA. It results in sequence specific degradation of mRNA and hence its depletion. Different biochemical, molecular, and pharmacological based assays have been standardized to study the inhibition of specific gene. Selective gene knockdown is able to decipher the various genes and pathways involved in various stages of cancer. RNAi is being extensively used to analyze the function of oncogenes, tumor suppressor genes; evaluate DNA damage and cell cycle pathways; apoptosis; angiogenesis and metastasis.

A detailed genome analysis of the human cell cycle, cell size, and proliferation using small interfering RNA was carried out. More than 95% of the protein-coding genes in the human genome were targeted to map the genes involved in progression of different phases of cell cycle. Quantitative fluorescence microscopy helped in identification of 1,152 genes critical in cell cycle. Depletion of these genes resulted in cell cycle arrest. These proteins were identified as kinases, phosphatase, and proteolytic proteins. The study also suggested that processes considered in regulating G1-S phase progression like receptor-mediated signaling, nutrient status, and translation also play important roles in the regulation of G2-M phase transition. This study also identified new roles of TNF-NF kappa B pathway in regulating G2-M which was not previously known. Moreover the 20% of these genes were found to be cancer-associated and 130 of them had increased mRNA levels in primary tumor sample as compared to normal tissue (Mukherji et al. 2006).

In another study, Knockdown of Rb was successfully carried out to identify its downstream targets using siRNA-mediated gene silencing, the microarray profiling coupled with systematic pathway analysis. Rb regulates the levels of E2F and its knockdown was successfully used to over-express E2F protein. It affected G1/S and G2/M transitions of the cell cycle, DNA replication and repair, mitosis, and apoptosis (Semizarov et al. 2004).

4.6. RNAi Based Cancer Gene Discovery

The process of specific gene knockdown to study the role of proteins in the particular pathway requires the standardization of large scale screening as well as analysis methods.

Many new target genes associated with the disease can be identified by large scale RNAi screens. One foremost role of the process is to identify the target which is required for the continued existence of tumor cells but is mostly redundant in normal cells. RNAi screens are very well suited to identify oncogenes critical for tumor cells survival. Once a cancer has been linked to a particular gene, novel ways of therapy can be designed. For example, HER2 (epidermal growth factor receptor) is over-expressed in breast cancer and Trastuzumab (Herceptin, Genentech/Roche); a humanized monoclonal antibody against (epidermal growth factor receptor) is being in clinical use for the treatment of breast cancer. Similarly imatinib (Gleevec, Novartis), targets BCR-ABL oncoprotein synthesized in chronic myelogenous leukemia (CML).

Various RNAi libraries used for screening and identifying genes are available in different formats (Table 5). They consist of chemically synthesized siRNAs or expressed short-hairpin RNAs (shRNAs) from suitable vector. Chemically synthesized siRNA libraries have high-transfection efficiencies, but are relatively expensive and only suited to study short term gene silencing effects. Vector based shRNA libraries provide long term and stable gene silencing effects and shRNA expression libraries further could be either constructed in plasmid based vector or viral vector. Plasmids expressing shRNA are stable but the cells should be compatible and be easily transfectable with plasmids. Viral libraries support wide repertoire of cells and have a high efficiency of infection. Amongst viral libraries, lentiviral libraries are especially useful to silence genes in the cells which are difficult to infect (Iorns et al. 2007).

Table 5 : Selected RNAi libraries.

Company/vendor	Details of library	Web Resource
Dharmacon	siRNA Libraries available in pre-defined (gene family and funcion), Duet (paired siGENOME and ON-TARGET *plus* SMARTpool reagents) and siGENOME Reverse Transfection Format for human and mouse	www.dharmacon.com/
Ambion	Silencer siRNA Libraries—for human, mouse, and rat cell systems	http://www.ambion.com
Cold Spring Harbor Laboratory	Library of retroviral vectors encoding shRNAs targeting all genes in human, mouse, rat and fruit fly genomes	http://codex.cshl.edu
Qiagen	genome-wide siRNA library targeted to 34 mRNA species	www.qiagen.com
Sigma-Aldrich, Open Biosystems	Human pooled shRNA library based on lentiviral vector	www.sigmaaldrich.com
Cellecta	Pooled Lentiviral shRNA libraries of up to 55,000 discrete sequences	http://www.cellecta.com/products-services/pooled-shRNA-libraries/

Libraries of endoribonuclease prepared siRNA (esiRNA) are obtained by in vitro processing of long dsRNA by RNAse III digestion to generate siRNA pools (Fig. 2) (Yang et al. 2002). They are also specific in silencing effects and their particular advantage over siRNA library is that they reduce non-specific off-target effects.

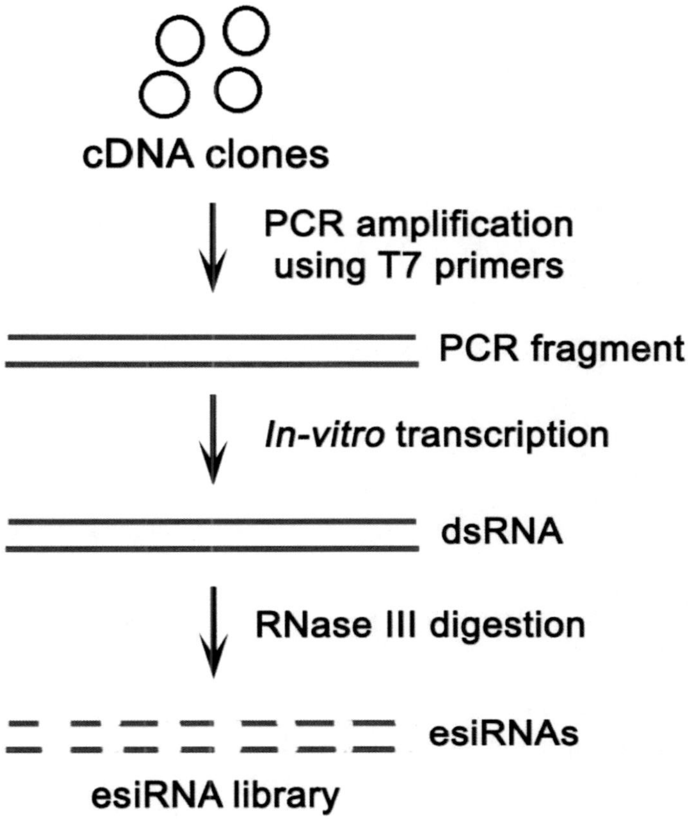

Fig. 2 : Generation of esiRNA library.

Two types of RNAi screening strategies are applied depending on the type of the assay (Fig. 3). In the first type, each gene is targeted separately in multi-well plates to determine the phenotypic change of interest. The cells in the multi-well plate are transfected with either naked siRNA or vector expressing shRNA. The end point of the assay could vary depending on the path to be analyzed and could range from determining viability of cells to image analysis. The new targets identified are reconfirmed and integrated with available data bases. Second approach employs the use of pooled shRNA libraries to transfect population of cells which are then screened for suitable phenotype. shRNAs in these cells can be PCR amplified and sequenced to identify the target gene of interest.

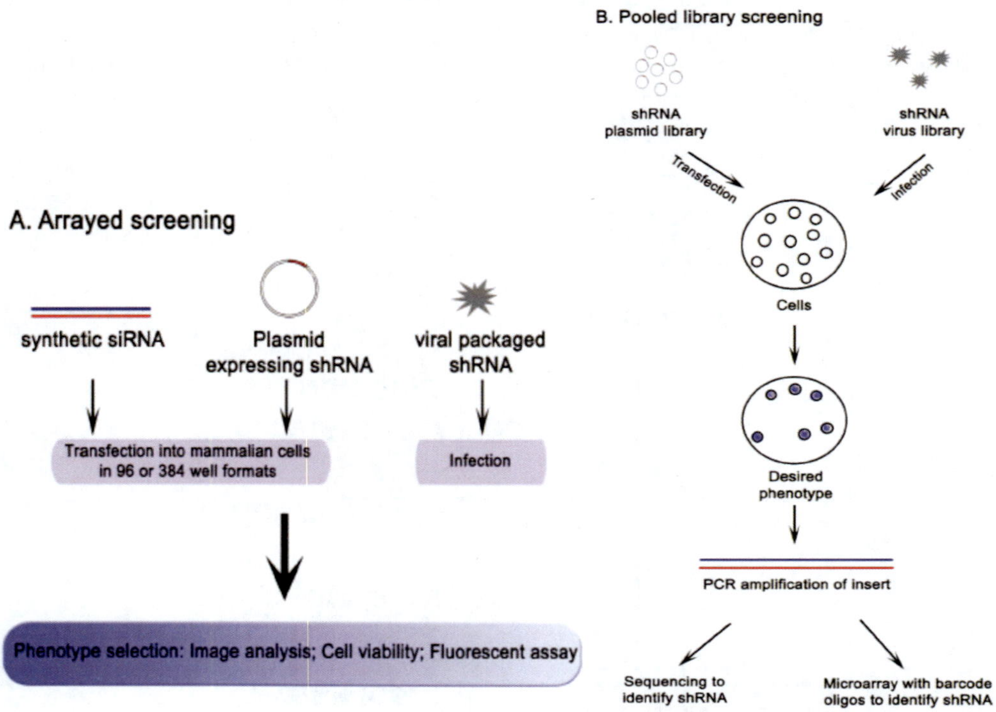

Fig. 3 : RNAi screening strategies. A. Arrayed screening: Synthetic siRNA, plasmid expressing shRNA or viral packaged shRNA can be used to transfect/infect cells and select the desired phenotype; B. Pooled library screening: Pools of plasmid/viral vectors expressing shRNA are used to transfect/infect cells. Cells are screened for desired phenotype and shRNA in the surviving cells is identified either by sequencing or barcode microarray analysis.

Moreover, these shRNA vectors can also have unique DNA sequence called barcode other than specific shRNA coding sequence. Cells are transfected with pooled shRNA library and divided into two populations: one population is taken as reference and the other is subjected to treatment. The cells exhibit differential response to treatment depending on specific shRNA harbored by these cells. These barcodes could be amplified by PCR, labeled with different fluorescent dyes and hybridized to DNA micro-array containing every barcode specific probe. Differential shRNA levels can be identified by colour change and correlated to specific phenotype change.

RNAi-barcode microarrays was used to identify genes involved in developing resistance to the anti-cancer cytotoxic drug Nutlin-3, which inhibits MDM2 and thus activates p53 (Brummelkamp et al. 2006). Additionally, the role of tumor suppressor PTEN in the PI3K pathway as a regulator of trastuzumab sensitivity was identified using shRNA screening (Berns et al. 2007).

RNAi screens to identify targets which are mandatory for the survival of tumor cell have resulted in identification of survivin in activated K-Ras cells (Sarthy et al. 2007) and glycogen synthase kinase 3β(GSK3β) and FBW7 (also known as FBXW7) as potential targets in Myc transformed cells (Rottmann et al. 2005).

RNAi screens were also used to identify novel tumour suppressor genes-transforming growth factor TGF β receptor II (*TGFBR2*); *PTEN*; and a novel tumour suppressor, RE1-silencing transcription factor (*REST*). REST was shown to restrict cells to grow in anchorage dependent manner. It has been shown to be frequently deleted in colorectal cancer cells (Westbrook et al. 2005). Thus identification and validation of target can lead to new drug discovery and development in a considerable less time as compared to other techniques.

4.7. RNAi as Therapeutic Tool

RNA interference technology and its role in targeting and down regulating the specific genes have shown a great promise as therapeutic tool for many diseases. Various growth and angiogenic factors/ their receptors, viral oncogenes or translocated oncogenes expressed in increased amounts to maintain the growth of the tumor are potential targets for RNAi based therapy. Biotechnology companies are putting in lot of effort in developing RNAi agents for therapy of various cancers (Table 6). Successful use of this technology holds a great promise in clinical treating medical conditions including cancer. FDA granted Orphan Drug Status for Quark's QPI-1002, siRNA based Prophylaxis for Delayed Graft Function (DGF) in Kidney Transplant Patients. DGF is a syndrome caused by ischemia and reperfusion injury which occurs after kidney transplant. In patients with DGF, the transplanted kidney does not function properly and requires intervention by dialysis. QPI-1002 is designed to temporarily inhibit the activity of the p53 gene, which is associated with apoptosis, and is believed to be critical in the ischemia and reperfusion injury process. RNAi-based inhibition of VEGF in cases of the wet form of macular degeneration has also reached an advanced stage of clinical trial.

Table 6 : RNA interference based drugs in various phases of clinical trials for carcinogenesis. [http://clinicaltrials.gov/]

S. No.	Company	Name of drug	Target	Disease/Indications
1	Calando Pharmaceuticals	CALAA-01	M2 subunit of ribonucleotide reductase	CancerSolid Tumor
2	Tekmira Pharmaceuticals Corporation	TKM-PLK1	PLK1 (polo-like kinase 1)	advanced solid tumor cancers
3	Alnylam Pharmaceuticals	ALN-VSP02	kinesin spindle protein (KSP) and vascular endothelial growth factor (VEGF)	solid tumors with liver involvement
4	Silenseed Ltd	siG12D LODER	KRASG12D	Pancreatic Ductal Adenocarcinoma Pancreatic Cancer
5	Silence Therapeutics AG	Atu027	KRASG12D	Advanced Solid Tumors
6	Duke University	Proteasome siRNA and tumor antigen RNA-transfected dendritic cells	Immunotherapy of Melanoma cells	Metastatic MelanomaAbsence of CNS Metastases

4.7.1. RNAi against oncogenic proteins

RNAi has been efficiently targeted against certain oncogenic factors such as Myc, a transcription factor and Ras, an important GTP binding protein, which are upregulated in cancers. RNAi studies have been able to successfully target mutated Ras and differentiate it from normal Ras (Fleming et al. 2005).

RNAi is able to target various translocations like BCR-ABL, growth and angiogenic factors like EGF, VEGF and human telomerase (hTR, hTERT). Carcinoembryonic antigen-related cell adhesion molecule 6 (CEACAM6), an intercellular adhesion molecule is overexpressed in human gastrointestinal cancer and has been linked to increased metastatic ability. CEACAM6-specific siRNA suppressed primary tumor growth by 68% versus control siRNA in mouse model (Duxbury et al. 2004).

Various DNA and RNA oncogenic viruses have been linked to cancer. These linkages are generally established by epidemiological data, serological data by identifying antibodies to viral antigens in patient's sera and by genetic studies. Human papillomavirus (HPV) is associated with cervical cancer. It is the second leading cancer among women. HPV pathogenesis has been directly linked to its oncogenes E6 and E7. They are expressed in epithelial layers and in both early as well as advanced stages of disease.

Gardasil is commercially available vaccine targeting HPV types 6, 11, 16, and 18 for the prevention of HPV related infections.

RNAi has been shown to be effective against cervical cancer by targeting E6 and E7 oncogenes in animal model (Yamato et al. 2008). Similarly the efficacy of RNAi against hepatocellular carcinoma (HCC) related cellular oncogenes or the hepatocarcinogenic hepatitis B virus (HBV) and hepatitis C virus (HCV) are being studied (Arbuthnot and Thompson, 2008). These results are promising but further clinical studies are required before RNAi based drugs can come to the market.

4.7.2. RNAi induced apoptosis

Cancer cells escape apoptosis which is the hallmark feature of cancer. As discussed earlier apoptosis is a complex process regulated by various activators and inhibitors. The defects in the levels of these regulatory proteins lead to interference in genetically programmed cell death. Various studies are underway to down regulate the anti-apoptotic proteins to promote apoptosis in cancer cells.

Bcl-2, anti-apoptotic protein, has been implicated in the carcinogenesis of a variety of solid tumors, including breast, lung, liver, gastric and matrix cancers. It has been targeted using siRNA to promote apoptosis in tumor cells. The significant reduction in tumor growth indicates it as a promising therapeutic agent for the treatment of malignant tumors induced by the *bcl-2* gene (Fu et al. 2005).

IAPs family regulates apoptosis independent of Bcl-2 family. They suppress apoptosis and thus the genes of the family are important targets. XIAP has been targeted by RNAi inside the cancer cells in animal models (Mohr et al. 2010).

Another member of IAPs, survivin belongs to inhibitors of apoptosis and is expressed in high amounts in fetal and embryonic organs and is not detected in differentiated tissues. Its levels have been detected in epithelium and endothelial cells during angiogenesis. Survivin expression has been found to be increased in various malignancies. Its higher levels also indicates high rate of tumor recurrence. siRNA was targeted against survivin mRNA and post-inhibition protein levels and inhibition rate were found to be 36.33% and 44.65%, respectively and furthermore apoptosis was also observed in human colon carcinoma cells (apoptosis rate: 17.24±2.13%) (Cai et al. 2009). Though level of inhibition achieved is limiting but at the same time the technology is promising.

4.7.3. RNAi targeting angiogenesis

The concept of "anti angiogenesis" as a potential novel anticancer therapy was first proposed by Judah M. Folkman (1971). He conceptualized that by inhibiting the vascularization of tumors it would deprive the malignant tissue of its oxygen and nutrient supply and thus inhibit tumor progression. It has turned into a reality and more than 10 drugs are in the market targeting various stages of vascularization of tumors (Table 7). The different strategies can target (a) angiogenic factors; (b) Extra cellular matrix degrading enzymes (Matrix metalloproteases); (c) Endothelial cells. These drugs will not kill the cancer cell directly but may keep the growth of tumor cells under check. The potential disadvantage could be that these drugs may need to be administered over a long period and thus can interfere with blood clotting, immune system and heart functions.

RNAi technology can also be used to target vascularization and factors responsible for angiogenesis to control the growth of cancer. Filleur and colleagues (2003) proved the concept that RNAi can be used successfully to inhibit VEGF in rat fibrosarcoma model. siRNA as alone or in combination with drugs have been tested in cell-line based or animal models for lung cancer (Ma et al. 2010), hepatocellular carcinoma (Arbuthnot and Thompson, et al. 2008) and various other types of cancer. These reports are very preliminary and a lot of hurdles/issues need to be addressed before RNAi based anti-tumor drugs can reach the market.

Table 7 : Currently available anti-angiogenic drugs approved by FDA. [http://www.angio.org/; http://www.cancer.gov/]

Category	Name	Company	Target	Indications
humanized Monoclonal antibody	Bevacizumab (Avastin)	Genentech	VEGF	Metastatic colorectal cancer (mCRC), non-small cell lung cancer (NSCLC), advanced breast cancer
chimeric monoclonal antibody	Cetuximab (Erbitux)	Bristol-Myers SquibbIm Clone	extracellular domain of epidermal growth factor receptor (EGFR)	mCRC, head and neck cancer
humanized IgG2 anti-EGFR monoclonal antibody	Panitumumab (Vectibix)	Amgen	EGFR	mCRC
humanized IgG1 monoclonal antibody	Trastuzumab (Herceptin)	Genentech	HER-2	Breast cancer
Small molecule TK inhibitor	Erlotinib (Tarveca)	Genentech OSIRoche	EGFR	NSCLC, pancreatic cancer
Small molecule	Sorafenib	BayerOnyx	VEGFR-1,	Advanced renal cell carcinoma,

Contd.

TK inhibitor	(Nexavar)		VEGFR-2, VEGFR-3, PDGFR-ß, and Raf-1	advanced hepatocellular carcinoma
Small molecule TK inhibitor	Sunitinib (Sutent)	Pfizer	VEGFR-1, VEGFR-2, VEGFR-3, PDGFR- ß, and RET	Advanced renal cell carcinoma, Gastrointestinal stromal tumor
A small molecule inhibitor	Temsirolimus (Torisel)	Wyeth	mTOR	Advanced renal cell carcinoma
proteasome inhibitor	Bortezomib (Velcade®)	Millennium	mechanisms are unclear	Multiple myeloma, mantle cell lymphoma (MCL).

5. Conclusion

RNA interference is the phenomenon that regulates the gene expression by small RNA molecules through inhibition of translation or modification of heterochromatin. Molecular changes at the genetic and epigenetic level result in disruption of pathways involved in cell growth and regulation. Both long and short ncRNAs have been shown to be critical mediators of cancer. MicroRNAs are a large family of short non-coding RNAs that are evolutionary conserved and negatively regulates the expression of their target genes at the posttranscriptional or posttranslational level. MiRNAs can act as tumor suppressor genes or oncogenes in human cancers. Rearrangements, deletions, amplification and mutations of miRNAs in cancer suggest their role in cancer pathogenesis.

Small-interfering RNAs are also short non-coding RNA molecules that bind their complementary mRNA and lead to its cleavage. These are 21-22 nucleotide long RNA duplex and require a perfect match to its complementary mRNA. The ability of exogenously introduced double stranded RNA to silence the specific gene has been demonstrated in cell culture and animal models and has become the method of choice to study the function of a particular gene. SiRNAs can be synthesized commercially with low production costs and selective gene knockdown can be used to identify critical genes and/or whole pathways involved in cancer. The efficacy of this approach depends on the selection of the target sequence and right siRNA design. In addition, siRNAs have favorable pharmacokinetic properties and can be taken up by a various tissues. However, siRNAs have a very short half life in the serum and delivery methods need to be standardized before RNAi reagents can be used as effective cancer therapy drugs.

The role of multiple genes and their interplay in mediating a particular action can be simultaneously or consecutively analyzed by RNAi based knockdowns. In addition, siRNA technology has been successfully applied for large-scale screens to identify new components in tumorigenesis and metastasis. RNAi has been effectively demonstrated to selectively silence the oncogenes associated with cancer *in vitro* and *in vivo*, establishing its potential as future therapeutic agent. It has also led to design novel therapeutic ways to block the expression of altered gene.

6. Acknowledgements

I thank Utkarsh Gupta for providing help to draw the figures.

References

Abdelrahim, M., Safe, S., Baker, C. and Abudayyeh, A. (2006) RNAi and cancer: Implications and applications. J RNAi Gene Silencing, 2:136-45.

Aguda, B.D., Kim, Y., Piper-Hunter, M.G., Friedman, A. and Marsh, C.B. (2008) MicroRNA regulation of a cancer network: consequences of the feedback loops involving miR-17-92, E2F, and Myc. Proc Natl Acad Sci U S A, 105:19678-83.

Aravin, A.A., Lagos-Quintana, M., Yalcin, A., Zavolan, M., Marks, D., Snyder, B., Gaasterland, T., Meyer, J. and Tuschl, T. (2003) The small RNA profile during Drosophila melanogaster development. Dev Cell, 5:337-50.

Arbuthnot, P. and Thompson, L.J. (2008) Harnessing the RNA interference pathway to advance treatment and prevention of hepatocellular carcinoma. World J Gastroenterol, 14:1670-81.

Bernard, D., Prasanth, K.V., Tripathi, V., Colasse, S., Nakamura, T., Xuan, Z., Zhang, M.Q., Sedel, F., Jourdren, L., Coulpier, F., Triller, A., Spector, D.L. and Bessis, A. (2010) A long nuclear-retained non-coding RNA regulates synaptogenesis by modulating gene expression. EMBO J, 29:3082-93.

Berns, K., Horlings, H.M., Hennessy, B.T., Madiredjo, M., Hijmans, E.M., Beelen, K., Linn, S.C., Gonzalez-Angulo, A.M., Stemke-Hale, K., Hauptmann, M., Beijersbergen, R.L., Mills, G.B., van de Vijver, M.J. and Bernards, R. (2007) A functional genetic approach identifies the PI3K pathway as a major determinant of trastuzumab resistance in breast cancer. Cancer Cell, 12:395-402.

Brummelkamp, T.R., Fabius, A.W., Mullenders, J., Madiredjo, M., Velds, A., Kerkhoven, R.M., Bernards, R. and Beijersbergen, R.L. (2006) An shRNA barcode screen provides insight into cancer cell vulnerability to MDM2 inhibitors. Nat Chem Biol, 2:202-6.

Cai, M., Wang, G., Tao, K. and Cai, C. (2009) Induction of apoptosis of human colon cancer cells by siRNA recombinant expression vector targeting survivin gene. J Huazhong Univ Sci Technolog Med Sci, 29:45-9.

Cho, W.C. (2007) OncomiRs: the discovery and progress of microRNAs in cancers. Mol Cancer, 6:60.

Chu, C.Y. and Rana, T.M. (2009) Structure and gene-silencing mechanisms of small noncoding RNAs. In Walter NG, Woodson SA, Batey RT. (Eds.). Non-protein coding RNAs, Springer.

Chumbalkar, V., Latha, K., Hwang, Y., Maywald, R., Hawley, L., Sawaya, R., Diao, L., Baggerly, K., Cavenee, W.K., Furnari, F.B. and Bogler, O. (2011) Analysis of phosphotyrosine signaling in glioblastoma identifies STAT5 as a novel downstream target of ÄEGFR. J Proteome Res, Jan 9. [Epub ahead of print]

Chung, S., Nakagawa, H., Uemura, M., Piao, L., Ashikawa, K., Hosono, N., Takata, R., Akamatsu, S., Kawaguchi, T., Morizono, T., Tsunoda, T., Daigo, Y., Matsuda, K., Kamatani, N., Nakamura, Y. and Kubo, M. (2011) Association of a novel long non-coding RNA in 8q24 with prostate cancer susceptibility. Cancer Sci, 102:245-52.

Duxbury, M.S., Matros, E., Ito, H., Zinner, M.J., Ashley, S.W. and Whang, E.E. (2004) Systemic siRNA-mediated gene silencing: a new approach to targeted therapy of cancer. Ann Surg, 240:667-74; discussion 675-6.

Eguchi, A., Meade, B.R., Chang, Y.C., Fredrickson, C.T., Willert, K., Puri, N. and Dowdy, S.F. (2009) Efficient siRNA delivery into primary cells by a peptide transduction domain-dsRNA binding domain fusion protein. Nat Biotechnol, 27:567-71.

Filleur, S., Courtin, A., Ait-Si-Ali, S., Guglielmi, J., Merle, C., Harel-Bellan, A., Clézardin, P. and Cabon, F. (2003) SiRNA-mediated inhibition of vascular endothelial growth factor severely limits tumor resistance to antiangiogenic thrombospondin-1 and slows tumor vascularization and growth. Cancer Res, 63:3919-22.

Fire, A., Xu, S., Montgomery, M.K., Kostas, S.A., Driver, S.E. and Mello, C.C. (1998) Potent and specific genetic interference by double-stranded RNA in Caenorhabditis elegans. Nature, 391:806-11.

Fleming, J.B., Shen, G.L., Holloway, S.E., Davis, M. and Brekken, R.A. (2005) Molecular consequences of silencing mutant K-ras in pancreatic cancer cells: justification for K-ras-directed therapy. Mol Cancer Res, 3:413-23.

Folkman, J. (1971) Tumor angiogenesis: therapeutic implications. N Engl J Med, 285:1182-1186

Fornari, F., Gramantieri, L., Giovannini, C., Veronese, A., Ferracin, M., Sabbioni, S., Calin, G.A., Grazi, G.L., Croce, C.M., Tavolari, S., Chieco, P., Negrini, M. and Bolondi, L. (2009) MiR-122/cyclin G1 interaction modulates p53 activity and affects doxorubicin sensitivity of human hepatocarcinoma cells. Cancer Res, 69:5761-7.

Friedman, J.M. and Jones, P.A. (2009) MicroRNAs: critical mediators of differentiation, development and disease. Swiss Med Wkly, 139:466-72.

Fu, G.F., Lin, X.H., Han, Q.W., Fan, Y.R., Xu, Y.F., Guo, D., Xu, G.X. and Hou, Y.Y. (2005) RNA interference remarkably suppresses bcl-2 gene expression in cancer cells in vitro and in vivo. Cancer Biol Ther, 4:822-9.

Galun, E. (2005) RNA silencing, World Scientific.

Goldsby, R.A., Kindt, T.J. and Osborne, B.A. (2000) Kuby Immunology, W.H. Freeman & Company.

Gomberts, B. (2002) Signal transduction, Academic Press.

Graf, F., Koehler, L., Kniess, T., Wuest, F., Mosch, B. and Pietzsch, J. (2009) Cell Cycle Regulating Kinase Cdk4 as a Potential Target for Tumor Cell Treatment and Tumor Imaging. J Oncol, 2009:106378.

Guo, S. and Kemphues, K.J. (1995) par-1, a gene required for establishing polarity in C. elegans embryos, encodes a putative Ser/Thr kinase that is asymmetrically distributed. Cell, 81:611-20.

Gupta, R.A., Shah, N., Wang, K.C., Kim, J., Horlings, H.M., Wong, D.J., Tsai, M.C., Hung, T., Argani, P., Rinn, J.L, Wang, Y., Brzoska, P., Kong, B., Li, R., West, R.B., van de Vijver, M.J., Sukumar, S. and Chang, H.Y. (2010) Long non-coding RNA HOTAIR reprograms chromatin state to promote cancer metastasis. Nature, 464:1071-6.

Hanahan, D. and Weinberg, RA. (2010) The hallmarks of cancer. Cell, 100:57-70.

Hao, Y., Crenshaw, T., Moulton, T., Newcomb, E. and Tycko, B. (1993) Tumour-suppressor activity of H19 RNA. Nature, 365:764-7.

Hwang, H.W. and Mendell, J.T. (2006) MicroRNAs in cell proliferation, cell death, and tumorigenesis. Br J Cancer, 94:776-80.

Inomata, M., Tagawa, H., Guo, Y.M., Kameoka, Y., Takahashi, N. and Sawada, K. (2009) MicroRNA-17-92 down-regulates expression of distinct targets in different B-cell lymphoma subtypes. Blood, 113:396-402.

Iorns, E., Lord, C.J., Turner, N. and Ashworth, A. (2007) Utilizing RNA interference to enhance cancer drug discovery. Nat Rev Drug Discov, 6:556-68.

Ji, P., Diederichs, S., Wang, W., Böing, S., Metzger, R., Schneider, P.M., Tidow, N., Brandt, B., Buerger, H., Bulk, E., Thomas, M., Berdel, W.E., Serve, H. and Müller-Tidow, C. (2003) MALAT-1, a novel noncoding RNA, and thymosin beta4 predict metastasis and survival in early-stage non-small cell lung cancer. Oncogene, 22:8031-41.

Kim, Y., Tewari, M., Pajerowski, J.D., Cai, S., Sen, S., Williams, J., Sirsi, S., Lutz, G. and Discher, D.E. (2009) Polymersome delivery of siRNA and antisense oligonucleotides. J Control Release, 134:132-40.

Kiuru-Kuhlefelt, S., Sarlomo-Rikala, M., Larramendy, M.L., Söderlund, M., Hedman, K., Miettinen, M. and Knuutila, S. (2000) FGF4 and INT2 oncogenes are amplified and expressed in Kaposi's sarcoma. Mod Pathol, 13:433-7.

Koh, M.Y., Spivak-Kroizman, T.R. and Powis, G. (2010) HIF-1alpha and cancer therapy. Recent Results Cancer Res, 180:15-34.

Kong-Beltran, M., Seshagiri, S., Zha, J., Zhu, W., Bhawe, K., Mendoza, N., Holcomb, T., Pujara, K., Stinson, J., Fu, L., Severin, C., Rangell, L., Schwall, R., Amler, L., Wickramasinghe, D. and Yauch, R. (2006) Somatic mutations lead to an oncogenic deletion of met in lung cancer. Cancer Res, 66:283-9.

Kutter, C. and Svoboda, P. (2008) miRNA, siRNA, piRNA: Knowns of the unknown. RNA Biol, 5:181-8.

Lee, J.S., Green, J.J., Love, K.T., Sunshine, J., Langer, R. and Anderson, D.G. (2009) Gold, poly(beta-amino ester) nanoparticles for small interfering RNA delivery. Nano Lett, 9:2402-6.

Lee, R.C., Feinbaum, R.L. and Ambros, V. (1993) The C. elegans heterochronic gene lin-4 encodes small RNAs with antisense complementarity to lin-14. Cell, 75:843-54.

Lee, Y.S. and Dutta, A. (2007) The tumor suppressor microRNA let-7 represses the HMGA2 oncogene. Genes Dev, 2121:1025-30.

Lin, C.J., Gong, H.Y., Tseng, H.C., Wang, W.L. and Wu, J.L. (2008) miR-122 targets an anti-apoptotic gene, Bcl-w, in human hepatocellular carcinoma cell lines. Biochem Biophys Res Commun, 375:315-20.

Lu, Y., Li, C., Zhang, K., Sun, H., Tao, D., Liu, Y., Zhang, S. and Ma, Y. (2010) Identification of piRNAs in Hela cells by massive parallel sequencing. BMB Rep, 43:635-41.

Ma, Y.P., Yang, Y., Zhang, S., Chen, X., Zhang, N., Wang, W., Cao, Z.X., Jiang, Y., Zhao, X., Wei, Y.Q. and Deng, H.X. (2010) Efficient inhibition of lung cancer in murine model by plasmid-encoding VEGF short hairpin RNA in combination with low-dose DDP. J Exp Clin Cancer Res, 29:56.

Malecová, B. and Morris, K.V. (2010) Transcriptional gene silencing through epigenetic changes mediated by non-coding RNAs. Curr Opin Mol Ther, 12:214-22.

Mao, C.P. and Wu, T.C. (2010) Inhibitory RNA molecules in immunotherapy for cancer. Methods Mol Biol, 623:325-39.

Meade, B.R. and Dowdy, S.F. (2008) Enhancing the cellular uptake of siRNA duplexes following noncovalent packaging with protein transduction domain peptides. Adv Drug Deliv Rev, 60:530-6.

Meng, F., Henson, R., Wehbe-Janek, H., Ghoshal, K., Jacob, S.T. and Patel, T. (2007) MicroRNA-21 regulates expression of the PTEN tumor suppressor gene in human hepatocellular cancer. Gastroenterology, 133:647-58.

Metzler, M., Wilda, M., Busch, K., Viehmann, S. and Borkhardt, A. (2004) High expression of precursor microRNA-155/BIC RNA in children with Burkitt lymphoma. Genes Chromosomes Cancer, 39:167-9.

Mohr, A., Albarenque, S.M., Deedigan, L., Yu, R., Reidy, M., Fulda, S. and Zwacka, R.M. (2010) Targeting of XIAP Combined with Systemic Mesenchymal Stem Cell-Mediated Delivery of sTRAIL Ligand Inhibits Metastatic Growth of Pancreatic Carcinoma Cells. Stem Cells, 28:2109-20.

Mukherji, M., Bell, R., Supekova, L., Wang, Y., Orth, A.P., Batalov, S., Miraglia, L., Huesken, D., Lange, J., Martin, C., Sahasrabudhe, S., Reinhardt, M., Natt, F., Hall, J., Mickanin, C., Labow, M., Chanda, S.K., Cho, C.Y. and Schultz, P.G. (2006) Genome-wide functional analysis of human cell-cycle regulators. Proc Natl Acad Sci U S A, 103:14819-24.

Napoli, C., Lemieux, C. and Jorgensen, R. (1990) Introduction of a Chimeric Chalcone Synthase Gene into Petunia Results in Reversible Co-Suppression of Homologous Genes in trans. Plant Cell, 2:279-289.

Nelson, E.C., Evans, C.P., Mack, P.C., Devere-White, R.W. and Lara, P.N. (2007) Inhibition of Akt pathways in the treatment of prostate cancer. Prostate Cancer Prostatic Dis, 10:331-9.

Oren, M. (1999) Regulation of the p53 tumor suppressor protein. J Biol Chem, 274:36031-4.

Pfeifer, G.P. and Rauch, T.A. (2009) DNA methylation patterns in lung carcinomas. Semin Cancer Biol, 19:181-7.

Pierotti, M.A., Schichman, S.A., Sozzi, G. and Croce, C.M. (2003) Oncogenes. In Bast, R.C., Kufe, D.W., Pollock, R.E., et al. (Eds) Holland-Frei Cancer Medicine. 6th Ed. BC Decker Inc.

Qi, L. and Gao, X. (2008) Quantum dot-amphipol nanocomplex for intracellular delivery and real-time imaging of siRNA. ACS Nano, 2:1403-10.

Rabinovich, G.A., Gabrilovich, D. and Sotomayor, E.M. (2007) Immunosuppressive strategies that are mediated by tumor cells. Annu Rev Immunol, 25:267-96.

Rinn, J.L., Kertesz, M., Wang, J.K., Squazzo, S.L., Xu, X., Brugmann, S.A., Goodnough, L.H., Helms, J.A., Farnham, P.J., Segal, E. and Chang, H.Y. (2007) Functional demarcation of active and silent chromatin domains in human HOX loci by noncoding RNAs. Cell, 129:1311-23.

Romano, N. and Macino, G. (1992) Quelling: transient inactivation of gene expression in Neurospora crassa by transformation with homologous sequences. Mol Microbiol, 6:3343-53.

Rottmann, S., Wang, Y., Nasoff, M., Deveraux, Q.L. and Quon, K.C. (2005) A TRAIL receptor-dependent synthetic lethal relationship between MYC activation and GSK3beta/FBW7 loss of function. Proc Natl Acad Sci U S A, 102:15195-200.

Sarthy, A.V., Morgan-Lappe, S.E., Zakula, D., Vernetti, L., Schurdak, M., Packer, J.C., Anderson, M.G., Shirasawa, S., Sasazuki, T. and Fesik, S.W. (2007) Survivin depletion preferentially reduces the survival of activated K-Ras-transformed cells. Mol Cancer Ther, 6:269-76.

Semizarov, D., Kroeger, P. and Fesik, S. (2004) siRNA-mediated gene silencing: a global genome view. Nucleic Acids Res, 32:3836-45.

Tam, O.H., Aravin, A.A., Stein, P., Girard, A., Murchison, E.P., Cheloufi, S., Hodges, E., Anger, M., Sachidanandam, R., Schultz, R.M. and Hannon, G.J. (2008) Pseudogene-derived small interfering RNAs regulate gene expression in mouse oocytes. Nature, 453:534-8.

Tseng, Y.C., Mozumdar, S. and Huang, L. (2009) Lipid-based systemic delivery of siRNA. Adv Drug Deliv Rev, 61:721-31.

Tuschl, T., Elbashir, S., Harborth, J. and Weber, K. (2003) The siRNA user guide. URL: http://www.rockefeller.edu/labheads/tuschl/sirna.html.

Van Themsche, C., Chaudhry, P., Leblanc, V., Parent, S. and Asselin, E. (2010) XIAP gene expression and function is regulated by autocrine and paracrine TGF-beta signaling. Mol Cancer, 9:216.

Westbrook, T.F., Martin, E.S., Schlabach, M.R., Leng, Y., Liang, A.C., Feng, B., Zhao, J.J., Roberts, T.M., Mandel, G., Hannon, G.J., Depinho, R.A., Chin, L. and Elledge, S.J. (2005) A genetic screen for candidate tumor suppressors identifies REST. Cell, 121:837-48.

Wilusz, J.E., Sunwoo, H. and Spector, D.L. (2009) Long noncoding RNAs: functional surprises from the RNA world. Genes Dev, 23:1494-504.

Yamato, K., Yamada, T., Kizaki, M., Ui-Tei, K., Natori, Y., Fujino, M., Nishihara, T., Ikeda, Y., Nasu, Y., Saigo, K. and Yoshinouchi, M. (2008) New highly potent and specific E6 and E7 siRNAs for treatment of HPV16 positive cervical cancer. Cancer Gene Ther, 15:140-53.

Yang, D., Buchholz, F., Huang, Z., Goga, A., Chen, C.Y., Brodsky, F.M. and Bishop, J.M. (2002) Short RNA duplexes produced by hydrolysis with Escherichia coli RNase III mediate effective RNA interference in mammalian cells. Proc Natl Acad Sci U S A, 99:9942-7.

Zhang, C. (2008) MicroRNomics: a newly emerging approach for disease biology. Physiol Genomics. 33:139-47.

Zhang, J.G., Wang, J.J., Zhao, F., Liu, Q., Jiang, K. and Yang, G.H. (2010) MicroRNA-21 (miR-21) represses tumor suppressor PTEN and promotes growth and invasion in non-small cell lung cancer (NSCLC). Clin Chim Acta, 411:846-52.

Zhang, Y., Li, M., Wang, H., Fisher, W.E., Lin, P.H., Yao, Q. and Chen, C. (2009) Profiling of 95 microRNAs in pancreatic cancer cell lines and surgical specimens by real-time PCR analysis. World J Surg, 33:698-709.

❑❑❑

CHAPTER – 24

Secondary Metabolites Production, Prospective and Applications

Indu Ravi[1], Garima Bhardwaj[2] and Kambaska Kumar Behera[2]
[1]IGNOU Regonal Centre, Jaipur- 302020, Rajasthan,India
[2]Department of Bioscience and Biotechnology, Banasthali University, Banasthali Rajasthan-304022
Email: kambaska@yahoo.co.in

Summary

Plants produce over 200,000 metabolites, which are important source of drugs and industrial materials. They have been traditionally classified as primary and secondary metabolites. The primary metabolites like sugars, amino acids and fatty acids that are needed for general growth and physiological development of plants are widely distributed in nature and are also utilized as food by man. Secondary metabolites are natural products, the absence of which does not result in immediate death, but in long term their impairment may lead to malfunction and reduced survivability. These plant secondary metabolites can be divided into groups that share the same core substructure, originated from the same biosynthetic pathways. Secondary metabolites play a major role in the adaptation of plants to the changing environment and in overcoming stress constraints. In this chapter we discussed the structure, properties, classification and functions of some of the important groups of plant secondary metabolites such as phenols, flavonoids, terpenoids and alkaloids. A brief overview of some of the important fungal secondary metabolites, their structures and importance has also been discussed. Biotechnological approaches which can be adopted to engineer plants for production of secondary metabolites have also been discussed.

1. Introduction

Metabolites are the intermediates and products of metabolism. The term metabolite is usually restricted to small molecules. The term "secondary" was introduced by Kossel in 1891, which implies that while primary metabolites are present in every living cell capable of dividing, the secondary metabolites are present only incidentally and are not of paramount significance for plant life. Secondary metabolites are chemicals produced by plants for which no role has yet been found in growth, photosynthesis, reproduction, or other "primary" functions. These chemicals are extremely diverse; many thousands have been identified in several major classes. Each plant family, genus and species produces a characteristic mix of these chemicals, and they can sometimes be used as taxonomic characters in classifying plants (Marcello Iriti and Franco Faoro, 2009).

Research in 'secondary metabolism' is about 50 years old. It began with the first biosynthetic studies by application of radioactively-labeled tracers. The history of secondary product research started about 200 years ago in 1806 with the first isolation of a natural product (*Principium somniferum*, i.e. morphine) by F. W. Serturner. It also advanced the first foundations of pharmaceutical industry and drug research. The first synthesis of a secondary product, indigo, by von Baeyer in 1886, provided a milestone in synthetic organic chemistry.

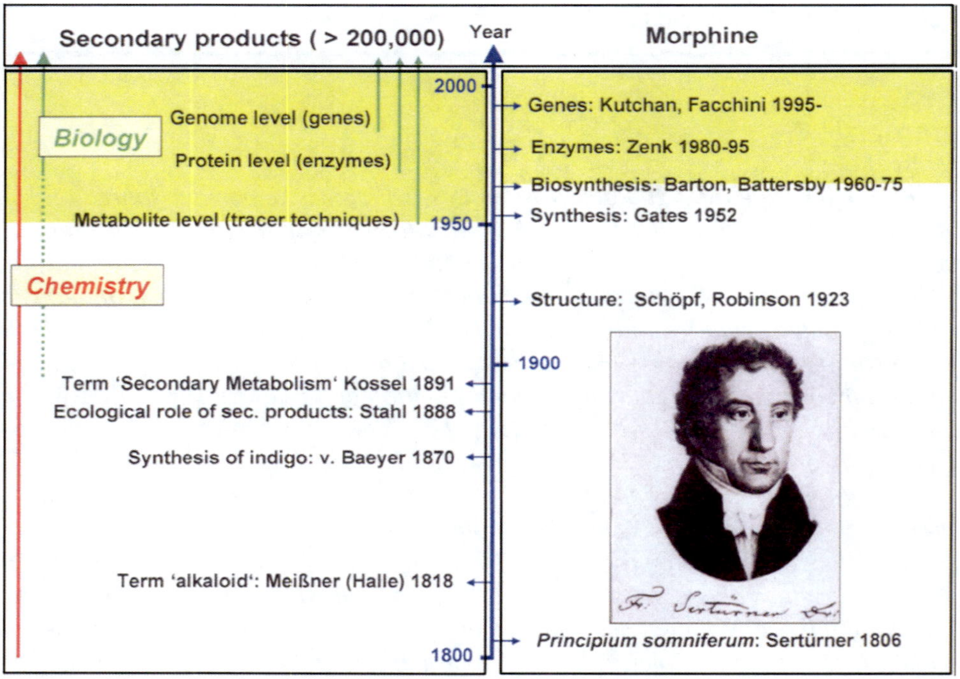

Fig.1. The history of secondary metabolite product research (Hartman, 2007).

A primary metabolite is directly involved in the normal growth, development and reproduction. A secondary metabolite is not directly involved in these processes, but usually has important ecological function. Typically primary metabolites are found across all species within broad phylogenetic groups and are produced using the same pathway in all species. Secondary metabolites, by contrast, are often species-specific (or found in only a small set of species in a narrow phylogenetic group), and without these compounds the organism suffers from only a mild impairment, lowered survivability, or else no phenotypic change at all.

The function or importance of these compounds to the organism is usually of an ecological nature as they are used as defenses against predators, parasites and diseases, for interspecies competition, and to facilitate the reproductive processes (coloring agents, attractive smells, etc). Since these compounds are usually restricted to a much more limited group of organisms, they have long been of prime importance in taxonomic research. Plant organisms being devoid of motility and immune system, have elaborated alternative defense strategies, involving the huge variety of secondary metabolites as tools to overcome stress constraints, adapt to the changing environment and survive.

In living organisms, secondary metabolites are usually considered not essential for growth and development, unlike the products of primary metabolism. This does not mean they are not important, as they relate plants with the components of their ecosystem, i.e., the physical environment (biotope) and the living community (biocenosis), thus resulting indispensable for the survival of the species. Functionally some secondary metabolites are as vital to plants as primary metabolites. Further, the division of metabolism in "primary" and "secondary" is somewhat unsatisfactory. In fact, some major chemical groups still classified as secondary metabolites, such as lipids, are required by all living organisms, as are some metabolites arising from a secondary pathway (i.e., giberellin A1 from the isoprenoid pathway) that can have a determinant endogenous regulator role.

2. Classification

Secondary metabolites can be classified by their chemical structure or physical properties into one or more of the following groups: alkaloids, terpenoids, polyketides, aliphatic, aromatic, and heteroaromatic organic acids, phenols, iridoids, steroids, saponins, peptides, ethereal oils, resins and balsams.

Classification of secondary metabolites based on their biosynthetic origin:

- Alkaloids: Hyoscyamine, Atropine, Cocaine, Morphine, etc.
- Terpenoids: Azadirachtin, Artemisinin, Steroids, Saponins
- Glycosides (heavily modified sugar molecules): Nojirimycin, Glucosinolates
- Phenols: Resveratrol
- Phenazines: Pyocyanin, Phenazine-1-carboxylic acid
- Polyketides: Erythromycin, Discodermolide

- Fatty acid synthase products: phloroglucinols
- Nonribosomal peptides: Vancomycin, Thiostrepton, Gramicidin, Bacitracin etc.
- Hybrids: Epothilone
- Polyphenols
- Ribosomal peptides

Secondary metabolites can also be classified on the basis of chemical structure (for example, having rings, containing a sugar), composition (containing nitrogen or not), their solubility in various solvents, or the pathway by which they are synthesized. A simple classification includes three main groups: the terpenes (made from mevalonic acid, composed almost entirely of carbon and hydrogen), phenolics (made from simple sugars, containing benzene rings, hydrogen, and oxygen), and nitrogen-containing compounds (extremely diverse, may also contain sulfur).

Table 1 : Classification of secondary metabolites on the basis of chemical structure.

Class	Example Compounds	Example Sources	Some Effects and Uses
Nitrogen-Containing			
Alkaloids	nicotine cocaine theobromine	tobacco coca plant chocolate (cocao)	interfere with neurotransmission, block enzyme action
Nitrogen-and Sulfur-Containing			
Glucosinolates	sinigrin	cabbage, relatives	
Terpenoids			
Monoterpenes	menthol linalool	mint and relatives, many plants	Interfere with neurotransmission, block ion transport, anesthetic
Sesquiterpenes	parthenolid	Parthenium and relatives (*Asteraceae*)	contact dermatitis
Diterpenes	gossypol	cotton	block phosphorylation; toxic
Triterpenes, cardiac glycosides	digitogenin	Digitalis (foxglove)	stimulate heart muscle, alter ion transport
Tetraterpenoids	carotene	many plants	antioxidant; orange coloring
Terpene polymers	rubber	Hevea (rubber) trees, dandelion	gum up insects; airplane tires
Sterols	spinasterol	spinach	interfere with animal hormone action

Contd.

Phenolics			
Phenolic acids	caffeic, chlorogenic	all plants	cause oxidative damage, browning in fruits and wine
Coumarins	umbelliferone	carrots, parsnip	cross-link DNA, block cell division
Lignans	podophyllin urushiol	mayapple poison ivy	cathartic, vomiting, allergic dermatitis
Flavonoids	anthocyanin, catechin	almost all plants	flower, leaf color; inhibit enzymes, anti- and pro-oxidants, estrogenic
Tannins	gallotannin, condensed tannin	oak, hemlock trees, birdsfoot trefoil, legumes	bind to proteins, enzymes, block digestion, antioxidants
Lignin	lignin	all land plants	structure, toughness, fiber

3. Structure and Properties of Secondary Metabolites:

3.1. Phenolic Compounds

Phenolic compounds consist of a large number of molecules of heterogeneous structure. Their common feature is the presence of at least one hydroxyl-substituted aromatic ring (C6) bearing one or more hydroxyl groups. They are mainly synthesized from cinnamic acid, which is formed from phenylalanine by the action of L-phenyloalanine ammonia-lyase (PAL) (EC 4.3.1.5), the branch point enzyme between primary (shikimate pathway) and secondary (phenylopropanoid) metabolism (Dixon and Paiva, 1995). Most phenolic compounds belong to the flavonoids. Lignin, the primary substance of wood, is the most common member of this group. This group also consists of a large number of molecules of heterogeneous structure. Their common feature is the presence of at least one hydroxyl-substituted aromatic ring system.

Phenols are acidic due to the dissociability of their -OH group. They are rather reactive compounds and as long as no steric inhibition due to additional side chains occurs, they form hydrogen bonds. Consequently, many flavonoids have intramolecular bonds. Another important feature is their ability to form chelate complexes with metals. Also, they are easily oxidized and, if so, form polymers (dark aggregates). The darkening of cut or dying plant parts is caused by this reaction. They have usually an inhibiting effect on plant growth. Among the phenylpropanol derivatives of lower molecular weight are a number of scents like the coumarins, cinnamic acid, sinapinic acid, the coniferyl alcohols and others. These substances and their derivatives are at the same time intermediates of the biosynthesis of lignin.

Table 2 : The Most Important Groups of Phenolic Compounds in Plants

Number of C-atoms	Basic skeleton	Class
6	C_6	simple phenols, benzoquinones
7	$C_6 - C_1$	phenolic acids
8	$C_6 - C_2$	acetophenone, phenylacetic acid
9	$C_6 - C_3$	hydroxycinnamic acid, polypropene, coumarin, isocoumarin
10	$C_6 - C_4$	Naphtoquinone
13	$C_6 - C_1 - C_6$	Xanthone
14	$C_6 - C_2 - C_6$	stilbene, anthrachinone
15	$C_6 - C_3 - C_6$	flavonoids, isoflavonoids
18	$(C_6 - C_3)_2$	lignans, neolignans
30	$(C_6 - C_3 - C_6)_2$	Biflavonoids
N	$(C_6 - C_3)_n(C_6)_n(C_6 - C_3 - C_6)_n$	lignins catecholmelanine (condensed tannins)

3.2. Flavonoids

In 1975, the number of identified flavonoids was estimated to be larger than 2000. Some important representatives and their biological significance are listed in the table below.

Table 3 : The Most Important Classes of Flavonoids and their Biological Significance

Class	No. of known members	Biological significance
Anthocyanin(s)	250	red and blue pigments
Chalcons	60	yellow pigments
Aurones	20	yellow pigments
Flavones	350	cream-coloured pigments of flowers
Flavonols	350	feeding repellents (?) in leaves
Dihydrochalcons	10	some taste bitter
Proanthocyanidins	50	astringent substances
Catechins	40	some have properties like those of tannins
Biflavonoids	65	anti-inflammatory activity
Isoflavonoids	15	oestrogen effect, toxic for fungi

The basic structure of flavonoids is derived from the C_{15} body of **flavone**. They differ from other phenolic substances in the degree of oxidation of their central pyran ring and also in their biological properties. Flavone is the generic name of a group of compounds produced from a skeleton of three ring systems called the flavone nucleus, with two side aromatic rings and a central oxygenated heterocycle. While some classes (the flavonones, for example) are colourless, the members of other classes (the anthocyanes, for example) are always coloured and known as pigments of flowers or other plant parts. Anthocyanes are normally red or yellow; their colour is pH-dependent.

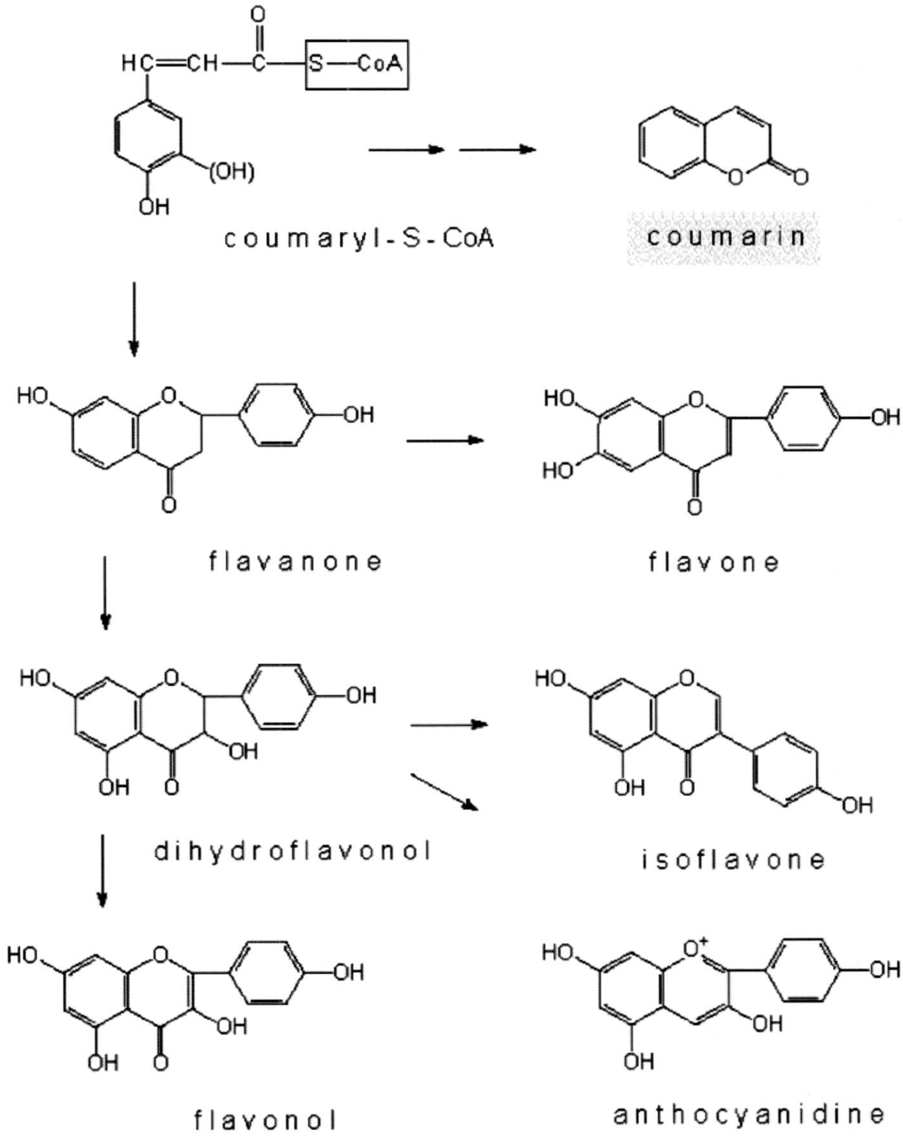

Fig. 2 : Structure of Flavonoids.

The variability of the flavonoids is largely based on the hydroxylation and methylation pattern of the three ring systems. They have therefore proven to be suitable traits for the study of the phylogenetic relations between higher plants. The quinones are another group of phenolic compounds. Some of its members function as co-factors. Accordingly, they do not actually belong to the secondary plant products but have to be counted among those of the basic metabolism. As has been mentioned before, phenolic compounds occur usually not unbound within plant tissues. They are mostly coupled to other molecules, often to glucosyl residues, but to sulphate- or acetyl-residues, too. One of the reasons may be that they are toxic when in a free state and are detoxified, at least partially, if coupled. Many low molecular weight compounds, for example thymol, are used in medicine as antiseptics due to their toxicity. Different types of bonds between flavonoids and a glycosyl residue lead to different derivatives that increase the range of flower colours. The glycosylation of flavonoids has an additional, ecologically important function that is connected with pest protection and protection against other animals. The function or importance of these compounds to the organism's development is usually of ecological nature as they are used as defense against predators for interspecies competition, and to facilitate the reproductive processes.

3.3 Terpenoids

In general the term 'terpene' is used to denote the compounds containing an intergral number of 5C units whether or not they contain other elements such as oxygen. Terpenoinds are a class of compounds constructed from 5-carbon building blocks, called **isoprene units** whose structural formula is,

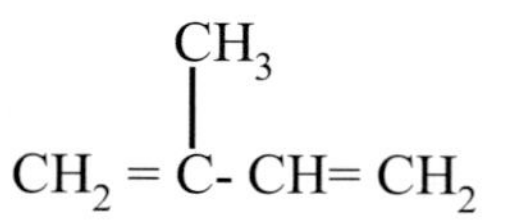

Isoprene units are biologically inert since the terpenoid biosynthetic enzymes are unable to recognize and use them for the synthesis of terpenoids. The biologically functional form of isoprene unit is **isopentenyl pyrophosphate**,

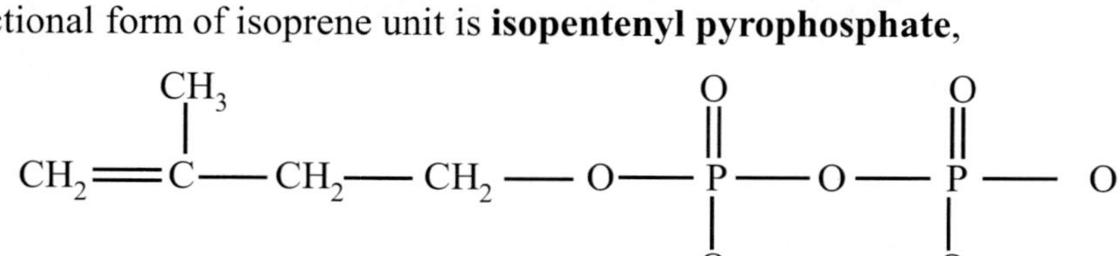

The terpenoids, are also known as isoprenoids because of the involvement of isoprene in their synthesis are classified according to the number of isoprene units they contain: hemiterpenes, monoteroenes, sesquiterpenes, diterpenes, triterpenes, tetraterpenes and polyterpenes which are respectively made up of one, two, three, four, six, eight and more than eight 5-carbon units.

Table 4 : Main classes of alkaloid precursors and derivatives (the example number refers to the corresponding alkaloid class).

Precursors	Alkaloid classes	Examples
Ornithine 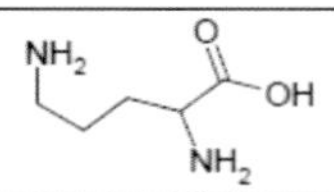	Pyrrolidine[1], Tropane, Pyrrozidine, Polyamines[2]	1. Cocaine, Atropine 2. Putrescine, Spermine, Spermidine
Leucine 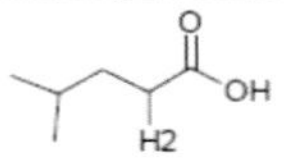	Pyrrole	
Lysine 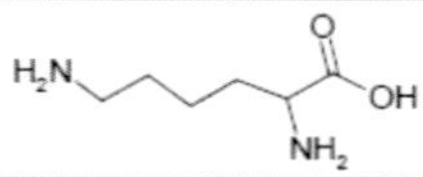	Pyperidine, Quinolizidine, Indolizidine	
Tyrosine 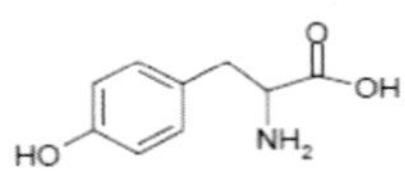	Isoquinoline Tetrahydroisoquinoline, Benzyltetrahydroiso-quinoline[1], Catechol-amines[2]	1. Morphine, Curarines, Papaverine 2. Noradrenaline, Adrenaline
Tryptophan 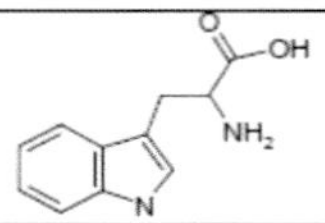	Indole[1], Carbolines, Quinoline[2], Pyrrolindole, Indolamines[3]	1. Vindoline,Catharantine 2. Quinine, Capthotecin 3. Melatonin, Serotonin
Histidine 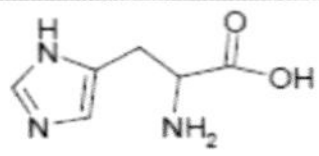	Imidazole	
Phenylalanine 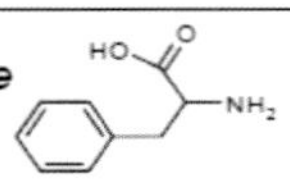		Ephedrine, Capsaicin
Anthranilic acid 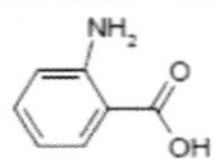	Quinazoline, Qunoline,Acridine	
Purine 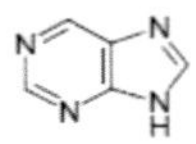		Theobromine, Theophylline, Caffeine
Geranylgeranyldiphosphate 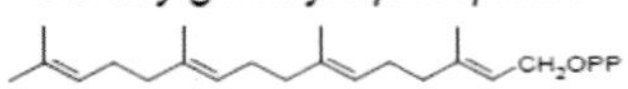	Terpenoidic	
Cholesterol 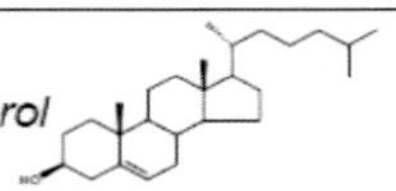	Steroidal	Solanin
Acetate 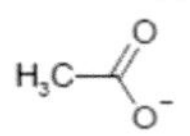	Piperidine	
Nicotinic acid 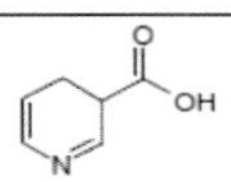	Pyridine	Nicotine

Isoprenoids (or terpenoids) represent the chemically and functionally most diversified class of low molecular mass lipids in plants, both primary and secondary metabolites. They include electron carriers (quinones), membrane constituents (sterols), vitamins (A, D, E and K), plant hormones (side chain of cytokinins, abscisic acid, gibberellins and brassinosteroids), photosynthetic pigments (chlorophyll, phytol and carotenoids) and essential oils.

3.4 Alkaloids

Many plants contain aromatic compounds called alkaloids. Chemically, they usually contain nitrogen in a heterocyclic ring of variable structure. This nitrogen frequently acts as a base, so many alkaloids are basic.

Alkaloids constitute a large number of phytochemicals of toxicological, pharmacological, nutritional and cosmetic interest. They are also of ecological importance for plants e.g. tropane alkaloids include cocaine and atropine, nicotine is a pyridine alkaloid, noradrenaline (or norepinephrine), adrenaline (epinephrine), papaverine, curarines and morphine arise from tyrosine. Melatonin and serotonin are indolamines, vindoline, catharantine (and their derivatives vincristine and vinblastine) are indole alkaloids, quinine and capthotecin are quinoline alkaloids and lysergic acid diethylamide (LSD) is an ergot alkaloid, all these arising from tryptophan. Histamine is an imidazole alkaloid, ephedrine and capsaicin derive from phenylalanine, solanin is a steroidal glycoalkaloid from cholesterol. Finally, purine alkaloids include theophylline, theobromine and caffeine, found in tea, cacao and coffee, respectively.

4. Biosynthesis of Secondary Metabolites

Different classes of secondary metabolites are synthesized by different metabolic pathways. Some of the important pathways by which these metabolites are synthesized in plants are given below.

4.1. Phenylpropanoid Pathway

The starting product of the biosynthesis of most phenolic compounds is shikimate. Among the phenylpropanol derivatives of lower molecular weight are the coumarins, cinnamic acid, sinapinic acid, the coniferyl alcohols and others. These substances and their derivatives are at the same time intermediates of the biosynthesis of lignin. Phenylpropanoids are a class of phenylalanine derivatives with a basic C6-C3 (phenyl-propane) skeleton (Fig. 3). In turn, the essential amino acid phenylalanine arises from the shikimate pathway, as well as the other aromatic amino acids tyrosine and tryptophan. Precursors of this pathway are phosphoenolpyruvate, from glycolysis, and erythrose 4-phosphate from pentose phosphate pathway, leading to two important intermediates, shikimic and chorismic acid. In further steps, after a branch point, phenylalanine and tyrosine are synthesized from prephenic and arogenic acid, whereas tryptophan from anthranilic acid.

The removal of an amino group from phenylalanine (deamination) via the step catalyzed by the enzyme phenylalanine ammonia-lyase (PAL), leads to cinnamic acid and, in turn the precursor of hydroxycinnamates after a series of hydroxylation of the benzene ring (Fig. 4). These compounds, including coumaric, ferulic and sinapic acids are reduced to the corresponding alcohols via aldehyde intermediates, namely coumaryl, coniferyl and sinapyl alcohol, collectively termed monolignols.

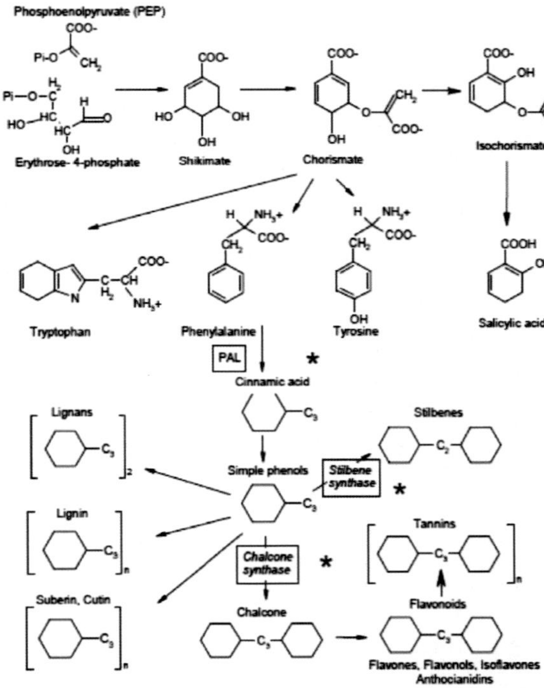

Fig. 3 : Aromatic amino acid biosynthesis from shikimate and phenylalanine derivatives of the phenylpropanoid pathway (Marcello and Franco, 2009).

Dimerization or polymerization of monolignols leads to lignans and lignin, respectively (Fig. 4). Lignification is a complex reaction in which peroxidases catalyze the polymerization of lignin units, consuming H_2O_2. Apposition of lignin in plant cell wall is a process occurring during the development of particular tissues, as well as in plant defence response, in order to strengthen the cell wall and protect the plasmalemma. Benzoic and hydroxybenzoic acids (C6-C1), such as salicylic acid (SA), a molecule importantly involved in systemic acquired resistance (SAR), represent another group of cinnamic acid derivatives, formed by cleavage of a C2 fragment from the phenylpropane structure (Fig. 4). Salicylic acid can also be synthesized directly from chorismateisochorismate by isochorismate synthase.

The metabolites mentioned till now can be collectively named simple phenols, to differentiate them from polyphenolic compounds. The latter have an additional benzene ring, arising from the cyclization of three malonic acid (C2) residues via stilbene synthase (STS) or chalcone synthase (CHS), to give stilbenes (C6-C2-C6) or flavonoids (C6-C3-C6), respectively (Fig. 3). Finally proanthocyanidins (PA), or condensed tannins, include oligo- and polymeric flavonoid derivatives (from flavanols catechins), with polymerisation degree ranging from 2 to 17 and more (Fig. 3). The numerous hydroxyl groups available in these molecules promote the formation of complexes with macromolecules, such as proteins and polysaccharides or metal ions.

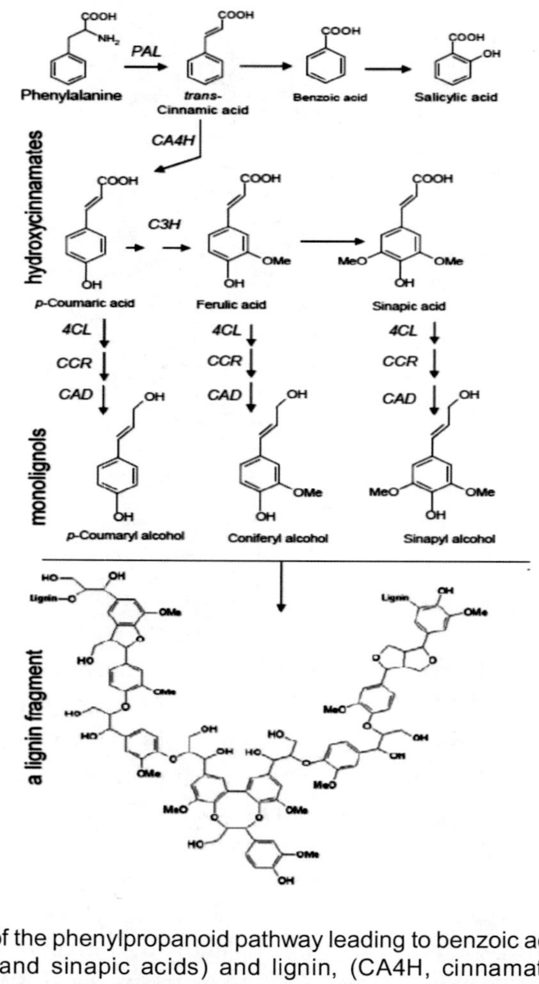

Fig. 4 : Main steps of the phenylpropanoid pathway leading to benzoic acids, hydroxycinnamates (coumaric, ferulic and sinapic acids) and lignin, (CA4H, cinnamate-4-hydroxylase; C3H, coumarate-3-hydroxylase; CAD, cinnamyl alcohol dehydro-genase CCR, cinnamoyl: CoA re-ductase; CL, 4-Coumarate: CoA ligase; PAL, phenylalanine ammonia-lyase ((Marcello and Franco, 2009).

4.2 Isoprenoid Pathway

Acetyl coenzyme A (CoA) represents the precursor for the isoprenoid biosynthesis (Fig. 5). Firstly, two molecules of acetyl CoA react to give acetoacetyl CoA and, then, with a further acetylCoA to produce â-hydroxy-â-methylglutaryl (HMG)-CoA. In plants, the same enzyme, HMG-CoA synthase, catalyses both reactions. The conversion of HMG-CoA into mevalonate, via HMG-CoA reductase, is the rate limiting enzyme of this pathway. Mevalonate kinase and mevalonate phosphate kinase phosphorylate, respectively, mevalonate and, then, mevalonate 5-phosphate, yielding mevalonate 5-diphosphate. Afterwards, mevalonate diphosphate decarboxylation, via mevalonate diphosphate decarboxylase, produces isopentenyl diphosphate (IPP), the five-carbon building block for the formation of isoprenoid chains. The enzyme IPP: dimethylallyl-PP isomerase converts IPP into dimethylallyl diphosphate (DMAPP), the acceptor for successive transfers of isopentenyl residues.

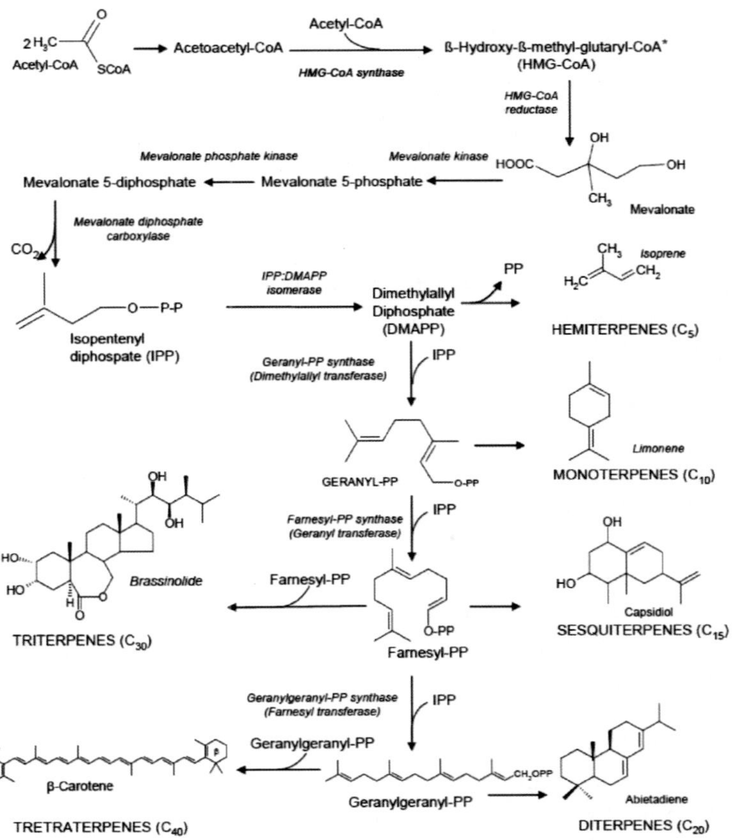

Fig. 5 : Isoprenoid pathway from acetyl-CoA (Marcello and Franco, 2009)

Hemiterpenes (C5), such as isoprene, originate from dimethylallyl-PP, upon the release of diphosphate. Differently, dimethylallyl-PP can condense with IPP, to form geranyl-PP via geranyl-PP synthase. In the same way, further chain elongation is attained by head to tail condensation of geranyl-PP to IPP, to produce farnesyl-PP via farnesyl-PP synthase. Analogously, geranylgeranyl-PP synthase catalyses the head to tail condensation of farnesyl-PP to IPP, thus yielding geranylgeranyl-PP.

Geranyl-PP is the precursor for the formation of monoterpenes (C10) or essential oils, including highly volatile open chain and cyclic compounds, such as menthol, limonene, geraniol, linalool and pinene. They are active in plant-microbe, plant-phytophagous and plant-plant interactions, due to their attractiveness or repulsiveness. Farnesyl-PP is the precursor for the synthesis of open chain and cyclic sesquiterpenes (C15), the largest group of isoprenoids, including essential oils and antibiotic compounds (phytoalexins).

Diterpenes (C20) derive from geranylgeranyl-PP, consisting of phytoalexins, plant hormones, the phytol side chain of chlorophylls, tocopherols and phylloquinone [18]. Additionally, triterpenes (C30) are synthesized from two molecules of farnesyl-PP (C15), by a reductive head to head condensation. The triterpene squalene is the precursor for sterols, important membrane constituents, via squalene synthase. Analogously, head to head condensation of two molecule of geranylgeranyl-PP (C20) leads to tetratepenes (C40), such as the carotenoids (carotene, lycopene) and xanthophylls (lutein, zeaxanthin, violaxanthin). Besides, isoprenoids are involved in protein prenylation, that is the synthesis of variously lengthened isoprenoid chains anchoring proteins in membranes, such as G proteins, ubiquinone, plastoquinone and cytochrome-a. Finally, natural rubber is a polyterpene, composed of over 1,000 isoprene units and deriving from polymerisation of geranylgeranyl-PP units.

4.3 Alkaloid Pathway

Alkaloids are a group of nitrogen-containing bases. Most of them are drugs. Only a few (like caffeine) are derived from purines or pyrimidines, while the large majority is produced from amino acids. A common pathway for alkaloid biosynthesis does not exist. A majority of alkaloids are amino acid derivatives grouped on the basis of their precursor and chemical structure. Therefore, the main groups include alkaloids arising from ornithine, leucine, lysine, tyrosine, tryptophan, histidine and phenylalanine, in addition to alkaloids arising from nicotinic (pyridine alkaloids) and anthranilic acid, acetate, isoprenoids and purine (Table 4). In turn, these classes include some minor divisions. Accordingly, pyrrole alkaloids arise from leucine; pyrrolidine, tropane and pyrrolizidine alkaloids from ornithine; piperidine, quinolizidine and indolizidine alkaloids derive from lysine; catecholamines, isoquinoline, tetrahydroisoquinoline and benzyltetrahydroisoquinoline alkaloids originate from tyrosine; indolamines, indole, carboline, quinoline, pyrrolindole and ergot alkaloids come from tryptophan and imidazole alkaloids from histidine. Anthranilic acid is the precursor of quinazoline, quinoline and acridine alkaloids, whereas isoprenoid alkaloids include mono- (geraniol), di- (geranylgeranyl-PP) and triterpene (cholesterol) derivatives (Figure 5).

4.4 Other Pathways for Synthesis of Secondary Metabolites.

Some secondary compounds are not only found in phylogenetically related species but also in species that are not directly linked. A wide range of metabolic compounds have developed in parallel in non-related species and are produced by different biosynthetic pathways.

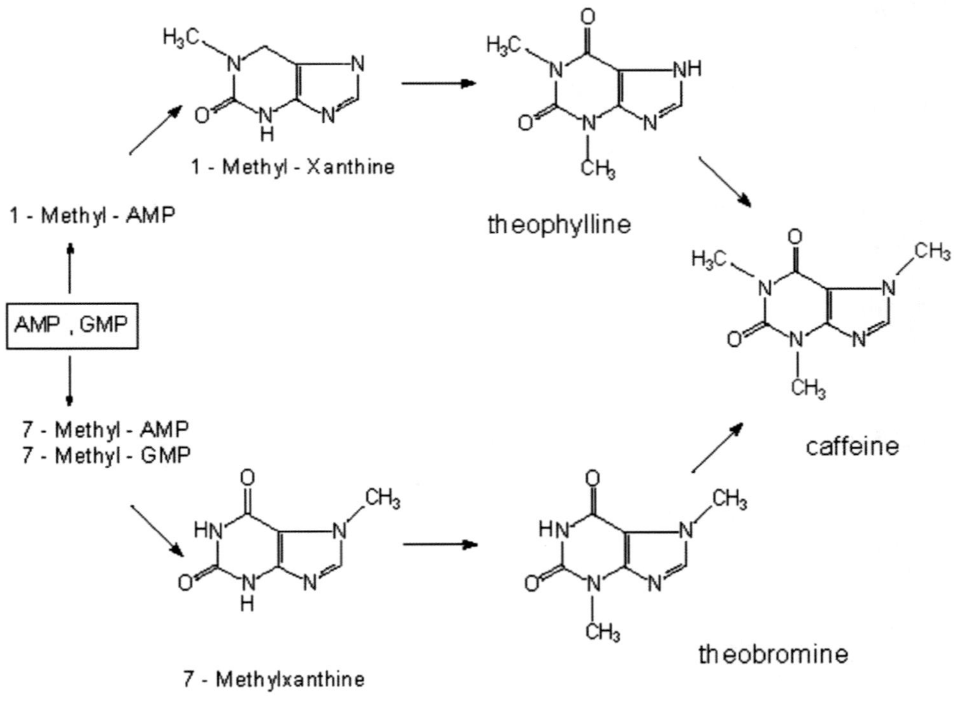

Fig.6 : Two different pathways for the synthesis of caffeine.

5. Fungal Secondary Metabolites

Filamentous fungi produce a diverse array of secondary metabolites that are not necessary for normal growth or development. Some secondary metabolites are exploited for their antibiotic and pharmaceutical activities, others are involved in disease interactions with plants or animals. The availability of fungal genome sequences has led to an enhanced effort at identifying biosynthetic genes for these molecules. However, the role of secondary metabolites in the fungi that produce them remains a mystery. Many of these fungi live saprophytically in the soil and such molecules may provide protection against other inhabitants in this ecological niche and interest in these compounds is considerable, as many natural products are of medical, industrial and agricultural importance. Some natural products are deleterious, while others are beneficial to humankind. Classes of fungal secondary metabolites include polyketides (e.g. aflatoxin and fumonisins), non-ribosomal peptides (e.g. sirodesmin, peramine and siderophores such as ferricrocin), terpenes (e.g. T-2 toxin, deoxynivalenol (DON)), indole terpenes (e.g. paxilline and lolitrems). There are a number of well known fungal secondary metabolites such as, penicillin, cephalosporins, ergotrate and statins etc. Some important group of fungal secondary metabolites are given below.

5.1. Aflotoxins

Aflatoxins are produced in members of the *Aspergillus parasiticus* group via the polyketide pathway. The pathway has around 20 steps, and the end products include a diversity of related compounds (bisfuranocoumarins) that can be readily converted one to another. Aflatoxin B1 is one of the most toxic compounds known. The toxin is formed commonly in plant materials held at relatively high moisture and temperature for long periods. Peanuts, corn and cotton are readily contaminated in the field. *The Aspergillus parasiticus* groups of fungi are common in soil. The fungi can colonise roots and spread through the plant. When harvested, contaminated seed will become toxic if not dried immediately and held in a dry form.

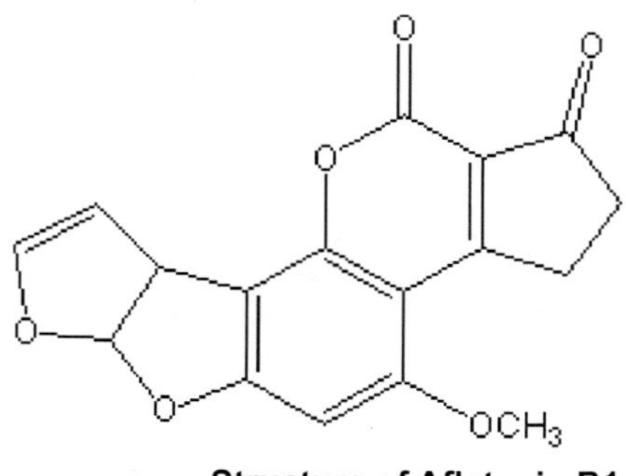

Structure of Aflatoxin B1

Aflatoxins are toxic and carcinogenic. At lower levels and following prolonged exposure, the toxins cause liver cancer in humans. Aflatoxin B1 is converted to Aflatoxin M (in milk) on passage through cows. Though less toxic, it does illustrate the potential damage caused by consumption of contaminated product. The function of aflatoxins in fungi is unknown. Toxicosis only takes place after consumption of the contaminated plant material. Animals tend to avoid contaminated feed, but as B1 is so highly toxic, even large animals can be killed by small, almost undetectable quantities.

5.2. Patulin

Patulin is a polyketide antibiotic synthesised on an acetate pathway. Its biosynthetic pathway is still unclear, but it appears that several alternate pathways may result in the same end product. Further, several related compounds may be produced from minor variations of the cultural conditions.

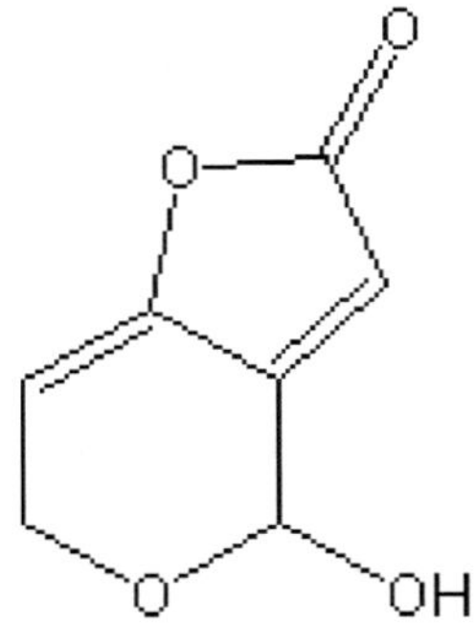

Structure of Patulin

Patulin was first thought to be a potential antibiotic. However, it is now also known to be a compound produced by *Penicillium expansum* in contaminated apples. Patulin is associated with storage rot of apples, and it is toxic to mammals. The role of patulin in rotting apples seems unclear, but it may be a general inhibitor of competitors, including mammals and microbes. However, the life cycle of apples tends to be decades, that of the fungus much shorter. The disparity in the length of the life cycle between organisms associated with fungi and the fungus is common.

5.3. Aromatic Compounds

Cyclic compounds can be synthesised via the polyketide or shikimic acid pathways. Zearalenone is one interesting example from this group. The compound regulates perithecium formation in the fungus. It also has an oestrogenic effect in mammals.

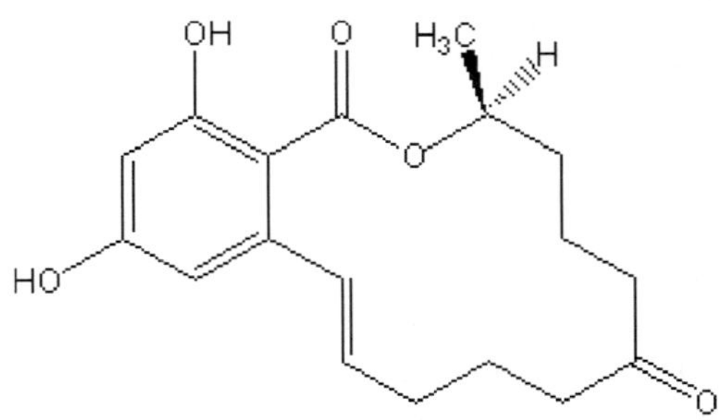

Structure of Zearalenone

The reproductive behaviour of pigs consuming feed contaminated with *Fusarium graminearum* may be modified. Consumption of high doses of zearalenone is commonly detrimental to pigs, but the compound has also been used deliberately at low doses as a growth supplement in sheep and cattle.

F. graminearum is a common hyphomycete found in soil. Some isolates are widely recognised as important plant pathogens in temperate and warm temperate climates. Thus contamination of feeds is likely to be widespread where there is poor control over humidity during storage. The main function of the compound appears to be its role in perithecium formation. The effect on mammals appears to be entirely coincidental.

5.4. Structure of Ergotamine, LSD and Ergometrine

Ergot alkaloids are synthesized from several pathways. Trypotophan from the shikimic acid pathway is attached to an isoprenoid moiety from the mevalonate pathway, and several amino acids from primary metabolism are added depending on the final product. Ergot alkaloids are produced as a complex mixture of related compounds from a branched pathway.

The activity of the alkaloids is as varied as the compounds. Their function in the fungus is unclear, but theories about coevolution of the fungi and their grass hosts are based on the benefit of the toxins in selecting grasses that deter herbivory. The role of the toxins in the fungi are, otherwise, almost inexplicable.

Given the importance of amino acids in the synthesis of alkaloids, it follows that nitrogen nutrition of the fungi is important. Carbon balance, carbon source, availability of precursors, and repression by excess of regulators such as ammonium are important both in industry (ergotamine) and in colonised grasses. However, the regulation of other secondary metabolites is as varied as their structures.

5.5. Plant Growth Regulators

Many pathogenic and benign fungi produce auxin, cytokinins, gibberellins and abscisic acid. In fact the gibberlellins were first found in the fungus *Gibberella fujikuroi*, a pathogen that causes tall, straggly growth of rice. The gibberellins are diterpenes produced by the mevalonate/ isoprenoid pathway. The functions of these compounds in fungi that colonise plants seem clear. Modification of host tissue enhances colonisation, releases nutrients for fungal metabolism and regulates host reproduction. The function of plant growth regulation in fungi found outside plants is unclear.

Some plant fungi appear to modify host production of growth regulators, resulting in alterations of host metabolism. For instance, initiation of AM in roots results in a slowing of root tip elongation and increase in lateral formation. The cause may be associated with a change in concentrations of auxins and cytokinins from the fungus or induced in the host, an increase in local concentration of phosphate due to the fungi, or a factor influenced by either.Increases in expression of plant hormones may be direct or indirect. Resolution of the identity and importance of the fungal compounds remains.

5.6. Toxins

In general, toxins associated with fruiting bodies are important because consumption of the fruiting body can result in poisoning. The principles of toxicosis are the same, though the topics of toxin fungi and toxic foods are commonly separated. In addition, toxins are produced from a myriad of pathways, and have enormously diverse effects. They may be produced at a different point in the life cycle of a fungus.

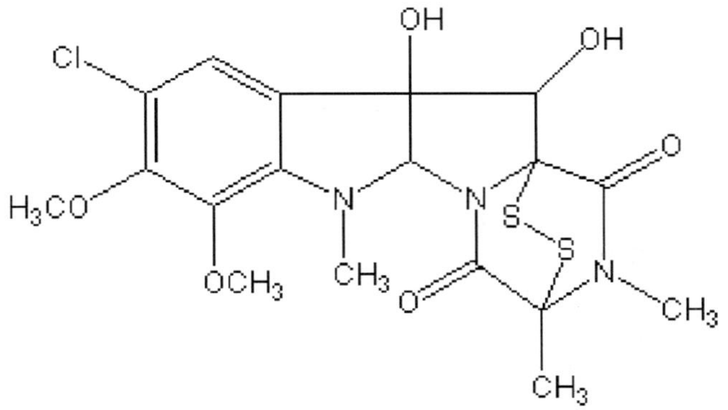

Structure of Sporodesmin

The function of toxins to fungi has been subject of much speculation. Colonisation (contamination) of organic materials is a prelude to the digestion of the material by the fungus. The production and expression of toxins is one mechanism the microbe has to protect the food, provided competitors detect the presence of the microbe and toxin. The conditioned response to the fungus thus reduces the consumption of the fungal substrate. The "detection" molecule may be other than the toxin.

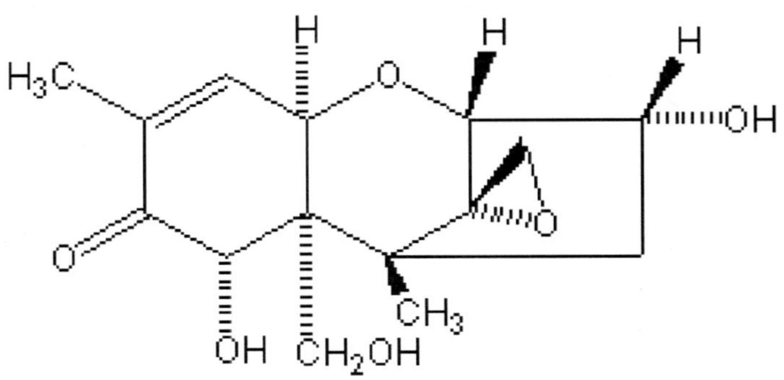

Structure of Vomitoxin

The plant defense theory predicts that plants allocate defenses only if they provide improved fitness. The presence of fungi in plants has been suggested to be due to their production of defense and deterrence compounds. Colonised plants would thus be relatively protected from herbivory and produce more offspring in the next generation. The fungi would be provided with a home and nutrients, and in a position to colonise the subsequent generation. Further, fungi are able to evolve new and more appropriate deterrents more rapidly than plants being under greater selection pressure, and having shorter generation times. The result is suggested to explain the relationship between grasses, which have few defenses, and the Balansioid fungi.

6. Engineering Secondary Metabolite Production in Plants

Genetic engineering of a secondary metabolic pathway aims to either increase or decrease the quantity of a certain compound or group of compounds (Facchini and Dixon, 2001). Recent achievements have been made in the metabolic engineering of plant secondary metabolism. Various pathways have been altered using genes encoding biosynthetic enzymes or genes encoding regulatory proteins. Other secondary metabolite pathways that have extensively been studied at the level of intermediates and enzymes mainly lead to pharmaceutically important products such as indole and isoquinoline alkaloids (Facchini and Verpoorte, 2000). Presently, a major constraint in engineering plant secondary metabolite production is that only a few genes of these pathways are known.

6.1. Metabolic Engineering of Plant Volatiles

Plants produce an amazing diversity of low molecular weight organic compounds known as secondary or specialized metabolites (Pichersky and Gang, 2000). More than 1% of these metabolites are lipophilic molecules with low boiling points and high vapor pressures at ambient temperature. They are mainly represented by terpenoids, phenylpropanoids or benzenoids, fatty acid derivatives and amino acid derivatives. These volatile compounds are released from leaves, flowers and fruits into the atmosphere and from roots into the soil. The contribution of volatiles to plant survival and overall reproductive success in natural ecosystems, and their impact on agronomic and other commercial traits, including yield and food quality, suggest that modification of volatile production through genetic engineering has the potential to improve cultivated plant species. Metabolic engineering requires a basic understanding of the biochemical pathways and the identification of the genes and enzymes involved in the synthesis of volatile compounds.

Volatile organic compounds (VOCs) are produced by a broad range of organisms, from bacteria to mammals, and they represent a vast chemical diversity. In plants, one of the preeminent roles of VOCs is their repellent or cytotoxic activity, which helps the plant deter its predators. Most studies on VOCs emitted by vegetative parts have been conducted in model plant species, and little is known about patterns of VOC emissions in diverse plant communities. Within the diverse array of naturally synthesized chemicals, volatile organic compounds (VOCs) are of interest. First, VOCs encompass a broad range of compounds that result from at least three distinct biosynthetic pathways: lipoxygenase, shikimic acid, and terpenoid pathways (Dudareva et al. 2006). Second, they represent one of the most diverse classes of secondary metabolites, with over 2,500 monoterpenes and 5,000 sesquiterpenes described so far (Wink 2006). The defensive function of VOCs has been studied intensively in some model plant species (Turlings and Wäckers, 2004) but not much is known about importance in complex plant communities such as tropical forests.

7. Biotechnological Approach for Production of Secondary Metabolites

Biotechnology offers an opportunity to exploit the cell, tissue, organ or entire organism by growing them in vitro and to genetically manipulate them to get desired compounds. The improvements in molecular biological research have given a new dimension to in vitro culture as well as for plant improvement, enhancing the yields of the product and resulting in multiple products or producing novel products from genetically engineered plants. There can be a number of biotechnological approaches for the production of plant derived secondary metabolites:

i. Plant Cell Culture as a Source of Secondary Metabolites

Plant cell cultures are an attractive alternative source to whole plant for the production of high-value secondary metabolites (Ravishankar et al., 1999; Dornenburg and Knorr, 1997). producing higher amount of secondary metabolites than in intact plants. There has been tremendous success in the production of high-value secondary metabolites such as shikonin from cell cultures of *Lithospermum erythrorhizon*, berberine from *Coptis japonica* and sanguinarine from *Papaver somniferum*. These are produced at industrial levels (table 5). However, there are still problems in the production of metabolites by cell cultures resulting from the instability of cell lines, low yields, slow growth and scale-up problems (Ravishankar and Venkataraman, 1993).

Table 5: High yields of secondary products (Ravishankar and Ramachandra Rao, 2000)

Product	Plant species	Yield (%D.W.)	Reference
Rosmarinic acid	*Saponaria officinalis*	36.0	Hippolyte et al. (1992)
Rosmarinic acid	*Coleus blumei*	21.4	Ulbrich et al. (1985)
Anthroquinones	*Morinda citrifolia*	18.0	Zenk et al. (1975)
Shikonin	*Lithospermum erythrorhizon*	12.4	Fujita (1988)
Berberine	*Thalictrum minus*	10.6	Kobayashi et al. (1988)
Jatrorhizine	*Berberis wilsonae*	10.0	Breuling et al. (1985)
Anthocyanins	*Perilla frutescens*	8.9	Zhong et al. (1994)
Berberine	*Coptis japonica*	7.5	Matsubara et al. (1989)
Diosgenin	*Dioscorea deltoidea*	3.8	Sahai and Knuth (1985)
Sanguinarine	*Papaver somniferum*	2.5	Park et al. (1992)
Serpentine	*Catharanthus roseus*	2.2	Zenk et al. (1977)

ii. Organ Cultures as a Source of Pharmaceuticals

Since production of secondary metabolites is generally higher in differentiated tissues, there are attempts to cultivate shoot cultures and root cultures for the production of medicinally important compounds.

iii. Immobilization of Plant Cells for the Production of Secondary Metabolites

Improvement in the secondary metabolite production of cell cultures is often associated with the organization and differentiation of plant cells. The concept of organization and differentiation led to the use of immobilization technology, which has long been used for microbes and enzymes.

iv. Biotransformations for the Production of Secondary Metabolites

Biotransformation can be defined as a process through which the functional groups of organic compounds are modified either stereo- or region specifically by living cultures, entrapped enzymes or permeabilized cells to a chemically different product.

v. Metabolic Engineering of Biochemical Pathways

It is envisaged that the use of plant cell culture approach for pharmaceutically important compounds can be made industrially applicable only upon the overproduction of metabolites or the production of novel compounds that too with the expression of more than one compounds of interest in a plant. This can be achieved by metabolic engineering.

8. Research Developments

The research developments which have taken place in the secondary metabolite production can be discussed considering the following aspects.

8.1 Mechanistic Aspects

In the first half of the 20th century proposals about the formation of secondary metabolites in plants were based mainly on analogy with reactions of organic chemistry and on comparison of structures rather than on biochemical evidence (Robinson, 1981). A number of important biosynthetic pathways were characterized at an enzymatic level in the 1970s and 1980s by the use of cell suspension cultures. Already in 1970, Grisebach and Hahlbrock had begun their successful research on the enzymatic characterization of the biosynthetic pathways of the general phenylpropanoids, as well as the branches leading to flavonoids (Hahlbrock, 1981) and lignins (Grisebach, 1981). Despite the highly successful application of cell cultures in biosynthetic research there are classes of secondary products that are not, or less efficiently, produced in dedifferentiated cell cultures. In the mid 1980s the state-of-the-art in biochemistry of plant secondary metabolism could be summarized as follows: secondary products are de novo synthesized from simple precursors of primary metabolism through an elaborated sequence of reactions catalyzed by specific enzymes. Metabolic channeling by organization of biosynthetic enzyme in metabolons (multienzyme complexes) appears to be a common feature of complex biosynthetic pathways of secondary metabolism. Examples are known from phenylpropanoids, terpenoids and alkaloids (Jorgensen et al., 2005). Two other general features of enzymes of secondary metabolism need to be mentioned: (i) they are often found in low concentrations, i.e. specific activities in crude protein extracts are often in the picokat-per-milligram-protein range, whereas those of enzymes of primary metabolism are generally three to five orders of magnitude higher and (ii) they lack feedback regulation as known from biosynthetic pathways of primary metabolism, e.g. amino acids and nucleotides. In the mid 1980s, the first genes encoding enzymes of plant secondary metabolism were cloned and functionally expressed.

8.2 Functional Aspects

Until the 1980s functional aspects of plant secondary metabolism were largely neglected by phytochemists and plant physiologists. In the 1950s, secondary metabolites were regarded as metabolic waste or detoxification products (Peach, 1950; Reznik, 1960). This view changed in the 1970s with the increasing biochemical knowledge of secondary metabolism. Secondary metabolites were now no longer considered as inert end products but as dynamic components of plant metabolism (Barz and Köster, 1980). Secondary metabolism comprises all interactions of plants with their biotic and abiotic environment. Since secondary metabolism is dispensable for growth and development, its components can be continuously modified and adapted to the demands of a continuously changing environmental selection pressure. Recently it was demonstrated that even flavonoids, which are generally regarded as typical secondary compounds with ecochemical functions, may possess primary functions. Flavonoids clearly affect auxin transport and developmental processes in *Arabidopsis* (Buer and Muday, 2004; Taylor and Grotewold, 2005), although it is not yet clear which flavonoids are specifically responsible for these effects.

Secondary metabolism comprises all interactions of plants with their biotic and abiotic environment. The inherent feature of secondary metabolism is its high genetic plasticity and diversity that guarantees flexible adaptations of plants to the demands of their continuously-changing environment. Herbivores or pathogens that during their co-evolutionary adaptations learned to cope with a certain defense barrier may now, as specialists, use a defense compound as a positive signal to recognize the host plant to which they are often socialized.

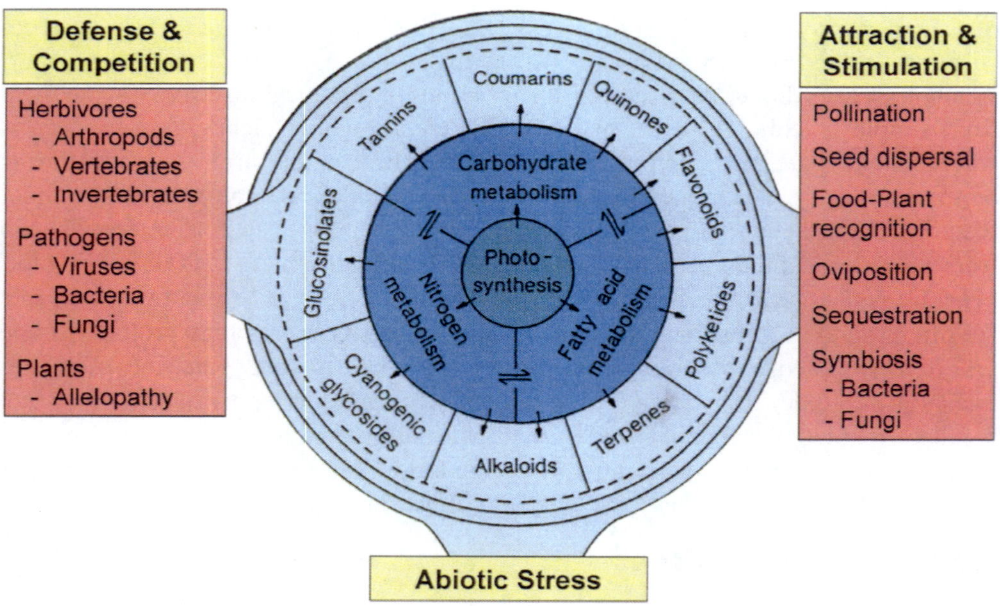

Fig. 7 : Ecological functions of plant secondary metabolism (Hartmann, 2007)

Plant metabolism exists at two functional levels. Primary metabolism covers all processes essential for growth and development whereas secondary metabolism is indispensable for the survival of the individual in its environment. Since secondary metabolism is dispensable for growth and development, its components can be continuously modified and adapted to the demands of a continuously changing environmental selection pressure. The two levels of plant metabolism:

Primary metabolism: Genes with high stringency controlling essential functions

1. Covers growth and development of the individual
2. Indispensable
3. Universal
4. Uniform,
5. Conservative

Secondary metabolism: Genes with high plasticity controlling functions that are under selection pressure of a continuously changing environment.

1. Covers interactions of the individual with its environment
2. Dispensable for growth and development
3. Indispensable for survival in the environment
4. Unique
5. Diverse
6. Adaptive

Sometimes a metabolite may include both primary and secondary functions. This is no contradiction and is usually easily explained through the evolutionary background. During evolution of the plant kingdom some secondary metabolites have been recruited for primary functions or attained both primary and secondary functions.

Compounds of Secondary Origin that Attained Primary Function

Diterpenoids	Gibberellines (phytohormone)
Sequiterpenoids	Abscisic acid (phytohormone)
Triterpenoids	Brassinosteroids (phytohormone)
Tertraterpenoids	Carotenoids, xanthophylls (photoprotection)
Flavonoids	Certain flavonoids (developmental regulators)
Benzoates	Salicylate (stress signal)

Compounds Which Attained Both Primary and Secondary Functions

Lignin	Cell wall strengthening and chemical defense
Canavanine	Chemical defense and seed nitrogen storage

Plants have evolved Three Major Strategies for Defense Against Pathogens and Herbivores:

(1) Constitutive accumulation of defense compounds, that is metabolites that are stored in the target tissue in concentrations needed to fulfill their functions. Most of the natural products that are isolated in substantial amounts from plants belong to this category.

(2) Constitutive accumulation of preformed defense compounds and 'ignition enzymes' well separated by spatial compartmentation. Upon de-compartmentation, e.g. by herbivore attack the 'ignition enzyme' liberates the active defense compound. Examples are the liberation of HCN from cyanogenic glycosides; the formation of mustard oils from the glucosinolates; the less-recognized formation of toxic monodesmosidic saponines from non-toxic bidesmosidic saponins stored in plant vacuoles; formation of coumarine from the respective ortho-glucoside; the release of vanilline from vanilloside in Vanilla fruits.

(3) Induced formation of defense compounds in response to microbial or herbivore attack. Classical examples are the phytoalexins as plant defense compounds that are specifically produced upon pathogen attack.

Whereas with defense strategies of (2) and (3) the causal principle is easily recognized, the efficiency of a constitutive defense can only be judged by the absence or presence of potential attackers.

8.3. Evolutionary Aspects

In the past about 10 years research, on plant secondary metabolism reached the gene level (Pichersky and Gang, 2000). First, fascinating insights into the creation of genetic diversity of secondary metabolism became visible and then ideas about the mechanisms of gene recruitment and diversification for novel functions arose. Gene duplication is assumed to be the major driving force for gene diversification and recruitment (Pichersky and Gang, 2000; Ober, 2005). If a gene that directs an essential function is duplicated, the duplicate is released from the stringent function of the mother gene; it may either be eliminated by inactivation (e.g., pseudogenization) or recruited for modified or new functions. There are at least two general possibilities how duplicate genes are recruited and functionalized: (i) continuous modification of its function during plant speciation, as consequence of which large gene families with rich functional diversity arise and (ii) the duplicate is recruited for a new stringent function in a novel biochemical environment, as consequence, a new single copy gene originates. Plant polyketide synthases (PKSs) (Austin and Noel, 2003) exemplifies the first possibility. Many other fascinating examples of functional diversification of genes are becoming visible including glucosyltransferases of secondary metabolism (Strack et al., 2003;

Gachon et al., 2005), O-methyltransferases of flavonoid metabolism (Ibrahim, 2005) and the enzymes catalyzing triterpene cyclization (Phillips et al., 2006).

9. Applications of Secondary Metabolites

Secondary metabolites play a major role in the adaptation of plants to the changing environment and in overcoming stress constraints. The function of secondary metabolites is as varied as their structure. Some appear to directly or indirectly benefit the plant or fungus, while the function of many others remains obscure. Secondary metabolites play an extremely important role in the functioning of fungi, providing one crucial element of their diversity. A broad array of protective functions is performed by secondary metabolites in both biotic and abiotic stress situations: antimicrobial, photoprotective, structure stabilizing, signalling. Diverse chemical characters can be determinants of various functions.

Plant secondary metabolism has multiple functions throughout the plant's life cycle. These functions can be classified as mediators in the interaction of the plant with its environment, such as plant–insect, plant–microorganism and plant–plant interactions. Moreover, several plant secondary metabolites are used for the production of medicines, dyes, insecticides, flavours and fragrances. Secondary metabolism is thus an interesting target for plant breeding. In addition, plants are a valuable source of a wide range of secondary metabolites, which are used as pharmaceuticals, agrochemicals, flavours, fragrances, colours, biopesticides and food additives.

Functions of Some Plant Secondary Metabolites:

1. Carotenoids

Carotenoids are organic pigments occurring in plants and are mostly found in red, orange and yellow fruits and vegetables. Other vegetables such as broccoli, spinach or curly kale also contain carotenoids. Carotenoids have antioxidative effects and prevent cancer. In addition to this they boost the immune system and reduce the risk of getting heart attacks.

2. Phytosterols

Phytosterols are found in plant foods such as sunflower seeds, sesame, nuts and soya beans. Phytosterols protect against colon cancer and lower cholesterol levels. Phytosterols are chemically similar to cholesterol and therefore they compete against each other for absorption in the body.

3. Saponins

Saponins are flavour additives, which are found in legumes and spinach. Saponins boost the immune system, lower the cholesterol levels in the blood and reduce the risk of getting intestinal cancer.

4. Glucosinolates

Glucosinolates are flavour additives, which are found in all types of cabbages, mustard, radish and cress. Glucosinolates prevent infections and inhibit the development of cancer.

5. Flavonoids

Flavonoids are organic pigments occurring in plants which give plants a red, violet or blue colour. Flavonoids have a particularly broad spectrum of efficacy. Flavonoids inhibit the growth of bacteria and viruses, protect the cells against the damages of free radicals, protect against cancers and heart attacks, have a repressive effect against inflammations and they influence blood coagulation.

6. Protease-inhibitors

Protease-inhibitors are found in plants that are rich in protein such as legumes, potatoes and wheat and they inhibit the decomposition of protein. Protease inhibitors protect the body against cancers and regulate the blood sugar levels.

7. Terpenes

Terpenes are plant flavours for e.g. the menthol in peppermint oil or the essential oils in herbs and spices. Terpenes decrease the risks of cancer.

8. Phytoestrogens

Phytoestrogens are natural plant hormones which are similar to the sexual hormones. Phytoestrogens are mostly found in wheat, legumes and wheat products. Phytoestrogens protect the body against hormonal dependant cancers such as breast, uterine and prostrate cancer.

9. Sulphides

Sulphides are compounds containing sulphur which are mostly found in plants that belong to the lily family such as onions, leeks, asparagus and garlic. Sulphides inhibit the growth of bacteria, lower cholesterol levels, protect the body from free radicals and have preventive effects against cancer.

10. Phytic acid

Phytic acid is found in wheat, legumes and flaxseeds. Phytic acid was considered undesirable for a long time because it binds trace elements such as iron and zinc and it also affects various digestive enzymes. However new studies have proved that phytic acid has an antioxidant effect in the large intestine.

Colour and scent performance by secondary metabolites, such as flavonoids, terpenoids, and other volatives can underlie attraction or repelling of insects and herbivores, while toxins can be involved in plant-plant allelopathic interactions. Water solubility and lack of electrical charge contributes to the compatible solutes being implicated in osmoregulation and protection of hydrophilic cellular sites, while hydrophobic molecules (carotenoids) are the best candidates to protect lipophilic surfaces such as membranes.

The apparent lack of primary function in the plant, and that many secondary metabolites have specific negative impacts on other organisms such as herbivores and pathogens, leads to the hypothesis that they have evolved because of their protective value. Many secondary metabolites are toxic or repellant to herbivores and microbes and help defend plants producing them. Production increases when a plant is attacked by herbivores or pathogens. Some compounds are released into the air when plants are attacked by insects; these compounds attract parasites and predators that kill the herbivores. Recent research is identifying more and more primary roles for these chemicals in plants as signals, antioxidants, and other functions.

Many secondary compounds have signalling functions e.g. plant hormones. They influence the activities of other cells, control their metabolic activities and co-ordinate the development of the whole plant. Other substances like flower colours serve to communicate with pollinators or protect the plants from feeding by animals or infections. Some plants, for example, produce specific phytoalexines after fungi infection that inhibits the spreading of the fungi mycelia within the plant. A number of substances are secreted and influence the existence of other species. Especially with algae but with fungi, too, the communication of the cells is stimulated or kept upright by extracellular substances. Many of them are antibiotic i.e. they inhibit the existence of competing species in the surrounding of their producer thus safeguarding its ecological niche. The mutual influence of plants by secretion is called allelopathy. It occurs not only in algae but also in higher plants.

Consuming some secondary metabolites can have severe consequences. Alkaloids can block ion channels, inhibit enzymes, or interfere with neurotransmission, producing hallucinations, loss of coordination, convulsions, vomiting and death. Some phenolics interfere with digestion, slow growth, block enzyme activity and cell division, or just taste awful. Most herbivores and plant pathogens possess mechanisms that ameliorate the impacts of plant metabolites, leading to evolutionary associations between particular groups of pests and plants. Some herbivores can store (sequester) plant toxins and gain protection against their enemies.

Secondary chemicals are important in plant use by humans. Most pharmaceuticals are based on plant chemical structures, and secondary metabolites are widely used for recreation and stimulation (the alkaloids nicotine and cocaine; the terpene cannabinol). The study of such plant use is called ethnopharmacology. Psychoactive plant chemicals are central to some religions, and flavors of secondary compounds shape our food preferences. The characteristic flavors and aroma of cabbage and relatives are caused by Nitrogen-and sulfur-containing chemicals, glucosinolates, which protect these plants from many enemies. The astringency of wine and chocolate derives from tannins. The use of spices and other seasonings developed from their combined uses as preservatives and flavorings. The nitrogen-containing alkaloids have been exploited as pharmaceuticals, stimulants, narcotics and poisons. Many of the promising pharmaceuticals are still extracted from their natural plant sources. Secondary metabolites are also being exploited in medicine and industry. Alkaloid biosynthesis has also been the target of intensive metabolic engineering, mainly to increase the amount of the target compounds but also to conduct the whole biosynthesis.

10. Future Perspectives

The mechanisms by which metabolites help the plant to resist stress remain largely unclear. While there is an increasing interest in above ground interactions, there is a strong need to study underground mechanisms and interactions of plants as well, because they have not been adequately addressed yet. The complementation of the ecological approach, with the understanding of the molecular basis of plant defense strategies, should be useful to fully comprehend the extent of the integration of these mechanisms. From the mechanistic point of view, the increasing knowledge of molecular gene information and the great impact of the genomic tools now available will advance the complete molecular characterization of secondary pathways as a base for their biotechnological application in medicinal and agricultural directions. The knowledge that secondary metabolites are functionally optimized under the ecological demands of plants (e.g., protection against animals and pathogens) should provide a valuable background. From the functional point of view, the powerful molecular tools now available provide a substantial base from which to explore the role of secondary metabolism in the diversity of plant–environment interactions. A pioneering attempt is the fusion of approaches of ecology and molecular biology by (Baldwin and his colleagues 2001), who are studying gene expression in natural populations of plants (e.g., *Nicotiana attenuata*). Microarray gene expression profiling will be a powerful tool to recognize changes of gene activity under environmental stress as, for instance, demonstrated for conifer defense upon attack of herbivorous insects (Ralph et al., 2006): among others the activities of genes of octadecanoid and ethylene signalling pathways, terpenoid biosynthesis, and phenolic metabolism were upregulated. The recent exciting first insights into the mechanisms of gene recruitment, diversification and new functionalization has brought secondary metabolism back to the forefront of biological research of biology.

Substantial progress in improving secondary metabolite production from plant cell cultures as well as economically important plants has been made within last few years. Advances in plant cell cultures could provide new means for the cost-effective, commercial production of even rare or exotic plants, their cells, and the chemicals that they produce. *In vitro* propagation of medicinal plants with enriched bioactive principles and cell culture methodologies for selective metabolite production is found to be highly useful for commercial production of medicinally important compounds. Knowledge of the biosynthetic pathways of desired compounds in plants as well as in cultures is often still vestigial, and strategies are needed to develop information based on a cellular and molecular level. New technologies will serve to extend and enhance the continued usefulness of higher plants as renewable sources of chemicals, especially medicinal compounds. Efforts in this field will lead to controllable and successful biotechnological production of specific, valuable and new plant chemicals.

Microbial secondary metabolites have been a tremendous resource for natural product drug discovery, and their usefulness will continue if the cryptic biosynthetic pathways can be accessed. The regulation of secondary metabolism in fungi is complex, involving multiple proteins and complexes that respond to various environmental and host stimuli. A greater understanding of the regulation of secondary metabolism in response to environmental stimuli will provide knowledge on the role of the product in the biology of the fungus. Future studies will uncover more information about the role of these molecules to the organisms which produce them.

References

Austin, M.B., Noel, J.P. (2003).The chalcone synthase superfamily of type III polyketide synthases. Nat. Prod. Rep. 20, 79–110.

Baldwin, I.T., Halitschke, R., Kessler, A., Schittko, U. (2001).Merging molecular and ecological approaches in plant–insect interactions. Curr. Opin. Plant Biol. 4, 351–358.

Barz, W., Koöster, J. (1980).Turnover and degradation of secondary (natural) products. In: Conn, E.E. (Ed.), the Biochemistry of Plants, vol. 7. Academic Press, New York, London, pp. 35–84.

Breuling, M., Alfermann, A.W., Reinhard, E. (1985).Cultivation of cell cultures of Berberis wisoniae in 20 L airlift bioreactors. Plant Cell Rep; 4:220–3.

Buer, C.S., Muday, G.K., (2004).The transparent testa4 mutation prevents flavonoid synthesis and alters auxin transport and the response of Arabidopsis roots to gravity and light. Plant Cell 16, 1191–1205.

Dixon, R., Paiva, N.L. (1995).Stress-induced phenylopropanoid metabolism. Plant Cell **7**, 1085.

Dixon, R.A. (2001).Natural products and plant disease resistance. Nature, 411:843-847.

Dornenburg, H., Knorr, D., (1997).Challenges and opportunities for metabolite production from plant cell and tissue cultures. Food Technol 51:47, 48, 50–52, 54.

Dudareva, N., Negre, F., Nagegodda, D. A., and Orlova, I. (2006).Plant volatiles: Recent advances and future perspectives. Crit. Rev. Plant Sci. 25: 417–440.

Edreva, A.M., Velikova V., Tsonev,T. (2007).Phenylamides in plants. Russ. J. Plant Physiol., 54, 287-301.

Facchini, P.J., (2001): Alkaloid biosynthesis in plants: biochemistry, cell biology, molecular regulation, and metabolic engineering applications. Annu Rev Plant Physiol Plant Mol Biol, 52:29-66.

Fujita, Y., (1988).Industrial production of shikonin and berberine. Applications of plant cell and tissue culture. Ciba Foundation Symposium 137. Chichester: Wileypp. 228–38.

Gachon, C.M., Langlois-Meurinne, M., Saindrenan, P. (2005).Plant secondary metabolism glycosyltransferases: the emerging functional analysis. Trends Plant Sci. 10, 542–549.

Grisebach, H. (1981).Lignins. In: Conn, E.E. (Ed.), the Biochemistry of Plants. Secondary Plant Products, vol. 7. Academic Press, New York, pp. 457–478.

Hahlbrock, K. (1981).Flavonoids. In: Conn, E.E. (Ed.), the Biochemistry of Plants. Secondary Plant Products, vol. 7. Academic Press, New York, pp. 425–456.

Hartman, T. (2007). From waste products to ecochemicals: Fifty years research of plant secondary metabolism. Phytochemistry 68:2831–2846.

Hippolyte, I., Marin, B., Baccou, J.C., Jonard, R. (1992).Growth and rosmarinic acid production in cell suspension cultures of *Salvia officinalis* L. Plant Cell Rep; 11:109–12.

Ibrahim, R.K. (2005).A forty-year journey in plant research: original contributions to flavonoid biochemistry. Can. J. Bot. 83, 433–450.

Jorgensen, K., Rasmussen, A.V., Morant, M., Nielsen, A.H., Bjarnholt, N., Zagrobelny, M., Bak, S., Møller, B.L. (2005).Metabolon formation and metabolic channeling in the biosynthesis of plant natural products. Curr. Opin. Plant Biol. 8, 280–291.

Kobayashi, Y., Fukui, H., Tabata, M. (1988).Berberine production by batch and semi-continuous cultures of immobilized Thalictrum cells in an improved bioreactor. Plant Cell Rep; 7:249–53.

Kossel, H., (1891).Über die chemische Zusammensetzung der Zelle.Arch Physiol Physiol Abt, Arch Anat Physiol: 181-186.

Marcello, I., Franco, F. (2009). Chemical Diversity and Defence Metabolism: How Plants Cope with Pathogens and Ozone Pollution. Int. J. Mol. Sci. 10, 3371-3399.

Matsubara, K., Yamada, Y., Kitani, S., Yoshioka, T., Morimoto, T., Fujita, T.(1989).High density culture Coptis japonica cells increases berberine production. J Chem Technol Biotechnol; 46:61–9.

Ober, D. (2005).Seeing double-gene duplication and diversification in plant secondary metabolism. Trends Plant Sci. 10, 444–449.

Park, J.M., Yoon, S.Y., Giles, K.L., Songstad, D.D., Eppstein, D., Novakovski, D., Friesen, L., Rosevear, I. (1992).Production of sanguinarine by suspension culture of Papaver somniferum in bioreactors. J Ferment Bioeng; 74:292–6.

Peach, K. (1950). Biochemie und Physiologie der Sekunda¨ren Pflanzenstoffe. Springer, Berlin Go¨ttingen Heidelberg.

Phillips, D.R., Rasbery, J.M., Bartel, B., Matsuda, S.P.T. (2006).Biosynthesic diversity in plant triterpene cyclization. Curr. Opin. Plant Biol. 9, 305–314.

Pichersky, E., Gang, D.R. (2000).Genetics and biochemistry of secondary metabolites in plants: an evolutionary perspective. Trends Plant Sci, 5:439-445.

Ralph, S.G., Yueh, H., Friedmann, M., Aeschliman, D., Zeznik, J.A.,Nelson, C.C., Butterfield, Y.S.N., Kirkpatrick, R., Liu, J., Jones,S.J.M., Marra, M.A., Douglas, C.J., Ritland, K., Bohlmann, J. (2006).Conifer defense against insects: microarray gene expression profiling of Sitka spruce (Picea sitchensis) induced by mechanical woundinh or feeding by spruce budworms (Choristoneura occidentalis) or white pine weevils (Pissodes strobi) reveals large-scale changes of the host transcriptome. Plant Cell Environ. 29, 1545–1570.

Ravishankar, G.A., Bhyalakshmi, N., Ramachandra Rao, S. (1999). Production of food additives. In: Ramawat KG, Merillon JM, editors. Biotechnology: secondary metabolites. New Delhi: Oxford IBH, pp. 89–110.

Ravishankar, G.A., Ramachandra Rao, S. (2000). Biotechnological production of phyto-pharmaceuticals. J Biochem, Mol Biol Biophys; 4:73–102.

Ravishankar, G.A., Venkataraman, L.V. (1993). Role of plant cell culture in food biotechnology: current trends, limitations and future prospects. In: Prakash J, Pierik RLM, editors. Plant biotechnology: commercial prospects and problems. New Delhi: Oxford IBH Press, pp. 255–74.

Reznik, H. (1960). Vergleichende biochemie der phenylpropane. Ergebnisse der Biologie 23, 14–46.

Robinson, T. (1981). The Biochemistry of Alkaloids. Springer, Berlin, Heidelberg, New York.

Sahai, O., Knuth, M. (1985). Commercializing plant tissue culture processes: economics, problem and prospects. Biotechnol Prog; 1:1–9.

Strack, D., Vogt, T., Schliemann, W. (2003). Recent advances in betalain research. Phytochemistry 62, 247–269.

Taylor, L.P., Grotewold, E. (2005). Flavonoids as developmental regulators. Curr. Opin. Plant Biol. 8, 317–323.

Turlings, T. C. J. and Wäckers, F. (2004). Recruitment of predators and parasitoids by herbivore-injured plants, pp 21–75, in R. T. Carde and J. G. Millar (Eds.). Advances in Insect Chemical Ecology. Cambridge University.

Ulbrich, B., Wiesner, W., Arens, H. (1985). Large scale production of rosmarinic acid from plant cell cultures of Coleus blumei Benth. In: Deus-Neumann B, Barz W, Reinhard E, editors. Secondary metabolism of plant cell culture. Berlin: Springer-Verlag. pp. 293–303.

Verpoorte, R., van der Heijden, R., Memelink, J. (2000). Engineering the plan cell factory for secondary metabolite production. Transgenic Res.9:323-343.

Wink, M. (2003). Evolution of secondary metabolites from an ecological and molecular phylogenetic perspective. Phytochemistry. 64:3–19.

Wink, M. (2006). Importance of plant secondary metabolites for protection against insect and microbial infections, pp 251–268, in M. Rai and M. Carpinella (Eds.). Naturally Occurring Bioactive Compounds. Elsevier, Amsterdam.

Zenk, M.H., EI-Shagi, E., Arens, H., Stockigt, J., Weiler, E.W., Deus, B. (1977). Formation of the indole alkaloids serpentine and ajmalicine in suspension cultures of *Catharanthus roseus*. In: Barz W, Reinhard E, Zenk MH, editors. Plant tissue culture and its biotechnological applications. Berlin: Springer-Verlag. pp. 27–43.

Zenk, M.H., EI-Shagi, H., Schulte, U. (1975).Anthraquinone production by cell suspension cultures of Morinda citrifolia. Planta Med Suppl; 79–101.

Zhong, J.J., Konstantinov, K.B. and Toshida, T. (1994). Computer-aided on-line monitoring of physiological variables in suspended cell cultures of *Perilla frutescens* in a bioreactor. J. Ferment Bioeng.**77**:445–7.

❑❑❑

CHAPTER – 25

Bioactive Molecules and Steps of DrugDiscovery

Upasana Singh
Department of Electrical and Computer Engineering, University of Alberta, 9107-116 street, Edmonton, Alberta, T6G 2V4, Canada
Email : upasana_rin@yahoo.com

Summary

The present era is witness of understanding the science of human diseases at molecular level. The advances of science in terms of Genomics, Proteomics, Metaboliomics, Lipidomics and bioinformatics have drawn an excellent pathway to discover more promising drugs. It's still a challenging task to reduce the time line to discover new drug, which is approximately 10-15 years. A drug discovery program requires billions of dollars but without the promise of success. After screening of thousands of compounds only one becomes successful in clinical trials and reaches to the market. Although there are lot many ways to approach drug discovery, nature has gifted many interesting compounds to fight the diseases. On the other hand isolation of pure compounds from natural products like bacteria found in soil or different parts of plants is itself a challenging area. The modern fashion of natural product isolation is guided by bioactivity. Hence a feasible assay to access the bioactivity has got a great importance.

Introduction

A drug in the market is the biggest achievement for a scientist associated with the discovery program. It takes approximately 10-15 years for a chemical to be a marketed as a drug. The drug discovery program starts with the process of concept i.e. evaluation of the target. In terms of drug discovery a target is a naturally existing cellular or molecular structure which participates in the pathogenesis of the disease of interest (Keiser et al.

2010; Pan et al. 2010). For example Rifampicin interferes directly with the bacterial machinery for transcribing ribonucleic acid (RNA) from DNA. Beta subunit of RNA polymerase is the target of Rifampicin (Cole, 1996). A target can be selected on the basis of established literature or a target can be new components which need extensive validation. The validation of a target needs to focus on few important points like rationales behind the selection and feasibility with scientific techniques to pursue the program. An assay is an integral part of a drug discovery program. An assay is a procedure to measure the activity of an enzyme or the growth of the cells. On the other way an assay is a procedure to measure the inhibitory activity of a drug against an enzyme or cell line. When an assay is on board, the next step is screening of the compounds. Screening is a procedure to identify the lead compound from a diverse chemical library. There are several criteria to consider a compound as a lead molecule. At this stage Structure Activity relationship (SAR), Selectivity, Potency, Bioavailability, Solubility etc. which are the 'drug like' properties is investigated. Once the compound fulfills the criteria of a lead, it goes to lead optimization step. At this stage, the efficacy of the compound is tested on the animal models (i.e. *in vitro* and *in vivo* pharmacology is studied) and pharmacokinetics is studied for the lead molecule. Now the compound is characterized for the evaluation of clinical candidate properties.

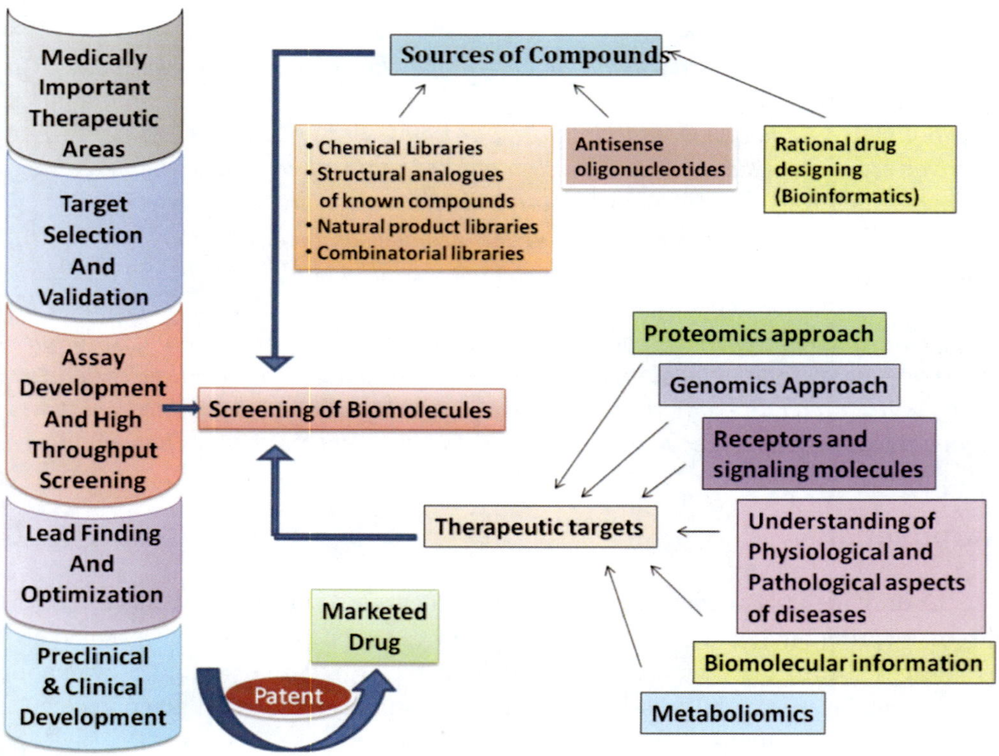

Legend –1 The flow chart of Drug discovery program and the role of natural product chemistry. A typical drug discovery program follows the critical path which is shown in the above diagram. The program starts with the identification of the target and ends with the marketed drug but very few of the compounds hit the market.

Biologically active pharmacophores are the natural products isolated from plants, fungi, bacteria, protozoans, insects and animals (Bindseil et al, 2001). Natural products were considered a valuable source of drug leads, and the testing of natural product extracts was widely practiced in the pharmaceutical industry. The overall objective in identifying lead compounds is to elaborate on tight binding inhibitors with good pharmacokinetic properties. In most cases, lead compounds are identified through successfully implementation of high-throughput assays and surveys of chemical diversity for compounds that inhibit the target selected (Strohl, 2000). The process of lead compound identification has been greatly enhanced by the advent of combinatorial chemical approaches to generating compound diversity. Approximately one-third of the top-selling drugs in the world are natural products or their derivatives. Moreover, natural products are widely recognized in the pharmaceutical industry for their broad structural diversity as well as their wide range of pharmacological activities (Strohl R, 2000). The wonder molecule discovered by Alexander Fleming in 1928 was derived from *Penicillium* fungi (Fleming, 1929). Penicillin was the first antibiotic effective against *syphilis* and *Staphylococcus* infections which was industrialized subsequently during World War II. The area of cancer therapy is highly dominated by the drugs which are originated from natural products. Apart from cancer therapy, the natural products originated drugs are dominating the area of infectious diseases. Beta-lactem or macrolides remained the core structure for infection causing agents where as adriamycin, taxol and vinblastine are the major anti tumor agents (Harvey, 1999). Statins are the top selling cholesterol lowering drugs which are derived from fungal metabolites. FK-506 (tacrolimus) and rapamycin (sirolimus) are immunosuppressive drugs and are natural products too. The term natural products also includes the secondary metabolites which are small molecules (mol wt. < 2000 amu) produced by an organism as byproduct and are not strictly necessary for the survival of the organism (Gullo and Hughes, 2005).

A significant development related to natural product science has been observed in the last few decades. The old fashion of natural product chemistry is set aside and new strategies are adapted.

1. The Old Flow chart of Natural Product Chemistry

a. Importance was given to the structure determination of the compounds from natural sources but not focused on the inhibitory activity.

b. Compounds were isolated and identified randomly from natural sources and *in vivo* biological testing was initiated.

c. Investigations related to chemotaxonomic studies were done

d. Ethnopharmacological and traditional information were used for the selection of organisms for isolation of compounds.

2. The Modern Strategies for Natural Product Chemistry

a. Emphasis was given on the bioactivity and the isolation and structure determination of the compound was initiated. This strategy was called as bioactivity guided purification of the lead molecule.

b. Natural product libraries were generated

c. Evaluation of the inhibitory activity of the compounds on cell culture, genetic manipulation, and strategies of natural combinatorial chemistry.

d. Bioactivity is given highest importance.

e. Introduction of modern strategies likes -

  i) Dereplication :- A natural product could be a mixture of actives in a single sample. The process to recognize and eliminate those active substances which are already studied is called dereplication. This stage of preliminary screening in the process of discovery of new pharmacologically active substances in mixtures of natural products is also called counter-screening.

  ii) Chemical fingerprinting - A unique pattern indicating the presence of a particular molecule, based on specialized analytic techniques such as mass- or x-ray-spectroscopy, used to identify a particular compound is called as chemical fingerprinting.

  iii) Metabolomics - The collection of all metabolites in a biological cell, tissue, organ or organism, which are the end products of cellular process, represents the metabolomics data. Metabolites are the intermediates and products of metabolism. In terms of metabolomics, a metabolite is usually defined as any molecule less than 1 kDa in size.

f. Organisms for natural product extraction were selected either randomly, using ethnopharmacological information, folkloric reputation, or traditional uses.

3. Historical Perspective of Natural Products towards Novel Drugs

The first member of the tetracycline group to be discovered is Chlortetracycline (Aureomycin) in the late 1940s by Dr. Benjamin Duggar, a scientist employed by Lederle Laboratories who derived the substance from a golden-colored, fungus-like, soil-dwelling bacterium named *Streptomyces aureofaciens* (Rishton, 2008). Chloramphenicol was the first antibiotic to be manufactured synthetically but was originally derived from the bacterium *Streptomyces venezuelae*, and introduced into clinical practice in 1949.

Streptomycin was first isolated on October 19, 1943 by Albert Schatz, from actinobacterium *Streptomyces griseus.* Erythromycin was isolated from metabolic products of a strain of "*Saccharopolyspora erythraea*" by Eli Lilly's research team, led by J. M. McGuire. Rifamycins were first isolated in 1957 from a fermentation culture of *Streptomyces mediterranei.* Lincomycin is a lincosamide antibiotic that comes from the actinomyces *Streptomyces lincolnensis (*Rishton G.M., 2008*).* Cephalosporin compounds were first isolated from cultures of *Cephalosporium acremonium* in 1948 and were effective against *Salmonella typhi*, the cause of typhoid fever, which had beta-lactamase. Vancomycin was first isolated in 1953 by from a soil sample collected from the interior jungles of Borneo by a missionary. The organism that produced it was eventually named *Amycolatopsis orientalis*. The original indication for vancomycin was for the treatment of penicillin-resistant *Staphylococcus aureus.* Amphotericin B originally extracted from *Streptomyces nodosus*, a filamentous bacterium, in 1955 at the Squibb Institute for Medical Research from cultures of an undescribed streptomycete isolated from the soil collected in the Orinoco River region of Venezuela. In the 1950s, an Italian research company, Farmitalia Research Laboratories, a new strain of *Streptomyces peucetius* was identified which produced a red pigment and an antibiotic was identified from this bacterium having significant activity against murine tumors. The compound was named as daunorubicin. However, the drug saw success in treating acute leukemia and lymphoma by 1967, it was established that daunorubicin could produce fatal cardiac toxicity (Rishton, 2008).

Along with the success of the drugs from natural products, some complications are also associated. Drugs developed from Natural products are mostly antibiotics although few exceptions are also there like statins (lovastatin, simvastatin, pravastatin) and immunosuppressive agents (cyclosporin, tacrolimus, rapamycin). The development of antibiotic resistance in pathogenic microorganisms like *Mycobacterium tuberculosis* create an urgent need for new antimicrobial agents as well as for those which are not affected by resistance mechanisms already present in the bacterial population (Singh et al, 2011). Targeting virulence is one of the alternative approaches to find new molecules to treat infections due to resistant bacteria. Novel strategies to identify these new antimicrobial agents are required (Singh et al, 2011). The development documented recently for anti-virulence molecules at the pre-clinical stages should allow new classes of molecules to progress as new antimicrobial agents with new mode of action.

4. Drawbacks Related to Natural Products

The Assay development part of drug discovery program has become more sophisticated. A considerable progress has been observed from cell based assays to pharmacologically relevant enzyme based assays, receptor based assays, protein-protein interactions and other biochemically oriented assay systems. The natural products extracts are considered as too dirty and too difficult to assay. To overcome this problem the concept of replica plate was introduced for High throughput assays.

5. Replica Plate

The natural product library consists of plant extracts and fractions, many of which are colored and interfere with the readings obtained in the assay plate (Singh, and Sarkar, 2006). It is a difficult task to identify the real inhibitors from the screening data. To get the actual readings in assay plates corrected for the color interference, the same screening is carried out with an identical plate map and other conditions except the addition of inhibitors. The inhibitors in the replica plate are added after the termination of the reaction. For presenting the data, the plate reader needs to have a template assigned with blank, control and sample for the respective wells mentioned in the plate map. After reading the plate, the values for each well are shown after deducting the average value obtained from blank wells. The reading from the replica plate is used to normalize the reading, obtained in the corresponding assay plate. The following formula is used to estimate the inhibition by the sample :-

$$\text{Percent inhibition} = 100 - \frac{[(S)_A - \{(C)_R - (S)_R\}]}{(C)_A} \times 100$$

Where :-

$(S)_A$ is the sample reading in the assay plate,

$(S)_R$ is the sample reading in the replica plate,

$(C)_R$ is the reading of the control in the replica plate,

$(C)_A$ is the reading of the control in the assay plate.

The modern paradigm of drug discovery is focused on the reduction of time line of drug discovery which is usually 8-14 years. On the other hand the time taken to progress from identification of Natural product as hit to knowing its chemical structure fairly long. So, synthetic compounds are preferred instead of natural products because their structure is known. A natural product often possesses structural complexity like multiple hydroxyl moieties, ketones and chiral centers.

6. Drug Like Properties of Compounds

Pharmacology describes the beneficial or adverse effect of a drug on living matter via biological activity or pharmacological activity. Usually a dose dependent activity was observed. If the pharmacophore behaves very uncommonly then the effects from beneficial to adverse was observed by the shifting of dose from low to high. Activity depends on the fulfillment of the 'ADME' criteria i.e. absorption, distribution, metabolism and excretion (Zhang and Wilkinson, 2007). The pharmacological activity depends basically on the toxicity of the compound. In the study of biomineralisation, bioactivity is often meant as the formation of calcium phosphate deposits on the surface of objects placed in simulated body fluid, a buffer solution with ion content similar to blood.

7. The Drug like Properties of a Compound Depends on the following Factors

A) **Lipinski's Rule of Five** (which describes the molecular properties of a drugs) Christopher A. Lipinski has introduced the rule of five in 1997 to evaluate the druglikeness of a compound. This rule is called as the rule of thumb to determine the druglikeness of a chemical compound that would make it an orally active drug in humans. The rule of five evaluates the important molecular properties of a drug for the pharmacokinetics in the human body as well as 'ADME' (Lipinski et al. 1997). The following criteria defines the rule of five –

i) Not more than 5 hydrogen bond donors (nitrogen or oxygen atoms with one or more (hydrogen atoms). Too many hydrogen bond donors can lead to low fat solubility and the penetration of the drug through the cell wall will be effected.

ii) Not more than 10 hydrogen bond acceptors (nitrogen or oxygen atoms). The drug has to be water soluble as it is transported in aqueous media like blood and intracellular fluid.

iii) A molecular weight under 500 daltons. More than 80 % of marketed drugs are small molecules and have molecular weight under 500 daltons.

iv) An octanol-water partition coefficient $\log P$ less than 5.To reach inside of the cell the drug has to be optimally soluble to both water and fat and get absorbed through the intestinal lining, as well as carried in aqueous blood to penetrate the lipid cellular membrane.

B) **QSAR (quantitative structure-activity relationship)** :– When biological activity or chemical activity is quantitatively correlated with a chemical structure then is defined as QSAR.

The relationship between the effect of the chemical (Three dimensional structure) on the biology of the target is defined as **Structure-activity relationship** (**SAR**). The functional group of the compound is responsible for the changes in the target molecules. Hence modification of the functional group affects the potency of the lead molecule (Zhang and Wilkinson, 2007).

C) Chemical Property

Chemical properties of a substance are the changes occurring in the internal structure during the chemical reaction. Examples of chemical properties Reactivity against other chemical substances, Heat of combustion, Enthalpy, Toxicity, Chemical stability, Flammability, Preferred oxidation state(s), Coordination number, Capability to undergo transformations, for example molecular dissociation, chemical combination, redox reactions under certain physical conditions in the presence of another chemical substance, preferred types of chemical bonds to form, for example metallic, ionic, covalent (Florence and Attwood, 2006).

D) **Physical property** – Physical property of a substance is a change in momentary state of a substance, where no change in the internal structure is observed. The mass, shape, size, type of bonds present and the structure of a substance are the physical properties. For example, the shape and size of a Nano - Particle is determined by High Resolution Transmission Electron Microscopy (Florence and Attwood, 2006).

E) **Molecular property** – Molecular properties include chemical properties, physical properties and structural properties but the SAR properties are not included in this category.

F) **Chemical structure** – This is the most important property of a compound or drug. This includes crystal structure, spatial arrangement of atoms, chemical bonds that hold the atoms together, orbital arrangement of the molecules and structural formula (Florence and Attwood, 2006).

8. Isolation and Purification of the Natural Products

A) Extraction

The selection of procedure for extraction of natural products depends on the source of the materials and the nature of the compound to be isolated (Sarker et al. 2005). The following steps are involved in an ideal extraction protocol-

i) Drying and grinding of plant material or homogenizing fresh plant parts (leaves, flowers, roots, seeds etc) or maceration of total plant parts with a solvents.

ii) Choice of solvents

   a. Polar extraction: water, ethanol, methanol and so on.

   b. Medium polarity extraction: ethyl acetate, dichloromethane and so on.

   c. Nonpolar: n-hexane, pet ether, chloroform and so on.

iii) Choice of extraction methods

   a. Maceration

   b. Boiling

   c. Soxhlet

   d. Supercritical fluid extraction

   e. Sublimation

   f. Steam distillation

B) Fractionation

A natural product extract is a mixture of number of compounds having similar polarities or molecular sizes. It is wise to make only few fractions initially because the compound of interest may spread in a number of samples instead of one or two. The methods commonly used for fractions are Vacuum liquid chromatography, column chromatography, size exclusion chromatography and solid phase chromatography.

C) Isolation

It is difficult to design an isolation protocol for a crude extract where the type of compound is totally unknown. A wrong selection of isolation method could result to lose of a potent molecule. If the target compound is known then the designing of isolation protocol depends on the nature of the compound of interest. The general features of the molecule that are helpful to ascertain the isolation include solubility (hydrophobicity or hydrophilicity), acid-base properties, charge, stability, and molecular size. In case of unknown compound a qualitative test is done for the presence of various types of compounds e.g. phenolics, steroids, alkaloids, flavonoids, etc. as well as analytical TLC or HPLC profiling.

The chromatographic techniques used in the isolation of various types of natural products are as follows –

i) Thin layer chromatography

ii) Preparative Thin layer chromatography

iii) Open column chromatography

iv) Flash chromatography

Modern chromatographic techniques

1. HPLC
2. Multiflash chromatography
3. Vaccum liquid chromatography
4. Chromatotron
5. Solid Phase extraction
6. Droplet countercurrent chromatography
7. HP TLC
8. Hyphenated techniques (e.g. HPLC-PDA, LC-MS, LC-NMR, LCMS- -NMR)

D) Quantification

The yield of the active compound is an important aspect of natural product isolation and purification. To estimate the recovery of the compound routine analytical methods can be used along with the standards.

9. Structure Determination

Basically, a compound is just a powder if its structure is not determined. Determination of structure is a very critical factor for the fate of the compound. The structure of a compound is the result of the coordinates of the atoms in a molecule. The methods for the determination of structure are Nuclear Magnetic Resonance (COSY, NOSY, HMBC & HMQC), infrared spectroscopy, Raman spectroscopy, electron microscopy, and x-ray crystallography (x-ray diffraction) (Walch et al. 2005).

The following are common methods for determining chemical structure:

- X-ray diffraction
- Proton NMR
- Carbon-13 NMR
- Mass spectrometry
- Infrared spectroscopy

The following are common methods for determining electronic structure:

- Electron-spin resonance
- Cyclic voltammetry
- Electron absorption spectroscopy
- X-ray photoelectron spectroscopy

10. Compound Libraries for High Throughput Screening

Experimental screening through chemical collection has been a major route for high throughput biological screening to discover novel compounds with interesting biological activity (Mishra et.al. 2008).

This activity include –

A) Compound Identification

Compounds are grouped into four classes

a) **Specialty sets (SS)** comprising known bioactives such as drugs, toxins, metabolites and others

b) **Natural Products** (NP) and derivatives from known and documented natural sources

c) **Targeted Libraries** (TL) subcategorized into protease, kinase, GPCR, ion channel, and nuclear receptor sets

d) **Diversity Compounds (DC)** all other compounds available for purchase or acquisition

B) Criteria for Submission of Compounds in a Chemical Library

For the development of a compound library at least 5 milligrams of the sample is required and its solubility is DMSO is essential (James et. al. 2008). The purity of the compound should be more than 90 % pure and should not overlap with the samples already present in the library.

The standard procedure to submit compounds to compound library are as follows

i) Submission of Data with the Compound

a. A unique identity is must along with the chemical structure of the compound. The unique identity should signify the structure (including salts, hydrates, etc.) and not the batch / lot / sample identifier.

b. The preferred data file is a Structure Data File (SDF). SDFs are generated by many applications, including CambridgeSoft ChemFinder, MDL ISIS and Isentris, and SciTegic Pipeline Pilot.

c. The alternate datafile is a tab delimited (.tdt or .txt) text file with the SMILES format. SMILES can be generated in Cambridge Soft Chem Draw and other applications.

d. The data file should include all known compound hazard information.

ii) Transfer of the Compound

a. **Samples** should be transferred into samples vials. The Repository do not accepts the samples delivered as solutions in a solvent or solvent mixture. The samples must be supplied between 5.00 and 20.00 mg. Samples must be purified by a method like reverse-phase HPLC that removes impurities.

b. **Acceptable Quality Control Data**. QC data must be supplied with the purified sample and must come from analysis conducted within 90 days. Machine-generated QC data-files must display the analysis date. Where a single UV wavelength is used to indicate LC purity, the UV wavelength must be in the range of 214 – 220 nm.

iii) Pub Chem Registration

The repository will register accepted samples in PubChem, the NCBI database of small molecules and biological data.

11. High Throughput Screening

HTS is automated process of testing ~ 10,000 samples/day against biological activity of targets. Traditionally screening has centered on simple, single protein compound interaction, 'mix and measure' primary assays that measure activity. HTS is considered as a point solution in bringing newest and fastest robotics in laboratory environment. Ultra HTS is now defined in the 1536 well formats that can screen 500,000 compounds per day. Now–a-days development of novel HTS assays and screening has been adapted for genomics, proteomics and metabolomics also (James et al. 2008).

The HTS assay system can be of 6 types such as :-

A. Growth inhibition assays

B. Intracellular growth inhibition assays

C. Surrogate assays of growth inhibition

D. Inhibition of target biochemical function

E. Binding to known structural targets

F. Computational design of inhibitors to known targets.

12. Basic Requirements for High Throughput Assay

a) Combination of modern robotics –An HTS robot can usually prepare and analyze many plates simultaneously, hence speeding up the data-collection process.

b) Data processing & control software –Data analysis soft ware are required with the HTS system for data management.

c) Liquid handling devices – Liquid Handler transfers liquid to deliver and mix all liquid reagents. Although includes removing liquids and plate washing in cost effective way

d) Sensitive detectors – Colorimetric, Fluorometric, Radiometric, NMR etc detection methods are also available with HTS system.

e) Screening protocol – Screening protocols can be developed in laboratory or standard assay kits for HTS are also available.

The main microplate format and working volumes

Microplate type	Format	Multiple of 96	Working volume
96	8x 12	1x96	100-300
384 regular volume	16x24	4x96	5-100
384 regular Low volume	16x24	4x96	1-10
1536	32x48	16x96	1-10
3454	48x72	36x96	0.2-3

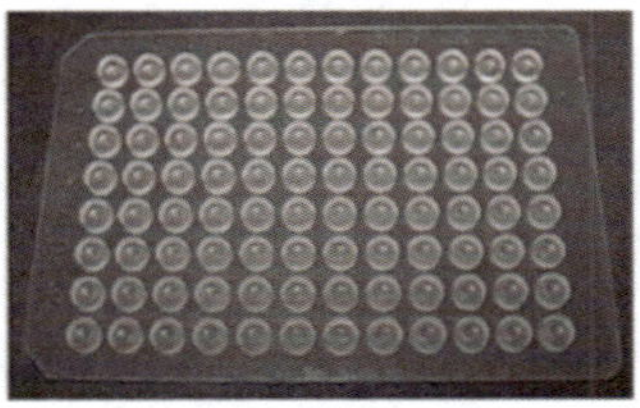

96 well plate

384 well plates

1536 well plates

13. Major factors for HTS Assay Development

The most appropriate assay technology Fluorescence, Radiometric, Colorimetric, NMR etc. based on target type,cost,ease of use. The most important factors to be considered are the optimum component concentrations for signal/Noise ratio, calculation of Z' factor to determine the robustness of the assay, the optimum incubation times (2-4hrs), linearity in activity should be maintained during incubation and the effect of standard inhibitors should remain same in plate format (Zhang et.al. 1999). Another factor to be considered is the plate format for the assay i.e. 96 well, 384 well up to 1536 well). In case of cell based assays the factors to be considered are best cell growth conditions, harvest method and passage number should be maintained properly.

Z'-factor Calculation

'Z' factor is one of the most important factors for High throughput assay development (Zhang et al, 1999).

$$\text{Z factor} = 1 - \frac{3\times(\sigma_p - \sigma_n)}{|\mu_p - \mu_n|}$$

Where :-

σ_p = Standard deviation of Positive control

σ_n = Standard deviation of negative control

μ_p = Signal in positive control

μ_s = Signal in nagative control

Z' Factor	Interpretation
1.0	Ideal. This indicates that the data have a huge dynamic range with negligible standard deviation. Z' can never be equal to 1.0 and absolutely can't be greater than 1.0.
Between 0.5 to 1.0	An Excellent assay
Between 0.0 to 0.5	A marginal assay
Less the zero	No difference between the positive control and negative control. Essentially useless assay.

13. Conclusion

The new drugs discovery picture has been changed dramatically in the past decade. Emphasis is now being placed on identifying "druggable targets". HTS is used as core activity to meet the challenges in rational drug discovery. Natural products have bright future prospects, because of the popularity of Ayurveda, Chinese medicines and Egyptian medicines are increasing day by day. Modern techniques like throughput activities are introduced to maintain the leadership. Innovation will remain the key driver of future therapy. The process of drug discovery has to experience a high level of complication. Failure is an inseparable part of drug discovery. The pre discovery process is related to understanding of the diseases and identification of the target molecules. At this level the contributing bodies are pharmaceuticals, government organizations, academics and non-profit making organizations. The discovery process contributes in identification of candidate drug and creation of new molecule. The goal of the preclinical stage is to determine the effectiveness and safety related issues. When a drug gets approval from Food and Drug Administration (FDA), it is called as Investigational New Drug (IND). Now the drug proceeds to clinical trials, although maximum failures of the drugs are observed at this stage only. A clinical trial is the stage, where the drug is tested on humans in three phases of trial. The first phase starts with testing on a small group of healthy volunteers and moving into larger number of volunteers. The FDA evaluates the data of the clinical trials and approves for the application on the patient. After all this exercise the drug has the good fortune to enter the market and then its manufacturing starts in the production unit. The pharmaceuticals file the patents to secure the area of their profit and pursue further research.

Acknowledged

"If at the end of my career, I can look back and know that something I did made a difference in one patient somewhere in the world, that'll be more satisfying and more gratifying than anything I can possibly imagine."

Jonathan Yingling, Ph.D., Eli Lilly and Company

Reference

Bindseil K.U., Jakupovic J., Wolf D., Lavayre J., Leboul J. and Pyl D.V.D. (2001). Pure compound libraries; a new perspective for natural product based drug discovery. DDT Vol. 6. 840-847.

Cole S T. (1996). Rifamycin resistance in mycobacteria. Res. Microbiol. 147: 48}52.

Chan J.A. and Hueso-Rodríguez J.A. (2002). Compound Library Management. High Throughput Screening Methods in Molecular Biology, Volume 190, 117-127

Fleming A. (1929). "On the antibacterial action of cultures of a penicillium, with special reference to their use in the isolation of B. influenzæ". Br. J. Exp. Pathol. 10 (31): 226–236.

Florence A.T. and Attwood D. (2006). Physicochemical Principles of Pharmacy, 4th Edition.

Gullo V.P. and Hughes D.E. (2005). Exploiting new approaches for natural product drug discovery in the biotechnology industry. Drug Discovery Today: Technologies Vol. 2.

Harvey A. L. (1999). Medicines from nature: are natural products still relevant to drug discovery? TiPS (Vol. 20) 196-198.

Keiser J. M., Irwin J.J., and Shoichet B.K. (2010). The Chemical Basis of Pharmacology. Biochemistry, 49, 10267–10276.

Lipinski C.A., Lombardo F, Dominy B.W. and Feeney P.J. (1997). "Experimental and computational approaches to estimate solubility and permeability in drug discovery and development settings". Adv. Drug. Del. Rev. 23: 3–25.

Mishra K.P., Ganju L., Sairam M., Banerjee P.K.and Sawhney R.C. (2008). A review of high throughput technology for the screening of natural products. Biomedicine & Pharmacotherapy 62 94-98.

Pan S.Y., Pan S., Yu Z.L., Ma D.L., Chen S.B., Fong W.F., Han Y.F. and Ko K.M. (2010). New Perspectives on Innovative Drug Discovery: An Overview. J. Pharm. Pharmaceut. Sci. 13, (3), 450 – 471.

Rishton G.M. (2008). Natural Products as a Robust Source of New Drugs and Drug Leads: Past Successes and Present Day Issues. The American Journal of Cardiology. Vol 101 (10A).

Strohl W. R. (2000). The role of natural products in a modern drug discovery program DDT Vol. 5, 39-41 (Editorial).

Singh U., Akhtar S., Mishra A., Sarkar D. (2011). A novel screening method based on menadione mediated rapid reduction of Tetrazolium salt for testing of anti-mycobacterial agents. Journal of Microbiological Methods Vol 84, (2), 202-207.

Singh U. and Sarkar D. (2006). Development of a Simple High-Throughput Screening Protocol Based on Biosynthetic Activity of *Mycobacterium tuberculosis* Glutamine Synthetase for the Identification of Novel Inhibitors. J. Biomo.l Screen. 11, 1035.

Sarker S. D., Latif Z. and Grey A.I. (2005). Natural products isolation Methods in Biotechnology Volume 20.

Walch N.B, Kuhan Till, Moskau D. and Zerbe O. (2005). Strategies and tools for Structure determination of Natural products using Modern methods of NMR Spectroscopy. Chemistry and biodiversity Vol 2 147-177.

Zhang M.O. and Wilkinson B. (2007). Drug discovery beyond the 'rule-of-five' Current Opinion in Biotechnology, 18:478–488

Zhang J.H.,Chung T.D.Y.and Oldenburg K.R. (1999). A simple statistical parameter for use in evaluation and validation of high-throughput screening assays. J. Biomol. Screen; 4:67-73.

❑❑❑

CHAPTER – 26

An Approach to Drug Discovery

Anupam Das Talukdar[1], Pankaj Chetia[1], Amitabha Bhattacharjee[2], Manabendra Dutta Choudhary[1]*
[1]Department of Life Science and Bioinformatics, Assam University, Silchar-788011
[2]Department of Microbiology, Assam University, Silchar-788011
Email : monichi_dc@writeme.com

Summary

Human being is dependent on nature and natural resources to combat diseases from time immemorial. Ayurveda describes the use of traditional herbs in medicinal preparations in ancient India. Such types of evidences are there in classical Chinese and Greek records also. Nowadays, modern tools are used to identify and purify the active principle from the herbal composition having the therapeutic effect. Bioinformatics tools are often used to screen the drug efficacy of natural products. Using computational and Bioinformatics tools, the efficacy of lead molecules can be screened. Moreover, drug targets can be identified using Bioinformatics and proteomics approach. QSAR and molecular docking tools are used to study the probable activity and mode of action of drug molecule. This approach has minimized the cost and time involved in the drug discovery process.

Introduction

The term "drug" refers to any substance that leads to desired physiological changes in the body. A drug may be a chemical compound or mixtures of chemical compounds or substances. Generally drugs may be of natural origin or synthetic. Natural products are being used as drugs from time immemorial. A natural product is a chemical compound or substance produced by a living organism, found in nature that usually has a pharmacological or biological activity for use in pharmaceutical drug discovery and drug design. A natural product can be considered as such even if it can be prepared by total synthesis.

Not all natural products can be fully synthesized and many natural products have very complex structures that are too difficult and expensive to synthesize on an industrial scale. These include drugs such as penicillin, morphine, and paclitaxel (Taxol). Such compounds can only be harvested from their natural source - a process which can be tedious, time consuming, and expensive, as well as being wasteful on the natural resource.

Records from as early as 2700 B.C. from China, traced to the Emperor Shennung, indicate the usefulness of plants for treating disease, and the Ebers papyrus, written in about 1550 B.C., includes many of the plants used in Egyptian medicine. Theophrastus (370-285 BC) began the scientific classification of plants, and Dioscorides *De Materia Medica* (77 AD) reported the uses, medicinal and otherwise, of over 600 plants. Ibn al-Baitar (1197-1248) listed over 1400 drugs and medicinal plants in his *Corpus of Simples*. In Europe, after the tenth century, much of the medicinal lore was based in the church, particularly the monastic orders, but by the 1500's, after the invention of the printing press, herbals available to the general public were popular, particularly in England. By the late 1700's, studies like William Withering's *An Account of the Foxglove and its Medicinal Uses* (1785) began to appear. These were based on case histories and described specific doses and gave administration instructions for herbal remedies. In the United States, before the advent of specific pharmaceuticals, herbal medicine was relied upon to treat many illnesses.

Drugs derived from natural products are usually secondary metabolites and their derivatives, and today those must be pure and highly characterized compounds. Until the late 1800's, organic chemistry was almost exclusively the study and use of natural products. The purpose of these compounds in the organisms and their formation was little understood or investigated, primarily due to the lack of appropriate techniques and structural theory. The natural products that were studied and used tended to be the compounds that occurred in the largest amounts, mostly in plants, and were most easily isolated in a pure, or sometimes not very pure, form by techniques such as simple distillation, steam distillation, or extraction with acid or base.

Many higher plants contain novel metabolites with antimicrobial and antiviral properties. However, in the developed world almost all clinically used chemotherapeutics have been produced by in vitro chemical synthesis. Exceptions, like taxol and vincristine, were structurally complex metabolites that were difficult to synthesize *in vitro*. Many non-natural, synthetic drugs cause severe side effects that were not acceptable except as treatments of last resort for terminal diseases such as cancer. The metabolites discovered in medicinal plants may avoid the side effect of synthetic drugs, because they must accumulate within living cells. Despite the potential limitations of natural products, these small molecules provide the source or inspiration for the majority of FDA-approved agents and continue to be one of the major sources of inspiration for drug discovery. In particular, these compounds are important in the treatment of life-threatening conditions (Newmann and Cragg, 2007).

Traditional Approaches of Drug Discovery from Natural Sources

Mankind's curiosity about taste, odour and cures from human and plant diseases have originated the Natural product chemistry. The term Natural product is commonly applied to material derived from plants, microorganisms and animals which are fine biochemical factories for the synthesis of both primary and secondary metabolites. Industrially, natural product serve as an essential source of oils, flavours, fragrance, resins, gums, dyes, pharmaceuticals and many other special products. The primitive man by trial and error method has acquired significant biological knowledge in determining the usefulness of plants and animals as food or medicine and is to be avoided for its poisonous and dangerous nature.

For centuries, China, India, Egypt and Greek led the world in the use of natural products in healing and folkloric medicine based on the use of plants and animals have played a significant role in reducing human sufferings and pain. From ancient times, people are well aware of extraordinary medicinal properties of garlic, cinchona etc, in curing many ailments. *Artemisia annual* L. (Compositae) was used in china for curing fever and malaria. In 1972, an anti malarial drug artemisinin was isolated and was found to be superior to other conventional ant malarial drugs. The biological effect of Ginseng (Paneax ginseng) was well recognized by the ancient Chinese. Its biological effects are mainly due to synergistic activity between saponins called ginsenosides and flavonoids. *Ginkgo balboa* is anther important folk drug used by the Chinese in improving mental alertness. The main constituents that are associated with Ginkgo are due to the activity of ginkgoloids and flavonoids.

Natural products have served as an important source of drugs since ancient times and about half of the useful drugs today are derived from natural sources. Chemodiversity in nature, e.g. in plants, microorganisms and marine organisms, still offers a valuable source for novel lead discovery, but rapid identification of the bioactive compounds of natural product mixtures remains a critical factor to ensure that this tool of drug discover can compete with recent developed technologies such as chemical compound libraries and high-throughput screening of combinatorial synthetic efforts. Rapid screening of natural product mixtures requires the availability of a library of reference of natural compounds and methods for simple identification of putative lead structural classes avoiding, to a large extent, the potential for false-positive results. The coupling of chromatographic methods such as high pressure liquid chromatography (HPLC) with diode array detection, mass spectrometry (MS) or nuclear magnetic resonance spectroscopy (NMR) or, and with, on-line bioactivity assays, is an important tool for high throughput screening of natural product mixtures (Vlietinck and Apers , 2001). The introduction of a dereplication step after extraction by using a reproducible preseparation method would enable the rapid elimination of false positives (Verpoorte, 1998). The effective use of automated procedures and databases in the isolation, identification and biological profiling of bioactive compounds from natural sources will be the best guarantee to the continued discovery of novel chemotypes from nature (Hook *et al.*, 1997).Structure-activity studies of these leads, preferentially combined with computergraphic model building, should result in molecules with optimal activity and bioavailability, fewer side effects and an acceptable therapeutic index and, consequently in good candidates, for the development to new drugs. (Vlietinck and Apers , 2001).

High performance liquid chromatography (HPLC) is used routinely in phytochemistry to 'pilot' the preparative isolation of natural products (optimisation of the experimental conditions, checking of the different fractions throughout the separation) and to control the final purity of the isolated compounds (Kingston, 1979; Hostettmann *et al.,* 1998). The development of hyphenated techniques related to this efficient separation technique in the past 20 years has provided powerful new tools such as LC/UVphotodiode array detection (LC/UV-DAD) (Huber and George, 1993), LC/mass spectrometry (LC/MS) (Niessen, 1999) and very recently LC/nuclear magnetic resonance (LC/NMR) (Spraul *et al.,* 1993). The combination of the high separation efficiency of HPLC with these different detectors has made possible the acquisition of on-line complementary spectroscopic data on an LC peak of interest within a complex mixture.

As crude plant extracts represent very complex mixtures containing up to hundreds of constituents, these new LC-hyphenated techniques have been rapidly integrated for the study of crude plant extracts (Wolfender *et al.,* 1998b).

At present, MS is one of the most sensitive methods of molecular analysis. Moreover, it has the potential to yield information on the molecular weight as well as on the structure of the analytes. Due to its high power of mass separation, very good selectivities can be obtained. LC/MS has been used in analytical chemistry since the mid '80s and now numerous LC/MS systems are available on the market and different LC/MS interfaces have been built up. These LC/MS interfaces must accomplish nebulisation and vaporization of the liquid, ionisation of the sample, removal of the excess solvent vapour, and extraction of the ions into the mass analyser (Vlietinck and Apers , 2001). Now the most commonly used LC/MS interfaces are electrospray (ES) (Whitehouse *et al.,* 1985) and atmospheric pressure chemical ionisation (APCI) (Bruins *et al.,* 1987).

Nuclear Magnetic Resonance (NMR) techniques have proved to be powerful tools for the structural determination of organic compounds since the beginning of their applications in chemistry. They are particularly useful in the field of natural products, where frequently the problem occurs of elucidating the structure of a completely novel compound, sometimes possessing unusual or unprecedented structural features. Early NMR spectrometers, equipped with permanent magnets working at low magnetic fields

(1.41 Tesla, corresponding to a 60 MHz carrier frequency for 1H nuclei), exclusively allowed the acquisition of 1H NMR spectra, affording structural information complementary to that available from chemical analysis or well-established physical methods like Ultraviolet (UV) or Infrared (IR) spectroscopies, or from the technique of Mass Spectrometry (MS), at that time routinely used by chemists. Nowadays modern NMR spectroscopy performed with cryospectrometers operating at high magnetic field (up to 21.14 Tesla, i.e. 900 MHz for 1H) and capable of executing a variety of sophisticated, multipulse and multidimensional experiments, has largely superseded the above methods. Thus, it has become the main choice for establishing chemical structures, particularly when X-ray crystallography is inapplicable. It requires only minute amounts of substance and frequently also allows the determination of stereochemical details, or even preferred conformation in solution. In addition, the variety of experiments available on modern NMR instruments has considerably reduced the time needed for the identification of known compounds, thus giving valuable help to the chemical knowledge of previously unexamined natural sources (Hostettmann and Wolfender 2001).

Bioactive Compounds

The term flavonoid refers to a class of plant secondary metabolites. Flavonoids are most commonly known for their antioxidant activity. Flavonoids are also commonly referred to as bioflavonoids in the media – these terms are equivalent and interchangeable, since all flavonoids are biological in origin.

Flavonoids are widely distributed in plants fulfilling many functions including producing yellow or red/blue pigmentation in flowers and protection from attack by microbes and insects. The widespread distribution of flavonoids, their variety and their relatively low toxicity compared to other active plant compounds (for instance alkaloids) mean that many animals, including humans, ingest significant quantities in their diet.

Tannins are astringent, bitter-tasting plant polyphenols that bind and precipitate proteins. The term tannin refers to the source of tannins used in tanning animal hides into leather; however, the term is widely applied to any large polyphenolic compound containing sufficient hydroxyls and other suitable groups (such as carboxyls) to form strong complexes with proteins and other macromolecules. Tannins have molecular weights ranging from 500 to over 20,000, Kda. Tannins are usually divided into hydrolyzable tannins and condensed tannins (proanthocyanidins).While hydrolyzable tannins and most condensed tannins are water soluble, some very large condensed tannins are insoluble.Tannins may be employed medicinally in antidiarrheal, hemostatic, and antihemorrhoidal compounds.

A steroid is a terpenoid lipid characterized by a carbon skeleton with four fused rings. Different steroids vary in the functional groups attached to these rings. Hundreds of distinct steroids have been identified in plants, animals, and fungi. Plant sterols are of two types- Phytosterols and Brassinosteroids. Saponins are glycosides of steroids, steroid alkaloids (steroids with a nitrogen function) or triterpenes found in plants, especially in the plant skins where they form a waxy protective coating. Some authors distinguish a third class of saponin, the alkaloid saponins. They dissolve in water to form a stable soapy froth; this isthought to be due to their amphiphillic nature. Saponins are believed to be useful in the human diet for controlling cholesterol . Saponins are also mild detergents and are used commercially as well as for research.

A reducing sugar is a type of sugar with an aldehyde or ketone group. This allows the sugar to act as a reducing agent, for example in the Maillard reaction and Benedict's reaction. Reducing sugars include glucose, glyceraldehyde, lactose, arabinose and maltose. All monosaccharides which contain ketone groups are known as ketoses, and those which contain aldehyde groups are known as aldoses. Significantly, sucrose is not a reducing sugar. It is in fact known as a non reducing sugar.

Benedict's reagent is used to determine if a reducing sugar is present. If it is a reducing sugar, the mixture will turn green/orange/red. Fehling's solution can also be used for the same purpose, as both contain copper (II) ions which are reduced to a brick red precipitate of copper (I) oxide when the solution is heated. A reducing sugar occurs when its anomeric carbon is free.

Marine Natural Products

Microorganisms have been the source of many valuable compounds in medicine, industry, and agriculture; most are derived from terrestrial habitats. After intensive studies on terrestrial microorganisms, consequent attentions have been focused on other ecosystems, especially those subject to extreme environmental conditions such as the desert, hot springs and the sea. Microorganisms living in "sea environments" are subject to high salinity, high pressure, poor nutrition, low temperature (or local high temperature), and complex relations with other different creatures. These factors have resulted in the development of unique metabolisms, which provide the opportunity to produce metabolites that differ from the terrestrial ones. Thus, sea microorganisms offer a wonderful resource for the discovery of new compounds and compounds with new activity. Although various bioactive compounds from marine microorganisms have been studied, antitumor agents and antibiotics remain the main research target in the last five years. Many new compounds with strong bioactivities have been isolated from marine microorganisms. Till date 62 antitumor agents, 38 antimicrobial, 11 enzyme inhibition and 5 antivirus agents has been isolated from marine fungi. From marine cyanobacteria the results are also impressive and the numbers are, 47,2, 4 respectively. From Bacterium 3 antitumor agents , 23 antimicrobial and 1 enzyme inhibition agents has been isolated so far (Xiaohong et al., 2008).

The world's oceans cover more than 70% of the earth's surface and contain over 200,000 invertebrate and algal species. These organisms live in complex communities and in close association with other organisms both macro- (e.g., algae, sponges, ascidians) and micro-organisms (e.g., nonfllamentous bacteria, fungi, actmomycetes). Some organisms derive their chemistry from dietary sources, while others synthesize the compounds de IZOVOS. Some compounds may be produced by associated microorganisms, while others may require an association between the host and microorganism to produce the compounds. The chemistry of any particular specimen can be affected by the habitat as well as by geographic and seasonal factors . In fact, the true biogenetic origin of most marme natural products remains a topic for debate within the marcine natural products community. Because of the diversity of marine organisms and habitats, marcine natural products encompass a wide variety of chemical classes, including terpenes, shikimates, polyketides, acetogenins, peptides, alkaloids of varying structures, and a multitude of compounds of mixed biosynthesis. In the past decade alone, the structures of over 5000 marine natural products have been published. In many instances, the class of compound present in the organism can be predicted based on the taxonomic classification of the source organism. Unfortunately, even knowing the class of compound does not always allow one to define a simple purification method. Series of marine natural products can often contain a variety of chemical functionalities (e.g., OH, OCH3, Et OAc, OSO,- Na^+). Each change in functionality has the potential to radically change the polarity of the compounds and therefore the method required for purification (Wright, 1998).

Chromatographic Techniques for Isolation of Pure Compound

TLC separations can also be used to select column chromatography conditions. TLC conditions that give a useful Rf value, i.e., compound separates from the majority of other components without staying at the origin or with the solvent front, can be approximately transferred to column chromatography. Identification of the target compound or on the TLC plate can be carried out by comparison with a standard, by chemical staining, or by an overlay assay carried out on top of the developed plate in the case of an unknown biologically active component, or by scraping off, extracting, and assaying portions of the adsorbent. However, it is not always straightforward to translate TLC systems to column systems. This is partly because TLC cannot be seen simply as a two dimensional column. Unlike column chromatography, TLC is a non equilibrium technique, which means that the conditions of the mobile and stationary phases are not constant throughout the plate but vary during the run and according to the position on the plate. This can lead to difficulties in trying to reproduce TLC separations in a column form. As a general rule, it is advisable to use a slightly less polar mobile phase for normal phase column chromatography (and slightly more polar for reverse phase columns) than that used to obtain a reasonable Rf value on TLC.

Armed with some knowledge of the nature of the compound and some idea of suitable general separation approaches, one can usually establish a successful isolation method much more quickly than if none of these preliminary tests is carried out.

Preparatory TLC (PTLC) Method

Preparatory TLC studies are generally performed for the separation or isolation of pure phytochemicals. In this method, at first the particular composition of the solvent, which results in the clear, distinct and uniform separation of individual compounds is determined by manipulating the solvent mixtures, on the basis of their polarity by trial & error method (generally solvents like petroleum ether (low polar), ethyl acetate(medium polar) and methanol (high polar) are used to compose the suitable solvent in variable concentrations). After determining the composition ratio of the solvents, nos. of TLC separations are studied. Since, for each TLC plate, the compounds are given to separate against the same composition of the eluent, therefore, the compound separation for each TLC plate becomes uniform and by identifying the same or similar compounds with expertized eyes on the basis of their extent of separation for each TLC observations, the portion of the separated compounds of same Rf value cut down and hence, the compound can be recovered from the silica gel and the compound mixture.

Column Chromatography

Column chromatography is very frequently used by organic as well as biochemists to purify liquids as well as solids. An impure sample is loaded onto a column of adsorbent, such as silica gel or alumina. An organic solvent or a mixture of solvents (the eluent) flows down through the column. Components of the sample separate from each other by partitioning between the stationary packing material (silica or alumina) and the mobile eluent .Molecules with different polarity partition to different extents, and therefore move through the column at different rates. The eluent is collected in fractions. Fractions are typically analysed by Thin Layer Chromatography to see if the separation of the compounds was successfull or not.

Packing a (Silica Gel) Column

1. A piece of wire is used to add a plug of cotton to the bottom of the column. There should be enough cotton that the silica does not fal out of the column. However, too much cotton or cotton packed too tightly will prevent the eluent from dripping at an acceptable rate.
2. The column is clamped to a ring stand and enough sand is added to fill the curved portion of the column.
3. A pinch clamp is placed on the tubing, then the column is packed up to 1/4 to 1/3 full with the intial eluent (The composition of eluent is often changed as the separation proceeds).

4. A slurry of silica is prepared in the initial eluent by pouring dry silica into a beaker of eluent.(a volume of silica gel, such as 20 ml is added, to approximately double the volume of eluent, i.e., 40 ml).
5. The slurry is poured into the column quickly and carefully. It is stirred and poured immediately to maximize the amount of silica that goes into the column instead of remaining behind in the beaker.
6. The pinch clamp is removed to allow the solvent to drip into a clean flask. The side of the column is tapped with a rubber stopper or tubing to help the silica settle uniformly.
7. A Pasteur-pipette is used to rinse any silica that is sticking to the sides of the column. The silica is allowed to settle, while the eluent continues to drip into the flask.
8. Once the silica is settled, sand is added to the top of the column (sand is heavier than silica). If the silica has not settled, the sand may sink into the silica instead of forming a layer on top of it.

Loading a Sample Onto a Column

9. The eluent is drained from the column, until no solvent remains above the surface of the sand.
10. Using a long Pasteur-pipette, the sample is carefully added to the column.
11. The eluent is drained from the column until no sample remains above the surface of the sand.
12. To rinse the container and pipette, 1 ml of the eluent is added to the sand. The eluent is drained from the column until no liquid remains above the surface of the sand.
13. The step-12 is repeated two or three times to completely transfer the sample onto the silica gel.

Eluting the Sample

14. Once the sample is rinsed into the silica, the eluent is carefully added to the top of the column. To avoid disturbing the top of the column, it is a good idea to carefully pipette out one or two inches of the solvent onto the column instead of pouring solvent directly onto the sand.

Analysis of Fraction

15. The fractions are analyzed by Thin Layer Chromatography to determine:-
    a. if the fraction contains more than one component and

b. if the fractions can be combined without affecting the purity of those fractions

Precautionary Measures

1. The success of the separation will be dependent on how well the column is packed and loaded. It is important to have level sand and silica. It is also important to carefully add the sample to the packed column.
2. The silica should not be allowed to dry out as the column progresses. Cracks will form within the silica column if it dries and compounds can fall down the cracks instead of partitioning between the mobile and stationary phases.

In silico Approaches for Newer and Effective Drug Molecule

Drug discovery process is a time taking and expensive process. The application of computational tools in drug discovery researches has significantly minimizes the time as well as the cost involved in the process. Bioinformatics tools are nowadays widely used in drug discovery researches. The databases like Protein data Bank, KEGG Pathway and KEGG Ligand, NCBI PubChem etc. are of immense use in modern drug discovery researches. Moreover tools like Drug Likeness screening, ADME/Tox screening, Quantitative Structure Activity Relationship, Molecular Docking etc. are used to predict and screen probable drug molecules. There are two approaches of *in silico* drug discovery process, viz. Ligand Based Drug Designing and Target Based Drug Designing.

Generally drug molecule acts on an enzyme or protein or any other specific receptor, resulting into a desired therapeutic effect. This enzyme or protein molecule is called a drug target. When the ligand (probable drug) is known and target is unknown, the approach used is called Ligand Based Drug Designing. And when the target is known and the ligand is unknown, the approach is called Target Based Drug Designing. The ligand binds to the drug target and inhibits the normal activity of the drug target. The *in silico* screening of target-ligand binding is called "docking". Docking study helps in understanding the fitness of two molecules to each other.

Drug Discovery

Drug discovery and development is an intense, lengthy and an interdisciplinary endeavor. Drug discovery is mostly portrayed as a linear, consecutive process that starts with target and lead discovery, followed by lead optimization and pre-clinical *in vitro* and *in vivo* studies to determine if such compounds satisfy a number of pre-set criteria for initiating clinical development. For the pharmaceutical industry, the number of years to bring a drug from discovery to market is approximately 12-14 years and costing upto \$1.2 - \$1.4 billion dollars. Traditionally, drugs were discovered by synthesizing compounds in a time-consuming multi-step processes against a battery of *in vivo* biological screens and further investigating the promising candidates for their pharmacokinetic properties, metabolism and potential toxicity. Such a development process has resulted in high attrition rates with failures attributed to poor pharmacokinetics (39%), lack of efficacy (30%), animal toxicity (11%), adverse effects in humans (10%) and various commercial and miscellaneous factors. Today, the process of drug discovery has been revolutionized with the advent of genomics, proteomics, bioinformatics and efficient technologies like, combinatorial chemistry, high throughput screening (HTS), virtual screening, *de novo* design, *in vitro*, *in silico* ADMET screening and structure-based drug. *In silico* methods can help in identifying drug targets via bioinformatics tools.

They can also be used to analyze the target structures for possible binding/ active sites, generate candidate molecules, check for their drug likeness , dock these molecules with the target , rank them according to their binding affinites , further optimize the molecules to improve binding characteristics

The use of computers and computational methods permeates all aspects of drug discovery today and forms the core of structure-based drug design. High-performance computing, data management software and internet are facilitating the access of huge amount of data generated and transforming the massive complex biological data into workable knowledge in modern day drug discovery process. The use of complementary experimental and informatics techniques increases the chance of success in many stages of the discovery process, from the identification of novel targets and elucidation of their functions to the discovery and development of lead compounds with desired properties. Computational tools offer the advantage of delivering new drug candidates more quickly and at a lower cost. Major roles of computation in drug discovery are; (1) Virtual screening & de novo design, (2) *in silico* ADME/T prediction and (3) Advanced methods for determining protein-ligand binding.

As structures of more and more protein targets become available through crystallography, NMR and bioinformatics methods, there is an increasing demand for computational tools that can identify and analyze active sites and suggest potential drug molecules that can bind to these sites specifically. Also to combat life-threatening diseases such as AIDS, Tuberculosis, Malaria etc., a global push is essential. Millions for Viagra and pennies for the diseases of the poor is the current situation of investment in Pharma R & D. Time and cost required for designing a new drug are immense and at an unacceptable level. According to some estimates it costs about $880 million and 14 years of research to develop a new drug before it is introduced in the market Intervention of computers at some plausible steps is imperative to bring down the cost and time required in the drug discovery process.

Researchers have invented computational tools to decode and rapidly determine whether natural compounds collected in oceans and forests are new -or if these pharmaceutically promising compounds have already been described and are therefore not patentable.These process will advances the speed of the process by which its possible to discover and describe new and biologically active molecules from organisms such as marine cyanobacteria, also known as blue-green algae. This, in turn, will accelerate the timeline for bringing new experimental therapies into clinical application.

In-silico screening of Natural products

***In-silico* screening** is the use of computational chemistry techniques to analyze large chemical databases in order to identify possible new drug candidates. Virtual screening techniques range from simple ones, based on the presence or absence of specific substructures, or match in calculated molecular properties, up to sophisticated virtual docking methods aimed at fitting putative ligand molecules into the target receptor site.

Drug likeness may be defined as a complex balance of various molecular properties and structure features which determine whether particular molecule is similar to the known drugs. These properties, mainly hydrophobicity, electronic distribution, hydrogen bonding characteristics, molecule size and flexibility and presence of various pharmacophoric features influence the behavior of molecule in a living organism, including bioavailability, transport properties, affinity to proteins, reactivity, toxicity, metabolic stability and many others.

The diversity of possible drug targets (of which each requires a different combination of matching molecular characteristics) is so enormous, that we do not believe that it is possible to find a common denominator for all of them and to express molecule drug-likeness by a single magic number. Simple count criteria (for example limits for **molecular weight, logP, or number of hydrogen bond donors or acceptors**) have also relatively limited applicability and are useful only to discard obvious non-drugs.

LogP (Octanol/water Partition Coefficient)

LogP is calculated by the methodology developed by Molinspiration as a sum of fragment-based contributions and correction factors. Method is very robust and is able to process practically all organic and most organometallic molecules.

Molecular Polar Surface Area TPSA

It is calculated based on the sum of fragment contributions. O- and N- centered polar fragments are considered. PSA has been shown to be a very good descriptor characterizing drug absorption, including intestinal absorption, bioavailability, Caco-2 permeability and blood-brain barrier penetration.

Molecular Volume

Method for calculation of molecule volume developed at Molinspiration is based on group contributions. These have been obtained by fitting sum of fragment contributions to "real" 3D volume for a training set of about twelve thousand, mostly drug-like molecules. 3D molecular geometries for a training set were fully optimized by the semiempirical AM1 method.

"Rule of 5" Properties

It is set of simple molecular descriptors used by Lipinski (1997) in formulating his "Rule of 5" (Joy et al., 1998). The rule states, that most "drug-like" molecules have logP <= 5, molecular weight <= 500, number of hydrogen bond acceptors <= 10, and number of hydrogen bond donors <= 5. Molecules violating more than one of these rules may have problems with bioavailability. The rule is called "Rule of 5", because the border values are 5, 500, 2*5, and 5.

Number of Rotatable Bonds - nrotb

This simple topological parameter is a measure of molecular flexibility. It has been shown to be a very good descriptor of oral bioavailability of drugs . Rotatable bond is defined as any single non-ring bond, bounded to non terminal heavy (i.e., non-hydrogen) atom. Amide C-N bonds are not considered because of their high rotational energy barrier (Joy et al., 1998).

Statistics in the Search of Newer Drug Molecule

Statistics has wide applications in drug discovery. In Quantitative Structure Activity Relationship, activity of an unknown molecule can be predicted studying the existing activity of similar compounds. Correlation and regression analysis play a major role in this. Here, biological activity can be expressed quantitatively as in the concentration of a substance required to give a certain biological response. Additionally, when physicochemical properties or structures are expressed by numbers, one can form a mathematical relationship, or quantitative structure-activity relationship, between the two. The mathematical expression can then be used to predict the biological response of other chemical structures.

QSAR's most general mathematical form is:

Activity = f(physiochemical properties and/or structural properties)

As the newer and efficient Bioinformatics and Chemoinformatics software are coming up, it has become easier to predict the physicochemical properties of any molecule. These physicochemical properties are more or less related to the biological activity of the molecule. In QSAR, these physicochemical properties are termed as

"Chemical Descriptors" : These descriptors are related to the structure and functional groups of the molecule and this is ultimately related to the biological activity of the molecule.

In QSAR, first a set of similar molecules with known activity are taken. Their descriptors are calculated using Chemoinformatics software. The correlation of each descriptor with its activity is calculated. And an equation is generated. This QSAR equation helps in predicting the activity of unknown similar molecule with known descriptors.

There are many practical purposes of a QSAR and these techniques are utilized widely in many situations. The purpose of in silico studies, therefore, includes the following:

1. To predict biological activity and physico-chemical properties by rational means.
2. To comprehend and rationalize the mechanisms of action within a series of chemicals.Underlying these aims, the purpose for developing these models include
3. Savings in the cost of product development (e.g. in the pharmaceutical, pesticide, personal products, etc.).
4. Predictions could reduce the requirement for lengthy and expensive animal tests.
5. Reduction (even, in some cases, replacement) of animal tests, thus reducing animal use and obviously pain and discomfort to animals.
6. Other areas of promoting green and greener chemistry to increase efficiency and eliminate waste by not following leads unlikely to be successful.

Applications of Qsar

The ability to predict a biological activity is valuable in any number of industries. Whilst some QSARs appear to be little more than academic studies, there are a large number of applications of these models within industry, academia and governmental (regulatory) agencies. A small number of potential uses are listed below :-

1. The rational identification of new leads with pharmacological, biocidal or pesticidal activity.
2. The optimization of pharmacological, biocidal or pesticidal activity.
3. The rational design of numerous other products such as surface-active agents, perfumes, dyes, and fine chemicals.
4. The identification of hazardous compounds at early stages of product development or the screening of inventories of existing compounds.

5. The designing out of toxicity and side-effects in new compounds.
6. The prediction of toxicity to humans through deliberate, occasional and occupational exposure.
7. The prediction of toxicity to environmental species.
8. The selection of compounds with optimal pharmacokinetic properties, whether it be stability or availability in biological systems.
9. The prediction of a variety of physico-chemical properties of molecules (whether they are pharmaceuticals, pesticides, personal products, fine chemicals, etc.).
10. The prediction of the fate of molecules which are released into the environment.
11. The rationalization and prediction of the combined effects of molecules, whether it be in mixtures or formulations.

The key feature of the role of in silico technologies in all of these areas is that predictions can be made from molecular structure alone.

Methods

Predictive models of all types are reliant on the data on which they are based, the technique to develop the model and the overall quality of the information including the item to be modelled. In silico models for the prediction of the properties and effects of molecules are no different. In almost all cases two types of information are required for a model (the effect to be modelled and descriptors on the chemicals) and a technique(s) to formulate the relationship(s). The data to be modeled are denoted as the X-matrix, the descriptors as the Y-matrix. From such a matrix various types of relationship may be obtained by statistical, or other, means. For instance, a structure–activity relationship will be formed for a categorical endpoint, e.g. active/non-active or toxic/non-toxic. In this case a molecular fragment or substructure is associated with an effect. A quantitative structure–activity relationship is based on a continuous endpoint, e.g. potency where activity (X) is a function of one or more descriptors (Y). To develop a SAR, as few as a single compound might be required – should there be a very firm basis (such as a well-established mechanism of action) for developing the relationship. For instance, if a compound is known to elicit a particular effect, and the structural determinant is recognized, that structural fragment can be extracted. This may be in the form of a "structural alert" which can be coded easily into software. Obviously, the greater the number of compounds with the same structural determinant demonstrating the same effect, the greater the confidence that can be demonstrated in the alert. The formation of SARs is usually appropriate for a qualitative (i.e. yes/no; active/inactive; presence of toxicity/absence of toxicity, etc.) endpoint. To develop a QSAR, a more significant number of compounds is required to develop a meaningful relationship. An often asked question is "how many compounds are required to develop a QSAR?". There is no direct and simple response to this question – other than "as many as possible". To provide some guide, it is widely accepted that between five and ten compounds are required for every descriptor in a QSAR (Topliss, 1972; Schultz, 2003) .This does suggest that a one descriptor regression-based QSAR could be developed on five compounds. This is possible, but is very reliant on issues such as data distribution and range. Ideally "many more" compounds are required to obtain statistically robust QSARs, with some modelling techniques being considerably more data hungry than regression analysis.

Sources of Some Essential Resources for (Q)SAR, are

Internet

There are obviously many Internet sites, wikis and blogs devoted to (Q)SAR, molecular modelling, drug design and predictive ADMET. Two of the most well established are

1. The homepage of the International Chemoinformatics and QSAR Society: www.qsar.org – this is a good starting place for those in the field of QSAR; it also contains excellent listings of upcoming meetings and resources.
2. The homepage of the Computational Chemistry List: www.ccl.net – this also contains excellent listings resources and freely downloadable software. *Journals* Papers relating to (Q)SAR are published in a very wide variety of journals from those in pure and applied chemistry to pharmacology, toxicology and risk assessment and as far as chemoinformatics and statistics. The following is a small number that is commonly used by the author; whilst the reader will hopefully find these suggestions useful, they are, by no means, an exhaustive list (see the resources section of www.qsar.org which lists over 250 journal titles).
3. *Chemical Research in Toxicology*
4. *Chemical Reviews*
5. *Journal of Chemical Information and Modeling*
6. *Journal of Enzyme Inhibition and Medicinal Chemistry*
7. *Journal of Medicinal Chemistry*
8. *Journal of Molecular Modelling*
9. *"Molecular Informatics (formerly QSAR and Combinatorial Science)"*
10. *SAR and QSAR in Environmental Research*

Books

There are many hundreds of books available in areas related to (Q)SAR. Again, the reader is referred to the resource section of www.qsar.org. A very short list is given below, clearly biased by the author's own interests and experience. Apologies are given for omission of other "favourite" or "essential" books that have not been listed.

1. Cronin MTD, Livingstone DJ (eds) (2004) *Predicting Chemical Toxicity and Fate*, CRC Press, Boca Raton, FL.
2. Helma C (ed) (2005) *Predictive Toxicology*, CRC Press, Boca Raton, FL.
3. Livingstone DJ (1995) *Data Analysis for Chemists – Application to QSAR and Chemical Product Design*, Oxford University Press, Oxford.
4. Todeschini R, Consonni V (2001) *Handbook of Molecular Descriptor*. Wiley, New York.
5. Triggle DJ, Taylor JB (series eds) (2006) *Comprehensive Medicinal Chemistry II – Volumes 1–8*. Elsevier, Oxford.

Software

It is well beyond the scope or possibility of this section to note individual software for use in (Q)SAR. Experienced QSAR practitioners will no doubt be familiar with many of the freely available and commercial packages available. For the novice, in addition to the resources listed on www.qsar.org and www.ccl.net.

Conclusion

New diseases are coming up everyday. To combat these diseases, the need of newer drugs is increasing at an exponential rate. Moreover, pathogens are developing resistance against existing drugs. So, it has become a major challenge for the drug researchers to design newer and effective drugs. Natural products have always been the most preferable source of drugs as they have lesser side effects. Application of Bioinformatics tools in natural product research as well as in the drug discovery research as a whole has revolutionized the entire process.

References

Newmann , D. J. and Cragg , G., 2007 . Natuaral products as drug over the past 25 years . *J. Nat. Prod.* 70(3): 461-477.

Lipinski, C. A., Lombardo, F. Dominy, B. W. and Feeney, P. J. (1997). Experimental and computational approaches to estimate solubility and permeability in drug discovery and development setting (review). *Advanced Drug Delivery Review* .23: 3-25.

Joy, P. P., Mathew, J. T. S. and Skaria, B. P.(1998).Medicinal Plants, Kerala Agricultural University. Pp-2-10.

Bruins, A.P., Covey, T.R. and Henion, J.T. (1987). Ion spray for combined liquid chromatography mass spectrometry/atmospheric pressure chemical ionisation mass spectrometry. *Anal. Chem.* 59:2642–2646.

Kingston, D.G.I. (1979). High performance liquid chromatography of natural products. *J. Nat. Prod.* 42: 237–260.

Hook, D.J., Pack, E.J., Yacobucci, J.J. and Guss, J. (1997). Approaches to automating the dereplication of bioactive natural products. The key step in high throughput screening of bioactive materials from natural sources. *J. Biomol. Screening.* 2: 145–152.

Huber, L. and George, S.A. (1993). *Diode array detection in HPLC*. New York, Marcel Dekker.

Hostettmann K. and Wolfender J.-L. (2001). Applications of Liquid Chromatography/UV/MS and Liquid Chromatography/ NMR for the On-line Identification of Plant Metabolites. In Corrado Tringali (Eds) *Bioactive Compounds from Natural Sources:Isolation,characterisation and biological properties.* Published by Taylor and Francis. London and New York. Pp.69-75.

Hostettmann, K., Marston, A. and Hostettmann, M. (1998). *Preparative Chromatography Techniques: Applications in Natural Product Isolation, 2nd. Ed*. Berlin, Springer.

Niessen, W.M.A. (1999) *Liquid Chromatography-Mass spectrometry (2nd ed)*. New York, Dekker.

Verpoorte, R. (1998). Exploration of nature's chemodiversity: the role of secondary metabolites as leads in drug development. *DDT* .3:232–238.

Vlietinck A.J. and S. Apers A. (2001).Biological Screening Methods in the Search for Pharmacologically Active Natural Products. In Corrado Tringali (Eds) *Bioactive Compounds from Natural Sources:Isolation, characterisation and biological properties.* Published by Taylor and Francis. London and New York. Pp.3-26.

Spraul, M., Hoffmann, M., Lindon, J.C., Nicholson, J.K. and Wilson, I.D. (1993). Liquid chromatography coupled with high field proton nuclear magnetic resonance spectroscopy: Current status and future prospects. *Anal. Proc.* 30: 390–392.

Wolfender, J.-L., Rodriguez, S. and Hostettmann, K. (1998b). Liquid chromatography coupled to mass spectrometry and nuclear magnetic resonance spectroscopy for the screening of plant constituents. *J. Chromatogr. A* 794: 299–316.

Whitehouse, R.C., Dreyer, R.N., Yamashita, M. and Fenn, J.B. (1985). Electrospray interface for liquid chromatographs and mass spectrometers. *Anal. Chem.* 57: 675–679.

Wright (1998). Isolation of Marine Natural Products. In Richard J. P. Cannell (Eds) Natural Products Isolation.Humana Press.Totowa, New Jersey. Pp. 368-370.

Xiaohong L, Fang XU, Changlun S, Zhigang S, Yongcheng L, and Wing L, Chan.(2008). Bioactive metabolites from marine microorganisms.In Atta-ur-Rahman (Ed.) Studies in Natural Products Chemistry,. Elsevier B.V. 35:197-206.

Topliss, J.G and Costello, R.J. (1972). Chance correlations in structure-activity studies using multiple regression analysis. J Med Chem 15:1066–1068.

Schultz, T. W, Netzeva, T.I, Cronin, M.T.D. (2003). Selection of data sets for QSARs: analyses of *Tetrahymena* toxicity from aromatic compounds. SAR QSAR Environ Res 14:59–81.Webliography:http://www.rdchemicals.com/drug-relevant-properties.html

❑❑❑

CHAPTER – 27

Drug Delivery Methodologies

B. Behera[1], S. S. Sagiri[1], T. Sudeep[1], V. Patil[1], V. Varghese[1], B. Biswal[1], K. Pal[1], S. Roy[2], S.S. Ray[1] and B. Nayak[3]
[1]Department of Biotechnology & Medical Engineering, National Institute of Technology, Rourkela-769008, Odisha, India.
[2]School of Pharmacy and Technology Management, SVKM's NMIMS, V L Mehta road, Vile Parle (W), Mumbai- 400056, India
[3]Department of Life Sciences, National Institute of Technology, Rourkela-769008, Odisha, India.
Email : pal.kunal@yahoo.com

Summary

The process of administration of a drug into the human body to cure a disease is regarded as drug delivery. The common routes of administration of the drugs include oral, nasal, inhalational, transdermal, dermal, ocular, rectal and vaginal. In the current chapter, an attempt has been made to discuss about the various methodologies which are being used to deliver the drug to the human patients.

1. Introduction

A drug may be defined as a bioactive agent, which may be used for the purpose of diagnosis, prevention or cure of a disease. The drug may elicit a desired action only when its concentration in the systemic circulation is within the therapeutic range. To maintain the therapeutic concentration in the systemic circulation, there is a need to deliver the drug at an appropriate rate at a specific site. The conventional delivery systems, though patient compliant methods for drug delivery, suffer from various disadvantages including quick delivery of the drug and non-specific release of drug in all environments. The rate of absorption of the drug from the specific site may also be altered by tailoring the properties of the delivery vehicles. The method of improving the bioavailability of the drug by delivering the drug at a specific site is regarded as spatial placement whereas when the bioavailability is increased by altering the rate of delivery of the drug is regarded as temporal delivery. From the above, it is quite evident that for targeting specific tissues or organs it would be wise to explore various routes of administration (e.g. topical, oral, intravenous, intramuscular and intranasal) depending on the site of action of the drug, pharmacokinetic parameters of the drug and the physiological conditions of the patient. This would help in improving the bioavailability of the drug at the specific site of action. As for example, the bioactive agents which are meant to release for a prolonged period of time into the systemic circulation are delivered as intramuscular depot while the bioactive agents meant for quick actions are delivered intravenously. Transdermal, buccal, sublingual, nasal, rectal and vaginal delivery systems may be used to avoid first-pass metabolism of the drugs. In this chapter, an attempt will be made to discuss about the various drug delivery methodologies employed to as a delivery system.

2. Oral Formulations

The oral delivery of the drugs is one of the oldest methods of drug administration. The oral formulations have been found to be one of the most patient compliant methods of delivery of drug, which may be attributed to the easy administration of formulation, and do not require any special training or employment of specialized health care professionals. After the formulation is administered orally, the drug may be absorbed from the various parts, *viz.* stomach, duodenum, jejunum and ileum, of the gastro-intestinal tract. Depending on the type of the formulation, the drug may be selectively released at a particular site thereby reducing the chances of either damaging the mucosa or the drug. As for example, the drugs penicillin G and insulin are degraded at a lower pH of the stomach and their release in the stomach may be avoided. Also, non-steroidal anti-inflammatory drugs, e.g. ibuprofen and nimesulide, are gastric irritant in nature and may precipitate gastric ulcer if the drug. Hence, care should be taken to avoid the release of the drug in the gastric environment. A sizable number of drugs currently being used in the pharmaceutical industries are highly potent. This explains the need of excipients for the formulation of the dosage form. A well formulated dosage form may improve the bioavailability of most drugs by 2-5 folds. The oral delivery systems have gained much importance which may also be accounted to the economical production of the dosage forms (Rathbone, Drummond, and Tucker). The oral dosage forms may be broadly classified into solid and liquid dosage forms. The solid dosage forms may further be classified as tablets, tablet triturates, capsules, pills, lozenges, cachets, troche, cone, powders and granules whereas the liquid dosage forms may be further classified as monophasic and biphasic liquid dosage forms. In the current section, an attempt will be made to discuss about the various available forms of solid and liquid dosage forms.

2.1. Solid Dosage Form

Tablets

Tablets, also known as compressed tablets, may be defined as small compressed masses containing one or more drugs along with suitable excipients like disintegrator and binder. They are generally circular in shape and may either be flat or biconvex in shape. The properties of the tablets may vary as per the need. Tablets may be formulated either by three methods: wet-granulation, dry-granulation and direct compression. The wet-granulation method involves the mixing of the drug(s) along with various excipients, including diluent and disintegrating agent, followed by the preparation of dough using an aqueous solution of binders. The dough, hence formed, is passed through a sieve so as to obtain granules. The wet-granules are then dried under suitable environment. These dried granules are then mixed with lubricants and compressed to form tablets (Fig. 1, Fig. 2). The dry-granulation method includes pre-compression of the drug-excipient mixtures. The pre-compressed mass is then reduced to certain size, which is then mixed with lubricants and subsequently compressed to obtain tablets. In the direct compression method, the crystalline drugs are mixed with various excipients and lubricants and are directly compressed to obtain tablets (Wang et al. 2010a).

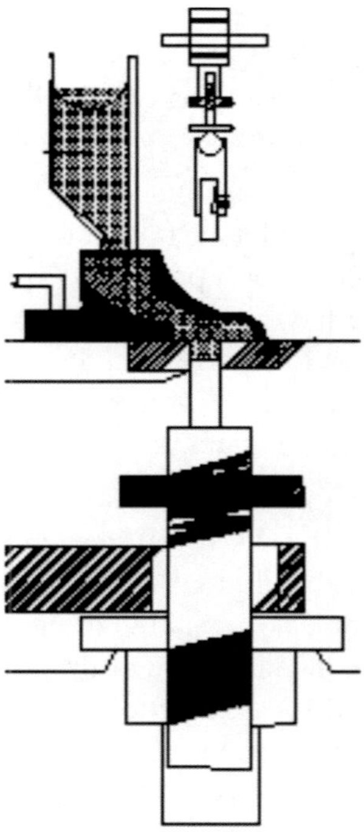

Fig. 1 : Schematic diagram of a tablet manufacturing unit (Redrawn from ,Rawlins 2006).

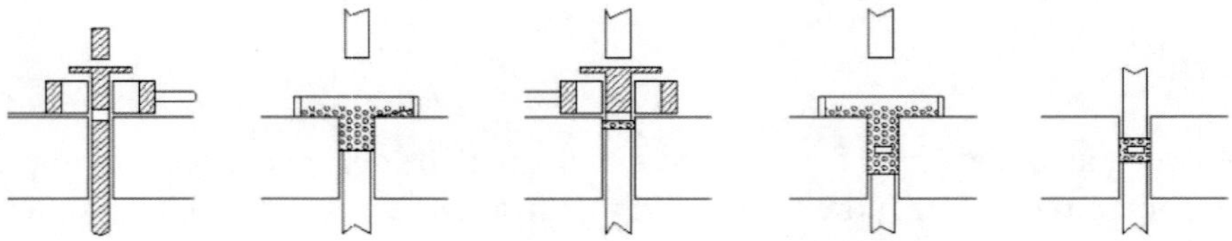

Fig. 2 : Stages of preparation of multilayered tablets (Rawlins 2006).

The tablets offer number of benefits which have lead to its their easy commercialization. The formulation and scaling up of tablets is very easy and economical which allows the manufacturing and marketing of the tablets at a cheaper price (van Santen et al. 2002). In addition to this, tablets have a longer shelf-life which may be accounted to the negligible moisture content in the tablets. If proper care is taken, it is feasible to develop batches with accurate dosing. The rate of release of the drug may be tailored by changing the composition of the formulation. All the above factors play an important role in wide spread use of the tablets in medication. Depending on the necessity, the properties of the tablets have been altered and may be categorized as rapid dissolution tablets, lozenges, solution tablet, effervescent tablets, sublingual tablets, chewing tablets, buccal tablets and implantable tablets. In the following paragraphs the various categories of the tablets will be discussed in brief.

If the drug is to be rapidly absorbed from the gastrointestinal tract, the tablet is required to disintegrate rapidly when the tablets are swallowed. These tablets are regarded as rapid dissolution tablets. The example of such drugs includes aspirin, paracetamol and nimesulide.

The second category of the tablets is not meant to be disintegrated. Rather they are meant to stay in the oral cavity for a prolonged period of time so as to form slow solution of drugs in saliva and have been named as lozenge tablets. These products have been designed to produce a local effect on the throat. Example of for mulations includes Vicks® and Halls® lozenges. The lozenges usually contain a very high proportion of sugar as excipients, which help in maintaining the integrity of the dosage form for a prolonged period of time.

Some tablets are meant to be dissolved very quickly in water so as to form a solution of the drug in water. The solution of the drug in water may be administered orally. The most common examples of this kind of tablets include biochemical tablets commonly used in homeopathic medication. The drug solution obtained by using these tablets may also be administered for external or parenteral delivery of the drugs. The term solution-tablet or solvella have been coined for the tablets which will lead to the formation of formulation for external use. If the tablets use a combination of an acid and a biocarbonate, which result in the release of carbon dioxide in the presence of water, for ready dispersion in water, then the tablets are regarded as effervescent tablets

(Rotthäuser et al. 1998). The common example of this includes dispersible Disprin® tablets. The tablets meant to be dissolved or disintegrated rapidly when kept under the tongue are classified as sublingual tablets. Glyceryl trinitrate is formulated as sublingual tablet.

Chewing tablets are the tablet dosage forms which are meant to be chewed (masticated) so as to break down the tablet mass. These tablets consist of higher amount of binders which enables the maintenance of the integral structure of the tablet. The drug commonly delivered by using chewing tablets is aluminium hydroxide, an antacid.

Buccal tablets are meant to be placed in the buccal cavity. These tablets may or may not contain mucoadhesive polymers. They are designed to release the drug for a prolonged period of time without being disintegrated. The drugs delivered by this method include ethisterone, theophyline (Boyapally et al. 2010) and diltiazam.

Some tablets are formulated for subcutaneous implantation into the human body. These tablets are known as implantable tablets. They may be formulated either by heavy compression or by fusion of the drug and the excipients. They are meant to be slowly absorbed from the site of implantation. Various steroids (Axelson et al. 1986), e.g. deoxycortone acetate and testosterone, have been formulated as implantable tablets.

Though tablets have gained much importance in medicine, they also suffer from certain constraints. The drugs to be used in higher quantities are difficult to formulate into tablets. Apart from this, the drugs with poor compressibility and are prone to degradation in the presence of moisture cannot be formulated as tablets. Drugs with poor wettability will take longer period of time for their dissolution and release from the tablets may not be uniform. Tablets do not provide ideal solutions for the formulation of liquid and semi-solid formulations. In the presence of gastrointestinal disorders, the residence time of the tablets gets decreased, which is not desirable. This leads to the reduced bioavailability of the drugs. Also, it is difficult to administer the tablets to the children and old aged patients.

Tablet Triturates

Tablet triturates, also known as moulded or compressed tablets, are obsolete now-a-days and are rarely used in medicine. They are small, cylindrical shaped molded tablets, which are meant to be completely soluble. The materials used for the preparation of moulded tablet triturates are usually moist and are formulated in a triturate mould whereas the compressed tablet triturates are usually prepared in the similar methods discussed in tablets. The compressed tablet triturates have rapidly soluble materials, *viz.* lactose, β-lactose, mannitol and dextrose, as the base. They have been further categorized either as dispensing tablets (potent drugs incorporated within powders and liquids which are never dispensed as a dosage form) or as hypodermic tablets (used for the preparation of solutions which are then injected).

Capsules

The capsules are solid-dosage forms which have the capability to encapsulate either solid (like powders and microparticles) or liquid/semi-solid drugs (Jones 2001). Gelatin and its products are used for the formulation of the capsule shells. Type A (derived from pork skin by acidic method) and type B (derived from bone and animal skins by alkaline method) have been extensively used for the production of capsules. The solid drugs are generally encapsulated within the hard-gelatin capsules and are hence referred to as dry-filled capsules. The liquid/ semi-solid drugs are encapsulated within soft-gelatin capsules, which derived its name due to the soft texture of the encapsulating gelatin sheet. This soft texture may be attributed to the presence of plasticizers like glycerol. The shells are tasteless and help in taste-masking and odour-masking of the medicaments. The preference by the patients is higher for capsules, hard and soft-gelatin capsules combined, as compared to the tablets. This is due to the relative ease of swallowing of the capsules in comparison with the tablets.

The hard-gelatin capsules consists of two parts, body and cap, which can slip over each other. The outer diameter of the body is slightly smaller than the internal diameter of the cap. The capsules are oblong shaped. The body is filled with the powder mixture or the microparticles, containing the medicament (Fig. 3). Subsequently, the cap is slipped over it to close the shells. The hard-gelatin capsules may also contain colourants and opacifying agents (Podczeck and Newton 1999). The shells may have 12-15 % of water content but this may vary depending on the storage conditions. Lower amount of water makes the shells brittle while higher amount of water makes the shells soft which ultimately leads to the loss of the shape.

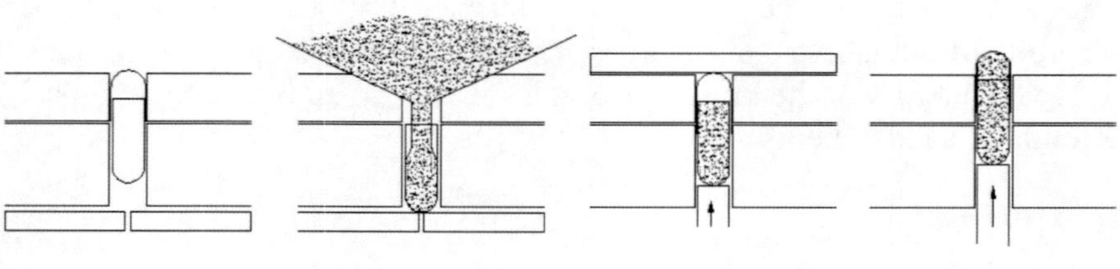

Fig. 3 : Stages of hard-gelatin capsule filling (Rawlins 2006).

The hard-gelatin shells are made by dipping the pegs of the stainless steel unit, having an array of uniform sized pegs protruding out of the unit, into gelatin solution. As the pegs are withdrawn from the gelatin solution, they are dried in kilns in which filtered air, with specific humidity. The dried shells are removed mechanically and collected. The body and the cap of the shells are prepared separately. The size of the shells may vary from 000 to 5 with 000 size shells being the largest and 5 size shells being the smallest. In general, 000 sized capsules may accommodate 600 mg and the 5 sized capsules may accommodate 30 mg powdered drug. The amount of the powder which may be accommodated in a capsule is also dependent on the density of the powder.

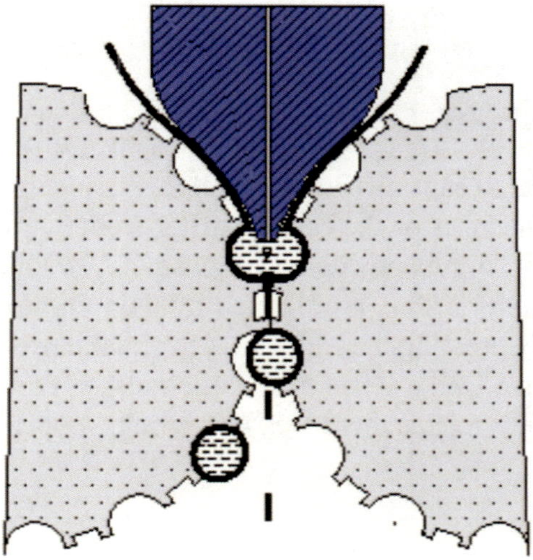

Fig. 4 : Stages of soft-gelatin capsule filling (Redrawn from (Rawlins 2006))

As compared with the hard-gelatin capsules, soft-gelatin capsules are soft and thicker and contain plasticizers like glycerine, sorbitol and similar polyols. The plasticizers help the shells to hold a higher amount of water and this necessitates the incorporation of the preservatives (e.g. methylparaben, propylparaben, sorbic acid) so as to inhibit the microbial growth on it. These are used to encapsulate liquid or semi-solid formulations whose external phase is apolar (Pissinati and Oliveira 2003). These capsules may be round, oval, oblong or suppository-shaped.

The soft-gelatin capsules are commonly prepared by rotary-die process (Fig. 4) (Cole et al. 2008). In this process, two gelatin ribbons are forced through rotary-dies moving counter-clock wise. At first, the gelatin ribbons are heat-sealed from the bottom, thereby forming a pouch-like structure, with the subsequent filling of the pouches with the liquid or semi-solid preparation using a metered dosage system and continuous heat-sealing of the edges. This results in the formation of a structure whose edges are joined by heat-sealing and encapsulating the medicament. In general, the sealing opens in 5 min and release the medicament in the gastrointestinal tract.

Pills

Pills are categorized as solid dosage forms which are intended for oral delivery of medicaments. They are usually small and round in shape. Drugs which have unpleasant taste may be formulated as pills. The use of the pills had decreased in last decade but their popularity is again increasing for the delivery of drugs not only by oral route but also by subcutaneous route. Conventionally, they are produced by converting the drug-excipient mixture into dough using glucose syrup as the binding solution. The dough is then rolled into a cylinder (pipe) and is subsequently cut into smaller pieces. These are then converted into small ball like structures, which are then coated with a sugar layer.

Cachets

Cachets are similar to capsules but the shells are made up of rice flour wafers, and are concave shaped. The sizes of the cachets differ from 3/4 to 1/8 inch diameter and may accommodate 0.2-1.5 g of powders (Simpson 1959). One of the wafers is filled with drug-excipient mixture. This is followed by the mointening of the edges of the wafers and subsequently pressing the edges so as to seal them off. Before administration, the cachets are immersed in water for few seconds so as to moisten the cachets. Moistening of the cachets makes them soft, elastic and slippery making them easier to swallow.

Cone

Cones are categorized as soild dosage forms whose name has been derived from the shape of the final product. Some authors have categorized them under the category of tablets. These formulations are meant to be inserted into the gingival cavity after the extraction of the tooth so as to release the antimicrobial drugs in the cavity for a prolonged period of time. They may also be devised to deliver astringents and anticoagulants to stop bleeding after the tooth extraction.

Powders

Powders can be defined as mixture(s) of fine particles. The powders are generally classified into five categories depending on their sizes: very fine (90-125 μm), fine (125-180 μm), moderately coarse (180- 355 μm), coarse (355-1000 μm) and very coarse (>1000 μm). Depending on the potency of the drug, the formulation may or may not contain excipients. A mixture of two or more drugs may also be dispensed as powders. The powder formulations may be intended either for internal or external use. In general, a good powder formulation is considered to have particles of uniform sizes. Presence of non-uniform particles may lead to the segregation of the powder particles, which may lead to non-uniform dosing of the drug(s). If the powders are dispensed in bulk then it is known as bulk powder whereas if they are dispensed as single doses then it they are known as divided powders. The divided powders may be dispensed in pouches made with cellophane, metallic foil or paper.

Granules

Granules may be defined as particles having sizes ranging from 2-5 mm. They are usually prepared by the methods already discussed in the tablets section.

2.2. Liquid Dosage Form

Liquid formulations consist of drug(s) either solubilised or dispersed in polar and/or apolar solvents. These formulations are easy to administer without any need of professionals or training and can be easily administered in young and old aged patients. The rates of absorption of the drugs are higher when they are administered as a solution. Apart from these, the dose to be administered may be easily tailored by changing the volume of the formulation. Like the two sides of a coin, liquid formulations too have some disadvantages including rapid reaction rate of the drugs when in solution which may render the drugs inactive and difficulty in carrying the containers. The liquid dosage forms may be categorized into two categories: monophasic liquid dosage forms and biphasic liquid dosage forms.

Monophasic Systems

The monophasic system may be regarded as homogenous mixture prepared by the dissolution of solid, liquid and/or gas in a liquid. The dissolution fluid is known as solvent while the component which is dissolved is known as solute. The monophasic dosage forms have been categorized as mixtures, elixers, syrups and paediatric drops.

The mixtures are categorized as liquid formulations which are administered in divided doses to the patients. The drug is dissolved, suspended or dispersed into a suitable solvent to give rise to a mixture. If the mixture is dispensed as a single dose, volume not exceeding 50 ml, then the formulation is known as draught. Mixtures generally have a short shelf-life as they are prescribed for acute clinical manifestations including cough, indigestion, diarrhoea, constipation and rheumatism. Depending on the type of solutes, the mixture have been classified in various groups: simple mixtures, mixtures containing diffusible solids, mixtures containing indiffusible solids, mixtures containing precipitate forming liquids and mixtures containing slightly soluble liquids.

Elixirs are defined as hydro-alcoholic preparations containing at least one bioactive agent in the formulation. Water and ethanol form the major components of the elixirs, which may contain glycerine, sorbitol, propylene glycol, flavouring agents, sugars and preservatives as addetives. Elixirs are optically clear products unlike various mixtures, which contain suspended oils or other vegetable components. Elixirs are generally used to for mulate the products containing potent drugs like antibiotics, antihistamines and sedatives. The bitter and nauseous taste can be masked with the help of flavouring agents. They have been categorized into non-medicated and medicated elixirs. The non-medicated elixirs are used as vehicle for the preparation of medicated elixirs e.g. aromatic elixirs, isoalcholic elixirs or compound benaldehyde elixirs. The medicated elixirs have been categorized as antihistaminic elixirs (e.g. chlarphenivamine maleate elixirs, diphenhydramine HCl elixirs), sedative and hypotonic elixirs (pediatric chloral hydrate elixirs), expectorant (terpin hydrate elixirs), miscellaneous (acetaminophen and paracetamol elixirs). Since the elixirs contain alcohol as one of the major component, they are generally stored in tightly closed amber coloured bottles to avoid the loss of alcohol by evaporation. They should be kept away from the direct heat and sunlight and be consumed within 7 days of preparation.

Syrups have been defined as formulations which are sweet, viscous and concentrated aqueous solutions of sucrose or other sugars in water or any other suitable vehicle. The syrups containing purified water as the vehicle are known as simple syrups. If the syrup contain various aromatic or pleasantly flavoured substances, the preparation is known as flavoured syrup. The syrups may be non-medicated or medicated. Simple syrup is a classical example of non-medicated syrup. It is prepared by dissolving 1 Kg of refined sugar in 500 ml of boiling distilled water with the subsequent addition of distilled water so as to make up the weight up to 1.5 Kg. The specific gravity of the syrup made by this method should be 1.33. The medicated syrup should contain at least one water soluble drug (e.g. diclofenac sodium). The presence of sugar in the syrups helps in preservation of the final formulation apart from the masking of the taste of the active agent. The sugar concentration in the syrup should be in between 65 and 67 % (w/w). The lower concentration of the sugar may result in the growth of the yeast and other microbes while a higher concentration may lead to the crystallization of the sugar as there is a change in the environmental temperature. The bacteriostatic effect of the syrups may be attributed to the osmotic effect exerted by the syrups on the microbes. The crystallization of the syrups may be avoided by the incorporating various polyhydric alcohols (e.g. glycerol and sorbitol) in the syrups. The syrups may contain the following excipients in the formulation: sugar polyols (e.g. glycerol and sorbitol), preservatives (e.g. parabens and bezoates), antioxidants (e.g. butylated hydroxytoluene and sodium metabisulfite), acids (e.g. citric acid → prevent the recrystallisation of sugar), buffering agents, chelating agents (e.g. sodium ethylenediaminetetraacetic acid), flavouring agents & flavour enhancers, colouring agents and ethyl alcohol (3-4% v/v). The syrups are meant to be stored in a closed container at temperatures not exceeding 30°C (Alanazi 2010).

Paediatric drops are the monophasic dosage forms meant for the administration of the drug to the children. In general, the dose administered to the children are less than 5 ml and may be attributed to the difficulty in the administration of the larger volume of formulations to the premature infants and children. These preparations contain potent medicaments, hence the precise measurement of the dose is important. This may be done either by using a dropper, which is accurately graduated in fractions of a millilitre or by using a pre-calibrated dropper.

Biphasic Systems

The biphasic liquid dosage forms contain two phases as in the case of emulsions and suspensions. These systems consist of either solid or liquid as the dispersed phase and a liquid as the continuous phase. If the dispersed phase is a liquid, then the formulation is regarded emulsion while the formulation containing solid as the dispersed phase is regarded as suspension.

Suspensions

Suspensions are defined as coarse dispersions containing dispersed insoluble solids in a liquid solvent. In general, the continuous liquid phase is either water or aqueous solutions. The sizes of the dispersed solids vary from 10-1000 μm. The suspensions have been administered either orally, injected intramuscularly or subcutaneously, inhaled or topically on the skin and in the ocular cavity. The suspensions may be categorized either as flocculated or deflocculated depending on the properties of the suspended particles (Fig. 5) (Lefebvre and Domenge 2002). If the particles occur as individual particles, the system is regarded as deflocculated suspension. These suspensions have shorter shelf-life and show a slower particle aggregation. The particles tend to aggregate by forming a chemical bridge in the flocculated suspensions. The aggregation of the particles may be increased by the addition of the flocculating agents. The flocs may be defined as light and fluffy conglomerates (Lester et al. 2010). Van der Waals forces are the predominant forces responsible for the formation of the flocs. Addition of counter ions in the ionic deflocculated solutions may result in the flocculation of the particles. The physical stability of the flocculated suspensions are better than that of the deflocculated suspensions but the bioavailability of the drugs when formulated as flocculated suspensions are lower than that of the formulations where the drugs have been suspended as deflocculated particles.

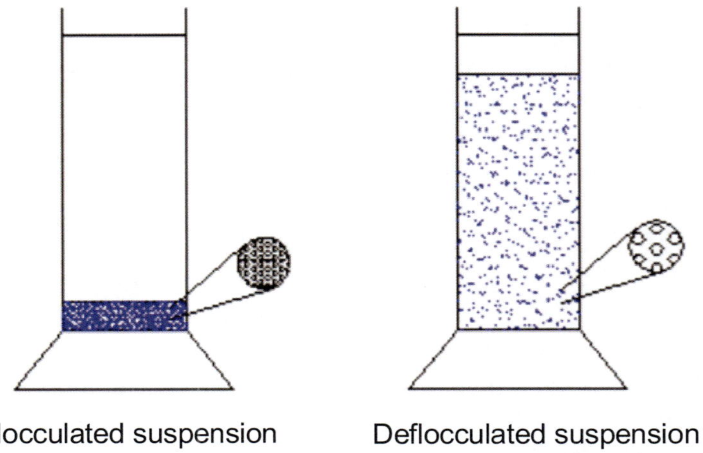

Fig. 5 : Characteristics of the flocculated and deflocculated suspensions (Swarbrick 2007).

The suspension has got various advantages which supports its candidature for formulating a product. The stability of a drug in solution form is a major concern as the drug molecules (e.g. procaine penicillin G) in the solution form may readily undergo chemical reaction and become inactive. The suspensions offer an effective taste and odour masking of the drugs by incorporating various sweetners and flavouring agents in the continuous phase. The drugs in the solid form act as a reservoir and help in improving the bioavailability of the drug by prolonging the absorption of the drug from the site of administration. Various water soluble polymers and inorganic compounds have been used to improve the dispersibility of the solutes in the continuous phase and are known as suspending agents. Table 1 enlists some commonly used suspending agents.

Table 1 : Suspending agents

Category	Example	Reference
Natural Polymers	Alginates, carageenan, xanthan gum, acacia, tragacanth, locust bean gum, guar gum	(Lu et al. 2003; Symes 1980; McHugh 1987)
Semi-synthetic Polymers	Carboxymethyl cellulose, microcrystalline celluslose, methyl cellulose, ethyl cellulose	(Bachmann et al. 1978)
Synthetic Polymers	Carbomers, polyvinyl pyrrrolidone, polyvinyl alcohol, poloxamer	(Dolan, Steelman, and Tumilowicz 1960)
Inorganic clays	Magnesium aluminium silicate, bentonite, hectorite	(Bhatty and Dollimore 1978)

Emulsions

Emulsions may be categorized as a heterogenous formulations consisting of two immiscible liquids, namely oil and water. These are regarded as colloidal system in which a liquid phase is dispersed in a continuous liquid phase, whose polarity is different from that of the dispersed phase (Pal 2010). The size of the inner phase may vary from 0.1-100 μm. Emulsions are generally considered as thermodynamically unstable system which may be attributed to the presence of increased interfacial free energy associated with them. This results in the inclination of the dispersed phase particles to coagulate with each other to form bigger particles so as to reduce the interfacial surface area and hence the interfacial free energy thereby resulting in the complete disruption of the emulsion (Tcholakova et al. 2006). Various compounds helps in reducing the interfacial free energy by forming a barrier layer amongst the two phases and are known as emulsifying agents. When formulated as a stable form, emulsions helps in masking the taste and odour of the drug incorporated in the dispersed phase and can help in the administration of the apolar drugs. Apart from this, the release of the drugs may be tailored by altering the composition of the emulsion. Drugs, prone to oxidation and hydrolysis, incorporated in the dispersed phase show limited chemical activity which improves the shelf-life of these drug formulations. Because of the above-mentioned properties, the principles of emulsification may be extended towards the formulation of creams and lotions for topical and transdermal applications.

The emulsions have been classified into four categories: simple emulsion, multiple emulsion, microemulsion and fine emulsions (Robins et al. 2002). The simple emulsions may be of two types, viz. oil-in-water or water-in-oil. They have only one interphase. The multiple emulsions have at least two interphases and the simplest form of multiple emulsion are oil-in-water-in-oil or water-in-oil-in-water. The multiple emulsions have been tried in sustained release formulations both orally and intramuscularly (Vladisavljevic et al. 2006). Microemulsions have the dispersed phase whose size is 0.01 μm (approx) are is not able to refract light. This results in the formation of clear/transparent solution. The microemulsions may become unstable, which can be attributed to the collisions of the particles due to the Brownian motion. The fine emulsions have dispersed phase size varying from 0.25-25 μm.

The preparation of the emulsion is dependent on the choice of the emulsifying agent. The property of an emulsifier is characteristic of hydrophilic-lipophilic balance (HLB). HLB scale is are an arbitrary scale dependent on the hydrophilicity and hydrophobicity of different component groups present in the emulsifying agent (Griffin 1949). The HLB scale varies from 0-20. Higher the HLB value, higher is the hydrophilicity of the emulsifying agent and the emulsifier promotes the formation of oil-in-water emulsion. On the contrary, emulsifier with lower HLB value promotes the formation of water-in-oil emulsion. The formulation of the multiple emulsion must contain both types of the emulsifiers to reduce the interfacial free energy from both the interfaces. Table 2 enumerates some commonly used emulsifiers and their HLB values.

The emulsion may be produced either by dry gum method, wet gum method and high pressure homogenization. In the dry gum method, proportion of volatile oil: water: gum and fixed oil: water: gum for the preparation of the primary emulsion is 2: 2: 1 and 4: 2: 1, respectively. In this method, the oil is triturated along with the gum. The trituration is continued with the addition of water till the cracking sound is produced marking the formation of the primary emulsion. To this primary emulsion, remaining of the water is added to make up the final volume. In the wet gum method, the gum is triturated along with the water. This is followed by the addition of oil, drop-by-drop, with continuous trituration to the mucilage (gum-water mixture) to form the primary. As the required volume of oil is added to form the primary emulsion, the volume of the emulsion is made up with the water to obtain the final volume. In the high pressure homogenization, the emulsifier, oil and water are mixed together to form a coarse emulsion. This coarse emulsion is subsequently passed through high-pressure homogenizer (Fig. 6) to produce emulsion with desired properties.

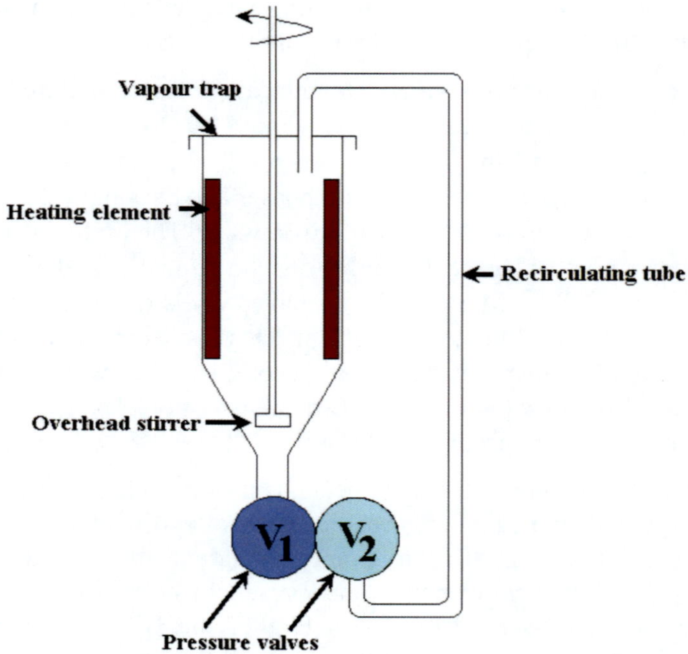

Fig. 6 : Schematic diagram of a two-valve pressure homogenizer

Table 2 : Some commonly used surfactant

Surfactant	HLB
Sorbitan trioleate	1.8
Propylene glycol monostearate	3.4
Sorbitan monostearate	3.4
Glyceryl monostearate	4.7
Sorbitan monolaurate	5.5
Polyoxyethylenesorbitan monooleate	15
Acacia	-
Gelatin	-
Bentonite	-

Tinctures

The word tinctures have been derived from the Latin word tingere, which means to dye or soak in colour. These are alcoholic or hydroalcoholic extracts or solutions of plants and their derivatives which act as a medicament. Due to this reason, the tinctures have been categorized under the class of galenicals. Apart from the use of hydroalcoholic solutions for the preparation of the tinctures, other solvents like vinegar, glycerol, ether and propylene glycol have also been tried (Cantor and Winicov 1979). The tinctures are generally prepared by the method of maceration and percolation. Various tinctures have been tried as adaptogens, anticatarrhals, astringents, carminatives, cholagogue, demulcents, and tonics (Bilia et al. 2007). The maceration technique involves incubation of the plant and its derivatives soaked into the solvent, kept in a stoppered container for certain duration and at a specific temperature. During the incubation period, the container is frequently agitated so as to promote maximum dissolution of the components into the solvent. After the incubation period, the solution is filtered out and stored in a well closed container. The percolation technique involves mixing of the ground drug along with the solvent so as to dampen the grounded drug and is subsequently transferred to a percolator for a period of 24 h to allow the percolation of the components. Most tinctures presently being used are simple, impure extracts made with alcoholic, hydroalcoholic and aqueous solutions. Some common examples of tinctures include tincture of *Cannabis sativa* tincture of benzoin, tincture of cantharides and tincture of ferric citrochloride.

Aromatic Waters

Aromatic waters are clear aqueous solution, saturated with volatile oils (e.g. rose oil, peppermint oil) and/or other aromatic or volatile substances e.g. camphor. The odour and taste of the aromatic water is derived from the drugs or volatile substances used to prepare the same. They may be prepared either by distillation, solution and alternate solution method. In the distillation method, coarsely ground odoriferous portion of the plant is placed still with sufficient purified water. The excess oil collected with the distillate rises to the top of the aqueous product and is removed. The remaining aqueous solution, saturated with volatile material is clarified by filtration. In the solution method, aromatic water is prepared by intermittently shaking 2 ml (if liquid) or 2 gm (if solid) of the volatile substance with 1000 ml of purified water in suitable container for a period of 15 min. The mixture is set aside for 12 h or longer to permit the excess oil and the solid substance to settle. The resulting solution is passed through a wetted filter paper and purified water was added as needed to bring the volume of the filtrate up to the prescribed quantity. They are categorized into two types, simple and concentrated aromatic waters. In the alternate solution method, volatile oil or suitably comminuted aromatic solid is thoroughly incorporated with talc and purified water is added to make up to 1000 ml. The resulting slurry is thoroughly agitated several times for a period of 30 min and subsequently filtered. Aromatic waters should be protected from light and hence stored in airtight and light resistant containers.

Linctuses

Linctuses are sweet, mucilaginous and viscous liquid preparations, which usually contain medicaments possessing demulcent, expectorant or sedative properties. They contain a high proportion of syrup except codeine linctuses, which contain chloroform water as diluent and glycerol as demulcent. Dilutions are prepared fresh and administered in small doses from 1/2 to 2 fluid drachms on the membranes of the throat. The dose volume is generally 5ml. On administration they form a temporary protective or remedial film over the mucous membrane of the throat and bronchial region thereby helping in alleviating the symptoms of cough. They are usually intended to be sipped and swallowed slowly without dilution to prolong the demulcent effect. Apart from vehicles, linctuses contain stabilisers, colouring agents, flavouring agents and preservatives. Since most linctuses contain syrup as the vehicle, they should be stored at constant temperature in a well closed containers. Sometimes invert syrup sugar or glycerol can also be used as vehicle.

3. Rectal Administration

Oral administration is the route of choice in the daily practice of pharmacotherapy. However, the potential for oral dosage form development is severely limited for drugs that are poorly absorbed in the upper gastrointestinal (GI) tract and unstable to enzymes. Some agents cause GI irritation or require doses in excess of 500 mg. Certain patient populations, notably children, the elderly, and those with swallowing problems, are often difficult to treat with oral tablets and capsules. Additionally, treatment of some diseases is best achieved by direct administration near the affected area. In these cases the rectal route may represent a practical alternative. The human rectum represents a body cavity in which drugs can be easily introduced and retained for suitable time to allow substantial release of the drug so as to improve the systemic/local bioavailability. Both systemic and local effect can be achieved through rectal route (Van Hoogdalem et al. 1991). Major class of therapeutic agents administered rectally include anticonvulsants, non-narcotic and narcotic analgesics, theophylline, antiemetics and antibacterial agents. Drugs inducing anaesthesia in children may also be administered rectally. Blood draining the lower part of the rectum largely bypasses the liver so that drugs showing a high first-pass metabolism when given orally are more effectively absorbed when administered rectally. This is because the rectum has lower haemorrhoidal vein and this vein absorbs about 70-80% of drug and bypasses the liver. The pH of rectum plays an important role in absorption of drug. The fluids that are present at rectum do not have buffer capacity, hence non ionized drugs are rapidly absorbed (De Boer et al. 1982). The barrier separating the rectum to blood is highly permeable to unionized drugs. The absorption of drug at the rectum depends on nature of the drug, diffusion of drug from rectum to site of action, amount of fluids secreted by the rectum and the thickness of the anorectal membrane. Drawbacks of rectal drug administration include the interruption of absorption by defecation and lack of patient acceptability.

Suppositories

Suppositories are defined as solid dosage forms intended for insertion in to body cavities such as rectum, vagina and urethra for systemic or local effects. These were first invented by Egyptians. Ancient people used wooden piece for making a suppository. The bioavailability of the drug from the suppository dosage form depends on the physicochemical properties of the drug as well as the composition of the base. The physical properties of the drug can thus be modified to a degree, as can the characteristics of the base selected for the delivery system (Hosny et al. 1996). They are available commercially in various shapes and in different sizes. The USP describes rectal suppositories to be tapered at one or both ends to resemble a torpedo shape weighing about 2 g each. Those intended for infants weigh about 1/2 the adult suppositories. Urethral suppositories are called bougies. They are pencil shaped and pointed at one extremity. Those intended for males are 100 - 150 mm long weighing about 4 g each. For females they are 60 - 75 mm in length weighing about 2 g. Suppositories employ vehicles that melt at body temperature or dissolve in the fluids present in the rectum or the vagina. Rectal suppository bases can be classified broadly into two groups, viz. fatty and water soluble or water miscible. The bases that are used for dissolution of drugs must be non-toxic and non-irritative. They should be compatible with drugs, have high melting and emulsifying properties, suitable for easy moulding into suppositories, should have acid value <0.2, saponification value of 200-245 and iodine value of <7.

The traditional cocoa butter or theobroma oil is immiscible with aqueous fluids. It is a naturally occurring triglyceride containing 40% unsaturated fatty acid. It satisfies many of the requirements for an ideal suppository base since it is innocuous, bland, unreactive and melts at body temperature. A major characteristic of theobroma oil is its polymorphism, i.e. its ability to exist in more than one crystal forms a phenomenon which has been attributed to the presence of high proportions of unsaturated fatty acids. The α, γ, β' and β are the four major crystal form of cocoa butter. Each of the different forms has different melting point as well as different drug release rate. Cocoa butter, however, possess several disadvantages which include rancidity during storage, melting in warm weather, liquefy when incorporated with certain drugs and isomerize to an undesirable lowered melting point. Substitutes available commercially include hard fat and hydrogenated vegetable oil such as coconut or palm kernel oil. A common method of producing fat intended for use as a suppository base involves interesterification.

Glycerinated gelatin is the hydrophilic base, which is usually used as a vehicle for vaginal suppositories. Those intended for rectal administration can be obtained by increasing the gelatin concentration. Glycerinated gelatin suppositories do not melt at body temperature but rather dissolve in the secretions of the cavities in to which they are inserted. They are prepared by molding. Another class of bases are the glycerine suppository bases used for cathartic action.

Water miscible bases also include the polyethylene glycol which improves the properties of the glycerinated gelatin and lipid type suppository bases. Suppositories of varying melting point can be prepared by blending polyethylene glycol of 1000, 4000 or 6000 MW. They can be prepared by both molding and cold compression method. Several non-ionic surfactants closely related to polyethylene glycol have been developed. Some of these are polyoxyethylene sorbitan fatty acid esters (Tween), polyoxyethylene stearates (Myrj) and the sorbitan fatty acid esters (Span and Arlacel). They may be used alone or in combination with other suppositories vehicle to obtain a varying melting ranges and consistencies.

Suppositories may be prepared by four methods namely hand molding, compression on a regular tablet press, pour molding and automatic molding. The hand molding method, involves rolling the well blended suppository base containing the drug into rod of desired length and diameter or in to balls of desired weight by hands. The rod shaped structures are then cut into pieces and the ends of those pieces are made into pointed tips. This is one of the simplest, practical, economical and oldest method for the manufacture of small number of suppositories. A more uniform and pharmaceutically elegant suppository can be prepared by compressing the blend using a wheel operated press. This method avoids heat and is also known as cold compression method. It avoids the sedimentation of insoluble solids in the suppository base, but it is too slow for large scale production. One of the major disadvantages of this technique while using fat-type bases is air entrapment. This unavoidable inclusion of air makes close weight control impossible and also favours oxidation of both base and the drug (Ibrahim et al. 1990).

The most common method employed in the manufacture of this dosage form is the pour molding method. Suppositories can be produced both on small and large scale using this technique. First, the base is melted preferably on a water bath and the drug is added to it by either emulsification or suspending methods. The blend is passed through chilled chrome or nickel plated molds and finally allowed to solidify. Automatic molding is similar to pour molding but automated machine is used instead of chilled molds. The molding operations like pouring, cooling and removal are performed by machine. All the operations are fully automated. This method results in production of 10,000 suppositories per h. Finished suppositories are then routinely inspected for their appearance and uniformity of the mix. To ensure uniformity to the labelled content they are assayed individually for active ingredients.

Melting range test is performed to check the physical and absorption characteristics. In order to ensure minimal breakage during package and shipping fragility test is performed.When suppositories are finally ready they should be stored at room temperature for 2 days. They are then packaged in boxes that hold them upright. Screw-capped glass containers are used for packing glycerine and glycerinated gelatin suppositories. Foiling is done in tin, aluminium, paper and plastic strips. Individually wrapped suppositories are usually packaged in slide, folding or set up boxes. A major advance in the manufacture of suppositories was the development of automated methods for molding suppositories directly in their wrapping material.

Pessaries

The term is derived from a greek word describing the small stones used for playing the game of draughts. They are exclusively used for their local action on vagina. Pessaries or vaginal suppositories are molded in to globular or oviform shape, or compressed on a tablet press in to modified conical shape weighing about 4 g each if prepared from oil of theobroma or cocoa butter and 10 g if made from glycerated gelatin. The large-scale manufacture of pessaries uses either the fusion method where the drug is dispersed in a molten base and the mixture poured into molds to solidify, or the cold compression method. There are mainly three different types of pessaries: therapeutic pessaries, pharmaceutical pessaries and occlusive pessaries. Therapeutic pessaries are similar to the outer ring of ear diaphragm. These are designed to support the uterus, vagina or rectum. The different uses of therapeutic pessaries include (a) treatment of prolapsed uterus (b) to perform abortions and (c) to treat urinary incontinence. These pessaries are fitted temporarily or permanently by the physician. These are used in conjunction with estrogen, because without it these pessaries may cause chronic damage to the rectal tissues. Pharmaceutical pessaries are used for the delivery of drugs into the rectum. These are used where action of drug is limited to surrounding areas of the rectum only. Generally they are used to treat inflammation or infection of the uterus.

Pessaries that are used for administration of contraceptive drugs like spermicide into the vagina are called occlusive pessaries (Vierhout 2004). The stem pessary is a type of occlusive pessary used for urinary administration of drug. Pessary can be readily inserted intravaginally and removed according to a predetermined schedule. A pessary set consists of pessary ring and sterilised gloves. By using the middle and index finger, the pessary ring is placed into the vagina. The average size of the ring is 4 cm. Once the ring is fitted, the patient is asked to walk for 1 h to avoid void. The patient with cystocele is treated by using ring passary with support (filled in center) while the patient with uterine prolapse is treated by using ring pessary without support (hallow). Once the pessary is inserted into patients, they should clean the pessary every 3-6 months. To prevent the infections that may occur in presence of pessary, estrogen must be applied vaginally for 2-3 times weekly. The patient has to contact the health care professional every 3-6 months to check for vaginal ulcers (Sulak et al. 1994). The pessaries may last upto 5 years .

Enemas

Enemas are one of the most ancient and widely used methods of introducing therapeutic substances into the body. They are rectal injections of liquid, either plain or containing drugs, into the rectum and colon to empty the lower intestine or to introduce food or medicine for therapeutic purposes. Those employed to evacuate bowel are known as evacuation enema. The later are known as retention enemas. They influence the general system by absorption, or to affect a local disease (Campieri et al. 1981) and possess anthelmintic, nutritive, sedative or stimulative properties. Radioopaque substances may be incorporated for roentographic examination of lower bowel. They are intended to be retained in the intestine and should not be used in larger quantities, usually more than 150 ml for an adult. Small volume preparations are known as microenema. Common vehicles for retention microenemas are small quantities of ethanol and propylene glycol. Sodium chloride, sodium bicarbonate, sodium mono hydrogen phosphate, and sodium dihydrogen phosphate are used to evacuate bowel. They should be labelled as "For rectal use only".

4. External Preparation

Medications are applied externally to the skin or inserted in to the body cavities for both local and systemic effects. They are available in solid, semisolid or liquid forms. Solid dosage form intended for the external application include the medicated topicals which are classified as ointments, creams, pastes, gels/jellies, dusting powders, poultices and plasters. Their common property is the ability to cling to the surface of application for reasonable duration before being washed off. This adhesion is due to their plastic rheologic behaviour which allows them to retain their shape and cling like a film to the skin.The skin is the largest multilayered organ. It covers a surface exceeding 20000 cm^2. Anatomically the human skin is divided in to three layers the epidermis, dermis and the subcutaneous fat layer. Its accessibility and the ability to maintain applied preparation intact for prolong period of time has resulted in its increasing route for many active ingredients, whether for local, systemic or regional effect (Denet et al. 2004). When a drug is applied topically, the drug diffuses out of the vehicle on to the surface tissues of the skin. Thereafter it follows either of the three modes of entry (a) through the follicular region (b) through the sweat ducts (c) or through unbroken stratum corneum. Once it crosses the stratum corneum there is no hindrance to penetration of the remaining layers of the epidermis and corium. The drug is readily transported into the circulation via capillaries. Topicals can also be applied to the mucous membrane. Preparation that are inserted in the body cavities also belong to this class. External liquid dosage form may further be classified in to lotions, liniments, paints, gargles, mouthwashes, ear drops, nasal solutions and inhalations. All external preparation are labelled as "For external use only".

Ointments

An ointment is a homogeneous, viscous, semi-solid preparation, that is intended for external application to the skin or mucous membranes. They are used as emollient for softening the skin or for the application of active ingredients to the skin for protective, therapeutic, or prophylactic purposes and where a degree of occlusion is desired (Yi-

Hung et al.1986) e.g. Paraffin ointment BP, Compound benzoic acid ointment, calamine ointment, Cetrimide emulsifying ointment etc. Some drugs are used in there micronized form e.g. coritcosteroids. The drugs are either dissolved or dispersed in the vehicle in the form of fine powder. The vehicles used for preparing ointment are termed as ointment bases. Choice of the vehicle is dependent on the formulation. They are classified in to four categories Oleaginous/ hydrocarbon bases, absorption bases, Water miscible bases and water soluble bases. Hydrocarbon bases forms thin film on the skin which are not absorbed through the skin.

These bases increases the skin hydrartion by preventing the loss of surface water .e.g. petroleum jelly, white wax, mixture of hard paraffin, soft paraffin & liquid paraffin. Hydrocarbon bases are classified as oleaginous bases along with those obtained from vegetable fixed oils or animal fats. Bases of this type include lard, benzoinated lard, olive oil and other oils. Petrolatum is a complex mixture of hydrocarbons, containing aliphatic, cyclic, saturated, unsaturated, branched and unbranched substances in varying proportions. It has a high degree of compatibility with variety of medicaments.

Absorption bases are non emulsified, anhydrous or hydrous and have a tendency to absorb additional water. Those that are anhydrous, absorb water to become W/O emulsion. Compared to hydrocarbon bases they are less occlusive and easier to spread, e.g. wool fat (anhydrous lanolin), hydrophilic petrolatum (aquaphor), wool alcohols, bees wax. The other class of absorption bases based on w/o emulsion, which are capable of absorbing more water as compared to its anhydrous weight. Hydrous wool fat (lanolin) is a classical example of this category.

Water miscible bases are o/w emulsions and are referred as creams. They are also known as water removable bases. Vanishing creams belong to this class. Advantages of this type of bases include (a) Miscibility with exudates from lesions, (b) high cosmetic acceptability, (c) easy removal from skin or clothing. Another class of ointment bases include the water soluble bases. They are prepared from low and high molecular weight macrogols (polyethylene glycol) .The low molecular weight macrogols are liquid while those having a high molecular weight are solid. water soluble bases are also known as greaseless ointment bases. Ointments are prepared by two methods (a) fusion method and (b) trituration method. In the former method, the ingredients are melted in a decreasing order and stirred thoroughly to ensure homogeneity until the mass cools down. This method is applicable for ointment base containing a number of solid ingredients. Most common and widely used method for extemporaneous preparation is the trituration method. The medicaments which are to be incorporated are generally insoluble in the base. The medicaments are reduced to fine powder and evenly distributed by grinding with a small amount of base, followed by dilution with gradually increasing amounts of base. Preservatives such as methyl and propyl paraben may be added to prevent contamination, deterioration and spoilage by bacteria and fungi since many of the components of these systems are substrates for these microbes. Ointments are stored in well closed container and in a cool place.

Pastes

A paste is a semisolid dosage form that contains a large proportion (20-50%) of solids like starch, zinc oxide, calcium carbonate, that are finely dispersed in a fatty or aqueous vehicle. They are stiffer as compared to ointments & creams. Proportion of solids is also more in pastes. They adhere well to the skin, forming a thick coat acting as an absorbent and protectant. Pastes are therefore used in the treatment of oozing lesions where they help in absorbing the serous secretions thus soothing the raw surface. Pastes are generally applied by means of wooden spatula or knife. Sometimes they retard the use of bandages & gauges. They might be applied as such on the skin or supporting or backing material like sterile gauge & then applied to the skin, e.g. lassar's paste (Zinc & Salicylic acid paste BP), magnesium sulphate paste, zinc & coal tar paste, triamcinolone acetonide dental paste etc. Based on the type of vehicle they are classified in to two types, Fatty pastes containing oleaginous bases e.g. Zinc oxide paste and those made from single phase aqueous gels e. g. carboxymethylcellulose sodium paste. Bases meant for formulation of pastes are classified in to Oleaginous/ hydrocarbon bases, water miscible bases and water soluble bases. The methods of preparation of pastes are similar to those of ointments.

Gels/Jellies

Gels are colloidal systems that contain insoluble inorganic drugs in their hydrated forms. They are characterized by a high degree of elasticity. Gels are sometimes referred as jellies. Jellies are water soluble bases prepared from natural gums such as tragacanth, pectin, alginates, or from synthetic derivatives of natural substances such as methyl cellulose and sodium carboxymethyl cellulose (Normand et al. 2003). They are used for medication, lubrication or contraceptive purpose. Gels like gelatin gel are also used as a suppository bases. They can also be used to administer the drug in other body cavities, e. g. phynylephrine hydrochloride nasal jelly. Gels can be classified as two phase gels or single phase gels. Gel mass of a two phase gel contain network of small discrete particles e.g. aluminium hydroxide gel. In 2 phase system, if the particle size is large then it is termed as magma e.g. bentonite magma. Gels of colloidal clays generally contain sodium bentonite. Their elastic deformability is limited and are similar to pastes in appearance. Their rigidity and yield value are independent of temperature. When subjected to stress above the yield value these gels break down to smooth and free-flowing sols. Single phase gels consist of organic macromolecule dispersed in a liquid in such a way that no boundaries are visible between them. They are prepared from synthetic macromolecules (carbomer) or from natural gums. Depending on the liquids gelled by means of suitable gelling agents, gels can also be classified as oleogels and hydrogels. Hydrophobic or oleogels contain liquid paraffin, with polyethylene or fatty oils gelled with colloidal silica or aluminium or zinc soaps (Osada et al. 2001). Hydrophiliic gels or hydrogels usually consist of water, glycerol, or propylene glycol gelled with suitable gelling agents like tragacanth, starch, cellulose, carboxyvinyl polymer, magnesium aluminium silicate etc (Tiwari et al. 2009).Gels are prepared by adding gelling agent to an aqueous solution of the drug. The mass is then triturated in a mortar until a uniform product is obtained. Gels are subjected to degradation, preservatives may be added to prevent the deterioration. They are stored in well closed containers in a cool place to prevent drying.

Dusting Powder

Powder intended for external use are termed as dusting powder .They are applied to the skin for antiseptic, antipruritic, astringent, antiperspirant, absorbent, protective and lubricant purposes. They may contain talc, zinc oxide, starch, kaolin, perfume & colour. The powder must be homogeneous and in fine form. The fineness of powders is expressed in terms of mesh size and thus the various ingredients used in the formulation are generally passed through sieve no 100 - 200. Particle size more than 150 μm (100-mesh), otherwise may lead to feeling of grittiness and lead to irritation of the skin. Dusting powders are of two types viz. medicals and surgical dusting powders. Medical dusting powders are intended for superficial skin conditions and sterility is rarely essential. However they must be free from pathogens These powders should not be applied to open wound or raw surfaces. The other class of dusting powders are the surgical powders. They are sterile preparations intended to be applied to major wounds, burns and on the umbilical cords of infants. They are also used in body cavities. Dusting powders are provided in specialized containers like sifter top containers or pressure aerosols or wide mouth containers to protect them from atmospheric condition. Brushes, puffs, or sterile gauge pads might be used for their application, e.g. zinc & salicylic acid dusting powder, dicophane dusting powder etc.

Plasters

Plasters are substances meant for external application to the skin. They adhere to the skin as by means of an adhesive material. The adhesive layer helps adhering the plastic backing with the skin. Proper balance of adhesive and cohesive forces between skin, backing material (like plastic, bandages, gauges) and plaster is necessary, so as to provide clean skin surface after removal of plaster. They provide protection and support to the area applied and help bringing the medications in the close contact with the skin. Some medicinal plasters are intended for long term applications to provide local or regional drug delivery. Such systems are the prototypes of transdermal delivery systems.

Liquid preparations intended for external application may be either in solution clear liquid, suspension containing dispersed particles or emulsions containing different phases. The choice of the formulation depends on variety of factors one of which is the type of drug and its solubility.

Liniments

Liniments are liquid preparation for external purpose. They may be applied with or without friction. Liniments containing non steroidal agents are used for their anti-inflammatory and analgesic effect (Wang et al. 2010b). They may be applied as paint with a soft brush. Because they are rubbed on to the affected area, they are also termed as embrocations. They may be massaged into the skin with considerable friction. Two types of vehicles are used in the formulation of liniments. High concentration of alcohol as in soap liniment and oil as in camphor liniment and methyl salicylate liniment. The former are majorly used as rubefacient, counterirritant and astringents .The later oily liniments are milder in action. They serve as protective coating. They should not be applied to broken or bruised skin. They are stored in tightly closed containers.

Lotions

Lotions are either semiliquid or liquid preparations that contain one or more active ingredients in an appropriate vehicles intended for application to the skin for their soothing effect. They differ from liniments by being aqueous rather than alcoholic or oleaginous in nature. Lotions containing suspended particles, evaporate when applied to the skin leaving a deposit of medicament on the surface. They are usually applied directly to the skin or the affected area without rubbing with the help of cotton wool or gauze. The particles are of colloidal dimension because of which they are more soothing to the inflamed area (Eccleston 1997). Lotions are prepared by triturating the ingredients to a smooth paste and then adding the remaining vehicle with triturating. High speed mixers and colloid mills produce better dispersions and therefore used on an industrial scale. They are used for local cooling, anaesthetic, soothing and protective purposes. They may contain either humectants such as glycerin, so that moisture is retained on the skin after application, or alcohol, which evaporates quickly imparting a cooling sensation to the skin. Antihistamines, benzocaine, calamine, betamethasone derivatives are common ingredients in an unofficial lotion. Lotion that are suspension or dispersion have a tendency to separate out on long standing therefore they must be labelled as "Shake well before use ".

Paints

Paints are solutions for application to the skin or mucosae usually with a soft brush. Skin paints contain a volatile solvent that evaporates quickly leaving a dry or resinous film of medicament. Paints that are used for mouth and throat infections are known as throat paints. The throat paints are applied using a throat brush. They contain a high content of glycerine, which due to the sticky nature adheres to the affected area and prolongs the action. All official paints are in solution form. Paints contain active ingredients meant for antiseptic, caustic, soothing, or stimulating properties. Those intended to remain in contact with a specified surface are usually prepared with collodion, glycerine and water. Those intended to be absorbed are prepared with oleic acid or fatty oils. Caustic substances are usually dissolved in vehicles such as distilled water, alcohol, or ethereal, whereas resinous substances, such as benzoin, storax, tolu balsam, or sandarac dissolved in ether, are employed as bases for medicated varnishes, and used for application to the skin and raw mucous surfaces. Paints containing dyes should be handled with precaution while dispensing to prevent spillage. All paints except throat paints are labelled "For external use only" and "Store in a cool place". They should not to be swallowed in large amounts.

Ear Drops

Ear drops are preparations in which the drug is dissolved or suspended is a suitable vehicle for local use in ear. The vehicles generally used are water, dilute alcohol, glycerine, propylene glycol. Ear drops are meant to be instilled into the ear with the aid of dropper for treating mild infection eg. cholroamphenicol, cleansing after infections eg. spirits, to soften the ear wax (xylene). They are dispensed in coloured, fluted bottles with the dropper. Sometimes plastic containers are also used. The labels contain "Not to be taken, store in a cool place and not to be diluted with water". To provide sufficient time for the preparation to act, it is necessary for the patients to remain on their side for a few minutes to prevent the drop flowing out of the ear.

Gargles

Gargles are aqueous solutions used for the treatment of mouth and throat infections They usually contain antiseptics, analgesics, astringents and have mild deodorant action. They are prepared by simple mixing. The volatile components are first dissolved in little alcohol.. This alcoholic solution is then mixed with talc or kaolin which acts as a suspending agent when the solution is added to aqueous solution and glycerol along with the drug. These solutions are either used directly or diluted with water if they are concentrated. Gargles may contain phenol or thymol as bactericide and potassium citrate for its mild astringent effect. This effect helps to restore the tone of a relaxed throat, stimulates the flow of saliva, which relives dryness. Gargles also contain flavouring and colouring agents. They are dispensed in clear fluted glass container and should not be swallowed in large amounts.

Mouthwashes

Mouthwashes are aqueous solution with a pleasant taste and odour. They are used for cleansing and deodorising the buccal cavity. They are refreshing, particularly for the bed ridden patients. Mouth washes contain four groups of exicipients viz., alcohol, surfactants, flavours and colouring agents. Alcohol is present in 10 - 20% concentration. It serves as a solubilising agent for some flavouring agents, helps in masking unpleasant taste and may function as a preservative. Humectants such as glycerin and sorbitol forms 5 - 20 % of the mouthwash. These agents increase the viscosity of the preparation. Non - ionic surfactants such as tween and span are used in the concentration of 0.1% - 0.5%. They help in removal of the debris by their foaming action and also help in the solublisation of flavouring agents. Anionic surfactants like sodium laurly sulphate may be used. Cationic surfactants such as cetylpyridinium chloride are used as antimicrobial agents. Flavours are used in conjuction with alcohol and glycerine to mask the bitter taste. Flavours must be safe to use. Peppermint, cinnamon, spearmint, menthol are some commonly used flavours in mouthwashes. In order to give a pleasant appearance to the product, colouring agents are used. Mouthwashes should be stored in a cool place

Nasal Solutions

Nasal drops are small volume aqueous solution that that are applied to the nasal passage in the form of nasal drops or sprays. Some preparations are available in the form of emulsions or suspensions. Nasal administration provides both a systemic and local effect. Nasal solutions are prepared so that they are similar to the nasal secretion in many aspects like the pH , toxicity and viscosity so that normal ciliary action is maintained. They are usually isotonic and slightly buffered to a pH of 6.8. In addition to antimicrobial preservatives, stabilizers similar to those in ophthalmic preparation are included in the formulation. Nasal solutions available commercially may also contain antihistamines, antibiotics, and drugs for asthama prophylaxis.

Inhalations

Inhalations are the preparation that deliver drug directly to the respiratory tract. For effective drug delivery, these systems should be capable of producing fine particles, usually <10 μm. Inhalation drug delivery has been traditionally used to treat respiratory diseases, but in recent times lung has been used as a portal for administering drugs to the systemic circulation. Lungs offer an attractive route for systemic drugs because of their large effective surface area. Drugs that are majorly administered through inhalation include steroids and cardiac agents. Many proteins are also administered via inhalations. Inhalations provide many advantages like rapid onset of action, bypassing of first pass effect, avoidance of degradation of drug in GI tract, tamper proof valve and container and require a low dose to minimize adverse reaction.

Three main dosage forms that have gained prominence are the metered dose inhalers (MDIs), dry powder inhalers (DPIs), and nebulizers. MDI contain a drug suspension in a propellant. They may also contain additives like surfactants and antioxidants. Earlier chlorofluorocarbons (CFC) were majorly used as propellants. Due to their effect on the ozone layer their usage wes curtailed by FDA. Nowadays liquefied hydrocarbons, non-CFCs and compressed gases are being used. The MDI consists of pressurized container and a valve. Dry powder inhalers are mixtures of micronized drug and a large particle size carrier (usually lactose). The main advantage of this system include product and formulation stability. The DPI device presents medication to the patient as a dry powder in a form that can be inhaled orally for delivery to the target lung tissues. The other class in this system is the nebulizer. This system delivers the therapeutic agent in the form of mist from aqueous or non- aqueous solution of drug using either compressed air or ultrasonication. Jet nebulizers or atomizers are most commonly used. Major disadvantage of this system is that they are bulky and non-portable.

Douches

Douches are aqueous solutions that are intended for rinsing body cavities. They function as cleansing or antiseptic agents. Douches are named according to the body cavities they are used in to E.g., nasal douche, pharyngeal douche, vaginal douche, eye douche. Those used in vagina and bladder are sometimes called as irrigation. Bladder douches should be sterile. Vaginal douches should be sterile if used after childhood or surgery. They contain antimicrobial agents such as benzalkonium chloride, parabens or chlorothymol, antipruritic and anaesthetic agents. Douches are dispensed in the form of powder with directions for dissolving in specified quantity of warm water.

5. Eye Preparations

Delivery of the drugs into the human eye has been used for the treatment of various clinical symptoms including conjunctivitis (inflammation of the conjunctiva). The formulations for the ocular delivery of the drugs should be sterile. Conventional ocular delivery systems include eye drops, eye lotions and eye ointments. All eye preparations are isotonic to the conjunctiva.

The eye drops are, in general, sterile saline solutions which may or may not contain any drug. The drops should have a pH in between 3.5 and 10 and should be isotonic to the lachrymal secretions. Any deviation from this results in the irritation of the cornea and conjunctiva. Eye drops have been used very frequently to administer drugs in the ocular cavity. Typical groups of drugs which have been successfully incorporated within the eye drops include steroids, antihistamines, non-steroidal anti-inflammatory drugs (NSAIDs) and prostaglandins. It has been found that with the increase in the viscosity of the eye drops there is a subsequent increase in the contact time of the eye drops with the conjunctival membrane and hence the improved residence time of the drug at the ocular site.

Eye drops may be dispensed either in glass or plastic bottles. Since the formulations for the eye drops are meant to be sterile, the containers should be have sterile packaging and should not allow the contamination of the formulation when it is being used. Many drugs are heat-sensitive and proper care should be taken during sterilization of the formulation. Generally, antimicrobial preservatives (e.g. phenylmercuric nitrate or acetate, benzalkonium chloride and chlorhexidine acetate) are present in the eye drops so as to inhibit the microbial growth. The eye drops are generally sterilized by filter sterilization method.

Eye lotions are the formulations used for cleansing the eyes as a first aid measure and do not contain any antimicrobial preservatives. They help removing the contaminants in the eyes thereby resulting the restoring the clear vision. They are basically solutions of normal saline dispensed in single use containers.

Eye ointments offer improved bioavailability for a drug due to the improved residence time of the drug in the ocular cavity. They may contain antimicrobial preservatives (e.g. cholorobutanols, paraben and organic mercurials), antioxidants and stabilizers. The formulations must not contain any particle greater than 90 µm, at most two particles of 50 µm and at most 20 particles of 25 µm. The restriction in the particle size of the drugs present in the ointment is, in order to avoid any damage to the ocular tissues. The only disadvantage of the eye ointments includes temporary loss of vision when the same is being administered in the eye. The eye ointments are filled into the sterilized collapsible metal or plastic tubes with nozzle shaped applicator. The capacity of the tubes is generally of 5g. If required, the eye ointments may be sterilized by ionization radiation after the formulations have been packaged into the final containers.

6. Advanced Techniques In Drug Delivery

Gels

The word gel has been derived from the word gelatos (meaning immobile). Gels may be defined as semi-solid formulations containing both solid and liquid components. The solid components form a 3-dimensional networked structure, which helps in immobilizing the liquid phase by the phenomenon of surface tension. Depending on the polarity of the liquid phase, the gels may either be categorized as hydrogels or organogels (Fig. 7). Hydrogels are defined as the gelled structures having polar liquid into it, whereas organogels contain apolar liquid into its structure. Gels have found extensive applications in controlled delivery systems, which may be administered by various routes including ocular region, ear cavity, orally, insufflatory, vaginal and rectal. The mechanisms of drug movement occur either by partitioning effect (Fig. 8) or by diffusion (Fig. 9) or by a combination of both. Partitioning of a drug deals with the distribution of the drug in the polar and apolar liquids when both are brought in contact with each other whereas the diffusion mechanism deals with the migration of solutes from a region of high concentration to a region of low concentration through a semi-permeable membrane or matrix.

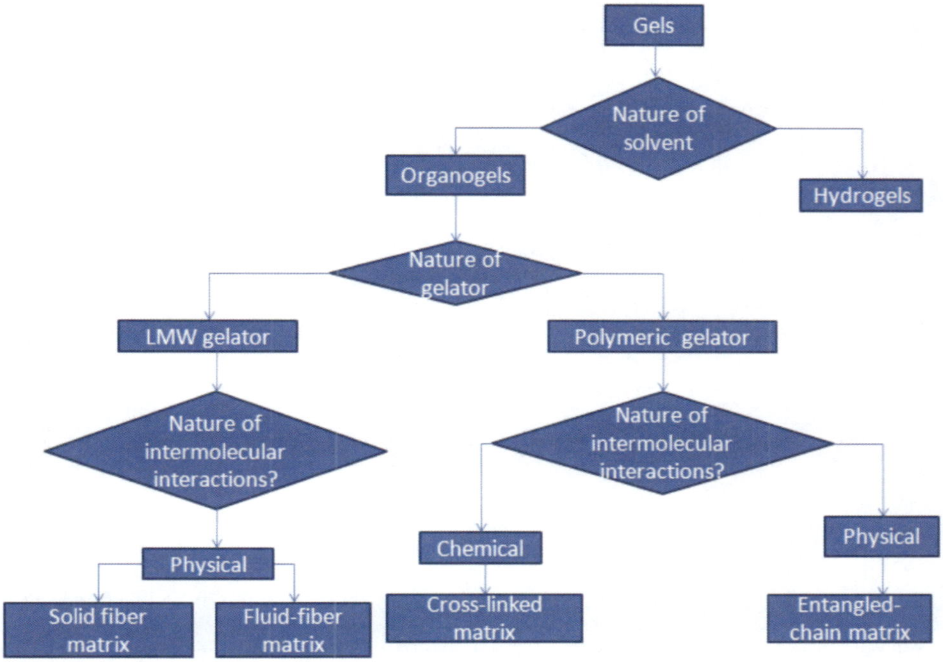

Fig. 7 : Types of gels, based on the interaction of the components

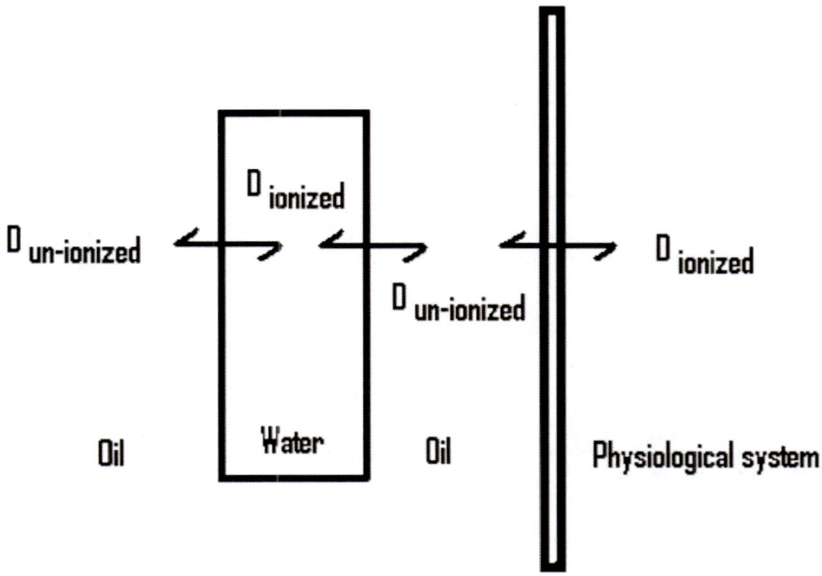

Fig. 8 : Partitioning of solutes from double emulsion to the physiological system

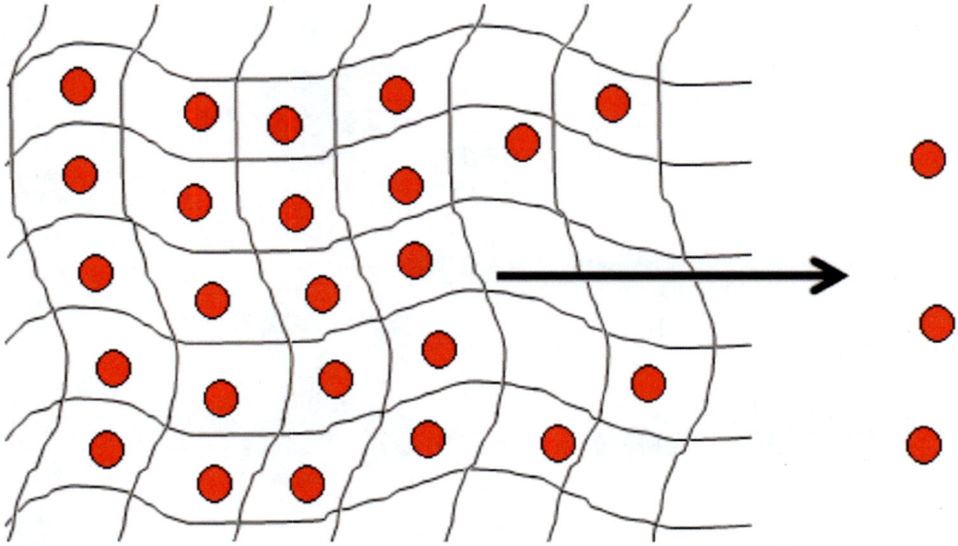

Fig. 9 : Diffusion of solutes from a region of high concentration to a region of low concentration

Hydrogels may be developed either from synthetic or natural polymers. Hydrogels developed from synthetic polymers are very stable (thermally and mechanically) in nature but they might be non-biocompatible depending on the method of synthesis or the chemistry of the polymers. Most of the biopolymeric hydrogels are inherently biocompatible but lacks mechanical and stability. The stability of these gels may be improved by crosslinking the biopolymers with multi-functional reagents, commonly known as cross-linkers. The most common example of cross-linkers includes glutaraldehyde, a dialdehyde which can react with both primary amines and hydroxyl groups in the presence of an acidic catalyst. It has been used extensively in various biomedical applications but the reagent is highly toxic and carcinogenic. The scientists are looking for the alternative natural crosslinkers which may be biocompatible in nature. This has led to the introduction of genipin, a natural crosslinker which is commonly used in food industries, in devising hydrogels. It reacts with primary amine groups and gives blue color pigmentation on reaction. Genipin is several thousand times less toxic as compared to glutaraldehyde and hence is gaining importance in devising various biomedical products including controlled delivery platforms. Similarly, various other naturally occurring natural crosslinkers (e.g. quinines and catechols) are also being tried to develop natural polymer based hydrogels. The above-mentioned cross-linkers result in the formation of covalent bond during the cross-linking procedure and hence are regarded as chemical crosslinkers. Any interaction amongst the polymers and the crosslinkers and/or drug resulting in the ionic, intermolecular hydrogel bonding and van der Wall's interactions are regarded as physical crosslinking method. The drug delivery vehicles devised using hydrogels may be either matrix type or reservoir type.

The matrix type delivery systems contain drug particles uniformly distributed throughout the polymer matrix whereas the reservoir type delivery systems contain reservoir (either solid or liquid) of a drug surrounded by a thin layer of hydrogel. Depending on the size of either of the delivery systems, they may further be classified as nano, micro and macro delivery systems. The size range of the nano-delivery system varies from 10-100 nm. These systems, due to their small size, can cross the blood-brain-barrier and hence can help in targeting the drug in the central nervous systems. The typical size of a micro-delivery system may vary from 1 μm to 1 mm. they have been used to entrap and encapsulate probacters and hydrophobic drugs. Macro-delivery systems may be used to develop inserts, implants, ocular patches, matrices for tissue engineering, transdermal and oral delivery systems.

Organogels may be developed by three mechanisms, viz., fluid-filled fiber mechanism, solid-fiber mechanism and polymerization reaction. The fluid-filled fiber mechanism deals with the formation of reverse-micellar tubular structures which get entangled with each other to form three dimensional networked structures thereby immobilizing the apolar solvent. The solid-fiber mechanism plays with the change in the solubility parameter of the organogelators with change in the temperature. The organogelators are solubilized in an apolar solvent at higher temperature followed by the lowering of the temperature. This results in the formation of solid tubular structures consisting of the organogelator molecules. The third mechanism deals with the formation of network structures from the monomers dissolved in the aploar solvent by undergoing a chemical reaction, forming covalent bonds. Unlike hydrogels, the field of controlled delivery using organogels is still in its novice stage and a lot of work needs to be done in this regard. Various organogels have been tried to formulate transdermal and topical formulations whereas only a handful of literatures were available for formulating oral delivery systems.

Solid Lipid nanoparticles

Solid lipid nanoparticles (SLNs), as the name suggest, are the particles made from the lipids whose size range varies from 50 nm to 1000 nm (Fig. 10). The lipids used to prepare the particles are solid at room temperature (e.g. stearic acid) (Kheradmandnia et al. 2010). The SLNs may be prepared by high pressure homogenization, microemulsion method, solvent emulsification-evaporation method and solvent emulsification-diffusion method.The high pressure homogenization technique (Shegokar et al. 2010) may be implemented either by hot or cold homogenization methods, depending on the thermo-sensitivity of the drugs. The drugs, which can bear the heating process, can be incorporated into the SLNs formulated by hot homogenization process whereas the heat-sensitive drugs may be incorporated into SLNs formulated by cold homogenization process. The microemulsion method employs the solidification of hot microemulsion of the lipids to prepare the SLNs. As per the solvent-emulsification-evaporation method, the lipids are dissolved in solvents like toluene and chloroform for the preparation of emulsion with the subsequent evaporation of the solvent under reduced pressure. This leads to the precipitation of the lipids as SLNs. The last method, solvent emulsification-diffusion method employs the preparation of emulsion using lipid solution in an apolar phase (saturated with water) and subsequent addition of excess water to the emulsion which results in the diffusion of the apolar phase into the aqueous phase. This results in the precipitation of the lipid molecules as SLNs (Subedi et al. 2009).

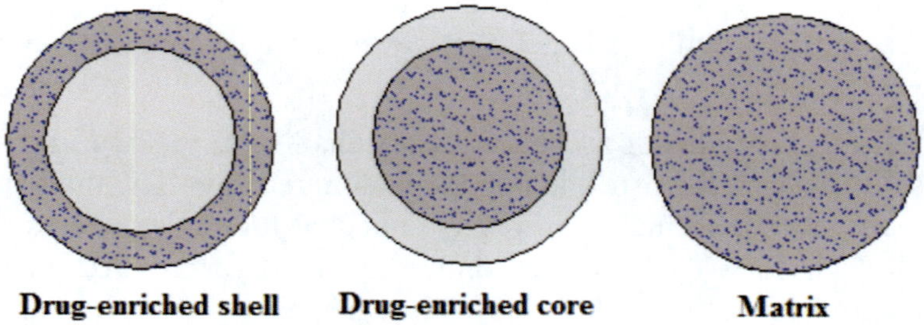

Fig. 10 : Models of distribution of drugs into the SLNs (modified from (Müller et al. 2002))

Dendrimers

Dendrimers may be defined as nano-particulate systems, which have got hyper-branched polymeric structures (Fig. 11). It has got a core on to which the hyper-branched structures are built up (Jang et al. 2009). The anionic dendrimers are more biocompatible as compared to the cationic dendrimers, as the later results in the leaking of the cell contents by interacting with the cell membrane. The cytotoxic effect of the dendrimers may be reduced by stealth technology, where the coating of the dendrimers is carried out using a biocompatible polymer like polyethylene glycol. This technology does not only help in reducing cytotoxicity but can also improve the residence time of the dendrimers in the physiological system by avaoiding their identification by the reticulo-endothelial systems(Svenson 2009).

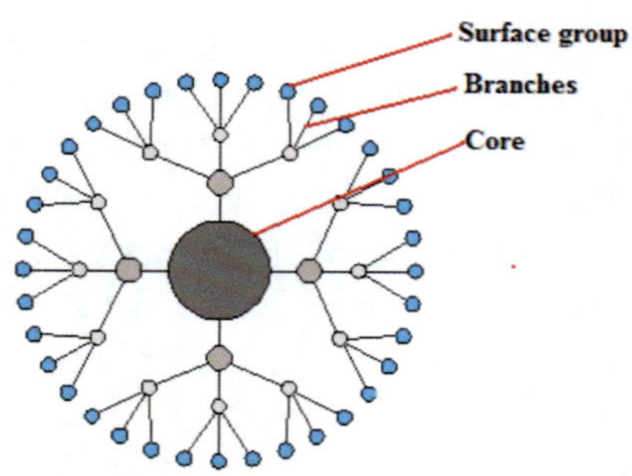

Fig. 11 : Schematic diagram of a dendrimer

Liposomes

Liposomes are nano-structured vehicles made up of phospholipid bilayer (Fig. 12). The liposomes may be tailored to carry both hydrophilic (within the core of the liposomes) and hydrophobic (within the palisade layer) compounds (Lasic 1998). The preparation of the liposomes is simple and can be carried out by sonication or freeze-drying method. The sonication method involves the dissolution of the phospholipids within the aqueous layer followed by the addition of the bioactive agents (Sudimack et al. 2002). The mixture, so obtained, is vortexed and subsequently sonicated to obtain liposomes. Vortexing results in the formation of large vesicles, whose sizes can be reduced by sonication. The freeze drying method involves the dissolution of the phospholipids within the hydro-alcoholic system, which are subsequently freeze-dried. The freeze-dried product is then hydrated, resulting in the spontaneous formation of the liposomes. Stealth technology may be employed to improve the residence time of the liposomes within the physiological system.

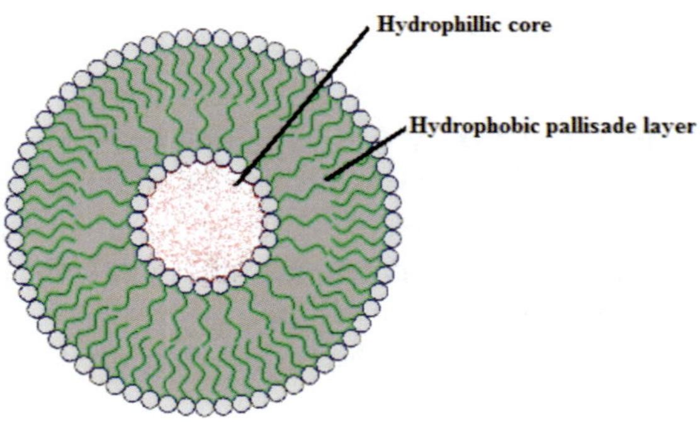

Fig. 12 : Schematic diagram of a liposome

Iontophoresis

Iontophoresis is a special kind transdermal delivery method which employs the use of electrical charge to deliver the drug into the systemic circulation across the skin barrier (Fig. 13). The technique facilitates the movement of ionic drugs under the influence of externally applied electrical potential and offers a simple non-invasive method to deliver the drug systemically. The system consists of three basic components, viz. a voltage source (which may produce either a DC voltage or voltages as per specially programmed microcontrollers), a reservoir for drugs and two electrodes (one active and one passive electrode). The polarity of the active electrode and that of the drug should be same. When the voltage is applied, the electrode forces the drug, of same polarity, through the skin into the systemic circulation. The passive electrode helps in completing the circuit and acts as the returning chamber for the counter-ions (sodium or chloride) from the physiological system. The amount of the drug penetrated into the systemic circulation, is considered to be proportional to the product of the current and duration of the current supply. The safe current limit has been set to 4 mA, which help avoiding any undue damage to the physiological system. The drugs which have been used for delivery using this technique include lidocaine, dexamethasone, hydrocortisone, penicillin and histamine.

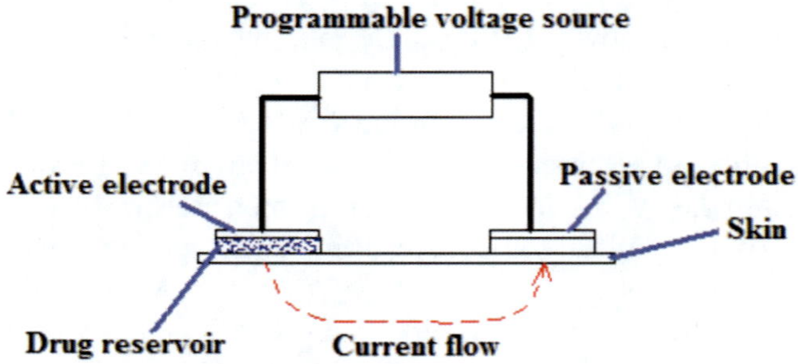

Fig. 13 : Schematic block diagram of an iontophoretic system

Sonophoresis

Like iontophoresis, sonophoresis is also a transdermal drug delivery system where an ultrasonic device is used to improve the permeation of the drugs through the skin to the systemic circulation (Fig. 14). The sonophoretic system helps increasing the kinetic energy of the drug molecules by a complex mechanism, the phenomenon of which is not totally understood till-date. It has been proposed that the permeation is enhanced either by cavitation, microstreaming and heating or their combination.

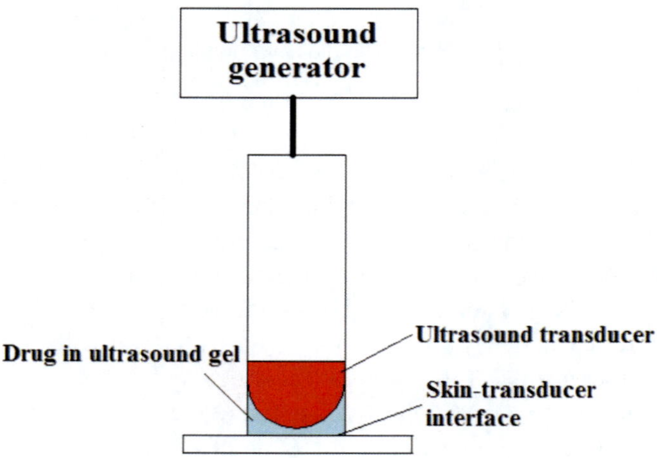

Fig. 14 : Schematic block diagram of a sonophoretic system

Conclusion

This chapter gives an overview on the different methods of drug delivery technologies, which have been tried to improve the bioavailability of the drug into the physiological system.

References

Alanazi, F. K. 2010. "Utilization of date syrup as a tablet binder, comparative study." Saudi Pharmaceutical Journal 18(2): 81-89.

Axelson, J. F., L. C. Zoller, J. E. Tomassone, and D. C. Collins. 1986. "Effects of Silastic progesterone implants on activity cycles and steroid levels in ovariectomized and intact female rats." Physiology & Behavior 38(6): 879-85.

Bachmann, E., E. Weber, M. Post, and G. Zbinden. 1978. "Biochemical effects of gum arabic, gum tragacanth, methylcellulose and carboxymethylcellulose-Na in rat heart and liver." Pharmacology 17(1): 39-49.

Bhatty, J. and A. Dollimore. 1978. "Studies on the suspending properties of aqueous solutions of some natural and synthetic hydrocolloids on china clay suspensions." Water Research 12(12): 1139-48.

Bilia, A. R., F. Eterno, M. C. Bergonzi, G. Mazzi, and F. F. Vincieri. 2007. "Evaluation of the content and stability of the constituents of mother tinctures and tinctures: The case of Crataegus oxyacantha L. and Hieracium pilosella L." Journal of Pharmaceutical and Biomedical Analysis 44(1): 70-78.

Boyapally, H., R. K. Nukala, P. Bhujbal, and D. Douroumis. 2010. "Controlled release from directly compressible theophylline buccal tablets." Colloids and Surfaces B: Biointerfaces 77(2): 227-33.

Campieri, M., G. Lanfranchi, G. Bazzocchi, G. Franzin, C. Brignola, A. Battocchia, and F. Sarti. 1981. "Treatment of ulcerative colitis with high-dose 5-aminosalicylic acid enemas." The Lancet 318(8241): 270-71.

Cantor, A. and M. Winicov. 1979. "Stabilized iodine tincture." Google Patents.

Cole, E. T., D. Cadé, and H. Benameur. 2008. "Challenges and opportunities in the encapsulation of liquid and semi-solid formulations into capsules for oral administration." Advanced drug delivery reviews 60(6): 747-56.

De Boer, A., F. Moolenaar, L. De Leede, and D. Breimer. 1982. "Rectal drug administration: clinical pharmacokinetic considerations." Clin Pharmacokinet 7(4): 285-311.

Denet, A., R. Vanbever, and V. Préat. 2004. "Skin electroporation for transdermal and topical delivery." Advanced drug delivery reviews 56(5): 659-74.

Dolan, M. M., R. L. Steelman, and R. R. Tumilowicz. 1960. "Carbopol 934: An improved suspending agent for insoluble test compounds." Toxicology and Applied Pharmacology 2(3): 331-37.

Eccleston, G. 1997. "Functions of mixed emulsifiers and emulsifying waxes in dermatological lotions and creams." Colloids and Surfaces A: Physicochemical and Engineering Aspects 123: 169-82.

Griffin, W. 1949. "Classification of surface-active agents by HLB." J. Soc. Cosmet. Chem 1(311): 97.

Hosny, E., S. Abdel-Hady, and K. El-Tahir. 1996. "Formulation, in-vitro release and ex-vivo spasmolytic effects of mebeverine hydrochloride suppositories containing polycarbophil or polysorbate 80." International Journal of Pharmaceutics 142(2): 163-68.

Ibrahim, S., T. El-Faham, S. Tous, and E. Mostafa. 1990. "Formulation, release characteristics and evaluation of ibuprofen suppositories." International Journal of Pharmaceutics 61(1-2): 1-7.

Jang, W., K. Kamruzzaman Selim, C. Lee, and I. Kang. 2009. "Bioinspired application of dendrimers: from bio-mimicry to biomedical applications." Progress in Polymer Science 34(1): 1-23.

Jones, B. E. 2001. "The filling of powders into two-piece hard capsules." International Journal of Pharmaceutics 227(1-2): 5-26.

Kheradmandnia, S., E. Vasheghani-Farahani, M. Nosrati, and F. Atyabi. 2010. "Preparation and characterization of ketoprofen loaded solid lipid nanoparticles made from beeswax and carnauba wax." Nanomedicine: Nanotechnology, Biology and Medicine.

Lasic, D. 1998. "Novel applications of liposomes." Trends in Biotechnology 16(7): 307-21.

Lefebvre, J.-L. and C. Domenge. 2002. "A comparative study of the efficacy and safety of fluconazole oral suspension and amphotericin B oral suspension in cancer patients with mucositis." Oral Oncology 38(4): 337-42.

Lester, D. R., M. Rudman, and P. J. Scales. 2010. "Macroscopic dynamics of flocculated colloidal suspensions." Chemical Engineering Science 65(24): 6362-78.

Lu, E.-X., Z.-Q. Jiang, Q.-Z. Zhang, and X.-G. Jiang. 2003. "A water-insoluble drug monolithic osmotic tablet system utilizing gum arabic as an osmotic, suspending and expanding agent." Journal of Controlled Release 92(3): 375-82.

McHugh, D. 1987. "Production, properties and uses of alginates." Production and Utilization of Products from Commercial Seaweeds. FAO. Fish. Tech. Pap 288: 58-115.

Müller, R. H., M. Radtke, and S. A. Wissing. 2002. "Solid lipid nanoparticles (SLN) and nanostructured lipid carriers (NLC) in cosmetic and dermatological preparations." Advanced drug delivery reviews 54(Supplement 1): S131-S55.

Normand, V., P. Aymard, D. Lootens, E. Amici, K. Plucknett, and W. Frith. 2003. "Effect of sucrose on agarose gels mechanical behaviour." Carbohydrate Polymers 54(1): 83-95.

Osada, Y., K. Kajiwara, and H. Ishida. 2001. Gels handbook: Academic Press London.

Pal, R. 2010. "Rheology of simple and multiple emulsions." Current Opinion in Colloid & Interface Science.

Pissinati, R. and W. P. Oliveira. 2003. "Enteric coating of soft gelatin capsules by spouted bed: effect of operating conditions on coating efficiency and on product quality." European Journal of Pharmaceutics and Biopharmaceutics 55(3): 313-21.

Podczeck, F. and J. M. Newton. 1999. "Powder filling into hard gelatine capsules on a tamp filling machine." International Journal of Pharmaceutics 185(2): 237-54.

Rathbone, M. J., B. K. Drummond, and I. G. Tucker. "The oral cavity as a site for systemic drug delivery." Advanced drug delivery reviews 13(1-2): 1-22.

Rawlins, E. 2006. Bentley's Textbook Of Pharmaceutics: Aitbs Publishers And Distributors.

Robins, M., A. Watson, and P. Wilde. 2002. "Emulsions--creaming and rheology." Current Opinion in Colloid & Interface Science 7(5-6): 419-25.

Rotthäuser, B., G. Kraus, and P. C. Schmidt. 1998. "Optimization of an effervescent tablet formulation using a central composite design optimization of an effervescent tablet formulation containing spray dried -leucine and polyethylene glycol 6000 as lubricants using a central composite design." European Journal of Pharmaceutics and Biopharmaceutics 46(1): 85-94.

Shegokar, R., K. Singh, and R. Müller. 2010. "Production & stability of stavudine solid lipid nanoparticles--From lab to industrial scale." International Journal of Pharmaceutics.

Simpson, J. M. 1959. "Drug resistance in out-patients treated with sodium PAS and isoniazid cachets." Tubercle 40(6): 453-56.

Subedi, R., K. Kang, and H. Choi. 2009. "Preparation and characterization of solid lipid nanoparticles loaded with doxorubicin." European Journal of Pharmaceutical Sciences 37(3-4): 508-13.

Sudimack, J., W. Guo, W. Tjarks, and R. Lee. 2002. "A novel pH-sensitive liposome formulation containing oleyl alcohol." Biochimica et Biophysica Acta (BBA)-Biomembranes 1564(1): 31-37.

Sulak, P., T. Kuehl, and B. Shull. 1994. "Vaginal pessaries and their use in pelvic relaxation." Obstetrical & Gynecological Survey 49(4): 256.

Svenson, S. 2009. "Dendrimers as versatile platform in drug delivery applications." European Journal of Pharmaceutics and Biopharmaceutics 71(3): 445-62.

Swarbrick, J. 2007. "Suspension." Encyclopedia of Pharmaceutical Technology. pp. 3603. New York: Informa Healthcare USA, Inc.

Symes, K. 1980. "The relationship between the covalent structure of the Xanthomonas polysaccharide (xanthan) and its function as a thickening, suspending and gelling agent." Food Chemistry 6(1): 63-76.

Tcholakova, S., N. Denkov, I. Ivanov, and B. Campbell. 2006. "Coalescence stability of emulsions containing globular milk proteins." Advances in colloid and interface science 123: 259-93.

Tiwari, A., J. Grailer, S. Pilla, D. Steeber, and S. Gong. 2009. "Biodegradable hydrogels based on novel photopolymerizable guar gum-methacrylate macromonomers for in situ fabrication of tissue engineering scaffolds." Acta biomaterialia 5(9): 3441.

Van Hoogdalem, E., A. De Boer, and D. Breimer. 1991. "Pharmacokinetics of rectal drug administration, Part I. General considerations and clinical applications of centrally acting drugs." Clinical pharmacokinetics 21(1): 11.

van Santen, E., D. M. Barends, and H. W. Frijlink. 2002. "Breaking of scored tablets: a review." European Journal of Pharmaceutics and Biopharmaceutics 53(2): 139-45.

Vierhout, M. 2004. "The use of pessaries in vaginal prolapse." European journal of obstetrics, gynecology, and reproductive biology 117(1): 4-9.

Vladisavljevic, G., M. Shimizu, and T. Nakashima. 2006. "Production of multiple emulsions for drug delivery systems by repeated SPG membrane homogenization: Influence of mean pore size, interfacial tension and continuous phase viscosity." Journal of Membrane Science 284(1-2): 373-83.

Wang, J., H. Wen, and D. Desai. 2010a. "Lubrication in tablet formulations." European Journal of Pharmaceutics and Biopharmaceutics 75(1): 1-15.

Wang, Y., L. Li, H. Li, Z. Zhu, L. Hua, F. Lei, M. Kheir, and L. Du. 2010b. "Transdermal permeation of geniposide in the herbal complex liniment in vivo and in vitro." International Journal of Pharmaceutics.

Yi-Hung, T., Y. Min-Yung, and N. Shun-Ichi. 1986. "Stability, anti-inflammatory effect and percutaneous absorption of indomethacin in absorption ointment formulations." International Journal of Pharmaceutics 28(1): 47-58.

❑❑❑

CHAPTER – 28

Enzyme
The Biological Messengers andits Kinetics

Biswanath Bhunia[1], Kambaska Kumar Behera[2], Soumya Sasmal[3] and Apurba Dey[4]
[1]Department of Biotechnology, Bengal College of Engineering and Technology, Durgapur-713212
Email : bbhunia@gmail.com
[2]Dept. of Bioscience and Biotechnology, Banasthali University, 304022 (Rajasthan)
[3]Department of Chemical Engineering, Indian Institute of Technology Guwahati
[4]Department of Biotechnology, NIT Durgapur

Summary

Substance that acts as a catalyst in living organisms, regulating the rate at which life's chemical reactions proceed without being altered in the process. Enzymes reduce the activation energy needed to start these reactions, without them, most such reactions would not take place at a useful rate. Because enzymes are not consumed, only tiny amounts of them are needed. Enzymes catalyze all aspects of cell metabolism, including the digestion of food, in which large nutrient molecules (including proteins, carbohydrates, and fats) are broken down into smaller molecules; the conservation and transformation of chemical energy and the construction of cellular materials and components. Almost all enzymes are proteins; many depend on a nonprotein cofactor, either a loosely associated organic compound (e.g., a vitamin) or a tightly bound metal ion (e.g., iron, zinc) or organic (often metal-containing) group. The enzyme-cofactor combination provides an active configuration; usually including an active site into which the substance (substrate) involved in the reaction can fit. Many enzymes are specific to one substrate. If a competing molecule blocks the active site or changes its shape, the enzyme's activity is inhibited. If the enzyme's configuration is destroyed (denaturation), its activity is lost. Most enzyme names end in -ase. Enzymes are chiral catalysts, producing mostly or only one of the possible stereoisomeric products (optical activity) and very sensitive to change in temperature, PH and other parameters of the medium. The fermentation of wine, leavening of bread, curdling of milk into cheese, and brewing of beer are all enzymatic reactions. The uses of enzymes in medicine include killing disease-causing microorganisms, promoting wound healing, and diagnosing certain diseases.

Introduction

Most of the enzymes are proteins that catalyze the biochemical reactions which increase or decrease the rates of chemical reactions. In enzymatic reactions, the molecules at the beginning of the process are called substrates, and they are converted into different molecules, called the products. Almost all processes in a biological cell need enzymes to occur at significant rates. Since enzymes are selective for their substrates and speed up only a few reactions from among many possibilities, the set of enzymes made in a cell determines which metabolic pathways occur in that cell. Like all catalysts, enzymes work by lowering the activation energy for a reaction, thus dramatically increasing the rate of the reaction. As a result, products are formed faster and reactions reach their equilibrium state more rapidly (Bender & Braubacher, 1973; Boyer, 1970).

Structurally enzymes are complex molecules which have a distinct three-dimensional structure. They typically consist of long strands of polypeptides all wrapped around each other and the three dimensional shape is important for its functioning. Their molecular weights range from about 10,000 to more than 1,000,000 KB. Like other proteins, enzymes consist of chains of amino acids linked together by peptide bonds. An enzyme molecule may contain one or more of these polypeptide chains. The sequence of amino acids within the polypeptide chains is characteristic for each enzyme and is believed to determine the unique three-dimensional conformation in which the chains are folded. This conformation, which is necessary for the activity of the enzyme, is stabilized by interactions of amino acids in different parts of the peptide chains with each other and with the surrounding medium. These interactions are relatively weak and may be disrupted readily by high temperatures, acid or alkaline conditions, or changes in the polarity of the medium. Such changes lead to an unfolding of the peptide chains (denaturation) and a concomitant loss of enzymatic activity, solubility, and other properties characteristic of the native enzyme. Enzyme denaturation is sometimes reversible (Copeland, 1996; Dixon & Webb, 1964). Some sources and catalysed reaction of important enzymes are shown in Table 1.

Characteristics of Enzyme

- Enzymes are Organic compound and acts as biocatalyst which is more power full than inorganic catalyst
- Enzymes are colloidal in nature, soluble in water, salt solution, glycerine etc.
- Enzymes are highly catalytic i.e., very small quantities of an enzyme can catalyse mesurable reaction

- Enzymes are very specific in their action i.e., a specific enzyme can acts on a specific substrates and change in to a specific products
- Enzymes do not initiate reaction but they speed up or retard the rate of chemical reactions
- Enzymes are not consumed, they retain their identity at the end of biochemical reaction, however sometimes they are damaged and inactivated due to degradation
- Some enzyme catalyzed reactions are reversible which help in metabolic process
- Boiling of enzymes destroy its activity, while freezing suspends its activity

Table : 1 Some Representative Enzymes, Their Sources, and Reaction Specificities

Enzyme	Sources	Reaction catalyzed
Pepsin	Gastric juice	Hydrolysis of proteins to peptides and amino acids
Urease	Jackbean, bacteria	Hydrolysis of urea to ammonia and carbon dioxide
Amylase	Saliva, pancreatic juice	Hydrolysis of starch to maltose
Phosphorylase	Muscle, liver, plants	Reversible phosphorolysis of starch or glycogen to glucose-1-phosphate
Transaminases	Many animal and plant tissues	Transfer of an amino group from an amino acid to a keto acid
Phosphohexose isomerase	Muscle, yeast	Interconversion of glucose-6-phosphate and fructose-6-phosphate
Pyruvic carboxylase	Yeast, bacteria, plants	Decarboxylation of pyruvate to acetaldehyde and carbon dioxide
Catalase	Erythrocytes, liver	Decomposition of hydrogen peroxide to oxygen and water
Alcohol dehydrogenase	Liver	Oxidation of ethanol to acetaldehyde
Xanthine oxidase	Milk, liver	Oxidation of xanthine and hypoxanthine to uric acid

Chemical Nature of Enzymes

Enzymes are special class of proteins that cause reactions to happen in the body without raising the temperature. They are considered organic catalysts. Since they are proteins and mostly are of two types (Fersht, 1999; Gutfreund, 1975):-

Simple Enzymes

I. Simple Enzymes are the enzymes which on hydrolysis, yields only amino acids only. Most of the digestive enzymes are simple enzymes (e.g., Pepsin, chymotrypsin, trypsin, Amylase etc.

II. Conjugated Enzymes or Complex Enzymes or Hollow Enzymes or Allosteric Enzymes:These are the type of enzyme which on hydrolysis, yields two parts.

a. Apoenzyme :- The Protein part of the Enzyme

b. Co-factor:- The non-protein part of the Enzyme

The catalytic activity of many enzymes depends on the presence of small molecules termed cofactors, although the precise role varies with the cofactor and the enzyme. Such an enzyme without its cofactor is referred to as an apoenzyme; the complete, catalytically active enzyme is called a holoenzyme.

Apoenzyme + cofactor = holoenzyme

Cofactors can be subdivided into two groups: metals and small organic molecules (Table 2). The enzyme carbonic anhydrase, for example, requires Zn2+ for its activity. Glycogen phosphorylase, which mobilizes glycogen for energy, requires the small organic molecule pyridoxal phosphate (PLP).

Cofactors that are small organic molecules are called coenzymes. Often derived from vitamins, coenzymes can be either tightly or loosely bound to the enzyme. If tightly bound, they are called prosthetic groups. Loosely associated coenzymes are more like co-substrates. Enzymes that use the same coenzyme are usually mechanistically similar.

Table 2 : Enzyme cofactors (Murray et al., 2009)

Cofactor	**Enzyme**
Coenzyme	
Flavin adenine nucletide	Monoamine oxidase
Nicotinamide adenine dinucleotide	Lactate dehydrogenase
Pyridoxal phosphate	Glycogen phosphorylase
Coenzyme A (CoA)	Acetyl CoA carboxylase
Biotin	Pyruvate carboxylase
Metal	
Zn^{2+}	Carbonic anhydrase
Zn^{2+}	Carboxypeptidase
Mg^{2+}	EcoRV
Mg^{2+}	Hexokinase
Ni^{2+}	Urease
Mn^{2+}	Superoxide dismutase

Active Sites of Enzyme

An enzyme usually contains one or more active sites, where reactions with substrates take place. An active site may comprise only a few amino acid residues; the rest of the protein is required for maintaining the three-dimensional integrity of the network. The specificity of enzymes for substrates varies from molecule to molecule. Many enzymes exhibit stereochemical specificity in that they catalyze the reactions of one conformation but not the other. For example, proteolytic enzymes catalyze only the hydrolysis of peptides made up of L-amino acids. Some enzymes are catalytically inactive in the absence of certain metal ions.

Mechanism of Enzyme Action

In the 1890s the German chemist Emil Fischer (1852-1919) proposed a lock-and key theory of enzyme specificity. According to Fischer, the active site can be assumed to have a rigid structure, similar to a lock. A substrate molecule then has a complementary structure and functions as a key (Figure 1). However, we now know that enzymes are flexible and that the shapes of the active sites can be markedly modified by the binding of substrate, as was postulated by Daniel E. Koshland, Jr., in 1958. The active sites of some enzymes assume a shape that is complementary to that of the transition state only after the substrate is bound. This process of dynamic recognition is called induced fit (Figure 2) (Murray et al., 2009).

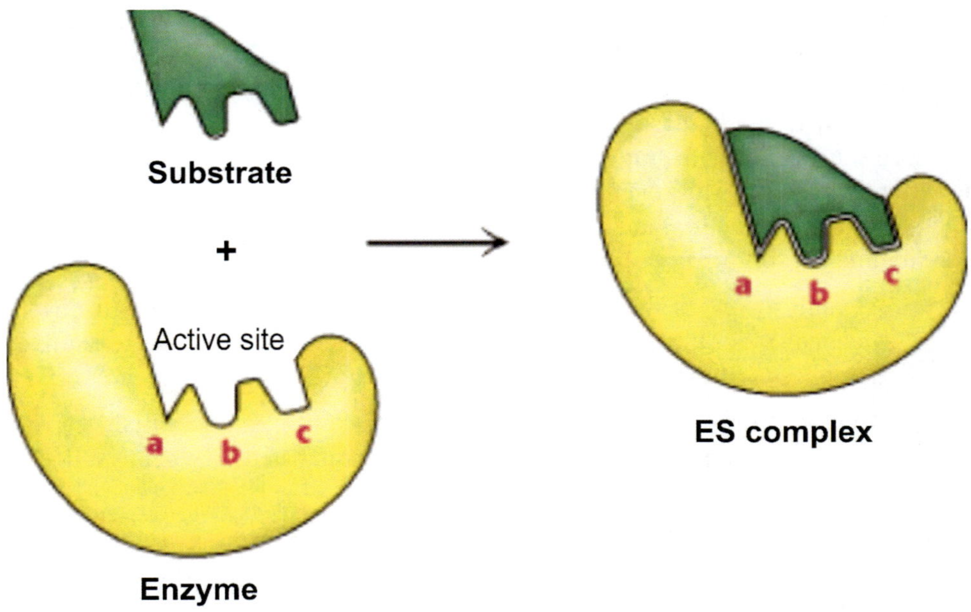

Fig 1 : Lock-and-Key Model of Enzyme-Substrate Binding.

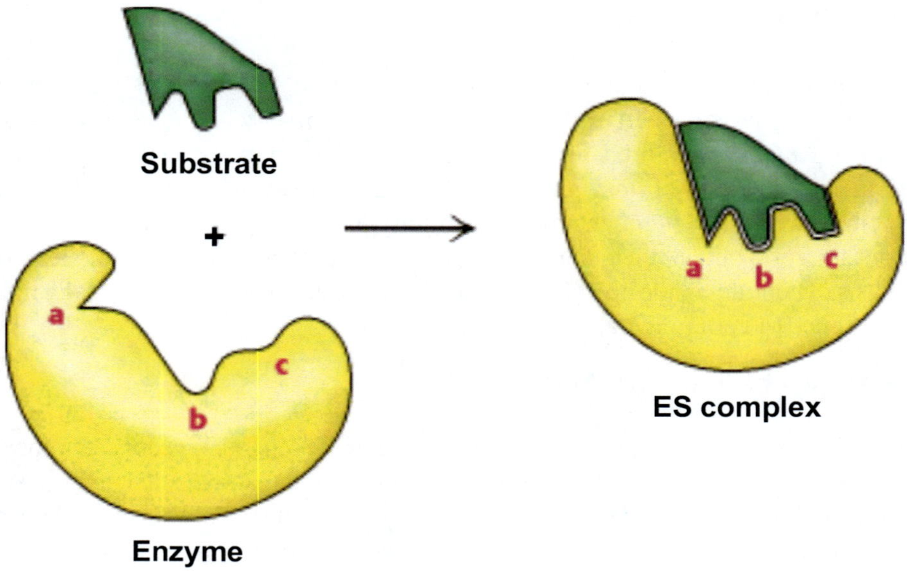

Fig. 2 : Induced-Fit Model of Enzyme-Substrate Binding..

How Enzyme Can Work

A catalyst is a substance that increases the rate of a reaction without itself being consumed by the process. A chemical reaction of substrate S to form product P goes through a transition state that has a higher free energy than does either S or P. The transition state is the most seldom occupied species along the reaction pathway because it is the one with the highest free energy. The difference in free energy between the transition state and the substrate is called the Gibbs free energy of activation or simply the activation energy (Murray et al., 2009). A reaction in which a catalyst is involved is called a catalyzed reaction, and the process is called catalysis. In studying catalysis, keep in mind the following characteristics:

1. A catalyst lowers the Gibbs energy of activation by providing a different mechanism for the reaction (Figure 3). This mechanism enhances the rate and it applies to both the forward and the reverse directions of the reaction.
2. A catalyst forms an intermediate with the reactant(s) in the initial step of the mechanism and is released in the product-forming step. The catalyst does not appear in the overall reaction.
3. Regardless of the mechanism and the energetics of a reaction, a catalyst cannot affect the enthalpies or Gibbs energies of the reactants and products. Thus, catalysts increase the rate of approach to equilibrium, but cannot alter the thermodynamic equilibrium constant.

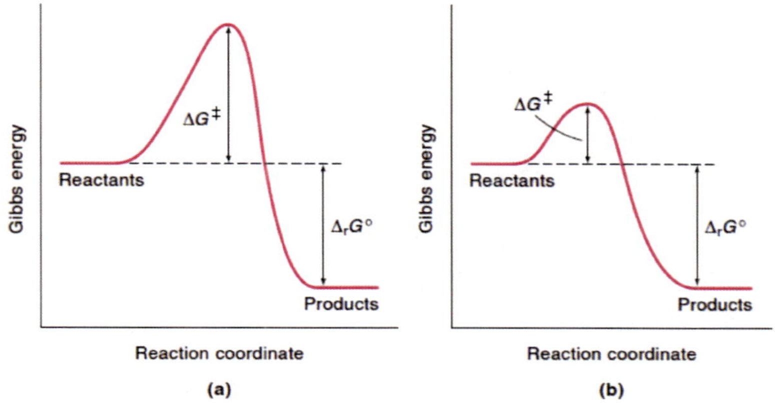

Fig. 3 : Gibbs energy change for (a) an uncatalyzed reaction and (b) a catalyzed reaction. A catalyzed reaction must involve the formation of at least one intermediate (between the reactant and the catalyst). The Δ_r G° is the same in both cases.

1. Concept of Basic Kinetics

The rate of a chemical reaction, as measured by the rate of disappearance of one of its reactants or appearance of one of its products, is in general proportional to the concentration of one or more of its reactants.

$$-\frac{dC_A}{dt} = k_r C_A C_B C_C .... \qquad (1)$$

This expression is called the "rate law" for the reaction, and kr is the "specific rate " or " rate constant." The concentration of a substance may, of course, appear in the rate law to a higher power than the first (Kraut, 1988). The substances whose concentrations appear in the rate law are not, in general, exactly identical with those which appear in the balanced equation for the reaction, so it is necessary to determine the rate law by experiment.

1.1. Zero-Order Reactions

Zero-order reactions (order = 0) have a constant rate. This rate is independent of the concentration of the reactants. The rate law is:

Rate = k, with k having the units of M/sec.

(M= Where M stands for concentration in molarity)

1.2. First-Order Reaction

For a first-order reaction, the rate of reaction is directly proportional to the concentration of one of the reactants.

Differential Rate Law

$$r = k[A]$$

The rate constant, k, has units of sec^{-1}.

1.3. Second-Order Reaction

For a second-order reaction, the rate of reaction is directly proportional to the square of the concentration of one of the reactants.

Differential Rate Law:

$$r = k[A]^2$$

The rate constant, k, has units of L mole-1 sec-1.

1.4. Mixed-Order or Higher-Order Reactions

Mixed-order reactions have a fractional order for their rate

e.g., $r = k[A]^{1/4}$

1.5. Integrated forms of the rate law

For rate laws involving only a single substance, the integrated forms of some of the simple rate laws are shown in Table 3:

Table 3 : Different order of reaction and rate laws

Zero order	$-\frac{dC}{dt} = k_0$	$\frac{C}{C_0} = 1 - \frac{k_0 t}{C_0}$
$\frac{1}{2}$ order	$-\frac{dC}{dt} = k_{1/2}C^{1/2}$	$\frac{C}{C_0} = \left(1 - \frac{k_{1/2}t}{2C_0^{1/2}}\right)^2$
First order	$-\frac{dC}{dt} = k_1 C$	$\frac{C}{C_0} = e^{-k_1 t}$
$\frac{3}{2}$ order	$-\frac{dC}{dt} = k_{3/2}C^{3/2}$	$\frac{1}{\left(1 + \frac{C_0^{1/2}k_{3/2}t}{2}\right)^2}$
Second order	$-\frac{dC}{dt} = k_2 C^2$	$\frac{C}{C_0} = \frac{1}{1 + C_0 k_2 t}$

2. Chemical Equilibrium

Chemical reactions are not fully irreversible and the transformation of the products back to the substrates is possible. Often we define a chemical reaction as

$$A + B = C + D$$

We know that equilibrium constant can be characterized by the concentration of the various substrates and products reaches towards equilibrium concentration. We can illustrate the rate equation through the following equation.

At the equilibrium, we have:

$$\vec{\upsilon} = \overleftarrow{\upsilon}$$

$$k_1 A_{eq} B_{eq} = k_{-1} C_{eq} D_{eq}$$

$$\frac{k_1}{k_{-1}} = \frac{C_{eq} D_{eq}}{A_{eq} B_{eq}} = K_{eq}$$

3. Catalyst

A catalyst is a substance that increases the rate of a reaction by participating chemically in intermediate stages of reaction and is ultimately liberated in a chemically unchanged form. Over a period of time, however, permanent changes in the catalyst-deactivation-may occur. Inhibitors are substances that slow down rates of reaction. The turnover ratio, the number of molecules converted per molecule of catalyst, can be in the millions. Many catalysts have s specific actions in that they influence only one reaction or group of definite reactions. The outstanding example is the living cell in which there are several hundred catalysts, called enzymes, each one favoring a specific chemical process (Daniel et al., 2001; Daniel et al., 2008; Daniel et al., 2007; Peterson et al., 2004). When a reaction can proceed by more than one path, a particular catalyst may favor one path over another and thus lead to a product distribution different from an un-catalyzed reaction. A catalytic reaction requires a lower energy of activation, thus permitting a reduction of temperature at which the reaction can proceed favorably. The equilibrium condition is not changed since both forward and reverse rates are accelerated equally. For example, a good hydrogenation catalyst also is a suitable dehydrogenation accelerator but possibly with a different most-favorable temperature (Klotz 1997; Perutz 1990).

4. Enzyme

The word enzyme comes from Greek: "in ferment". As early as the late 1700s and early 1800s the digestion of meat by stomach secretions and the conversion of starch to sugars by plant extracts and saliva were observed.

Studying the fermentation of sugar to alcohol by yeast, Louis Pasteur came to the conclusion that this fermentation was catalyzed by "ferments" in the yeast. In 1897, yeast extracts were used to ferment sugar independent from the living yeast cells by Eduard Buchner, proving that the fermentation was caused by molecules that functioned even outside the cells. He called these molecules enzymes (Segal 1975; Walsh 1979; Wyman & Gill, 1990).

4.1. Enzymatic Nomenclature

Many enzymes have common names that provide little information about the reactions that they catalyze. For example, a proteolytic enzyme secreted by the pancreas is called trypsin. Most other enzymes are named for their substrates and for the reactions that they catalyze, with the suffix "ase" added. Thus, an ATPase is an enzyme that breaks down ATP, whereas ATP synthase is an enzyme that synthesizes ATP.

To bring some consistency to the classification of enzymes, in 1964 the International Union of Biochemistry established an Enzyme Commission (EC) to develop a nomenclature for enzymes. Reactions were divided into six major groups numbered 1 through 6 (Table 4). These groups were subdivided and further subdivided, so that a four-digit number preceded by the letters EC for Enzyme Commission could precisely identify all enzymes (Bender & Braubacher, 1973; Boyer, 1970; Copeland, 1996; Murray et al., 2009).

Table 4 : Nomenclature of enzymes (Copeland, 1996; Murray et al., 2009)

Subclass	Name
EC 1	Oxidoreductases
EC 1.1	Acting on the CH-OH group of donors
EC 1.2	Acting on the aldehyde or oxo group of donors
EC 1.3	Acting on the CH-CH group of donors
EC 1.4	Acting on the CH-NH2 group of donors
EC 1.5	Acting on the CH-NH group of donors
EC 1.6	Acting on NADH or NADPH
EC 1.7	Acting on other nitrogenous compounds as donors
EC 1.8	Acting on a sulfur group of donors
EC 1.9	Acting on a heme group of donors

Contd.

EC 1.10	Acting on diphenols and related substances as donors
EC 1.11	Acting on a peroxide as acceptor
EC 1.12	Acting on hydrogen as donor
EC 1.13	Acting on single donors with incorporation of molecular oxygen (oxygenases)
EC 1.14	Acting on paired donors, with incorporation or reduction of molecular oxygen
EC 1.15	Acting on superoxide radicals as acceptor
EC 1.16	Oxidising metal ions
EC 1.17	Acting on CH or CH2 groups
EC 1.18	Acting on iron-sulfur proteins as donors
EC 1.19	Acting on reduced flavodoxin as donor
EC 1.20	Acting on phosphorus or arsenic in donors
EC 1.21	Acting on X-H and Y-H to form an X-Y bond
EC 1.22	Acting on halogen in donors
EC 1.97	Other oxidoreductases
EC 2	Transferases
EC 2.1	Transferring one-carbon groups
EC 2.2	Transferring aldehyde or ketonic groups
EC 2.3	Acyltransferases
EC 2.4	Glycosyltransferases
EC 2.5	Transferring alkyl or aryl groups, other than methyl groups
EC 2.6	Transferring nitrogenous groups
EC 2.7	Transferring phosphorus-containing groups
EC 2.8	Transferring sulfur-containing groups
EC 2.9	Transferring selenium-containing groups
EC 3	Hydrolases
EC 3.1	Acting on ester bonds
EC 3.2	Glycosylases
EC 3.3	Acting on ether bonds
EC 3.4	Acting on peptide bonds (peptidases)
EC 3.5	Acting on carbon-nitrogen bonds, other than peptide bonds
EC 3.6	Acting on acid anhydrides
EC 3.7	Acting on carbon-carbon bonds
EC 3.8	Acting on halide bonds
EC 3.9	Acting on phosphorus-nitrogen bonds
EC 3.10	Acting on sulfur-nitrogen bonds
EC 3.11	Acting on carbon-phosphorus bonds
EC 3.12	Acting on sulfur-sulfur bonds

EC 3.13	Acting on carbon-sulfur bonds
EC 4	Lyases
EC 4.1	Carbon-carbon lyases
EC 4.2	Carbon-oxygen lyases
EC 4.3	Carbon-nitrogen lyases
EC 4.4	Carbon-sulfur lyases
EC 4.5	Carbon-halide lyases
EC 4.6	Phosphorus-oxygen lyases
EC 4.99	Other lyases
EC 5	Isomerases
EC 5.1	Racemases and epimerases
EC 5.2	cis-trans-Isomerases
EC 5.3	Intramolecular isomerases
EC 5.4	Intramolecular transferases (mutases)
EC 5.5	Intramolecular lyases
EC 5.99	Other isomerases
EC 6	Ligases
EC 6.1	Forming carbon-oxygen bonds
EC 6.2	Forming carbon-sulfur bonds
EC 6.3	Forming carbon-nitrogen bonds
EC 6.4	Forming carbon-carbon bonds
EC 6.5	Forming phosphoric ester bonds
EC 6.6	Forming nitrogen-metal bonds

4.2. Mechanism of Enzymatic Catalysis

Enzymes are extremely selective catalysts. Not only do they catalyze one type of reaction only (e.g. hydrogenation), but they will perform it with only one specific molecule (e.g. ethylene but not propene). The selectivity of some enzymes is so high that they are capable of discriminating between left- and right-handed isomers. It is believed that selectivity is controlled mainly by the size, shape, and chemical nature of the cleft. An example of an enzymatic reaction is the first reaction of the glycolysis, catalyzed by the enzyme hexo-kinase (Fig.4).

Glucose + ATP $\rightarrow$ Glucose-6-phosphate + ADP

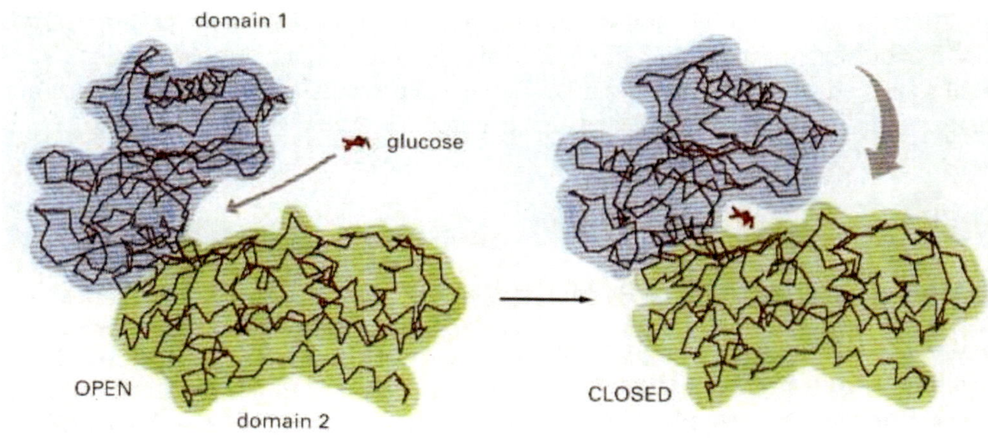

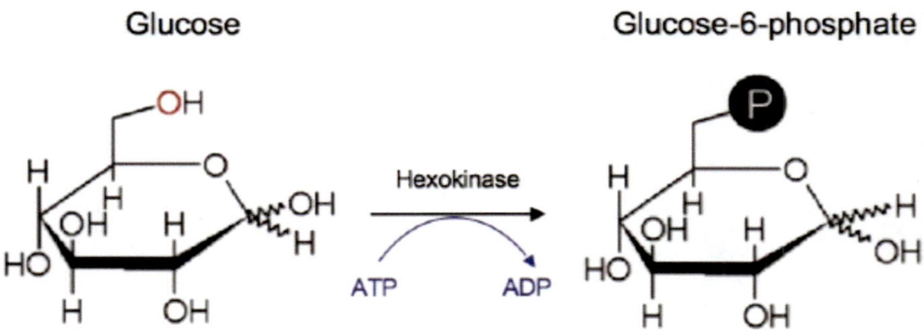

Fig. 4 : Action of Hexokinase on in active glucose molecule

Enzymes are substrate specific and it facilitates the reaction by reducing the activation energy as illustrated in the Fig 5. The mechanism of chymotrypsin is illustrated in Fig 6. Having developed the basic equations of enzyme kinetics, we shall now consider some reactions catalyzed by chymotrypsin, a digestive enzyme. Aside from its important role in digestion, chymotrypsin catalysis is significant for being the system whose study provided the first evidence for the general existence of covalent enzyme- substrate complexes. Chymotrypsin is one of the serine proteases, a family of protein-cutting enzymes that includes trypsin, elastase, and subtilisin. It has a molar mass of 24,800 daltons, 246 amino acid residues, and one active site (containing the serine residue) per molecule. Chymotrypsin is produced in the mammalian pancreas, where it takes the form of an inactive precursor, chymotrypsinogen. Once this precursor has entered the intestine, it is activated by another enzyme, trypsin, to become chymotrypsin. In this way, it avoids self-destruction before it can digest food. The enzyme can be prepared in highly purified form by crystallization. Table 5 represents a comparison of enzyme-catalyzed reactions and their uncatalyzed counterparts.

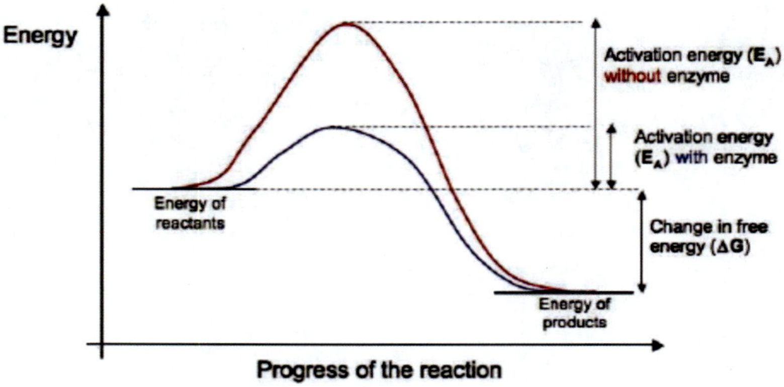

Fig. 5 : Enzyme and Activation energy

Table 5 : Comparisons of Enzyme-Catalyzed Reactions and Their Unanalyzed Counterparts

Reaction	Enzyme	Uncatalyzed Rate, vu (sec-1)	Catalyzed Rate, ve (sec-1)	ve/vu
$CH_3–O–PO_3^{2-} + H_2O \rightarrow CH_3OH + HPO_4^{2-}$	Alkaline phosphatase	1 x 10^{-15}	14	1.4 x 10^{16}
$H_2N–\overset{O}{\overset{\|\|}{C}}–NH_2 + 2H_2O + H+ \rightarrow 2NH_4^+ + HCO_3^-$	Urease	3 x 10^{-10}	3 x 104	1 x 10^{14}
$R–\overset{O}{\overset{\|\|}{C}}–O–CH_2CH_3 + H_2O \rightarrow RCOOH + HOCO_2CH_3$	Chymotrypsin	1 x 10^{-10}	1 x 10^2	1 x 10^{12}
Glycogen + Pi → Glycogen + Glucose-1-P (n (n - 1)	Glycogen phosphorylase	<5 x 10^{-15}	1.6 x 10^{-3}	>3.2x 10^{11}
Glucose + ATP → Glucose-6-P + ADP	Hexokinase	<1 x 10^{-13}	1.3 x 10^{-3}	<1.3x 10^{10}
$CH_3CH_2OH + NAD^+ \rightarrow CH_3\overset{O}{\overset{\|\|}{C}}H + NADH + H^+$	Alcohol dehydrogenase	<6 x 10-12	2.7 x 10-5	>4.5 x 106
$CO_2 + H_2O \rightarrow HCO_3^- + H^+$	Carbonic anhydrase	10^{-2}	10^5	1 x 10^7
Creatine + ATP → Cr-P + ADP	Creatine kinase	<3 x 10^{-9}	4 x 10^{-5}	>1.33x 10^4

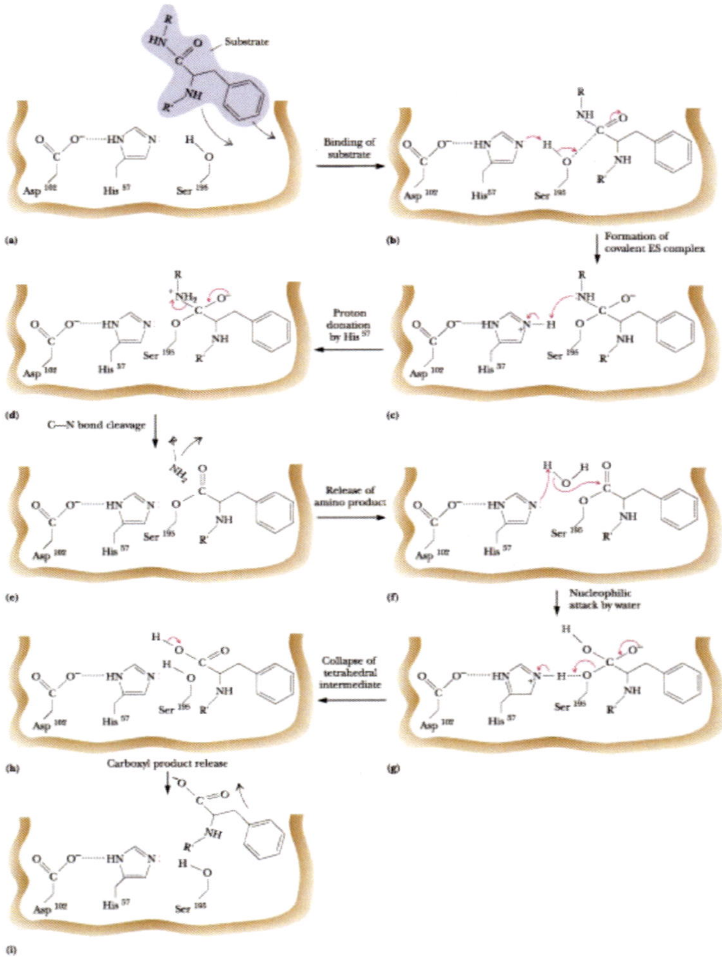

Fig. 6 : A detailed mechanism for the chymotrypsin reaction

4.3. Working Principle of Enzyme

An enzyme is a protein with catalytic properties. As a catalyst, an enzyme lowers the energy of activation of a reaction (Ea), thereby increasing the rate of that reaction without affecting the position of equilibrium-forward and reverse reactions are affected to the same extent.

Based on Linus Pauling's views, Joseph Kraut eloquently pointed out that "an enzyme can be considered a flexible molecular template, designed by evolution to be precisely complementary to the reactants in their activated transition-state geometry, as distinct from their ground-state geometry. Thus an enzyme strongly binds the transition state, greatly increasing its concentration, and accelerating the reaction proportionately.

This description of enzyme catalysis is now usually referred to as transition state stabilization."

$$E+S \xrightarrow{K_{14}^{\pm}} E+S^{+} \xrightarrow{k_u}$$
$$ES \xrightarrow[K_e^{+}]{} \tag{2.1}$$

The upper pathway represents the un-catalyzed reaction; the lower pathway represents the enzyme-catalyzed reaction. Four equilibrium constants can be written for the scheme 2.1

$$K_S = \frac{[E][S]}{[ES]} \qquad K_t = \frac{[E][S^{\neq}]}{[ES^{\neq}]}$$

$$K_{e=}^{\neq} = \frac{[ES^{\neq}]}{[ES]} \qquad K_{u=}^{\neq} = \frac{[E][S^{\neq}]}{[E][S]} \tag{2.2}$$

The ratio of the equilibrium constants for conversion of substrate from the ground state to the transition state in the presence and absence of enzyme is related to the ratio of the dissociation constants for ES and ES‡ complexes:

$$\frac{K_e^{\ddagger}}{K_u^{\ddagger}} = \frac{[ES^{\ddagger}]/[ES]}{[E][S^{\ddagger}]/[E][S]} = \frac{[E][S]/[ES]}{[E][S^{\ddagger}]/[ES^{\ddagger}]} = \frac{K_s}{K_t} \tag{2.3}$$

Absolute reaction rate theory predicts that the rate constant of a reaction (kr) is directly proportional to the equilibrium

$$k_r = \kappa v K^{\ddagger} \tag{2.4}$$

Relative changes in reaction rates due to enzyme catalysis are given by the ratio of reaction rates for the conversion of substrate to product in the presence (ke) and absence (ku) of enzyme

$$\frac{k_e}{k_u} = \frac{\kappa_e v_e K_e^{\ddagger}}{\kappa_u v_u K_u^{\ddagger}} = \frac{\kappa_e v_e K_s}{\kappa_u v_u K_t} \tag{2.5}$$

The magnitudes of the enzymatic rate acceleration, ke/kn, can be extremely large, in the range 1010 to 1014. Considering that it is unlikely that the ratio ?e?e/?n?n differs from unity by orders of magnitude (even though no data exist to support this assumption), we can rewrite Eq. (2.5) as .

$$\frac{k_e}{k_u} \approx \frac{K_s}{K_t} \tag{2.6}$$

The ratio Ks/Kt must therefore also be in the range 1010 to 1014.This important result suggests that substrate in the transition state must necessarily bind to the enzyme much more strongly than substrate in the ground state, by a factor roughly equal to that of the enzymatic rate acceleration. Equation (2.6) provides a conceptual framework for understanding enzyme action. For example, one can address the question of how good an enzyme can be. Identifying ke with kcat, Eq. (2.5) can be rewritten as

$$\frac{k_{cat}}{K_s} = k_u \frac{\kappa_e v_e}{\kappa_u v_u} \frac{1}{K_t} \tag{2.7}$$

The ratio kcat/Ks (M^{-1} s^{-1}) is the second-order rate constant for the reaction of free enzyme with substrate. The magnitude of this rate constant cannot be greater than the diffusion coefficient of the reactants. Thus, a perfectly evolved enzyme will have increased strength of transition-state binding (i.e., decreased Kt) until such a diffusion limit is reached for the thermo-dynamically favored direction of the reaction.

Table 6 : Difference between equilibrium and kinetic studies (Copeland, 1996)

Procedure	**Equilibrium studies**	**Enzyme kinetic studies**
Time dependence	Time independent rapid	Time dependent directional studies equilibrium progression
Constants	Thermodynamic constants: Dissociation (association) constants	Kinetic constant: Michaelis constant, maximum velocity
Detection principal	Free and bound components chemically identical	Substrate and product chemically different
Detection sensitivity	Dependent on macromolecule concentration	Dependence on product formation
Macromolecule/Enzy-me amounts required	Macromolecule and ligand in in comparable amounts [E]~[A]	Catalytic enzyme amounts [E]<<[A]
Purity requirements	Macromolecule present in high purity	High enzyme activity, no requirement for high purity if no disturbing influences

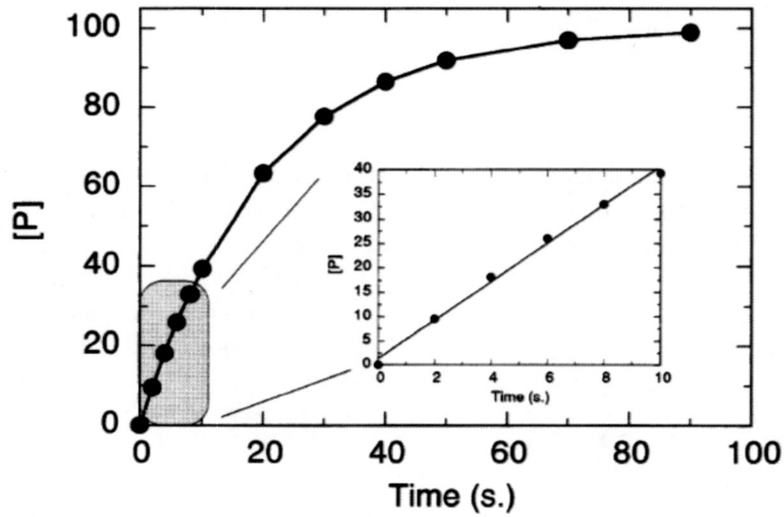

Fig. 7. Reaction progress curve for the production of product during an enzyme-catalyzed reaction. Inset highlights the early time points at which the initial velocity can be determined from the slope of the linear plot of [P] versus time.

5. Order of Reaction for Enzymatic Reaction

Reactions as we normally consider them, such as A (reactants) P (products) represent overall reactions that specify only the stoichiometric relationships between reactants and products. The rate or velocity (v), of such a reaction will dependent on concentrations of reactants (and possibly products) and can be written

v = d[P]/dt = - d[A]/dt = k[A]n[B]m

Where k is a proportionality constant (i.e., rate constant) and the sum of exponents, m + n is called the order of the reaction. Most reactions are either first order (sum of exponents = 1) or second order. Some may also be zero or third order. Mixed order is also possible, in which case the sum of exponents is not an integer.

The overall reaction cannot give us the rate equation because the overall mechanism tells us nothing about the mechanism by which reactants are converted to products. The reaction mechanism gives a description of how reactants are converted to products at the molecular level, and consists of one or more elementary reactions, which, in the case of a complex mechanism consisting of more than one elementary step, generally show the presence of reaction intermediates (I), all of which eventually cancel out, leaving only reactants and products.

The number of molecules involved in a given elementary reaction is called the molecularity of that reaction. Elementary reactions can be uni-molecular (one molecule involved in given elementary reaction), bimolecular (two molecules collide with each other in a given elementary reaction), etc.). Commonly, one finds uni-molecular and bimolecular reactions. Ternary elementary reactions are very rare. What commonly happens is that for a given mechanism.

Consisting of one or more elementary reactions, one of these elementary reactions is the rate determining step and determines the overall rate. Unimolecular reactions that determine the overall rate give rise to first order reaction rates, and bimolecular reactions give rise to second order reactions.

Note that the rate equation above is a differential equation. The integrated form of the rate equation depends on the order of the equation, which in turn depends on the molecularity.

Consider a first order reaction corresponding to a unimolecular process:

$$d[A]/dt = -k[A]$$

To obtain the integrated form of the rate equation, we separate variables and integrate:

$$d[A]/[A] = - kdt$$

$$[A] = [A]e\text{-}kt$$

6. Steady State Approximation

6.1. Generation of Initial Rate Equation

In 1956, King and Altman devised a systematic method, applicable to random as well as sequential binding order mechanisms, for deriving initial velocity rate equations. This manual technique uses rate constants and reactant concentrations obtained from all valid pathways of a mechanism in the formulation of an initial velocity rate equation. As the complexity of the mechanism increases so does the difficulty in deriving the rate equation even with the technique of King and Altman. Utilizing linear graph theory concepts, the method developed by Lam and Priest generates all valid patterns for an enzyme mechanism. An enzymatic mechanism is described as a connected graph with numbered branches and numbered nodes, corresponding to reaction steps and enzyme forms, respectively. Patterns can be generated from the graph by applying Wang algebra principles in a stepwise procedure. These patterns are then used to generate the initial velocity equations.

6.2. Generation of the Constraint Equation

The value of the rate constants belonging to cycles within a random binding mechanism is restricted as a result of the law of micro reversibility. The law of micro reversibility, also known as the principle of detailed balance, requires forward and reverse fluxes in a time independent state to be equal, consider an entire reaction for a mechanism at steady-state conditions (Lienhard, 1973). The concentrations of its intermediate enzyme complexes remain constant. If a cycle within the mechanism is at equilibrium, with no net change occurring for its reactant concentrations, then all enzyme complex concentrations for that cycle remain constant. For each of these enzyme complexes in the cycle, the change of its concentration over time, i.e., the forward flux minus the reverse flux of that reaction step is equal to zero. It then follows that the forward flux is equal to the reverse flux in a cycle.

If an intermediate is always present in amounts much less than those of the reactants (other than the enzyme) the rate of change of its concentration is much smaller than that of the reactants. This condition is ensured whenever, as is usually the case in enzyme-catalysed reactions, the concentration of substrate is much higher than that of the enzyme; it is not necessary for the amount of intermediate to be small compared with the amount of enzyme. For example, in the scheme

$$E + A \xrightarrow[K_{-1}]{K_1} E + Z$$

If [EA] is always much less than [A] the following equation is obeyed to a good approximation:

$$\frac{d[EA]}{dt} = k_1[E][A] - k_{-1}[EA] - k_2[EA] = 0$$

The intermediate EA is said to be in a steady state. The use of this approximation to obtain an overall rate expression is known as the steady-state treatment or the steady-state approximation.

At the very beginning of the reaction the concentration of EA in the above scheme is rising from zero to its steady-state value. The steady-state approximation is not valid during these early times and the kinetics are known as pre-steady-state kinetics or transient-phase kinetics. The transient phase of an enzyme-catalysed reaction usually occupies a very brief period of time (usually a small fraction of a second), and special techniques must be used for investigating this phase of the reaction

The rate of reaction of an enzyme-catalyzed reaction is not defined during the transient phase, because there is not a one-to-one stoichiometry between the reactants. In the steady state a one-to-one stoichiometry is established and the rate of reaction can be defined. This rate, extrapolated back to zero time, is called the initial rate and given the symbol v0. The subscript 0 is normally omitted when no other kinds of rate are at issue, i.e. when extended time courses are not being analysed.

The ratio [E] [A]/[EA] at equilibrium is called the substrate dissociation constant and given the symbol KsA When only one substrate is in question the qualifier A may be omitted, and when it is included its location is a matter of typographical convenience. The substrate dissociation constant should not be confused with the Michaelis constant KmA. , to which it bears no necessary relationship in general. In the example given above KsA is equal to the ratio k-1/k1.

7. Enzyme kinetics

Enzyme kinetics describes the rate of change of reactant concentrations in a chemical reaction. A simple, one-substrate reaction can be described by Equation 1, where E is the enzyme, S is the substrate, and P is the product:

$$E + S \underset{k_2}{\overset{k_1}{\rightleftarrows}} ES \underset{k_4}{\overset{k_3}{\rightleftarrows}} E + P \qquad (1)$$

Initially there is no product, so the rate constant k4 is zero. Under steady-state conditions, the concentration of ES stays the same while the concentrations of the reactants and products are changing; that is, the rate of ES formation is the same as the rate of breakdown (Equation 2).

$$k_1[E][S] = (k_2 + k_3)[ES] \qquad (2)$$

Equation 2 can be rearranged and simplified by defining the Michaelis constant, Km (Equation3):

$$K_m = \frac{k_2 + k_3}{k_1} \qquad (3)$$

When the entire enzyme is bound to substrate (i.e., substrate concentration is at saturation levels) and is thus in the ES state, the rate of the reaction is at a maximum or Vmax. Combining Km with Vmax, Michaelis and Menton derived a mathematical equation (Equation 4) to describe the relationship between the initial velocity (or reaction rate) and substrate concentration:

$$V_0 = \frac{V_{max}[S]}{K_m + [S]} \qquad (4)$$

Figure 4 illustrates a typical plot for a one-substrate enzyme-catalyzed reaction. At relatively low substrate concentration, vo increases as the substrate concentration increases. As vo approaches

Vmax the reaction rate is constant and independent of substrate concentration.

Rearranging the Michaelis-Menton equation to the form in Equation 5 makes it apparent that the Michaelis constant, Km, has the same units as substrate concentration or moles/liter. It should also be apparent that when Km is the substrate concentration, the initial velocity, vo, is equal to one-half the maximal velocity, Vmax. (That is, using Eqn. 5, when Km = [S], Vmax/vo = 2.).

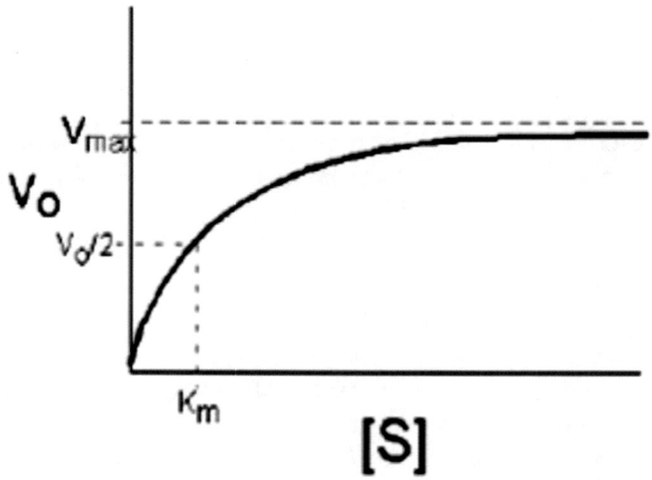

Fig. 8 : Michaelis menton plot

$$K_m = [S]\left(\frac{V_{max}}{V_0} - 1\right)$$

7.1. Briggs-Haldane Approach (Briggs and Haldane, 1925)

The change of the intermediate concentration with respect to time is assumed to be negligible, that is, d(CES)/dt = O. This is also known as the pseudo-steady-state (or quasi-steady-state) assumption in chemical kinetics and is often used in developing rate expressions in homogeneous catalytic reactions.

7.2. Significance of Km

1. Km can give an approximation value of the affinity towards substrate; the lower the value of Km, the greater the affinity.
2. Km value is constant and unique for a particular enzyme. The numerical value reveals to find out the origin of the enzyme.
3. Characteristics of ligand and their effect on enzyme reflect on the numerical value of Km.

4. To design the enzymatic assay conditions to determine the value of Vmax Because the substrate concentration should me greater than the Km value.
5. The best substrate is that where the Vmax/ Km ratio is highest, thus we choose the right substrate for a reaction.

7.3. Significance of K_{cat}

The value of Kcat is sometimes referred to as the turnover number for the enzyme, since it defines the number of catalytic turnover events that occur per unit time. The units of K_{cat} are reciprocal time (e.g., min^{-1}, s^{-1}). Turnover numbers, however, are typically reported in units of molecules of product produced per unit time per molecules of enzyme present.

In the laboratory we can easily determine the turnover number as Kcat by measuring the reaction velocity under conditions of [S]>>Km so that v approaches Vmax. The rate of enzyme turnover under most physiological conditions in vivo, however, is very different from our laboratory situation. In vivo, the concentration of substrate is more typically 0.1-1.0Km. When [S]<<Km we must change our expression for velocity to :-

$$\upsilon = \frac{k_{cat}}{K_m}[E][S]_f$$

(Since [S] < K < here, the free enzyme concentration is well approximated by the total enzyme concentration [E]; thus we used this term in above equation in place of [E]_ .) Recalling our definition of K_, we note that:

$$\frac{k_{cat}}{k_m} = \frac{k_{cat}k_1}{k_{-1} + k_{cat}}$$

Although the values of Km and Vmax can be estimated from plots like that in Figure 4, it is easier and more accurate (without using nonlinear regression) to use a plot of a linearized form of the Michaelis-Menton equation (Leytus et al., 1983). The reciprocal of the Michaelis-Menton equation (Equation 6) is the most common linearization and the concomitant plot (Figure 5) is referred to as a double reciprocal or Lineweaver-Burk plot.

$$\frac{1}{v_0} = \frac{K_m}{V_{max}}\frac{1}{[S]} + \frac{1}{V_{max}}$$

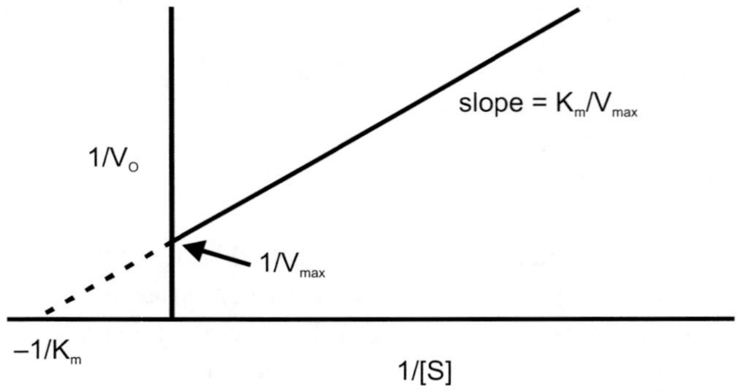

Fig. 9. Lineweaver-Burk (double reciprocal plot).

By comparing Equation 6 to the general equation for a straight line (y = a+ bx), it is clear that the slope of the line is Km/Vmax and the y-intercept is 1/Vmax. The x-intercept is equal to -1/Km. It is important to note that a shortcoming of using a double-reciprocal plot is that at low substrate concentrations (high values for 1/[S]) small deviations in the initial velocity (vo) can badly skew the regression line through the data points. For this reason other linearized forms of the Michaelis-Menton equation, for example Equation 7, used to produce an Eadie-Hofstee plot, may be used.

$$V_0 = \frac{V_0}{[S]} + V_{max} \qquad (7)$$

As Equation 7 indicates, the "independent variable" (x-axis) actually comprises the true independent variable ([S]), as well as the dependent variable (vo), so the axes are not independent of each other. Generally, this is to be avoided. The values for Km and Vmax can be determined using either linearized form. The calculated values will not be identical but they should not be very different.

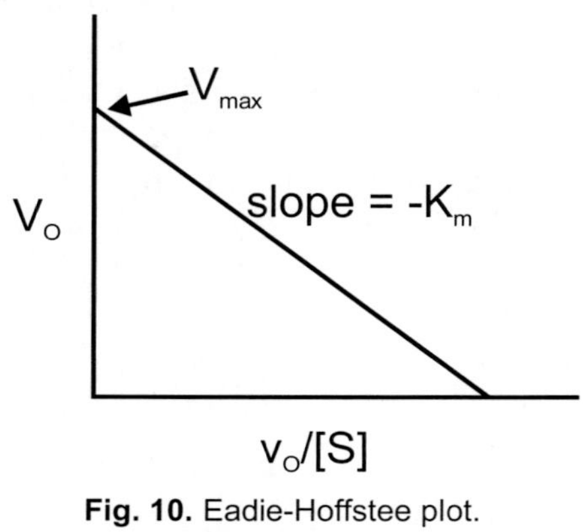

Fig. 10. Eadie-Hoffstee plot.

8. Enzyme Kinetics of multiple substrates

The introduction of alternative substrates in enzyme processes is an important approach in the study of living materials and their transformations, as the induced perturbations allow a more detailed understanding of the normal initial states (Ault, 1974). In particular, alternative substrates are employed to bring out the features of enzyme active centres due to their selective and competitive reactions.

8.1. Rapid Equilibrium Random BI BI

The following discussion is based on the concise rules and nomenclature introduced by Cleland. Substrates are termed in the sequence of their binding to the enzymes as A, B, C, products in the sequence of their release P, Q, R. Different enzyme states are named E, F, G. With substrates and products the enzyme forms transitory complexes, EA, EP, etc., that decay in a unimolecular step releasing substrate or product. Catalytic reactions occur from central complexes which are set in parentheses, e.g. (EAB), to distinguish them from transitory complexes. Central complexes cannot bind further substrate or product as all sites are already occupied; rather, they release substrate or product in uni molecular steps. As steady-state kinetics provides no information on the catalytic conversion of substrate into product (and vice versa) on the enzyme molecule, only one central complex for both states before and after catalysis (EAB-EPQ) is defined (Klotz, 1976).

The number of substrates participating in the forward and of products participating in the reverse reaction will be expressed by the terms uni, bi, ter, quad for 1, 2, 3, and 4, respectively, and these terms are added to the name of the reaction mechanisms. Two substrates being condensed into one product is a bi uni mechanism. Mechanisms where all substrates must bind before product can be released are called sequential mechanisms. The binding of substrates may be either random or ordered. In a ping-pong mechanism, product is already released before all substrates are bound. The enzyme here exists in two or more forms, modified by groups of the substrate. In iso mechanisms the enzyme isomerizes into two or more stable conformations. The reaction equations are presented in a schematic form (Vol'kenshtein et al., 1971). The progressing reaction is symbolised by a horizontal line, the reaction coordinate. The various enzyme states appearing during the reaction course are indicated below this line. Substrate binding is indicated by vertical arrows on the reaction coordinate, dissociation of product by vertical arrows from it. Rate constants are ascribed on the sites of the arrows, the forward rate constant on the left and the reverse on the right. Alternatively, the binding constant is indicated. The reaction schema for an ordered BI BI mechanism in the conventional style:

$$E + A \xrightarrow[K_{-1}]{K_1} EA$$

$$EA + \xrightarrow[K_{-2}]{K_2} EAB - EPQ \xrightarrow[K_{-3}]{K_3} EQ + P$$

$$EQ \xrightarrow[K_{-4}]{K_4} E + Q$$

Will be Transformed Accordingly

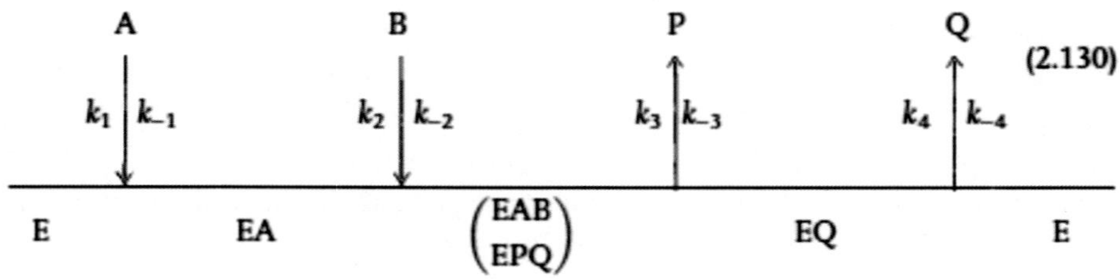

(2.130)

In the following the most frequent multi-substrate mechanisms are presented and methods to analyze and distinguish between them will be discussed. Finally methods for deriving complex rate equations are introduced.

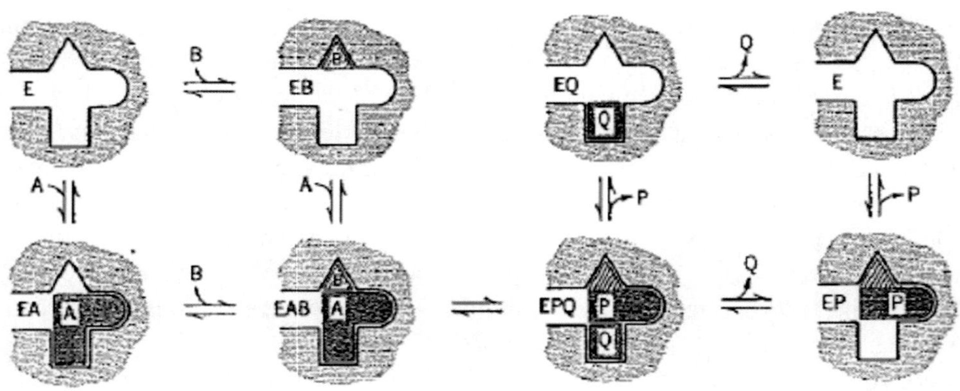

Fig. 11 : A schematic representation of BI BI mechanism

8.2. Ping Pong BI BI Mechanism

A third possible mechanism involves a transfer of phosphate from MgATP to the enzyme, followed by the release of MgADP before the glucose binds and picks up the phosphate. This type of mechanism is called Ping Pong because the enzyme oscillates between two stable forms, E and F. The reaction sequence is:

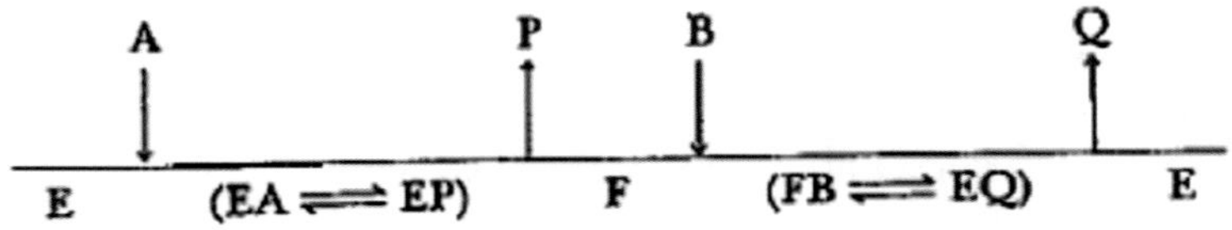

where A=MgATP, P=MgADP, B=glucose, and Q=glucose-6-phosphate. FB represents enzyme-phosphate-MgADP. Enzyme species F represents the stable (possible covalent) enzyme phosphate. FB represents enzyme-phosphate-glucose, and EQ represents enzyme-glucose-6-phosphate. A steady-state treatment yields:

$$\frac{v}{V_{max}} = \frac{[A][B]}{K_{mB}[A]+K_{mA}[B]+[A][B]} \quad (61)$$

When rearranged to show A as the varied substrate at different fixed concentrations of B, the equation becomes:

$$\frac{\upsilon}{V_{max}} = \frac{[A]}{K_{mA}+[A]\left(1+\frac{K_{mB}}{[B]}\right)} \quad (62)$$

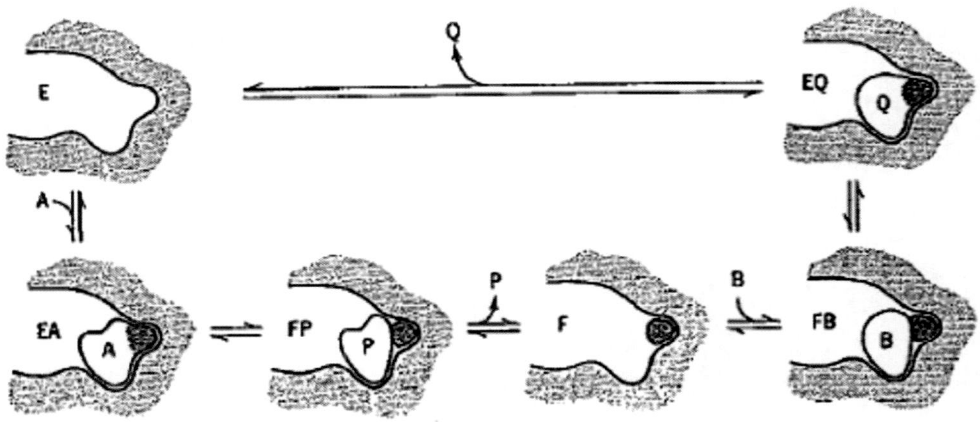

Fig. 12. Schematic representation of ping pong BI BI mechanism

Since the KmA term in the denominator is not multiplied by any factor, the family of reciprocal plots obtained at different fixed B concentrations will be parallel. A symmetrical equation results when [B] is varied at different fixed a concentrations. The reciprocal plots and replots for varied [A] are shown in Figure 12.

Hexokinase does not yield parallel reciprocal plots, so the Ping Pong mechanism can be discarded. However, initial velocity studies alone will not discriminate between the rapid equilibrium random and steady- state ordered mechanisms. Both yield the same velocity equation and families of intersecting reciprocal plots, other diagonostic procedures must be used (e.g., product inhibition, dead-end inhibition, equilibrium substrate binding, and isotope exchange studies).

Table 7. Abscissa and ordinate intercepts of secondary plots for bi-substrate reactions, (Boyer, 1970)

Primary plot		**Secondary plot**					
Designation of axis		**Designation of axis**		**Intercepts**			
Y	**X**	**Y**	**X**	**Random/ordered mechanism**		**Ping-pong mechanism**	
				Ordinate	**Abscissa**	**Ordinate**	**Abscissa**
$1/v$	$1/[A]$	Sl_A	$1/[B]$	K_{mA}/V	$-K_{mA}/K_{iA}K_{mB}$		
		Ab_A	$1/[B]$			$1/K_{mA}$	$-1/K_{mB}$
		Or_A	$1/[B]$	$1/V$	$-1/K_{mB}$	$1/V$	$-1/K_{mB}$
$1/v$	$1/[B]$	Sl_B	$1/[A]$	K_{mB}/V	$-1/K_{iA}$		
		Ab_B	$1/[A]$			$1/K_{mB}$	$-1/K_{mA}$
		Or_B	$1/[A]$	$1/V$	$-1/K_{mA}$	$1/V$	$-1/K_{mA}$
$[A]/v$	$[A]$	Sl_A	$1/[B]$	$1/V$	$-1/K_{mB}$	$1/V$	$-1/K_{mB}$
		$1/Ab_A$	$1/[B]$			K_{mA}	$-1/K_{mB}$
		Or_A	$1/[B]$	K_{mA}/V	$-K_{mA}/K_{iA}K_{mB}$		
$[B]/v$	$[B]$	Sl_B	$1/[A]$	$1/V$	$-1/K_{mA}$	$1/V$	$-1/K_{mA}$
		$1/Ab_B$	$1/[A]$			K_{mB}	$-1/K_{mA}$
		Or_B	$1/[A]$	K_{mB}/V	$-1/K_{iA}$		
$v/[A]$	v	Sl_A	$1/[B]$			$-1/K_{mA}$	$-1/K_{mB}$
		$1/Ab_A$	$1/[B]$	$1/V$	$-1/K_{mB}$	$1/V$	$-1/K_{mB}$
		Or_A	$1/[B]$	K_{mA}/V	$-K_{mA}/K_{iA}K_{mB}$		
$v/[B]$	v	Sl_B	$1/[A]$			$-1/K_{mB}$	$-1/K_{mA}$
		$1/Ab_B$	$1/[A]$	$1/V$	$-1/K_{mA}$	$1/V$	$-1/K_{mA}$
		Or_B	$1/[A]$	K_{mB}/V	$1/K_{iA}$		

9. Heterogeneous Enzyme Kinetics

Enzyme kinetic principles were developed based on the assumsion that enzyme and substrate both are in same state that is homogeneous state. In practical many enzymes are present which acts on heterogeneous mode, e.g. celulase enzyme which digests cellulose materials in heterogeneous mode (Albery & Knowles, 1987; Navarro-Lozano et al., 1995). The enzyme is soluble in water but the substrate is insoluble. The enzymes reacts with substrate or catalyze the system by means of adsorption followed by de-sorption from the substrate, we can imagine this phenomena with the help of classical thermodynamic collision theory and free fickian diffusion principle. The classical Michelis Menten equation (MM) can not consider such parameters. To overcome such limitations in classical Michelis Menten equation there are two factors were added viz., adsorption co efficient and de-sorption co efficient.

9.1. Fractal Michaels Kinetics

Fractal scheme for an enzyme (ellipsoid) acting on chained substrates (dashed curve) Fig 7.The enzyme binds to the chain through its active site tunnel. After cleaving off a substrate unit (bar), the enzyme slides along the chain (in the! direction) for the next catalytic cycle

Fig. 13. Illustration of fractal michaelis kinetics phenomena (Segal, 1975)

We know,

$$E + S \underset{K_2}{\overset{K_1}{\rightleftarrows}} ES \underset{K_4}{\overset{K_3}{\rightleftarrows}} E + P \tag{1}$$

We can re derive it in the following format

$$\frac{v}{k_2[E]} = \frac{[ES]}{[ES]+[E]} = \frac{[ES]/[E]}{[ES]/[E]+1} = \frac{[S]}{K_m+[S]}$$

$$k_2[E]dt = -d[S] - K_m \frac{d[S]}{[S]},\ k_2[E]\int_0^r dt$$

$$= \_ = -\int_{[s]}^{[s]} d[S] - K_m \int_{[s]}^{[s]} d\ \text{In}[S],\ k_2[E]t = [P] - k_m\ \text{In}\left(1 - \frac{[P]}{[S]}\right)$$

$$\frac{[E]_a}{[E]_b} = \frac{t_b}{t_u}$$

ta and tb are the time needed to reach the same [P], from the same [S], when the initial enzyme concentrations are [E]a and [E]b, respectively.

Taking log we got

$$\frac{\log([E]_a / [E]_b)}{\log(t_b / t_a)} = 1$$

9.1.1. Formalism

Formalism of MM equation can be done in two way time variant and time invariant

9.1.2. Time-variant formalism

In this case the general fractal formalism of replacing a rate constant k with a time-dependent "rate coefficient" g,

Where, g = kt-f [0<f<1]

There fore the velocity of product formation can be written as,

9.1.3. Time-invariant formalism

In this case, the rate constant becomes time invariant, we can write such phenomena in terms of reaction velocity is

$$\nu = -\frac{d[S]}{dt} = k_2[ES]^{h_{ES}}; h_{ES} > 1$$

9.1.4. Jammed" Michaelis kinetics

"Jammed" scheme for enzymes (ellipsoids) acting on substrate chains dashed lines) that are packed orderly with defined spacing. Being 'oversized" in comparison to the inter-chain space, the enzymes anchored on adjacent chains may jam each other.

The reaction velocity can be written in such case as,

$$\upsilon = \left(1 - \frac{[E]}{j[S]}\right) k_2[ES] = k[ES]$$

10. Enzyme Inhibitors

Enzyme inhibitors are molecules that interact in some way with the enzyme to prevent it from working in the normal manner. There are a variety of types of inhibitors including: nonspecific, irreversible, reversible - competitive and noncompetitive. Poisons and drugs are examples of enzyme inhibitors (Leytus et al., 1983).

10.1. Competitive Inhibition

Any chemical substance which has a molecular structure that closely resembles a substrate can reduce or inhibit the activity of an enzyme. Such an inhibitor is called competitive inhibitor. This situation is comparable to a lock jammed by a key almost similar to the original one. In the presence of a competitive inhibitor, it takes a higher substrate concentration to achieve the same velocities that were reached in its absence. So while Vmax can still be reached if sufficient substrate is available, one-half Vmax requires a higher [S] than before and thus Km is larger. Lineweaver-Burk plot (Fig 8) displays these results.

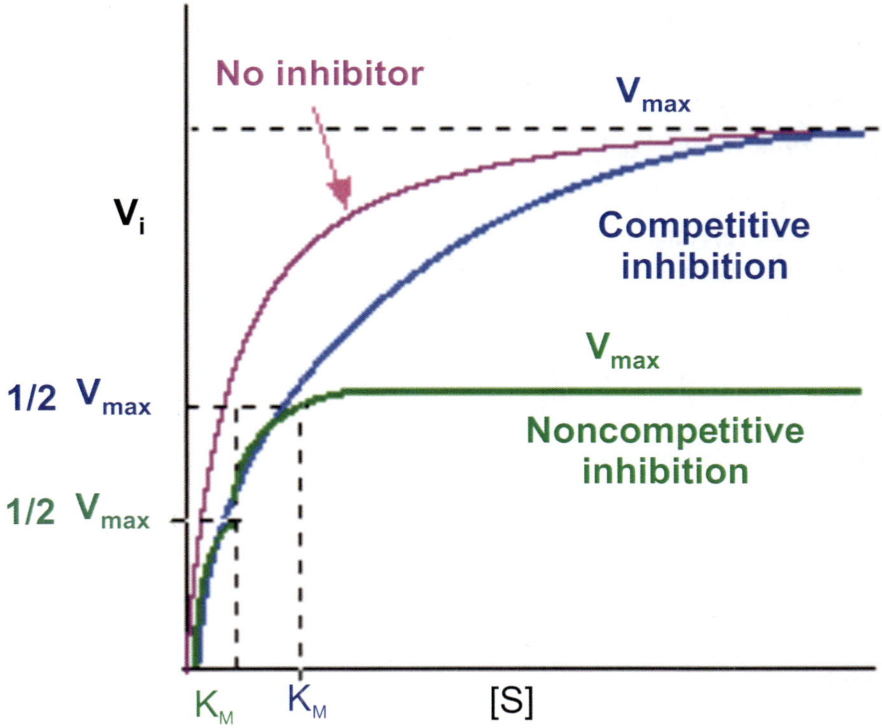

Fig.14. Line weaver plot for competitive inhibition

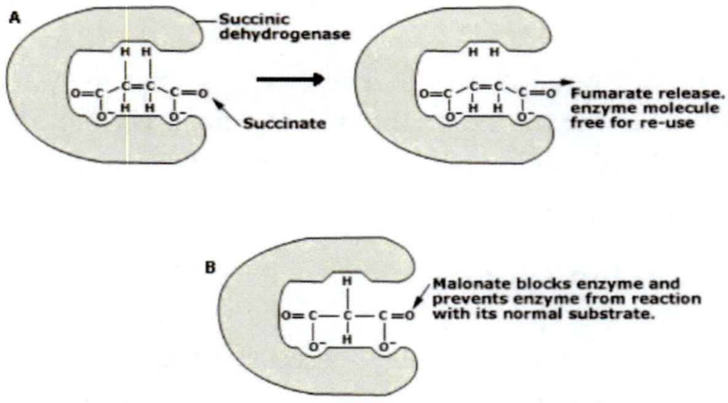

Fig. 15. An example of competitive inhibition (a) the action of the enzyme inhibition succinic dehydrogenase on substrate (b) competitive of the enzyme by malonate (Bender & Braubacher, 1973)

10.2. Noncompetitive Inhibition

Chemical substances such as cyanides or phosphide can inhibit the action of repiratory enzymes. They are not similar to the substrate molecule and as such do not compete with it. However, such inhibitor may become attached any site on the enzyme, other than the substrate binding site.

With noncompetitive inhibition, enzyme molecules that have been bound by the inhibitor are taken out of the game so

1. Enzyme rate (velocity) is reduced for all values of [S], including
2. Vmax and one-half Vmax but
3. Km remains unchanged because the active site of those enzyme molecules that have not been inhibited is unchanged.

This Line weaver-Burk (Fig 16) plot displays these results.

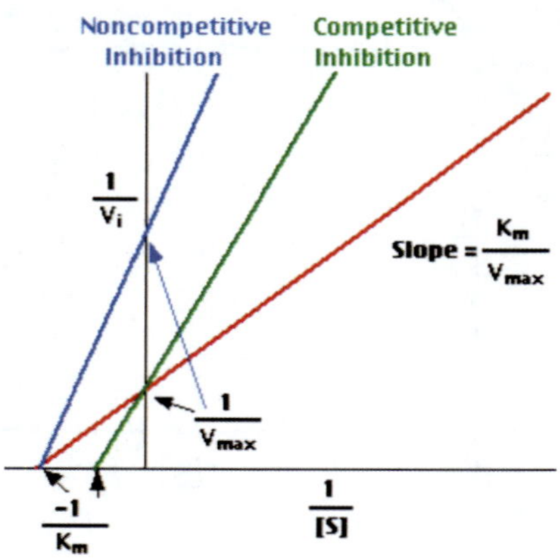

Fig.16 : Line weaver plot for non competitive inhibition

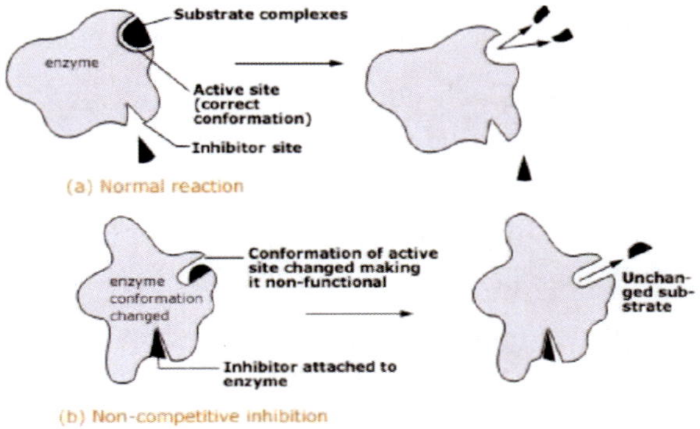

Fig. 17 : Mechanism of non competitive inhibition

10.3. Allosteric or Feedback Inhibition

The activity of some enzymes, particularly those involved in metabolic pathways, is controlled by a self-regulating mechanism. Some specific substance, most often the product itself, acts as an inhibitor. Such an inhibitor binds to an enzyme at a specific site and modifies the active site of the enzyme. This prevents the binding of substrate molecule. Such sites on the enzymes are called allosteric sites and such enzymes are called allosteric enzymes.

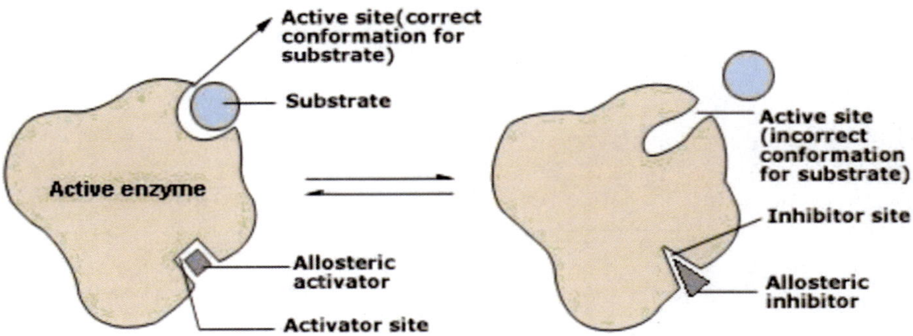

Fig. 18. Schematic representation of allosteric enzyme activity

11. Inhibition Mechanism

Common feedback inhibition mechanisms,

(1) The basic feedback inhibition echanism, where the product (P) inhibits the committed step (A->B).

(2) Sequential feedback inhibition. The end products P1 and P2 inhibit the first committed step of their individual pathway (C->D or C->F). If both products

are present in abundance, all pathways from C are blocked. This leads to a buildup of C, which in turn inhibits the first common committed step A->B.

(3) Enzyme multiplicity. Each end product inhibits both the first individual committed step and one of the enzymes performing the first common committed step.

(4) Concerted feedback inhibition. Each end product inhibits the first individual committed step. Together, they inhibit the first common committed step.

(5) Cumulative feedback inhibition. Each end product inhibits the first individual committed step. Also, each end product partially inhibits the first common committed step.

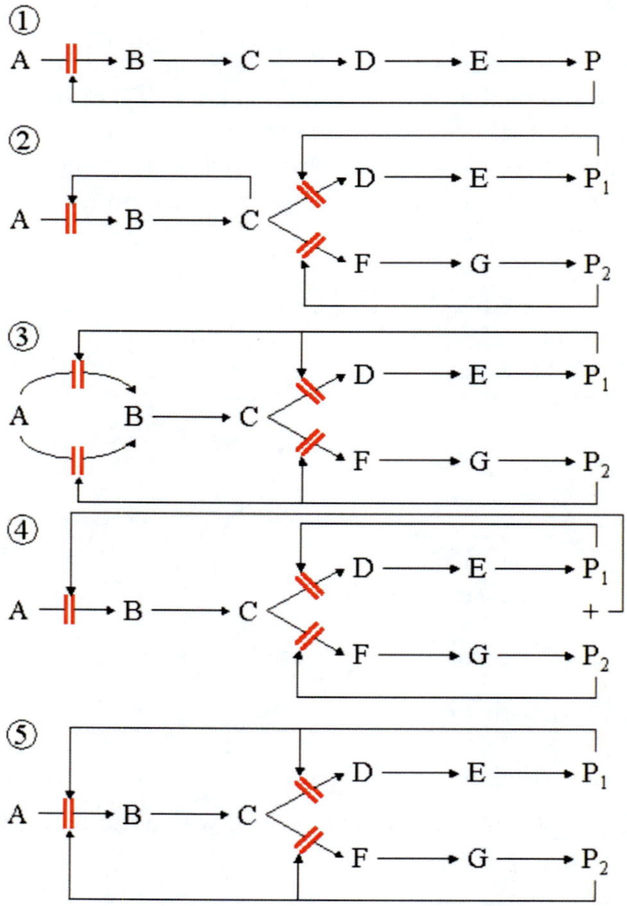

Fig. 19 : Feedback inhibition mechanism

Table 8 : Different type of inhibitors and their kinetic effect (Dixon & Webb, 1964)

Inhibitor Type	Binding Site on Enzyme	Kinetic effect
Competitive Inhibitor	Specifically at the catalytic site, where it competes with substrate for binding in a dynamic equilibrium- like process. Inhibition is reversible by substrate.	Vmax is unchanged; Km, as defined by [S] required for ½ maximal activities, is increased.
Noncompetitive Inhibitor	Binds E or ES complex other than at the catalytic site. Substrate binding unaltered, but ESI complex cannot form products. Inhibition cannot be reversed by substrate.	Km appears unaltered; Vmax is decreased proportionately to inhibitor concentration.
Uncompetitive Inhibitor	Binds only to ES complexes at locations other than the catalytic site. Substrate binding modifies enzyme structure, making inhibitor- binding site available. Inhibition cannot be reversed by substrate.	Apparent Vmax decreased; Km, as defined by [S] required for ½ maximal activity, is decreased

12. Quantum Mechanical Methods for Enzyme Kinetics

Three aspects are emphasized, namely the treatment of the electronic structure of atoms involved in the chemical process in an enzyme, the incorporation of quantum vibration free energies into the calculation of the potential of mean force, and the use of multidimensional tunneling methods for estimating the transmission coefficient.

13. Enzyme Kinetics and Bioinformatics

Information about kinetics is obligatory for the assessment of any enzyme used in several fields of biological researches like, clinical diagnosis, pharmaceutical research or drug development, as it endows with essential information about behavior of an enzyme in different situations. For example, it is essential in areas of drug development for understanding pharmacokinetics, interactions within the formulation, and for stability testing. High throughput screening is needed to establish the metabolic stability, kinetic parameters and prediction of metabolic clearance, and the information can be used to develop in vitro models that aid in the initial design. Such pharmacokinetic information can be scaled up to in situ models of the enzyme activity in model organs (Warshel & Florian, 1998).

Certain enzymes might have unaltered levels in the blood of an individual with a particular disease; they might have an altered activity. Knowledge of Km and Vmax and methods for their accurate determination are therefore required for diagnosis.

To know the exact mechanism of drug metabolism, interaction among different drugs is important to design a drug. This process is not only time consuming but also expensive and sometime fatal towards human civilization. The knowledge of bioinformatics facilitate and gear up the research on medicine at lower cost as such kind of expensive and vital researches can be done in situ virtually with computer (Agarwal, 2005).

The major thrust area in this field id determination of kinetic parameters. There are two types of enzymatic kinetic experiments transient (or pre-steady state) methods, which are of special interest if detailed mechanistic information is required; and steady-state methods, which are easier, but may lead to less detailed mechanistic information. Steady-state reaction rates can be analyzed in two manners initial rate analysis, using the traditional Lineweaver-Burk plots for example; and progress curve analysis .are employed Method for Inferring Kinetics and Network Architecture (MIKANA) ,infers biochemical pathway mechanisms from time course data using a global non-linear modeling technique to identify the elementary reaction steps which constitute the pathway (Warshel, 2003). Bayesian experimental design method the designs are based around prior distributions of parameter estimates and their variance, rather than on chosen single-point values. Importantly, they incorporate all previous scientific knowledge, which can be from a variety of sources, and not just knowledge of working designs and statistics) are also employed now a days (Garcia-Viloca et al., 2004).

References

Agarwal, P.K. 2005. Role of protein dynamics in reaction rate enhancement by enzymes. J Am Chem Soc, 127(43), 15248-56.

Albery, W.J., Knowles, J.R. 1987. Energetics of enzyme catalysis. II. Oversaturation, case diagrams, reversible and irreversible behaviour. J Theor Biol, 124(2), 173-89.

Ault, A. 1974. An introduction to enzyme kinetics. J Chem Educ, 51(6), 381-6.

Bender, M.L., Braubacher, L.J. 1973. Catalysis and Enzyme Action. McGraw-Hill, New York.

Boyer, P.D. 1970. The Enzymes. Academic Press, New York.

Copeland, R.A. 1996. Enzymes: A Practical Introduction to Structure, Mechanism, and Data Analysis. VCH Publishers, New York.

Daniel, R.M., Danson, M.J., Eisenthal, R. 2001. The temperature optima of enzymes: a new perspective on an old phenomenon. Trends Biochem Sci, 26(4), 223-5.

Daniel, R.M., Danson, M.J., Eisenthal, R., Lee, C.K., Peterson, M.E. 2008. The effect of temperature on enzyme activity: new insights and their implications. Extremophiles, 12(1), 51-9.

Daniel, R.M., Danson, M.J., Eisenthal, R., Lee, C.K., Peterson, M.E. 2007. New parameters controlling the effect of temperature on enzyme activity. Biochem Soc Trans, 35(Pt 6), 1543-6.

Dixon, M., Webb, E.C. 1964. Enzymes. Academic Press, New York.

Fersht, A. 1999. Structure and Mechanism in Protein Science. W. H. Freeman, New York.

Garcia-Viloca, M., Gao, J., Karplus, M., Truhlar, D.G. 2004. How enzymes work: analysis by modern rate theory and computer simulations. Science, 303(5655), 186-95.

Gutfreund, H. 1975. Enzymes: Physical Principles. John Wiley & Sons, New York.

Klotz, I.M. 1976. Free energy diagrams and concentration profiles for enzyme-catalyzed reactions. J Chem Educ, 53(3), 159-60.

Klotz, I.M. 1997. Ligand-Receptor Energetics. John Wiley & Sons, New York.

Kraut, J. 1988. How do enzymes work? Science, 242(4878), 533-40.

Leytus, S.P., Peltz, S.W., Mangel, W.F. 1983. Adaptation of acyl-enzyme kinetic theory and an experimental method for evaluating the kinetics of fast-acting, irreversible protease inhibitors. Biochim Biophys Acta, 742(2), 409-18.

Lienhard, G.E. 1973. Enzymatic catalysis and transition-state theory. Science, 180(82), 149-54.

Murray, R., Rodwell, V., Bender, D., Botham, K.M., Weil, P.A., Kennelly, P.J. 2009. Harper's Illustrated Biochemistry. Tata McGraw Hill New Delhi.

Navarro-Lozano, M.J., Valero, E., Varon, R., Garcia-Carmona, F. 1995. Kinetic study of an enzyme-catalysed reaction in the presence of novel irreversible-type inhibitors that react with the product of enzymatic catalysis. Bull Math Biol, 57(1), 157-68.

Perutz, M. 1990. Mechanisms of Cooperativity and Allosteric Regulation in Proteins. Cambridge University Press, New York.

Peterson, M.E., Eisenthal, R., Danson, M.J., Spence, A., Daniel, R.M. 2004. A new intrinsic thermal parameter for enzymes reveals true temperature optima. J Biol Chem, 279(20), 20717-22.

Segal, I.H. 1975. Enzyme Kinetics. John Wiley & Sons, New York.

Vol'kenshtein, M.V., Magarshak Iu, B., Stefanov, V.E. 1971. [Diagrams of enzyme reactions catalyzed by isoenzymes]. Dokl Akad Nauk SSSR, 197(5), 1193-5.

Walsh, C. 1979. Enzymatic Reaction Mechanisms. W. H. Freeman, San Francisco.

Warshel, A. 2003. Computer simulations of enzyme catalysis: methods, progress, and insights. Annu Rev Biophys Biomol Struct, 32, 425-43.

Warshel, A., Florian, J. 1998. Computer simulations of enzyme catalysis: finding out what has been optimized by evolution. Proc Natl Acad Sci U S A, 95(11), 5950-5.

Wyman, J., Gill, S.J. 1990. Binding and Linkage: Functional Chemistry of Biological Macromolecules. University Science Books, Sausalito, CA.

❑❑❑

CHAPTER – 29

Pharmacogenomics A Way to Personalized Medicine

Pramod Katara
Department of Bioscience and Biotechnology, Banasthali University - 304022, India
Email : pmkatara@gmail.com

Summary

One of the most seductive lures of the genomic revolution is the promise of personalized medicine. The rapid identification of tens of thousands of human genes and hundreds of thousands of DNA variations that might influence disease susceptibility has spawned a new field Pharmacogenetics and Pharmacogenomics deal with the genetic basis underlying variable drug response in individual patients. The traditional pharmacogenetic approach relies on studying sequence variations in candidate genes suspected of affecting drug response i.e. single or few genes. On the other hand, pharmacogenomic studies encompass the sum of all genes, i.e., the genome. Numerous genes may play a role in drug response and toxicity, introducing a daunting level of complexity into the search for candidate genes. The high speed and specificity associated with newly emerging genomic technologies enable the search for relevant genes and their variants to include the entire genome. These new technologies have essentially spawned a new discipline, termed pharmacogenomics, which seeks to identify the variant genes affecting the response to drugs in individual patients. Moreover, pharmacogenomic analysis can identify disease susceptibility genes representing potential new drug targets. All of this will lead to novel approaches in drug discovery, an individualized application of drug therapy, and new insights into disease prevention. Current concepts in drug therapy often attempt treatment of large patient populations as groups, irrespective of the potential for individual, genetically-based differences in drug response. In contrast, pharmacogenomics may help focus effective therapy on smaller patient subpopulations which although demonstrating the same disease phenotype are characterized by distinct genetic profiles. Whether and to what extent this individual, genetics-based approach to medicine results in improved, economically feasible therapy remain to be seen.

Introduction

It is well recognized that different individuals shows variable drug response (VDR) to the same drug. Some patients' shows a life threatening adverse reaction while others fail to shows an expected therapeutic level. Intermediate responses between above two extreme cases are also possible (Shastry, 2005). These differences are often greater among members of a population than the same individuals at different times (Vesell, 1989). The existence of large population differences with small intrapatient variability is consistent with inheritance as a determinant of drug response; these inherited determinants generally remain stable throughout a person's lifetime. It is estimated that genetics can account for 20 to 95 percent of variability in drug metabolism, disposition and effects (Kalow et al., 1998) Although many nongenetic factors influence the effects of medications, including age, organ function, concomitant therapy, drug interactions, and the nature of the disease, there are now numerous examples of cases in which interindividual differences in drug response are due to genetic variants in genes encoding drug-metabolizing enzymes, drug transporters, or drug targets (Evans, 1999; McLeod et al. 2001). Thus outcome of medication is often unpredictable, ranging from beneficial effects to lack of efficiency to serious adverse effects.

Clinical observations of inherited differences in drug effects were first documented in the 1950s, (Kalow, 1956; Carson et al. 1956) giving rise to the field of pharmacogenetics, and later pharmacogenomics. Although the two terms are synonymous for all practical purposes, pharmacogenomics uses genome-wide approaches to elucidate the inherited basis of differences between persons in the response to drugs (figure1).

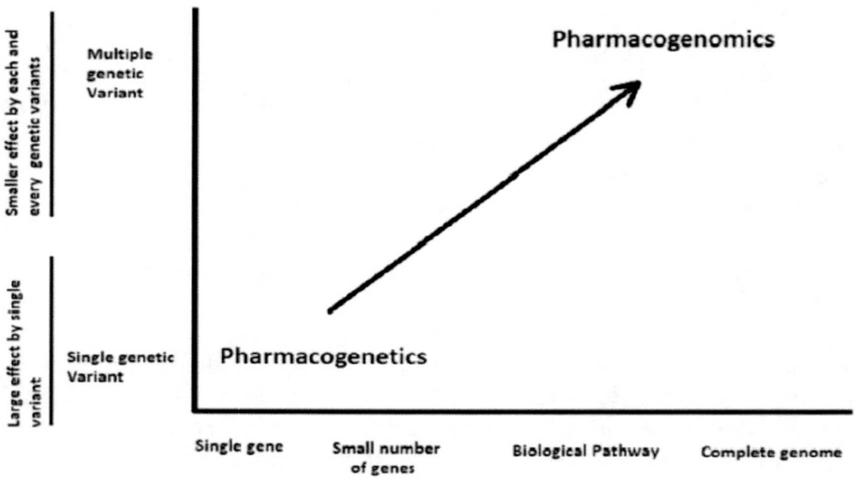

Fig. 1 : Transition of Pharmacogenetics into Pharmacogenomics

For more than 5 decades, studies of pharmacogenetics have provided ample examples of causal relations between genotypes (single gene or small number of genes)

and drug response to account for phenotypic variations of clinical importance in drug therapy. The convergence of pharmacogenetics and human genomics in recent years has dramatically accelerated the discovery of new genetic variations that potentially underlie variability in drug response, giving birth to pharmacogenomics. In addition to the rapid accumulation of knowledge on genome-disease and genome-drug interactions, there arises the hope of individualized medicine (Ma and Lu, 2011).

Concepts of Pharmacogenomics

Pharmacogenomics is the study of how an individual's genetic inheritance affects the body's response to drugs. The term comes from the words pharmacology and genomics and is thus the intersection of pharmaceuticals and genetics. Pharmacogenomics holds the promise that drugs might one day be made for individuals and adapted to each person's own genetic makeup. Environment, diet, age, lifestyle, and state of health all can influence a person's response to medicines, but understanding an individual's genetic makeup is thought to be the key to creating 'personalized medication' with greater efficacy and safety. Pharmacogenomics combines traditional pharmaceutical sciences such as biochemistry with annotated knowledge of genes, proteins, and polymorphisms (SNPs).

The successful completion of the Human Genome project has provided an opportunity to understand drug efficacy and toxicity in post-genomic era. More than 1.4 million single-nucleotide polymorphisms were identified in the sequencing of the human genome (Sachidanandam et al. 2001), with over 60,000 of them in the coding region of genes. Some of these single-nucleotide polymorphisms have already been associated with substantial changes in the metabolism or effects of medications and some are now being used to predict clinical response (Evans et al. 2001; McLeod and Evans, 2001). Because most drug effects are determined by the interplay of several gene products that influence the pharmacokinetics and pharmacodynamics of medications, including inherited differences in drug targets (e.g., receptors) and drug disposition (e.g., metabolizing enzymes and transporters).

Genetic Polymorphism/Variants and VDR

Mutation is defined as any change in a DNA sequence away from normal. This implies there is a normal allele that is prevalent in the population and that the mutation changes this to a rare and abnormal variant. A mutation in the DNA sequence consider as genetic polymorphism when it present in at least 1% of the population. DNA sequences can be altered at many levels, from major alterations in the genome to minor changes in the sequence of nucleotides in a particular gene. A few types of major DNA abnormalities, including missing or extra copies of a chromosome and gross breaks and rejoinings (translocations), can be detected by microscopic examination. Most changes in DNA, however, are too subtle to be detected by this technique and require specific molecular analysis. Molecular DNA abnormalities, or mutations, are responsible for many inherited diseases, such as cystic fibrosis, sickle cell anemia, and Huntington's disease, and may predispose an individual to cancer, major psychiatric illness, and other complex diseases. Yet most of the abnormalities in DNA sequences that are passed from generation to generation have no visible clinical impact and are detected only by generalized analysis of sequencing.

A polymorphism is a stable difference in DNA sequence at the same locus (a specific position in the genome) among individuals. The differing DNA sequences at a locus are known as alleles. For a given gene, there may be sequence differences in the promoter, coding, noncoding, or untranslated region. For example, a known DNA sequence in one individual may differ in others by one or more base pairs. These differences in base pairs may result in different amino acids; intron and exon splice sites, or stability of an mRNA, thus responsible for genetic variants in among the population. Single-nucleotide polymorphisms (SNPs) are the most frequent type of genetic variation in the human genome. Considerable evidence suggest that genetic variants are involved in genes, controlling (1) Drug metabolism, (2) Drug transport, (3) Disease susceptibility and (4) Drug targets and determine drug efficacy and toxicity. Their presence in such genes affects their functionality and, most of the time, responsible for variable drug response (VDR). Some this variability in drug responses is responsible for adverse drug reaction (ADR), and may account for death of patients (Shastry, 2006).

SNP and Its Role in VDR

A major contributor to the variation in genetic profile are the single nucleotide polymorphisms (SNPs), which are highly abundant throughout the genome, and both current and future methodologies have the potential to screen millions of SNP genotypes in one analysis. SNPs represent genetic variations in individuals; in order to be called SNPs, they must occur in at least 1% of population. With the completion of the human genome project, about 12 million true SNPs have been identified to date (SEARCH Collaborative Group, 2008). SNPs can be present in the coding part (cSNPs) but also in the non-coding part of the genome. Some SNPs render a CYP isoenzyme inactive, and people with the variant CYP gene can have impaired drug metabolism. This causes an increased plasma level of the drug at ordinary dosages, increasing the risk of adverse effects. Similarly, some SNPs increase the activity of a CYP isoenzyme, resulting in very rapid metabolism of drugs and failure to reach therapeutic levels at ordinary dosages—a problem that might be misinterpreted as poor compliance. Polymorphisms in genes encoding CYP isoenzymes that metabolically activate chemicals may explain the differences seen in individual susceptibility to carcinogenicity and toxicity from several chemical agents (Rioux, 2000). Studies also reported the presence of SNPs in drug transporters specifically in ABC-transporter gene families. Testing and identifying SNPs would help to establish a personalized medication, to avoid some side effects and to administer the right drug and the right dose (Jain, 2009).

HapMap and Pharmacogenomics: The HapMap is the haplotype map (HapMap) of the human genome, which will describe the common patterns of human genetic variation. It is a key resource for researchers to find genetic variants affecting health, disease and responses to drugs and environmental factors. The information related to HapMap is freely available to researchers around the world (The International HapMap Consortium, 2010). Groups of closely linked alleles that tend to be inherited together can be used to map human disease genes very accurately (Thorisson, 2005). All our chromosomes come in pairs, one in each pair inherited from each parent. While each chromosome of a pair contains the same genes in the same order, the sequences are not identical. For example, there are single nucleotide polymorphisms (SNPs) approximately every 1000 nucleotides. It is therefore possible to distinguish sequence variants that come from our mother and our father. These are termed maternal and paternal alleles.

The ability to distinguish between maternal and paternal alleles allows human disease genes to be mapped by linkage analysis. In germ cells, which produce eggs or sperm, the maternal and paternal chromosomes pair up and exchange segments of DNA, a process called recombination. After recombination, the chromosomes contain a mixture of alleles from each parent. Recombination will occur frequently between DNA sequences that are a long way apart but only rarely between sequences that are close together. Therefore, by measuring the frequency of recombination between the disease gene and other DNA sequences whose location is already known, the position of the disease gene can be established.

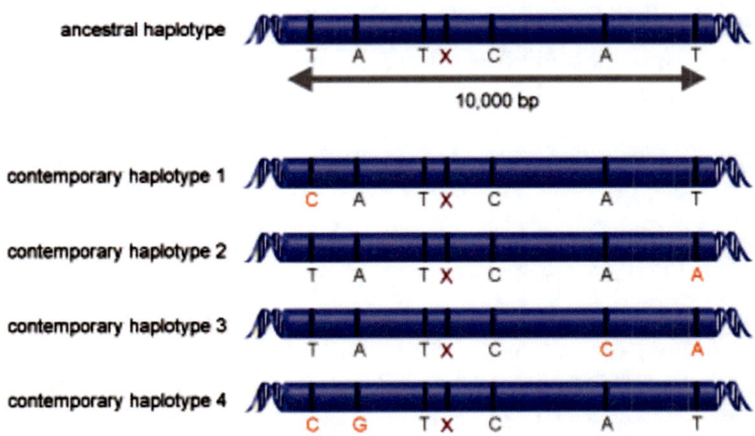

Fig. 2 : Haplotype mapping: A new mutation (X) arises in the proximity of six single nucleotide polymorphisms, with the ancestral haplotype signature TATCAT. Over several generations, the haplotype signature may be eroded by recombination. For example, contemporary haplotype 1 was produced by recombination between the first and second SNPs. The new alleles are shown in pink. However, the smallest conserved haplotype signature in all patients carrying the disease allele places the disease between SNPs 3 and 4. This technique provides a candidate region of about 10 000 bp, which is smaller than most human genes.

Another consequence of recombination is that blocks of sequences on the same chromosome tend to be inherited together, a phenomenon known as linkage disequilibrium. Such groups of alleles, which are rarely separated by recombination, are known as haplotypes. In the human genome, haplotypes tend to be approximately 60 000 bp in size and therefore contain up to 60 SNPs that travel as a group.

Haplotypes can be exploited for the fine mapping of disease genes. The principle of haplotype mapping is shown in the Figure 2. A new mutation responsible for a genetic disease always enters the population within an existing haplotype, which is termed the ancestral haplotype. Over several generations, recombination events may occur within the haplotype but the disease allele and the closest SNPs still tend to be inherited as a group. If this haplotype can be identified in a group of patients with the disease, typing the alleles within the haplotype allows a conserved region to be identified, which pinpoints the mutation responsible for the disease. Due to the abundance of SNPs, this technique has the potential to map genes very accurately. There is therefore much interest in developing a haplotype map of the entire human genome (Twyman, 2003).

CYP450 and Its Role in VDR

The most widely studied group of proteins displaying pharmacogenetic variability is the drug metabolizing enzymes and specifically, the cytochrome P450s (CYP). The CYP450 represents a large group of enzymes located mainly in the microsomes of the endoplasmic reticulum of liver. A small part of these enzymes can be found in the intestinal epithelium, lungs or other organs. These enzymes share few characteristics: all are bound to membranes within a cell (which provide the term of cyto) and contain a heme pigment (the origin of the term chrome and P) that absorbs light at a wavelength of 450 nm when exposed to carbon monoxide.

CYP450 is a drug metabolizing enzymes and well known for their oxidative degradation of endogenous chemicals present in the diet, and large range of Xenobiotics, from drug (i.e Morphine, tacrolimus, cyclosporine etc.) to environmental carcinogens (Gardiner and Begg, 2006). The CYP450 is involved mainly in the first phase of drug metabolism but is catalyzing a large number of chemical reactions on an almost unlimited variety of substrates and conjugates. These properties, added to the fact that P450s occur in families and subfamilies with different degrees of sequence similarity make them the ideal starting materials for genetic research and testing (He et al. 2005).

Table 1 : Drugs metabolized by various CYP450, and other enzymes.

S.No.	Enzyme	Drug
1	CYP1A2	Imipramine, Tacrine, Propranolol
2	CYP2C9	Cyclosporine, Nefazodone, Losartan
3	CYP2C19	Omeprazole, Lansoprazole
4	CYP2D6	Desipramine, Amitriptyline, Imipramine, Metoprolol, propranolol
5	CYP3A	Amitriptyline, Cyclosporine, Erythromycin, Imipramine, Losartan, Midazolam, Nefazodone, Omeprazole, Tacrolimus, Lovastatin, Triazolam
6	Glucuronosyl transferase	Labetalol, Morphine, Naloxone
7	TPMT	6-mercaptopurine
8	*S*-methyltransferase	Caprtopril

CYP450 isoforms: Various isoforms of CYP450s are reported. (An isoform is a CYP enzyme variant that derives from one particular gene). They are classified according to the similarities of their amino-acid sequences. Such classification allows division of CYP isoforms into:

- Families - CYP *families* contain genes that have at least a 40% sequence homology. There are at least 74 CYP families, but only about seventeen of these have been described in man. Families are numbered - for example CYP2, CYP21.
- Subfamilies - Members of a *subfamily* must have at least a 55% identity. About thirty subfamilies are well characterized in man. Subfamilies are identified by a letter, and thus we get CYP3A, CYP2D.
- Individual genes - There are fifty or so genes important in man. Individual genes are identified by a number, for example CYP2D6.
- Genetic variant (*allele*): An *allele/variant* is one of two or more forms of a gene. Individual variant is identified by a number following by asterisk'*' for example CYP2C19*3.

CYP450 shows variable drug response that is largely due to the differences in CYP450 enzyme activity, caused by polymorphic variations between population, age or environmental factors such as diet. All CYP450 phenotypes/genotype can be divided in four classes: Ultra-rapid metabolisers, Extensive metabolisers, intermediated metabolisers and Poor-metabolisers (Table 2).

Table 2 : Alterations in metabolic activity caused by CYP450 variants enable individuals to be classified into four main groups.

S.No.	Metabolic Type	Metabolism of Drug	Metabolism of pro-drug
1	Ultra-rapid Metabolisers (UM)	Rapid metabolism so that circulating drug level rapidly decreases below therapeutic level, so no benefit at standard doses.	Rapid conversion of pro-drug. Toxic levels and adverse affects at standard doses.
2	Extensive Metabolisers (EM)	'Normal' metabolism and optimal therapeutic benefit at standard drug doses	'Normal' metabolism and optimal therapeutic benefit at standard drug doses
3	Intermediate Metabolisers (IM)	Decreased drug metabolism. Susceptible to toxic accumulation of drug and side effects at standard drug doses.	Inadequate transformation of pro-drug into active metabolite. Reduced benefit at standard doses.
4	Poor (PM)	Limited or no drug metabolism so that drug accumulates to toxic levels. Serious risk of adverse events at standard doses.	Inability to transform pro-drug into active metabolite so little or no therapeutic benefit.

Till date 57 CYP450 enzymes are reported, only a few of them are responsible for abnormal/adverse pharmacological response, as 6 of them are responsible for the metabolization of over 90% of all known drugs: CYP1A2, CYP2C9, CYP2C19, CYP2D6, CYP3A4 and CYP3A5. Each enzyme is termed as isoform since each derives from a different gene.

Clinical relevance of genetic variability in CYP genes depends on a variety of factors like: patient's clinical status, drug's therapeutic index, number of enzymes involved in the drug's metabolism, smoking, and other concomitant administered drugs (Ward et al. 2008). The frequency of variant alleles of CYP families varies among populations according to the race and ethnic background (Xie et al. 2001). For instance, there are 78 reported variants of CYP 2D6 that are associated with adverse drug reactions. Many of these polymorphic genes encode inactive enzymes. These inactive enzymes may produce adverse drug reactions among patients because of their poor metabolic activity (e.g. risperidone adverse effect) (de et al. 2005). Similarly, several inactivating genetic polymorphisms have been reported in another member of the CYP family namely CYP2C19 (CYP2C19*2 and CYP2C19*3), which are also associated with adverse drug reactions. This enzyme is responsible for the metabolism of proton pump inhibitors (e.g. omeprazole and lansoprazole). Approximately 2–4% of white and 4% of African Americans have poor metabolism (Wilkinson 2005). Additionally, CYP2C9*2 and

CYP2C9*3 alleles reduce the clearance of warfarin and increase the risk of bleeding and CYP2C9*13 allele is associated with reduced metabolism of lornoxicam (Voora et al. 2005; Guo et al. 2005). Similarly, CYP2C8 plays a role in the disposition of therapeutic drugs (Totah et al. 2005).

Table 3 : Clinically Important genetic polymorphism of drug metabolism that influence drug response.

S.No.	Enzyme	Frequency of polymorphism	Drug	Drug effect
1	UDP-glucouronosyle -transferase	10-15%	Irinotecan	Diarrhoea, myelosuppression.
2	CYP2C9	14-28%	Warfarin Tolbutamide	Haemorrhage, Hypoglycaemia
3	CYP2D6	5-10% (Poor metabolisers) 1-10% (Ultra-rapid metabolisers)	Antiarrhythmics Antidepressants	Proarrhythmic and other toxic effects, Toxicity in poor metabolisers, inefficacy in Ultrarapid metabolisers
4	CYP2C19	3-6% (Whites) 8-23% (Asians)	Omeprazole Diazepam	Higher cure rates when given with clarithromycin, Prolonged sedation
5	Dihydropyrimidine dehydrogenase	0-1%	Fluorouracil	Neurotoxicity, myelotoxicity
6	Plasma pseudo-cholinesterase	1-5%	succinylcholine	Prolonged apnoea
7	N-acetyltransferase	40-70% (Whites) 10-20% (Asians)	Sulphonamides Amonafide	persensitivity, Mylelotoxicity
8	Thiopurine methyl transferase	0-3%	Mercaptopurine, Azothioprine	Myelotoxicity

ABC Transporter and their Role in VDR

Many different drug transporters are expressed in various tissues, such as the epithelial cells of the intestine and kidney, hepatocytes, and brain capillary endothelial cells (van Aubel et al. 2000; Gao and Meier, 2001) All Transporters have been classified as primary and secondary transporters. Secondary transporters are driven by an exchange or co-transport of intracellular and/or extracellular ions (Dresser et al., 2001; Lee et al., 2001a) The driving force for primary active transporters like ATP-binding cassette (ABC) transporters, such as MDR, MRP, and BCRP, is ATP hydrolysis (Lee et al. 2001a; Schinkel and Jonker, 2003).

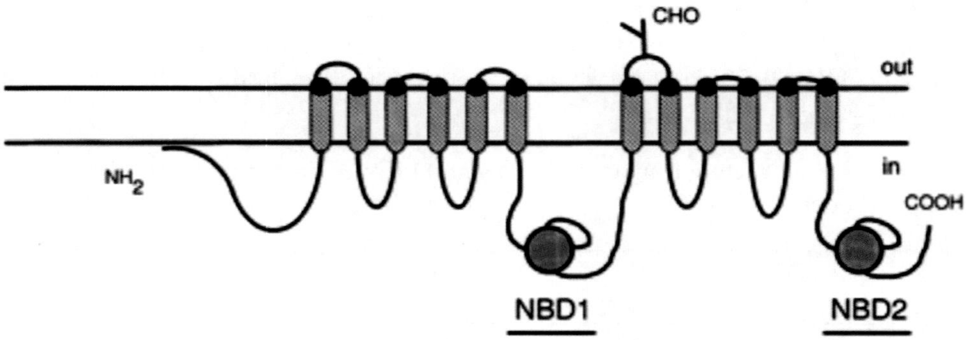

Fig. 3 : Two-dimensional membrane topology model for P-glycoprotein

ATP binding cassette (ABC) transporters are found all living organisms from microbes to humans (Dean et al. 2001). The wild-spread presence of these proteins with a relatively conserved structure and function suggests a fundamental role. Typically, they transport ligands across cellular lipid membranes, which are critical for most aspects of cell physiology, including the uptake of nutrients, drug and elimination of cytotoxin and drug-substrate. In Human 49 ABC transporters are known, mutations in such transporters, are resulted in genetic disorders/ Adverse response including a bleeding disorder and a number of eye and liver diseases (Jacquemin, 2000; Albrecht et al. 2005), all of which are caused by the failure to export a specific ligand across a lipid bilayer (HollandI,2003). The normal function of some human ABC transporters is to secrete from cells cytotoxic compounds (dietary cytotoxics and therapeutic drugs). These transporters (P-glycoprotein, BCRP, and MRP1) are highly expressed in the gut, liver and kidneys where they restrict the bioavailability of administered drugs. P-glycoprotein and BCRP in particular are also expressed in the epithelia of sensitive tissues [for example, the brain and placenta] and in stem cells (Jonker et al.,2000; Verdon et al., 2003), where they perform a barrier function (Sarkadi et al. 2006). There are also a few atypical ABC proteins.

Genetic variability in drug transporters plays a role in the resistance of malignant cells to anticancer agents (Sharom, 2008). For instance, polymorphism in the ABC-binding cassette (ABC) gene may affect the function and expression of proteins (Lepper et al. 2005). This may cause certain drug induced side effects and uncertainty in treatment efficacy. One notable example is that in certain patients the reduced rate of methotrexate metabolism produced a severe methotrexate overdosing and nephrotoxicity. This defect is attributed to the heterozygous mutation (R412G) in the highly conserved amino acid arginine of the ABCC2 gene (Hulot et al. 2005), which encodes the human multidrug resistant protein-2 (MRP2). Interestingly, this mutated region is associated with substrate affinity and hence the mutant protein has a reduced rate of methotrexate elimination. In some other cases, a long-term use of methotrexate induces pancytopenia (which is determined by white blood cells and platelet counts (Lim et al. 2005). However, it is also known that polymorphisms always need not have to produce functionally defective proteins. For example, in the multidrug resistant gene (MDR1), certain polymorphisms may not have any effect on the drug response (Sills et al., 2005). However, this could be due to nonsignificant statistical power. Thus, pharmacological studies on drug transporters (another target) are beneficial for predicting patients who are at risk in some cases at least.

Table 4: ABC transporters with known drug substrates, GS-X represents glutathione conjugates of drugs.

S.No.	Common Name	Substrates	Normal Location
1	P-gp, MDR1	Neutral and cationic organic compounds	Intestine, Liver, Kidney, Blood-brain barrier
2	MRP1	GS-X and other conjugates, organic anions	Widespread
3	MRP2,cMOAT	GS-X and other conjugates, organic anions	Liver, Kidney, Intestine
4	MRP3, MOAT-D	GS-X conjugates anti-folate, bile acids.	Pancrease, Kidney, Intestine, Liver, Adrenal
5	MRP4,MOAT-B	Nucleoside analogs, methotrexate	Prostate, Testis, Ovary, Intestine, Pancrease, lung
6	MRP5,MOAT-C	Nucleoside analougs, cyclic nucleotides, organic anions	Widespread
7	MRP6,MOAT-E	Anionic cyclic pentapeptide	Liver, Kidney
8	MXR, BCRPABC-P	Anthracyclines, Mitixantrone	Placenta, Intestine, Breast, Liver

Applications of Pharmacogenomics

Pharmacogenetics could be used to *'rescue' existing drugs* that have been withdrawn from use because of adverse drug reactions in a number of people. Retrospective genotyping of clinical trial subjects could identify the genetic make-up of the often small proportion of patients who suffered adverse reactions. Alternatively, a subgroup of individuals may be identified who, for genetic reasons, respond well to the drug and do not suffer side effects. In either case, the drug could be placed on the market with the proviso that specific genetic tests are administered prior to drug prescription.

The clinical diagnosis of many diseases may eventually incorporate pharmacogenetic testing, using *pharmacogenomics markers*. Test results could assist clinicians to identify disease subtypes and the most effective treatment for the disorder.

Table 5 : Example of pharmacogenetic markers for drug responses.

S. No.	Polymorphic Genes	Drugs	Drug function	Pharmacogenetic Effects
1	CYP2C9	Warfarin	Anticoagulant	Hemorrhage in poor metabolizers
2	CETP (Cholesterol ester transport protein)	Pravastatin	Used for lowering cholesterol and preventing cardiovascular disease	-Responder with TaqB1 genotype-Non-responder with TaqB2 genotype
3	UGT1A1 (UDP-glucuronosyl transferase 1A1)	Irinotecan	For treatment of cancer	Severe diarrhea in poor metabolizers
4	Beta2-Adrenergic receptor	Albuterol	To prevent difficulties in breathing i.e Asthama	-Responder with Arg-16 genotype-Non-responder with Gly-16 genotype
5	5-Lipoxygenase	Zileuton	For treatment of Asthama	-Responder with wild-type 5-LO promoter-Non-responder with variant 5-LO promoter

The process of drug research and development may be significantly altered by pharmacogenomics. Pharmacogenomic approaches would focus on the identification of genetically determined drug targets involved in disease and genetic polymorphisms associated with treatment response. Such research could assist pharmaceutical companies to develop more effective drugs with fewer side effects.

Subjects' eligibility to participate in clinical trials may be decided by the results of pharmacogenetic tests. Pharmaceutical companies may exclude some subjects on the basis of their genotype, e.g. excluding those individuals with a genetic predisposition to adverse drug reactions or who may not respond to treatment. As a result, clinical trials and drug approval may become quicker and more efficient to perform, and will make effective use of patents.

'Pharmacogenetic profiles' that catalogue both a patient's genetic predisposition's to diseases, and their predicted responses to different drugs could be used in clinical medicine. This information could be relevant to clinical decisions, informing physicians of recommended dosages, possible adverse reactions and potential drug interactions. It may even become possible to 'personalise' medicine, creating drugs that suit each individual, based on their genotype

Barriers to Pharmacogenomics Progress

Pharmacogenomics is a developing research field that is still in its infancy. Several barriers will have to be overcome before many pharmacogenomics benefits can be realized, some of such barriers are as follows: 1). **Complexity of finding gene variations that affect drug response** - Single nucleotide polymorphisms (SNPs) are DNA sequence variations that occur when a single nucleotide (A,T,C,or G) in the genome sequence is altered. SNPs occur every 100 to 300 bases along the 3-billion-base human genome, therefore millions of SNPs must be identified and analyzed to determine their involvement (if any) in drug response. Further complicating the process is our limited knowledge of which genes are involved with each drug response. Since many genes are likely to influence responses, obtaining the big picture on the impact of gene variations is highly time-consuming and complicated. 2) **Limited drug alternatives** - Only one or two approved drugs may be available for treatment of a particular condition. If patients have gene variations that prevent them using these drugs, they may be left without any alternatives for treatment. 3) **Disincentives for drug companies to make multiple pharmacogenomic products** - Most pharmaceutical companies have been successful with their "one size fits all" approach to drug development. Since it costs hundreds of millions of dollars to bring a drug to market, will these companies be willing to develop alternative drugs that serve only a small portion of the population. 4) **Educating healthcare providers** - Introducing multiple Pharmacogenomics products to treat the same condition for different population subsets undoubtedly will complicate the process of prescribing and dispensing drugs. Physicians must execute an extra diagnostic step to determine which drug is best suited to each patient. To interpret the diagnostic accurately and recommend the best course of treatment for each patient, all prescribing physicians, regardless of specialty, will need a better understanding of genetics (McKinnon et al. 2007).

Ethics

Ethical issues in Pharmacogenomics tend to involve the general concern that the ability to diagnose a genetic disorder before any treatment is available does more harm than good to the patient. Pharmacogenetic testing have many benefits, both for individuals and drug companies. It could identify people who are susceptible to adverse drug reactions, and could also identify those who are unlikely to benefit from a particular drug. It would also make possible the resurrection of some older drugs that are safe and effective for most people, but have been taken off the market because a few people had serious reactions. Companies would be able to reduce the size of clinical trials, creating greater efficiency (Wertz, 2003).

Although genetic testing for drug responses appears to have many benefits, there are also potential harms, some major issues that may cause negative impact of pharmacogenomics:

Self-image: Once genetic information has been obtained, patients may be required to disclose the information to insurers and employers. Additionally, the tests may reveal information about family members. It should also be made clear that the test results could make the patient ineligible for some treatments for their condition. Finally, patients should appreciate that knowledge of their own genetic make-up could adversely affect their self-image, particularly if the test results are 'negative'(i.e. suggest the individual is unlikely to respond to therapy).

Patient privacy: is another issue raised by the application of pharmacogenetic testing to clinical treatment. As medical records, the results of pharmacogenetic tests should remain confidential. However, the familial nature of such information may require more stringent safeguards. Additionally, clinical trials involving pharmacogenomic drugs may include the storage and databasing of participants' DNA. If the information were used in future protocols without explicit consent, it would raise questions about access to information and privacy, as well as autonomy and informed consent.

Discrimination: Individuals could suffer stigmatisation and discrimination as a result of pharmacogenetic testing. Although pharmacogenetic testing may not directly test for disease susceptibility genes, testing for a number of different polymorphisms may result in an individual being labelled as a 'non-responder'. Such patients could potentially suffer similar insurance and employment discrimination to those individuals identified as being as high risk of developing a genetic disease.

Racism: Pharmacogenomics and pharmacogenetics raise fears about racism because they focus on individual and population differences in genetic make-up. If disadvantageous polymorphisms are associated with a particular ethnic group, members of that group could suffer stigmatisation. As a result, individuals from that ethnic group may be denied access to certain medical treatments and may also suffer insurance or employment discrimination. If this situation arose it may discourage people from ethnic minorities from participating in pharmacogenetic and pharmacogenomic research (Rothstein et al. 2002).

Regulation

At present, no countries have regulations specifically relating to pharmacogenomics. Should these techniques become widely used; a number of regulatory issues will arise. Pharmacogenomic drugs and pharmacogenetic tests will require new labelling and prescription guidelines. Prescription guidelines for such drugs will need to include information on effective drug dosage for each genotype and will require the use of an appropriate pharmacogenetic test. This will in turn require regulation of pharmacogenetic tests. Even if the results of pharmacogenetic testing suggest that an individual may respond adversely to, or derive no therapeutic benefit from a drug, the physician and/or patient may still wish to use it. Thus regulations may also need to cover the permissibility of 'off label' uses of pharmacogenomic drugs (Pendergast, 2008, Kelley et al. 2011.).

Conclusion and Future Aspects

The newly emerging genomic technologies enable the search for relevant genes and their variants to include the entire genome and help in the search for candidate genes. Moreover pharmacogenomic analysis can identify the disease susceptibility genes representing potential new drug targets. The current concept of drug therapy is to apply treatment for a large population group whereas application of pharmacogenomic will help individual application of drug therapy or on smaller patient sub population. But whether this individualized medicine will lead to improved and economically feasible therapy is yet to be seen. These new approaches to drug discovery and treatment will steer the pharmaceutical and diagnostic industries closer together. It holds the promise that drugs might one day be tailor-made for individuals and adapted to each person's own genetic makeup. Environment, diet, age, lifestyle, and state of health all can influence a person's response to medicines, but understanding an individual's genetic makeup is thought to be the key to creating personalized drugs with greater efficacy and safety. In the future, researchers have the potential to subdivide each disease according to genetics, not symptoms. Specific diagnoses may be based on the molecular mechanisms involved rather than clinical presentation. Molecular mechanism differences will subdivide patient groups with common diseases like hypertension, diabetes, and cancer. Health care professionals will use genetic tests to predict how a disease will progress and the therapeutic response to anticipate. Genetically defining patient populations will help improve outcomes and genetic prognostics will revolutionize treatment and improve cost effectiveness. Pharmacogenomics is already making an impact in a wide array of disease states and drug therapy; it will eventually become part of standard patient management in selecting and monitoring drug therapy. Pharmacogenomics will definitely help us to sharpen our medical and pharmaceuticals tools. Drugs will become more precise and efficient and the risk of toxic side effects will be reduced. But at the same time increasing amounts of information and databases will be collected, which may be put to a variety of uses.

Refrences

Albrecht, C., McVey, J. H., Elliott, J. I., Sardini, A., Kasza, I., Mumford, A. D.,Naoumova, R. P., Tuddenham, E. G., Szabo, K. and Higgins, C. F. (2005) A novel missense mutation in ABCA1 results in altered protein trafficking and reduced phosphatidylserine translocation in a patient with Scott syndrome. Blood 196: 542-549.

Carson, P. E., Flanagan, C. L., Ickes, C. E. and Al-ving, A. S. (1956) Enzymatic deficiency in primaquine-sensitive erythrocytes. Science 124:484-485.

de Leon, J., Susce, M. T., Pan, R. M, Fairchild, M., Koch, W. H. and Wedlund, P. J. (2005) The CYP2D6 poor metabolizer phenotype may be associated with risperidone adverse drug reactions and discontinuation. J Clin Psychiatry 66(1):15-27..

Dean, M., Rzhetsky, A. and Allikmets, R. (2001) The human ATP-binding cassette (ABC) transporter superfamily. Genome Res 11(7):1156-66.

Dresser, M. J., Leabman, M.K. and Giacomini, K.M. (2001) Transporters involved in the elimination of drugs in the kidney: organic anion transporters and organic catio*n transporters.* J Pharm Sci 90*:* 397-421.

Evans, W. E., Relling and M. V. (1999) Pharmacogenomics: translating functional genomics into rational therapeutics. Science 286:487-49.

Gao, B. and Meier, P. J. (2001) Organic anion transport across the choroid plexus. Microsc Res Tech 52: 60-64.

Gardiner, S. J. and Begg, E. J., (2006) Pharmacogenetics, drug-metabolizing enzymes, and clinicalpractice, Pharmacol Rev 58(3), 521-90.

He, P., Court, M. H., Greenblatt, D. J. and Von Moltke, L. L. (2005) Genotype-phenotype associations of cytochrome P450 3A4 and 3A5 polymorphism with midazolam clearance in vivo. Clin Pharmacol Ther 77: 373–387.

Holland, I. B., Cole, S. P. C, Kuchler, K. and Higgins, C. F. (2003) ABC. *Proteins: From Bacteria to Man*. London: Academic Press.

Hulot, J. S., Villard, E., Maguy, A., Morel, V., Mir, L, Tostivint, I. et al. (2005) A mutation in the drug transporter gene ABCC2 associated with impaired methotrexate elimination. Pharmacogenet Genomics 15: 277–285.

Jacquemin, E. (2000) Progressive familial intrahepatic cholestasis. Genetic basis and treatment. Clin. Liver Dis 4:753-763.

Jain, K. K. (2009) Basic technologies for developing personalized medicine in: Textbook of personalized medicine, Springer 15-19.

Jonker, J. W., Smit, J. W., Brinkhuis, R. F., Maliepaard, M., Beijnen, J. H., Schellens, J. H. and Schinkel, A. H. (2000) Role of breast cancer resistance protein in the bioavailability and fetal penetration of topotecan. J Natl Cancer Inst 92: 1651–1656.

Kalow, W., Tang, B. K. and Endrenyi, I. (1998) Hypothesis: comparisons of inter- and intra-individual variations can substitute for twin studies in drug research. Pharmacogenetics 8:283-289.

Kalow, W. (1956) Familial incidence of low pseudocholinesterase level. Lancet 2:576-576.

Kelley, R.K, Van Bebber, S.L., Phillips, K.A. and Venook, A.P. (2011) Personalized medicine and oncology practice guidelines: a case study of contemporary biomarkers in colorectal cancer. J Natl Compr Canc Netw 9(1):13-25.

Lee, G., Dalla, S., Hong, M. and Bendayan, R. (2001a) Drug transporters in the central nervous system: brain barriers and brain parenchyma considerations. Pharmacol Rev 53: 569-596.

Lepper, E. R., Nooter, K., Verweij, J., Acharya, M. R., Figg, W. D. and Sparreboom, A. (2005) Mechanism of resistance to anticancer drugs: the role of the polymorphic ABC transporters ABCB1 and ABCG2. Pharmacogenomics 6: 115–138.

Lim, A. Y., Gaffney, K. and Scott, D. G. (2005) Methotrexate-induced pancytopenia: serious and under reported? Our experience of 25 cases in 5 years. Rheumatology 44: 1051–1055.

Ma, Q. and Lu, A. Y. (2011) Pharmacogenetics, pharmacogenomics, and individualized medicine. Pharmacol Rev 63(2):437-59.

McKinnon, R. A., Ward, M. B. and Sorich, M. J. (2007) A critical analysis of barriers to the clinical implementation of pharmacogenomics Ther Clin Risk Manag 3(5):751-9.

McLeod, H. L. and Evans, W. E. (2001) Pharmacogenomics: unlocking the human genome for better drug therapy. Annu Rev Pharmacol Toxicol 41:101-12

Pendergast, M. K. (2008) Regulatory agency consideration of pharmacogenomics. Exp Biol Med (Maywood) 233(12):1498-503.

Pistoi. (2002) Facing your genetic destiny, part II. *Scientific American.*

Vesell, E. S. (1989) Pharmacogenetic perspectives gained from twin and family studies. Pharmacol Ther 41:535-552.

Rioux, P. P. (2000) Clinical trials in pharmacogenetics and pharmacogenomics: methods and applications. Am J Health Syst Pharm 57(9):887-98.

Rothstein, M. A. and Epps, P. G. (2002) Ethical and legal implications of pharmacogenomics. Nature Reviews Genetics 2(3):228-31.

Sachidanandam, R., Weissman, D., Schmidt, S. C., et al. (2001) A map of human genome sequence variation containing 1.42 million single nucleotide polymorphisms. Nature 409:928-933.

Sarkadi, B., Homolya, L., Szakacs, G. and Varadi, A. (2006) Human multidrug resistance ABCB and ABCG transporters: participation in a chemoimmunity defense system. Physiol Rev 86: 1179–1236.

Schinkel, A. H. and Jonker, J. W. (2003) Mammalian drug efflux transporters of the ATP binding cassette (ABC) family: an overview. Adv Drug Deliv Rev 55: 3-29.

SEARCH Collaborative Group, Link, E., Parish, S., Armitage, J. et al, (2008) SLCO1B1 variantsand statin-induced myopathy - a genomewide study, N Engl J Med 359(8), 789-99.

Sharom, F. J. (2008) ABC multidrug transporters: structure, function and role in chemoresistance. Pharmacogenomics 9(1):105-27.

Sills, G. J., Mohanraj, R., Butler, E., McCrindle, S., Collier, L., Wilson, E. A. et al. (2005) Lack of association between the C3435T polymorphism in the human multidrug resistance (MDR1) gene and response to antiepileptic drug treatment. Epilepsia 46: 643–647.

Shastry, B. S. (2005) Genetic diversity and new therapeutic concepts. J Hum Genet. 50(7):321-8. Epub.

The International HapMap Consortium. (2010) Integrating common and rare genetic variation in diverse human populations. Nature 467, 52-58.

Thorisson, G. A., Smith, A.V., Krishnan, L. and Stein, L. D. (2005) The International HapMap Project Web site. Genome Research, 15:1591-1593. Twyman R. Haplotype mapping. Wellcome trust-2003 (http://genome.wellcome.ac.uk/doc_WTD020781.html)

Van Aubel R. A., Masereeuw, R. and Russel, F. G. (2000) Molecular pharmacology of renal organic anion transporters. Am J Physiol Renal Physiol 279: F216-F232.

Verdon, G., Albers, S. V., Dijkstra, B. W., Driessen, A. J. and Thunnissen, A. M. (2003) Crystal structures of the ATPase subunit of the glucose ABC transporter from Sulfolobus solfataricus: nucleotide-free and nucleotide-bound conformations. J Mol Biol 330: 343–358.

Ward, M. B., Sorich, M. J. and McKinnon, R. A. (2008) Cytochrome P450 Part 2: Genetics of Inter-Individual Variability. J Pharm Pract Res 38:226-9.

Wertz, D. C. (2003) Ethical, social and legal issues in pharmacogenomics. The harmacogenomics Journal 3:194–196.

Wilkinson, G. R. (2005) Drug metabolism and variability among patients in drug response. N Engl J Med 352.

Xie, H. G., Kim, R. B., Wood, A. J. J. and Stein, C. M. (2001) Molecular basis of ethnic differences in drug disposition and response. Ann Rev Pharmacol Toxicol 41: 815–850.

❑❑❑

CHAPTER – 30

Molecular Evolution and Evolutionary Computing Tools

Arun Kumar Sharma and Kambaska Kumar Behera
Department of Bioscience and Biotechnology, Banasthali University, Rajasthan,302240 India
Email : kambaska@yahoo.co.in

Summary

Molecular evolution is the process of evolution at the scale of DNA, RNA, and Protein. Molecular evolution emerged as a scientific field in the 1960s as researchers from molecular biology, evolutionary biology, and population genetics sought to understand recent discoveries on the structure and function of nucleic acids and protein. It covers two broad areas of study : the evolution of macromolecules, and reconstruction of the evolutionary history of genes and organisms. The evolution of macromolecules refer to the characterization of the changes in the genetic material (DNA or RNA sequences) and its products (Proteins of RNA molecules) during evolutionary process, and to the rates and patterns with which such changes occur, and also the mechanism responsible for such changes. The second area also known as "molecular phylogenetics", deals with the evolutionary history of organisms and macromolecules as inferred from molecular data and the methodology of tree reconstruction.

Introduction

Molecular evolution emerges as the results of fusion of population genetics, evolutionary biology, and molecular biology. However, a few scientific disciplines have significant contribution in advancement of molecular evolution, are computational biology and bioinformatics. In order to understand the basics of molecular evolution, brief information about such scientific disciplines are given bellow-

Population genetics is the study of allele frequency distribution and change under the influence of the four main evolutionary processes: natural selection, genetic drift, mutation and gene flow. It also takes into account the factors of population subdivision and population structure.

Molecular biology is the branch of biology that deals with the nature of biological phenomena at the molecular level through the study of DNA and RNA, proteins, and other macromolecules involved in genetic information and cell function.

Evolutionary biology is a sub-field of biology concerned with the origin of species from a common descent and descent of species, as well as their change, multiplication and diversity over time.

Computational biology is an interdisciplinary field that applies the techniques of computer science, applied mathematics and statistics to address biological problems. The main focus lies in the development of computational and statistical data analysis methods and in developing mathematical modeling and computational simulation techniques.

Bioinformatics is the application of statistics and computer science to the field of molecular biology. Major research efforts in the field include sequence alignment, gene finding, genome assembly, drug design, drug discovery, protein structure alignment, protein structure prediction, prediction of gene expression and protein-protein interactions, genome-wide association studies and the modeling of evolution.

The lack of integration between evolutionary concepts and molecular analysis is the major problem for the beginnere researchers working in area of molecular evolution. This chapter is designed to boost up beginners and advance researchers working on molecular evolution by providing various evolutionary concepts and also computational software particularly, what types of molecular data and evolutionary analysis can be done.

Mutation is the Basis of Molecular Evolution

Mutation is the fundamental basis of molecular evolution or in other word we can say molecular evolution came into existence after discovery of DNA physical structure and subsequently mutations. In molecular biology and genetics, mutations are changes in nucleotide of DNA sequence of a cell's genome or the DNA or RNA sequence of a virus. Mutations are caused by a variety of artificial and natural way such as radiation, viruses, transposons and mutagenic chemicals, as well as errors that occur during meiosis or DNA replication. Mutations arises by means of natural way can have role in molecular evolution or is the basis to study molecular evolution.

Point Mutation

It is a type of mutation that causes the replacement of a single base nucleotide with another nucleotide of the genetic material, DNA or RNA. Point mutations can be categorized by many way :--

Nature of Base Replacement

Transitions: It denotes replacement of a purine base with another purine or replacement of a pyrimidine with another pyrimidine.

Transversions: It denotes replacement of a purine with a pyrimidine or vice versa.

Functional Basis

Nonsense mutations: code for a stop, which can terminate the protein synthesis.

Missense mutations: code for a different amino acid

Silent mutations: code for the same amino acid

Common categorization used in molecular evolution

Synonymous: code for same amino acid

Non-synonymous: code for different amino acid

Genetic Polymorphism

A population is monomorphic at a locus if there exists only one allele at the locus. A locus is said to be polymorphic if two or more alleles coexist in the population. However, if one of the alleles has a very high frequency, say 99% or more, then none of the other alleles are likely to be observed in a population sample unless the sample size is very large. Thus for practical purposes, a locus is commonly defined as polymorphic only if the frequency of the most common allele is less than 99%. This definition is obviously arbitrary, and in the literature, one may find thresholds other than 99%.

Measures of Genetic Diversity

Genetic diversity is the variation of heritable characteristics present in a population of the same species. It serves an important role in evolution by allowing a species to adapt to a new environment and to fight off parasites. Genetic diversity in the populations of an organism is measured using several parameters such as nucleotide diversity $Pi_{(TT)}$, haplotype diversity (Hd) and index of genetic differentiations (Fst) and heterozygosity (*He*). Details of each genetic diversity measure methods are given below-

Nucleotide Diversity (π)

Nucleotide diversity is defined as the average number of nucleotide differences per site between any two randomly chosen sequences. Nucleotide diversity is denoted by term Pi (π) and is expressed as-

Measures of Nucleotide Diversity

$\theta = 4N_e\mu$ N_e = effective population size; μ = mutation rate per generation

$\hat{\theta} = S/a_1$ S = proportion of segregating sites
$a_1 = (1 + 1/2 + 1/3 + ... + 1/n-1)$
n = number of sequences

$\pi = \frac{n}{n-1}\sum_{ij} \hat{x}_i\hat{x}_j \pi_{ij}$ $\hat{x}_i$ = frequency of the ith type of DNA sequence
π_{ij}=proportion of different nucleotides between the ith and j^{th} type of DNA sequence

Under the neutral theory of molecular evolution and the infinite-sites model and assuming random sampling from a random mating population and no recombination:

$\theta = 4N_e\mu = \pi$

Nucleotide diversity provides an estimate of pattern of molecular diversity across the stretch of given DNA sequences, stretch of DNA under selection and functional constrains.

Haplotype Diversity (Hd)

Haplotype diversity is also denoted as gene diversity and is defined as the probability that two randomly chosen haplotypes are different in the samples. Haplotype diversity is estimated as-

$$\hat{H} = \frac{n}{n-1}(1 - \sum_{i=1}^{k} p_i^2)$$

Where n: number of gene copies in the samples, k: number of haplotypes, P_i : sample frequency of the i-th haplotype and *H*: haplotype diversity.

Haplotype diversity can assess evolutionary history of genetic trait, gene flow and population structure etc.

Diversity Fixation Index (Fst)

Fixation index (F_{ST}) is a measure of population differentiation based on genetic polymorphism data, such as Single nucleotide polymorphisms (SNPs) or microsatellites. F_{ST} used for comparing the genetic variability within and between populations, and is frequently used in the population genetics.

Several definitions of $\mathbf{F}_{ST}$ have been used, all measuring different but related quantities. A common definition given by Hudson et al.(1992) is:-

$$F_{ST} = \frac{P_{Between} - P_{Within}}{P_{Between}}$$

Where $\Pi_{Between}$ and Π_{Within} represent the average number of pairwise differences between two individuals sampled from different ($\Pi_{Between}$) or the same (Π_{Within}) population. The average pairwise difference within a population can be calculated as the sum of the pairwise differences divided by the number of pairs.

F_{ST} detect population structure, mating pattern in population, population differentiations, and gene flow.

Heterozygosity (*He*)

Heterozygosity defined as the average probability that a pair of alleles randomly obtained from the population is different. Heterozygosity ranges between 0 and 1, with high values (i.e. those close to 1) reflecting high genetic diversity levels of that population.

The heterozygosity (*He*) is calculated as-

$He= [n/ (n-1)][1-\Sigma pi2]$,

where n is the number of isolates analyzed and pi is the frequency of the i-th allele in the population.

Pattern of genetic diversity across the chromosome, selective sweep events, and bottleneck effect

Molecular Clocks

In the comparative studies of hemoglobin and cytochrome 'c' protein sequences among different species, the rates of amino acid replacement were observed to be similar among various mammalian lineages. Therefore, first time Zuckerkandl and Pauling (1965) proposed that for a given protein, the rate of molecular evolution is approximately constant over time in all lineages and is called molecular clock. It is used to estimate the time of occurrence of events called speciation or radiation. The molecular data used for such calculations is usually nucleotide sequences for DNA or amino acid sequences for proteins. In molecular evolution, molecular clock use to detect speciation and divergence events. Molecular clock hypothesis assumes that mutation rates among species is constant over time, and using this assumption, molecular clock works and determines divergence time between different taxa or species. For example, let us assume that the rate of non-synonymous substitution for the α chain of hemoglobin is 0.56×10^{-9} substitution per site per year, and that a-globins from rat and human differ by 0.093 substitutions per site. Then, under the molecular clock hypothesis, the divergence time between human and rat lineages is estimated to be approximately $0.093 / (2 \times 0.56 \times 10^{-9}) = 80$ million years ago.

Neutral Theory and Neutrality Test

Neutrality test signify whether mutation-evolving process is random or non-random (Neutral theory) (Kimura 1983). Neutral theory of molecular evolution speculates that genetic variations found in the population are neutral with respect to each other i.e. the variation had no associated fitness differential (identical relative fitness). The null hypothesis of neutral evolution states that mutation-evolving process is random. Deviation (D) to neutrality can be determined by two well-known tests i.e. Tajima and Fu & Li D test. Details of both D tests are given below.

Tajima's D Test

Tajima's D test is based on the differences between the number of segregating sites and the average number of nucleotide differences (Tajima 1983;1989).Tajima's test is based on the neutral model prediction that estimates of S/a1, and of k, are unbiased estimates of θ,

Where-

S: is the total number of segregating sites.

$a1 = \Sigma(1 / i)$ from i =1 to n-1

n: the number of nucleotide sequences

k: is the average number of nucleotide differences between pairs of sequences.

$\theta = 4Nu$ (for diploid-autosomal; N and u are the effective population size, and the mutation rate per DNA sequence per generation, respectively).

Fu and Li's, D Test

Fu and Li's, D* tests are based on the neutral model prediction that estimates of η /a1, (n -1) ηs / n, and of k, are unbiased estimates of è,

Where-

η is the total number of mutations

$a1 = \Sigma(1 / i)$ from i =1 to n-1

n: the number of nucleotide sequences

s: the total number of singletons (mutations appearing only once among the sequences).

k: the average number of nucleotide differences between pairs of sequences.

$\theta = 4Nu$ (for diploid-autosomal ; N and u are the effective population size, and the mutation rate per DNA sequence per generation, respectively) and 2Nu for haploid organism.

The D* test statistic is based on the differences between η_s, the number of singletons (mutations appearing only once among the sequences), and η (the total number of mutations) (Fu & Li 1993).

Natural Selection

Natural selection is defined as the differential reproduction of genetically distinct individuals or genotypes within a population. Differential reproduction is caused by differences among individuals in such traits as mortality, fertility, fecundity, mating success, and the variability of the offspring. Natural selection is predicted on the availability of genetic variation among individuals in characters related to reproductive success. Selection may lead to changes in allele frequencies over time. However, changes in allele frequency are not only determined by natural selection and genetic drift may also play an important role in changing frequency of alleles. The **fitness** of a genotype denotes measures of individual's ability to survive and reproduce.

Types of Selection

Mutations occurring in genic fraction of the genome may or may not alter the organism's phenotype. In such case, they may or may not affect the fitness of the organism that carries the mutation.

Negative Selection

Most new mutations arising in a population reduce the fitness of their carriers. Such mutations are called deleterious, and they will be selected against and eventually removed from the population. This type of selection is called negative or purifying selection.

Neutral

Mutation leads no effect over the fitness of an organism. Some time, a new mutation may be as fit as the best allele in a population, such mutation is selectively neutral and its fate is not determined by selection.

Positive Selection

In rare cases, a new mutation may arise that increases the fitness of its carriers. Such mutation is called advantageous, and it will be subjected to positive selection.

Selection Over Codon

Positive selection is operating on a gene can be detected by comparing the relative abundance of synonymous and non-synonymous substitutions within the gene sequences.

For a pair of sequences, this is done by first estimating the number of synonymous substitutions per synonymous site (d_S) and the number of non-synonymous substitutions per non-synonymous site (d_N), and their variances: Var (d_S) and Var (d_N), respectively. With this information, we can test the null hypothesis that HO: $d_N = d_S$ using a Z-test:

$$Z = (d_N - d_S) / \text{SQRT} (\text{Var} (d_S) + \text{Var} (d_N))$$

The level of significance at which the null hypothesis is rejected depends on the alternative hypothesis (HA):

HO: $d_N = d_S$

HA: (a) $d_N \neq d_S$ (test of neutrality).

(b) $d_N > d_S$ (positive selection).

(c) $d_N < d_S$ (purifying selection).

Molecular Phylogenetics

A phylogenetic tree is a branching diagram showing the inferred evolutionary relationships among various biological species based upon similarities and differences in their physical and/or genetic characteristics. Molecular phylogenetics uses DNA/RNA/Protein sequences or mini/microsatellite distance data to gain information on an organism's evolutionary relationships and is expressed as phylogenetic tree. Before describing molecular phylogenetics, a brief information about several phylogenetic terms are given bellow-

Rooted phylogenetic tree: It is a tree that show the most recent common ancestor of all the entities at the leaves of the tree.

Unrooted trees: It illustrates the relatedness of the leaf nodes without making assumptions about ancestry at all.

Dendrogram: It is a diagrammatic representation of a phylogenetic tree.

Cladogram: It is a tree formed using cladistic methods. This type of tree only represents a branching pattern, i.e., its branch lengths do not represent time.

Phylogram: It is a phylogenetic tree that explicitly represents number of character changes through its branch lengths.

Chronogram: It is a phylogenetic tree that explicitly represents evolutionary time through its branch lengths.

Types of Phylogenetic Tree Reconstruction Methods

Brief description of each phylogenetic tree reconstruction method is given below.

Unweighted Paired-group Method with Arithmetic Means (UPGMA)

UPGMA is an easy approach and assumes a constant evolutionary rate (mutation rate) over the time for all lineages under study. According to the UPGMA tree reconstruction assumptions, approximately a linear relation exists between evolutionary distance and the divergence time.

Neighbor-Joining Method (NJ)

Neighbor-Joining tree reconstruction method is based on the minimum evolution principle and it provides an approximate algorithm for finding the shortest tree (Saitou & Nei 1987).This is accomplished by sequentially finding neighbors that minimize the total length of the tree. Unlike UPGMA method, NJ method can incorporate different rates of evolution in different lineages. Both UPGMA and NJ are called distance matrix methods that depend on the measure of genetic distance between all of the pairs of taxa under examination.

Minimum Evolution method (ME)

In the ME method, distance measures that correct for multiple hits at the same sites are used, and a topology showing the smallest value of the sum of all branches (S) is chosen as an estimate of the correct tree.

Parsimony Method

Under parsimony, the preferred phylogenetic tree is the tree that requires the least evolutionary change to explain some observed data. Parsimony is part of a class of character-based tree estimation methods which use a matrix of discrete phylogenetic characters to infer one or more optimal phylogenetic trees for a set of taxa.

Maximum Likelihood Method (ML)

The most popular alternative phylogenetic method is maximum likelihood phylogenetic inference. Maximum likelihood analysis functions much like parsimony analysis, where trees are scored based on a character dataset, and the tree with the best score is selected. Maximum likelihood is a parametric statistical method, in that it employs an explicit model of character evolution. Such methods are potentially much more powerful than non-parametric statistical methods like parsimony, but only if the model used is a reasonable approximation of the processes that produced the data.

Bayesian Phylogenetic Inference

Bayesian phylogenetic analysis uses Bayes' theorem, which relates the posterior probability of a tree to the likelihood of data, and the prior probability of the tree and model of evolution. However, unlike parsimony and likelihood methods, Bayesian analysis does not produce a single tree or set of equally optimal trees. Bayesian analysis uses the likelihood of trees in a Markov chain Monte Carlo (MCMC) simulation to sample trees in proportion to their likelihood, thereby producing a credible sample of trees.

Distance Matrix Methods

Non-parametric distance methods were originally applied to phenetic data using a matrix of pairwise distances. These distances are then reconciled to produce a tree. The distance matrix can come from a number of different sources, including measured distance or morphometric analysis, various pairwise distance formulae applied to discrete morphological characters, or genetic distance from sequence, restriction fragment, or allozyme data.

Tree-Building Methods Can be Assessed on the Basis of Several Criteria

Efficiency, power, consistency, robustness, and falsifiability are the major parameter on that basis one can select tree-building method.

Limitation

Although phylogenetic trees produced on the basis of sequenced genes or genomic data in different species can provide evolutionary insight, they have important limitations. They do not necessarily accurately represent the species evolutionary history. The data on which they are based is noisy; horizontal gene transfer, hybridization between species that were not nearest neighbors on the tree before hybridization takes place, convergent evolution, and conserved sequences, can confound the analysis.

Assessing Tree Topology

When there are more than a few taxa, then there are a very large number of tree topologies. To determine the statistical support to various parts of a tree becomes critical. The most common approach is to determine the level of statistical confidence in a particular node of the tree by calculating bootstrap values (Felsenstein 1981; 1985).

Bootstrapping is an approach that re-samples the original data to get significance levels. The bootstrap belongs to a class of methods called re-sampling techniques because it estimates the sampling distribution by repeatedly re-sampling data from the original data set. Generally, 500-1000 bootstrapping is good for inferring phylogeny confidence levels. In general sense bootstrap is a statistical technique that is widely used to assess confidence limits in phylogenies

Coalescence Theory

Under this theory, all alleles in a sample are assumed to derive from a common ancestor. Therefore, it is possible to estimate the coalescent time of two alleles into one common ancestor. After one coalescent event, a sample of N alleles is reduced to N-1. Time is considered to move backwards in the model until the time to a common ancestor for the entire sample is estimated. By modeling this stochastic process, a distribution of genealogies emerges, enabling the estimation of parameters of interest, such as the Time to the Most Recent Common Ancestor (TMRCA).

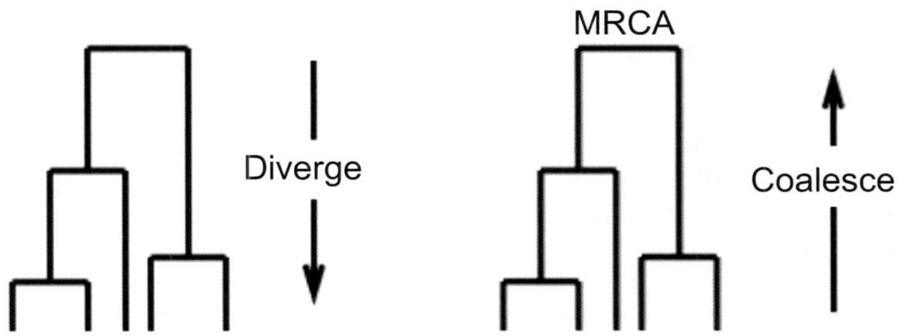

Fig. 1 : Phylogenetic tree showing coalescence and divergence

TMRCA Estimates of the Molecular Age

The age of most recent common ancestor of an organism can be determined using following equation:

$dS = 2\mu_s \text{ x } Tw$

Where dS is synonymous substitution rate at synonymous site, $\mathbf{i}_s$ is synonymous mutation rate per generation and T*w* is TMRCA of species under study

Synonymous mutation rate can be determined using following equation:

$Ks = 2\mu_s \times Tb + d_s$

Where, Ks is the divergence at synonymous sites between species (species under study and its closely related taxa), d_S is the nucleotide diversity (polymorphism) within species (species under study), μ_s is synonymous mutation rate per generation and T*b* is time of the divergence between species.

Effective Population Size Estimates (*Ne*)

Effective population size (*Ne*) defined as "an ideal population of size *N* in which all parents have an equal expectation of being the parents of any progeny individuals". *Ne* denotes the actual reproductive individuals in a population (Wright 1931).

Effective population size of an organism id determined by following equation:

$\theta = 4Ne\theta$ (for diploid organism) and

$\theta = 2Ne\mu$ (for haploid and mitochondrial genome)

Where θ (theta) is the mutation parameter, *Ne* effective population size and μ is the mutation rate per site per year.

Population Bottleneck

A population bottleneck (also called genetic bottleneck) is an evolutionary event in which a significant percentage of a population or species is killed or otherwise prevented from reproducing (by means of natural or artificial force). Population bottlenecks increase genetic drift, as the rate of drift is inversely proportional to the population size i.e. larger the population size minimizes effect of genetic drift. The reduction in a population's dispersal leads, over time, to increased genetic homogeneity. Severe population bottlenecks can also markedly increase inbreeding due to the reduced pool of possible mates.

Detecting genetic diversity can clearly demonstrate the signature of population bottleneck. Populations that have experienced a recent reduction in their effective population size exhibit a co-relative reduction of the allele numbers and gene diversity at polymorphic loci. However, the allele numbers are reduced faster than the gene diversity. Thus, in a recently bottlenecked population, the observed gene diversity is higher than the expected equilibrium gene diversity.

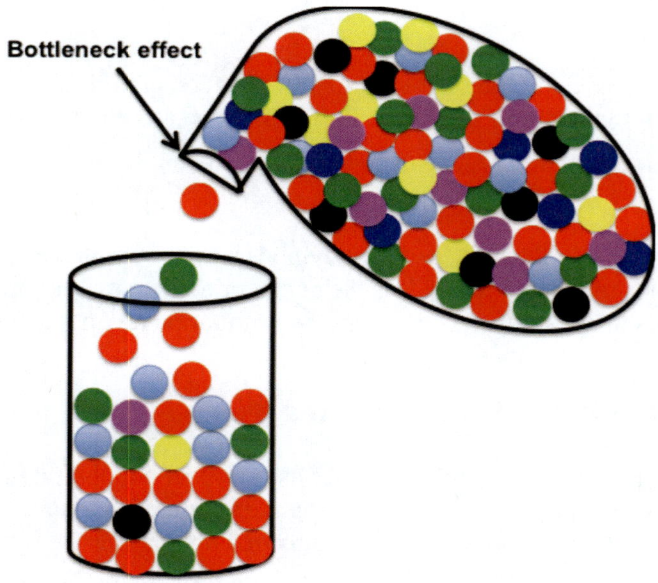

Fig. 2 : Diagrammatic representation of bottleneck effect

Founder Effect

The loss of genetic variation occurs when a new colony is established by a very small number of individuals from a larger population. Both, population bottleneck and founder effect reflect reduction of genetic diversity/variation in the population.

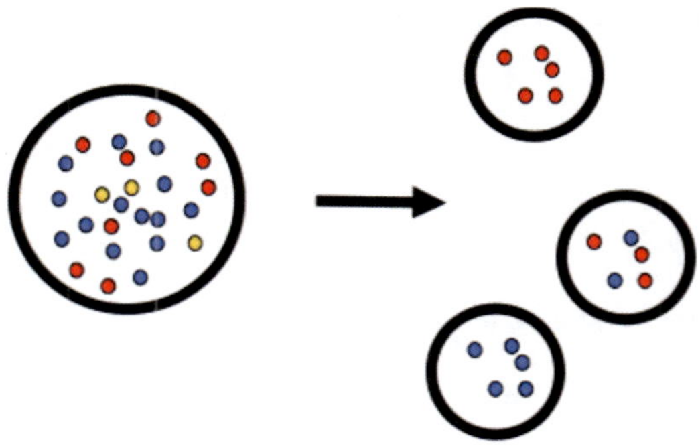

Fig. 3 : Diagrammatic representation of founder effect

Selective Sweep

A selective sweep is the reduction or elimination of genetic variation around a mutation that arises due to positive natural selection. Natural selection favor individuals that have a higher fitness as a result the frequency of newly mutated variant increases in the population. As its frequency increases, neutral and nearly neutral genetic variation linked to the new mutation also become more prevalent. This phenomenon is called genetic hitchhiking. One of the very good examples of selective sweep is drug resistance gene of *Plasmodium falciparum*. The selected mutant *dihydrofolate reductase* gene known to confer drug resistance (Anti-folate drug) show a clear drop of genetic variation **(Figure 4).** Reduction of genetic diversity is the characteristic features of both selective sweep and bottleneck events. In case of selective sweep, genetic diversity declined to a specific region of chromosome (local effect) whereas in bottleneck effect reduction of genetic diversity is throughout the genome (global effect).

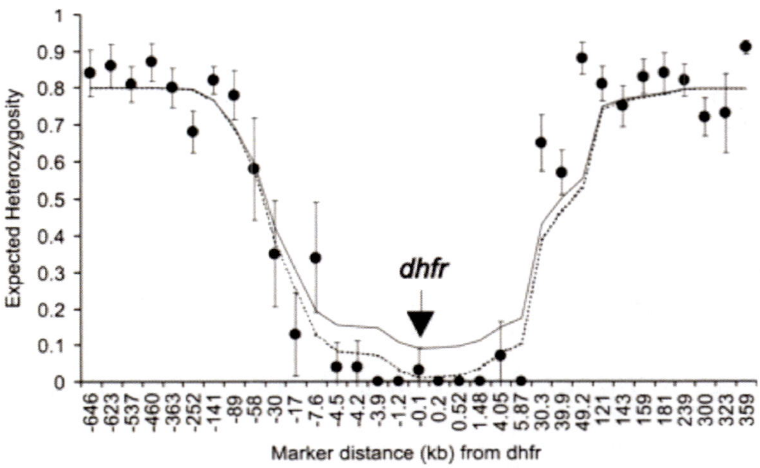

Fig. 4 : Selective sweep at *dhfr* gene of *Plasmodium falciparum*

Software Used in Molecular Data Analysis

This section describe various software being extensively use in analysis of raw data such as DNA sequence editing and microsatellite typing as well as evolutionary analysis using raw data to understand the various evolutionary facts and hypothesis. Below is brief description of such evolutionary tools.

DNA Lasergene version 7.00 (*http://www.dnastar.com/web/index.php*)

This is highly useful, license version and commercially available software used for DNA/Protein/RNA sequence analysis. DNA Lasergene software has specialized modules for specific function, such as primer designing (Primer Select), DNA sequence editing (EditSeq), multiple DNA/Protein sequence alignment and phylogenetic tree construction by Neighbor-Joining (N-J) methods (Meg Align) and contig formation (SeqMan). DNA sequence identity and dis-similarity among isolates was also carried out using MegAlign module.

MEGA Version 4.0 (*http://www.kumarlab.net/publications*)

Molecular Evolutionary Genetic Analysis (MEGA) software (Kumar et al., 2008; Kumar et al. 1994; Tamura et al. 2007) is user-friendly and highly applicable in evolutionary genetic analysis. MEGA provides many tools such as multiple DNA sequence alignment, DNA sequence editing with viewing chromatogram simultaneously, phylogenetic analysis using different methods (Neighbor-Joining, Minimum evolution, Unweighted paired-group method with arithmetic means and Maximum Parsimony), Neutrality test (Tajima's D), positive and negative selection over codons and genetic distance. Software uses only DNA sequence data for analysis.

DnaSP Version 4.0 (*http://www.ub.es/dnasp*)

DNA sequence polymorphism (DnaSP) software (Rozas et al. 2003) is user-friendly, freely available, and highly applicable in evolutionary genetic analysis. DnaSP offers many analysis tools such as identification of DNA sequence polymorphism (synonymous and non synonymous), genetic diversity measures (nucleotide diversity, haplotype diversity, theta and average number of nucleotide difference between sequence pair), neutrality tests (Tajima's D test and Fu & Li's D test), recombination, linkage disequilibrium, population size changes, population sub-division, divergence analysis between populations and gene flow. It uses only DNA sequence data for analysis.

Arlequin Version (*http://cmpg.unibe.ch/software/arlequin3*)

Arlequin is freely available, population and evolutionary genetics software package. Arlequin offers very useful population genetic analysis tools such as population genetic structure, haplotyping, multi-locus system, genotype assignment, linkage disequilibrium, neutrality test, population comparisons and differentiations, mismatch distributions, nucleotide diversity and F*st* (F-statistics). All types of data such as DNA sequence, microsatellite, allele frequency, and RFLP are used by Arlequin software.

Muscle (*http://www.phylogeny.fr/*)

This is html webpage based online tool to study sequence alignment, phylogenetic tree re-construction. It has two modules, one for non-specialist (non-biologist), and other for specialist (biologist). For non-specialist, all parameter are default value and single click option is available to construct phylogenetic tree, whereas in specialist module, one can custom all parameter according to their requirement.

Network Version 4.5.0.0 (*http://www.fluxus-technology.com/*)

This is freely available software, used to inferred evolutionary history of haplotypes of given gene of an organism. Network program is used to create the haplotype networks (phylogenetic network) of a gene. Network program provides evolutionary relationship of haplotypes and their evolutionary path. DNA/protein sequences, minisatellites, and microsatellites data are used by Network program.

Bottleneck Version 1.2.02 (*http://www.ensam.inra.fr/URLB*)

Bottleneck software is, freely available and used to detect the signature of recent bottleneck in the population and to detect the signature of stable population of a species in a specific geographical region. Minisatellites and microsatellites data types are used by Bottleneck software. Two tests were carried out for bottleneck analysis, one is heterozygosity deficiency and another is mode shift analysis. Brief description of both the tests is given below.

Population 1.2.30 beta (*http://www.cnrs-gif.fr/pge*)

Population is very useful, freely available, and user-friendly software for population genetic analysis. Population software provides several tools such as genetic distance of individuals and populations, phylogenetic analysis (N-J and UPGMA) of individual and population, allelic frequencies, population structure and F*st* analysis. Only allele size data (mini and microsatellites) is used by population software.

Microsatellite Analyzer version 4.00 (*http://i122server.vu-wien.ac.at*)

Microsatellite analyzer (MSA) is highly applicable software for population genetic analysis, and is freely available. MSA offers several analysis tools such as allele count, allelic frequencies, heterozygosity, global F*st*, and various genetic distances. MSA is specialized for analysis of large sample sizes and multi-locus data. Microsatellites and minisatellites data are used by MSA software.

Conclusion

Over the past few decades, major advances in the field of molecular biology, coupled with advances in genomic technologies, have led to an explosive growth in the biological information generated by the scientific community. This deluge of genomic information has, in turn, led to an absolute requirement for computerized databases to store, organize, and index the data, and for specialized tools to view and analyze the data. Bioinformatics can be viewed as *the use of computational methods to make biological discoveries* (Altman et al.2001).It is an interdisciplinary field involving biology, computer science, mathematics, and statistics to analyze biological sequence data, genome content and arrangement, and to predict the function and structure of macromolecules. The ultimate goal of the field is to enable the discovery of new biological insights as well as to create a global perspective from which unifying principles in biology can be derived (Baldi and Brunak1998).

Molecular evolution and evolutionary computing tools has emerged as a research topic in both natural and computational evolution. It is a notion put forward to investigate the fundamental mechanisms that enable a system to evolve. A number of hypotheses have been proposed in modern biological research based on the examination of various mechanisms in the biosphere for their contribution to evolvability. Therefore, it is intriguing to try to transfer new discoveries from Biology to and test them in Evolutionary Computation (EC) systems, so that computational models would be improved and a better understanding of general evolutional mechanisms is achieved.

References

Altman, R. B., Valencia, A., Miyano,S. and Ranganathan, S.(2001). "Challenges for intelligent systems in biology," IEEE Intell. Syst 16(6):14–20.

Baldi, P. and Brunak, S. (1998). Bioinformatics: The Machine learning approach. Cambridge, MA: MIT Press.

Felsenstein, J.(1981).Evolutionary trees from DNA sequences: a maximum likelihood approach. J. Mol. Evol. 17: 368-376.

Felsenstein, J.(1985). Confidence limits on phylogenies: an approach using the bootstrap. Evolution 39: 783-791.

Fu, Y.X., and Li, W.H. (1993). Statistical tests of neutrality of mutations. Genetics, 133, 693-709.

Kimura, M. (1983). The Neutral Theory of Molecular Evolution. Cambridge University Press, Cambridge, MA.

Kumar, S., Nei, M., Dudley, J.and Tamura, K. (2008). MEGA: a biologist-centric software for evolutionary analysis of DNA and protein sequences. Brief Bioinform 9: 299-306.

Kumar, S., Tamura, K. and Nei, M. (1994). MEGA: Molecular Evolutionary Genetics Analysis software for microcomputers. Comput. Applic. Biosci., 10:189-191.

Rozas, J., Sanchez-DelBarrio, J.C., Messeguer, X., Rozas, R., (2003). DnaSP, DNA polymorphism analyses by the coalescent and other methods. Bioinformatics 19: 2496–2497.

Saitou, N.and Nei, M.(1987).The neighbor-joining method: a new method for reconstructing phylogenetic trees. Mol Biol Evol 4: 406-425.

Tajima, F. (1989). Statistical method for testing the neutral mutation hypothesis by DNA polymorphism. Genetics 123: 585–595

Tajima, F.(1983). Evolutionary relationship of DNA sequences in finite populations. Genetics 105:437-460.

Tamura, K., Dudley, J., Nei, M. and Kumar,S.(2007).MEGA4: Molecular Evolutionary Genetics Analysis (MEGA) software version 4.0. Molecular Biology and Evolution 24:1596-1599.

Wright S. (1931) Evolution in mendelian populations. Genetics, 16, 97–159.

❑❑❑

CHAPTER – 31

Fungal Nano-Biotechnology Present Statusand Future Perspectives

Kambaska Kumar Behera
Department of Bioscience and Biotechnology, Banasthali University
Rajasthan,302240 India
Email : Kambaska@yahoo.co.in

Summary

Nanotechnology is an interdisciplinary areas of science and technology which refers broadly to a field of applied science and technology whose unifying theme of control of matter is on the molecular scale that integrates material science and biology. Due to their advantages over non-biological systems, several research groups have explored the use of biological systems for the synthesis of nanoparticles. Among the different microbes used for the synthesis of nanoparticles, fungi are efficient candidates for fabrication of metal nanoparticles both intra-and extracellulary. The nanoparticles synthesized using fungi present good polydispersity, dimensions and stability. The potential applications of nanotechnology and nanoparticles in different fields have revolutionized the health care, textile and agricultural industries and they are described in present chapter.

Introduction

Nanotechnology, the science of fabricating and putting into use nanoscale devices or particles for solving mankind's new emerging problems is being heralded as the science that will have a major impact on everything we do in our day to day lives. From intelligent shape and colour changing nanomaterials to nanoscale robots that will perform extremely specific microsurgery inside the human body, the applications of nanotechnology are immense and very diverse at the same time. A researcher working in this new but rapidly developing area has just to unleash his imagination to develop newer applications that are novel and have crucial applications in almost all spheres of human activity. The convergence of microbial biotechnology with this emerging field of science holds great promise in drug delivery platforms for diseases such as cancer; biosensor development for microbial pathogen detection in food and environmental samples and as nanoscale cell factories for a variety of enzymes and other materials used in industry(Wright, 2000; 2005). The first step in nearly all major nano-biotechnology platforms is the development of nano-particles/materials that have unique properties. Synthesis of nano-particles using abiotic chemical or physical processes have limitations in regards to their eco-friendliness and biosafety for human or any biological applications in living organisms. To overcome these limitations, biogenic nanoparticles synthesized with the help of biological system such as fungi, is now being taken up by researchers worldwide. Microorganisms, such as bacteria, molds, yeasts and viruses, in the living environment are often pathogenic and cause severe infections in human beings. There is a pressing need to search for new antimicrobial agents from natural and inorganic substances (Kim et al.1998; Cho et al.2005). Among the microbes endophytic fungi that reside inside the tissues of nearly all plants are comparatively an untapped source of novel genes and metabolites. Endophytic fungi has unique habitats inside the plant tissues and can also be used as efficient microbial cell factories. Fungal biologists are now slowly unraveling some of the endless per mutations and combinations of gene expression exhibited by endophytic fungi and trying to establish their relevance to human welfare. Focusing the inherent bio-processing ability of the fungal endophytes to generate safer, eco-friendly and cost-effective nanomaterials is thus, a window by which socially acceptable and sustainable nanotechnology based platforms can be evolved. This chapter is therefore an attempt to summarize the current state of research and development in this emerging area of fungal endophyte mediated synthesis of nanoparticles/materials. Issues related to biosafety and some possible directions for the future are also discussed at the end, primarily to ignite scientific thought among researchers and policy makers for a more sustainable use of endophytic fungal nano-biotechnology for a better tomorrow (Ingle et al.2008).

History of NanoScience

The dawn of the journey into the nano-world can be traced back to 1959, when physicist Richard Feynman hypothesized that manipulation of individual atoms might be possible. In a talk titled "There's Plenty of Room at the Bottom," at the annual meeting of the American Physical Society in 1959 at the California Institute of Technology, he challenged the scientific community, asking his colleagues "What would happen if we could arrange the atoms one by one the way we want them [?]". Feynman hypothesized that atoms and molecules could be manipulated like building blocks. Though his enthusiasm ultimately proved contagious, it was not until the 1980s that the idea became a reality with the invention of the scanning tunneling microscope (STM). The STM is widely described as the first nanotechnology-related invention because the microscope could take advantage of the wave-like properties of electrons to indirectly image and manipulate individual atoms. The first "proof-of-principle" that atoms could be precisely positioned took place in 1989 when IBM scientists manipulated 35 xenon atoms to form the letters IBM. The word nanotechnology is based on the prefix "nano" which is from the Greek word meaning "dwarf". The term nanotechnology was first coined by Norio Taniguchi in 1974 to describe technology at the atomic level in semiconductor circuits. In more technical terms, the word "nano" means 10^{-9}, or one billionth of something. A sheet of paper is about 100,000 nanometers thick; a single gold atom is about a third of a nanometer in diameter. Dimensions between approximately 1 and 100 nanometers are known as the nanoscale. Unusual properties can emerge in materials manufactured at the nanoscale including catalytic, electrical, magnetic, mechanical, optical, and thermal properties that differ in important ways from the properties of conventionally scaled materials. For instance, most people are familiar with gold it is yellow, it melts at about 1000°C, and it is non-reactive (Kowshik et al.2002). On the nanoscale, things look quite different. For instance, at 5 nm, gold is expected to melt at 830°C, and at 2 nm, it would melt at just 350°C. Even the magnetic properties of a material can change aluminum is not usually magnetic, but it becomes magnetic when individual atoms form a cluster with a diameter of less than 1 nm, and at 20–30 nm, aluminum will explode. Similarly, at the nanoscale an insulating material may change into a conductor, and a previously insoluble material may become soluble (Kim et al.2008; 2009).

Nanotechnology can therefore be broadly defined as the understanding and control of matter at dimensions between approximately 1 and 100 nanometers, where unique phenomena enable novel applications. Encompassing nanoscale science, engineering technology, Nanotechnology involves imaging, measuring, modeling, and manipulating matter at the nanoscale. It involves (1) the research and technology development at the atomic, molecular, or macromolecular levels, on a scale of about 1–100 nm; (2) the creation and use of structures, devices, and systems that have novel properties and functions due to their size; and (3) the ability to control or manipulate on the atomic scale. The basic 'building blocks' or units upon which Nanotechnology is based are the nanomaterials that can be positioned and manipulated for a variety of different applications(Duran et al.2005;Gade et al.2008). Fullerenes, nanotubes, nanowires, semiconductor nanocrystals, nanoparticles and organic molecules are some examples of such types of nanomaterials. These 'building blocks' can further be grouped into two categories. These are: inorganic nanomaterials and organic nanomaterials. Examples of inorganic nanomaterials are semiconductor nanocrystals (Quantum dots), metals, oxides, glass and clay. Similarly, deoxyribonucleic acid (DNA), proteins and dendrimers are examples of organic nanoparticles (Mann, 1993; Mann and Ozin, 1996). In order to use these basic building blocks for generation of useful nanodevices, nanotechnologists employ computational tools as well as laboratory tools to model, observe, and position nanomaterials (Pal et al.2007; Falkiewicz-Dulik and Macura, 2008).Computational nanotechnology involves designing and modeling nanomaterials and devices with the help of computers and specific software packages. The laboratory tools frequently used in nanotechnology are scanning tunneling microscope (STM) used to view nanomaterials, atomic force microscope (AFM) used to position nanomaterials and nanotweezers that enable researchers to grab nanomaterials and position them using a movable platform (Kowshik et al.2002).

The two fundamentally different approaches to nanotechnology are graphically termed as 'top down' and 'bottom up'.' Top-down' refers to making nanoscale structures by machining and etching techniques, whereas 'bottom-up', or molecular nanotechnology, applies to building organic and inorganic structures atom-by-atom, or molecule-by-molecule. Approaches in "top-down" category are referred to as different forms of lithography. Photolithography, which uses different kinds of electromagnetic radiation, is currently used to manufacture computer chips and other microelectronic devices. Photolithography is not; however, an effective tool for fabricating structures with features below 100 nanometers (Sanghi and Verma, 2009). E-beam lithography, a technique that employs beams of electrons to write, can produce some high-resolution nanostructures. Soft lithographic techniques, such as printing, molding, and embossing, involve physical or chemical deformation of the functional material to yield the desired structure. A new approach termed as "dip pen lithography" which is modeled after a traditional ink pen has also been developed by Chad Mirkin's laboratory at Northwestern University, USA. In this technique, different types of molecules can be placed on a nano-sized probe. Water molecules between the probe and a gold substrate act as a bridge over which the molecules are transferred from the probe to the gold surface. This technique has been used to write nano-sized words. Approaches in the "bottom up" category describe techniques for coaxing atoms and molecules to form nanomaterials (Luo et al.2005). One bottom up technique, referred to as "positional assembly," involves using a probe to move atoms into certain arrangements. Although positional assembly allows control over individual atoms, it is time-consuming, cannot presently be used to create complex nanostructures, and does not represent an efficient means for commercial production. Another bottom up approach is described as "chemical self-assembly." Different atoms, molecules, or nano-scale objects are mixed together and spontaneously organize into stable, well-defined structures. Because self-assembly methods are based on chemical reactions, they are simple and relatively inexpensive (Magudapathy. 2001).However, they do not offer the precision necessary for constructing designed, interconnected patterns that top down approaches might. Keeping these constraints in view, the development of reliable experimental protocols for the synthesis of nanomaterials over a range of chemical compositions, sizes, and high monodispersity is one of the most challenging issues in current nanotechnology (Fortin et al.2000).

Nanomaterials and Their Applications

Nanomaterials are arrangements of matter with dimensions of less than 100 nanometers that exhibit unique characteristics due to their size. Nanomaterials are at the leading edge of the rapidly developing field of nanotechnology. Nanomaterials have been used commercially since the 1930s. An example is silver in photographic film. Another important nanomaterial that has been produced in large volumes for a long time is carbon black. Nanomaterials can also be produced unintentionally, for example in the formation of ultra fine particles in combustion, or in natural processes. On a sunny day, new formation of organic nanoparticles takes place through photo-oxidation of organic substances that are emitted from forests. Other natural processes that lead to the new formation of nanoparticles are volcanic eruptions. Minerals that are formed through weathering and through redox chemical processes also occur naturally at nanoscale. Humic substances are present in very large quantities on earth and also occur at nanoscale. They also interact with nanominerals. In addition, all living organisms are made up of biological macromolecules, which, like many viruses, can be regarded as a natural form of advanced nanoparticles. Nanoparticles are tiny versions of the conventional materials. These include the diverse atomic elements of periodic table, both benign and toxic. Nanomaterials use in traditional process and products can provide resource efficiency and other ecological benefits like source reduction, pollution reduction, eco-friendly coatings like self-cleaning, resistance to heat, UV etc. A key criterion for something to be considered a nanomaterial is finding the property of the material like size-dependency, e.g., the tenability of optical absorption of metals by varying its thickness. Making a material nanoscopic might make it more absorptive under some relevant conditions (e.g., absorbing light), which in turn could have some unusual effects.

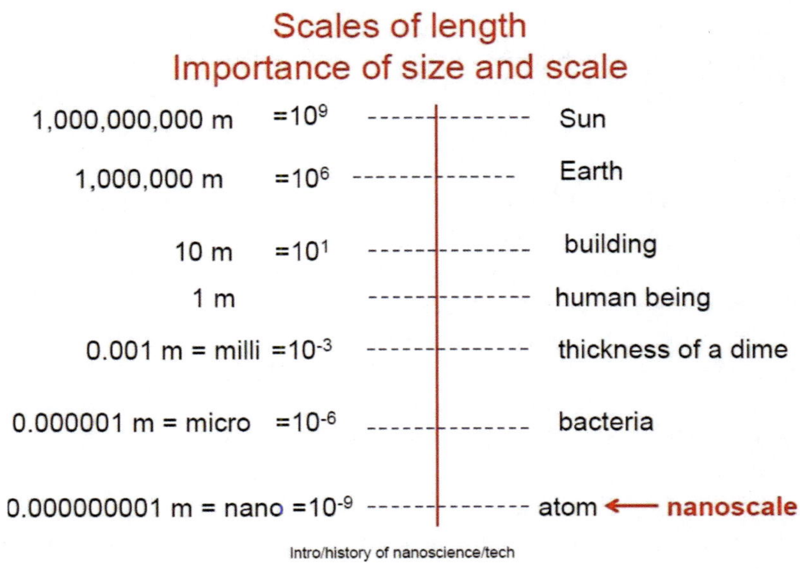

Fig.1 : Scales of length, Importance of size and scale of Nanoparticles

For example, cadmium selenide crystals at 1.5 nm will appear yellow, at 4 nm it will appear red. Manufactured nanomaterials, which are 100 nanometers (0.1 micrometers) or less in diameter, are already used in commercial products in powders, sprays, and coatings that are being used in a variety of products, including sunscreens, automobile parts, tennis rackets, scratch-proof eyeglasses, stain-repellent fabrics, self-cleaning windows, and many more.

Biogenic Nanoparticles and Their Advantages

Current chemical methods for the synthesis of nanoparticles are energy intensive and rely upon toxic chemicals; are outdated, costly and produce hazardous wastes. These facts not only make them unsuitable for any biomedical application, but also make them environmentally unsafe. On the other hand, the physical methods of nanoparticles synthesis such as sputter deposition, laser ablation or cluster beam deposition are often difficult to achieve. Moreover, these methods are time consuming and still under development. Among the biological systems, the synthesis of metal nanoparticles using different plants like Alfalfa, *Aloe vera*, *Cinnamomum camphora*, *Capsicum annum*, *Medicago sativa*, *Brassica juncea*, *Azadiracta indica* and *Cymbopogon flexuosus* has been demonstrated successfully by many nano-biotechnologists.

Mycogenic Nanoparticles

Fungi are eukaryotic, spore-producing, achlorophyllous organisms that are usually filamentous, having branched somatic structures with hyphae surrounded by cell walls. Fungi can accumulate metals by physicochemical and biological mechanisms including extracellular binding by metabolites and polymers, binding to specific polypeptides, and metabolism-dependent accumulation. Fungal species such as *Fusarium oxysporum* can readily synthesize metal nanoparticles extracellularly using high-levels of secreted proteins and/or enzymes that not only stabilize the particles but allows for an improved yield over an intracellular one(Ahmad et al.2003, 2005). An extracellular synthesis has obvious advantages over an intracellular process, when it comes to downstream processing, since there would be very little handling of the fungal biomass. In contrast, with an intracellular synthesis, sophisticated instruments would be required to isolate the nanoparticles from the biomass and into the cell-free filtrate (Birla et al.2009). These attributes make fungi the best choice among the biological systems for the synthesis of nanomaterials. In addition, fungal mycelia can withstand flow pressure, agitation and other conditions in bioreactors or other chambers compared to that of plant materials and bacteria (Anil kumar et al. 2007; Shivshankar et al. 2003). The extracellular secretions of reductive proteins in case of fungi are more in comparison to other biological systems, and can be easily handled in downstream processing. Mycogenic nanoparticles are mostly precipitated outside the cell and are devoid of unnecessary cellular components, making them suitable for direct use in various applications without further processing. Fungal systems have already been explored for the synthesis of metal nanoparticles of silver, gold, zirconium, silica, titanium, iron (magnetite) and platinum. Further, the methodology for extracellular synthesis of nanoparticles using fungi is extremely cost-effective and less time consuming (Bansal et al.2003)

Generation of Nanoparticles Using Endophytic Fungi

Endophytes are microbes that reside within the internal living tissues of higher plants as endosymbionts (Balaji et al.2009). Endophytic fungi are now considered as potential sources for novel drugs and enzymes. Keeping the inherent advantages of using a fungal system for the production of nanomaterials, endophytic fungi are now being slowly employed as myconanofactories for the synthesis of nanoparticles (Basavaraja et al.2009; Bhainsa and D'Souza 2006).A list of the use of endophytic fungi for the production of different nanoparticles is given in Table 1. However, it must be kept in mind that most of the technologies to generate harmless and ecologically safe nanoparticles from endophytic fungi in particular, are in a developing stage. As the technology to produce nanoparticles from endophytic fungi is being refined, the possibilities to generate different types of nanomaterials are limitless.

Biosynthesis of Silver Nanoparticles

The reductase is responsible for the reduction of Ag^{+} ions and the subsequent formation of silver nanoparticles and was also reported in the strain of *F. oxysporum* and it was pointed out that this reductase was specific to *F. oxysporum*. However, *Fusarium moniliforme*, did not result in the formation of silver nanoparticles, neither intracellularly nor extracellularlybut contained intra and extra cellular reductases in a similar fashion as *F. oxysporum* (Ahmad, et al.2003; Klittich and Leslie,1988). This is an indication that probably the reductases in this kind of *Fusarium* are important for Fe (III) to Fe (II) but not to Ag (I) to Ag (0). Moreover, in *F. moniliforme* anthraquinones derivatives were not detected unlike the case of *F. oxysporum*. Both *fusarium* were alike in the production of naphthaquinones (Medentsev and Alimenko ,1998) but differed in the production of anthraquinones.

Probably, in our case, Ag (0) reduction was mainly due to a conjugation between the electron shuttle with the reductase participation as shown in (Fig.1).

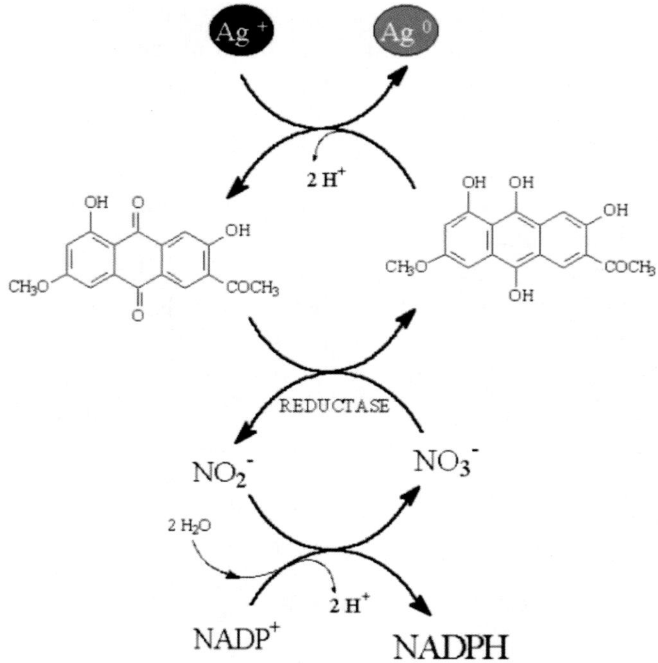

Fig. 2 : Hypothetical mechanisms of silver nanoparticles biosynthesis

Biosynthesis of Gold Nanoparticles

The synthesis of gold nanoparticles in aqueous medium using *Terminalia catappa* (Almond) leaf extract as the reducing and stabilizing agent. On treating chloroauric acid solutions with *Terminalia catappa* (TC) leaf extract rapid reduction of chloroaurate ions is observed leading to the formation of highly stable gold nanoparticles in solution. TEM analysis of the gold nanoparticles indicated that they ranged in size from 10 to35 nm with average size of 21.9 nm (Ankamwar.2010).

Steps of Biological Synthesis of Gold Nanoparticles

The broth used for the reduction of Au^{3+} ions to Au^0 was prepared by taking 10 g of thoroughly washed and finely cut *Terminalia catappa* leaves in a 500 mL Erlenmeyer flasks with 40 mL of sterile distilled water and boiled it for 15 min. In a typical experiment, 0.2 mL of broth was added to 50 mL of 10^{-3} M aqueous chloroauric acid ($HAuCl_4$) solution. Within an hour (50 minutes) cherry red solution was obtained and studied and measured through viz. **UV-Vis spectroscopy studies, X-ray diffraction (XRD) measurement, Fourier transforms infrared (FTIR) spectroscopy measurements and TEM measurements**

Table 1 : Endophytic fungi used in the synthesis of nano-particles.

Name of the endophytic fungi	Host Plant species	Nanoparticle synthesized	Intracellular / Extracellular	Size range (nm)/ shape
Aspergillus clavatus	*Azadirachta indica*	Gold	Intracellular	20-35nm Nanotriangles (Verma.2010)
Colletotrichum sp.	*Pelargonium graveolens*	Gold	Extracellular	8 -40 nm, Spherical(Arya. 2010)
Amylomyces rouxii strain KSU-09	*Phoenix dactylifera*	Silver	Extracellular	5-27 nm, pherical(Musarrat et al.2010)
Verticillium sp.	*Taxus* sp.	Gold and Silver	Intracellular	2-20nm,Variable (Sastry et al. 2003)
Fusarium oxysporum (A plant pathogen)	*Cymbopogon flexuosus*	Gold and Silver	Extracellular	20-25nm,Variable (Duran et al. 2005)
Alternaria alternata	*Amaranthus cruentus*	Silver	Extracellular	20–60 nm (Gajbhiye et al. 2009)
Penicillium fellutanum	*Rhizophora annamalayana*	silver	Extracellular	20-60 nm (Kathiresan et al, 2009)

Nano-Biotechnology Bubble Burst: Are Nanoparticles Safe?

The small size of nanoparticles can give them greater access to body tissues and organs than larger particulates. Once within the body, the larger surface area of nanomaterials per unit of mass makes them more chemically reactive than their normal-scale counterparts, and therefore more likely to interact with biological molecules. This is particularly likely when the size / shape or other characteristics of a nanomaterial mimics biologically active components of the body's systems, such as DNA-binding proteins, immune molecules and cell components. Studies indicate that some nanomaterials may interact with DNA, cause inflammation and oxidative damage. Particulate matter less than 10μm (10,000 nm) has been linked to increased lung cancer and cardiopulmonary disease. While all particulate air pollution is hazardous, smaller inhaled particles have long been known to be more damaging to body tissues than larger particles, inducing inflammation and tissue damage. The risks are especially high among individuals with pre-existing heart and lung ailments, including asthma and chronic obstructive pulmonary disease, suggesting that millions of people with these conditions may be more vulnerable to the hazards of inhaled nanomaterials. Carbon nanotubes have also been shown to cause irreversible and progressive lung damage in animal models. Nano-cerium oxide particles that are added to diesel fuel to decrease toxic diesel emissions and increase fuel efficiency, if inhaled can induce cell damage that is likely to lead to cancer. Nanomaterials may also pose risks to the environment. Nanoscale silver ions act as an antimicrobial biocide, a substance that can kill or inhibit the growth of microbes. Nanoscale silver is used in numerous commercial products including washing machines, clothing, kitchen utensils, wound dressings, and food storage containers from which, either during normal use of the product or through degradation after the product is discarded, nanoscale silver ions may find a way into the waste water outlet. Once these ions are released into the environment, their biocidal activity may wipe out the populations of beneficial microbes, there by leading to harmful disturbances in key ecosystems and ecological food webs.

Applications of Nanoparticles

Medicine and Health Care

A brief overview of nanoparticle synthesis using microorganisms such as bacteria, fungi, actinomycetes, and yeast has been reported. The antimicrobial activity of silver compounds is well known. Colloidal silver, however, was only discovered at the beginning of the 20th century and with the growing resistance of microorganisms to antibiotics has renewed the interest in biological effects of this form of silver (Mandal et al. 2006). Several researchers have studied the antimicrobial activity of silver nanoparticles against bacterial strains like *E. coli*, *Staphylococcus aureus*, *P. aerugenosa* and also antimicrobial activity of textile fabrics and wound dressings impregnated with silver nanoparticles has been successfully tested against several common pathogens, namely methicillin resistant *Staphylococcus aureus* (MRSA), *Enterococcus faecalis*, *Pseudomonas aeruginosa, Escherichia coli*, *Enterobacter cloacae,Proteus vulgaris, Klebsiella pneumoniae* and *Acinetobacter baumannii* (Rai et al. 2009b). Similarly Morones et al. (2005) have evaluated that the concentration of silver nanoparticles that prevents bacterial growth was different for each type of bacteria. *P. aeruginosa* and *V. cholera* were found to be more resistant to silver nanoparticles than *E. coli* and *Salmonella* typhus. Above $75\mu g\ ml^{-1}$, however, no significant growth was shown by any of the bacteria and silver nanoparticles have potential to become a new generation of antimicrobials (Rai et al. 2009). Elechiguerra et al. (2005) studied the interaction of metal nanoparticles with viruses and demonstrated that silver nanoparticles undergo a size-dependent interaction with HIV-1; the nanoparticles of 1-10 nm attached to the virus. The regular spatial arrangement of the attached nanoparticles, the center-to-center distance between nanoparticles, and the exposed sulfur-bearing residues of the glycoprotein knob suggested that, through preferential binding, the silver nanoparticles prevented the HIV-1 virus from binding to host cells.

Copper nanoparticles also inhibited the growth of *E. coli*, *Salmonella choleraesuis*, *Salmonella typhimurium*, *Staphylococcus aureus* and *Saccharomyces cerevisiae*.'Stealth coatings, combined with targeted molecules were expected to transform nanoparticles into 'magic bullets', which has been the ultimate goal of medical research. Although there is still a long way to go in this endeavour, the first anti-tumour and anti-fungal drug using simple delivery systems are already being marketed. Nanoparticles were for targeted oral drug delivery to the inflamed gut tissue in inflammatory bowel disease and found that this strategy of local drug delivery showed improvement compared with existing colon delivery devices. Gelperina et al. (2005) reported the use of metal nano particles for the drug delivery in tuberculosis patients due to technological advantages of nanoparticles such as high stability, high carrier capacity, feasibility of incorporation of both hydrophilic and hydrophobic substances, and of variable routes of administration, including oral application and inhalation as well as controlled (sustained) drug release. Pandey and Khuller (2007) studied nanoparticle for the development of oral drug delivery system and suggested that nano-encapsulation might be useful for developing a suitable oral dosage form for streptomycin and for other antibiotics that are otherwise injectable. Nanocrystals can be used as carrier-free colloidal drug delivery systems with a particle size between 100 and 1000 nm, as they provide viable drug delivery strategies that overcome drug solubility problems (Gao et al. 2008). Gold nanoparticles have unique optical and chemical properties that make them ideally suited for a number of applications in bio-nanotechnology, including optical probes and targeted drug delivery (Glomm 2005).

Nanoparticles have been used recently to improve the present imaging techniques for *in vivo* diagnosis of biomedical disorders. Iron oxide nanoparticles, are currently being used in patients for both diagnosis and therapy leading to more effective medication with lesser side effects (Gao et al.2008).Optical and electron beam contrasting qualities of gold colloids provide excellent detection with electron microscopy, for techniques to label target molecules through conjugation protocols, for protein A, avidin, streptavidin, glucose oxidase, horse radish peroxidase and IgG as immunoblotting, flow cytometry, and hybridization assays (Nam et al. 2003). A unique, sensitive and highly specific immunoassay system based on the aggregation of gold nanoparticles that are coated with protein antigens in the presence of their corresponding antibodies was also developed (Thanh and Rosenzweig 2002).The interaction of nanoparticles with biomolecules and microorganisms is an expanding field of research.

Agriculture

Agriculture is the backbone of most developing countries, with more than 60% of the population reliant on it for their livelihood. Biosynthesis of nanoparticles by plant extracts is under exhaustive research. Rickman et al. (1999) reported on the application of nanotechnology in precision farming to increase crop yields (outputs) while minimizing input (fertilizers, pesticides, herbicides). Precision farming made use of computers, global satellite positioning systems, and remote sensing devices prepared by using different nanomaterials to measure highly localized environmental conditions, in order to determine whether crops were growing at maximum efficiency. Nanopesticides and nanoherbicides were being used extensively in agriculture with several industries making formulations which contain 100–250 nm nanoparticles that are more soluble in water thus increasing their activity. Other companies employed nanoemulsions of nanoscale particles, which could be either water or oil-based and contained uniform suspensions of pesticide or herbicide nanoparticles of 200–400 nm (Rickman et al. 1999).

Textile Industry

Nanoparticles possess a large surface area-to volume ratio as well as a high surface energy, the use of nanoparticles could provide high durability for treated fabrics that lead to an increase in durability of the textile functions. Silver nanoparticles have been used to produce "self-cleaning" or "anti-odour"clothes, furniture textiles or automotive interiors, household products such as kitchen cloths, sponges, towels, antibacterial wound dressings, patient dresses, bed lines or reusable surgical gloves and masks (Lee et al. 2007), protective face masks, suits against biohazards, cosmetic products, tooth brushes, ultra hydrophobic fabrics with potential applications in the production of highly water repellent materials, upholstery and sportswear (Ramaratnam et al. 2008). Table 2 shows some of the applications of metal nanoparticles synthesized from different fungi.

Future Perspectives

The fungi are now well established as suitable system for the synthesis of different kinds on nanomaterials with great potential and promise for advanced diagnostics and biosensors, targeted drug delivery and smart drugs and immuno-isolation therapies. In the future, molecular machine systems and medical nanorobots will allow for immediate diagnosis and elimination of pathogenic bacteria, chromosome replacement and individual cell surgery *in vivo* and the efficient augmentation and improvement of natural physiological functions. It is anticipated that nanotechnology will be a major nano force and continue to evolve and expand in many areas of applied medical sciences, including diagnostics (nanosensors),development of artificial muscles and limbs, lab-on-a chip technology for more efficient drug discovery, electronics, nanomachines, nanorobots for in vivo treatment, electronics and agriculture.

Table 2 : Applications of metal nanoparticles synthesized by different fungi

Name of fungi	Nanoparticle synthesized	Application	Reference
Saccharomyces prombe	CdS	Electric diode	Kowshik et al. (2002)
Usnea longissima	Bioactive nanoparticles	Antifungal activity	Shahi and (Usnic acid) Patra (2003)
Fusarium oxysporum	Silver	Textile Fabrics	Duran et al. (2007)
F. oxysporum PTCC 5115	Silver	Response surface methodology to study the effect of physical parameters on synthesis of nanoparticles	Karbasian et al. (2008)
F. acuminatum	Silver	Antibacterial activity	Ingle et al. (2008)
Aspergillus niger	Silver	Antibacterial activity	Gade et al. (2008)
Trichoderma viride	Silver	Vegetable and Fruit preservation	Fayaz et al. (2009)
Rhizopus oryzae	Gold	Water hygiene management	Das et al. (2009)
Alternaria alternata	Silver	Antifungal activity	Gajbhiye et al. (2009)
A. niger	Silver	Wound healing activity	Sundaramoorthi et al. (2009)
Phytophthora infestans	Silver	Antibacterial activity	Thirumurugan et al. (2009)

Is Evolutionary Systems Biology the Answer? (Future Direction)

Ayurveda (Indian traditional system of medicine) has always emphasized on a holistic approach towards life and health. Ayurveda treats each and every individual patient as a unique biological system that has gone wrong due to the imbalance of three elemental energies. The entire science of nanotechnology and nanoparticles is based on the concept of nanoscale manipulation and synthesis of matter to yield novel materials and their subsequent applications. Keeping into view the health and environmental safety issues related with the use of the nanomaterials, we feel that studies on the evolutionary systems biology of living organisms holds the key to successful application of nanomaterials to cure diseases and make them safe for human use. This approach, which is slowly gaining momentum, deals with understanding of a living organism/ system in totality with an evolutionary perspective. One example of this can be the chloroplasts of plants, which are thought to be of microbial origin. Evolution of green plants or for that matter any living organism leads to the incorporation of new genes that may or may not lead to visible phenotypes.

In a similar fashion nanoparticles may find a way into the biological systems for which they are being primarily generated. The multitude of effects of such an invasion and possible integration of nanomaterials within existing biological systems would no doubt be difficult to study by present day state-of the art toxicological screening. The Holy Grail for biology now, is therefore, to understand the interplay of numerous sub-systems that go into the making of a complex biological system like the human body itself. A better understanding of the biological system as a whole, will perhaps equip us with more tools to engineer nanoparticles that will not only be safe but also ecologically friendly and thus sustainable in a longer time frame.

Conclusion

Nanoparticles are increasingly used in a wide range of applications in science, technology and medicine. Since they are produced for specific purposes which cannot be met by larger particles and bulk material. Myconano-biotechnology has scopes for solving some of the major problems faced by humans in today's world. However, as with all powerful tools of science, this rapidly evolving technology needs to be handled with care. A study on the human health and environmental risks associated with the use of mycogenic nanoparticles is important to keep the promise of green nanotechnology intact. Presently nanoparticles synthesized from endophytic fungi have not yet been given proper attention by companies working in the field of nanotechnology. However, as demands for green nanotechnology grows and more research on the viability of scaling up the laboratory based proof of concepts to the industrial level come up, it can be envisioned that commercially viable products based on endophytic fungi generated nanomaterials will start becoming visible in consumer markets worldwide.

Though nanotechnology has been hailed as the "next Industrial Revolution" and promises to have substantial impacts into many areas of our lives. These impacts will be manifest through many of the novel applications that nanotechnology will enable; these applications will take advantage of features that are only realized through nanoscale manipulations even these technological advances has many ethical and social questions will, or have been, raised (Allhoff et al.2007;Brayner, 2008).

References

Ahmad A, Mukherjee P, Senapati S, Mandal D, Khan MI, Kumar, R. and Sastry, M.(2003). Extracellular biosynthesis of silver nanoparticles using the fungus *Fusarium oxysporum*. Colloids and Surfaces B: Biointerfaces 28, 313–318.

Ahmad. A., Senapati, S., Khan MI, Kumar, R. and Sastry, M. (2005).Extra-/intracellular biosynthesis of gold nanoparticles by an alkalotolerant fungus, Trichothecium sp. J. Biomed Nanotechnol 1:47–53.

Allhoff, F., Lin, P., Moor, J.and Weckert, J.(2007). Nanoethics: Thesocial and ethical dimensions of nanotechnology. Wiley,Hoboken, NJ

Anil kumar, S., Abyaneh, M.K., Gosavi, S.W., Kulkarni, S.K., Pasricha, R.,Ahmad, A. et al. (2007).Nitrate reductase-mediated synthesis of silver nanoparticles from AgNO3. Biotechnol Lett 29:439-45.

Ankamwar,Balaprasad.(2010).Recent Trends in Nanoscience and Nanotechnology (Nanoparticle and Nanostructured Materials)"VDM Publishing House Ltd, Germany (2010).

Balaji, D.S., Basavaraja, S., Deshpande, R., Mahesh, D.B., Prabhakar,B.K. and Venkataraman, A.(2009). Extracellular biosynthesis offunctionalized silver nanoparticles by strains of *Cladosporium cladosporioides* fungus. Coll Surf B Biointerfaces68:88–92.

Bansal, V., Rautaray, D., Bharde, A., Ahire. K., Sanyal, A., Ahmad, A., et al. (2005). Fungus-mediated biosynthesis of silica and titania particles. J. Mater Chem 15:2583-9.

Basavaraja, S., Balaji. S.D., Legashetty, A., Rasab, A.H. and Venkatraman,A.(2008). Extracellular biosynthesis of silver nanoparticlesusing the fungus *Fusarium semitectum*. MaterRes Bull 43:1164–1170.

Bhainsa, K.C. and D'Souza, S.K. (2006). Extracellular biosynthesis ofsilver nanoparticles using the fungus *Aspergillus fumigatus*. Coll Surf B Biointerfaces 47:160–164

Birla, S.S., Tiwari, V.V., Gade, A.K., Ingle, A.P., Yadav, A.P. and Rai, M.K.(2009). Fabrication of silver nanoparticles by Phoma glomerata and itscombined effect against *Escherichia coli, Pseudomonas aeruginosaand Staphylococcus aureus*. Lett Appl Microbiol 48:173-9.

Brayner, R. (2008).The toxicological impact of nanoparticles. Nanotoday 3, 48–55.

Cho, K.H., Park. J.E., Osaka, T.and Park, S.G.(2005). The study of antimicrobial activity and preservative effects of nanosilver ingredient. Electrochim Acta 51:956-60.

Duran, N., Marcato, P.D., Alves, O.L., D'Souza, G.and Esposito, E.(2005). Mechanistic aspects of biosynthesis of silvernanoparticles by several *Fusarium oxysporum* strains. J Nanobiotechnol 3:8–14.

Elechiguerra, J.L., Burt, J.L., Morones, J.R et al.(2005). Interactionof silver nanoparticles with HIV-1. J Nanobiotechnol 3:6.

Fortin, D.and Beveridge, T.J.(2000).Mechanistic routes toward biomineral surfacedevelopment. In: Baeuerien E, editor. Biomineralization. From biologyto biotechnology and medical applications. Weinheim: Wiley-VCH p. 7.

Falkiewicz-Dulik, M.and Macura, A.B.(2008).Nanosilver as substance biostabilising footwear materials in the foot mycosis prophylaxis. Mikologia Lekarska 15:145-50.

Gade, A.K., Bonde, P.P., Ingle, A.P., Marcato, P., Duran, N. and Rai, M.K.(2008)..Exploitation of Aspergillus niger for synthesis of silver nanoparticles. JBiobased Mater Bioenergy 2:243-7.

Gao, L., Zhang, D.and Chen. M. (2008). Drug nanocrystals for theformulation of poorly soluble drugs and its application as a potential drug delivery system. J Nanopart Res 10:845–862.

Gelperina, S., Kisich, K., Iseman, S.D., Heifets, L.(2005).Thepotential advantages of nano particle drug delivery systems in chemotherapy of tuberculosis. Am J Respir Crit Care Med 172:1487–1490

Glomm, W.R .(2005). Functionalized gold nanoparticles forapplications in bionanotechnology. J Dispers Sci Technol26:389–414.

Ingle, A., Gade, A., Pierrat, S., Sönnichsen, C.and Rai, M.(2008) Myco-synthesis of silver nanoparticles using the fungus Fusarium acuminatum and its activity against some human pathogenic bacteria. Curr Nanosci 4:141-144.

Kim, T.N., Feng, Q.L., Kim, J.O., Wu, J., Wang, H.and Chen ,G.C., et al.(1998).Antimicrobial effects of metal ions (Ag+, Cu2+, Zn2+) in hydroxyapatite.J Mater Sci Mater Med 9:129-34.

Kim, K.J., Sung, W.S., Moon, S.K., Choi, J.S., Kim, J.G.and Dong, G.L.(2008). Antifungal effect of silver nanoparticles on dermatophytes. J Microbiol Biotechnol 18:1482-4.

Kim, K.J., Sung, W.S., Moon, S.K., Choi, J.S., Kim, J.G.and Dong, G.L.(2009).Antifungal activity and mode of action of silver nano-particles on *Candida albicans*. Biometals 22:235-42.

Klittich, C.J.R. and Leslie, J.F.(1988). Nitrate reduction mutants of *Fusarium-moniliforme* (*Gibberella-fujikuroi*) Genetics.118:417–423.

Kowshik, M., Vogel, W., Urban, J., Kulkarni, S.K.and Paknikar, K.M.(2002).Microbial synthesis of semiconductor PbS nanocrystallites. Adv Mater 14:815-518.

Lee, H.Y., Park, Y.K., Lee, Y.M., Kim, K. and Park, S.B.(2007). A practical procedure for producing silver nano coated fabric and its antibacterial evaluation for biomedical applications.Chem Commun 2959–2961

Luo, L., Yu, S., Qian, S.and Zhou, T.(2005).Large-scale fabrication of flexible silver/cross-linked poly (vinyl alcohol) coaxial nanoscale by a facial solutionapproach. J Am Chem Soc 127:2822-3.

Magudapathy, P., Gangopadhyay, P., Panigrahi, B.K., Nair, K.G.M.and Dhara, S.(2001). Electrical transport studies of Ag nanocrystallites embedded in glass matrix. Physics B 299:142-6.

Mandal, D., Bolander, M.E., Mukhopadhyay, D., Sarkar, G.and Mukherjee, P.(2006). The use of microorganisms for the formation of metal nanoparticles and their application. Applied Microbiology and Biotechnology 69: 485–492.

Mann, S.(1993). Molecular tectonics in biomineralization and biomimeticmaterials chemistry. Nature 365:499-505.

Mann, S.and Ozin, G.A.(1996).Synthesis of inorganic materials with complex form. Nature 382:313-8.20.

Medentsev, A.G. and Alimenko, V.K. (1998). Naphthoquinone metabolites of the fungi. Phytochemistry. 47:935–959.

Morones, J.R., Elechiguerra, J.L., Camacho, A., Holt, K., Kouri, J.B.,Ramirez, J.T.and Yacaman, M.J. (2005). The bactericidal effect of silver nano particles. Nanotechnology 16:2346–2353.

Nam, J., Thaxton, C.S.and Mirkin, C.A. (2003). Nanoparticle based biobarcodes for the ultrasensitive detection of proteins.Science 301:1884–1886.

Pal, S., Tak, Y.K.and Song, J.M.(2007). Does the antibacterial activity of silvernanoparticles depend on the shape of the nanoparticle? A study of thegram-negative bacterium Escherichia coli. Appl Environ Microbiol 73:1712-20.12.

Pandey, R.and Khuller, G.K. (2007). Nanoparticle based oral drugdelivery system for an injectable antibiotic-streptomycin.Chemotherapy 53:437–441.

Pandey,R. and Khuller, G.K. (2006). Nanotechnology based drug delivery system(s) for the management of Tuberculosis. Indian Journal of Experimental Biology 44:357-366.

Rai, M., Yadav, A., Bridge, P.and Gade, A.(2009). Myconanotechnology: A new and emerging science. In Applied mycology (Eds Rai and Bridge), CAB International, pp. 285-267

Ramaratnam, K., Iyer, S.K., Kinnan, M.K., Chumanov, G., Brown, P.J. and Luzinov, I. (2008). Ultra hydrophobic textiles using nanoparticles:lotus approach. J Eng Fibers Fabrics 3:1–14.

Rickman, D., Luvall, J.C., Shaw, J., Mask, P., Kissel, D. and Sullivan, D.(1999). Precision agriculture: changing the face of farming.[www. ghcc. msfc. nasa. Gove /precisionag/].

Sanghi, R., Verma P.(2009). Biomimetic synthesis and characterisation of proteincapped silver nanoparticles. Bioresource Technol 100:501-504.

Shivshankar, S., Ahmad, A.and Sastry, M.(2003).Geranium leaf assisted biosynthesisof silver nanoparticles. Biotechnol Prog 19:1627-31.

Thanh, N.T.K.and Rosenzweig,Z.(2002).Development of anaggregation-based immunoassay for anti-protein a usinggold nanoparticles. Anal Chem 74:1624–1628.

Wright, G.D.(2005). Bacterial resistance to antibiotics: enzymatic degradationand modification. Adv Drug Deliv Rev 57:1451-70.

Wright, G.D.(2000). Resisting resistance: new chemical strategies for battling superbugs. Chem Biol 7: 127-132.

❑❑❑

CHAPTER – 32

Lac Cultivation
An Overview to IndianContext

Kambaska Kumar Behera[1] and Bishnu Prasad Dash[2]
[1]Department of Bioscience and Biotechnology, Banasthali University, Rajasthan-304022, India
[2]Department of Bioscience and Biotechnology, Fakir Mohan University, Nuapadhi, Balasore- 756020, India
Email : kambaska@yahoo.co.in

Summary

Lac is a resinous exudation from the body of female scale insect Laccifer lacca Kerr. belongs to family Lacciferidae of phylum Arthropoda, class Insecta and order Hemiptera. The lac insect lives on native trees in India, China, Thiland, Burma/Myanmar and Malaysia. In India it is chiefly grown on trees like Kusum, "Palas", and "Ber. It is also purposely cultured on shoots of these tree species. Lac cultivation is an alternative agricultural profession and provides subsidiary source of income for a large number of farmers mainly in Jharkhand, Chhattisgarh, West Bengal, Odisha, North-East states and other parts of the country. India, which is the highest producer of lac, contributes around 55% of the total world requirement followed by China and Thiland. The country exports around 80-90% of its production. Jharkhand is the 'Lac State of India' which alone contributes about 59% of the national production followed by Chattisgarh (23%), West Bengal (12%), while Odisha, Gujarat, Maharastra, Uttar Pradesh, Andhra Pradesh and Assam are the minor producers. The income form lac cultivation is about 8% of their total agricultural income. And most of the lac growing families are among the poorest of the poor in the tribal dominated state of the country. The cultivation is done extensively in the remote interior pockets of the different state. The average yield of stick lac per tree of kusum, ber and palas is 10 kg, 6 kg and 4 kg respectively. India fetches approximately Rs.120 to 130 crores of foreign exchange through export of lac every year. The present review, deals with the information concerning the general aspect of Lac, Life cycle and steps of cultivation, importance of lac with Indian and international status .

Introduction

Lac is one of the most valuable gifts of nature to mankind and is unique material of animal origin.The natural, resinous complex(Lac) is being secreted from the tiny, primitive soft bodied insect, *Laccifer lacca* through a large number of glands situated all over its body. It is also a unique material which thrives on the tender twigs of specific host trees *viz.*, palas (*Butea monosperma*), ber (*Zizyphus mauritiana*), *kusum* (*Schleichera oleosa*), *Flemingia semialata*, *Ficus spp.* etc.(Pal 2009).The raw lac is the source of three valuable, natural and renewable products *i.e.* resin, dye and wax. *Rangeeni* and *Kusmi* are the two strains of lac insect which are classified based on preference of the insect for specific host plants (Srinivasan 1956; Roonwal et al.1958). Lac cultivation is an important source of income for livelihood of the tribal population of India. Besides, it has high potential for generating employment for both men and women in forest and subforest areas of Jharkhand, Chhattisgarh, Madhya Pradesh, West Bengal, Maharashtra, Odisha and parts of Uttar Pradesh, Andhra Pradesh, Gujarat and NEH region particularly in the off agricultural seasons(Sengupta 1972).Recent research carried out by Indian Lac Research Institute in Ranchi have revealed that income generated from lac cultivation is next only to cultivation of paddy. The income form lac cultivation is about 8% of their total agricultural income. And most of the lac growing families are among the poorest of the poor. The cultivation is done extensively in the remote interior pockets of the state (Varshney and Srivastav 1989). Further, it has also been found that about half of the total lac-host trees are still lying un-exploited for lac cultivation in lac growing areas. There are also areas where in spite of large number of host trees, lac cultivation is not carried out at all. If these lac hosts could be utilised for cultivation of lac, it would not only increase the national production of lac and add to the income of the farmers but also help prevent indiscriminate felling of trees for fuel and timber purposes which indirectly help the biodiversity of conservation process(Chauhan and Teotia 1973; Sharma and Ramani 1999).Lac insect agro-ecosystems have been highlighted for their conservation potential, but management to intensify lac production reduces biodiversity and may restrict ecosystem services (Chen et al. 2008). The relationship between lac insects and their host trees (Bhagat 1988; Chen & Wang 2007a; Chen et al. 2003; Chen et al. 2004a, b; Chen et al.2005a, b; Chen et al. 2006a, b; Kher & Lakra 1988; Mishra et al. 2000; Saha & Jaipuriar 2000;Subbarayudu & Ram 1997; Varshney & Srivastava1989) farming systems and ecology of lac insects which include lac production have seldom been described also in this chapter.

Lac Production Status in India

The Indian Council of Agricultural Research report **on Lac Statistics at a glance (2010)** reported on the basis of survey in the markets of different lac producing districts and states, the estimated national production of sticklac during 2009-10 was approximately 16,495 tons(Pal et al.2010).Jharkhand state ranks first followed by

Chhattisgarh, Madhya Pradesh, West Bengal and Maharashtra. These five states contributed around 95 per cent of the national lac production. Contribution of Jharkhand in national lac production was 41.98 per cent followed by Chhattisgarh (30.28 %), Madhya Pradesh (14.49 %), West Bengal (5.18 %) and Maharashtra (3.15 %). Regarding share of different crops, *jethwi* ranks 1st (42.55 %) followed by *baisakhi* (34.62 %), *aghani* (14.40 %) and *katki* (8.43 %) in national lac production. In the year 2009-10, the production of *baisakhi* and *jethwi* crop has increased by around 11 per cent and 33 per cent respectively while production of *katki* and *aghani* crop has decreased by 73 per cent and 39 per cent respectively in comparison to last year production. At national level the production of lac was around 4 percent lesser than that of last year. Similarly the country's production of lac was 23,229 tonnes in 2006-07 (Saha and Jaipuria 2000; Pal et al. 2007). There is decrease in total lac production in India during the period of 2009-10 as compared to 2006-07(Pal et al.2010).

Table 1 : Lac production in India during 2009-10 (in tons)

Sl.No.	Name of the State	Name of lac crop(in tons)				Total Production
		Baisakhi	Jethwi	Katki	Aghani	
1	AndhraPradesh	15	4	15	1	35
2	Assam	40	0	60	0	100
3	Chhattisgarh	2470	1550	340	635	4995
4	Gujarat	10	200	10	25	65
5	Jharkhand	385	5000	140	1400	6925
6	Madhya Pradesh	1910	130	205	145	2390
7	Maharashtra	355	0	165	0	520
8	Meghalaya	5	0	5	0	10
9	Uttar Pradesh	175	25	0	0	200
10	West Bengal	275	200	335	45	855
11	Total	5710	7019	1390	2376	16,495

Lac Production Status in Other Country

Being a high remunerative crop lac is also grown in other country viz. Thailand, Indonesia, parts of China, Myanmar, Philippines, Vietnam, Cambodia etc. Thailand is the second largest lac exporter after India with a production of about 35% of the world market (Chuntanaparb 1985). In China, lac was traditionally produced on the upper part of the mountain slopes, from 900-1500 m in altitude (Chen et al.2010). Similarly Lac has been cultivated in about three hundred hectares of land annually in Bangladesh from which is able to provide 180 tons of crude lac (Alam and Sarker 2000).

Different Forms of Lac

Based on the methods of collection and processing various forms of lac known in commercial basis such as:

Stick Lac:- It is raw lac, obtained by scraping the lac encrustation from the dry twigs, cut down before emergence of the new insects. It contains dead bodies of the lac insects, bark of the host plant, dried leaves, dust and other extraneous impurities

Seed lac : The semi refined product obtained after crushing, sieving and winnowing of sticklac, followed by washing and drying is called seedlac. It contains impurities such as sand and insect debris etc.

Shellac : It is the refined from of lac available as thin flakes. It is obtained by stretching the heat softened seed lac, freed from infusible materials. It is mainly the lac of commerce and is graded based on its colour and wax content.

Button Lac : It is another form of heat-purified lac, where the molten resin is cast in to button shaped cakes instead of being drawn in to sheets, as in manufacture of shellac. The cakes are largely used for bonding mica splitting into micanite.

Bleached Lac : A major portion of lac consumed in the world today is in the form of bleached lac or white lac which is obtained through treatment alcoholic solution of seed lac with activated charcoal by bleaching a sodium carbonate solution of seed lac with sodium hypochlorite solution.

De-waxed Lac : It is another form of lac in which wax has been removed is known as De-waxed Lac. It forms clear solution with sprit and is extensively used in furniture polishing, coating tablets in pharmaceutical industry, etc.

Life Cycle of *Laccifer lacca*

The insect survives on the succulent green branches of the host plants and completes it's life cycle within six months. During the life cycle it secretes reddish brown gelatinous substance around the branches of host plants, which gives stick lac. On an average three hundred thousand insects produce one Kg of lac resin. The lac insect starts its life as a minute boat shaped, red-coloured larvae, 0.5 mm long and 0.25 mm broad. The larvae emerges in large numbers, generally twice a year, in June and November from the lac cells of the female insects and crawl over the surface of twigs and branches of the host plants they infest. A healthy female produces 300-1000 larvae. After a brief period,the larvae emerge from the cell in search of suitable places for settlement; and larval emergence may continue for several weeks. The proportion of males to females is generally 1:3. Once settled the larvae do not move again. A week after settling, the larvae starts secreting lac from glands distributed under the cuticle all over the body, except the mouth parts, the two breathing pore and the anus. The larva gets encased due to its own secretions, which increases in size with growth of insects. The larva moults thrice before reaching to maturity. The duration of each larval stage depends on temperature, humidity and host plant species.

Distinctive Feature of Male and Female

The sex is readily recognized by the shape of cell even in the early stages of development. In case of male, the growth is more on longitudinal axis, whereas in female, the growth is more in vertical axis. During pre-pupal and pupal stages, the larvae does not feed, and mouth become aborted. After the pupal stage, the adult male emerges by pushing open the operculum. The female larvae becomes swollen and assumes the form of a roundish bag which completely occupies the space inside the lac cell. After the third and final moult, the female is sexually mature and is fertilized by males which has a life only of 62-92 hours after emergence. Lac secretion by the female continues and the size of the insect as well as lac cell increases rapidly. The female continues to secrete lac until the eggs are laid. As the time for egg laying is reached, the body of female contracts on the ventral side, gradually vacating a space in the Iac cell in which it is enclosed. Simultaneously the wax filaments are secreted and shed in vacated space for providing dry dressing and cushion for the incoming larvae. The anal tubercle is withdrawn inside for egg laying. These eggs hatch into larvae immediately after laying. Temperature plays an important role in egg laying and larval emergence. The larvae inside the lac cell becomes inactive below 20°C but their capacity to produce lac is not reduced. '

Strains of Lac Insects

The commonest and the most widely occurring species of lac insect in India is *Laccifer lacca* (Kerr), which produces the bulk of commercial lac. Fourteen species of the genus have been recorded in India, but much confusion prevails regarding their exact identity, and it is possible that many of them are only strains or varieties of *Laccifer lacca.* Some of the species found in India occur also in other countries. *L. albizziae*, for instance, is found in Ceylon. *L. chinensis* is the chief source of commercial lac in Thailand. Other countries in which *Laccifer* spp. Occur include Burma, Cambodia, Laos, North Vietnam, China, Formosa, Indonesia, Philippines, Malaya, Pakistan, Nepal, Sikkim and Bhutan (Saha and Jaipuria 2000; Kher and Lakra.1988).

Strains of lac insect : There are mainly two strains of the lac insect, commonly recognized in India, the *Rangeeni* and the *Kusumi*. The lac crops raised from them are named after the months in which they are harvested. Each strain completed its life cycle twice a year, but he seasons of maturity differ considerably, thus in the case of *Rangeeni* strain, the *Katki* crop is harvested in October-November from inoculations made in June-July. Similarly the *baisakhi* crop is obtained in June-July from plants inoculated in October-November. Usually as a general practice the *baisakhi* crop is harvested earlier in April or May as *ari* i.e., in immature state of development. Likewise in the case of

Kusumi strain, the *aghani* crop is harvested in January-February from plants inoculated in June-July, where as the *jethwi* crop is obtained in June-July from January-February infection. In Karnataka State, the *Rangeeni* strain completed three cycles in 13 months on jallari (*Shorea talura*), though the same. Mysore strain when grow on jallari in northern India, has only two life cycles in a year. Likewise, *kusumi* brood raised in northern India when inoculated on kusum trees in Tamil Nadu State, completes its life-cycle in different months. Equable climatic conditions in the south seem to be responsible for these variations (Subbaray and Ram 1997; Bhagat 1988).

Host Plants

There are allover a hundred species of plants on which lac insets have been recorded in India, but from the point of view of large scale lac production, the following are important host plant for different strains (Subbaray and Ram 1997).

A. *Rangeeni* Strain: There are Different Host plants for *Rangeeni* Strain such as

1. Palas, dhak (*Butea monosperma*): The plant is very common throughout the greater part of India, extending in North-west Himalayas as far as Jhelum; in the outer Himalayas it ascends up to 900 m. and in the hills of Southern India up to 1,200 m. It is the commonest lac host in the country(Roonwal et al.1958).
2. Ber, beri, kul (*Ziziphus mauritiana*): The Ber plant is also found throughout the greater part of India, either wild or naturalized, ascending up to 1,500 m in the Himalayas. It is one of the major hosts in most lac growing tracts. It is the only important lac host in Murshidabad and Malda districts of West-Bengal and in Hoshiarpur district of Punjab.
3. Ghont, Kat-ber (*Ziziphus xylopyra*): The plant species is fairly and widely distributed in India especially in Rajasthan, Uttar Pradesh, Madhya Pradesh and the Deccan Peninsula from Konkan southwards. It is an important lac host in Madhya Pradesh.
4. Porho, khunia, jahrphali (*Ficus cunia*): It is a plant of potential importance found throughout the greater part of India, especially on the sides of ravines. It is a valuable plant for preserving broodlac in hot summers. It is utilized for lac cultivation mostly in Chota Nagpur and Assam and other parts of India
5. Barh, bargad (*Ficus bengalensis*) and Peepal (*Fiscus religiosa*): These hosts plants are particularly valued in the hotter parts of the country as they are able to carry broodlac through the summer months. The plant species are found throughout India.
6. Arhar (*Cajanus cajan*): Extensively cultivated throughout India, it is an important lac host in Assam hills where it is grown as a biennial or triennial

crop. In the plains, where it is usually an annual crop, it is of little value for lace cultivation. *Laccifer chinensis*, which is common in Assam, is probably the species that thrives on it.

7. Tapria siris (*Albizzia lucida*): Distributed in the sub-Himalayan tract from Nepal eastwards and in Assam.It is another important lac host in Assam and has given good results in Chota Nagpur. It grows well in several areas in northern India.

Other host plants employed to a small extent in the cultivation of *Rangeeni* strain in different parts of India are: panjan, sandan (*Ougeinia oojeinensis*) for *baisakhi* crop; khair (*Acacia catechu*) for *ari* (premature cutting) of *baisakhi* and pansaura (*Grewia serrulata*) both for *baisakhi* and *katki* crops.

B. *Kusumi* Strain

Kusum (*Schleichera oleosa*): The plant species is mostly found in the dry forests of the sub-Himalayan tract from Sutlej Eastwards and throughout central and Southern India. Almost the entire *Kusumi* lac is grown on this host in Chota Nagpur, Madhya Pradesh, Odisha and parts of Karnataka and Tamil Nadu States.

Khair (*Acacia catechu*): Common in most parts of India, it has been successfully employed in Chota Nagpur for carrying *aghani* crop. It remains leafless for a considerable period during the summer, and it is most important for lac cultivation that it should occur in association with kusum so that the latter may be used as host for the jethwi crop. During favourable seasons, when extra broodlac is available, cultivators in Chota Nagpur inoculate khair trees and harvest a satisfactory crop.

The two strains of lac insects complete their life in India twice a year, but the season of maturity differs considerably. Each healthy Rangeeni female lac insect in its life time produces about 0.029 gms of lac and each Kusumi female : about 0.069 gm.

Ecology and Management of Lac-Hosts

Since lac insect is dependent entirely upon the host-plant on which it lives and feeds, it will be realized that successful cultivation of lac depends upon proper use of the lac insect and the host-plants.

Ecological Conditions: For the successful cultivation of lac certain ecological conditions are necessary for the host-plants. Three important conditions as summarized by Srinivasan (1956) are as follows:-

1. Lac does best in open areas where there is plenty of circulation of air around the host-plants. In the list of hosts there are several species whose habitat is in conditions where free circulation of air is not possible. When trying out new hosts, care should be taken to confined the trials to only those plants that are found occurring in the open or in situations where free circulation of air is assured. Lac cultivation should not be tried in deep forests.

2. Another point needing emphasis is that lac cultivation should not be tried in areas liable to forest fires, e.g., scattered trees in grass land. Where the habitat of any of the host-plants comes under this category, cultivation should be avoided unless protection from fire can be absolutely assured.
3. Species which are deciduous for long periods during the hot months will not be suitable as host-plants during the summer crop season. They can be taken up for cultivation only for the crops which mature in the cold seasons provided alternative host-plants are available in the vicinity to take on the summer crops.

Direct Sowing and Nursery Planting: Lac-hosts may be propagated either by direct sowing at the site required or by first raising seedlings in nurseries and latter on transplanting them to the site required. Glover (1937) recommends the following:

Direct sowing for:

Khari (*Acacia catechu)*

Kusum: *Schleichera oleosa*

Nursery sowing and transplanting for:

Palas: *Butea monospema.*

Ber: *Zizyphus mauritiana*

Ghont: *Zizyphus xylopyra*

Manuring: To ensure the healthy establishment of trees, suitable manuring is highly advisable. Once the trees have become well established, manuring need not be continued.Glover (1937) has recommended the use of the following manure which should be dug to an area of about 1meter (or 3 feet) radius and a depth of about 15-20 cm (or 6-9 inches) around each plant. If artificial manure is not available, natural manure (well rotted cowdung) may be used.

Composition of Manure Quantity Required Per Plant

Potassium sulphate	28.4 gms
Ammonium sulphate	28.4 gms
Concentrated super phosphate	450 gms

Soil and Water : The soil should not be deficient in calcium, if deficient lime may be added to make up for the deficiency. Care should also be taken to avoid water-logging of young seedlings as that causes mortality. At the same time water should be conserved for the dry months by such devices as contour ridging, terracing etc.

Age of Host: The age at which a host-plant is suitable, i.e., mature enough to take on its first lac infection varies from species to species. It also depends upon the climatic conditions of the area as well as the methods employed for the management of the plant which determine its vigour and vitality to withstand the drain imposed upon it by the lac infection. For the major host-plants, the following may be found to be the suitable age:-

Sl.No.	Plant Species	Age when first suitable for Lac infection
1	Khair:*Acacia catechu*	About 8 years
2	Ber:*Zizyphus mauritiana*	About 18 years
3	Ghont:*Zizyphus xylopyra*	About 18 years
4	Palas:*Butea monosperma*	8-10 years
5	Kusum:*Schleichera oleosa*	About 15 years

In places where Ber is the only host available a more or less similar practice of keeping the trees continuously under infection is followed. But at other places where Ber trees occur along with Palas the former are in variably explored in the *Baisakhi* season only. Being known to be a poor carrier of brood lac through the summer season, Ber is put to use for production of lac and not brood crop. The trees are heavily infected with brood obtained from the Katki crop on the Palas and the harvest is reaped as Ari or immature in April-May before, yet the lac insects start dying in large numbers as a result of the summer heat and drought. Thus, the Ber has to depend on Palas for brood lac and where lac fails on Palas, it naturally results in the Ber trees going with little or no infection at all that year. With Kusum which is ordinarily a large-sized tree, the general practice is again to keep it under continuous inoculation, the cultivation being confined to different sectors of the same tree from season to season.

Lac Cultivation

Lac cultivation is initiated from broodlac, which is the twig of host tree carrying lac encrustation, with larvae about to emerge from mother cells. The twigs of the host are cut, bundled and tied at convenient places on a fresh host plant, so that the emerging larvae swarm and settle on nearby succulent shoots. Traditional methods of lac cultivation, however, have not proved always reliable in providing sustained supplies of healthy broodlac for raising two crops a year. Optimum yield of lac from a given quantity of broodlac is also not always obtained, and fluctuations in lac production have been the rule in the past. The proper management of the host stock, use of pest-free brood, time and manner of inoculating host plants and control of pests and predators of the lac insect are all important for ensuring good yields of lac. In recent yeasrs, much value information of lac culture has emerged as result of researches at the Indian Lac Research Institute at Namkum, Jharkhandr. Improved techniques of cultivation have been evolved and efforts are being made to popularize them among lac growers.

Preservation of broodlac: Successful lac cultivation depends on the continuous availability of broodlac in the required quantities. Since different host plants react differently to weather conditions, the broodlac survives better on some host species than on others. This is especially true in the case of hosts during hot summers. In order, to ensure a steady supply of broodlac, it is advisable to have a large variety of trees under lac wherever possible rather than restrict cultivation to one host. A shortage of broodlac periodically occurs in areas where only one host, Ber or Palas, is exploited for lac production. These two hosts account for 80% of the annual lac production in India and being deciduous, they remain leafless for a long period during summer, thus exposing the lac insect to heat, drought and hot winds. In order to overcome this difficulty in areas where only Palas or Ber is available resort may be had to partial pruning of Ber and partial defoliation of Palas in October-November just before inoculating the plants. As a result, the trees put forth new leaves which last during hot months and provide shade to the maturing *baisakhi* brood. The second way of preserving adequate brood during the hot summer months is to select only hosts which normally have green leaves during summer for raising the *baisakhi* crop. Certain *Ficus* spp., notably *F. cunia* and *F. lacor*, and a few leguminous plants like *Albizzia lucida* and *Ougeinia oojeinensis* have proved valuable for this purpose.

Preventive measures for insect pests of lac are:-(i) Immersion of freshly harvested sticklac, not wanted for brood, as well as phunki lac (i.e., broodlac after larval emergence is complete) in running or deep stagant water (ii) Scraping of lac from twigs immediately after harvesting and killing larvae and pupae of the pests by burning, crushing, drowning or by fumigation with carbon bisulphide (10cu.ft. of space) before storage (iii) Avoiding cultivation of early and late maturing varieties of lac, at least for brood purposes, in the same locality to prevent the speread of pests (iv) Encasing of broodlac for inoculation in 60-80 mesh wiregauze baskets of 30 cm x 7 cm in size. The last method is particularly recommended for areas where lac cultivation is being introduced for the first time. The baskets permit free exit to lac larvae but exclude enemy insects.

Proper management of host plants with a view to ensure their vitality and vigour helps to reduce damage by parasites. Parasites and Predators emerge from lac cells in large numbers towards the close of the emergence of lac insect larvae. The practice of leaving broodlac on trees even after swarming is complete, therefore, favours the spread of enemy insects. To reduce the risk to the minimum, removal of *phunki* lac from inoculated trees should be completed as early as possible, i.e., within 2-3 weeks of the date of swarming. For the same reason, self or natural inoculation of host trees should be avoided except where absolutely necessary.

Harvesting and Yield: The instruments used for harvesting lac crops are those normally used for pruning; in fact, if harvesting is properly done there is no necessity for pruning the hosts. Lac required for brood purposes is cut a few days before larval emergence, as broodlac meant for dispatch over long distances to other lac growing centres is cut a week or so before emergence. For despatch to outside stations by motor, rail or air, broodlac is packed in bamboo baskets with a capacity of 15-20 kg.

Encrustations of lac from excess crop, which are not required for brood are separated from the twigs by scraping. The scraped material, green lac or sticklac is spread thinly (10-15 cm deep) in a covered and well-ventilated place and periodically raked until dry. Broodlac twigs employed for inoculating fresh host plants are collected after the completion of swarming, and encrustations from them are also scraped out and dried. The sticklac thus obtained is then bagged for storage or for the market. Lac should not be stored in the green condition as fermentation sets in and fungus growth is favoured and lumps are formed and the quality of lac deteriorates. Long storage of even dried lac affects the solubility and other properties of lac and reduces its market value. The yield of lac is 2 ½ - 3 times the weight of broodlac employed for inoculating the host but in a healthy crops, it may be as high as 5-7 times. The yield of scraped lac is usually about one-third of the harvested material (which includes twigs) in the case of *Rangeeni* crops and about one-half in the case of *Kusumi* crops(Chen et al.2005b).

Method of cultivation: Cultivation of Lac is fairly a simple process and considering the limited labour required for it, the yield is fairly remunerative to the cultivators. But to get a fair return cultivation should be made in a systematic and scientific method. Very often the cultivators are deprived of a good crop due to lack of knowledge in the cultivation of lac. The main operations involved in the cultivation of lac are:-

1. **Pruning:** The bulk of the lac crop is grown on host trees kusum, Palas and Ber. The nature and extent of host plants and the time of the year at which pruning is to be done have an important effect on bearing and the production of feeding ground for the lace insect and total production. During pruning, care is to be taken not to damage the general health of the host trees(Pal 2009).

2. **Coupe system:** If a same tree is continuously inoculated its vitality and vigour suffers, and progressively yields become lower. So, it is necessary that the host plants should be rested periodically. The coupe system of cultivation provides the maximum use of host plants resources consistent with their vigour and well being. Available host trees are divided into blocks or coupes, the size and number of which are determined by the kind of trees and the strains of the lac insect. Each coupe is inoculated according to the schedule.

3. **Inoculation :** Brood lac should be cut just before the emergence of larvae. This is determined visually by the appearance of yellow spots on the encrustations at the anal region of healthy females. The spots gradually increase in size and become orange. When the spots are about a 1/4 the size of lac cell, the twigs are cut. Emergence of larvae will start after about five days. Branches with healthy brood lac should be cut of a convenient length of 15-30 cm and tied with strings to new shoot of the host trees either singly or in small bundles. Longitudinal inoculation allows maximum contact between brood and host.

4. **Preservation of brood lac :** Successful lac cultivation depends on the continuous avail-ability of brood lac in the required quantities. To ensure a steady supply of brood lac, it is always advisable to have a large variety of trees under lac wherever possible. One way to preserve adequate brood lac in summer months is to have host plants which normally have green leaves in summer.
5. **Alternation of broods :** If brood lac from a particular host is used year after year in the same locality and on the same host species, the quality and production is likely to go down. Hence, it is always better to exploit more than one host, e.g., Khair, which is more widely distributed and is a host for both Rangeeni and Kusumi strains and can successfully be alternated with Kusum.
6. **Control of parasites and predators of lac insect :** Several species of parasitic wasps lay their eggs in or on the body of the lac insect. The grubs on hatching feed on the lac insects and mature adults emerge from the lac cell by cutting circular holes.The damage due to parasites is usually 5-10 percent. Main predators are two moths, *Eublemma anabilis*, and *Holocera pulverea*, and three species of Chrysopa (Lacewings). The caterpillars of two moths tunnel through and eat away lac encrustations as well as insects, whereas larvae of *Chrysopa sp*. suck the body fluids of lac insects. The damage is about 40 per cent

Disease Management of Lac are

1. Immersion of freshly harvested stick lac, not waited for, brood, in running or deep stagnant water.
2. Scrapping of lac from twigs immediately after harvesting and killing larvae and pupae of the pests by burning, or fumigating with carbon di-sulphide before storage.
3. Avoiding cultivation of early and late varieties of lac in the same locality.
4. Encasing of brood lac for inoculation in 60-80 mesh wire gauge baskets. The baskets allow only larvae of lac but not the insect enemies.

Ensuring Beneficial Insects for Lac Insect

Some insect are indirectly beneficial to lac cultivation in a way, that they keep off the parasites enemy insects. Some ants act as body guard for lac insect. Beneficial insects mostly belong to the group of chaleids, braconoids, ichneumtids and bethylids. These parasitize, the eggs or larvae of *Eublemma anabalis* and *Holocera pulverea*.

Harvesting

After four months of inoculation, according to season, the insects will have ceased to emerge and the lac-covered shoots (stick-lac) are ready for gathering. These are then cut, a liberal allowance being left on trees for serving as brood lac. After being dried in the sun for a about week, the lac is scrapped off with a knife or crushed by a grinding mills then dried in the shade. It is afterwards ground soaked in water for 24 hours and thoroughly washed. The washed granular material which is of a pale-orange colour is the seed lac of commerce. This is heated, then formed into thin sheets (shellac).The latter is often blended, according to grades required or uses to which it is applied, with a small percentage of pine-resin and arsenic. The arsenic being used for the purpose of improving the colour. The trade grades are: T.N. (The standard grade), fine-orange, garnet-lac, Tongue-lac etc. Lac should not be stored in the green condition as fermentation sets in and fungus growth is reported.

Yield and Costs (Trade)

No accurate data regarding yield and cost are available; much depends on locality and host:-plant. Cultivation etc. involves little labour. Once the host trees are established, the crop requires little attention, while wild trees or those grown for shade or boundary lines, may be utilized as lac-bearers. Two crops a year may be obtained one ("Baishakhi" or Summer crop) taking about 4 months between May to August and another ("Katiki" or winter crop) taking 4 months from October to January. Some trees e.g. Kusum, bears a good crop only every second or third year(Mishra et al.1999; 2000).

The annual yield of stick-lac per year per tree may be only 1 to 1.5 kg or as much as 8-16 kg in case of well maintained trees. In the central provinces of India, an average of 340 kg per hectare is said to be often obtained from "Palas" Butea trees; taking into consideration of 275 to 290 trees per hectare. Prices fluctuate considerably, the best grades commanding quite a high price in the market (Mishra and Sushil 2000).

Table 2 : Cost of Cultivation of Lac in Different host Plant (Rs.) (Pal et al. 2009)

Sl.No.	Particulars	Palas (50 hosts)	Ber (50 hosts)	Kusum(10 hosts)
1	Gross Return	11,701.50	28,875.00	44,170.50
2	Cost of Cultivation	3,532.60	7,961.30	11,042.00
3	Net Return	8,168.90	20,913.70	33,128.50
4	Input out put ratio(BC ratio)	3.31	3.63	4.0
5	Family labour income	8,924.90	22,908.70	34,388.50
6	Farm Business income	9,008.90	22,992.70	34,472.50

Structure and Composition Lac

The major constituent of sticklac is the resin (70-80%), other constituents are: sugars, proteins, and soluble salts, colouring matter, wax, sand, woody matter, insect bodies and other extraneous matter, and volatile oil in traces. The colouring matter Lac dye is described as the scarlet pigment present in the live, pre-emergent insects (*Laccifer lacca*) which develops a resinous cocoon, known as stick lac on the twinges of over 160 host trees. The crude lac (Stick lac) contains 1 per cent dye pigment. The dye stuff is obtained by aqueous extraction of sticklac. From ancient times, lac dye has been employed in India as a skin cosmetic and as a source for dyeing wool and silk, while dyeing of leather with lac was traditionally followed in China. Colour of the dye can be modified with appropriate choice of mordants from violet to brown. Lac dye is based on anthraquinoid type of structure and composed of laccaic acid and erythrolaccin. Laccaic acid is water soluble compound whereas erythrolaccin is water insoluble. The laccaic acid is the mixture of five closely related compounds i.e., laccaic acid A, B, C, D and E. During the extraction of lac dye the amount of dye depends on the percentage of laccaic acid present in stick lac (Gupta and Prasad 1993).

Lac dye is usually extracted from the by products of shellac industry. Stick lac is made into shellac by thoroughly washing in plain water. The water soluble dye pigment, laccaic acid is taken for further extraction of lac dye. Lac dye is acidic in nature, soluble in water and other solvents such as methyl alcohol, amyl alcohol, acetone, acetic acid, formic acid, but insoluble in ether, chloroform and benzene. Since lac dye is acidic in nature can safely be applied to protein substrates such aswool, silk and other animal hair (Indian Lac Research Institute 2006).

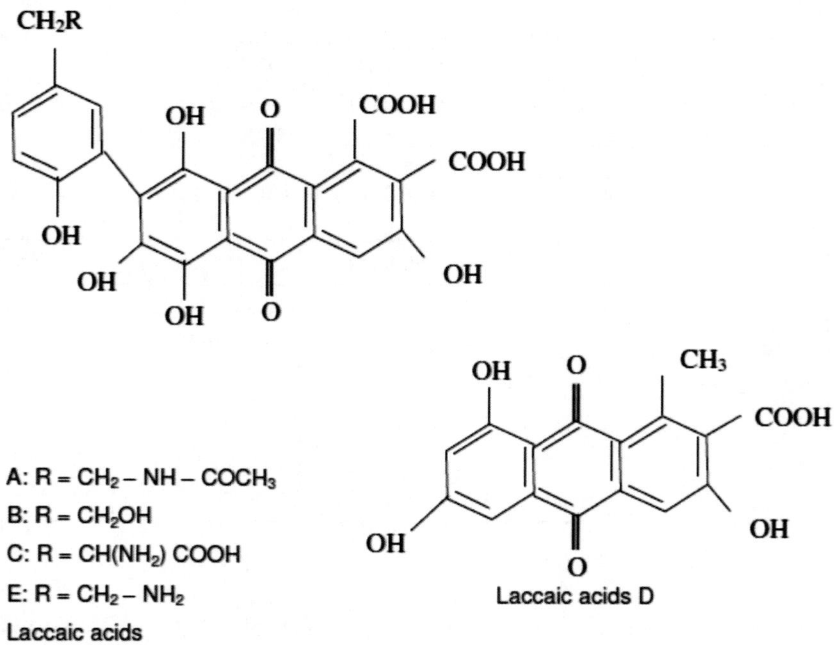

Structure of laccaic acid

The crude dye is fractionated by column chromatography into five fractions, designated as laccaic acid A,B,C,D and E respectively, of which A and B are the major constituents. Laccaic acid A contains 2.5% nitrogen and has the molecular formula $C_{26}H_{19}NO_{12}$. Laccaic acid B has no nitrogen and its molecular formula is $C_{24}H_{16}O_{12}$. It is characterized as alcohol corresponding to laccaic acid A, where -OH replaces -$NHCOCH_3$. Laccaic acid C, crystallized from methanol is the first naturally occurring anthraquinone,

Extraction of Lac Dye

There are various method proposed by different scientist for extraction of Lac dye which are discussed below:-

Protocol-I: The crushed stick lac was washed thoroughly with water. The solution was allowed to settle, and the supernatent liquid is taken off. The PH of the solution was brought to 8 by addition of caustic soda and then treated with calcium chloride (with agitation). The precipitated dye was filtered and washed to make it free from impurities (Prasad and Agarwal 1989).

Protocol-2: Lac dye can be obtained by extracting stick lac with water and sodium carbonate solution followed by precipitating with lime (Paul et al. 1996).

Protocol-3: Stick lac was crushed into pieces, wetted with water and kept over night. The wet pieces of stick lac were churned. The water soluble dye driven out during churning. The solution was filtered, and the extracted dye was found to be 50 to 98 percent pure (Patra 1998).

Protocol-4: Lac dye is an important by- product of lac industry; when stick lac is crushed and washed with water, the water soluble dye nearly 1.0 per cent dissolves and is thrown off as a lac effluent. The waste effluents of lac processing units when treated with mineral acids, the lac effluent precipitates out as a solid sludge which is filtered. The filtrate is treated with calcium carbonate to obtain calcium salt of lac dye. The calcium salt is converted into sodium salt which is then passed through cation exchange resin column. The dilute solution is concentrated to give pure lac dye crystals which are washed with cold water and sun dried (Anonymous 2006).

Properties of Lac Dye

Gupta and Prasad (1993) conducted a study on "Lac dye-A potential material for colouring food and proteinous fibres." Lac dye is non-toxic, soluble in methyl alcohol, amyl alcohol, acetone, acetic acid and formic acid but insoluble in ether, chloroform and benzene. It dissolves slowly in ethyl alcohol. The aqueous solution is orange-red where as alkaline solution is reddish violet. It also dissolves in concentrated, sulphuric acid with a beautiful carmine red colour. The study on "Lac as a natural dye" was carried out by Agrwal (1997). Lac dye is acidic in nature melts above 23°C. Dissolves 50 to55 percent and 98 to 99 percent in coldwater and boiling alcohol, respectively.

Importance and Uses of Lac

The major constituent of stick-lac is the resin (70-80%), other constituents present are sugar, proteins, and soluble salts (2-3 %); colouring matter (1-2%); wax (4-6%) sand, woody matter, insect bodies and other extraneous matter (8-12%) including presence of volatile oil in traces. The colouring matter in lac includes a water-soluble red-dye, laccaic acid, and an alkali and spirit soluble yellow dye, erythrolaccin (Anonmous 2006). The natural resin, wax and dye, all of which are safe for different applications especially, related to human consumption and contact. Being natural, easily available, non-toxic in nature, soluble in common solvents, very good film forming ability, adhesive and moulding properties, lac resin finds application in diverse areas viz. Surface coating (furniture polish, paint, varnish), printing ink, handicraft, jewellery, gasket cement, sealing wax, micanite, leather finishing, insulating varnish etc. Some of the emerging areas where lac resin and dye find usage are coating of fruits, vegetables, synthesis of fine chemicals (aleuritic acid, isoambrettolide etc.), coating of candy and chocolate and lac dye as food colour in beverages and sausages etc(Ghosh et al.1964;2001). Previously lac's chief use was the manufacture of gramophone records but now it has been replaced by synthetic substitutes (Prasad and Agarwal 1989). Similarly Lac dye was formerly used for dying silk, wool and feathers but has now been replaced by synthetic dye-stuffs. Lac is also used as a coat for metal ware to prevent tarnishing and for preserving archeological and zoological specimens. Lac is the principle ingredient of sealing wax, manufacture of glazed paper, printing and water proofing inks, nail polishes, dental plates, ammunition, bangles, wax crayons and optical frames. It is used as an ingredient of varnishes, polishes, finish lacquers etc. for protective and decorative purposes. Aqueous varnishes or emulsions are used for leather dressing and leather finishing. Lac is also used for stiffening and finishing felt, straw and silk hats. Lac is used in electrical industry in the form of insulating varnishes and moulded insulators. It also possess very good adhesion to mica. However, the biggest consumption of lac is in the surface coating industry (Gupta and Prasad 1993).

Production and Trade

Lac is grown in India mostly as a subsidiary crop by small village farmers and the individual production often ranges from a few pounds to a maund. The average annual production of lac in India during past 30 years is over 11 lakh maunds (md) (Paul et al.1996; Mishra et al.2000; Pal 2009).

Baisakhi Crop: This crop accounts for 66% of the total annual production and receives particular attention from growers and trade alike. Failure of this crop is not as frequent as that of others. The maximum annual production of baisakhi crop was 11,13, 000 md. (1939-40) and the minimum, 3,65, 350 md. (1943-44). Bihar accounts for the major share of the production (45%) followed by Madhya Pradesh (29%) and West Bengal (18%).

Katki Crop: This crop constitutes 25% of total annual production. During the period 1930-60, the maximum production was 5, 01, 000 md. (1956-57) and the minimum,1,39, 250 md.(1932-33).As the katki crop supplies brood for the next baisakhi crop, an increase for decrease in the former is generally followed by similar variations in the baisakhi crop. Bihar contributes 32%, Madhya Pradesh 39%, Assam 4% and West Bengal9% to the total production of katki crop. Assam produces only the katki **Lac** crop, while this crop also occupies an important position in Madhya Pradesh.

Jethwi Crop: The jethwi crop, on an average, constitutes only 2.6% of the total annual production. Bihar accounts for 40% of the total jethwi crop and Madhya Pradesh, for 34%. This crop supplies broodlac to the aghani crop, the prospects of which are eagerly awaited by the trade, because of the high prices that it fetches in the market. Since the period of development of jethwi crop falls during the summer, it is particularly liable to failure.

Aghani Crop: This crop accounts for 6.8% the total annual production, of which 44% is produced by Bihar and 23% by Madhya Pradesh. During the period 1930 to 1960, the highest production was 3,28, 250 md. (1936-37) and the lowest, 38,500 md. (1952-53).

Marketing : The producers of Lac usually sell their crops in the nearest rural market (haat) to dealers or agents of manufacturers. Rural markets, small or large, are scattered throughout the lac growing areas and number over 600 in Bihar, 850 in Madhya Pradesh, 125 in West Bengal and 300 in Assam. With the approach of the cropping season, small amounts of rather immature lac start coming into the market; then comes the main flush of crop spread over nearly two moths, followed by decreasing quantities. In normal years, there is usually a carry over of sticklac fo all kinds, but mostly form baisakhi and katki crops. The quantity carried over depends upon the sizeof the crop and the export during the year, and it has a marked influence on the price of the new crop. The cultivator keeps himself informed about price fluctuations, but has little holding capacity (Rao and Singh 1990).

Exports: At present, lac is primarily an export commodity and the internal consumption forms only 5-8% of the total production. Lac is exported mostly in the form of seedlac, shellac and buttonlac. Small quantities of sticklac, kirilac and refuse lac are also exported. Shellac and buttonlac constitute 58-72% of the export, seedlac 22-34% and sticklac 0.2-1.3% (Rao and Singh 1990; Pal et al.2010).

Imports : Till recently India used to import lac from Burma and Thailand. Burma, in the pre-war years, produced annually 54,000 md. of sticklac, all of which was exported to India. During the war and also in subsequent years, Burma did not export any lac. Thailand also exported all its sticklac to India in pre-war years. Since the war, however, the pattern of lac trade with Thailand has completely changed, and India is no longer the only country to which lac is exported by that country (Rao and Singh 1990; Pal et al.2010).

Reference

Agarwal, S.C.(1997).Lac as a natural dye. The Indian Tex. J., 107(10): 26-28.

Anonymous.(2006).Lac dye a potential material for proteinous fibres. Indian Lac research Institute, Exhibition on dissemination of appropriate technology for hand loom sector pp. 41-45.

Bhagat, M.L. (1988). Field studies on the initial mortality of lac insect *Kerria lacca* Kerr. in relation to host plants. Indian Forester 114: 339-341.

Chauhan, N. S. and Teotia, T. P. S. (1973).A new variant in the lac insect, Kerria lacca (Kerr). Entomol. Newsl 3:33–34.

Chen, Y.L., Chen, Y.Q., Li, Q., Zhang Y. and Zhou X.Y.(2008). Preliminary study on the spider community in *Kerria* spp. ecosystem. Journal of Fujian College of Forestry 28: 179-183. (in Chinese).

Chen, Y.Q. and Wang, S.Y. (2006a). Growth of *Schleichera oleosa* infested by *Kerria lacca*. Entomological Knowledge 43: 549-552. (in Chinese).

Chen, Y.Q. and Wang, S.Y. (2006b). The influence on plants nutrient contents parasitized by *Kerria lacca*. Entomological Knowledge 43: 691-695. (in Chinese).

Chen, Y.Q. and Wang, S.Y. (2007a). Effects of host plants on natural population of Kerria yunnanensis (Hemipterae: Kerridae). Chinese Journal of Applied Ecology 18(4): 761-765. (in Chinese).

Chen, Y.Q., Chen X. M.and Li, K.(2004b).Preference of lac insect to host branch in foraging. Forest Research17: 159-166. (in Chinese).

Chen, Y.Q., Chen X.M. and Li, K. (2003). A probe into the mortality of lac insect population and improvement of mortality formula. Forest Research 16: 135-140.(in Chinese).

Chen, Y.Q., Chen X.M., Li, K., Shi, L. and Chen,Z.Y.(2005b). A study on population dynamics of lac insect and itslac secreting. Journal of Northwest Forestry University 20: 170-174. (in Chinese).

Chen, Y.Q., Chen X.M., Li, K., Shi,L.and Chen,Z.Y.(2005a). A primarily study on relationship between mineral salt contents of host trees and Kerria lacca. Chinese Journal of Ecology 24: 523-527. (in Chinese).

Chen, Y.Q., Chen X.M.and Li, K.(2004a). A primarily study on relationship between amino acid content of host trees and *Kerria lacca*. Forest Research 17: 362-367. (in Chinese).

Chen,Y., Li, Q., Chen,Y .,Wang, S. and Yang, Y.(2010). Lac-production, arthropod biodiversity and abundance, and pesticide use in Yunnan Province, China. Tropical Ecology 51(2): 255-263.

Ghosh, A.K., Mukherjee, M. and Sengupta, S.C.(1964). Reclamation of lac dye from lac effluents. Res. Ind. 9(5): 129-131.

Ghosh, S.S., Sengupta, D. and Thangavelu, K. (2001).Application of lac dye on Tasar silk. Indian Silk 40(4): 27-29.

Gupta, D.C. and Prasad, K.M.(1993).Lac dye a potential material for colouring food andproteinous fibres. Regional Workshop on Natural Dyes, pp.156-159.

Kher, S.and Lakra. P.K.(1988). Preliminary observations on some recorded and unrecorded host plants and natural enemies of lac insect, Kerria lacca (Kerr.) in Haryana. Haryana Agricultural University Journal of Research 18: 131-133.

Mishra, Y. D. and Sushil, S. N.(2000).A new trivoltine species of genus Kerria Targioni–Tozzeti (Homoptera: Tachardidae) thriving on Schleichera oleosa (lour.) Oken from eastern India. Orient. Insects 34: 215–220.

Mishra, Y. D., Sushil, S. N., Bhattacharya, A., Kumar, S., Mallick, A. and Sharma, K. K(1999). Intra specific variation in host plants affecting productivity of Indian lac insect, *Kerria lacca* (Kerr). J. Non-Timber For. Prod 6: 14–117.

Mishra, Y. D., Sushil, S. N., Kumar, S. and Bhattacharya, A. (2000).Variability of lac productivity and related attributes of Kerria spp.(Homoptera: Tachardidae) on ber (*Ziziphus mauritiana*). J. Entomol.Res 24:19–26.

Mishra, Y.D., Sushil, S.N., Kumar S. and Bhattacharaya, A.(2000). Variability in lac productivity and related attributes of Kerria spp. (Homoptera:Tachardiidae) on ber (Zizyphus *mauritiana*). Journalof Entomological Research 24: 19-25.

Pal, G.(2009). Impact of ScientificLac Cultivation Training on Lac Economy-A Study in Jharkhand. Agricultural Economics Research Review 22: 139-143.

Pal,G., AK Jaiswal, A.K. And Bhattacharya, A (2010).Lac Statistics at A Glance. Indian Institute of Natural Resins and Gums (ICAR) Ranchi- 834010, Jharkhand.

Patra, S.K.(1998). Potential of lac dye. The Indian Tex. J., 108(12): 40-41.

Paul, R., Jayesh, M. and Naik, S.R.(1996).Natural dye: classification, extraction and fastness, properties. Tex. Dyer and Prin., 29(2): 16-23.

Prasad, K.M. and Agarwal, S.C.(1989).Insect dyes an industrial perspective, compendiam of the first national seminar on natural dyes, Lucknow pp.33-34.

Rao, A.R. and Singh, P.(1990). Lac cultivation and marketing. Indian Forest 116(5): 459-463.

Roonwal, M. L., Raizada, M. B., Chatterjee, R. N. and Singh, B.(1958). Descriptive account of the host plants of the lac insect, *Laccifer lacca* (Kerr) and the allied plants in the Indian region (Part 1 & 2),Indian Lac Cess Committee, Ranchi p.140.

Saha, S.K.and Jaipuriar S.K.(2000). Variability and interrelationship of biological parameters and resin characteristics in lac insect, *Kerria lacca* (Kerr.). Shashpa 7: 17-20.

Sengupta, S.C.(1972). Twenty five years of research in lac. Indian Farm., 22(7): 13-23.

Sharma, K. K. and Ramani, R.(1999). An update on synoptic catalogue of lac insects (Homoptera: Tachardidae). J. Bombay Nat. Hist.Soc 96: 438–443.

Sharma, K. K., Ramani, R. and Mishra, Y. D.(1997).An additional list ofthe host plants of lac insects, Kerria spp. (Tachardidae: Homoptera).J. Non-Timber For. Prod 4:151–155.

Sharma, K.K., Jaiswal, A.K. and Kumar, K.(2006).Role of lac culture in biodiversity conservation : issues to stake and conservation strategy. Curr. Sci., 91(7): 894-898.

Srinivasan, M. M.(1956).Lac, its cultivation and manufacture in India,Chapter IV – Host plants of the lac insect. Indian Forest 82: 180–193.

Subbarayudu, B.and Ram R.L.(1997). Distribution of host plants of the lac insect, Kerria lacca (Kerr.). Journal of Entomological Research 21: 187-192.

Varshney, R.K. and Srivastava P.N.(1989).Amino acid and carbohydrate constituents of the India lac insect, *Kerria lacca* (Homoptera: Tachardiidae) at different stages of development. Annales de la Societe Entomologiquede France 25: 380-382.

❑❑❑

FSC
www.fsc.org
MIX
Papier | Fördert gute Waldnutzung
FSC® C083411

Kolibri 360 GmbH
Ferdinand-Jühlke-Straße 7
99095 Erfurt, Deutschland
produktsicherheit@kolibri360.de